Holger Kohl · Günther Seliger · Franz Dietrich ·
Sebastián Mur
Editors

AF173132

Sustainable Manufacturing as a Driver for Growth

Proceedings of the 19th Global Conference
on Sustainable Manufacturing,
December 4–6, 2023
Buenos Aires, Argentina

 Springer

Editors
Holger Kohl
Technische Universität Berlin
Berlin, Germany

Günther Seliger
Technische Universität Berlin
Berlin, Germany

Franz Dietrich
Technische Universität Berlin
Berlin, Germany

Sebastián Mur
Instituto Tecnológico de Buenos Aires
Buenos Aires, Argentina

ISSN 2195-4356   ISSN 2195-4364 (electronic)
Lecture Notes in Mechanical Engineering
ISBN 978-3-031-77428-7   ISBN 978-3-031-77429-4 (eBook)
https://doi.org/10.1007/978-3-031-77429-4

This Springer imprint is published by the registered company Springer Nature Switzerland AG
The registered company address is: Gewerbestrasse 11, 6330 Cham, Switzerland

If disposing of this product, please recycle the paper.

Preface

We are pleased to publish a collection of papers presented at the 19th Global Conference on Sustainable Manufacturing (GCSM), held on December 4–6, 2023, in Buenos Aires, Argentina. The conference is annually sponsored by the International Academy for Production Engineering (CIRP), committed to excellence in the creation of sustainable products and processes. GCSM 2023 was jointly organized by the Institute for Machine Tools and Factory Management (IWF) at the TU Berlin and Fraunhofer Institute for Production Systems and Design Technology (IPK) together with the Instituto Technológico de Buenos Aires (ITBA).

The GCSM 2023 brought together more than 130 attendees form 30 countries providing a global forum of academics, researchers, and specialists from universities, research institutes, and industry from across the globe, working on topics related to sustainable manufacturing. A unique feature of the GCSM conference series is its integration of industrial engineering perspectives, sustainable manufacturing applications in emerging and developing countries, as well as education and workforce development for advancing sustainable manufacturing. Plenary keynote speeches by experienced personalities from academics and industry, paper sessions presentations, and workshops of student teams from different countries offered new insights and chances for exchange of ideas. The conference featured eleven keynote speakers who shared recent advances in cutting-edge research and industry practices; these prominent and internationally recognized experts elaborated how technologies in the product, process, and system domains can enable sustainable manufacturing.

This volume documents almost 100 contributions presented at GCSM 2023 in 21 sessions held over three days. The proceedings are organized according to the conference program, classified into four broad categories as: Sustainable Processes, Sustainable Manufacturing Systems, Sustainable Manufacturing Products, and Crosscutting Topics in Sustainable Manufacturing. The papers cover a variety of topics in these areas related to modelling and simulation of manufacturing processes, product design for sustainability, metrics for sustainability assessment, energy efficiency in manufacturing, strategies and business models, as well as education and workforce development for sustainable manufacturing. All papers published in these proceedings have been reviewed by experts from the international scientific committee.

In addition to keynotes and paper sessions, two sessions on student projects were included in the GCSM 2023 program to further its mission, by involving the younger on the challenges of sustainable manufacturing. Students from different countries exchanged their perspectives on how to tackle the "Sustainable Development Goals" of the United Nations, by presenting and discussing specific projects.

Holger Kohl
Franz Dietrich
Günther Seliger
Sebastián Mur

Lecture Notes in Mechanical Engineering

Series Editors

Fakher Chaari, *National School of Engineers, University of Sfax, Sfax, Tunisia*

Francesco Gherardini ⓘ, *Dipartimento di Ingegneria "Enzo Ferrari", Università di Modena e Reggio Emilia, Modena, Italy*

Vitalii Ivanov, *Department of Manufacturing Engineering, Machines and Tools, Sumy State University, Sumy, Ukraine*

Mohamed Haddar, *National School of Engineers of Sfax (ENIS), Sfax, Tunisia*

Editorial Board Members

Francisco Cavas-Martínez ⓘ, *Departamento de Estructuras, Construcción y Expresión Gráfica Universidad Politécnica de Cartagena, Cartagena, Spain*

Francesca di Mare, *Institute of Energy Technology, Ruhr-Universität Bochum, Bochum, Germany*

Young W. Kwon, *Department of Manufacturing Engineering and Aerospace Engineering, Graduate School of Engineering and Applied Science, Monterey, USA*

Tullio A. M. Tolio, *Department of Mechanical Engineering, Politecnico di Milano, Milano, Italy*

Justyna Trojanowska, *Poznan University of Technology, Poznan, Poland*

Robert Schmitt, *RWTH Aachen University, Aachen, Germany*

Jinyang Xu, *School of Mechanical Engineering, Shanghai Jiao Tong University, Shanghai, China*

Lecture Notes in Mechanical Engineering (LNME) publishes the latest developments in Mechanical Engineering—quickly, informally and with high quality. Original research or contributions reported in proceedings and post-proceedings represents the core of LNME. Volumes published in LNME embrace all aspects, subfields and new challenges of mechanical engineering.

To submit a proposal or request further information, please contact the Springer Editor of your location:

Europe, USA, Africa: Leontina Di Cecco at Leontina.dicecco@springer.com
China: Ella Zhang at ella.zhang@cn.springernature.com
India, Rest of Asia, Australia, New Zealand: Swati Meherishi
at swati.meherishi@springer.com

Topics in the series include:

- Engineering Design
- Machinery and Machine Elements
- Mechanical Structures and Stress Analysis
- Automotive Engineering
- Engine Technology
- Aerospace Technology and Astronautics
- Nanotechnology and Microengineering
- Control, Robotics, Mechatronics
- MEMS
- Theoretical and Applied Mechanics
- Dynamical Systems, Control
- Fluid Mechanics
- Engineering Thermodynamics, Heat and Mass Transfer
- Manufacturing Engineering and Smart Manufacturing
- Precision Engineering, Instrumentation, Measurement
- Materials Engineering
- Tribology and Surface Technology

Indexed by SCOPUS, EI Compendex, and INSPEC.

All books published in the series are evaluated by Web of Science for the Conference Proceedings Citation Index (CPCI).

To submit a proposal for a monograph, please check our Springer Tracts in Mechanical Engineering at https://link.springer.com/bookseries/11693.

Committees

International Organizing Committee

Prof. Dr.-Ing. Holger Kohl	International Chairman
Prof. Dr.-Ing. Franz Dietrich	Local Chairman
Prof. Dr.-Ing. Günther Seliger	Founding Chairman
Valentin Eingartner	Organizational Team
Maxim Mintchev	Organizational Team
Nwaoyibo Donatus Junior	Organizational Team

National Organizing Committee

Ing. Sebastián Mur	National Chairman
Prof. Dr. Leopoldo de Bernárdez	National Academic Advisor
Prof. Dr. Giampaolo Campana	Latin American Academic Advisor
María Pía Mendoza	Local Organizational Team
Clara Weppler	Local Organizational Team

International Scientific Committee

Prof. Ahmed Abu Hanieh	Birzeit University
Dr. Samy Abu Salih	Braude, Karmiel
Dr. Feri Afrinaldi	Andalas University
Prof. Tülin Aktin	İstanbul Kültür Üniversitesi
Dr. Elita Amrina	Andalas University
Prof. Afif Aqel	Birzeit University
Prof. Fazleena Badurdeen	University of Kentucky
Prof. Dirk Bähre	Universität des Saarlandes
Prof. Peter Ball	University of York
Prof. Wahidul Biswas	Curtin University
Prof. Daniel Braatz	Universidade Federal de São Carlos
Prof. Erhan Budak	Sabancı Üniversitesi
Prof. Laszlo Cser	Budapesti Corvinus Egyetem
Prof. Pedro Filipe Cunha	Escola Superior de Tecnologia de Setubal
Dr. Roy Damgrave	University of Twente

Local Scientific Committee

Dr. Eliana Agaliotis	Instituto Tecnológico de Buenos Aires
Prof. Leopoldo De Bernardez	Instituto Tecnológico de Buenos Aires
Prof. Diego Cafaro	Universidad Nacional del Litoral
Prof. Giampaolo Campana	Università di Bologna
Dr. Leonel Chiacciarelli	Instituto Tecnológico de Buenos Aires
Prof. Antonio Corradi	Università di Bologna
Dr. Gabriela Corsano	Consejo Nacional Investigaciones Científicas y Técnicas
Prof. Sebastián D'Hers	Instituto Tecnológico de Buenos Aires
Prof. Daniel Martinez Krahmer	Instituto Nacional de Tecnología Industrial
Prof. Raúl Marino	Instituto Tecnológico de Buenos Aires
Dr. Úrsula María Montoya Rojo	Instituto Tecnológico de Buenos Aires
Prof. Sebastián Mur	Instituto Tecnológico de Buenos Aires
Prof. Daniel Ryan	Instituto Tecnológico de Buenos Aires
Prof. Cecilia Smoglie	Instituto Tecnológico de Buenos Aires
Prof. Olga Timoteo	Universidad Cayetano Heredia
Dr. Dannisa Chalfoun	Instituto Tecnológico de Buenos Aires
Dr. Nancy Lis García	Consejo Nacional Investigaciones Científicas y Técnicas

Contents

Education

Life Cycle Thinking

Design and Innovation

Technical Valuation

Social Valuation

Remanufacturing

Repair and Maintenance

Factory Planning and Production Management

Data and Simulation

Data and Learning

Technical Processes

Additive Manufacturing

Energy Generation and Efficiency

Materials and Resource Efficiency

Student Sessions

Business Models and Regional Integration

Olive Sector Integrated Artificial Intelligence and Modern Technologies: Model for Palestine

Ahmed Abu Hanieh$^{(\boxtimes)}$ (ID) and Afif Akel Hasan (ID)

Birzeit University, Al Marj Street 1, P.O. Box 14, Birzeit, Palestine
ahanieh@birzeit.edu

Abstract. Olive sector in Palestine suffers from serious problems starting from cultivation process passing through oil extraction process ending with final treatment. Most of the processes depend on old traditional ways away from modern technologies. The main focus of this paper is to put forward some basis and roadmap for implementing artificial intelligence and new technological components in olive sector processes. The hilly nature of Palestine prevents using heavy duty mechanical instruments for harvesting. Most of olive trees are planted on hilly terrains hence requiring the use of light weight hand-held harvesting tools. Soil moisture and nutrition elements can be determined through sensors to decide the necessity to add fertilizers. Olive fruits can be tested to check the oil content and quality before harvesting and extraction process. The final produced virgin oil can be turned to extra virgin oil by implementing further treatment processes where modern technologies are implemented. Modern technologies can also be integrated in post processes of olive sector like touristic wood works, soap manufacturing, olive fruit pickling, and other oil-based food products.

Keywords: Olive sector · Artificial Intelligence · Modern Technologies · Olive oil · Supply chain · Food technologies

1 Introduction

Worldwide olive oil market exceeded $12000 million in 2022 and is predicted to exceed $21000 million in 2030 (https://www.verifiedmarketresearch.com/product/olive-oil-market/). Spain is the largest producer of olive oil contributing to 42% of the world production. Meanwhile 80% of production comes from Mediterranean countries. (https://www.researchandmarkets.com/reports/5715833/olive-oil-market-size-global-forecast-2023?utm). Around 10 million trees exist in Palestine producing around 177 ton of olives and around 39000 ton of extracted olive oil in 2019 [1]. Olive sector in Palestine was worth $60–90 million in 2021. Around 800000 to 100000 people are involved in this sector economy during cultivation, harvesting and production, and over 15% of working women are in the sector (https://www.aljazeera.com/news/2021/10/14/infographic-palestines-olive-industry) [2].

Olive tree and olive oil are in rooted in Palestine heritage and history, many of the villages and towns names are related to this sector [3]. According to Abu Hanieh et al. in [4],

© The Author(s) 2025
H. Kohl et al. (Eds.): GCSM 2023, LNME, pp. 3–10, 2025.
https://doi.org/10.1007/978-3-031-77429-4_1

olive tree symbolizes pride cultural heritage and rooting of people in the land. In addition, it forms a main ingredient in the traditional diet and cuisine in Palestine. The products of olive tree are used as a base for many local industries including soap, firewood, pickled olives, souvenir and some pharmacological products. Olive sector involves the planting and cultivation activities, the harvesting, extracting and other industrial process. This sector traditionally is a labor extensive sector and women as main player in the harvesting and traditional pickling and soup manufacturing. Modernization of the sector will improve its quality and its economic value. The use of AI, automation, control, sensing and machine learning in the olive sector have been investigated by many researchers worldwide, some of this recent literature is reviewed below.

Ordukaya and Bekir in [5] used two methods for oil quality control; one is based on electronic nose EN and the other on machine learning algorithm. They concluded that both methods are cheaper and faster than normally used chemical methods. Muhammad Waleed et al. in [6] devised an automatic method for detection & enumeration of live trees, the algorithm is based on RGB images of SIGPAC viewer. Alkelani and Awad in [7] used machine learning and data mining methods developed and algorithm to investigate the impact of climate factors on olive oil production as well as predicting the future production and oil yield in Palestine. Abu-Khalaf in [8] employed an electronic nose EN based on chemical sensors to check the quality of olive oil. He found that such EN could be used to predict olive acidity accurately. According to Panagiotis Christias and Mariana Mocanu: in order to predict company potential profits from olive farming developed a machine learning framework, then applied it to Hellenic Island in Crete [9]. Rapa and Ciano in [10] reviewed lifecycle analysis studies for olive oil supply chain; they concluded that largest impact on environment comes from using fertilizers and pesticides. Alsayat, and Ahmadi in [11] investigated olive oil companies' performance as affected by improvements in the supply chain. They developed a model based on AI and machine learning and found that implementing IoT in this sector would improve companies' performance considerably.

Mechanization of the cultivation and harvesting processes are being introduced in Palestine however at low level and slowly. AI, IoT, modern IT have been investigated in many of the olive oil producing countries especially in the Mediterranean region, however little being investigated for this sector in Palestine. In this paper a model to be developed to improve the productivity, profitability and sustainability of olive sector supply chain in Palestine by employing AI, automation and IT tools.

2   Olive Sector Supply Chain

Olive oil is one of the most important products in Palestine. The sector is considered one of the main income sources of life for many families. Virgin olive oil is used in most of the Palestinian traditional dishes and pickled olive is used as starter dish in house and in restaurant. The annual crop fluctuates every year where the total production quantity decreases to half the quantity every other year. This raises the necessity for storing, packaging to facilitate the exportation of this product [12].

In general, most of olive trees in Palestine are planted since a long time ago and they can live for thousands of years. Some olive trees in Palestine have been planted during the existence of Roman Empire in the region and some are from the Islamic centuries. On one hand, the need arises to plant new olive trees when people try to exploit new lands. On the other hand, there will be a need to substitute the burnt and damaged trees by Israeli settlers with new ones.

The lifecycle and supply chain of olive oil is shown in Fig. 1. This figure shows that the supply chain starts with cultivation process where the existing trees are served by ploughing, irrigation, fertilizing, pruning and harvesting. The picked olive fruits are transported to the oil extraction presses (factories) where the fruits are pressed and mixed to extract the oil in pure phase. In most of the cases, the oil is packed in 16-L cans directly after this stage and sent to be used in houses and restaurants. This can fit for local use but for export requirements there is a need to make further treatment which constitutes the third stage in the supply chain. There are other industrial applications for olive oil like pharmaceutical or cosmetic industries, soap manufacturing, canned food and other implementations.

Most of the mentioned stages in the supply chain are carried out in traditional ways without or with the minimum level of technology. The lack of using modern technologies reduces the produced quantity, increases the use of unskilled human labor, reduces the dependence on machines and modern intelligence, reduces the quality of product and could increases the cost of production. This raises the necessity to investigate using modern Artificial Intelligence, big data, Internet of Things and other modern technologies in olive oil production process.

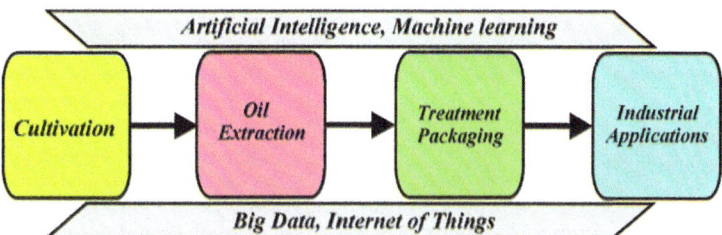

Fig. 1. General supply chain of olive oil

The detailed components of the olive supply chain are shown in Fig. 2. Where, Fig. 2 (a) depicts the main inputs and outputs of the cultivation phase. Farmers still use animals for the ploughing process between olive trees because most of the trees are planted on mountains and hilly areas and it is difficult for a machine tractor to go through the trees without damaging their branches. Fertilizing is usually done by adding animal manure to the trees every year, in few cases farmers use chemical fertilizers as well. Most of old olive trees depend on rainfall irrigation in winter only, hence without being irrigated during summer, this affects negatively the quantity of the production and reduces the percentage of oil in olive fruits. Only newly planted trees are irrigated regularly to help them dig their roots in soil and it stops after a few years. Weeding is required to get rid of bad weeds. Harvest of olive fruits is usually done manually where families go together in a yearly celebration shape to pick olives in groups. This is a beautiful tradition but lacks technicality and needs to be replaced by technological methods.

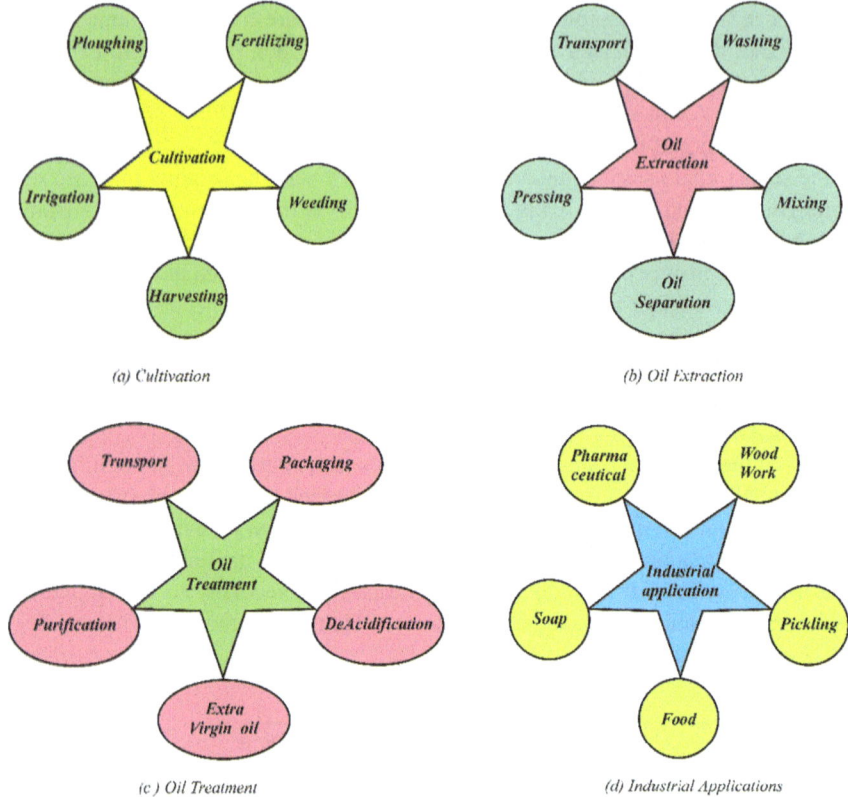

(a) Cultivation

(b) Oil Extraction

(c) Oil Treatment

(d) Industrial Applications

Fig. 2. Detailed supply chain of olive sector

Figure 2(b) shows details of the oil extraction process where harvested olive fruits are transported oil extraction press (factory) using animals, tractors and light weight trucks. In press, the fruits are washed using water, pressed and mixed, then oil is separated from waste pomace to be filled in special cans. Pressing process is done by two ways; the first is the old traditional ways using big heavy stones to crash fruits into small pieces and the paste is put under manual presses to extract oil, the other way is by using modern machines to mill the fruits and centrifugal turning to separate oil.

In order to obtain better extra virgin olive oil product, Fig. 2(c) shows the treatment processes required after oil extraction. Oil is transported from presses to modern oil treatment factories where oil is purified and deacidified to reduce its acidity and cope with the international standards. Extra virgin oil is then packed in small and big containers according to the required usage. It is worth mentioning that this stage does exist in Palestine and there is a high need for it to qualify products to international markets.

The last phase shown in Fig. 2(d) represents the industrial applications of olive oil. Olive oil can be used in pharmaceutical cosmetic products manufacturing knowing its healthy and rich content. Oil is also used widely in Palestine to produce soap. One of the main problems is that all soap factories are very elementary and use old non-smart techniques, and this shows that they need to be modernized. Oil can also be used in canned food like beans, salads, tomato paste and others. Olive fruits can be pickled using salt and vinegar where they can be served as starter dishes in kitchen and restaurants.

3   AI and Modern Technologies in Olive Sector

As mentioned early, most of the processes and techniques used in the olive oil supply chain in Palestine are traditional ones and need to be improved and developed to ame-liorate the production quality and quantity from one side and to reduce the cost and increase the added value from the other side.

Looking at the cultivation shown in Fig. 3(a), ploughing techniques can be improved by using modern self-driving tractors. These tractors must be smart and programmed for obstacle avoidance technique because they will work between unarranged trees. Fertilizing process needs to check the Nitrogen level in soil and thus, Nitrogen sensors can be installed near each tree to check its requirements of fertilizers. Irrigation techniques can be added to improve the product where this process needs to add humidity and temperature sensors. For newly planted trees, smart sprinkles or drip irrigation can be added. Weeding can be done by controlled drones and lasers to reduce efforts done in this process. Manual harvest can be replaced by smart vibrating machines and hand-held pneumatic or electric picking tools.

The AI modern techniques that can be used in oil extraction phase are shown in Fig. 3(b). Transportation of harvested fruits on hills and mountains can be done by using rails with smart carriages to place crops near to transportation trucks. Washing process should be accompanied by a recycling technique of water to reduce the waste. The amount of produced oil can be improved by using smart controlled and modern press machines to control and reduce the amount of oil wasted in pomace during separation process. Image processing tests can be used during mixing and separation as well to improve the product. Pomace can be pressed in small molds to be used as fuel in winter.

Oil treatment phase can be improved by adding AI smart techniques as presented in Fig. 3(c). If the press is close to the treatment plant, then Automatic Guided Vehicles (AGV) can be used for the transportation process. Acidity tests and image processing can be used for the deacidification and purification techniques. Humidity and temperature monitoring sensors are needed to control the product during packaging and storage phase. Freezing test is used to check the extra virgin olive oil.

Figure 3(d) shows that smart controlled machines and smart production lines can be used in pharmaceutical and food production applications. Computer Numerical Control (CNC) and Direct Numerical Control (DNC) techniques can be used in wood working and soap manufacturing factories to reduce time and improve products. Other techniques are required in pickling factories to test brine concentration for example. Table 1 shows the detailed description of AI techniques.

4   Conclusions and Recommendations

In this paper current practices in olive sector process are examined in Palestine in addition to the some of the modern IT tools investigated by researchers in large and intensive olive sector counties. A model is developed to implement such modern techniques in the olive

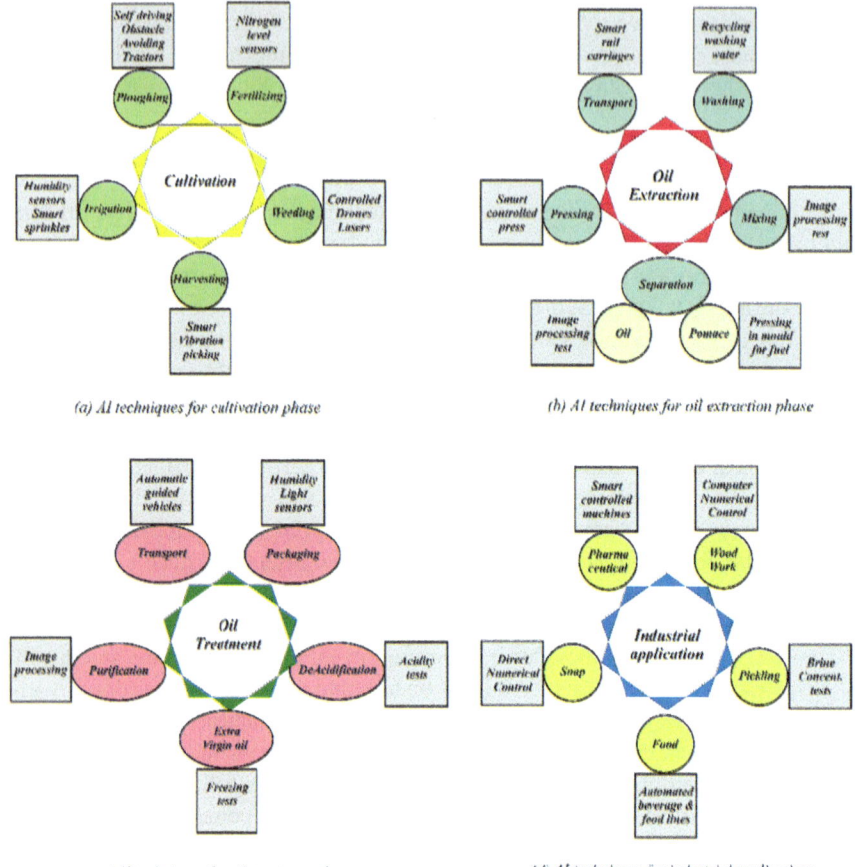

(a) AI techniques for cultivation phase

(b) AI techniques for oil extraction phase

(c) AI techniques for oil treatment phase

(d) AI techniques for industrial applications

Fig. 3. AI techniques implemented in the stages of olive sector supply chain

sector in Palestine hence increasing its profitability. Suggested recommendations for this purpose include; the use of light weight hand-held harvesting tools, implementing sensors to decide the necessity to add fertilizers, and need for irrigation. Check the oil content and quality through modern sensing instruments before harvesting and extraction process. Apply modern technologies for extra virgin oil treatment processes and olive-based products.

Table 1. Details of the AI techniques used in olive sector

Phase	Process	AI Technologies
Cultivation	Ploughing	Self-driving and obstacle-avoiding tractors
	Fertilizing	Nitrogen level sensor
	Irrigation	Temperature sensors Humidity sensors Smart sprinkles
	Weeding	Smart controlled drones Lasers
	Harvesting	Vibration machines Pneumatic picking tool
Oil Extraction	Transport	Smart rail and carriage
	Washing	Recycling
	Pressing	Smart controlled press
	Mixing	Image processing
	Separation	Image processing
Oil treatment	Transport	Automatic Guided Vehicle
	Packaging	Humidity, light sensors
	Purification	Image processing
	De-acidification	Acidity tests
	Extra virgin oil	Freezing test
Industrial applications	Pharmaceuticals	Smart controlled machines
	Soap manufacturing	Direct Numerical Control
	Wood Working	Computer Numerical Control
	Pickling	Brine concentration test
	Food	Automated beverage and food production lines

References

1. Palestinian Central Bureau of Statistics (2019) State of Palestine. Publish Date: 18-08-2021. https://www.pcbs.gov.ps/statisticsIndicatorsTables.aspx?lang=en&table_id=902. Last accessed 10 May 2023
2. Olive oil market size and forecast. Report ID: 16740. Published Date: Nov 2022. https://www.verifiedmarketresearch.com/product/olive-oil-market/. Last accessed 10 May 2023
3. Sharkawi M (2019) Introducing olive culture in Palestine. AFRAT, ICP, PCR and Tétraktys-European Union. University of Lorraine-Nancy
4. Abu Hanieh A, Karaeen M, Hasan A (2020) Model for sustainability of olive sector in Palestine. Procedia Manuf 43:269–276
5. Ordukaya E, Karlik B (2017) Quality control of olive oils using machine learning and electronic nose. J Food Qual 2017. https://doi.org/10.1155/2017/9272404

6. Waleed M, Um T-W, Aftab K, Umair K (2020) Automatic detection system of olive trees using improved K-means algorithm. Remote Sens 12(760). https://doi.org/10.3390/rs12050760

7. Alkelani M, Awad M (2021) Prediction of olive oil productivity using machine learning decision tree algorithm. Int J Emerg Technol 12(1):13–18

8. Abu-Khalaf N (2021) Identification and quantification of olive oil quality parameters using an electronic nose. Agriculture 11(674). https://doi.org/10.3390/agriculture11070674

9. Christias P, Mocanu M (2021) A machine learning framework for olive farms profit prediction water. 13(3461). https://doi.org/10.3390/w13233461

10. Rapa M, Ciano S (2022) A review on life cycle assessment of the olive oil production. Sustainability 14(654). https://doi.org/10.3390/su14020654

11. Alsayat A, Ahmadi H (2023) Workers' opinions on using the internet of things to enhance the performance of the olive oil industry: a machine learning approach. Processes 11(271). https://doi.org/10.3390/pr11010271

12. Haddad M (2023) Infographic: Palestine's olive industry. 14 Oct 2021. https://www.aljazeera.com/news/2021/10/14/infographic-palestines-olive-industry. Last accessed 10 May 2023

Urban Agrifood Circularity: Exploring Consumable and Capital Micro Circular Production Loops

Peter Ball[1]([✉]) [iD], Ehsan Badakahshan[2] [iD], Joseph Bell[7], Nicola Holden[3] [iD],
Jens Jensen[4] [iD], Ifeyinwa Kanu[5] [iD], Lydia Smith[6] [iD], and Xiaobin Zhao[7] [iD]

[1] School for Business and Society, University of York, York YO10 5ZF, UK
peter.ball@york.ac.uk
[2] Department of Management, Sheffield Hallam University, Sheffield S1 1WB, UK
[3] Department of Rural Land Use, Scotland's Rural College, Edinburgh EH9 3JG, UK
[4] Scientific Computing, Science and Technology Facilities Council, Swindon SN2 1SZ, UK
[5] IntelliDigest Ltd., Edinburgh Business School Incubator, Edinburgh EH14 4AS, UK
[6] Innovation Hub, NIAB, Lawrence Weaver Road, Cambridge CB3 0LE, UK
[7] Wasware Ltd., Future Business Centre, Cambridge CB4 2HY, UK

Abstract. The circular economy concept is typically applied at scale, especially for high value products and materials. The dominant manufacturing model is for large global factories with research and practice independent of agriculture. This paper challenges the dominant "big is beautiful" ethos and explores how agricultural and industrial production can operate at local, urban scale with wastes circulating for consumable and capital production. The research case is a UK city where food wastes could be used for food production and beverage production waste could be used to produce building materials. The research explores the industrial symbiosis engineering challenge of small-scale waste conversion and the digital challenge of identifying and measuring waste flows for conversion. In considering waste conversions through local, distributed manufacture this paper also tackles the digital challenge of how to source local, small volume material flow data for optimization. Future potential research avenues of micro manufacture as well as digital twins are discussed.

Keywords: circular economy · urban manufacturing · distributed manufacturing · industrial symbiosis · digital

1 Introduction

Production has become increasingly global [1]. Responding to the sustainability imperative and digitally enabled production advances [2] there is interest in alternative production configurations, including local production. Technological advances could make small scale production efficient and economically feasible [3] using urban space [4]. Manufacturing and agriculture have been organized and optimized separately which misses opportunities for better resource use and lower environmental impact.

© The Author(s) 2025
H. Kohl et al. (Eds.): GCSM 2023, LNME, pp. 11–19, 2025.
https://doi.org/10.1007/978-3-031-77429-4_2

Technical production is now removed from urban centers [5] and whilst urban centers are expanding [6] production is remote with residents increasingly detached from their food [7] and product origins. With the growing interest in urban food production, coupled with advances in controlled environment agriculture (CEA) opportunities exist to exploit advances in technology, urban vacancies, and localized wastes.

This paper draws on exploratory and experimental research on the feasibility of local urban circularity. Local here refers to distances of hundreds of meters rather than throughout a town or city. The research considers how agrifood waste flows can be captured using digital technologies then modeled for nutrients to grow more food or create building products to build structures, i.e., create capital. Secondary publicly accessible data or primary data can be modeled to optimize flows and experiments can test the feasibility of those flows.

This paper first presents available knowledge in urban circularity from peer reviewed literature. The knowledge gaps identified lead on to methods to consider the gaps in digital and physical flows. Virtual models and physical prototypes are then described in the models section. The results from these experimental steps are presented and discussed for the advances as well as the emergent challenges. The paper then concludes with the contributions of the research and potential for future work.

2 Literature Focus

This review builds on earlier work to explore (1) urban space for productive operations, (2) small-scale manufacturing, potentially in urban settings and (3) circularity, especially hyper local. Starting with keyword searches for these three groups, particular focus was given to whether theories exist for urban operations, whether micro manufacturing has been applied for low value, non-additive operations and how circularity has been used for low value, low volume, local operations.

Urban space is increasingly for residential dwellings and retail with manufacturing on the periphery [5]. Few authors [e.g. 8] consider exploiting it for productive and clean activities rather than further residential development. Locating manufacturing in urban areas [9] holds potential but could be challenging if higher knowledge or skills are required to fulfil requirements [10], if the technology levels are high. Authors [e.g. 11, 12] see opportunities for food production (including controlled environments such as vertical farms) and link this to circularity which is discussed later. Whilst authors see opportunities for the deployment of manufacturing and agriculture in urban areas, there is a dearth of literature to guide design and implementation. This is particularly so as a system rather than specific technology advance.

Geographically distributed production can be located close to local markets [3] with small scale production, potentially mobile, meeting individual requirements rapidly at low cost [13]. Micro-factories, especially modular, can be small, self-sufficient, and enabled by digital technologies [10]. When related to food, producers "need a well-designed network of scalable and modular manufacturing systems in their geographically distributed production sites." [14]. Authors show the potential for small-scale localized production, however, when considering the implementations, most work considers high-value processes or products, especially additive.

Exploiting urban food waste circularity has potential [11] but lacks literature on the trade-offs between wastes being used for further food production (retained in the food system) or for technical applications, e.g. building products (lost from the food system but offsetting other carbon emissions). Finally, research on short conceptual circularity loops [15] is not matched by research on the physically short circularity loops.

Modular, small-scale, urban operations come with speculation of better sustainability credentials through productive, localized, circular, customized delivery but performance evidence is scant. An opportunity exists to contribute to exploring farm-to-fork agrifood circularity at a local, urban level with manufacturing as a single system.

3 Methods

From the gaps identified, there are three areas of focus: data flows, flows of nutrients from wastes and flows of building materials from wastes. Nutrients from food business operators (FBO) wastes will be considered for supply to a controlled environment agriculture (CEA) vertical container farm for plant growth. Wastes from beer production will be tested for bio resin extraction to make bio board for construction. Here manufacturing is not solely a technical operation but part of an agricultural system which could advance sustainability by transforming the overall system not just optimizing technical manufacture.

Data capture from a combination of real data and simulated data inferred from norms expected from business operations was input into two computational models. As this was testing feasibility, the data accuracy was not a major consideration and assumptions could be tested through sensitivity analysis. One computational (simulation) model considered wastes and product logistics, the other (optimization) model considered how perishable food stocks can be timely stored and used.

For the nutrient extraction from food waste for food production nutrients, quality was more important than cost and delivery. Desk research evaluated quality (both safety and matching nutrient composition to farm requirements). The focus was whether digester output from restaurant food waste aligned with farm requirements and whether safety risks existed. This determined how much of the nutrients available locally could be used locally and what level of external input was still required.

Finally, the building materials from production waste used lab experimentation to synthesize bio board samples. Spent grain can be used for bio resin extraction, so this tested if a particular brewery waste was suitable for bio resin to create bio board.

4 Models

Three areas were investigated for an urban conceptual model using the UK city of York given its diversity of resource flows. The 200m x 200m area had restaurants, bars, hotels, student accommodation, a brewery, and a vertical farm (Fig. 1). The farm is in a shipping container constructed entertainment complex, SPARK:York with production and consumption and a density of valuable resource flows close by. Some resource flows could be accessed live, others could be inferred from public waste data or estimated from geospatial mapping.

Fig. 1. Sketch map of the Piccadilly area in the center of the city of York.

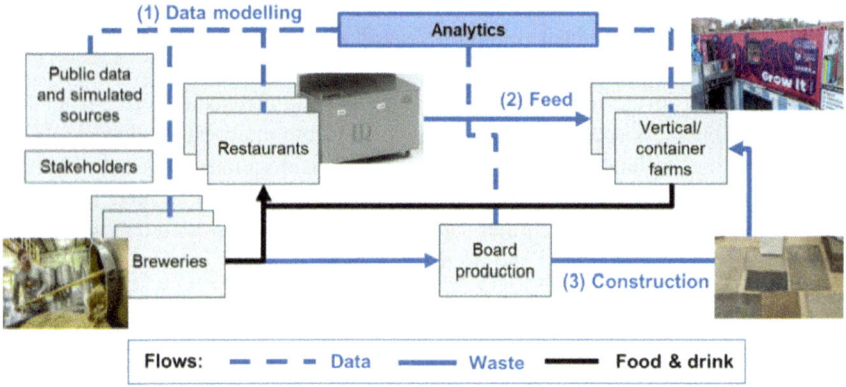

Fig. 2. The urban flows modeled based on the center of the city of York (Source: Authors).

A conceptual model (Fig. 2) captured production and consumption, flows of product and 'waste' as well as potential data flows. The brewery and container farm supply local businesses with beer and leafy greens (black solid flow lines in Fig. 2). Local restaurants could capture their waste foods in an iDigest digester [16] (domestic appliance size) to create nutrients for crop growth (the waste flow "(2) Feed" shown in blue). The brewery could supply its spent grain 'DDGS' waste to Wasware [17] to process into materials for creating structures such as a vertical container farm (the waste flow "(3) Construction" shown). Finally, data capture (and estimations) from businesses could feed analytics to improve the system (flow "(1) Data" shown in dashed blue lines). Two models were created to maximize use and value extraction: waste flow logistics between businesses and optimization of perishable food stocks.

5   Results and Evaluation

The outcomes for (1) data flows, (2) feed flows and (3) construction flows are reported here. First, data and data flows were captured or created to build two models. Two types of data were utilized: static data for position, capacity, operating hours, etc. and dynamic data for demand, production, waste, storage, etc. From this a spreadsheet implementation [18] of material use optimization was chosen for simplicity and flexibility for assessing the storage, use and quality decay for perishable wastes according to customizable profiles. The modeling demonstrated localized flows of waste input, waste utilization and loss of waste due to age. This gave an understanding of potential waste utilization levels. The second logistics model for the system level flows required the complexity handling of AnyLogistix [19]. It incorporated multistage flows from production and consumption including processing to new forms for local use. Given this was to test the concept, previous modeling with live data from the vertical farm through the operating system API was not utilized. The outcome was the concept evaluation of time-based, varying flow volumes through the production assets. Better input data will be needed to evaluate actual performance.

The second area considered nutrients available from IntelliDigest's digestors for safe feedstock for vertical farm crop production. We considered whether nutrients could be used in hydroponic crop production (growth in soil-less substrates). The analysis considered the acceptable minimum and maximum values of plant nutrient requirements for two common CEA target crops (e.g., lettuce, herbs) for a range of macro- and micro-nutrients (ammonia, potassium, sodium, etc.) alongside three active vertical farm operations for the same nutrients. Using these target and typical nutrients it is possible to match requirements with the iDigest nutrient outputs. This evaluates how food waste digester outputs (e.g., within restaurants) can be ultimately directed to farms and returned for further crop production.

The third area of investigation was the creation of bio board. These are shown in Fig. 3. Spent grain was collected from the York brewery immediately after its use in brewing. This was then processed into bio resin and for bio leather or mixed with chippings, compressed, and cured to produce bio board.

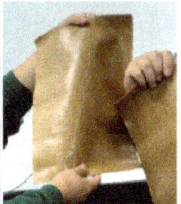

Fig. 3. Brewery spent grain used to create bio resin, bio board and bio leather (Source: Authors)

6   Discussion

The research themes were investigated for their individual and collective feasibility for urban agrifood circularity. This section considers how sustainable manufacturing research can build on this. The following section details future research opportunities.

Firstly, simulation modeling and digital twins are new for urban circularity but not for production or supply chain. Further opportunities arising are how to exploit circular supply chain modeling to evaluate and optimize system performance using public data and data inferred from public mapping. Conceptual feasibility was considered here but not aspects of economic (cost effectiveness), social (employment, engagement) or environment (carbon, best use of agrifood wastes, physical impacts).

The second area investigated was nutrient flows. This drew on historical data and desk evaluation to assess the likely feasibility of using digested food wastes as nutrient feed to minimize the municipal waste collection and reduce import of nutrients, especially by capturing the waste as early as possible before it degrades. Cost and quality need to be compared to current given fertilizer cost and demand for high quality food with increased micro and macro nutrients. Given vertical farm providers optimize nutrient supply for crop and farm operation, the locally processed digestate from iDigest would need to be optimized prior to use in vertical farms.

Lastly for construction waste flows, the research demonstrated several advances. First, spent grain from a local brewery had the necessary composition for multiple material applications. Second, materials could be made to viable quality levels in small scale manual production. Third, the apparent material properties lent themselves to two vertical farm requirements: paneling or shelving. Relating back to earlier, potential local sourcing and local use would ideally be satisfied through local production. Viable production scale is needed. This would be evaluated for its local footprint (physical and environmental). It would be valuable to know if production take place in a permanent or mobile facility.

This research was focused on the feasibility of digital and physical flows. What has not been considered are aspects of efficiency, economics, social impact, environmental evaluation. Scale and efficiency are considered synonymous, however, efficient large systems are not implicitly effective [20]. There is unfinished work here to assess whether sustainable manufacturing would be enhanced through localized, distributed manufacturing. Additionally, no environmental assessments were possible within the scope of work but are needed to understand advantages and disadvantages at system scale and uncover whether there are otherwise hidden negative effects at scale.

7   Conclusion and Future Research

This paper has considered urban agrifood circularity using a conceptual model of a small UK urban area. The research considered how agrifood waste could be retained and used for value adding production activities at small scale rather than being aggregated and transported away from the city, potentially for low value uses. The research considered three areas: capture and use of data for local flow optimization, processing of food waste into nutrients for the operation of controlled environment agriculture (vertical farming) and processing of waste into bio board for capital infrastructure.

There are contributions to both knowledge and practice. The research outcomes demonstrate feasibility in circulating resources in hyper-local urban settings to minimize transport, exploit high value properties and create opportunities through urban agricultural production, technical production and consumption. There is a 'small is beautiful' challenge to the dominant high efficiency, high volume, regional or global production with significant material, waste, employee and product travel. This learning contributes to directions for sustainable manufacturing research. Contributions from a practice perspective, are on better understanding of how wastes can be used for high value consumables (nutrients for growing food) and capital products (high value building products) that could offset higher carbon material use. This learning challenges the best circularity route for materials and whether flows should be maintained as biological nutrients (renewable) and/or substitute technical nutrient (finite) flows. Opportunities exist for outreach, training and consultancy, to optimize these flows: to educate the public and policy makers to foster understanding for the circular economy, and train and support the participants in optimizing flows. Finally, there are opportunities for discovering new uses of materials to further reduce waste.

There are four areas for potential future research. The first area is modeling. This research used historic and simulated data. There is potential for scaled models using live data. The scale would be presented by at urban area containing multiple business that are for consumption as well as production. Metrics are needed, especially sustainability, to evaluate impacts. Potential extensions are how to cope with very high volume of some data sources and hard to obtain data as well as whether models could progress through levels of sophistication from digital models to digital shadows (live data inputs) to digital twins (live data inputs and decision outputs).

The second area from a manufacturing perspective would be the collection, processing and quality of nutrient flows. Onsite robotic bio-upcycling through the iDigestors onsite enhances the recovery of bio-nutrient from food waste using enzymatic bio-catalysis. Previously work on gem lettuce showed the bio-nutrients are superior in value and balance compared to conventionally sourced nutrients leaving a gap in research for suitable pre-processing for the vertical farm environment.

The third area to investigate is how laboratory material production can be scaled. This is both a general production challenge as well as a challenge to understand if bulky building products can be produced in small scale urban production facilities practically and economically. Additional research would test materials for strength, durability, insulation, etc. hence suitability for construction.

The fourth area for research is evaluation. Here the physical and data flows and analysis were evaluated but not the impact of the new system or the relative impact compared to the current production and consumption system including using urban space and reassigned material. Evaluation of the combined carbon, employment, economic and social impacts to uncover the positive and negative effects of the changes suggested at scale is needed. Cost and impact of circular flow optimization could focus on reducing cost or CO_2e or waste, but these can be time varying and mutually conflicting goals (Reducing waste by refrigeration versus cost by doing nothing).

Acknowledgements. We thank UK's UKRI STFC Food Network+ Extension (ST/T002921/2) for funding the "Circular urban vertical farming: Data, models and optimisation of waste flows" project and Prof Katherine Denby and Dr Alana Kluczkovski from University of York and FixOurFood, Transformations to Regenerative Food Systems UKRI BBSRC BB/V004581/1 project for their contributions and vertical container farm access.

References

1. Cheng Y, Farooq S, Johansen J (2015) International manufacturing network: past, present, and future. Int J Op. Prod. Mgt 35(3):392–429
2. Schroeder A, Bigdeli AZ, Galera Zarco C, Baines T (2019) Capturing the benefits of industry 4.0: business network perspective. Prod Planning Cont 30(16):1305–1321
3. Kleer R, Piller FT (2019) Local manufacturing and structural shifts in competition: market dynamics of additive manufacturing. Int J Prod Econ 216:23–34
4. Matt DT, Orzes G, Rauch E, Dallasega P (2020) Urban production. A socially sustainable factory concept. Comp Ind Eng 139:105384
5. Qui R, Xu W, Zhang J (2015) The transformation of urban industrial land use: a quantitative method. J Urban Manage 4(1):40–52
6. Li X et al (2021) Global urban growth between 1870 and 2100. Commun Earth Environ 2:201
7. Veldhuizen LJ, Giller KE et al (2020) Missing middle: connected action on agriculture and nutrition across global, national and local levels. Glob Food Sec 24:100336
8. Celanie A, Ciaramella A, Dettwiler P (2016) Identification of vacant space; a prerequisite for industrial and societal development. In: CIB world building congress, vol 1
9. Srai J, Kumar M, Graham G et al (2016) Distributed manufacturing: scope, challenges and opportunities. Int J Prod Res 54(23):6917–6935
10. Zadra R (2021) Modular microfactories, back to the future: emerging topics for long-term resilience in manufacturing. World Manufacturing Forum, Italy
11. Lehmann S (2012) Optimizing urban material flows and waste streams in urban development principles of zero waste and sustainable consumption. Sustainability 3:155–183
12. Avgoustaki DD, Xydis G (2020) Indoor vertical farming in the urban nexus context: business growth and resource savings. Sustainability 12(5):1965
13. Matt DT, Rauch E (2013) Design of a network of scalable modular manufacturing systems to support geographically distributed production of mass customized goods. CIRP Conf Intell Comp Mfg Eng Procedia CIRP 12:438–443
14. Srai JS, Graham G, Hennelly P, Phillips W, Kapletia D, Lorentz H (2020) Distributed manufacture: a new form of localised production? Int J Ops Prod Mgt 40(6):697–727
15. Bocken NMP et al (2016) Product design and business model strategies for a circular economy. J Ind Prod Eng 33(5):308–320
16. iDigest, IntelliDigest waste food processor. https://intellidigest.com/. Accessed 2023/07/07
17. Wasware. https://www.was-ware.com/. Last accessed 2023/07/07
18. STFC. Source code model simulation. https://github.com/stfc/YorkCircular. 2023/07/07
19. AnyLogistix. Supply chain software. Accessed 2023/07/07. https://www.anylogistix.com/
20. Schumacher EF (1975) Small is beautiful. Abacus, London

Ecosystem Collaboration for Sustainability: Learnings from the Food System

Cadence Hsien[1,2(✉)] ⓘ and Steve Evans[1]

[1] Institute for Manufacturing, Department of Engineering, University of Cambridge, Cambridge CB3 0FS, UK
ljh95@cam.ac.uk

[2] Singapore Institute of Manufacturing Technology, Agency for Science, Technology and Research, 2 Fusionopolis Way, #08-04, Innovis 138634, Singapore

Abstract. Individual firm and ecosystem level efforts are required to meaningfully improve sustainability across the food system. While there has been extensive research on collaboration for sustainability in the food system, these studies focus on working along the supply chain. As the food system is complex with relations and effects spanning across supply chains and industries, another approach is for firms to look beyond their traditional supply chain to tackle sustainability at their ecosystem level. In this regard, there is a need to investigate how firms collaborate with partners to influence sustainability in their ecosystem. In this paper, we present findings of an exploratory study with interviews of 17 firms across the food system that work with partners in their ecosystem for sustainability. We found that informal collaboration among partners is common but occur differently at certain stages of innovation and development and happens even when sharing of benefits are not yet agreed upon. The findings suggest how individual firms can support ecosystem change and how they can work with partners to develop and propagate innovative sustainability solutions in the food system.

Keywords: sustainability · ecosystem · collaboration

1 Introduction

The food system is one of the most important sectors in the world and contributes significantly to global sustainability concern like carbon emissions [1] and biodiversity loss [2] and is linked to all 17 United Nation's sustainable development goals [3]. Due to the interconnectedness of the system, firms have to work with partners and stakeholder beyond their organizational boundaries to meaningfully improve sustainability across the food system. This has been done through efforts and standards to progress towards sustainable supply chain management through supplier environmental performance management and cooperation with suppliers for sustainable products, among others [4]. Studies have also been conducted to study collaboration among firms in the food supply chain [5]. However, as the food system is complex with relations and effects across supply chains and industries, firms need to look beyond their traditional supply

© The Author(s) 2025
H. Kohl et al. (Eds.): GCSM 2023, LNME, pp. 20–27, 2025.
https://doi.org/10.1007/978-3-031-77429-4_3

chain to tackle sustainability at their ecosystem level. A distinction between supply chain and ecosystem is the nature of the relationships among actors. Relationships in supply chains are frequently bilateral and hierarchically controlled, differentiating it from the multilateral interdependencies in ecosystems [6, 7]. In addition, conventional supply chains frequently has contractual relationships that formalizes the value exchange process while members in business ecosystems need to align their visions and coordinate efforts and resources to ensure that their outcomes of their investments are synergistic for the ecosystem [8]. This sparks the question of how collaboration for sustainability occurs at the ecosystem-level.

As research on food systems often focus on supply chain and few look at ecosystem-level collaboration for sustainability, this topic is underexplored and there is a lack of understanding of how firms in the food system collaborate for sustainability at the ecosystem-level and what collaborative activities they take. This study aims to investigate how firms collaborate with partners to improve sustainability in their ecosystem and propagate innovative sustainability solutions. Through an exploratory study of 17 food-related businesses, we identify characteristics of firms and collaborative activities they take for sustainability in their ecosystem.

2 Background

The concept of ecosystem in business management was introduced by Moore [9] where firms are viewed as part of a business ecosystem that spans across a variety of industries and coevolve around new innovation through competition and cooperation. More recently, Thomas and Autio [10] reviewed work on ecosystem in management and conceptualised ecosystems as "a community of hierarchically independent, yet interdependent heterogeneous participants who collectively generate an ecosystem output", and consensus among scholars is that business ecosystems are especially relevant to the pace of innovation today, where firms need to collaborate with ecosystem members to create and deliver value they cannot deliver by themselves [10, 11]. The basis of collaboration in an ecosystem is the agreement of a shared business objective and that actors can identify mutually beneficial partnerships [12]. Through collaboration, firms can benefit from resource sharing [13] and develop ideas for competitive advantage.

Sustainability researchers have made use of the ecosystem concept to emphasize how businesses need to work at the ecosystem level to create and deliver sustainability value that an individual company cannot deliver by themselves [14]. In the context of sustainable business ecosystem for the circular economy, Konietzko et al. [15] reviewed studies on circular oriented innovation and developed a set of principles for circular ecosystem innovation to achieve circularity as a common outcome. The set of principles are categorized into the broad themes of collaboration, experimentation, and platformization. Of the themes, collaboration has been identified to be the most represented, suggesting its importance. A list of collaboration principles was developed to support ecosystem development; however, the principles assume a multi-firm orchestrated effort to develop circular innovation at the ecosystem-level and does not include specific actions individual firms can take to support ecosystem-level effects. There is therefore a need to develop further understanding of what collaborative and collaboration-supporting actions individual firms can take for ecosystem-level sustainability.

3   Research Method

To investigate how businesses in the food system collaborate at the ecosystem-level for sustainability, we conducted an exploratory study based on interviews of businesses in the food system who act on sustainability and collaborate with partners at their ecosystem-level for sustainability. We conducted theoretical sampling [16] to identify potential businesses where we could observe the phenomenon of interest. The criteria for selection include businesses in the food system that collaborate with other firms for sustainability or have a sustainable value proposition that requires collaborating with partners to fulfil. To collect representative data across the food system, the sample included businesses in different roles across the supply chain and also included firms that are not typically considered as part of the food supply chain, as ecosystems include firms that may be cross industry [9].

Data was collected from a total of 17 food-related businesses with different roles across the food supply chain who are acting to make their ecosystem more sustainable. Despite the country where the businesses operate from, in many of the businesses operate regionally or globally. Data collection consisted of semi-structured interviews with strategy or sustainability leaders in the company who have experience working with stakeholders to implement business changes for sustainability. The interviews aimed to understand the attitudes the businesses have towards sustainability and how they work with businesses in their ecosystem for sustainability. Questions for the semi-structured interview are listed in Table 2. The questions were designed to be open to prompt discussions and allow exploration of the topics during the interview [16]. Supplementary secondary data was collected from company website, public presentations and documents, and media sources (Table 1).

All interview recordings were transcribed, and secondary data collected before and after the interviews for triangulation to reduce bias. Upon collecting the data, the data was analysed using content analysis to identify themes from the data.

4   Results and Discussion

Although the businesses have different sustainability goals and strategies, many of them exhibit certain firm characteristics and are involved in varied collaborative activities and ecosystem-supporting activities. The following key findings were seen by interviewees to be instrumental to ecosystem success.

4.1 Firm's Characteristics

Leader's Mindset. Firms driving sustainability have a leader's mindset. They want to be technological or thought leaders in their industries and push the boundaries of sustainability, frequently seeing this as a competitive strategy. This involves investing in research and development (R&D) for new sustainability solutions and educating their customers about it. For example Firm F conducts R&D of food by-product upcycling and educate their customers about it by testing and communicating about it in their restaurant. Firm O is building a digital product where they are several years ahead of

Table 1. Businesses interviewed in exploratory study

Firm	Type of business in the food system	Venture type	Country	Interviewee
A	Farmer & food producer	SME	Italy	Owner
B	Food producer	Start-up	France	Chief Sustainability Officer and Co-Founder
C	Food producer	SME	Ireland	Commercial Director
D	Food producer	Start-up	UK	CEO
E	Food producer	MNC	Sweden	Head of Circular Innovation
F	Foodservices	SME	Singapore	Managing Director and Co-Owner
G	Foodservices	SME	Singapore	Chief Growth Officer and Co-owner
H	Foodservices	MNC	Singapore	Lead, Corporate Responsibility
I	Packaging	MNC	UK	Sales manager
J	Packaging	Start-up	Chile	Circular Economy Partnerships
K	Packaging	Start-up	Singapore	CEO
L	Packaging	Start-up	UK	CEO
M	Packaging	Start-up	Singapore	CEO
N	Technology provider	Start-up	USA	Chief Impact Officer
O	Technology provider	Start-up	Sweden	CEO and Co-Founder
P	Food waste recycling	SME	Singapore	Managing Director
Q	Food waste recycling	SME	UK	Commercial Director

the market and have to educate their customers about how the market will evolve and how to take advantage of it. They understand that it involves taking their customers on a journey as the market and sustainability solutions regarding sustainability evolves.

Ecosystem Perspective. Firms have an ecosystem perspective of their business environment, seeing relationships with other members of their ecosystem beyond that of a transactional supplier-customer relationship. Though not always explicitly described by interviewees as having an ecosystem view, firms exhibit understanding of their interdependencies with members of their ecosystem, express interest in the wellbeing of the ecosystem, and have a long-term mindset to promote the longevity and prosperity of the ecosystem. Firm A described their interdependence with a partner, who supplies an innovative sustainable packaging, by sharing that they actively promote and educate about their supplier's innovative sustainable packaging so that more will start using it, leading to higher demand for the sustainable packaging in the ecosystem and making

Table 2. Interview questions for the semi-structure interviews

Data to be collected	Interview questions
Attitudes towards sustainability	• What does your company do and how does it relate to sustainability? • Tell me about the sustainability journey of your company
Working with ecosystem members for sustainability	• Which are the groups of stakeholders you engage with regarding sustainability? • How do you work with them? • Are there any partners that are playing a big part to allowing you achieve your sustainability goals and how do you engage with them? • How do you see your role in the ecosystem?

the sustainable packaging more cost effective for themselves. In describing their approach to improving the sustainability of their value chain, the interviewee from Firm E shared that they want to "[make] sustainability not only a nice to have or an add-on, but actually an integral part of our business that also economically benefits [us] and the other stakeholders because from our perspective this is the only way to make it actually long term viable." The firms' characteristics of having an ecosystem perspective and understanding of how they are interdependent result in them wanting to take actions to support the wellbeing of their ecosystem.

4.2 Collaborative Activities

Informal Co-investment in Innovation. Collaborative activities among ecosystem members in the form of informal co-investment is common and occur differently across stages of innovation and deployment. At an early stage of innovation, we found that R&D co-investment is loose, where there is no formal contract to define the outcome of the investment and how the benefits will be shared, but an understanding of interdependences and commitment for R&D momentum is crucial. This is observed in the collaboration between Firms F, P, and several other partners to upcycle food by-product. In this case, Firm F sought out technology partners who shared their vision of upcycling food by-product, and the collaboration was agreed upon without an R&D or commercial agreement. Both firms described their investment to be based on goodwill and emphasised exploring the R&D opportunity instead of getting into a contract prematurely, which may cause the R&D to be inflexible. This works because the firms understand that they have unique resource configuration [5] and are interdependent (e.g. technology, knowledge, raw material, market access), and have to continue collaboration for subsequent R&D and marketing stages. Despite the informal collaboration, the interviewee from Firm F described a motivation factor that seeing each other's commitment to the collaboration supports each other and provides a sense of "journeying together". This self-reinforcing activity fuels the collaboration despite an absence of a formal agreement.

They expressed a common understanding that a commercial agreement will be required in later stages when they are able to capture value from the outcome of the R&D.

At innovation pre-deployment stage, firms co-invested due to the expectation of mutual benefit during innovation deployment. In a pre-deployment collaboration between Firms A and L, the firms collaborated to test a sustainable packaging and make it more usable. This happened despite no formal agreement or monetary exchange for the process, but with both parties understanding the mutual value in this collaboration.

Co-marketing. Co-marketing as a collaborative activity is commonly seen among firms at the deployment and scaling stage of a sustainable innovation and is observed in majority of the firms in the study. Among start-ups and SMEs in our data, it is common to co-market at tradeshows and champion each other to gain acceptance by the wider business environment and customers about the sustainable innovation. This tells a more holistic story of their sustainability efforts where they work with sustainable firms in the ecosystem to have a greater impact on sustainability and brings the collaborative advantage of taping on each other's network.

4.3 Ecosystem-Supporting Activities

With a common goal to support the sustainability transition of the food system, firms collaborate informally by participating in ecosystem-supporting activities.

Invest in the Wellbeing of Their Ecosystem. Firms frequently invest in sustainability education that support their business but also increases the level of sustainability awareness in their ecosystem. Firms A, B, D, I, and L include education about sustainable alternate packaging in their marketing which have contributed to the acceptance of such alternate packaging as being able to represent good quality product, paving the way for wider consumer acceptance of sustainable alternate packaging. This may be especially necessary for firms pushing the boundaries of sustainability and introducing ideas and innovation not commonly understood in the market. By doing so, they also pave the way for other businesses with a similar sustainable value proposition.

Another form of investment into the ecosystem is that individuals in the firms contribute to an informal support network to further sustainability in the broader ecosystem by sharing knowledge and information. Interviewees from Firms B, C, and H share about the informal support networks that they contribute to where they share about their sustainability experiences, ideas, and suppliers. Firm B described sharing about their B Corp journey and provide advice for peers who aim to apply for a B Corp certification, supporting the sustainability journey of firms in the ecosystem. In these examples of informal support network, which may be hosted informally on chat application like whatsapp or telegram, other members in the network are like-minded individuals with an interest in sustainability and may be in similar or adjacent industries. Such informal networks develop as like-minded individuals seek for support among peers to learn about and inspire each other about sustainability. Firms with individuals in such networks not only contribute sustainability knowledge to the ecosystem but also gain a support network and a sustainability knowledge resource.

Take Actions That Have Cascading Effects. As leaders in sustainability, firms believe that they are able to influence their ecosystem or the broader business community. One way they do this by aiming to inspire and set an example for their peers. From Firm C, "what we're really trying to do is bring a focus so that we can be a leading sustainable light in Ireland…I do believe that with brands such as ourselves, it will help inspire this more so in Ireland." This is can be seen as an influential activity as interviewees described seeking out and taking inspiration from the broader business environment, for example Firm B taking inspiration from Patagonia and Hiut Denim who take sustainability actions in another industry.

In addition, larger firms like Firms G and H expressed that they attempt to influence their suppliers by making procurement requests, creating demand for sustainable products while influencing their suppliers about the possibility of sustainable business practices and opportunities. This differs from having a sustainable supplier procurement policy as the market for sustainable products may not sufficiently mature to be specified in a sustainable supplier procurement policy. Firm H shared that they request to buy the ugly food from their SME suppliers to prevent food waste and that "we are trying to be the first mover for that for a lot of our SME vendors…we show our willingness to work with them to reduce their waste to help them become more sustainable." This helps their suppliers realise that there is demand for ugly food which they can offer to other customers, creating a cascading effect of increasing supply for sustainable products in the market. This has been observed mainly in non-competitive circumstances, though cascaded effects may benefit competitors.

5   Conclusion

The findings of this exploratory study provide insights on how firms in the food system approach collaboration at an ecosystem level for sustainability. We found that it is important that firms have a leader's mindset and an ecosystem perspective, understanding how they are interdependent with other ecosystem members and take action to work with partners to develop and propagate sustainability solutions in the food system. The results suggest that informal collaborative activities between closely linked ecosystem partners and loosely linked firms at a broader ecosystem level are common and these activities are self-reinforcing by creating support for participating firms. A limitation is the limited number of firms due to the exploratory nature of the study and future research can investigate these insights with a larger number of businesses across the food system.

References

1. Crippa M, Solazzo E, Guizzardi D, Monforti-Ferrario F, Tubiello FN, Leip A (2021) Food systems are responsible for a third of global anthropogenic GHG emissions. Nat Food 2(3):3. https://doi.org/10.1038/s43016-021-00225-9
2. Benton TG, Bieg C, Harwatt H, Pudasaini R, Wellesley L. Food system impacts on biodiversity loss
3. United Nations. Food Systems Summit × SDGs. United Nations. https://www.un.org/en/food-systems-summit/sdgs. Accessed 13 June 2023

4. Seuring S, Müller M (2008) From a literature review to a conceptual framework for sustainable supply chain management. J Clean Prod 16(15):1699–1710. https://doi.org/10.1016/j.jclepro.2008.04.020
5. Dania WAP, Xing K, Amer Y (2018) Collaboration behavioural factors for sustainable agri-food supply chains: a systematic review. J Clean Prod 186:851–864. https://doi.org/10.1016/j.jclepro.2018.03.148
6. Adner R (2017) Ecosystem as structure: an actionable construct for strategy. J Manag 43(1):39–58. https://doi.org/10.1177/0149206316678451
7. Jacobides MG, Cennamo C, Gawer A (2018) Towards a theory of ecosystems. Strat Mgmt J 39(8):2255–2276. https://doi.org/10.1002/smj.2904
8. Moore JF (2006) Business ecosystems and the view from the firm. Antitrust Bull 51(1):31–75. https://doi.org/10.1177/0003603X0605100103
9. Moore JF (1993) Predators and prey: a new ecology of competition. Harvard Bus Rev
10. Thomas LDW, Autio E (2020) Innovation ecosystems in management: an organizing typology. In: Oxford research encyclopedia of business and management. Oxford University Press. https://doi.org/10.1093/acrefore/9780190224851.013.203
11. Adner R (2006) Match your innovation strategy to your innovation ecosystem. Harvard Bus Rev: 12
12. Radziwon A, Bogers M, Bilberg A (2017) Creating and capturing value in a regional innovation ecosystem: a study of how manufacturing SMEs develop collaborative solutions. IJTM 75(1/2/3/4):73. https://doi.org/10.1504/IJTM.2017.085694
13. Ahuja G (2000) Collaboration networks, structural holes, and innovation: a longitudinal study. Adm Sci Q 45(3):425–455. https://doi.org/10.2307/2667105
14. Hsieh Y-C, Lin K-Y, Lu C, Rong K (2017) Governing a sustainable business ecosystem in Taiwan's circular economy: the story of spring pool glass. Sustainability 9(6):1068. https://doi.org/10.3390/su9061068
15. Konietzko J, Bocken N, Hultink EJ (2020) Circular ecosystem innovation: an initial set of principles. J Clean Prod 253:119942. https://doi.org/10.1016/j.jclepro.2019.119942
16. Eisenhardt KM (1989) Building theories from case study research. Acad Manag Rev 14(4):532–550

Business Model Design in Context of Circular Economy

Benjamin Gellert[✉], Henry Nicolai Buxmann, and Ronald Orth

Fraunhofer Institute for Production Systems and Design Technology, Pascalstr. 8-9, 10587
Berlin, Germany
benjamin.gellert@ipk.fraunhofer.de

Abstract. In dynamic environments, businesses not only look for sustainable technology solutions, but also for new business model opportunities that management can apply as they move to a more sustainable future. Being embedded in the process of analyzing the status quo and identifying potential improvements of sustainability management, this article focuses on the development of a catalog with circular economy-based business model patterns to provide an overview of universally applicable implementation measures for business model development. For this purpose, four steps were conducted: An extensive literature research to identify a comprehensive collection of sustainable business model patterns with specific emphasis on circular economy, reduction of identified patterns to exclude redundancies and ambiguities, derivation of the influence of these patterns on the elements of the Business Model Canvas—taking into account their impact on environmental sustainability, and a final enrichment with further details for practical application. This includes a definition, benefits, barriers, implementation steps as well as filterable categories for the specific selection of business model patterns. The result is a catalog which provides a valuable resource for businesses looking to adopt more sustainable business practices.

Keywords: circular economy · sustainability · biological transformation · bioinspiration · catalog · business models · business model patterns · business model canvas

1 Introduction

Societies and enterprises around the world are currently facing enormous challenges. These include climate change in particular, which can be seen as a major driver of a necessary structural change towards a greenhouse gas-neutral and resource-efficient economy [1, 2]. Enterprises have to adapt to changing climatic conditions and weather patterns on the one hand, and shoulder the impact on global supply chains with increasing resource scarcity on the other hand [3].

The problems of scarcity, the climate crisis, but also environmental pollution have their origin among others in the currently prevailing system of linear economy [4]. There, mainly non-renewable, fossil energies and primary raw materials are used in production

© The Author(s) 2025
H. Kohl et al. (Eds.): GCSM 2023, LNME, pp. 28–36, 2025.
https://doi.org/10.1007/978-3-031-77429-4_4

processes with high green-house gas emissions, and are then subsequently used and disposed as non-recyclable waste [5]. Despite these conditions and the fact that with a circular economy (CE) the European Union could reduce CO_2 emissions by about 83% by 2050, only 7.2% and thus only a small part of the global economy is circular according to the Circularity Gap Report 2023 [6, 7]. This paper focuses on providing practical ways to strengthen the CE at enterprise level.

2 Concepts

2.1 Circularity and Biological Transformation

The transformation process towards CE is based on value cycles (reuse, refurbishment, recycling) and networks (sharing, leasing, repairing), largely avoiding the use of fossil raw materials and radically reducing the volume of waste [8]. In order to reach a sustainable state of circularity, research has increasingly focused on the concept of circular bio-economy. Bio-economy focuses on the development and implementation of innovative processes and products as well as the development of new business models and value creation networks, which in total concentrate on a circular use of raw materials in production processes under economic, environmental and social aspects. Hence, the bio-economy is a description of the economic framework conditions of a sustainable industry [9].

A sustainable industry, however, also needs technological innovations [10]. This connection is made in the Biological Transformation (BT), which is understood as the increasing use of materials, structures and principles of living nature in technology with the aim of sustainable value creation [11]. In the context of value creation, this systematic application of knowledge about biological processes leads to an increasing integration of production, information and biotechnology with the potential to profoundly change future products, manufacturing processes and organizations. In this context, BT proceeds in three complementary core principles [12]:

1. *Bioinspiration*: The transfer of biological phenomena to the design of products and manufacturing technologies (e.g. lightweight construction, swarm intelligence, closed-loop principles).
2. *Biointegration*: The integration of biological materials, structures, processes into production systems (e.g. substitution of chemical processes by biological ones).
3. *Biointeraction*: Generation of new, biointelligent value creation and production technologies through the interaction of technical, informational and biological systems (e.g. human-technology interfaces, biocomputing).

The research project BioFusion 4.0, led by Fraunhofer IPK in Berlin, deals with the goal of understanding the interdependencies between the aforementioned principles of BT and their interactions with production, services and work. In cooperation with various technology and application partners, solutions are being developed for bioinspired, biointegrated and biointelligent value creation in the application areas of mechanical and systems engineering (with a focus on automotive and electrical industries) and the waste disposal industry.

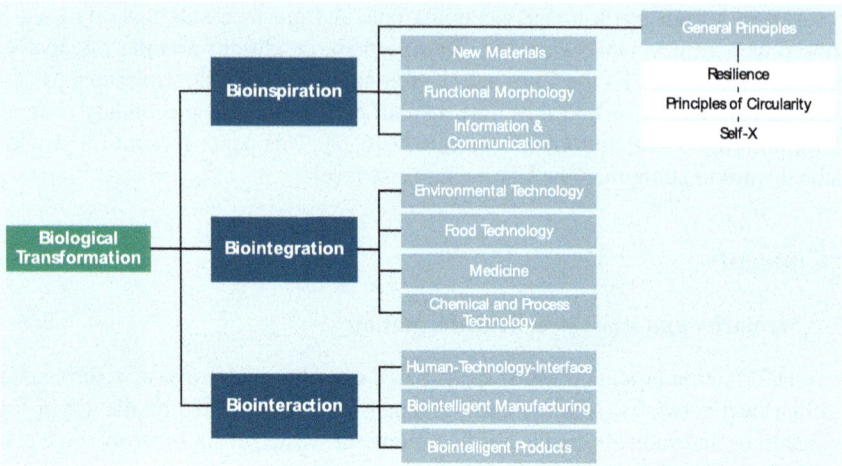

Fig. 1. The taxonomy of Biological Transformation in manufacturing (according to [13]).

Central aspect of the initial research question was the specification of a taxonomy for BT based on the three core principles (Fig. 1).

The different principles of bioinspiration, biointegration and biointeraction were each assigned three to five subcategories according to the taxonomy development methodology of Nickerson [14]. Consequently, specific solutions or principles of BT were assigned to each of these subcategories. The degree of transfer of biological factors to technology increases, starting with bioinspiration and ending with biointeraction. In the BioFusion 4.0 project, innovative technology solutions are being developed across all core principles and are tested in a demonstrator. In this paper, the focus is on business models of BT, which are based on existing processes in nature and therefore mapped in the general principles of bioinspiration (resilience, principles of circularity, self-x).

2.2 Business Model Generation

Essentially, a business model describes how an enterprise functions and how it designs its business [15]. The focus is on the creation, retention and distribution of a value proposition by the enterprise and which aspects on the business side as well as on the market side are required for it. This includes the key resources, activities, partners and costs on one side and the customer relations, customer segments, revenues as well as distribution channels on the other side [16].

The definition of a business model is usually carried out for an existing enterprise or an enterprise to be developed and is therefore very enterprise-specific. Transferability to other entities is often not possible due to individual characteristics every enterprise possesses.

Business model patterns, however, are general ways in which business models' function. As global design templates, these patterns are independent of industries and organizational sizes and are consequently defined in a universal manner. By skillful design and adaptation, suitable business model patterns can be applied in any organization [17].

In order to make enterprises resilient to the increasing global challenges at strategic business model level and at the same time mitigate the effects of the resource-consuming linear economy, specific opportunities for action are required as to how the transformation process towards CE can succeed. Bioinspired business model patterns offer generally valid solutions in this respect.

So far, however, there is a lack of a comprehensive overview of business model patterns that show bioinspired business model development options on the way to a circular and resilient economy. Therefore, a business model pattern catalog was developed and embedded as an instrument in a practical process model for business model development in the sense of CE. In the following, the process model will be explained, and the catalog will be derived and presented in extracts.

3 Overview on Circular Business Model Development Process

The process model follows a three-stage structure: analysis, design and implementation (see Fig. 2).

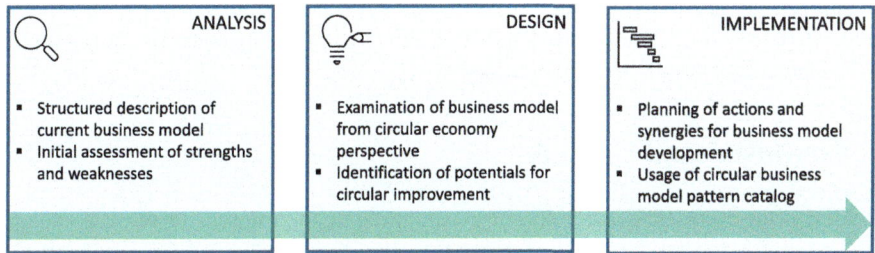

Fig. 2. Process Model.

In the analysis, the current business model (status quo) of the enterprise is first assessed using the Business Model Canvas (BMC), following Osterwalder and Pigneur [16]. At the same time, an initial examination of strengths and weaknesses can provide an early view of optimization opportunities. In the second stage, the business model is examined from a CE perspective in order to detect potential for improvement and possible applications of circular principles. This step is decisive for the design of the BT at business model level. The final stage is implementation, in which measures for a circular business model development are planned and their impact on the enterprise is evaluated. As an overview of these measures, the circular business model pattern catalog (CBMPC) was developed by Fraunhofer IPK.

4 Results

4.1 Circular Business Model Pattern Catalog

The foundation of the CBMPC was laid with an initial literature research of business model patterns. The result was a longlist of relevant patterns and technical solutions across the entire taxonomy of BT. In the course of focusing on business model patterns of CE, all listed items were then analyzed in expert groups with regard to their character (technology vs. pattern), redundancies and classification in bioinspiration. The result was a refined list of 35 biologically transformed circular business model patterns.

To obtain categorization possibilities on the one hand and further information on the applicability in enterprises of the patterns on the other hand, the catalog was enriched with the following aspects (see Table 1). The purpose is to offer filter functions that allow a specific selection of business model patterns, which are suitable for the individual conditions in enterprises based on the previously determined CE potentials.

The following chapter uses an example business model pattern to show how this framework is implemented.

4.2   Selected Example from Circular Business Model Catalog

Table 2 depicts an excerpt from the CBMPC providing an in-depth view of the circular business model pattern Take-Back-Management. The information contained follow the frame shown in Table 1. Based on a joint analysis from Fraunhofer experts the impact that the introduction of the business model pattern has on the different BMC elements of an enterprise is illustrated in Table 2. Herby the impacts were analyzed from a circular perspective. The abbreviations [I+] and [I−] indicate if the impact of the pattern on the BMC element would be positive or negative.

5   Conclusion and Outlook

To address the crisis of resource shortages and mitigate their impact on climate change, it is imperative for enterprises to embrace the principles of CE. The presented circular business model pattern catalog (exemplary illustrated through the pattern 'Take-Back-Management') offers practical solutions at a general business level, providing actionable strategies for a circular future. Within the BioFusion 4.0 project, this catalog serves as a tool integrated into the process model for the Biological Transformation of enterprises, where it is implemented and continuously refined. The primary objective to provide an overview of how circular business model patterns can impact entities is given. However, as future development progresses, enriching the catalog with more detailed pattern profiles and causal loops will enhance the visibility of interactions within business operations. This helps to map other participants along the value chain and to include their contribution in the context of circular business model development. Furthermore, to measure the impact from a business perspective, the inclusion of quantitative sustainability Key Performance Indicators (KPIs) alongside current qualitative impact relations can augment the value of the catalog.

Table 1. CBMPC-Framework

Definition	Brief description of respective circular business model pattern
Influence on Business Model Canvas elements	Positive/negative influence (regarding environmental sustainability) of business model patterns on BMC elements [16], modified: key resources, key activities, key partners, value proposition, customer segments, customer relations, competition, cost structure, revenue streams
Benefits for the enterprise	Benefit categories such as resource reduction, differentiation through sustainable product offering, extended product lifespan, quality improvement, higher margins, etc.
Barriers during implementation	Barrier categories such as cost increase, dependency on key partners, cost and effort increase, customer resistance, etc.
Implementation steps	General steps for implementation of business model pattern
R-strategy	Ten R-strategies according to Kirchherr et al. [18]: Refuse, Rethink, Reduce, Reuse, Repair, Refurbish, Remanufacture, Repurpose, Recycle, Recover
Value creation cycle	Value creation stages according to Lacy et al. [4]: Design, Procurement, Production, Logistics, Marketing & Sales, Use Phase, End of Life, Recycling (Reverse Logistics)
CBM typology	Circular Business Model Framework according to Lacy et al. [4] and Försterling et al. [19]: Circular Inputs, Resource Recovery, Life-Cycle Extension, Product Service Systems, Collaboration Platforms
Sustainable business model typology	Types of sustainable business models according to Ahrend et al. [20]: Eco-Effectiveness, Eco-Efficiency, Sharing Economy

Table 2. Business Model Pattern: Take-Back-Management (TBM).

Business Model Pattern	Take-Back-Management
Definition	Take Back Management describes the return of old, discarded products by the manufacturer/seller for the reuse/resale of installed materials in terms of enterprise's own material cycles. The mandatory return of the system at the end of the leasing period ensures the proper recycling, re-usage or disposal of the products/materials
Value Proposition	[I+] Takeover of waste disposal for customers [I+] Sustainable product range using reused materials
Customer Segments	[I+] Outreach to environmentally conscious target groups
Customer Relations	[I+] Customer loyalty through mandatory return → potential new purchase
Revenue Streams	[I+] Revenue from the sale of returned materials
Competition	No impact
Key Resources	[I+] Knowledge about use of own products/materials
Key Activities	[I+] Analysis of the usage behavior possible [I−] Additional cost for setting up the infrastructure for traceable take-back
Key Partners	[I+] Independence from raw material suppliers [I+] Establishment of new, sustainable partnerships for drop-off facilities
Cost Structure	[I+] Reduction of material cost [I−] Cost for take-back [I−] Cost for disposal/recycling
Benefits for the enterprise	Cost reduction, establishment/expansion of key partnerships, differentiation through sustainable product offering, information gain, customer group expansion, strengthening of customer loyalty, revenue increase
Barriers during implementation	Cost increase, effort increase
Implementation steps	1. Definition of TBM-responsibilities, 2. Set-up of external/internal take-back infrastructure (delivery, disassembly, material reprocessing, etc.), 3. Return analysis (disposal, recycling, sale), 4. Informing customers
R-Strategies	Remanufacture, Repurpose, Recycle
Value Creation Cycle	End of Life, Recycling (Reverse Logistics)
Circular Business Models	Resource Recovery
Type	Eco-Efficiency

Acknowledgement. This paper was written as part of the research project "Biological Transformation 4.0: Further development of Industrie 4.0 by integrating biological principles (BioFusion

4.0)", funded by the German Federal Ministry of Education and Research (BMBF) and supervised by the Project Management Agency Karlsruhe (PTKA). The responsibility for the contents of this publication lies with the authors.

References

1. IPCC (2023) Summary for policymakers In: Climate change 2023: synthesis report. A report of the inter-governmental panel on climate change. In: Core Writing Team, Lee H, Romero J (eds) Contribution of working groups I, II and III to the sixth assessment report of the intergovernmental panel on climate change
2. Dowbiggin A (2021) Climate risk and business: new challenges for organizations. Springer International Publishing AG, Cham
3. Bardt H, Chrischilles E, Mahammadzadeh M (2012) Klimawandel und Unternehmen. Wirtschaftsdienst 92:29–36. https://doi.org/10.1007/s10273-012-1347-6
4. Lacy P, Long J, Spindler W (2020) The circular economy handbook: realizing the circular advantage, 1st edn. Palgrave Macmillan UK; Imprint Palgrave Macmillan, London
5. Ellen MacArthur Foundation (2019) Completing the picture: how the circular economy tackles climate change. https://ellenmacarthurfoundation.org/completing-the-picture. Accessed 22 June 2023
6. Circle Economy (2023) The circularity gap report 2023
7. Ellen MacArthur Foundation (2015) Growth within: a circular economy vision for a competitive Europe. https://ellenmacarthurfoundation.org/growth-within-a-circular-economy-vision-for-a-competitive-europe. Accessed 22 June 2023
8. Ellen MacArthur Foundation (2015) Towards a circular economy: business rationale for an accelerated transition. https://ellenmacarthurfoundation.org/towards-a-circular-economy-business-rationale-for-an-accelerated-transition. Accessed 22 June 2023
9. Buller J, Daschner R, Grimm L, Hofer M, Hüsing B, Krayer J et al (2022) Zirkuläre Bioökonomie für Deutschland: Eine Roadmap der Fraunhofer-Gesellschaft zur Umsetzung der Bioökonomie in Deutschland. Berlin
10. Byrne G, Dimitrov D, Monostori L, Teti R, van Houten F, Wertheim R (2018) Biological-isation: biological trans-formation in manufacturing. CIRP J Manuf Sci Technol 21:1–32. https://doi.org/10.1016/j.cirpj.2018.03.003
11. Neugebauer R (2019) Biologische transformation. Springer, Berlin, Heidelberg
12. Bauernhansl T, Brecher C, Drossel W-G, Gumbsch P, ten Hompel M, Wolperdinger M (eds) (2019) Biointelligenz: Eine neue Perspektive für nachhaltige industrielle Wertschöpfung: Ergebnisse der Voruntersuchung zur biologischen Transformation der industriellen Wertschöpfung (BIOTRAIN). Fraunhofer Verlag, Stuttgart
13. Berkhan M, Kremer G, Riedelsheimer T, Lindow K, Stark R (2023) Taxonomy for biological transformation principles in the manufacturing industry. In: Kohl H, Seliger G, Dietrich F (eds) Manufacturing driving circular economy. GCSM 2022. Lecture notes in mechanical engineering. Springer, Cham. https://doi.org/10.1007/978-3-031-28839-5_109
14. Nickerson RC, Varshney U, Muntermann J (2013) A method for taxonomy development and its application in information systems. Eur J Inf Syst 22:336–359. https://doi.org/10.1057/ejis.2012.26
15. Osterwalder A, Pigneur Y, Tucci CL (2005) Clarifying business models: origins, present, and future of the concept. CAIS. https://doi.org/10.17705/1CAIS.01601
16. Osterwalder A, Pigneur Y (2013) Business model generation: a handbook for visionaries, game changers, and challengers. Wiley, New York

17. Gassmann O, Frankenberger K, Choudury M (2017) Geschäftsmodelle entwickeln: 55 innovative Konzepte mit dem St. Galler Business Model Navigator, 2nd edn. Hanser, München
18. Kirchherr J, Reike D, Hekkert M (2017) Conceptualizing the circular economy: an analysis of 114 definitions. Resour Conserv Recycl 127:221–232. https://doi.org/10.1016/j.resconrec.2017.09.005
19. Försterling G, Orth R, Gellert B (2023) Transition to a circular economy in Europe through new business models: barriers, drivers, and policy making. Sustainability 15:8212. https://doi.org/10.3390/su15108212
20. Ahrend KM (2022) Geschäftsmodell Nachhaltigkeit: ökologische und soziale Innovationen als unternehmerische Chance. Springer, Berlin, Heidelberg. https://doi.org/10.1007/978-3-662-65751-5

From Talk to Action: A Diagnosis of Drivers of Sustainable Logistics Practices in LATAM, a Collaborative Study Between Industry and Academia

Teresa Brandi[1](✉), Heidi Romero[2], Agatha Clarice da Silva-Ovando[3], M. Ileana Ruiz-Cantisani[4], Jocabed Becerra Soliz[3], and Leonardo Fuentes Pereira[3]

[1] Instituto Tecnológico de Buenos Aires, Iguazú 341, C1437 Buenos Aires, Argentina
tbrandi@itba.edu.ar
[2] Instituto Tecnológico de Santo Domingo, Av. Los Próceres 49, Santo Domingo, República Dominicana
[3] Centro de Operaciones Logísticas, Universidad Privada Boliviana, Av. Capitán Victor Ustáriz km. 6.5, La Paz, Bolivia
[4] Tecnologico de Monterrey, Av. Eugenio Garza Sada 2501, 64700 Monterey, Mexico

Abstract. Supply chains are estimated to produce more than 50% of the world's CO_2 (World Economic Forum, 2020). As climate change becomes more evident, raising awareness about sustainable supply chains is becoming a priority. This study focuses on the survey results on sustainable supply chains conducted in 2022 through a collaboration between universities in Latin America. More than a dozen universities collected a sample of over 400 surveys across ten countries. We applied the Partial Least Squares (PLS) regression model to respond to how the combination of a set of drivers and the characteristics of a company influence the level of commitment to sustainability, measured in terms of their existing measures, practices, and policies. As a result, we identified the level of understanding of sustainability for companies of different sizes and sectors and the drivers of businesses to design, implement, and control the outcomes of sustainable practices in their operations.

Keywords: Sustainability · Sustainable logistics · Sustainability drivers · Latin America

1 Introduction and Literature Review

The current situation of continual ecosystem degradation has drawn attention to environmental preservation from the scientific community, industry, and society as a whole. The parties have developed many techniques to address this issue, ranging from large-scale programs to smaller, more particular solutions. Environmental management, waste minimization, reverse logistics, green purchasing, sustainable supply chains, and sustainable operations are just a few examples [1]. There are numerous reasons why businesses should use sustainable practices. Despite the multiple benefits associated with these

© The Author(s) 2025
H. Kohl et al. (Eds.): GCSM 2023, LNME, pp. 37–45, 2025.
https://doi.org/10.1007/978-3-031-77429-4_5

applications, public perception remains one of the most prevalent motivators for pursuing sustainability [2]. Industry leaders need more time to be ready to invest in holistic solutions to improve and streamline their operations' environmental impact. As a result, despite numerous studies demonstrating the positive effects that a well-planned sustainable operation may give to an organization, numerous initiatives continue to be explored on a surface level [2–4].

Many studies have defined companies' drivers to implement sustainable practices in their operations. Authors agree that there are both external and internal factors that trigger the implementation of green initiatives. Among the elements found, the most important are operation cost, quality, and efficiency, customer or stakeholder satisfaction [4–6]. However, the authors also recognize the need to comply with local norms and governmental regulations as a big trigger. Williamson et al. [4] classify companies through the type of environmental activities implemented (active or passive actions) and their motivation (regulation or business performance). It was found that even though many companies could see the improvement in business performance as motivation, most wouldn't start these activities voluntarily [4, 6, 7]. Usually, internal and external stakeholders play a significant role in generating the need for these improvements, such as employees, suppliers, customers, and the community, among others [6, 7]. Further, Damert et al. [6] find that even the external stakeholders' pressure to implement sustainable practices may come to identify potential risks or opportunities. However, despite their motivation, the firm size seems relevant for a company to decide on these implementations [8]. Additionally, different works discuss the role of ethical, cultural, and relational concerns in implementing sustainable practices in operations and how corporate leadership, market strategies, and the work environment influence or discourage such practices [7, 9, 10].

Laari et al. [10] approach further applies green supply chain initiatives among companies, suppliers, and customers. Authors generated profiles of companies according to their size, market strategy, and tier. The study found four main clusters regarding companies and their suppliers: Low collaboration and low monitoring; Average collaboration and average monitoring; Average collaboration and high monitoring; and finally, High Collaboration and high monitoring. Most manufacturers and trading companies (around 38%) fell into average collaboration and monitoring, showing low commitment to improvement in those initiatives. Companies implementing market differentiation as their primary strategy are likely to adopt green practices; however, the effects sought by those are quite small.

There are still many challenges for companies of all sizes to implement green practices into their operations. Not only the noticeable changes in their work environment and process may be perceived as an obstacle, but there are still many technological, financial, and workforce-related issues to be adjusted. Further, companies must rely on collaboration and the generation of public policies that allow and favor these implementations [11, 12]. Usually, to ensure a more sustainable and robust performance, companies opt to apply standard norms such as ISO 14000 or other local laws, overseeing the detail of operations and just seeking an overall approach. It may be confusing "where to start" due to all the potential initiatives in this field: green manufacturing, lean operations, green lean, green supply chain management, reverse logistics, and circular economy, among other trends [8].

As a result, we identified a need for companies to develop long-term and engaging efforts to embrace incorporating sustainable operations into their processes truly. To better understand what factors would lead to such results, this article works on three main hypotheses, as presented.

H1: The drivers for green operations in companies influence the type of sustainable practices applied.

H2: The type of sustainable practice applied impacts the adoption of policies and evaluation criteria to ensure robustness in implementation.

H3: The influence of drivers for green operations in companies in sustainable practices applied differ among companies with different profiles in terms of size and sector.

The remainder of this article is organized as follows. Section 2 describes the methodology used in this study, ranging from the survey to the PLS analysis. Section 3 presents the study's findings. Following that, in Sect. 4, we discuss our main findings, and emphasize our work's primary conclusions.

2 Methodology

The study applies a two-phase methodology. Initially, the data collection was carried out through a comprehensive survey at the regional level, which has been analyzed up to now, applying only descriptive statistical tools. However, through the experience and knowledge generated by the survey, authors have identified potential connections between the drivers and the practices implemented by companies in the region [13]. This idea has been reinforced after a literature review on other similar experiences, which also found such relationships according to the type of driver, the source of motivation (internal or external), and the company profile. In the second stage, seeking to explain the previous idea, the authors used Partial Least Squares Structural Equation Modeling (PLS-SEM) to identify cause-effect relationships in path models with latent variables [14]. The framework used for this model is presented in Fig. 1.

2.1 Survey Design and Data Collection

Under the supervision of the Instituto Tecnologico de Buenos Aires, a "Sustainable logistics" survey was issued in 2014 to diagnose the degrees of application of sustainable practices in the logistics of local enterprises and promote awareness about its environmental impact. By 2022, 13 institutions from 10 different countries collaborated to depict a landscape of sustainable practices in all-sized businesses in Latin America and the Caribbean. The survey focused on better understanding sustainable practices, the learning process, and the knowledge contained within regional businesses. The surveys were disseminated through direct emails, social networks, and direct invitations by each higher education institution in their respective countries. Both individual and joint initiatives were held to increase the spread of the tool in the region. After the dissemination

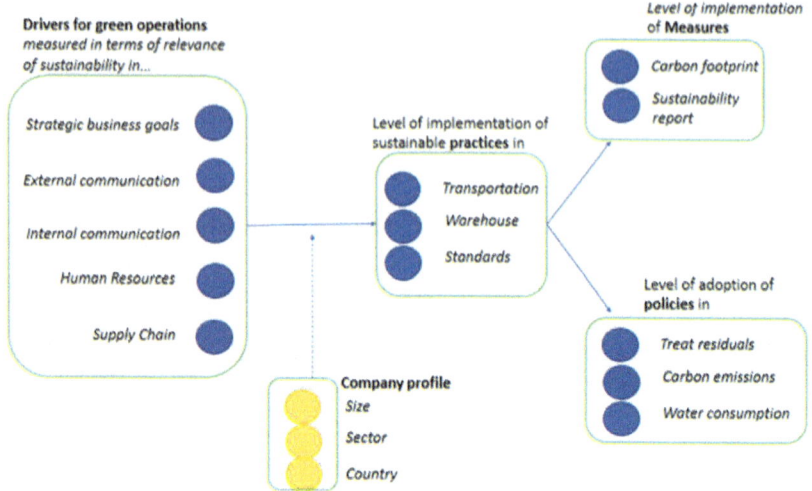

Fig. 1. Conceptual model to explain the hypotheses derived from literature and practice.

Table 1. Constructs, indicators, and survey questions

Constructs	Survey question and indicators	Type	
Drivers	What is the relevance of sustainability in the following aspects? R1 = Strategic business goals R2 = External communication R3 = Internal communication R4 = Operation/production R5 = Human Resources R6 = Supply chain	Categorical (1–4) 1 = not considering 2 = somewhat taken under consideration 3 = taken under consideration 5 = fully integrated	
Metrics	MEDP1 = Does the company measure the carbon footprint? MEDP2 = Does the company issue a sustainability report?	Categorical MEDP1 (0–2) 0 = not measured 1 = ongoing implementation to measure 2 = measured	Categorical MEDP2 (0–3) 0 = no report 1 = ongoing report implementation 2 = report implemented but not comprehensive 3 = comprehensive report implemented

(continued)

Table 1. (*continued*)

Constructs	Survey question and indicators	Type
Practices	PRACL1 = Select the best practices of sustainability in transportation implemented PRACL2 = Select the best practices in sustainability for warehouses implemented PRACL3 = Select the standards implemented	Categorical PRACL1 Transportation (0–15) PRACL2 Warehouse (0–12) PRACL3 Standards (0–13)
Policies	POL1 = Does the company treat residues over the law requirement? (Solids, liquids, and gasses) POL2 = Does the company have a plan to reduce carbon emissions? POL3 = Does the company have a plan to reduce energy and or water consumption?	Categorical POL1 (0–3) POL2 (0–2) 0 = no plan 1 = plan is currently under implementation 2 = comprehensive plan implemented POL3 (0–2)

period, the results were cleaned and processed using Python to obtain a comprehensive descriptive analysis of the data.

The survey consists of four parts: First, the questions are directed to understand the company's profile (size, activity type, location). Then, the survey questions all company policies and practices regarding warehousing and transportation operations. Finally, the questions are directed toward if and how companies measure the implementation of those practices. From this structure, we have defined the levels of constructs used for the model.

As shown in Table 1, the construct Practices contain three variables PRAC1, PRAC2, and PRAC3. In this survey section, companies were given a list of best practices and instructed to select the currently implemented ones. The variable PRACL1 Transportation is the number of best practices established; there are 15 best practices listed, while PRACL2 Warehouse has 12 best practices listed. PRACL3 is the number of industry standards the company has selected from 13 alternatives. In the constructed labeled Policies, the variable POL1 is the number of residues that receive further treatment than the law requirement (gas, solid or liquid). POL3 Counts if the company plans to reduce water or energy.

2.2   Partial Least Squares (PLS) Regression Model

Partial Least Squares (PLS) regression models have been used to answer how the combination of a set of drivers and a company's attributes influences the level of commitment to sustainability in terms of the level of implementation of sustainability practices, measures, and the adoption of policies. The benefits of this method include carrying out studies that reestimate complex models with several paths, small sample sizes, and variables without making assumptions about the distribution of the data [13]. Given all the previous arguments, it can be understood that the PLS-SEM is adequate and adapts to the objectives and the exploitative nature of the investigation.

A structural model consisting of four variables (drivers, practices, measures, and policies) derived from the conceptual model has been described using SmartPLS 4.0. The tool allows us to conduct analyses to enhance the prediction in the econometric models to build managerial implications. Due to the nature of the research and the result to be reached through this paper, it is necessary to elaborate and validate a conceptual model by analyzing the information collected through a survey.

3   Results

The survey was carried out between June and November 2022; it was addressed to companies from 7 Latin American and Caribbean countries. The responses were 474, 187 from the Dominican Republic, 115 from Argentina, and the rest from Peru, Bolivia, Uruguay, Colombia, and Mexico. Almost half of the respondents were from large companies; the rest were evenly split between medium, small, and micro firms.

The first step in the analysis phase consisted of evaluating the degree of significance of the conceptual model. Using the PLS-SEM method, we used Cronbach's alpha reliability coefficient to test the model significance and the integral consistency of the variables defined in the model: drivers, practices, measures, and policies. All estimated values of Cronbach's Alpha of the four independent variables are greater than 0.6, indicating the model is significant when the scale is closer to 1.

The second step was to examine the coefficients of the variables in the measurement model to evaluate the constructs' validity. In Fig. 2, we observed that all the coefficient values are above 0.5, which indicates that they adequately explain a statistically significant part of the variation of its construct, so we can conclude that the construct validity is confirmed.

Additionally, the latent variables' variance inflection factor (VIF) values are lower than 5, concluding that there are no collinearity problems between the independent variables in the model. Then, we explored the heterogeneity in the data, considering that we have identified different companies' profiles regarding company size and sectors in which companies operate. A marginal data analysis was conducted using filtered data sets in which we compared the results between large, medium, small, and micro firms. Concerning company size, the results are consistent regarding the external outer model's validity, and no significant differences are identified in the inner model. Additionally, we analyze the effects of 7 different sectors displayed in Table 2.

The results show a more substantial relation between measures and practices in most sectors and lower drivers and policies. Specifically, the analysis revealed less variance

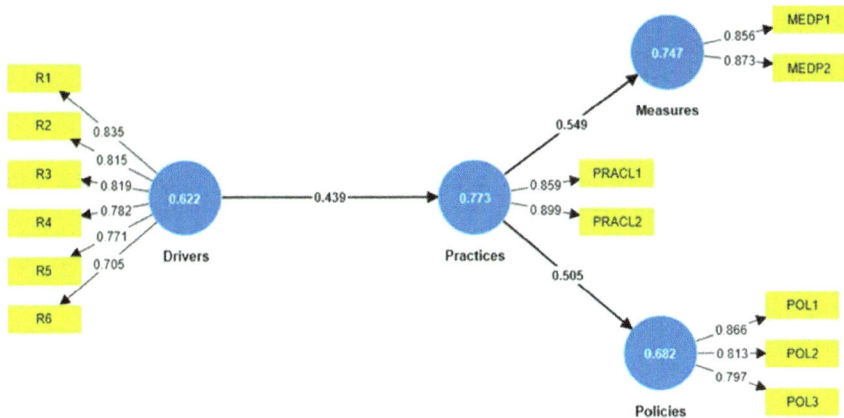

Fig. 2. Structural model

Table 2. Average Variance Extracted (AVE) comparison per sector.

Average Variance Extracted (AVE)				
Sector	Drivers	Measures	Policies	Practices
1	0.653	0.626	0.658	0.693
2	0.689	0.760	0.613	0.804
3	0.783	0.790	0.742	0.910
4	0.644	0.710	0.697	0.689
5	0.536	0.798	0.707	0.791
6	0.640	0.590	0.565	0.804
7	0.580	0.752	0.696	0.788

for all the sectors in the significant relation mentioned before; this implies that these sectors have better responsiveness to these two factors in their performance and behavior. Therefore, it is observable that the sector type moderates the connection between measures and practices in the study context.

4 Discussion and Conclusions

Theoretically, the first contribution of this work is the creation of the conceptual framework that leads to this model. The framework was based on the learning after years of executing the annual sustainability survey and supported by literature research. The second contribution was testing the validity of our model with empirical data, applying the PLS-SEM method. Practically, our contribution relies on identifying the drivers for implementing sustainable practices in the industry, which will be relevant for decision-makers in establishing strategies and policies to motivate these practices.

During the execution of the model, the hypotheses H1 and H2 raised were confirmed. Based on the results, we found a direct relationship between the drivers and the sustainable practices, meaning that the processes relate to the approaches the companies implemented. The most significant positive relationship is found with the transportation and warehousing practices. Regarding the profile (size or sector) of the companies and their relationship with sustainable practices, it could be observed based on the results that it does not have an impact that directly influences the implementation of sustainable actions or policies, thus rejecting the third hypothesis raised in the article. What is understood is that regardless of the size of the company, the application of the practices is not compromised.

Given this result, it would be recommended to continue with a more in-depth analysis, feeding the model and the information in a more specific way directly related to the profile of the companies and how this is related to sustainability practices. Something important that could not be identified within the model but within the literature and analysis of information is the relationship of the motivation variable to predict the implementation of sustainable practices in the industry. Conceptually it is understood that the motivation behind companies' policies and actions can be a good predictor of outcomes. A company exclusively motivated by image might be inclined to more visible actions. In contrast, companies with the primary motivation are cost savings might focus solely on cost and neglect to measure or communicate a reduction in environmental impacts.

The survey implemented included a question regarding a company's motivation to implement sustainable practices, however, the PLS-SEM method could not include this data because of how this question is currently structured. For future research, it is essential to redefine this question to allow the outcoming variables to include this aspect in the model. Furthermore, it is interesting to point out that industry standards did not fit the model as was expected. Further research should analyze whether this variable is irrelevant or if there is a problem with how it is measured within the survey.

References

1. D'Agostini M, Tondolo VAG, Camargo ME, Dullius AIS, Tondolo RRP, Russo SL (2017) Relationship between sustainable operations practices and performance: a meta-analysis. Int J Product Perform Manage 66(8):1020–1042. https://doi.org/10.1108/IJPPM-11-2015-0168
2. Aguinis H, Glavas A (2012) What we know and don't know about corporate social responsibility: a review and research agenda. J Manag 38(4):932–968. https://doi.org/10.1177/0149206311436079
3. Baughn CC, (Dusty) Bodie NL, McIntosh JC (2007) Corporate social and environmental responsibility in Asian countries and other geographical regions. Corp Soc Responsib Environ Mgmt 14(4):189–205. https://doi.org/10.1002/csr.160
4. Williamson D, Lynch-Wood G, Ramsay J (2006) Drivers of environmental behaviour in manufacturing SMEs and the implications for CSR. J Bus Ethics 67(3):317–330. https://doi.org/10.1007/s10551-006-9187-1
5. Bin H et al (2020) Crowd-sourcing a way to sustainable urban logistics: what factors influence enterprises' willingness to implement crowd logistics? IEEE Access 8:225064–225075. https://doi.org/10.1109/ACCESS.2020.3044921
6. Damert M, Feng Y, Zhu Q, Baumgartner RJ (2018) Motivating low-carbon initiatives among suppliers: the role of risk and opportunity perception. Resour Conserv Recycl 136:276–286. https://doi.org/10.1016/j.resconrec.2018.05.002

7. El Akremi A, Gond J-P, Swaen V, De Roeck K, Igalens J (2018) How do employees perceive corporate responsibility? Development and validation of a multidimensional corporate stakeholder responsibility scale. J Manag 44(2):619–657. https://doi.org/10.1177/014920631556 9311
8. Piyathanavong V, Garza-Reyes JA, Kumar V, Maldonado-Guzmán G, Mangla SK (2019) The adoption of operational environmental sustainability approaches in the Thai manufacturing sector. J Clean Prod 220:507–528. https://doi.org/10.1016/j.jclepro.2019.02.093
9. Arias-Arévalo P, Martín-López B, Gómez-Baggethun E (2017) Exploring intrinsic, instrumental, and relational values for sustainable management of social-ecological systems. E&S 22(4):art43. https://doi.org/10.5751/ES-09812-220443
10. Laari S, Töyli J, Ojala L (2017) Supply chain perspective on competitive strategies and green supply chain management strategies. J Clean Prod 141:1303–1315. https://doi.org/10.1016/j.jclepro.2016.09.114
11. Maxner T, Dalla Chiara G, Goodchild A (2022) Identifying the challenges to sustainable urban last-mile deliveries: perspectives from public and private stakeholders. Sustainability 14(8):4701. https://doi.org/10.3390/su14084701
12. To WM, Tang MNF (2014) The adoption of ISO 14001 environmental management systems in Macao SAR, China: trend, motivations, and perceived benefits. Manage Environ Qual Int J 25(2):244–256. https://doi.org/10.1108/MEQ-01-2013-0002
13. Hair JF Jr, Sarstedt M, Hopkins L, Kuppelwieser VG (2014) Partial least squares structural equation modeling (PLS-SEM): an emerging tool in business research. Eur Bus Rev 26(2):106–121. https://doi.org/10.1108/EBR-10-2013-0128
14. Busu C, Busu M (2021) Research on the factors of competition in the green procurement processes: a case study for the conditions of Romania using PLS-SEM methodology. Mathematics 9(1):1. https://doi.org/10.3390/math9010016

Aligning Strategic Priorities in Manufacturing and Sustainable Operational Practices in Firms: Exploration and Evidence in North-East Colombia

Juan Felipe Reyes-Rodríguez[(⊠)] ⓘ, María Nikolle del Cairo-Jiménez, and Liza María Martínez-Zúñiga

Universidad Industrial de Santander, Bucaramanga, Colombia
jfreyrod@uis.edu.co

Abstract. Recently, issues such as global warming and ecosystem deterioration have demanded corporate actions towards a responsible orientation to operations. Particularly, manufacturing firms can exhibit such a responsible orientation through the implementation of sustainable operational practices (SOP). Such practices can also pave the way to consolidate the different strategic priorities in manufacturing (SPM) that drive business value. Based on a sample of 412 manufacturing firms in Bucaramanga, Colombia, the purpose of this study is to characterize different portfolios of SOP as well as profiles of firms based on their SPM. Results evidence three forms of SOP, related to: (i) product design and transformation, (ii) supply chain, and (iii) reverse logistics. Furthermore, sampled firms exhibit four main SPM: (i) natural environment and social responsibility, (ii) flexibility, (iii) innovation and efficiency, and (iv) quality and customer satisfaction. Results show that there are remarkable differences regarding SOP related to product design and transformation as well as supply chain across profiles of SPM. Notwithstanding, the extent of implementation of reverse logistics takes place regardless of their SPM. The study contributes to an increased understanding of how firms adopt a more responsible behavior. Furthermore, the study provides evidence concerning the articulation of such organizational response and the priorities to capture value in manufacturing.

Keywords: Corporate sustainability · Sustainable operational practices · strategic priorities in manufacturing · manufacturing firms

1 Introduction

This study addresses the need for businesses to care for the natural environment and society, beyond generating wealth for shareholders. In manufacturing companies, concrete actions related to resource use, waste generation, and their impact on individuals and the biosphere are crucial. These actions, termed sustainable operational practices (SOP) align with strategic priorities of the firm to achieve market advantages [1].

© The Author(s) 2025
H. Kohl et al. (Eds.): GCSM 2023, LNME, pp. 46–55, 2025.
https://doi.org/10.1007/978-3-031-77429-4_6

This study aims to explore the characterization of SOP in manufacturing firms and their alignment with strategic priorities in manufacturing (SPM). Thus, the guiding research questions are: 1) How are SOP characterized in manufacturing firms? 2) How do these practices align with SPM in such firms?

To achieve these purposes, a sample of 412 manufacturing companies in the Metropolitan Area of Bucaramanga, Colombia, is used. The study structure comprises a literature review to position the research conceptually, followed by a description of the methodological considerations, including data collection, measurements, and analysis procedures [2, 3]. The study ends with the presentation and discussion of the results, along with the derived conclusions and recommendations from the analysis.

2   Literature Review

Literature review is based on searching terms such as sustainable operational practices, environmental/sustainable initiatives, environmental/sustainable actions. Further terms included manufacturing strategies, manufacturing priorities and competitive priorities.

2.1   Sustainable Operational Practices

SOP comprise planning, production, supply, and logistics activities with a sustainable perspective in manufacturing [2]. SOP are implemented throughout the product life-cycle stages, including product design, production, supply chain management, and reverse logistics.

In product design, SOP focus on eco-design to create environmentally friendly products manufacturing. They involve making environmentally conscious decisions during new product development, estimating environmental consequences, and promoting green innovation [2, 4]. During production, SOP integrate environmental goals to reduce emissions, waste, and resource consumption. Innovative technologies may be used to mitigate environmental risks in the production process [4].

SOP extend to the supply chain, encouraging environmentally and socially responsible practices among suppliers and customers. This involves joint actions to address design, raw material selection, packaging, and logistics based on environmental considerations. Finally, SOP in reverse logistics focus on managing waste through recycling, remanufacturing, and proper disposal, with a view of achieving maximum recovery of value and create market opportunities [2, 4].

2.2   Strategic Priorities in Manufacturing

Strategic deployment in firms involves translating corporate strategy into functional priorities [5]. Manufacturing strategies contribute to corporate goals through SPM to ultimately boost competitiveness. Initially, four SPM were identified in literature: cost, quality, flexibility, and delivery. Over time, additional priorities emerged, including after-sales service, time-to-market, and environmental protection. Recent studies validate six relevant manufacturing priorities: cost, quality, reliability, flexibility, environmental protection, and social well-being [1, 3].

2.3 Integration of Perspectives

The implementation of SOP in manufacturing firms is expected to align with SPM so as to ultimately enhance environmental and social performance. The literature suggests integrations of SOP on SPM like cost, quality, flexibility, and delivery when the focus is on pollution prevention [6]. However, disaggregating SOP into social and environmental practices reveals a more significant impact on SPM related to environmental practices [7]. Further, the alignment between SOP and SPM is influenced by the financial situation of firms [8]. Environmental performance, when combined with lean manufacturing, can find better compatibility as a SPM itself [3]. Nevertheless, literature lacks consensus, underscoring the need to understand SOP's disaggregation and its alignment with SPM [3, 7].

3   Research Methods

The study focuses on manufacturing companies in the Metropolitan Area of Bucaramanga (MAB), Santander, Colombia. The region prioritizes resource-efficient management for increased productivity and competitiveness.

An online questionnaire was used to collect data from a final sample of 412 manufacturing firms operating in the MAB. The response rate was 6.123%, with 67.48% of respondents being general managers, partners, and/or owners of the companies. The final sample predominantly consisted of small businesses (96.84%). In terms of activities, sampled firms engaged mainly in manufacturing food, beverages, textiles, and other products (52.18%). Data collection occurred in 2022.

The questionnaire development involved adapting 23 items for SOP measurement from previous studies [9, 10]. Items assessed the level of implementation of SOP in product design, production transformation, supply and distribution, as well as reverse logistics, using a 7-point Likert scales. SPM were measured by asking respondents to rank 10 priorities from most important to least important in manufacturing for competitiveness in their markets. Items related to SPM were also adapted from previous literature [1, 5]. Structural variables included firm size (number of full-time employees) and economic activity (NACE codes 10-32).

The instrument, with respective items and scales, underwent review by academic experts in sustainability and questionnaire development. Adjustments were made based on feedback to obtain the final version for distribution to companies.

After data cleaning, exploratory principal component analysis (EPCA) determined SOP's structure and internal consistency using Cronbach's alpha. Cluster analysis identified company profile groups based on SOP. Another cluster analysis classified firms' SPM. ANOVA compared mean differences in SOP and SPM among profile groups.

4   Results

EPCA on SOP items revealed three factors, explaining 65.246% variance (see Table 1).

The three extracted factors were labeled as product design and production (12 items), supply chain (8 items), and reverse logistics (3 items). Loading values were larger

Table 1. Exploratory principal component analysis of SOP

Items—Sustainable operational practices (SOP)	MEAN (S.D.)	Factor 1: Product design and production	Factor 2: Supply chain	Factor 3: Reverse logistics
Production processes reduce waste generation and toxic emissions	5.374 (1.597)	**0.825**	0.206	0.057
Design of products with the aim of using the least amount of materials possible	5.527 (1.497)	**0.815**	0.010	0.107
Production processes reduce water and energy consumption	5.367 (1.581)	**0.806**	0.197	0.106
Design of products with the aim of having reusable and/or recyclable packaging	5.415 (1.516)	**0.796**	0.080	0.168
Replacing toxic substances and inputs with environmentally friendly ones	5.466 (1.574)	**0.788**	0.298	-0.006
Design of products with the aim of avoiding or reducing the use of hazardous and/or toxic materials or inputs	5.651 (1.530)	**0.778**	0.234	0.063
Design of products with the aim of consuming the least amount of energy possible	5.490 (1.599)	**0.774**	0.193	0.034
Design the packaging of products to be easy to recycle. Reuse. Remanufacture and/or decompose	5.422 (1.497)	**0.770**	0.162	0.203
Design of products to be easy to recycle. Reuse. Remanufacture and/or decompose	5.330 (1.616)	**0.762**	0.320	0.137
Design of products with the aim of reducing the use of their packaging	5.318 (1.653)	**0.717**	0.190	0.264

(*continued*)

Table 1. (*continued*)

Items—Sustainable operational practices (SOP)	MEAN (S.D.)	Factor 1: Product design and production	Factor 2: Supply chain	Factor 3: Reverse logistics
Production processes recycle waste to be treated and reused	5.595 (1.653)	**0.702**	0.110	0.302
Use of clean technologies in production processes to reduce pollution and/or resource consumption	5.226 (1.614)	**0.700**	0.373	0.062
Audits of the environmental management of our suppliers	3.762 (1.678)	0.091	**0.842**	0.109
Supplier selection based on environmental criteria	3.922 (1.727)	0.174	**0.831**	0.118
Provide suppliers with design specifications aligned with environmental attributes	4.350 (1.708)	0.193	**0.743**	0.264
Suppliers required to use environmentally friendly packaging (e.g., recyclable. Biodegradable. etc.)	4.333 (1.781)	0.249	**0.741**	0.182
Reuse of the energy from processes to transfer it as a resource to another company	3.646 (1.705)	0.131	**0.692**	0.176
Suppliers required to implement environmental management systems	4.541 (1.703)	0.392	**0.667**	0.050
Use of environmentally friendly transportation when distributing products	4.485 (1.75)	0.247	**0.666**	0.198
Return of products to suppliers to be recycled. Reused or remanufactured	4.080 (1.877)	0.163	**0.657**	0.381

(*continued*)

Table 1. (*continued*)

Items—Sustainable operational practices (SOP)	MEAN (S.D.)	Factor 1: Product design and production	Factor 2: Supply chain	Factor 3: Reverse logistics
Collection of products purchased by customers to recycle or reuse them	4.420 (1.92)	0.088	0.366	**0.790**
Collection of packaging returned by customers to recycle or reuse it	4.862 (1.887)	0.190	0.289	**0.765**
Repairing. Refurbishing. or remanufacturing component parts from returned or defective/damaged products	5.107 (1.779)	0.209	0.190	**0.650**
Cronbach's alpha		0.950	0.909	0.760

Explained total variance: 65.246%. Orthogonal varimax rotation method
Loading values of the items associated with each factor are in boldface

than 0.600 (Kaiser-Meyer-Olkin measure was 0.948, and Bartlett's test was significant). Cronbach's alpha was larger than 0.700 for each factor.

Two-stage cluster based on the summated scales of the factors formed 4 firm groups. ANOVA in Table 2 compared SOP mean values among groups, including the number of employees.

Table 2. Firm profile groups based on SOP

Sustainable operational practices (SOP)/variables	Profiles of postures towards sustainability				ANOVA F
	Advanced	Emphasis on design and production	Emphasis on market environment	Laggards	
Product design and production	**6.223**	5.515	4.821	2.809	210.423*
Supply chain	**5.173**	3.202	3.532	2.004	211.359*
Reverse logistics	**5.716**	2.714	5.071	2.373	279.809*
Firm size (number of employees)	15.480	7.210	13.080	8.030	0.670
Number of firms	197	63	118	34	

*$p < 0.001$

Cluster analysis identified four firm groups: "Advanced" (N = 197) with consistently higher SOP values, "Laggards" (N = 34) with consistently lower SOP values, and two groups with specific emphases: one (N = 63) on product design and production, and the other (N = 118) on supply chain and reverse logistics. Statistically significant differences in SOP types were found within groups but not for the case of firm size.

Then, two-stage cluster analysis based on the 10 SPM resulted in four company groups. Table 3 shows ANOVA results for SPM and firm size.

Table 3. Firm profile groups based on SPM

Strategic priorities in manufacturing (SPM)/variables	Profiles of strategic priorities in manufacturing (SPM)				ANOVA F
	Env. and social issues	Flexibility	Efficiency and innovation	Quality and customer satisfaction	
Manufacturing cost	5.260	5.556	**8.687**	7.698	46.464*
Product quality	9.082	2.370	9.133	**9.265**	228.101*
Delivery speed and timing	5.260	4.482	**7.520**	6.704	34.581*
Flexibility in production	2.712	**7.482**	6.033	4.327	67.273*
New product development/level of innovation	5.795	5.815	**5.933**	4.469	12.022*
Environmentally friendly products and processes	**7.123**	6.111	4.927	2.864	110.366*
Social responsibility	**6.096**	5.111	3.967	2.568	65.359*
Conformity to customer specifications	5.616	5.037	4.247	**6.648**	31.712*
Customer service (post-sale and/or technical support)	5.041	5.741	2.713	**6.562**	103.533*
Variety/scope of product range offered	3.014	**7.296**	1.840	3.895	53.870*
Firm size (number of employees)	8.840	11.930	13.010	14.820	0.294
Number of firms	73	27	150	162	

*$p < 0.001$

The analysis revealed four distinct profiles of companies based on SPM. Group 1 prioritizes environmental and social aspects, while Group 2 focuses on flexibility. Group 3 emphasizes efficiency and innovation, and Group 4 prioritizes quality and customer satisfaction. ANOVA results in Table 3 show significant differences in SPM among the groups, but not in the number of employees.

Finally, Differences in SOP among SPM profiles were evaluated through ANOVA (see Table 4). Statistically significant differences were found in both product design/production SOP and supply chain SOP.

Table 4. SOP according to profiles of SPM

Sustainable operational practices (SOP)	Profiles of strategic priorities in manufacturing (SPM)				ANOVA F
	Env. and social issues	Flexibility	Efficiency and innovation	Quality and customer satisfaction	
Product design and production	**5.819**	5.506	5.548	5.137	5.832*
Supply chain	**4.654**	4.560	4.362	3.633	14.415**
Reverse logistics	4.872	4.901	4.929	4.621	1.183
Firm size (number of employees)	73	27	150	162	

$^{*}p < 0.01$; $^{**}p < 0.001$

The highest mean values for SOP were found in the profile of SPM on environmental and social issues ($N = 73$), while the lowest were in the quality and customer satisfaction profile ($N = 162$). The flexibility ($N = 27$) and efficiency and innovation ($N = 150$) profiles also showed notable SOP values.

Further analysis included the comparison of SOP based on firm size and economic activity, considered as control variables. In both cases, statistically significant differences were only evidenced concerning reverse logistics SOP. In terms of size, medium-sized firms had higher mean value (mean $= 6.111$, ANOVA F $= 3.483$, $p < 0.05$). Based on economic activity, higher mean values were evidenced in firms that manufacture petrochemical, pharmaceutical and plastic products (mean $= 5.505$, ANOVA F $= 5.449$, $p < 0.05$).

5 Conclusions and Implications

The results show that environmental and social SPM align with SOP in product design, production, and the supply chain, contributing to improved sustainability performance in firms. SOP focused on product design and production underpin SPM related to flexibility, efficiency, innovation, and customer satisfaction. Efficiency and innovation SPM also align with SOP at the product design and production level, driving resource use optimization and greening products. Further, quality and customer satisfaction SPM align with SOP related to product design and production, emphasizing continuous improvement and conformity to standards. However, none of the characterized SPM show strong alignment with SOP in reverse logistics, indicating potential challenges in this alignment.

Managers should assess sustainability integration in their operations and align SOP accordingly, while regulatory bodies should provide incentives for SOP implementation, especially in reverse logistics. The study enriches the fields of operations management and strategy by understanding the alignment between SOP and SPM in manufacturing firms.

Acknowledgement. This study is part of the project entitled "Evaluation of Sustainable Operational Practices and their Relationship with Manufacturing Strategies in Companies in the Metropolitan Area of Bucaramanga," funded by Universidad Industrial de Santander, code 2987.

References

1. Macchi M, Savino M, Roda I (2020) Analysing the support of sustainability within the manufacturing strategy through multiple perspectives of different business functions. J Clean Prod 258
2. D'Agostini M, Tondolo VAG, Camargo ME, Dullius AIS, Tondolo RRP, Russo SL (2017) Relationship between sustainable operations practices and performance: a meta-analysis. Int J Product Perform Manag 66(8):1020–1042
3. Queiroz GA, Filho AGA, Costa Melo I (2023) Competitive priorities and lean–green practices—a comparative study in the automotive chain' suppliers. Machines 11(1):50
4. Tondolo VAG, D'Agostini M, Camargo ME, Tondolo RRP, Souza JL, Longaray AA (2021) Sustainable operations practices and sustainable performance: relationships and moderators. Int J Product Perform Manag 70(7):1865–1888
5. Longoni A, Cagliano R (2015) Environmental and social sustainability priorities. Int J Oper Prod Manag 35(2):216–245
6. Jabbour CJC, Maria da Silva E, Paiva EL, Almada Santos FC (2012) Environmental management in Brazil: is it a completely competitive priority? J Clean Prod 21(1):11–22
7. Galeazzo A, Klassen RD (2015) Organizational context and the implementation of environmental and social practices: what are the linkages to manufacturing strategy? J Clean Prod 108:158–168
8. Fura B (2022) The role of financial situation in the relationship between environmental initiatives and competitive priorities of production companies in Poland. Risks 10(3):52
9. Sundram VPK, Bahrin AS, Othman AA, Munir ZA (2017) Green supply chain management practices in Malaysia manufacturing industry
10. Ahmadi-Gh Z, Bello-Pintado A (2022) Why is manufacturing not more sustainable? The effects of different sustainability practices on sustainability outcomes and competitive advantage. J Clean Prod 337:130392

Towards Circular Business Models in the Punching Industry: Leveraging Smart Sensor Technology for Sustainable Manufacturing Processes

Daniel Wörner[✉], Lukas Budde, and Thomas Friedli

Institute of Technology Management, University of St. Gallen, 9000 St. Gallen, Switzerland
daniel.woerner@unisg.ch

Abstract. The punching industry (PI) remains largely dependent on unsustainable resource use in its manufacturing process, consuming finite materials. Evolving market needs for increased environmental sustainability in the PI causes punching companies to reevaluate their existing manufacturing processes. Considering this trend, this research explores the prerequisites for transitioning to circular business models (CBMs), a subset of sustainable business models, by using smart sensor technology (SST) supporting a sustainable manufacturing process (SMP). Prior research has shown that using different sensors pertaining punching tools and machines is prevalent. Standard parts in the punching tool itself address these challenges of an imprecise set-up of the punching tool on the punching machine resulting in process improvements and efficiency enhancements. Yet, no similar data-driven punching tool concepts have been installed highlighting the novelty. To address this gap, this study examines how data can be gathered and successfully utilized to support SMP fostering CBMs in the PI based on data-driven punching tools. Mixed methods are applied in a practice oriented project, involving a Swiss-based manufacturer. The results demonstrate several key drivers supporting SMPs enabled by data-driven punching tools. Future research should involve a larger number of interviews and further field testing to increase maturing SST data-driven punching tools.

Keywords: Punching industry · punching tool · smart sensor technology · circular business model · sustainable manufacturing process

1 Introduction

Sustainable development has globally been gaining increasing importance for the past decades [1]. Constantly changing market requirements, deteriorating profit margins, and rising environmental awareness have caused producing companies to rethink their existing manufacturing processes, e.g., punching process (PP) (see Fig. 1). The current economic business models are based on a "take-make-dispose" system and rely on continuous growth and extraction of resources and therefore perpetrate profound negative

© The Author(s) 2025
H. Kohl et al. (Eds.): GCSM 2023, LNME, pp. 56–64, 2025.
https://doi.org/10.1007/978-3-031-77429-4_7

impacts on the natural environment and resource availability [2]. Nearly 25% of annual metal production never enters a product but is scrapped through the supply chain illustrating the significant potential for metal punching technologies to reduce metal waste [3]. The need for greater sustainability, i.e., as caused by material waste, can be considered an important driver for business model innovation [4]. Research indicates that digital technology is a promising lever to activate these drivers to shift production and consumption to a circular economy (CE) to improve process efficiency and accuracy [5]. CBMs form a subset of SBM [6]. CBMs can be operationalized by implementing CE-strategies, e.g., "narrowing" (i.e., optimizing natural resources), "slowing" (i.e., extending the product's life), "closing" (i.e., using the product after its use phase) [7], supported by "regenerating" (i.e., considering material input and output flow), and "informing" (i.e., utilizing technological support systems, e.g., data generation with SST) [8]. The concept of SBM considers the efforts of companies to promote sustainability activities and is yet often perceived as a source of competitive advantage [9]. SBMs holistically represent business models that incorporate proactive multi-stakeholder management providing monetary and non-monetary value in a long-term perspective [6]. Sustainable manufacturing processes (SMP) are broadly defined to include all aspects that aim to improve process metric types regarding environmental impact, energy consumption, economic cost, worker safety, and waste management [10] and serve as a basis and provider of technological elements to foster a CE [2, 6]. The rising usage of sensors in production processes enables access to larger amounts of digitizable information, making resource efficiency measurable and potential savings usable [11]. Smart sensor technology (SST) can provide a solution to these challenges by collecting inputs from the physical environment and using built-in computing resources to perform predefined functions upon detection of specific input and then process data before passing it on [12]. The prospective impact of new technologies on CBM development in terms of value creation is still largely unexplored, hindering the implementation of successful CBMs [13], and the transition of existing industries, e.g., to a state of extending product life cycles through an integrated life cycle management (LCM) [14]. The development of reliable, cost-effective sensors to measure representative process variables is a key technical challenge [15], resulting in a significant gap in transferring theoretical considerations into practical results within the punching industry (PI). To address this shortage, the following research question is posed: "*How can data be gathered and successfully utilized to support sustainable manufacturing processes in the punching industry based on data-driven punching tools?*".

2   Punching Machines and Tools

Mechanical sensors have become indispensable in modern PP [16]. Tool monitoring controls the sequence of the PP over the entire punching cycle and provides both protection of the tool and the punching machine against overloading while indirectly detecting gross changes in the quality of the punched parts. The monitoring can be executed at multiple locations, i.e., in the tool, in the punching machine, and on the punched part [12]. Force measurements with strain gauges or piezo sensors and distance measurements with optical sensors are suitable for measuring in the tool itself. The forces are measured at defined positions in the machine ram, a tool plate, or a special pressure plate in the force flow of individual punches or punch groups. The force spreads conically downwards and can be measured in this cone [17].

Despite technical progress largely being driven by manufacturing technology improvements, most manufacturing processes in the PI are based on human knowledge neglecting the full optimization potential. Hence, many processes involving punching tools cannot be directly improved since the processes tend to be highly human knowledge driven. Advances in SST could provide solutions addressing this issue. It is an industry-wide phenomenon that traditionally the design process in sheet metal forming has been performed by a few key industry professionals with high experience and craftsmanship [18]. Considering Industry 4.0, a large variety of tasks can be supported in punching production, e.g., quality assurance and production planning [19]. Digitalization enables new business potential for the manufacturing industry, causing many companies to develop data-based business models in pilot projects [20]. The PI is at the lever to accelerate SMP and LCM (see section 1), but only a few firms have the in house know-how for developing innovative sensor-monitored punching parts. In conclusion, no measurement unit of the bending in the guide elements has been developed yet [17].

This paper focuses on such innovative SST detecting an incorrect installation position of the punching tool during a tool change and supports the ongoing production process, which can lead to increased wear, reduced component quality or tool damage [21]. This subject becomes increasingly relevant due to its positive impact on the SMP, particularly regarding environmental impact, manufacturing costs, energy consumption, and waste management [10]. Aside from literature research, this study follows a practice-based project perspective by gaining insights from a globally operating Swiss-based manufacturer operating in the PI.

3   Research Method

As the study considers qualitative aspects, the data collection follows a mixed methods approach by Eisenhardt [22]. Qualitative data is collected by conducting 12 semi-structured interviews with relevant actors across the PI ecosystem reaching from blue to white collar workers, e.g., punching tool users, punching tool service provider, punching tool manufacturers, supplier of punching parts, and punching machine manufactures. The interview findings are contextualized according to the approach by Braun and Clarke [23]. Besides the interviews, we worked with an associated technology research partner, which designed the innovative SST, to explore the potential of this innovative technology during a field test. To strengthen the data collected scientifically, relevant literature and information provided by the interviewees, e.g., firm reports, firm corporate websites, and white papers, were examined. Iterative cross-checks and repetitive examination of the obtained data led to robust analyses. Information is clustered with analytical tools such as open-ended interview coding.

4   Findings

Punching defines a non-cutting manufacturing process that allows the short-term production of highly precise and complex workpieces with a rather optimum material utilization and relatively low scrap percentage. Due to the large volumes and potential of process optimization it gains increasing importance in shift towards SMP [17].

Figure 1 illustrates the main activities carried out during the set-up process of a punching tool on a punching machine while highlighting the key differences without and with the usage of SST. The interview findings revealed that the set-up process until mass production can be divided into six consecutive steps. Initially, (1) a new customer request of punching parts is received. Subsequently, (2) the request must be set up correctly on the punching machine. In case A, without the use of SST, (3.1) a manual visual inspection of the installation position is followed by (4) quality acceptance testing in an iterative process. In a next step, (5) the final quality control of the parts, (3D sensor technology etc.) completes the punching machine setup process (and marks the end of the static setup process). Subsequently, the (6) start of mass production marks the transition to the dynamic production process, where out challenges are posed, i.e., heat generation and bending of the punching parts. Steps (3.1) and (4) include numerous time-consuming iterations and intensive knowledge-based decisions with currently no SST support (see Fig. 1 [A]). Without any SST multiple iteration loops are needed given external characteristics concerning product and machine parameters (see Fig. 2) and cannot be overcome since an imprecise contact surface leads to suboptimal punching results. Case B, in contrast, utilizes SST (see Fig. 1 [B]), allows an automatic inspection of the installation position to the machine operator, minimizing such iterations. This approach highly reduces the time needed of an otherwise purely manual based inspection of the installation position, passing the (4) acceptance test much quicker. By reducing the time and jobs involved of individual stages in the manufacturing process (i.e., reducing required tools and lubricants), using efficient metal forming technologies (i.e., reducing material waste), and reducing overall energy consumption, the SST contributes inherently to a positive environmental impact. Thus, SST positively influences SMPs supporting the transition towards CBMs and associated advantages in the PI [24].

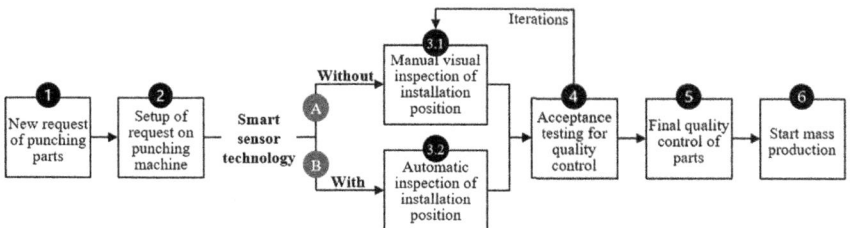

Fig. 1. Visualization of main activities performed (1) to (5) in the set-up process of a punching tool on a punching machine until mass production starts (6) without [A] and with [B] smart sensor technology.

As existing knowledge is scattered across levels and actors (see Fig. 2), technologies in the tool itself bears high potential for improvement, an SST within the standard parts of a punching tool can reduce the error rate of the installation position. In addition, data recording can assist especially less trained operators to setup the punching machine, as the SST closes the lack of knowledge through its provided feedback. Monitoring the highly dynamic PP during (6) mass production illustrates additional potential, serving as a basis for future SST improvements, e.g., high production speeds and the resulting high temperatures can cause a displacement of the upper and lower tool part. Such a displacement is currently not accurately measured.

The overall PI ecosystem is illustrated in Fig. 2. On a micro perspective, there are four actors influencing the PI on a firm level, although it is possible that some actors are embodied by the same company. The tool constructor (TC) mainly covers the creation of the prototype design and the development of the new punching tools. The tool manufacturer (TM) is responsible for the implementation of the design proposals and the verification of design and layout. The tool user (TU) receives the tool from the TM or the TC and uses it for punching production. In addition, the tool service (TS) is responsible for the provision and performance for specific machine or tool parts. TS must decide which actions to take, while a data driven decision support is often not available. As such, SST makes the CE-strategy "informing" more actionable, which in turn supports the transition toward CBMs. To leverage the full potential, it is essential that all actors are contributing their efforts to attaining a successful SMP, starting from the design of the tool to the optimal resource-efficient usage. At an aggregated layer, the activities performed by the four actors have significant implications on LCM, tool, and machine design as well as for human resources. These are triggered by the key drivers on the second layer affecting the PP, exceeding the firm-level and substantially have an influence on each other. Interview informants mentioned several key drivers affecting the PP in multiple ways but with varying relevance based on the product related market environment, i.e., high precision manufactures tend to have a higher need for an SST solution. Product-related parameters influencing the PP include the complexity and material of the product, as well as the scalability of the SST solution and the lot size. In addition, existing machine-related parameters such as cutting clearance, modular construction, exposure to forces, self-precision and the existing technology must be considered and planned for during the PP. Profitability calculations are influenced by the holistic system, consisting of the product characteristics and the machine parameters. Only if both are in sufficient condition, the negative influences regarding the product and manufacturing costs, caused by the key drivers, can be minimized and the PP can be optimized, attributing toward the "narrowing" CE-strategy. On a meso perspective, we are highlighting relevant actors on a PI supply chain-level, namely the suppliers of data-driven punching tools enabled by SST can optimize the SMP. Starting with the customer ordering of the punching parts, data supports the decisions of the TC regarding prototype design gathered with SST. It complements lessons learned from prior design tools by assessing used materials, end-of-life recovery of products, machine efficiency, and conceptual aspects including environmental impact [25]. Such data collection also accelerates the tool improvement, design, and development process of a TC, supporting to optimize the efficiency and reliability of the punching solution while reducing CO_2 emissions, energy consumption, and possibly tool manufacturing costs [5]. Using SST allows determining the OEE and to maximize profitability calculations. Using different data sources, from internal firm processes to external supply chain partners and customers, a data-driven platform utilizing SST enables data collection with embedded sensors to potentially measure real-time data and report information with minimal human interaction [26]. Data sharing among the actors increases information availability, supports data analysis [5], and enables a better traceability concerning the root cause analyzes of prior failures, e.g., within the SMPs. It is important to highlight, that the quantitative impact on increased efficiency and reduced emissions have to be further elaborated. As such the novel solution has yet to be monitored in the PP over time, e.g., in terms of CO_2 reduction or energy savings.

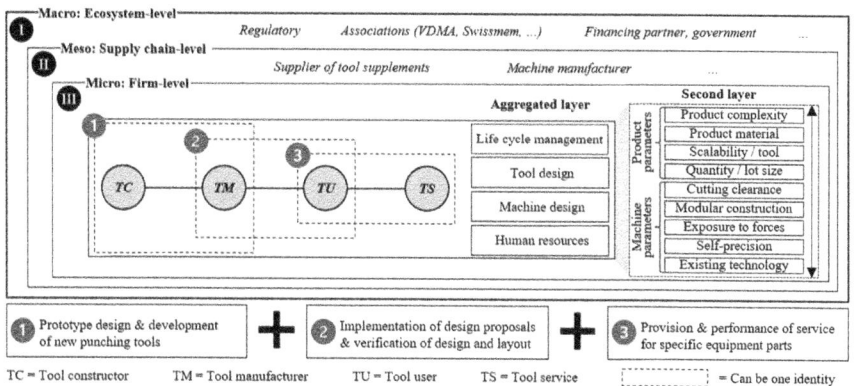

Fig. 2. : Illustration of the key actors in the punching industry (PI) across three levels (micro, meso and macro) influencing the sustainable manufacturing process on a firm level towards circular business models.

In sum, the interview data revealed that although a variety of sensors already exist measuring different parameters regarding the punching tool and the punching machine (see section 2), little measurement is performed in the tool itself. Yet, given the current trends towards SMPS fostering CBMs, there is a strong need for such a SST. The need and acceptance of such SST highly depends on the product related market environment, e.g., a higher need for firms producing high precision products, i.e., for electrical engines compared to thrust washers used in combustion engines.

5 Conclusion

The production of dimensionally accurate stampings with minimal material and processing costs in combination with the thin nature of sheet metal has always been a complicated process [21]. Our findings reveal that both product and machine have to be holistically considered in order to enable a SMP in the PI. As such, data-driven punching tools enabled by SST are of particular interest if the machine has an older construction year, as they occur relatively often due to the long usage of punching machines. Along the main activities performed (1) to (5) in the set-up process of a punching tool (see Fig. 1), digitalization supports the collection, integration, and analysis of data, as well as process automation and can be leveraged by the key players at the firm-level (see Fig. 2) in the PP. Data and SST serve as a direct enabler of SMPs fostering CBMs within the PI by exploring the technology-based solution of data-driven punching tools creating significant positive environmental impacts on the punching ecosystem. This work's data collection follows a mixed methods approach that is focused on primarily Swiss-based companies. The boundary conditions and limitations of our approach need to be considered when internalizing and generalizing our findings. First, we studied leading firms with the prevailing know-how, market conditions and business environment in the DACH region. The potential for SST is high but remains limited due to the relatively high average of new machines (high precision) and the currently sufficient percentage of skilled labor on a firm level. Extending our research on a global scale, especially China, India and partly north America should be considered for the application of our data-driven tool given its shortage of skilled labor and high percentage of old machines with less accuracy [27]. The SST based solution enables a simple and non-knowledge-based set-up of the punching tool demonstrating a great impact on overcoming skill shortage among on-site operators. Second, optimizing the whole PP to support a SMP other factors must also be considered, e.g., tool surface roughness [28] or intelligent tool design [29] contribute to reductions in mass, maximum stress, maximum displacement of tools, and process automation [30]. SST can positively contribute to process improvements and efficiency enhancements supporting current trends of SMPs fostering CBMs and CE in the PI. Additional data generation is an important lever to promote LCM and to determine the quantitative impact. We suggest exploiting the full potential, other opportunities, and interlinkages on top of the suggested SST should be further investigated, e.g., additional field testing in manufacturing processes, usage of data analytics concerning process and tool design, digital platform across PI actors, further scalability of IoT solutions in the PP.

References

1. Bjørnbet MM, Skaar C, Fet AM, Schulte KØ (2021) Circular economy in manufacturing companies: a review of case study literature. J Clean Prod 294:1–14. https://doi.org/10.1016/j.jclepro.2021.126268
2. Jawahir IS, Bradley R (2016) Technological elements of circular economy and the principles of 6R-based closed-loop material flow in sustainable manufacturing. Procedia CIRP 40:103–108. https://doi.org/10.1016/j.procir.2016.01.067
3. Metal forming and sustainability. [Media Releases & Ad Hoc News]. Accessed 08 Feb 2023. https://www.ictp2017.org/sites/www.ictp2017.org/files/industry-workshop-3.pdf
4. Foss NJ, Saebi T (2017) Fifteen years of research on business model innovation: how far have we come, and where should we go? J Manag 43(1):200–227. https://doi.org/10.1177/0149206316675927
5. Liu Q, Trevisan AH, Yang M, Mascarenhas J (2022) A framework of digital technologies for the circular economy: digital functions and mechanisms. Bus Strateg Environ 31(5):2171–2192. https://doi.org/10.1002/bse.3015
6. Geissdoerfer M, Vladimirova D, Evans S (2018) Sustainable business model innovation: a review. J Clean Prod 198:401–416. https://doi.org/10.1016/j.jclepro.2018.06.240
7. Bocken NMP, de Pauw I, Bakker C, van der Grinten B (2016) Product design and business model strategies for a circular economy. J Ind Prod Eng 33(5):308–320. https://doi.org/10.1080/21681015.2016.1172124

8. Kirchherr J, van Santen R (2019) Research on the circular economy: a critique of the field. Resour Conserv Recycl 151
9. Bocken NM, Short SW, Rana P, Evans S (2014) A literature and practice review to develop sustainable business model archetypes. J Clean Prod 65:42–56. https://doi.org/10.1016/j.jcl epro.2013.11.039
10. Haapala K, Zhao F, Camelio J, Sutherland J, Skerlos S, Dornfeld D, Jawahir I, Zhang H, Clarens A (2014) A review of engineering research in sustainable manufacturing. In: International manufacturing science and engineering conference. https://www.researchgate.net/Haa pala_MSEC2011-50300_MEDLCEpaper_final#pf4
11. Neligan A, Baumgartner RJ, Geissdoerfer M, Schöggl J (2022) Circular 8 disruption: digitalisation as a driver of circular economy business models. Bus Strat Environ: 1–14. https://doi.org/10.1002/bse.3100
12. Schuster A (2000) Überwachungssysteme, Pressenautomatisierung und Sensorik. In: Hochleistungswerkzeuge der Stanztechnik. Technische Akademie Esslingen
13. Rosa P, Sassanelli C, Terzi S (2019) Towards circular business models: a systematic literature review on classification frameworks and archetypes. J Clean Prod 236:1–17. https://doi.org/10.1016/j.jclepro.2019.117696
14. Biloslavo R, Bagnoli C, Massaro M, Cosentino A (2020) Business model transformation toward sustainability: the impact of legitimation. Manag Decis 58(8):1643–1662. https://doi.org/10.1108/MD-09-2019-1296
15. Lim Y, Venugopal R, Ulsoy AG (2008) Advances in the control of sheet metal forming. IFAC Proc Vol 41(2):1875–1883. https://doi.org/10.3182/1001
16. Zhang G, Li C, Zhou H, Wagner T (2018) Punching process monitoring using wavelet transform based feature extraction and semi-supervised clustering. Procedia Manuf 26:1204–1212. https://doi.org/10.1016/j.promfg.2018.07.156
17. Kolbe M (2020) Stanztechnik: Grundlagen – Werkzeuge – Maschinen, 13th edn. https://doi.org/10.1007/978-3-658-30401-0
18. Jonsson C, Stolt R, Elgh F (2020) Stamping tools for sheet metal forming: current state and future research directions. https://doi.org/10.3233/ATDE200087
19. Fluck E (2019) Technische Handbücher. OTTO Vision Technology GmbH Jena
20. Favoretto C, Mendes G, Filho M, de Oliveira M, Ganga G (2022) Digital transformation of business model in manufacturing companies: challenges and research agenda. J Bus Industr Market 37(4):748–767. https://doi.org/10.1108/JBIM-10-2020-0477
21. Satpute D, Chavan G, Musale S (2020) A literature review on optimization of stamping tool. Int J Emerg Technol Innov Res 7(12):24–26. http://www.jetir.org/papers/JETIR2012006.pdf
22. Eisenhardt KM (1989) Building theories from case study research. Acad Manag Rev 14(4):532–550. https://doi.org/10.2307/258557
23. Braun V, Clarke V (2006) Using thematic analysis in psychology. Qual Res Psychol 3(2):77–101. https://doi.org/10.1191/1478088706qp063oa
24. Cioffi R, Travaglioni M, Piscitelli G, Petrillo A, Parmentola A (2020) Smart manufacturing systems and applied industrial technologies for a sustainable industry: a systematic literature review. Appl Sci 10:1–22. https://doi.org/10.3390/2897
25. Rosa P, Sassanelli C, Urbinati A, Chiaroni D, Terzi S (2020) Assessing relations between circular economy and industry 4.0: a systematic literature review. Int J Prod Res 58:1662–1687. https://doi.org/10.1080/2019.1680896
26. Zacharaki A et al (2020) Toward a new era of refurbishment and remanufacturing of industrial equipment. Front Artif Intell 3:570562. https://doi.org/10.3389/frai.2020.570562
27. Arias LM, Artola G, Porto I (2021) Hot stamping research scenarios form the last decade. Mater Proc 3(26):1–11. https://doi.org/10.3390/IEC2M-09245

28. Sigvant M et al (2019) Friction in sheet metal forming: influence of surface roughness and strain rate on sheet metal forming simulation results. Procedia Manuf 29:512–519. https://doi.org/10.1016/j.promfg.2019.02.169
29. Karen I, Kaya N, Öztürk F (2015) Intelligent die design optimization using enhanced differential evolution and response surface methodology. J Intell Manuf 26:1027–1038. https://doi.org/10.1007/s10845-013-0795-1
30. Wróbel I, Skowronek A, Grajcar A (2022) A review on hot stamping of advanced high-strength steels: technological-metallurgical aspects and numerical simulation. Symmetry 14(969):1–17. https://doi.org/10.3390/sym14050969

Challenges of Low-Carbon Transition of Manufacturing Industry in Developing Countries and Research on the Construction of Green Manufacturing System in China

Yizhi Song[1], Benxiao Yang[1], Xiaqing Liu[1], Jianhua Zhang[1], Frida Li[2(✉)],
Xinyi Tong[2], and Jerome Feldman[2]

[1] Service-Oriented Manufacturing Institute (Hangzhou) Co., Ltd., Hangzhou, China
songyizhi@isom.org.cn
[2] Global Alliance of Innovators E.V., Berlin, Germany
yrli33@yahoo.com

Abstract. Resource and environmental issues are common challenges faced by mankind, and it is a common choice for global economies to promote green growth and implement a new green deal. Based on the current situation that developing countries are in the stage of high input, high consumption and high emission development, this paper analyzes the challenges of resource and energy efficiency, technology, and environmental impacts, and takes China as an example to study and analyze the background of the establishment of the green manufacturing system, the overall strategic framework and implementation strategy, as well as the effectiveness of the current stage. This can provide a reference model for industrial green low-carbon transition and sustainable manufacturing.

Keywords: Manufacturing · Production · Sustainable development · Management · Environment

1 Challenges of Low-Carbon Transformation

83% of the global population lives in developing countries [1]. The industrial development and structure upgrading of developing countries play a pivotal role in global ecological and economic development. However, developing countries face serious challenges in low-carbon transformation, which are mainly reflected in:

The Effects of International Political Situation: The Russian-Ukrainian conflict and the impact of the Corona epidemic intertwined and superimposed, bringing great uncertainty to the international situation. The current international situation of geopolitical tensions and camp confrontation has seriously slowed down the world economic recovery, affecting developing countries in the development of sustainable manufacturing technology exchanges and green economy cooperation.

© The Author(s) 2025
H. Kohl et al. (Eds.): GCSM 2023, LNME, pp. 65–72, 2025.
https://doi.org/10.1007/978-3-031-77429-4_8

High Energy Consumption and High Pollution: Developing countries are mostly at the low end of the industrial chain, and often rely on the exploitation and utilization of water, oil and gas, and mineral resources. Insufficient environmental protection in the processes has further led to increased ecological vulnerability and seriously affected the sustainable development of the region.

Lack of Sustainable Manufacturing Technology: Due to the unsmooth financial chain for the transformation of scientific and technological achievements, the lack of technological innovation talents, the separation of technological research and industrial system etc. have led to insufficient capacity of sustainable manufacturing technology in most developing countries.

Low Productivity, Profit-Driven : The real productivity gap between developing and developed countries has further widened, reaching 1:18 by 2021.The weak recovery of emerging markets will cause inequality between countries and further aggravate inflation [2]. The higher profits of highly polluting enterprises make the lack of motivation for manufacturing transformation and the lack of punishment in some countries, resulting in high pollution, crime, and other social problems.

Insufficient Technical Capacity for Digital Transformation : Small and medium-sized enterprises in developing countries are in the exploration stage of digital transformation, facing difficulties such as a lack of digital talent, a weak foundation for data collection, and a low level of technology application. The difficulties of digital transformation will inevitably affect the transformation of manufacturing industries.

While developing countries are striving to improve the quality of life, the existing challenges and crises will come to a head if the conventional way of development are followed. To break this cycle new innovative and far-reaching solutions are urgently required [3].

2 China's Approach to Meet the Challenges

As the world's most populous developing country, China has developed rapidly. However, its crude development model with the characteristics of "high input, high consumption, high emissions" has brought about huge issues. Resource shortage, environmental pollution and ecological damage have become the bottleneck that restricts the sustainable development of its manufacturing industry [4].

As of 2000, China began to take energy saving and emission reduction as an important part of industrial upgrading. In 2005, the Chinese government issued the "Outline of the National Medium- and Long-Term Scientific and Technological Development Plan (2006–2020)", which listed green manufacturing as one of the three major directions for the development of manufacturing [5].

In the "Twelfth Five-Year Plan" period (2011–2015), the specific plans focused on industrial energy conservation, clean production, comprehensive utilization of resources and environmental protection equipment have been vigorously implemented and achieved certain results. However, in the face of the new requirements of the "Thirteenth Five-Year Plan" (2016–2020), the problem of unsystematic, unbalanced, and uncoordinated green development of industry is becoming increasingly prominent. In

2015, in China's national strategy, "Made in China 2025", the need to strive to build a green manufacturing system that is highly efficient, clean, low-carbon and recycling was explicitly put forward.

For the 14th Five-Year Plan period (2021–2025), China has formulated the "14th Five-Year Plan for Green Industrial Development", which puts forward the overall goal, sub-targets and specific tasks on green industry. The purpose is to strengthen the construction of a green manufacturing support system to lay the foundation for the industrial carbon dioxide peaking by 2030.

3 Green Manufacturing System and Its Development Strategy

China's manufacturing community believes that green manufacturing is a sustainable manufacturing model in the new era, which considers production efficiency, as well as environmental impact and resource efficiency. It strives to achieve high resource utilization efficiency, low energy consumption, and low pollution to the environment in the whole life cycle of product design, production, packaging, transportation, consumption, and end-of-life treatment [6].

3.1 China's Green Manufacturing System

According to the concept of the Chinese Ministry of Industry and Information Technology, the national green manufacturing system consists of four parties of main body and the green manufacturing standard, which is the basic framework for assessing, guiding, and promoting the development of the green manufacturing architecture, as shown in Fig. 1.

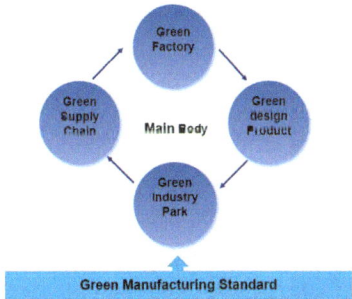

Fig. 1. Main body and layout of China's green manufacturing architecture

Green products are energy-saving, water-saving, low-pollution, renewable and recyclable products, whose production, use and disposal processes take full account of the impact on the resources and environment. It is the ultimate embodiment of green technology application.

Green factory refers to the factory with the smallest negative impact in the whole production and manufacturing process and the highest efficiency of resource utilization, whose production mode is environmentally friendly, capable of clean production, intensification of land use.

Green industry park is a manufacturing park that embodies the concept of resource conservation in terms of infrastructure, ecological environment as well as planning, spatial and temporal layout, industrial chain design, energy/resource utilization, and operation and management. It should focus on product manufacturing and energy supply, with the proportion of industrial added value exceeding 50%.

Green supply chain consists of the enterprises in different positions of the industrial chain, whose procurement, production, marketing, recycling, and logistics etc. work together to enhance the efficiency and performance, and to realize the full utilization of resources with minimal impact on the environment.

Green manufacturing standard architecture consists of 39 groups of standards in seven categories: comprehensive foundation, green products, green factories, green enterprises, green industry parks, green supply chains and green evaluation and services. They are the foundation of green manufacturing implementation [7].

Based on the standards and evaluation indicators, the relevant management departments have formulated a series of operable documents, which can evaluate the product, factory, industry park and supply chain in detail. For example, the green factory evaluation system includes seven level 1 indicators and 25 level 2 indicators. The total score sums up to 100 points. Each evaluation element of green factory and its weight are shown in Table 1 [8]. The general indicator is one vote negative item, and it does not occupy score in Table 1.

Most of the green attribute indicators are quantitative indicators, such as waste gas, wastewater and physical pollution, etc., and the detection and calculation standards of the indicators are relatively sound. Some of the resource and energy evaluation indicators require industry-based data, such as material consumption, use of hazardous substances, material utilization, water saving, energy saving, etc. At present, these standards in this regard are still unsound, which affects, to a certain extent, the credibility of green evaluation. The green management indicators are all qualitative evaluation indicators. There are few evaluation indicators with standard basis, which is the weak link of green evaluation.

3.2   China's Green Manufacturing Development Strategy

China's green manufacturing action and green manufacturing strategy have adopted a systematic methodology with multiple approaches. In addition to industrial structural transformation and energy efficiency improvement, the following strategies have been adopted for green manufacturing:

Establishing policy system support for industrial green development: The industrial administration departments have formulated, revised and completed more than 500 industrial green development standards and documents.

Table 1. Green factory evaluation elements

No	Level 1 Indicators	Level 2 Indicators	Count	Mandatory Option		Selectable Option		Weight
1	Infrastructure	Infrastructure	3	4	12	5	9	20%
		Lighting		2		3		
		Facilities		6		1		
2	Management System	General requirements	4	2	4	2	5	15%
		Environmental management system		1		1		
		Energy Management System		1		1		
		Social responsibility		0		1		
3	Energy Resources Inputs	Energy Inputs	3	1	6	5	9	15%
		Resource Inputs		3		2		
		Procurement		2		1		
4	Products	Eco-design	5	1	3	2	9	10%
		Hazardous Substances Utilization		1		1		
		Energy Saving		1		1		
		Carbon reduction		0		3		
		Recycling		0		2		
5	Environmental Emissions	Air pollution	5	1	5	1	5	10%
		Water pollution		1		1		
		Solid waste		1		0		
		Noises		1		0		
		Greenhouse Gases		1		3		
6	Achievements	Site Scorching	5	3	12	3	12	30%
		Raw material harmlessness		1		1		
		Purification of production30%		3		3		
		Resourcefulness of waste		3		3		
		Energy Decarbonization		2		2		
	In Total		25					100%

Establishing green manufacturing demonstration projects: The advanced green manufacturing models are selected from the national, provincial and municipal levels. They play a demonstration-led role in creating green factories, green industrial parks, and green supply chains.

Green manufacturing system integration: By forming consortiums, a great number of manufacturing system integration projects will be carried out among the important industries like machinery, electronics, chemicals, food, textiles, home appliances, etc. They are focused on solving the industry's weakness in green design and process, and insufficient collaboration between upstream and downstream enterprises.

Promoting integrated service providers: At the national level, green manufacturing solution service providers will be cultivated for providing integrated services for the green manufacturing transformation.

Carrying out fiscal, tax and financial support policies: Comprehensively utilize fiscal, tax, financial and other policies to build the capacity of green finance to support the development of green industry. By the end of 2022, the balance of local and foreign currency green loans reached 22.03 trillion RMB, and the stock of green bonds reached 1.5 trillion RMB [9].

Intelligent support for green development: Continuously accelerates the large-scale green data centers and other digital infrastructure for the green and low-carbon transformation. China has implemented the strategy of "Industrial Internet + Green Manufacturing" creating typical application scenarios of intelligent energy management and fostering the promotion of "Industrial Internet + Green Low Carbon" solutions.

Building bridges for deepening international cooperation: Through bilateral and multilateral mechanisms such as China-EU, China-France, China-Britain, China-Korea, China-Brazil, the Belt and Road Cooperation Initiative, and the BRICS Partnership, the experience-sharing and technology exchanges have been carried out.

4 Effectiveness of China's Green Manufacturing

4.1 Demonstration of Green Manufacturing System

According to relevant information in January 2022, since the Ministry of Industry and Information Technology (MIIT) launched the construction of the green manufacturing architecture in September 2016, the sixth batch of green manufacturing lists have been released, including a total of 3,159 green design products, 2,783 green factories, 223 green industry parks, and 296 green supply chain management demonstration enterprises.

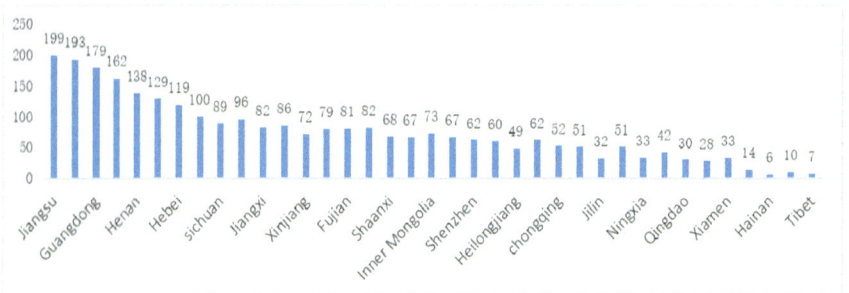

Fig. 2. Regional Distribution of Green Factories

The top five categories of products with the highest number of green products are household detergents, household refrigerators, printed circuit boards, construction materials and composite fertilizers. Green factories come from 19 provinces and cities, and the top three regions with the largest number of green factories are province Jiangsu, Guangdong and Henan, as shown in Fig. 2. The average utilization rate of solid waste disposal in the green industrial parks assessed exceeds 95% and the level of energy and resource utilization has improved significantly.

4.2 Green Manufacturing System Integration Program

Industrial leading enterprises connected their upstream and downstream enterprises, third-party service companies and research institutes to form a consortium, which undertakes the task of green manufacturing system integration, focusing on technology, mode, standard application and innovation by taking demand and problem as a compass [10].

Until now there are a total of 368 green manufacturing system integration projects in three categories: the key process breakthrough projects (57%), design platform building and application (34%) and green supply chain construction (9%) [11].

Thousands of upstream and downstream enterprises of key industries, third-party service companies and research institutions have participated in system integration projects. They built green design information databases, design evaluation tools and data share platforms and intelligent equipment etc.

4.3   Green Manufacturing System Solution Providers

In 2019 and 2020 the open bidding announcement for green manufacturing system solution providers were tendered. Seven main tendering fields of system integration and application include green key process, manufacturing common technology and equipment, product green design and manufacturing, factory digitalization and green updating, manufacturing supply chain, end-product resource utilization and development of data base capacity enhancement [12]. Twelve green manufacturing system solution providers were selected in 2019, and another 75 were added in 2020. The green manufacturing system solution providers enable China's industrial green transformation with a strong service capability.

5   Insights and Future Development Outlook

By setting up initiatives, governments help their economies to adapt to changing environments or initiate a change itself [3]. China's green manufacturing and industrial transformation can be seen in its distinctive features. These can also serve as references for other developing countries.

- **Enhancement of overall coordination and implementation capacity**: Led by government departments and industry organizations, the active bodies establish the evaluation standards, select models of products, factories, industry parks, and supply chains to demonstrate and encourage others. At the same time promotes the joint research and sharing of key technological innovations in green and smart manufacturing.
- **Support manufacturing enterprises with service**: give them training about Environmental Social Governance (ESG), guide them to establish the carbon foot printing system as well as lead them to carry out the Science-Based Targets Initiative (SBTi) certification etc. so that they can integrate into the international green and low-carbon industrial chain.

For the future, China should strengthen the international cooperation and exchange of green manufacturing and intelligent manufacturing. China's evaluation standards in green manufacturing system, especially the evaluation indexes of green supply chain need to be improved. It is necessary to supplement the basic data of the industry, strengthen the data tracking of the whole supply chain, and improve the credibility of the green evaluation.

References

1. United Nations Conference on Trade and Development, Now 8 billion and counting: Where the world's population has grown most and why that matters. Available at https://unctad.org/data-visualization/now-8-billion-and-counting-where-worlds-population-has-grown-most-and-why (retrieved 15 Nov 2022)

2. United Nations, Global Sustainable Development Report 6 2023 7 8 9 10 Advance, Unedited Version 2023. Available at https://sdgs.un.org/sites/default/files/202306/Advance%20uned ited%20GSDR%2014June2023.pdf
3. Kohl H, Schliephack W, Muschard B (2020) Increasing challenges for sustainability for manufacturing industry based on global, national, and technological initiatives. In: 17th global conference on sustainable manufacturing, Procedia manufacturing, 43, 293–298
4. Shi D (2018) Theory and practice of china's industrial green development-along with policy options for deepening green development in the 19th national congress. J Fin Econ 1:3–11
5. Huajun C (2020) Status and future development strategy of green manufacturing research. China Mech Eng 2:135–144
6. Ministry of Industry and Information Technology of the People's Republic of China, Interpretation of "Made in China 2025". Available at https://wap.miit.gov.cn/ztzl/lszt/zgzz2025/zcjd/art/2020/art_5183349830704af084057f04ef30ad7f.html (retrieved Oct 2022)
7. State Council Information Office of the People's Republic of China, Ministry of Industry and Information Technology Interpretation of the Green Manufacturing Standard System Construction Guidelines, Beijing 2016
8. Sun X (2022) Green factory series: what does the green factory evaluation indicator include? Available at https://www.chndaqi.com/news/305037.html (retrieved 11 Oct 2022)
9. State Council Information Office of the People's Republic of China, China's green loans to maintain high growth in 2022. Available at https://www.gov.cn/xinwen/2023-02/03/con tent_5739935.htm (retrieved Feb 2023)
10. Ministry of Industry and Information Technology (2016), Circular of the ministry of finance ministry of industry and information technology on organizing green manufacturing system integration work, Beijing
11. Beijing Volcano Power Network Technology Co., Ministry of Industry and Information Technology (2016) Green manufacturing system integration project announcement (List). Available at 2016 "Green manufacturing system integration project announcement (list)-Polaris energy storage network (bjx.com.cn)" (retrieved Nov 2022)
12. China Recruitment International Bidding Co (2019) Bidding announcement for green manufacturing system solution providers. Available at http://cntcitc.com.cn/smore.htmlcontentId= 0455818d40414b118022b425be294bfa. (retrieved July 2022)

Risk Assessment for Small Manufacturing Businesses in Soweto

Takalani Tshabalala[✉]

University of Johannesburg, 55 Beit St, Doornfontein, Johannesburg 2028, South Africa
tshabalalat@uj.ac.za

Abstract. A risk assessment is tool that assists businesses to identify, analyse and evaluate various types of risks that their severity. The manufacturing sector serves as the backbone of the social and economic development in South African, Soweto. Risks assessments have been conducted in various sectors but not the Soweto manufacturing businesses. It is therefore imperative that this study focuses on this very important sector to assist it in being sustainable. A risk assessment was conducted for 120 manufacturing businesses in Soweto. The study identified 5 types of risks namely the financial, operational, business, legal and third-party risk. The businesses were categorised based on the outcome of the production, and a risk rating tool was used to analyse and evaluate the level of risks that exists. The risk rating tool categorised the risks in terms of low-risk, medium-risk, and high-risk. The tools' outcome was the total risk type per manufacturing business type and the risk values per risk type. The study found that financial risk is a predominant risk that affects all the manufacturing businesses that were engaged, and that food, glass, textile, and wood manufacturers are the businesses impacted most.

Keywords: Risk Assessment · Manufacturing businesses · Risk rating

1 Introduction

A risk assessment in an organisation is defined as a process of identifying potential risks that may cause harm in organisation and analyzing the depth of these hazards [1]. Risk assessment is generally concerned with three phases, namely, Risk identification, Risk analysis and Risk Evaluation [2, 3]. The identification, analysis and evaluation of risk is conducted primarily through the engagement with business owners as they would have the advanced knowledge on the operations management of the business [4]. The purpose of a risk assessment is to ensure that the working environment is safe to work in, and that all measures are in place to ascertain the worker's safety. A risk assessment ensures that business is well prepared for the any potential hazards that may be present in the business environment [1].

The manufacturing industry is regarded as the pillar of social and economic development. The sector contributes largely the level employment in South Africa and facilitates the alleviation of poverty in the country. The manufacturing sector constitutes 12% of

H. Kohl et al. (Eds.): GCSM 2023, LNME, pp. 73–80, 2025.
https://doi.org/10.1007/978-3-031-77429-4_9

the total GDP in South Africa [5]. Soweto is a township in Johannesburg that is stricken by poverty and unemployment [6]. The manufacturing sector in Soweto is faced by many challenges including lack of funding, large competition, and negative economic impacts. It is therefore imperative that a risk assessment of these manufacturing businesses is conducted. Literature indicates that risk assessments are predominant within various sectors, including the banks and telecommunications, however, there has been no available data indicating a risk assessment conducted specifically in the manufacturing businesses in Soweto [2, 7, 8]. A risk assessment in this regard, will assist in identifying the type of risks that exists within the manufacturing sector in Soweto, analyse and evaluate these risks. The risk assessment will further allow these businesses to mitigate the potential risks and eventually achieve sustainability within the local manufacturing industry.

This study aims to conduct a risk assessment for the manufacturing businesses in Soweto. The identification of risk will include classifying the type of risks that exist in businesses. Analysis of risk will be done through assigning values to each risk type, and finally the evaluation of risk will be concluded by calculating the total risk values per business manufacturing type.

The following sections are included in this research paper; Sect. 2 focuses on literature review, Sect. 3 discusses the methodology used, Sect. 4 addresses the results, Sect. 5 explains the recommendations and Sect. 5 provides the conclusion of the research.

2 Literature Review

According to [9] there are specific tools that are suitable for a risk assessment in manufacturing sector, specifically in Small Medium Enterprises. Oduoza [10] recommends that the application of the generic risk management system that is a five step process, is mostly applicable. This system includes;

a. Establishing the context of the risk and the potential hazard exposure
b. Risk assessment which involves risk identification, risk analysis and evaluation (is it worthwhile, cost benefit analysis).
c. Risk treatment which could be; reduce, avert, transfer, mitigate, retain the risk or if it is an opportunity to exploit, share, enhance, or ignore it.
d. Monitor and Review
e. Communicate and consult

Islam and Tedford [11] does however argue that depending on the manufacturing operation, the sequence of risk management activities is broken down into three sections.

(1) Identification of key risk indicators (KRI) from potential risk factors that underpin enterprise performance in that sector. This is achieved using algorithms such as Bayesian Belief Network (BBN), Probabilistic Neural Network (PNN), Analytical Hierarchy Process (AHP).
(2) The key risks indicators that could impact on cost, safety, quality and time management in the manufacturing process are critical.
(3) Different combinations of the key risk indicators interact and could impact on one or more of the performance indicators

The safety risk can be evaluated based on the intermediate distance between the human and any moving object that may cause an injury [12]. For the calculation of the distance values, among the entities, the coordinates of the robot's joints, which correspond to a certain frame timestamp, are also imported into the distance calculation algorithm. Jones and Kumar [13] Based on the assembly process specifications, different control, safety and operator support strategies have to be implemented in order for the human safety and the overall system's productivity to be ensured. The variability of assembly processes (screwing, insertion, fixing etc.), components (small/big/flexible parts), robotic equipment (single arm, dual arm, overhead) and station layouts, dictates different approaches, in terms of human-robot interaction. Operational risk assessment according to [14] is illustrated as a "conditional view" function that is responsible of the provision of a remaining useful life (RUL) prediction of a health managed component. This conditional view provides a basis for operational risk estimation, which in turn serves to identify maintenance actions that can be deferred. The risks that are most prevalent in the manufacturing sector are human risk factor, organisational risk factors, technology risk factors, economic, political, financial risk factors, and Environmental risk factors [15] however [16] found the following risks exist in manufacturing industries, Cost Related Risk Factors, Time Related Risk Factors, Quality Related Risk Factors and Safety Related Risk Factors. This evidences that each organisation faces various types of risks and that includes the manufacturing sector.

3 Risk Assessment Method

Risk is rated on the impact on the business, which can be economical or reputational, and its likelihood of occurring shortly. This is the common pattern of risk across businesses. The impact of risk is rated as follows [17];

- Low: A low-rated event is one with little/no impact on the business activities and the reputation of the firm.
- Medium: An event resulting in risks that can cause an impact but not a serious one is rated as medium.
- High: A major event that can cause reputational and economic damage, resulting in huge business and client base losses.

The risk-rating formula is based on its recurrence, which can change depending on the type of business that is being considered [17]. For example, for a fast-food company, a frequent likelihood rating will be something that can happen every day, whereas, for an investment bank it would be something that happens frequently.

1. Frequent
2. Likely
3. Rare

A total of 120 small manufacturing businesses were identified in Soweto for the risk assessment. The businesses were classified under the following manufacturing categories: Food, Plastic, Metal, Textile, Chemical, Wood, Electronics, Cut and Sew Apparel, Glass. The businesses were assessed under five types of risks [1], namely; Financial Risk, Operational Risk, Business risk, Legal risk and Third-Party risk.

Table 1. Risk type rating

	Financial Risk	Operational Risk	Business risk	Legal risk	Third-Party risk	Total Manufacturing business type
Food	(3)	(3)	(3)	(3)	(3)	15
Plastic	(3)	(2)	(3)	(1)	(1)	10
Metal	(3)	(3)	(3)	(3)	(3)	15
Textile	(3)	(2)	(3)	(2)	(3)	13
Chemical	(3)	(2)	(1)	(2)	(1)	9
Wood	(3)	(3)	(3)	(3)	(3)	15
Electronics	(3)	(2)	(1)	(1)	(3)	10
Cut and Sew Apparel	(3)	(3)	(1)	(1)	(1)	9
Glass	(3)	(3)	(3)	(3)	(2)	14
Total per risk type	27	23	21	19	20	

High risk = (3) Medium risk = (2) Low risk = (1)

A risk rating tool [18] was provided to the businesses to identify the type of risk that exists withing their businesses and rate the severity of the risk. The risk rating tool was adopted from literature [1, 2] as a means to assess the business risks. Five risk types were assessed based on each manufacturing businesses (see Table 1). The risks were coded (3) for a high risk, (2) for a medium risk and (1) for a low risk. A business category that reports a total between 0 and 5 is ranked as a low total risk, 6–10 is ranked as medium total risk, and 11–15 is ranked as high total risk.

4   Results and Discussion

A total of 120 manufacturing businesses participated in the study. The largest contributor of all manufacturing businesses is the food manufacturers in Soweto with a total of 38 businesses that were engaged. The second largest type of businesses in the metal manufacturers with a total of 30 businesses. 16 and 12 businesses were categorised as the wood and plastic manufacturers.

The least number of manufacturers that engaged in the study include 9 businesses in the cut-and-sew apparel, 6 businesses in textile, 4 businesses in textile, 3 businesses in the electronic equipment manufacturers and 2 businesses in glass manufacturing.

The classification of manufacturing businesses in Soweto is new, although the business directory includes most of these business operations, however no document has yet classified these businesses under the above specific categories. There is a clear domination of the food and metal manufacturers in Soweto compared to all other types of manufacturing businesses.

4.1   Risk Totals Per Business Type

The results indicated that the businesses categorised under food and wood manufacturing had the highest risks in all five categories. Glass and textile categorised businesses had the second highest risk, showing only a medium risk in operational, Legal, and third-party risk items. The plastic, electronics, chemical and cut-and sew apparel categorised businesses have reported a medium to low risk under Operational, Business, Legal and Third-party risks.

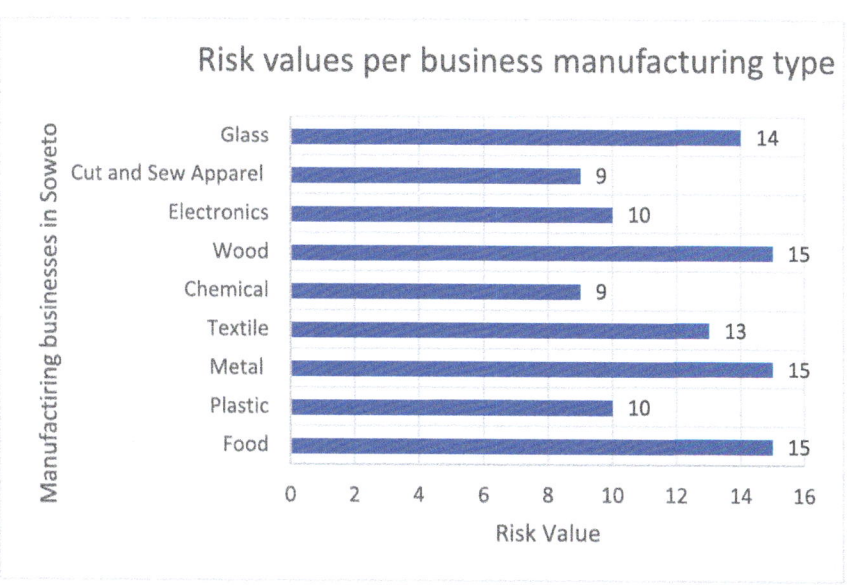

Fig. 1. Risk values per business manufacturing type. 0–5 = Low risk, 6–10 = Medium risk, 11–15 = high risk

It can be concluded, as indicated in Fig. 1, that businesses within the manufacturing sector in Soweto are facing difficulties based on the large rating of risks that experiencing. The food, metal, textile, wood, and glass businesses are experiencing the most risks withing their working environment. To achieve business sustainability, it is important that these businesses adopt mitigating strategies for the identified risk to allow for long-term survival in the business environment.

4.2   Risk Values Per Risk Type

Financial risk has dominated as the highest risk type amongst all manufacturing businesses concerned. The impact of financial risk is directly related to the struggling South African economy, and therefore poses as the highest threat in the Soweto manufacturing sector. Operational rand business risks, reported to be the second and third risk factors that are affecting the sector. With a small margin, third-party and legal risk were the last two lowest risk factors reported.

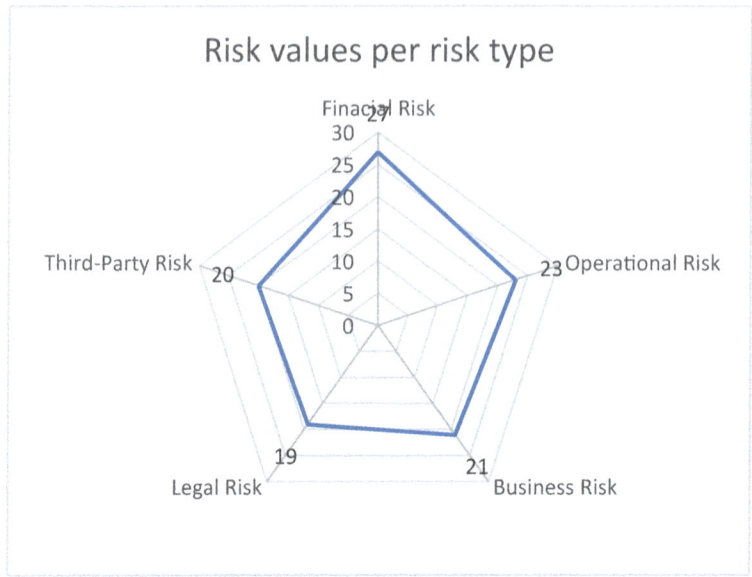

Fig. 2. Risk values per risk type

Third party risk is expected to increase as South African faces energy power issues. The country currently faces high levels of load-shedding, ensuring that businesses are without power for longer hours. The impact of third-party risks is expected to be reported by businesses in the last term of the 2023.

All five risks identified pose a threat to the sustainability and development of local manufacturing businesses in Soweto, see Fig. 2, however a mitigation of the financial risk would have a direct reduction in the operational and business risk. This would allow the business to apply resources to reduce the legal and third-party risks.

5   Conclusion

A risk assessment tool was used to analyse five types of risks that exist within the manufacturing companies in Soweto. The study found it evident that the financial, business, and operational risk pose a challenge for most of the businesses in manufacturing. Manufacturing sustainability will only be achieved if mitigation strategies are adopted to reduce the risks and their impact. The major risk that is a bigger threat to the survival of manufacturing businesses, specifically the food, metal, wood, and glass manufactures, is the financial risk, it is estimated that the sustainability of these businesses will not be possible.

A risk appetite and risk tolerance need to be conducted in future research. This will enable the development of a practical risk mitigation strategy that would allow the success and survival of manufacturing businesses.

References

1. Aven T (2016) Risk assessment and risk management: review of recent advances on their foundation. Eur J Oper Res 253(1):1–13
2. Fenton N, Neil M (2018) Risk assessment and decision analysis with Bayesian networks. CRC Press
3. Wangen G (2017) Information security risk assessment: a method comparison. Computer 50(4):52–61
4. Zio E (2018) The future of risk assessment. Reliab Eng Syst Saf 177:176–190
5. Bank SAR (2020) Occasional bulletin of economic notes (no. 10407). South African Reserve Bank
6. Davis A (2015) Harnessing entrepreneurial potential in Soweto as a catalyst for inclusive growth. New Vision Public Affairs 7:22–35
7. Malekipirbazari M, Aksakalli V (2015) Risk assessment in social lending via random forests. Expert Syst Appl 42(10):4621–4631
8. Van Greuning H, Bratanovic SB (2020) Analyzing banking risk: a framework for assessing corporate governance and risk management. World Bank Publications
9. Oduoza CF (2020) Framework for sustainable risk management in the manufacturing sector. Procedia Manuf 51:1290–1297
10. Islam MA, Tedford D (2012) Implementation of risk management in manufacturing industry—an empirical investigation. People 2(3):258–267
11. Covello VT, Mumpower J (1987) Risk analysis and risk management: an historical perspective. Risk Anal 5(2):103–120
12. Jones T, Kumar S (2007) Comparison of ergonomic risk assessments in a repetitive high-risk sawmill occupation: saw-filer. Int J Ind Ergon 37:744–753
13. Michalos G, Makris S, Tsarouchi P, Guasch T, Kontovrakis D, Chryssolouris G (2015) Design considerations for safe human-robot collaborative workplaces. Procedia CIrP 37:248–253
14. Papachatzakis P, Papakostas N, Chryssolouris G (2007) Condition based operational risk assessment an innovative approach to improve fleet and aircraft operability: maintenance planning. In: 1st European air and space conference, Berlin, Germany, pp 121–126
15. Tchankova L (2002) Risk identification-basic stage in risk management. Environ Manag Health 13(2/3):290–297
16. Michael MH, Paul N (2006) Strategic risk management using complementary assets: organizational capabilities and the commercialization of human genetic testing in the UK. Res Policy 35:355–374
17. Crouhy M, Galai D, Mark R (2001) Prototype risk rating system. J Bank Finance 25(1):47–95
18. Mulder P, Kousky C (2023) Risk rating without information provision. AEA Pap Proc 113:299–303

Criteria-Driven Comparison of Hydrogen, Charcoal and Liquefied Petroleum Gas as Cooking Fuels Based on a Systematic Literature Review

Lukas Sturm[1]([✉]), Semih Severengiz[1], Dhanashri Satish Salokhe[2], and Gaurav Bhatia[2]

[1] Bochum University of Applied Sciences, Sustainable Technologies Laboratory, Am Hochschulcampus 1, 44801 Bochum, Germany
lukas.sturm@hs-bochum.de
[2] Technische Universität Berlin, Global Production Engineering, Pascalstraße 8-9, 10587 Berlin, Germany

Abstract. Green Hydrogen could be used as a sustainable cooking fuel, especially in countries of the global south in which cooking with carbon-based, or biomass fuels is prevalent. This paper presents a criteria based assessment of the sustainability of cooking with hydrogen in comparison to charcoal and liquefied petroleum gas. The research methodology adopts a systematic literature review to provide a comprehensive overview of the research. Based on this, a morphological analysis is conducted to display possible hydrogen cooking scenarios. A chosen hydrogen scenario is then subjected to a sustainability assessment and compared with the aforementioned conventional cooking methods. The findings of this paper reveal that while hydrogen presents itself as a potential alternative, it is currently not a more sustainable option. Despite its lower climate change impact when compared to liquefied petroleum gas and charcoal, hydrogen falls short in several other sustainability aspects, such as investment costs, marine and freshwater eutrophication or safety concerns related to the usage of the energy source.

Keywords: Green Hydrogen · Cooking · Sustainability Assessment

1 Introduction

Cooking with carbon-based or biomass fuels such as kerosine, charcoal or firewood is unsustainable due to two major reasons. It is very harmful to human health and it contributes to global warming. Those cooking fuels are predominantly used in developing economies and affect the health of almost 3 billion people every year, largely due to the particulate matter, carbon monoxide and hydrocarbons emitted during their combustion [1]. In sub-Saharan Africa alone, 80% of the population relies on biomass for energy use in their homes, 95% of which is for cooking needs [2]. Combustion of solid biomass (e.g. charcoal, firewood, waste) is accountable for 25% of global black carbon emissions,

H. Kohl et al. (Eds.): GCSM 2023, LNME, pp. 81–88, 2025.
https://doi.org/10.1007/978-3-031-77429-4_10

estimated to have the second largest radiative forcing behind CO2 [3]. In contrast to those conventional cooking fuels, hydrogen presents the potential for decarbonization and the ability to mitigate air pollution within households. Existing research mainly concerns the technical feasibility of cooking with hydrogen, highlighting the compatibility of stoves with hydrogen as fuel [4, 5]. Little research deals with the sustainability of cooking with hydrogen. A few studies concern with the ecological or economic impact of hydrogen [1, 3], leaving out the social perspective. Therefore, this paper summarizes the applications of cooking with hydrogen and assesses the sustainability in comparison to existing alternatives. The first part of the paper adopts a systematic literature review (SLR) to provide a comprehensive overview of the existing research. Based on this, a morphological analysis is conducted to display possible hydrogen cooking scenarios. The selected hydrogen cooking scenario is then subjected to a criteria-driven sustainability assessment and compared with conventional cooking methods. The assessment encompasses various criteria, including climate change impact, investment costs and safety considerations. Through this research, we aim to systematize existing literature related to cooking with hydrogen and shed light on the potential of cooking with hydrogen as a sustainable alternative to conventional cooking methods.

2   Methodology

In this work, a SLR was conducted to obtain an overview of the existing literature on cooking with hydrogen, to analyse it and extract specific information on the possible technical implementation of hydrogen cooking technologies, as well as to collect data for a robust criteria-driven sustainability assessment of the cooking scenarios considered. The SLR is based on various standards and guidelines summarized by Snyder [6]. The databases Scopus and Web of Science were analysed. The search string was designed to cover all papers dealing with the topic of cooking with hydrogen. Accordingly, the keyword "hydrogen cook*" plus variations in combination with the keyword "stove" plus variations was used to retrieve the existing literature. Additionally, to cover papers dealing with sustainability aspects the keywords "Sustainability Assessment AND Cook*" plus variations were used. The query for the defined keywords was narrowed down to the article's title, abstract and keywords. In total, a sample of 172 publications for analysis after excluding duplicates were acquired. The selection and analysis of publications followed the procedure proposed by Tranfield et al. [7] (see Table 1).

Table 1. Procedure for the selection and analysis of the publications.

No.	Step	Number of papers after review step
1	Title and abstract reading	42
2	Introduction and conclusion reading	32
3	Full paper reading	30

Using the 30 chosen publications, a comprehensive content analysis was conducted to extract data on existing hydrogen cooking scenarios and subsequently assess the sustainability of both hydrogen and conventional cooking approaches. A synthesis of the extracted data was performed and is presented in Sect. 3. For the following cooking scenarios sufficient data for the sustainability assessment were available (see Table 2).

Table 2. Cooking scenarios compared in the criteria-driven sustainability assessment.

Scenario	Description
Charcoal	Charcoal bought, stored and burned for cooking
LPG	Liquefied petroleum gas (LPG) bought, stored in cylinders and burned for cooking
Green Hydrogen	Hydrogen self-produced by solar mini-grid and proton exchange membrane electrolyser, stored in cylinders and burned for cooking

The sustainability assessment was based on a criteria catalogue divided into ecological impacts, economic impacts and social acceptance. For the ecological impacts data of Life Cycle Assessments (LCA) according to the guidelines in ISO 14040/44 [8, 9] were retrieved. Regarding the economic impacts, the study primarily focused on costs related to the investment, operation, and maintenance of cooking systems. The investment costs were calculated taking into account the expenses for the stoves and, in the hydrogen scenario, also for the solar mini-grid, electrolyser and hydrogen storage. For the operation costs, we only considered the expenditures related to the fuels used in each scenario. Since little data about the social acceptance of the scenarios were found, a survey was conducted to complement the existing data. The survey involved 209 participants, mainly from India and Ghana.

3 Results

3.1 Systematic Literature Review

Based on the 30 publications filtered by the SLR a morphological analysis was conducted presenting possible scenarios for cooking with hydrogen and referencing the sources from which the data originated (see Table 3). Due to the data availability we chose the in Table 2 defined "Green Hydrogen" scenario for comparison which is highlighted in italicized value in the morphological box (MB). Table 4 presents the analysed publications and the data extracted for the criteria-driven sustainability assessment.

3.2 Criteria Based Sustainability Assessment

In order to assess the ecological sustainability of the cooking scenarios, criteria 1–10 were synthesized (see Table 4). As can be seen, the hydrogen scenario is the best option regarding four criteria: climate change, ozone depletion, fossil fuel depletion and photochemical oxidants formation (the latter jointly with LPG). The greatest difference is noticed for the criterion climate change. However, hydrogen performs worst with

Table 3. MB depicting hydrogen cooking scenarios found in analysed publications. The chosen scenario for the sustainability assessment is highlighted in italicized value. Fuel: Hydrogen cooking fuel, Sorc: Hydrogen energy source, Prod: Hydrogen production technology, Stor: Hydrogen storage technology, Stov: Stove, PEM: Proton exchange membrane electrolyser, AEL: Alkaline electrolyser, CGC: Compressed gas cylinder, MHC: Metal hydride cylinder, CHCC: Catalytic hydrogen combustion cooker, DHCC: Direct hydrogen combustion cooker, H2: Hydrogen.

Parameter	H_2 Cooking Scenario Criteria					Source
	Fuel	Sorc	Prod	Stor	Stov	
H_2	X					1, 5, 10–17
H_2 and methane						4, 10, 18, 19
H_2 and natural gas						20, 21
H_2 and LPG						1, 3
PV		X				1, 3, 11–16, 18
Wind						11, 14
PEM			X			1, 5, 15, 16
AEL						3, 12
CGH				X		1, 3, 15
MHC						5, 11, 14, 16
CHCC						5, 11, 12, 17
DHCC					X	11, 13, 14, 16, 17, 19
Gas stove working with H_2						3, 4, 10, 15, 18

regard to metal depletion. The remaining criteria show different trends. For instance, hydrogen performs better than charcoal for nine out of ten criteria. The biggest difference is seen for photochemical oxidants formation. Furthermore, the hydrogen scenario requires significantly less primary energy. Relative to LPG, hydrogen has lower fossil fuel depletion, ozone depletion and climate change contribution. For six other criteria hydrogen performs worse than LPG, namely terrestrial acidification, freshwater and marine eutrophication, primary energy demand, metal depletion and human toxicity.

To assess the economic sustainability of the cooking scenarios, we synthesized criteria 11–13. Hydrogen emerges as the most favourable option for operation cost. However, in this scenario significant investment costs are incurred for the acquisition of the solar mini-grid and the hydrogen system, leading to a much higher investment compared to other scenarios where only the cooker incurs these costs. Consequently, the maintenance costs in the hydrogen scenario are significantly elevated as well. In summary, while hydrogen proves to be the best option for operation cost due to its lack of ongoing costs, it falls short in two out of three criteria, primarily due to the substantial investment and maintenance costs associated with the required technical equipment.

To assess the social acceptance of the cooking scenarios, we synthesized the results of the aforementioned survey, focusing on criteria 14–17, which provide a summary of the data extracted from the responses. Hydrogen is considered the cleanest cooking fuel

Table 4. Publications' analysis for sustainability assessment. Detailed information on the criteria can be found in the sources indicated. The values given correspond to average values from the publications indicated. Dim: Dimension, SC: Scenario Charcoal, SL: Scenario LPG and SGH: Scenario Green Hydrogen, CC: Climate Change, TA: Terrestrial acidification, FE: Freshwater eutrophication, ME: Marine eutrophication, PED: Primary energy demand, OD: Ozone depletion, POF: Photochemical oxidants formation, FFD: Fossil fuel depletion, MD: Metal depletion, HTP: Human toxicity potential, InvCost: Investment cost, OpCost: Operation cost, MaC[1]ost: Maintenance Cost.

Dim	No	Criteria	Measured Unit	Value SC	Value SL	Value SGH	Source
Environmental Impact	1	CC	g CO_2-eq/MJ	248	365	4	1, 3, 22, 23
	2	TA	g SO_2-eq/MJ	7.5	2.2	3.5	1
	3	FE	mg P-eq/MJ	84	4	22	1, 22, 23
	4	ME	mg N-eq/MJ	24	7.5	16	1, 22, 23
	5	PED	MJ/MJ	0.109	0.020	0.036	1
	6	OD	mg CFC-11-eq/MJ	0.013	0.026	0.007	1, 22, 23
	7	POF	g NMVOC-eq/MJ	4.2	0.2	0.2	1
	8	FFD	g oil-eq/MJ	14.65	55	11	1, 23
	9	MD	g Fe-eq/MJ	2.95	2.55	13.40	1, 23
	10	HTP	g 1,4-DB-eq/MJ	158	11.5	52	1, 22, 23
Economic Impact	11	InvCost	USD	10.11	45.25	9222	3, 13, 17
	12	OpCost	USD/kg	0.34	0.97	0[1]	13, 16, 17, 24–26
	13	MaCost	USD/year	1	1.40	275.16	3, 27
Social Acceptance	14	Inaccessibility	Perception of inaccessibility	1.52[2]	1.41	2.59	Survey
	15	Danger	Perception of danger	2.32	1.69	1.95	Survey
	16	Indoor air pollution	Perception of indoor air pollution	2.80	1.48	1.52	Survey
	17	Unfamiliarity	Unfamiliarity with energy source in %	8.19	8.17	34.8	Survey

in terms of indoor air pollution, alongside LPG. However, there are certain drawbacks associated with hydrogen. Participants perceive it as less accessible than charcoal and LPG, and it is relatively unfamiliar as a cooking fuel to most respondents. Furthermore,

---

[1] The authors assumed that the operating costs in SGH are negligible.

[2] Based on the survey options, responses were quantified. Criteria 14-16 were rated on a scale of 3 to 1. The higher the value, the higher the perceived danger/air pollution/accessibility.

when comparing hydrogen to LPG, hydrogen falls short in 3 out of 4 criteria. Specifically, LPG is perceived to be safer, more accessible, and enjoys higher familiarity among people. In conclusion, while hydrogen has some positive attributes, such as being considered clean, it faces challenges in terms of accessibility and familiarity, and it scores lower than LPG on 3 out of 4 social acceptance criteria.

Summarizing the results of each criterion of the social, economic and environmental impact respectively acceptance, it is evident that hydrogen underperforms in the areas of social acceptance and economic impacts compared to both alternatives. In the area of environmental impact hydrogen excels in reducing greenhouse gas emissions and ozone depletion and performs significantly better than the charcoal scenario in most of the considered criteria. However, compared to LPG it performs worse in 6 out of 10 environmental criteria.

4 Discussion and Conclusion

Through this research, our primary objective was to systematize existing literature related to cooking with hydrogen, outline various potential scenarios, and assess the sustainability based on several criteria in comparison to conventional cooking methods. The results of the sustainability assessment show significant differences between cooking with hydrogen, LPG and charcoal. Hydrogen is the best option for 5 criteria: Operation cost, climate change, ozone depletion, fossil fuel depletion and photochemical oxidants formation (the latter jointly with LPG). However, it is also the worst option for the criteria investment cost, maintenance cost, unfamiliarity with the energy source, inaccessibility of the energy source and metal depletion. The selected hydrogen cooking scenario was found to be more sustainable in 12 out of 17 considered criteria compared to cooking with charcoal. On the other hand, it is worse than LPG for 11 out of 17 criteria. The exception to this are the criteria operation cost, climate change, ozone depletion and fossil fuel depletion. A key reason for these differences is the still emerging green hydrogen economy, resulting in significant production costs for off-grid renewable hydrogen systems. In contrast, the market for LPG and charcoal is well-established, as is their familiarity and acceptance as cooking fuels. Furthermore, while the hydrogen scenario primarily excels in reducing greenhouse gas emissions due to the exhaust-free combustion, its overall environmental performance is compromised by extensive infrastructure requirements.

This research does come with certain limitations, primarily concerning the sustainability assessment process. The comparison in this paper does not include weighted criteria. However, a prioritization of criteria would be helpful for a qualified evaluation. The prioritization could be conducted through pairwise comparisons by experts, based on stakeholder input. The prioritisation of criteria can be also useful as a basis for promoting technical solutions through public incentives and as an aid for investment decisions by companies. For this reason, the compiled results are more useful as indications for further research. The findings show that empirical research, e.g. through real laboratories and experiments, would be necessary to collect reliable social and economic data. Acceptance research should also be conducted after people have gained experience with hydrogen as cooking fuel, since most people have no experience or even access to hydrogen so far. To make a long-term alternative possible investment costs for cooking with green hydrogen would have to drop rapidly and familiarity with hydrogen would have to be significantly increased. Considering the environmental impacts, it could be justified to replace LPG and charcoal with hydrogen if climate change in particular is prioritized. However, the mitigation of climate change would be at the expense of other environmental impacts that should be taken into account.

References

1. Schmidt Rivera XC, Topriska E, Kolokotroni M, Azapagic A (2018) Environmental sustainability of renewable hydrogen in comparison with conventional cooking fuels. J Clean Prod 196:863–879. https://doi.org/10.1016/j.jclepro.2018.06.033
2. International Energy Agency (2022) Africa energy outlook 2022
3. Schöne N, Dumitrescu R, Heinz B (2023) Techno-economic evaluation of hydrogen based cooking solutions in remote African communities—the case of Kenya. Energies 16(7). https://doi.org/10.3390/en16073242
4. Liu X, Zhu G, Asim T, Mishra R (2023) Combustion characterization of hybrid methane hydrogen gas in domestic swirl stoves. Fuel 333. https://doi.org/10.1016/j.fuel.2022.126413
5. Fumey B, Stoller S, Fricker R, Weber R, Dorer V, Vogt UF (2016) Development of a novel cooking stove based on catalytic hydrogen combustion. Int J Hydrog Energy 41(18):7494–7499. https://doi.org/10.1016/j.ijhydene.2016.03.134
6. Snyder H (2019) Literature review as a research methodology: an overview and guidelines. J Bus Res 104:333–339. https://doi.org/10.1016/j.jbusres.2019.07.039
7. Tranfield D, Denyer D, Marcos J, Burr M (2004) Co-producing management knowledge. Manag Decis 42(3/4):375–386. https://doi.org/10.1108/00251740410518895
8. International Organization for Standartization. ISO 14040: environmental management—life cycle assessment principles and framework. International Organization for Standardization, Geneva
9. International Organization for Standardization. ISO 14044: environmental management and life cycle assessment and requirements and guidelines. International Organization for Standardization, Geneva
10. Palacios A, Bradley D (2022) Hydrogen and wood-burning stoves. Philos Trans R Soc Math Phys Eng Sci 380(2221). https://doi.org/10.1098/rsta.2021.0139
11. Großmann U-P, Lehmann J, Menzl F (2000) Non-stationary hydrogen cooker with portable hydride storage and catalytic hydrogen burner. Int J Hydrog Energy 25(1):87–90. https://doi.org/10.1016/S0360-3199(99)00008-7
12. Hernández Y et al (2018) Development and characterization of an ecological hydrogen stove. J New Mater Electrochem Syst 21(1):33–36. https://doi.org/10.14447/jnmes.v21i1.519
13. Onwe CA, Rodley D, Reynolds S (2020) Modelling and simulation tool for off-grid PV hydrogen energy system. Int J Sustain Energy 39(1):1–20. https://doi.org/10.1080/14786451.2019.1617711
14. Suha Yazici M, Yavasoglu HA, Eroglu M (2013) A mobile off-grid platform powered with photovoltaic/wind/battery/fuel cell hybrid power systems. Presented at the 8th international journal of hydrogen energy, pp 11639–11645. https://doi.org/10.1016/j.ijhydene.2013.04.025
15. Topriska E, Kolokotroni M, Dehouche Z, Potopsingh R, Wilson E (2015) The application of solar-powered polymer electrolyte membrane (PEM) electrolysers for the sustainable production of hydrogen gas as fuel for domestic cooking. Renew Energy Serv Mankind 1:193–203. https://doi.org/10.1007/978-3-319-17777-9_18

16. Topriska E, Kolokotroni M, Dehouche Z, Novieto DT, Wilson EA (2016) The potential to generate solar hydrogen for cooking applications: case studies of Ghana, Jamaica and Indonesia. Renew Energy 95:495–509. https://doi.org/10.1016/j.renene.2016.04.060
17. Mukelabai MD, Wijayantha KGU, Blanchard RE (2022) Hydrogen for cooking: a review of cooking technologies, renewable hydrogen systems and techno-economics. Sustainability 14(24):16964. https://doi.org/10.3390/su142416964
18. Fang Z, Zhang S, Huang X, Hu Y, Xu Q (2023) Performance of three typical domestic gas stoves operated with methane-hydrogen mixture. Case Stud Therm Eng 41. https://doi.org/10.1016/j.csite.2022.102631
19. Jones DR, Dunnill CW (2021) On the initiation of blow-out from cooktop burner jets: a simplified energy-based description for the onset of laminar flame extinction in premixed hydrogen-enriched natural gas (HENG) systems. Fuel 294. https://doi.org/10.1016/j.fuel.2021.120527
20. Zhao Y, McDonell V, Samuelsen S (2021) Corrigendum to 'influence of hydrogen addition to pipeline natural gas on the combustion performance of a cooktop burner' [Int J Hydrogen Energy 44 (2019) 12239–12253]. Int J Hydrog Energy 46(17):10586–10588. https://doi.org/10.1016/j.ijhydene.2020.12.136
21. Soroka BS, Pyanykh KY, Zgurskyi VO (2022) Mixed fuel for household gas-powered appliances as an option to replace natural gas with hydrogen. Sci Innov 18(3):10–22. https://doi.org/10.15407/scine18.03.010
22. Cimini A, Moresi M (2022) Environmental impact of the main household cooking systems—a survey. Italian J Food Sci 34(1):86–113. https://doi.org/10.15586/ijfs.v34i1.2170
23. Aberilla JM, Gallego-Schmid A, Stamford L, Azapagic A (2020) Environmental sustainability of cooking fuels in remote communities: life cycle and local impacts. Sci Total Environ 713:136445. https://doi.org/10.1016/j.scitotenv.2019.136445
24. Afrane G, Ntiamoah A (2012) Analysis of the life-cycle costs and environmental impacts of cooking fuels used in Ghana. Appl Energy 98:301–306. https://doi.org/10.1016/j.apenergy.2012.03.041
25. Gujba H, Mulugetta Y, Azapagic A (2015) The household cooking sector in Nigeria: environmental and economic sustainability assessment. Resources 4(2):412–433. https://doi.org/10.3390/resources4020412
26. IRENA (2017) Biogas for domestic cooking: technology brief. International Renewable Energy Agency, Abu Dhabi
27. Aemro YB, Moura P, De Almeida AT (2021) Inefficient cooking systems a challenge for sustainable development: a case of rural areas of Sub-Saharan Africa. Environ Dev Sustain 23(10):14697–14721. https://doi.org/10.1007/s10668-021-01266-7

Education

Bridging the Educational Gap in Circular Design and Engineering—An Educational Concept and Case Study from Austria

Selim Erol$^{(\boxtimes)}$ ⓘ and Roman Hörbe ⓘ

University of Applied Sciences Wiener Neustadt, Johannes Gutenberg-Straße 3, 2700 Wiener Neustadt, Austria

`selim.erol@fhwn.ac.at`

Abstract. Despite promising developments towards a circular economy, a recent study by EUROSTAT concluded that only 12% of EU's economy is circular. One cause is suspected in particular in the fact that the scientific knowledge relating to the design and development of sustainable and recyclable products and processes has not arrived sufficiently in industrial practice. Educational offers for practitioners in circular design and engineering are still sparse and taken up hesitantly by companies in this context either. To close this educational gap, we have developed a particular training program for learning of Circular Design and Engineering principles. The development of the program was publicly funded and involved five industry partners from furniture and interiors industry. In this paper, we describe the underlying educational concept and lessons learned from a particular application of the concept in the Austrian furniture and interiors industry.

Keywords: Circular Economy · Circular Engineering · Engineering Education

1 Introduction

The global consumption of resources is constantly increasing [1]. The associated environmental impact and climate change represent the greatest challenges of the 21st century for the economy, society and politics. However, both on political and practical level measures have been taken to tackle these challenges. In Austria, for example, the amendments to the Waste Management Act have been enacted, as a legal basis for recycling-oriented products and production.

However, the circular material use rate, according to EUROSTAT [2] increased from 10.8% in 2020 to 12.3% in 2021, setting Austria just above the EU average, but still at a very low level. Austria is significantly above average when it comes to waste generation [3]. One cause is suspected in the fact that scientific knowledge related to the design and development of sustainable and recyclable products and processes has not arrived sufficiently in industrial practice. A study by the Federal Environment Agency found that although ecological product design is on the agenda of many companies, it is not yet seen as an inherent part of product development [4]. This fact is also stated by

H. Kohl et al. (Eds.): GCSM 2023, LNME, pp. 91–99, 2025.
https://doi.org/10.1007/978-3-031-77429-4_11

Federal Ministry of Climate Protection and Environment. In order to enable the transfer of know-how to the relevant stakeholders, it is key to provide modular and scalable concepts which are easily accessible. The proposed action plan covers integration of circular economy (CE) basics in tertiary education as well as development of training programs—especially workshops and seminars—for Austrian companies [5].

A survey of the degree of implementation of a sustainable circular economy in Austrian production industry has shown awareness on the strategic level, primarily geared towards compliance and efficiency. Radical changes at product level, in production and in business models are still rare. The development of competencies in circular design and engineering does not seem to be on the agenda of Austrian companies either [6].

In this paper, we present results from "CircularPro", a project aimed at developing and conducting a training program for the furniture and interiors industry, which enables firms to bring the concept of CE from a strategic to the operational level in product development and production. In the following sections we provide an overview on the background motivation and related activities (Sect. 2). In Sect. 3 we explain both target group and the development process we followed. In Sect. 4 we describe the evaluation results from the first term we conducted the training and lessons learned.

2 Background and Related Work

The Austrian Federal Ministry for Climate Action and Environment (BMK) defined a national Circular Economy Strategy to achieve a climate neutral and sustainable CE by 2050. Products shall be designed in a way that the lifetime is extended, the amount of recycled materials is increased and to possibility for recycling or repair at the end of the lifecycle of the product is enhanced [7].

In order to achieve the set goals until 2050, several actions were formulated, with a major point being education and training for personal [7]. The fact that the majority of required resources for a product are defined in the design phase, highlights the importance of a proper education and training on circular design and engineering [8].

Several initiatives already have started to develop respective educational offerings. The Research Group for Ecodesign of the TU Wien developed a training program on CE, tailored to the Austrian construction sector (circular design_Bau) [9]. The INTERREG project EcodesignCircle [10] by German Environment Agency aimed at strengthening awareness and practical application of the circular design approach. The main output of the project was a toolkit of methods to guide circular design but did not develop educational formats.

The relevance of education in the field of CE has already been addressed in scientific papers. Romero-Luis et al. [11] concluded that scientific literature addressing CE in education is limited and in an early stage. Also different didactic approaches like serious games [12], collaborative projects [13], constructive alignment and problem-based learning [14] can be found In literature. However, existing educational concepts and programs are mainly tailored towards students in primary, secondary or tertiary education. Considering the fact, that firms are in need to take measures towards more circular and sustainable products, practitioners in product development need practice oriented and job compatible educational offerings. This is where our approach comes in.

3 Approach

The basic methodological approach we used for the development of a training program for learning circular design and engineering methods and tools ("CircularPro") can best be described as participatory educational design. According to Janssen et al. [15] participatory design in education is preferable when sustainable improvement of quality of teaching and engagement in learning is in focus. Especially, in educational settings with new complex contents and participants with different educational and industry backgrounds are involved, participatory design has advantages over top down approaches where learners are not involved in the design of an educational offer [16].

3.1 Target Group

For developing CircularPro we could build partly upon our experience from developing existing study programs at our University such as a bachelor program in Sustainable production and CE and a master program in Eco-Design. However, the target of this project is to close the educational gap in practice of product design and engineering regarding principles of CE. Therefore, the target group is practitioners in the field without prior education in sustainable and CE. As the incorporation of circularity principles involves the whole product life-cycle we targeted not only designers and engineers but also marketers and executives. As an industry sector we have chosen the furniture and interiors sector in Austria. Austria is among the most densely forested countries in Europe, with forest covering about half the area of Austria. It serves as raw material for an important value chain in Austria—the timber industry. As part of the timber industry, the furniture industry in Austria plays an important role in exports [2]. To gain representative input from the chosen industry sector we selected five industry partners differing in firm size and product portfolio. Firm size varied between 3 and 280 employees, product portfolio covers furniture for private living, office furniture and public space furniture.

3.2 Development Process

The development of the CircularPro training program involved several steps that were carried out in a participative and iterative manner. First, we narrowed the scope of the training program to competencies and skills in the application of circular design and engineering principles, methods and tools. Choosing furniture and interiors industry as the target industry (step 1) helped a lot in providing a starting point for subsequent refine-ment of learning targets towards concrete competencies and skills (step 2). For this step we invited participating firms to an online workshop. We prepared an interactive white board allowing for collaborative brainwriting (collecting, grouping and prioritizing of ideas) regarding desired learning targets and contents. While most ideas where collected in the workshop itself, we let participants add ideas for 4 weeks afterwards to allow reflection and maturing of ideas and needs. After that period, we prepared a preliminary structure of learning modules covering the consolidated ideas. This preliminary concept was again presented to participating firms to collect feedback. Having identified require-ments regarding content and time we started to search for experts covering the different topics. Experts for general (non-industry specific topics) were recruited from our scientific staff. For industry specific topics we invited industry experts. The requirements from industry representatives were presented to experts and again we collected feedback and refined contents. As a result, experts started the development and online provision of learning resources. Learning resources were again discussed in a workshop to sharpen contents. To provide a consistent format and a single point of access for participants we prepared templates for developing learning materials and a Moodle course to manage and access these resources.

The training was finally held at our university campus in the City of Wiener Neustadt. Part of the modules were held in a seminar room, those modules with more hands-on exercises were conducted in the university's fab lab. The workshop lasted for a total of five days, where each day was dedicated to an individual module. Altogether, a total of 9–15 persons attended the workshop each day (Table 1).

Table 1. Development process for CircularPro training program

Step		Method and tools	Participative	Iterative
1	Scoping of main learning targets and target group	Workshop, Brainstorming, Miro board	Yes, with core team	Yes
2	1st Refinement of learning targets, competencies and skills and content	Workshop, Miro board	Yes, with firms	Yes
3	Definition of learning modules and time schedule	Workshop, Word, Powerpoint	Yes, with firms	Yes
4	Selection of experts (trainers) for modules	Expert pool of university and institute	No	No
5	2nd Refinement of learning targets, competencies and skills and content	Workshop	Yes, with experts	Yes
6	Development and provision of learning resources	Cumulative and individual feedback, Moodle	Yes, with experts	Yes
7	Conduction of training program	Presentation, Group Exercises, Case-studies	n/a	n/a
8	Evaluation of training program	Questionnaires, open discussion	n/a	n/a

4   Results

Based on the previously mentioned approach, an educational concept was developed consisting of five modules, each module itself consisting of two parts.

Module 1 of the training program is aimed towards teaching the participants the idea and principles of CE. Part 1 of the module focuses on the concept of CE, applicable strategies for implementing and the legal framework that surrounds CE in Europe. By presenting several case-studies of successful implementations in the industry, participants are able to learn from others, reflect on the applicability in their own company and engage in discussions. Part 2 has a very strong focus on Life Cycle Assessment (LCA) as a method and tool to determine environmental impact along the whole lifecycle of a product. At the end of module 1, the participants are able to name, understand and explain the principles of CE. The participants are capable to design a project road map for the implementation of CE in their own companies and can name tools that aid the implementation (e.g. LCA).

In module 2, the focus is entirely on circular materials. Several, commonly used, materials such as metals, polymers and composite materials are explained and compared in terms of technical and ecological characteristics. For a more hands-on experience, the module is supported by a material library containing 50 samples of certified circular and sustainable materials. At the end of module 2, the participants know the basic characteristics of materials regarding their circular and technical properties and their potential for circular furniture and interiors.

Module 3 focuses on circular design. Participants are presented methods that enable them to make decisions in the development process of products, that support circularity in a comprehensive way. Participants are introduced into the Circular Design Rules Methodology, developed by the Institute of Design Research Vienna (IDRV) and learn its application within a group project that is conducted in part 1 of the module. In part 2, participants learn more about the communication aspect of circular design.

Part 1 of module 4 is dedicated to the repairability of products. Legal aspects and basic requirements for the repairability of products are explained. Participants have the opportunity to reflect on the repairability of their own products and learn basic principles to improve repairability of their products. Part 2 of the module focuses on recyclability. Participants learn about disposal- and recycling processes and how circular design and engineering affect this process.

In the final module, participants reflect on their own products and business model in regard CE. The participants engage in the serious game 'Make it Circular' [17], and develop new business models and services for their own companies. Part 2 of the module expands the circular design aspect to the entire value chain and how this affects the design of environmentally sustainable products and processes by using Circular-Design-Toolkit [18] (see Fig. 1).

In order to assess the success of the training program, participants received a questionnaire at the end of each part to rate different aspects of the training program. Each aspect could be rated on a scale from 1 (very dissatisfied) to 6 (very satisfied). The result of the survey can be seen in Fig. 2. In this graph a score of 100 would represent all participants rating every module with the highest possible score (very satisfied) and a score of 0 would represent every module being rated at the worst possible score (very dissatisfied).

Overall the feedback was positive, with modules 1, 4 and 5 receiving the most positive feedback. Participants noted, that the theoretical input in module 1 was very valuable for them and the part could have been extended to another module, especially the introduction to the LCA software. Modules 4 and 5 were very interactive and geared

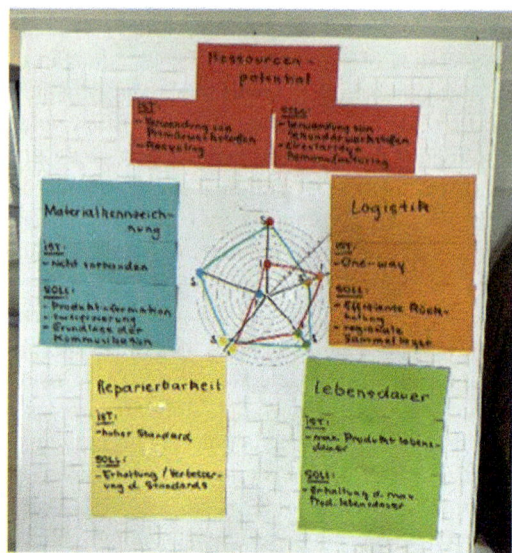

Fig. 1. Radar chart developed in module 5

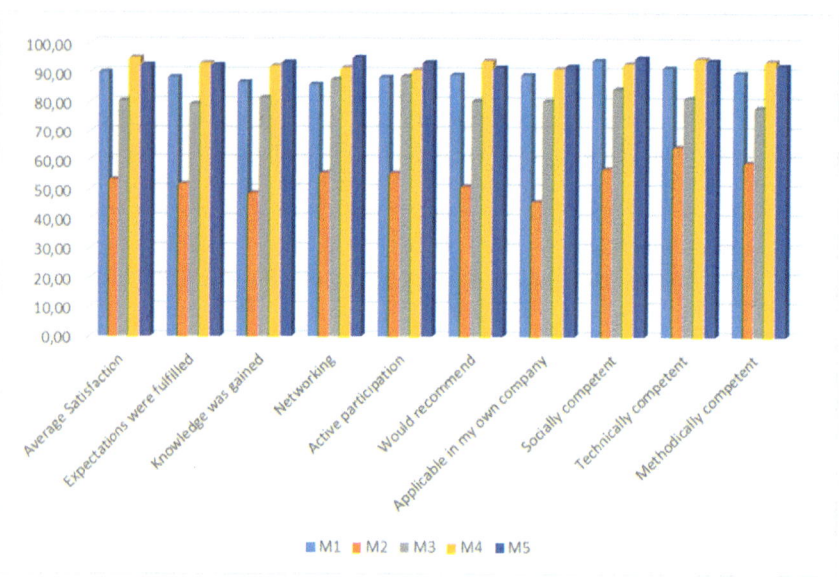

Fig. 2. Summary of results of the participant survey

towards practical learning. The participants appreciated the variety of the modules and the possibility to apply theoretical in group exercises.

Module 2, received a largely negative feedback. It was concluded, that the focus was too general and too theoretic. A detailed discussion and evaluation on sustainability and circularity properties of materials particularly interesting for furniture and interiors would have been beneficial. In addition, participants remarked that interactive exercises would have been appropriate to enable participants to apply theoretical input.

In module 3 some theoretical inputs from modules 1 and 2 were repeated and the module involved many interactive sessions. Participants provided the feedback, that more distinct theoretical input on circular design options would have been beneficial, hence the feedback was not as overwhelmingly positive.

Summarizing the feedback from the participants, the training program, was perceived adequate in addressing educational needs of participating firms. Still, several conclusions for improvements can be drawn: (1) the concept needs a strong focus on the target industry, (2) in order to spike the participants interest, their particular products and problems need to be addressed, e.g. by modules explicitly dedicated to develop individual circular design and engineering strategies, (3) the theoretical input on circular materials should be reduced and focus more on the target industry, (4) Designing and engineering circular products must be strongly connected to sustainability goals. Basic knowledge of sustainability is a prerequisite to learn circular design and engineering.

5 Conclusion

In this paper, we described our approach in developing a modular and scalable training program to learn principles, methods and tools of circular design and engineering. The training program was tested in an initial term with five firms from Austrian furniture and interiors industry. By evaluating the feedback provided by the participants of the training program, it was possible to confirm the relevance of the concept for the participating firms.

In order to further refine the concept, it will be necessary to apply lessons learned from the first term into an updated training program and collect further feedback. To investigate the long-lasting effects of the training, participating companies will need to be interviewed regarding the impact of the training program on product development processes and projects.

It was not yet possible to assess the scalability and feasibility of the modular design of the training program, due to a limited number of participants and the focus on a single industry. In order to be able to finally assess scalability and feasibility for larger audiences and other industry sectors, the training program will be continually offered on additional terms addressing also other industries.

Acknowledgement. This project was funded by the Austrian Research Promotion Agency (FFG) and was carried out as part of the "Innovationscamps 2022" program.

References

1. OECD (2019) Global material resources outlook to 2060—economic drivers and environmental consequences. OECD
2. EUROSTAT (2023) Circular material use rate
3. EUROSTAT (2023) Waste generation per capita. Accessed 16 Dec 2022
4. Umweltbundesamt (2017) Ökologisches Design als Qualitätskriterium in Unternehmen stärken. Accessed 9 July 2023
5. BMK (2022) Österreich auf dem Weg zu einer nachhaltigen und zirkulären Gesellschaft: Die österreichische Kreislaufwirtschafsstrategie. Wien
6. Schöggl J-P, Stumpf L, Rusch M, Baumgartner RJ (2022) Die Umsetzung der Kreislaufwirtschaft in österreichischen Unternehmen – Praktiken, Strategien und Auswirkungen auf den Unternehmenserfolg (in De;de). Österr Wasser- und Abfallw 74(1–2):51–63. https://doi.org/10.1007/s00506-021-00828-3
7. BMK (2023) Bundes-Abfallwirtschaftsplan 2023: Teil 1
8. Kaiser O (2018) VDI ZRE Kurzanalyse Nr. 21: Ressourceneffizienz in der Holzmöbelindustrie. Berlin
9. Technische Universität Wien. IDEA: Innovationscamp circular_design BAU [online]. Available https://www.tuwien.at/mwbw/ikp/klft-ecodesign/ecodesign/forschungsprojekte/bewusstseinsbildung/idea. Accessed 30 June 2023
10. Interreg Baltic Sea Region. Ecodesign as driver of innovation in the BSR—Interreg Baltic Sea Region [online]. Available https://interreg-baltic.eu/project/ecodesign-circle/. Accessed: 9 July 2023
11. Romero-Luis J, Carbonell-Alcocer A, Gertrudix M, Del Carmen Gertrudis Casado M (2021) What is the maturity level of circular economy and bioenergy research addressed from education and communication? A systematic literature review and epistemological perspectives. J Clean Prod 322
12. Whalen K, Berlin C, Ekberg J, Barletta I, Hammersberg P (2018) 'All they do is win': Lessons learned from use of a serious game for circular economy education. Resour Conserv Recycl 135:335–345
13. Williams M, McDonough M, Edge S (2021) Interdisciplinary circular economy design education through local and regional partnerships. PLATE Product Lifetimes Environ 9:432–436
14. Kirchherr J, Piscicelli L (2019) Towards and education for the circular economy (ECE): five teaching principles and a case study. Resour Conserv Recycl 150
15. Janssen F, Könings K, van Merriënboer J (2017) Participatory educational design: how to improve mutual learning and the quality and usability of the design? Eur J Educ 52(3):268–279
16. Cober R, Tan E, Slotta J, So H, Könings K (2015) Teachers as participatory designers: two case studies with technology-enhanced learning environments. Instr Sci 43(2):203–228
17. Acatech. Make it circular! Zirkuläre Geschäftsmodelle im Unternehmen spielerisch kennenlernen - acatech [online]. Available https://www.acatech.de/publikation/. Accessed 9 July 2023
18. Ecodesign Circle. Circular design toolkit [online]. Available https://circulardesign.tools/. Accessed 9 July 2023

Concept of an Integrated Information and Education Platform for Sustainable Development

Anna M. Nowak-Meitinger⑩, Mrunali K. Arute, Devarsh P. Upadhyay, and Roland Jochem⁽✉⁾ ⑩

Technische Universität Berlin, Straße Des 17. Juni 135, 10623 Berlin, Germany
{owak,Roland.jochem}@tu-berlin.de

Abstract. To meet the growing demand for sustainable products, companies are increasingly striving to manufacture products aligned with the United Nations' (UN) 17 Sustainable Development Goals (SDGs). To communicate this sustainability to customers, companies often use sustainability seals to ensure that the environmental, social and economic impacts during production meet certain criteria. However, the increasing number of these seals awarded by independent organizations often creates confusion rather than clarity due to the lack of transparent criteria and award procedures. Consequently, these seals overlap the usefulness which creates an ambiguity in the consumer's mind while making a purchase. This paper introduces the concept of an information platform, which provides consumers with information about product sustainability, considering the entire product lifecycle (PLC). It aims to collect and evaluate data from different product groups to enable consumers to make purchasing decisions based on their criteria and values. Research was conducted to assess consumers' access to consolidated information on sustainable product development and manufacturing. The identified gap led to the presentation of the multi-level model of the digital platform. The concept benefits consumers, companies and society by facilitating information and education about sustainability aspects distinguishing companies from the competition in the production of consumer-oriented and sustainable products.

Keywords: sustainable products · product lifecycle · sustainable manufacturing · digital platform · 17 SDGs

1 Introduction

Recently, companies considered sustainability the fifth attribute of decision-making for production systems, alongside cost, time, quality and flexibility [1]. Sustainability is a central issue in society and represents a major challenge. Users are responsible for the sustainable use of products, while manufacturers are responsible for producing sustainable products [2]. The 17 SDGs of the UN call on all industrialised and developing countries to act in a global partnership for a sustainable future [3]. According to the sustainable product policy and eco-design guidelines, sustainable products should become

H. Kohl et al. (Eds.): GCSM 2023, LNME, pp. 100–108, 2025.
https://doi.org/10.1007/978-3-031-77429-4_12

the norm in the EU [4]. New regulations will be introduced to make almost all physical goods sold in the EU more environmentally friendly, circular, and energy efficient at every stage of their life cycle, from design to use, repurposing, and end-of-life [4]. The question arises regarding how sustainability is defined and can be increasingly implemented and communicated transparently. Sustainability seals can be helpful in terms of providing orientation for consumers and customers and setting incentives for manufacturing companies and service providers [5]. In 2014, there were already more than 1000 different labels in Germany, awarded by independent institutions, which refer to sustainability [5]. This large variety of existing sustainability labels is confusing and therefore only moderately helpful for consumers. Furthermore, customers lack awareness of these labels and information about the sustainability of products.

The aim of this research is to develop a concept for an information platform that gives required information about the sustainability of the products while transparently evaluating the PLC against the 17 SDGs. It will also allow users to learn more about certain sustainability aspects. As a result, consumers can compare various products on the same platform to make an informed and smart decision for sustainable products.

2 Theoretical Background

The concept of the information platform for sustainability is based on the 17 SDGs with the perspective of the whole PLC. Both basic concepts are briefly introduced.

2.1 Sustainable Development in Manufacturing Industry

The 17 SDGs have been developed by UN, such as the elimination of poverty and hunger, creating decent work, responsible consumption and production, innovation and infrastructure and economic prosperity and many more [3]. Therefore, global partners are trying to implement the SDGs in various industries to promote sustainable products and raise awareness of sustainability worldwide and sustain themselves in the competitive market for a long time. As a result, there is an increasing demand for environmentally friendly products among consumers. Companies must actively pursue proactive eco-efficiency and eco-effectiveness strategies in production to achieve social, economic and environmental benefits [6]. However, complex sustainability criteria and their targets make it difficult for manufacturers and users to maintain an overview when manufacturing and purchasing products. For many consumers, it is challenging to understand the content of sustainability labels, and the trustworthiness of voluntary communications is often questioned [7].

2.2 Product Lifecycle

Different approaches for PLC exist, such as market-oriented, manufacturer-oriented, and user-oriented [8], with the latter being the focus of this paper. The generic PLC defined in [8] consists of five stages: ideation, definition, realisation, use/service and disposal/recycling/retirement. The product idea and product description are developed in the ideation and definition phase. The product acquires its final shape in the realization phase in order to be sold to consumers. In the use/support phase, consumers use the product and the company provides services. In the final phase, the product is no longer valuable to consumers and can be disposed of/recycled/repaired by users or manufacturers [8]. To produce more effective and sustainable products, companies should consider the complete PLC and sustainability aspects in the product development stage [2] and adopt more ways to produce sustainable products, such as the adoption of natural capitalism, the blue economy and the concept of regenerative design [9]. To implement sustainability throughout PLC and achieve the UN's 17 SDGs, a circular economy is essential, which also decreases dependency on virgin resources for production. With the help of the standard ISO 14040/14044, the life cycle assessment (LCA) is performed to evaluate the PLC in regard to its impact on the environment [10–12].

3   State of the Art

Extensive research has been conducted to identify the state of the art and to find similar initiatives in the EU. The main finding was that there is a large amount of information available on product sustainability, but at the same time there is no platform that provides a consolidated and comparable overview.

3.1   Informative Initiatives and Projects on Sustainable Products

The legal framework of sustainable product policy and eco-design guidelines aims to align all products manufactured or sold in the EU with technical standards for sustainability, which facilitates the use of the Eco-design approach to establish product-level requirements that promote energy efficiency, circularity, and overall reduction of environmental and climate impacts [4]. According to this, the aim of the European Digital Product Passport (DPP) is to increase the transparency of the manufacturing processes and the origin of the resources used [13]. DPPs are planned to be available for customers and consumers in the form of QR codes to promote a circular economy considering the whole PLC and give information about the environmental sustainability of the products [13, 14]. A DPP shall assist consumers and businesses in making informed purchasing decisions. The framework supports repairs and recycling and enhances transparency about the environmental impact of a product throughout its life cycle [13]. DPP will gather important data on environmental sustainability aspects and provide product information that can be useful for the planned information platform.

The Sustainable Platform [15] is a privately-operated, chargeable database that provides access to key and transparent corporate ESG (Environmental, Social and Governance) and SDG performance. It provides sustainable portfolio construction and benchmark sustainability for leading investors and advisors for identification, management of risk and engagement ratings with independent company research to make strategic decisions [15]. Since this platform is not free of charge and the evaluation refers to the company as a whole rather than to individual products or product groups, it does not meet the requirements of an information platform for every consumer.

The Sustainability Code (Deutscher Nachhaltigkeitskodex, DNK) assists organizations in developing strategies and reporting on sustainability by giving them insightful feedback [16, 17]. It consists of 20 sustainability criteria mainly divided into strategy, process management, environment, and society to make the company's activities more transparent and visible on the platform. This sustainability code is free and organization-oriented to meet corporate social responsibility (CSR) reporting obligations [16, 17], however the platform is not consumer-oriented.

The German Council for Sustainable Development (RNE) is hosting the platform Sustainability Shopping Basket (dt.: Der nachhaltige Warenkorb) as an orientation guide for making everyday life more sustainable [18]. This free platform offers various decision-making assistance on all forms of consumption and sustainable alternatives. Additionally, it provides a summary of reliable seals and product labelling [19]. The main objective of the platform is to spread knowledge and awareness about sustainable choices in public sector procurement. It highlights the fact that sustainable consumption means making wise choices, occasionally buying less, and fundamentally keeping the social and environmental impacts of goods and services in mind [18]. This platform does not refer specifically to PLC stages or single SDGs.

The Sustainability Scanner [19] is a conceptual app designed to help consumers make sustainable purchasing decisions by providing information about a product's carbon footprint and the company's sustainability practices by scanning the barcode of products [19]. This concept is close to the concept presented in this paper, but leaves open how the required information is collected and presented.

Consumers and customers also receive certain information about products on companies' websites. However, this information is not necessarily verified by a third-party organization, nor is it structured and provided in a standardised, comparable form. In addition, it is not practical to access and read each website individually.

3.2 Research Gap and Requirements

The conclusion that can be drawn from the above studies is that none of the existing platforms offers consolidated information for specific companies and their product groups, or even allows direct product comparison. In order to bridge this gap, a new platform should be developed, which meets the following defined requirements for users: free access, trustworthiness, consumer friendliness, comprehensibility for many user groups, holistic information on PLC and the 17 SDGs, information on sustainability activities of the manufacturers, comparability of product groups and specific products to support decision making. The platform should also meet the following requirements for companies: assessment and presentation of sustainable development activities, knowledge sharing and benchmarking, and connecting existing platforms, information systems and websites from the manufacturers' point of view. Finally, the goal is to meet the general requirement of promoting education in sustainable development by providing applied information on the 17 SDGs and linking them to the PLC.

4   Conceptual Design of the Information Platform

For the methodological development of the platform, the V-model approach according to guideline VDI 2206 is adopted. The design phase includes platform requirements, functionality, and data collection, followed by the implementation phase and the verification and validation phase [19, 20]. This paper aims to present the first steps of the concept and design phase, including platform requirements and functions as well as the planned data collection for the platform. Further work will be focused on the implementation and test phase.

4.1   Functional Framework of the Information Platform

The concept of the web-based information platform provides an information and data-sharing model between users and companies that assists decision making and knowledge sharing to create transparency and foster sustainable production and consumption. The platform will include interactive visualizations to facilitate use and provide deeper insight. Figure 1 shows the basic framework of the platform from user and functional perspective. The platform consists of three levels, while each level provides deeper information about sustainability criteria of specific products.

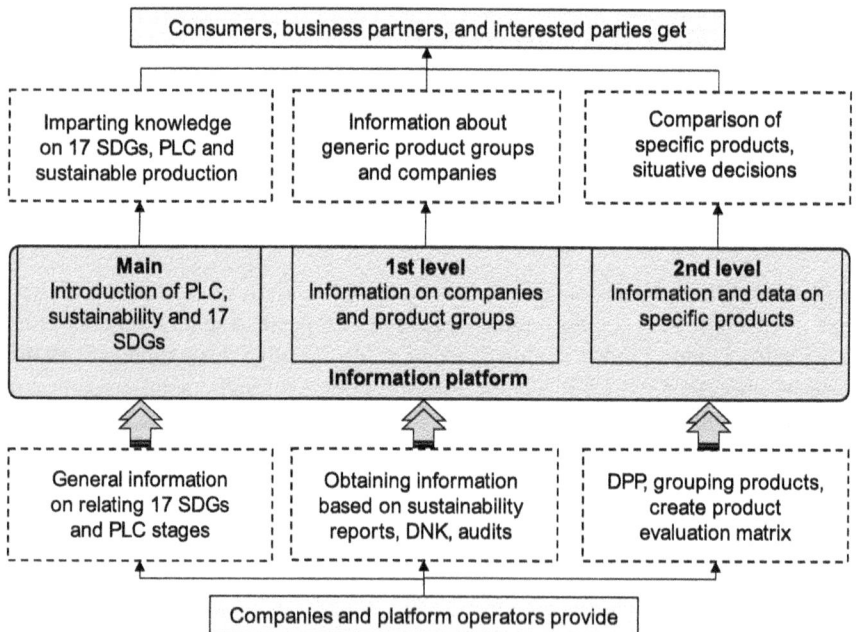

Fig. 1. Functional framework of information platform

The main page offers insights into the 17 SDGs and the concept of PLC, which will be the result of a theoretical and practical analysis. Users find detailed information about the 17 SDGs and their targets, the entire PLC from the ideation phase to the end-use phase, the importance of sustainability, and the contribution and progress of sustainability, as well as how to improve society through initiatives. In addition, links to official information sites, e.g., Sustainability Shopping Basket and DNK should be provided to broaden the information base.

In level 1, the SDGs are mapped to PLC and explained using product groups. Users receive detailed information and data about different companies and their products. The particular product-related processes are carried out individually in each company. Therefore, reference should be made to a generic PLC, i.e., what the products contain, how they are produced, what resources are used, and how they are transported to customers. Information on the use and disposal or recycling should also be accessible.

The final level of the platform provides a detailed overview of product-specific sustainability indicators and facilitates comparison between products. This is offered using a product evaluation matrix to help users making decisions according to their personal values and situational requirements. This can lead to different trade-offs depending on the situation. The complexity of the sustainability criteria often makes a quantitative comparison unfeasible, thus the criteria are compared qualitatively and evaluated by the users themselves.

4.2   Data Acquisition and Collection for the Information Platform

The platform will be filled with information on the 17 SDGs, the PLC, sustainability criteria, company data, manufacturing processes and specific product information. The basic data will be provided by the companies in order to be collected and verified by the platform operators. DPPs in particular provide individual product data in this context. Furthermore, manufacturers should provide among others sustainability reports, LCA results, ISO 14001 [22] and related environmental management standards.

Data can then be provided for the evaluation and comparison of products, e.g., material origin, use of renewable raw materials, type of energy used for production and transport, as well as repair and recycling concepts. The aim is to develop a scientifically based set of criteria for a product evaluation matrix in line with the environmental, social and economic dimensions of sustainability and the DNK. This matrix shall be used for the evaluation process and lead to high trustworthiness and knowledge transfer through transparency and acceptance while supporting sustainable decision making.

Providing all the data needed for the platform is challenging for companies due to the complexity of PLC processes and networks that involve many stakeholders, e.g., suppliers, manufacturers, retailers, users and recyclers who need to share and analyse data collaboratively [23]. In the context of circular economy, data sharing models and data governance models are currently being developed and tested to support the creation of comprehensive product information, such as Product Circularity Data Sheet (PCDS) and DPP [24, 25]. The criteria matrix for the information platform will be based on such existing data sharing models and visualized in a user-friendly way. The platform provider, as an independent organization, should ensure that all information is reviewed and updated at least annually by automatically incorporating newly revised data.

4.3 Impacts on Users and Companies

The presented platform enables users to follow a systematic approach to sustainable decision making. In addition, the platform provides companies with the opportunity to share knowledge on sustainability good practices and compare the implementation of sustainability activities across the PLC, as well as showcase their own sustainable development. This establishes sustainable production as a predominant and competitive area and promotes environmentally conscious business and consumption.

5 Conclusion

The connection between sustainability and PLC is important in the context of sustainable production. The presented information platform serves as a bridge for the information gap between consumers and manufacturers. Until now, consumers have lacked the means to evaluate and compare sustainability practices used in specific products, while manufacturers have had difficulty highlighting their sustainable manufacturing practices. The proposed platform aims to address these challenges by providing freely accessible, user-friendly, trustworthy and comprehensible resources to identify the level of sustainable production in different products. As a visualized source of information, this platform solves the problem of limited availability of production and manufacturing information to consumers. The aim is to establish scientifically based criteria for assessing and comparing sustainability aspects, to promote trustworthiness, and to facilitate the dissemination of knowledge through transparency. Consequently, this initiative enables consumers to make informed and reflected purchasing decisions and allows companies to improve the sustainability of their products through benchmarking, which provides a competitive advantage.

References

1. Salonitis K, Ball P (2013) Energy efficient manufacturing from machine tools to manufacturing systems. Procedia CIRP 7:634–639
2. Steinbach T, Parthey F, Weidmann M, Anderl R (2023) Development of a potential analysis for the introduction of sustainable digitization solutions. In: Kohl H, Seliger G, Dietrich F (eds) Manufacturing driving circular economy. GCSM 2022. Lecture notes in mechanical engineering. Springer, 559–566
3. UN, United Nations (2015) Department of economic and social affairs sustainable development, the 17 Goals. https://sdgs.un.org/goals. Last accessed on 2023/06/08
4. European Commission. Sustainable product policy & eco-design. https://single-market-economy.ec.europa.eu/industry/sustainability/sustainable-product-policy-ecodesign_en. Last accessed on 2023/06/08
5. Revermann C, Petermann T, Poetzsch M (2014) Chancen und Kriterien eines allgemeinen Nachhaltigkeitssiegels, Endbericht zum TA-Projekt, Büro für Technikfolgen-Abschätzung beim Deutschen Bundestag (TAB), TAB-Arbeitsbericht Nr. 163
6. Peralta ME, Soltero V (2021) Chapter 1—sustainable manufacturing: needs for future quality development. In: Gupta K, Salonitis K (eds) Handbooks in advanced manufacturing, sustainable manufacturing. Elsevier, 1–28

7. Turunen LLM, Halme M (2021) Communicating actionable sustainability information to consumers: the shades of green instrument for fashion. J Clean Prod 297:1–10
8. Stark J (2020) Product lifecycle management, vol 1, 4th edn. Springer, Switzerland
9. Has M (2022) Business and sustainability, is part of sustainable products: life cycle assessment, risk management, supply chains, eco-design, Berlin. De Gruyter, Boston
10. Quernheim N, Winter S, Arnemann L, Wolff S, Anderl R, Schleich B (2023) Concept for the evaluation and categorization of sustainability assessment methods and tools. In: Kohl H, Seliger G, Dietrich F (eds) Manufacturing driving circular economy. GCSM 2022. Lecture notes in mechanical engineering. Springer, 721–728
11. International Standard Organization: Environment management-Life cycle assessment-Principles and framework (ISO 14040:2006 + Amd 1:2020)
12. International Standard Organization. Environmental management—life cycle assessment—requirements and guidelines (ISO 14044:2006 + Amd 1:2017 + Amd 2:2020)
13. European Commission. The digital product passport. https://hadea.ec.europa.eu/calls-propos als/digital-product-passport_en. Last accessed on 2023/06/08
14. Adisorn T, Tholen L, Götz T (2021) Towards a digital product passport fit for contributing to a circular economy. Energies 14:2289. https://doi.org/10.3390/en14082289
15. Sustainable Platform. https://www.sustainableplatform.com. Last accessed 2023/06/21
16. The sustainability code. https://www.deutscher-nachhaltigkeitskodex.de/en-gb/. Last accessed on 2023/07/21
17. Yvonne Z (2016) Der Deutsche Nachhaltigkeitskodex. Eine erste Bilanz, is a part of CSR und Nachhaltige Innovation. Springer, Berlin, Heidelberg, 55–67. https://doi.org/10.1007/978-3-662-57697-7_6
18. German Coucil for Sustainable Development/Deutscher Nachhaltigkeitsrat. Der Nachhaltige Warenkorb, https://www.nachhaltigkeitsrat.de/projekte/der-nachhaltige-warenkorb. Last accessed on 2023/06/08
19. Petersen E. Sustainability scanner. https://www.emmajpetersen.com/sustainability-scanner. Last accessed on 2023/06/21
20. VDI/VDE 2206 (2020) Development of cyber-physical mechatronic systems. Beuth Verlag
21. Ali A (2017) ISO 26262 functional safety standard and the impact in software lifecycle. J Univ Appl Sci 1(11)
22. ISO 14001:2015. Environmental management systems—requirements with guidance for use (ISO 14001:2015)
23. Piétron D, Hofmann F, Jaeger-Erben M (2023) Die digitale circular economy, Zirkuläre Daten-Governance für eine Ressourcennutzung von der Wiege zur Wiege, Friedrich-Ebert-Stiftung [online] Available https://library.fes.de/pdf-files/a-p-b/20544.pdf
24. Mulhall D, Ayed A-C, Schroeder J, Hansen K, Wautelet T (2022) The product circularity data sheet—a standardized digital fingerprint for circular economy data about products. Energies 15(9):3397
25. Piétron D, Staab P, Hofmann F (2023) Digital circular ecosystems: a data governance approach. Gaia 32(Supplement 1):40–46

Innovating Blended Learning Model for Professional Education in Manufacturing

Peter Krajnik[1]([⊠]) [iD], Philipp Hoier[1] [iD], and Sampsa Laakso[2] [iD]

[1] Department of Industrial and Materials Science, Chalmers University of Technology, 412 96 Gothenburg, Sweden
`krajnik@chalmers.se`
[2] Department of Mechanical and Materials Engineering, Faculty of Technology, University of Turku, Joukahaisenkatu 3, 20520 Turku, Finland

Abstract. Advancements in technology and the increasing prevalence of digitalization in industry require a new approach to professional education. The primary objective is to enhance the skills of working professionals, ensure content is relevant to industry needs, increase learner engagement, and optimize learner and instructor efficiency. To achieve these goals, a new methodology is proposed, utilizing constructive alignment, outcome-based education, and blended learning strategies. This approach incorporates asynchronous digital learning, synchronous online lectures, and interactive debriefing sessions, providing an engaging blend of self-paced learning and active, instructor-led experiences. Evaluation results show an improved course structure and a positive learning experience, despite initial implementation challenges. While not exclusively designed for sustainable manufacturing education, this approach offers innovative learning pathways and has the flexibility to integrate specific modules, such as those related to sustainability. Based on evaluation feedback and measurable learner outcomes, ongoing refinements to this model suggest a promising shift in the approach to professional education within the manufacturing sector.

Keywords: Blended Learning · Professional Education · Manufacturing · Sustainability

1 Introduction

Professional education is critical in the manufacturing sector, a field characterized by continuous innovation and the integration of new technologies. These advances are driving significant changes in manufacturing practices, requiring a workforce skilled in the use of these new tools and methods [1]. Skills play a significant role in the economic growth of a society, on the innovation process, as well as on social inclusion. The global manufacturing industry currently faces an acute talent gap, due to a lack of adequately trained working professionals. Addressing this gap necessitates a deeper understanding

H. Kohl et al. (Eds.): GCSM 2023, LNME, pp. 109–117, 2025.
https://doi.org/10.1007/978-3-031-77429-4_13

of how technological advancements in manufacturing are generating new job roles, and the subsequent modifications needed in educational and training programs. Navigating the difficulties related to the supply and demand of skills requires a transformation in the educational models within the manufacturing sector [2]. The International Academy for Production Engineering (CIRP) has promoted the concepts of Teaching and Learning Factories (TLFs) to enhance practical, industry-oriented education and foster academia-industry collaboration [3]. However, TLFs face challenges in scalability and resource intensity [3].

Professional education is closely connected to sustainable development and supports its economic pillar through efficient and accessible educational services. As advancements in science and technology facilitate new industrial innovations, blended learning has become a fundamental tool for organizations to promptly adapt and offer relevant and sustainable learning experiences. While this paper acknowledges the importance of incorporating blended learning to advance sustainable development, its primary focus is to provide effective, and easily implementable training modules that can seamlessly integrate sustainable manufacturing concepts and learning content.

The development of professional education in the Swedish manufacturing industry is a reflection of wider trends in technical training. The discussed initiative began in 2006, with a focus on "Professional Education in Metal Cutting," in partnership with major industrial players such as Volvo Group, Scania, Seco Tools, and Siemens. The training program had multiple goals: (i) to increase interest in technical training among younger industry professionals, (ii) to adapt to fast-paced technological changes, (iii) to introduce new concepts, and (iv) to sustain critical machining skills for the Swedish manufacturing industry. The training program was designed to be versatile and not confined to specific job roles, e.g., machine operators, workshop technicians, production engineers. The flexibility in the modules enabled customization of content to suit different participant mixes, with the option to adjust the balance between theoretical learning and discussing real-world case studies. Although drawn from experiences with large Swedish companies, the program's adaptive design implies that it can be potentially applied in comparable industrial settings worldwide.

Due to the Covid-19 pandemic, all in-person training was suspended in 2020, necessitating the training to be converted entirely to an online format. When restrictions eased, certain components of instruction, such as labs or workshops, were reintroduced. Notably, there was no complete return to the traditional in-person model. The rationale behind this is multifold: companies acknowledged the cost savings from reduced travel, while training providers saw distinct advantages to online instruction. For instance, enlisting remote industry experts for guest lectures. The new mandate focused on designing, developing, implementing, and evaluating blended learning pathways for professional education—to guide future instructional delivery while creating a marketplace that integrates both existing and new learning modules into training programs.

2 Methodology

A new methodology is proposed for the development of blended learning pathways in professional education. The methodology has its theoretical basis in outcome-based education [4], constructive alignment [5, 6], while using Bloom's taxonomy [7] to identify the cognitive levels in the intended learning outcomes. Collective learning [8, 9] and activity theory [10, 11] are included in the model because experience has shown that professional training benefits from interaction between the participants, and this is often the most positive feedback from trainees in the course. In addition, the methodology is refined based on the best practices collected from research literature on online learning [12, 13], blended learning [14, 15], flipped classroom methodology [16, 17], lifelong learning [18, 19], and assessment of online courses [20, 21].

The blended learning approach in this concept involves combining asynchronous (self-paced) and synchronous (real-time) activities, as well as mixing independent and interactive learning activities. The design methodology is illustrated in Fig. 1, it starts by identifying the intended learning outcomes based on the subject matter, lifelong learning goals and the requirements of different stakeholders. The intended learning outcomes are divided into three groups, i.e., cognitive levels, using the modified Bloom's taxonomy [22]: (i) Remember and Understand, (ii) Apply and Analyze, (iii) Create and Evaluate. This method adds an additional cognitive level—knowledge construction—which serves as a reminder that the most important learning outcome is students' active learning after their formal education. Students need to take control and responsibility for their own learning. After the learning outcomes are defined in the cognitive categories, appropriate learning activities are designed for each of them.

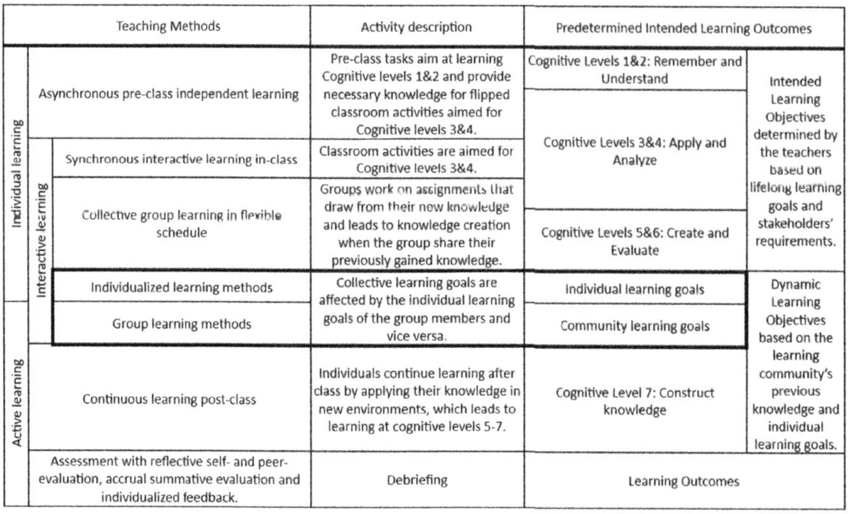

	Teaching Methods	Activity description	Predetermined Intended Learning Outcomes	
Individual learning — Asynchronous pre-class independent learning		Pre-class tasks aim at learning Cognitive levels 1&2 and provide necessary knowledge for flipped classroom activities aimed for Cognitive levels 3&4.	Cognitive Levels 1&2: Remember and Understand	Intended Learning Objectives determined by the teachers based on lifelong learning goals and stakeholders' requirements.
Interactive learning — Synchronous interactive learning in-class		Classroom activities are aimed for Cognitive levels 3&4.	Cognitive Levels 3&4: Apply and Analyze	
Collective group learning in flexible schedule		Groups work on assignments that draw from their new knowledge and leads to knowledge creation when the group share their previously gained knowledge.	Cognitive Levels 5&6: Create and Evaluate	
Active learning — Individualized learning methods		Collective learning goals are affected by the individual learning goals of the group members and vice versa.	Individual learning goals	Dynamic Learning Objectives based on the learning community's previous knowledge and individual learning goals.
Group learning methods			Community learning goals	
Continuous learning post-class		Individuals continue learning after class by applying their knowledge in new environments, which leads to learning at cognitive levels 5-7.	Cognitive Level 7: Construct knowledge	
Assessment with reflective self- and peer-evaluation, accrual summative evaluation and individualized feedback.		Debriefing	Learning Outcomes	

Fig. 1. Design methodology for blended learning in professional education (*collective learning highlighted with thicker borders*).

Asynchronous independent activities are intended for learning the Cognitive Level 1 outcomes, complemented with synchronous flipped classroom activities. Students learn from assigned materials or their own sources before the first synchronous class, depending on their pre-existing knowledge of the topic. The first live synchronous session is designed to activate the knowledge the students have learned prior to class through discussions, group assignments, and exercises. The in-class synchronous activities, targeting Cognitive Levels 1–4, are designed to create initial affinity among students and ease the transition to subsequent group stages.

The group stage includes both asynchronous activities and independent learning, depending on how the groups have agreed to distribute the work, as well as synchronous activities (within a group) to prepare for the final live session, where groups give and receive feedback. The duration of the group stage is intentionally long (9 days) to accommodate the flexible schedules of participants.

The final live session provides a debriefing of the group learning outcomes for the whole class, which helps to align the individual group outcomes in the broader context of the entire learning pathway. Participants in the learning pathway are assessed through self- and peer-evaluation. Independent learning is evaluated using methods such as unproctored online exams, and students receive feedback from the teacher/instructor during the debriefing session.

3 Implementation

The design of the professional courses allows for flexibility to accommodate the primary work commitments of the trainees, offering either four-module (3 ECTS) or six-module (4.5 ECTS) options. Each module is organized over a two-week period (see Fig. 2) and can be illustrated as follows:

- Pre-learning: Before the start of each module, students/trainees engage in pre-learning learning activities, such as viewing short videos or working through digital learning nuggets. These activities are designed to familiarize trainees with the module's topic, adhering to the flipped classroom model, and ensuring they are ready to actively participate in the upcoming live lecture sessions.
- Synchronous instructions: Trainees participate in live online sessions using video conferencing tools, interacting primarily with the instructor(s) in real time. Chat is a supplemental feature that helps trainees ask questions or share comments. These online sessions not only deliver the instructional content, but also facilitate interaction and group discussion among course participants, for example by using breakout rooms for more focused discussions.
- Asynchronous learning period: Trainees are given two weeks to gain a deeper understanding of the subject matter at their own pace. Trainees have access to screencasts recorded during live lectures, readings, and lecture handouts. They also engage in individual or group assignments, using conferencing or digital tools for quizzes.
- Debriefing session: The live debriefing sessions provide an opportunity for trainees to reflect on their learning experience. Instructors provide feedback on learning objectives and assignments, facilitating an open dialogue about the module's subject. While structured, these debriefings are also designed to encourage sharing of experiences in

an informal setting. For example, participants are encouraged to share challenges they have encountered in their respective companies—or to disagree with the instructors.

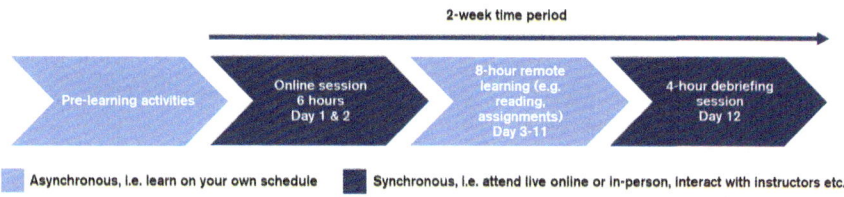

Fig. 2. Learning pathway example for blended professional education.

In the context of the courses discussed, a module refers to a distinct unit of content devoted to a specific topic, similar to a chapter in a book. For example, in a Metal Cutting course, modules such as "Machinability", and "Cutting tools" each house their respective pre-learning activities, lectures, assignments, assessments, and debriefings.

4   Evaluation

Two courses were developed using the proposed methodology. They covered diverse aspects of metal cutting and were conducted in 2021 and 2022 for working professionals in manufacturing industry, ranging from bearing, tooling to automotive sectors. Course 1 consisted of two-week modules, while Course 2 utilized shorter, one-week modules. Each course had 12–15 participants.

Feedback from trainees was collected during each course implementation. First, interviews with randomly selected course representatives were conducted halfway through the course, to provide timely and informal feedback that allowed for potential adjustments to the remainder of the course. Second, an anonymous survey was sent to all participants at the end of the course, to solicit trainees' opinions and perceptions of their learning experience. In the survey, learners were asked to rate their level of agreement with multiple statements using a five-point Likert scale. In addition, open-ended questions were also included in the survey to receive more specific feedback. Some noteworthy comments are summarized in Table 1.

High average scores were observed for all statements in Table 1. The overall impression of the course, the course structure, and the teaching received particularly high scores—indicating a positive perception of the format and delivery of the courses.

More specific comments from the course participants further emphasized the positive perception of the course structure, as reflected in comments 1 and 2 (in Table 2). The two-week asynchronous period between lectures and debriefing was particularly well-received, as it provided sufficient time provided to explore the topics in depth through individual and group assignments (Comment 2). The group assignments also had the positive effect of allowing networking among the trainees in an online format (Comment

Table 1. Average scores of five-level Likert scale course evaluation. The levels are: 1. Strongly disagree, 2. Disagree, 3. Neutral, 4. Agree, 5. Strongly agree. Response rates were 92% (Course 1 2021), 33% (Course 1 2022), and 47% (Course 2 2022).

Statement	Average ratings		
	Course 1 (2021)	Course 1 (2022)	Course 2 (2022)
The course structure is appropriate to reach the intended learning outcomes	4.64	4.75	4.29
The teaching worked well	4.55	4.50	4.43
The assessment (e.g., assignments) tested whether I had reached the intended learning outcomes	4.45	4.75	4.00
The organization and teaching of this course have been designed and executed so that everyone can feel included, welcome and seen/heard	4.64	4.75	4.29
What is your overall impression of the course? (1. Very poor, … 5. Very good)	4.82	4.75	4.57

6). In contrast, the one-week asynchronous period was not well received, as participants found it difficult to fit into their work schedules (Comment 3). Another aspect praised by trainees was the online format of the course (Comments 4–6), in particular its time efficiency.

Despite the largely positive feedback, some common challenges associated with online teaching were identified. These include the issue of passive participants (Comments 7–9), which highlights the importance of incorporating interactive elements, such as interactive discussions or group assignments, in both synchronous and asynchronous sections of the course.

5   Conclusions

The transformation of professional education in the Swedish manufacturing sector was marked by a shift from traditional in-person training to a fully online model. At the core of this change was a course design methodology that incorporated constructive alignment, outcome-based education, and blended learning. Positive feedback was received for the learning experience, but challenges were identified, particularly the creation of resource-intensive digital pre-learning materials. The online format was appreciated for its efficiency, flexibility, and access to recorded lectures. A two-week module structure was preferred for deeper engagement with the topics. Networking was facilitated by group assignments, even in an online environment. However, scheduling issues were

Table 2. List of relevant responses to open-ended questions in course surveys.

Comments on course structure	
1	*"I liked the way with modules. And also the assignments, both group and individuals. And that we ended with discussions on Friday."*
2	*"[...] The time between lecture and debrief was perfect to work on the assignments and to dive deeper into the topic, and the debriefings were very good for reviewing the learnings and for discussing open questions." (Course 1)*
3	*"Earlier courses with wider spans felt easier to manage than 6 weeks of stuffing. Managing group assignments in a couple of days was more of a stress than an aid in this, even though the discussions were meaningful." (Course 2)*

Comments on online-format	
4	*"For me the structure was just right. Having the course online allowed more time for me to go away and do some self-learning to catch up on areas that i wouldn't have been familiar with before joining the course."*
5	*"[...] I preferred the virtual aspect more than i thought i would and actually think it was better than face to face. I don't think it would have attended if it hadn't been online. [...]"*
6	*"Got to know new persons and did not expect that from a remote course. So this was a learning as well that remote can replace physical course. Also time effective, if physical I would probably not have assigned."*

Comments on teaching, learners' active participation, and learner-instructor interaction	
7	*"The teaching was mostly very good. I would only encourage to have more (very short-5 min) discussions between lectures to prompt students to participate and not just listen."*
8	*"This is a challenge, some hesitate much more than others before taking part of the discussions or assignments."*
9	*"It might be a good idea to set "rules" or "expectation" regarding camera use for the students, maybe specify that during set points such as for in-class discussion it is expected to switch on your camera to enable interaction, perhaps after/between different segments where discussion is wanted. I feel this would improve the discussions and make it easier for the lecturers as well."*

reported with one-week modules. The need for more interactive elements to engage passive participants was recognized. To address this, interactive features need to be integrated into both pre-learning content and live-streamed labs, complemented by the use of breakout rooms in live sessions. This highlights the need for further refinement of the blended learning approach presented.

Acknowledgment. The authors are grateful for the opportunity to extend the testing of blended learning beyond Sweden—reaching Austria, Spain, and the Czech Republic. This was made possible due to the funding provided by the EIT Manufacturing program for the project "22194—Hybrid Learning Paths for Professional Education in Manufacturing".

References

1. Martinez W (2018) How science and technology developments impact employment and education. Proc Nat Acad Sci (PNAS) 115/(50):12624–12629
2. Chryssolouris G, Mavrikios D, Mourtzis D (2013) Manufacturing systems: skills & competencies for the future. Procedia CIRP 7:17–24
3. Abele E et al (2017) Learning factories for future oriented research and education in manufacturing. CIRP Ann 66(2):803–826
4. Spady WG (1994) Outcome-based education: critical issues and answers. American Association of School Administrators, Arlington, VA
5. Biggs J (1996) Enhancing teaching through constructive alignment. High Educ 32:347–364
6. Biggs JB, Tang CS (2011) Teaching for quality learning at university: what the student does. McGraw-Hill Society for Research into Higher Education & Open University Press, Berkshire, England
7. Bloom BS, Engelhart MD, Furst EJ, Hill WH, Krathwohl DR (1956) Taxonomy of educational objectives: the classification of educational goals. Longman, New York
8. Gurnee H (1937) Maze learning in the collective situation. J Psychol 3:437–443
9. Gurnee H (1939) Effect of collective learning upon the individual participants. Psychol Sci Public Interest 34:529
10. Yasnitsky A, der Veer R (2016) Revisionist revolution in Vygotsky studies: the state of the art. Routledge, New York
11. Engeström Y (2015) Learning by expanding, 2nd edn. Cambridge University Press, New York
12. Moore JL, Dickson-Deane C, Galyen K (2011) E-learning, online learning, and distance learning environments: are they the same? Internet Higher Educ 14:129–135
13. Dumford AD, Miller AL (2018) Online learning in higher education: exploring advantages and disadvantages for engagement. J Comput High Educ 30:452–465
14. Hrastinski S (2019) What do we mean by blended learning? TechTrends 63:564–569
15. Horn MB, Staker H (2014) Blended: using disruptive innovation to improve schools. Wiley, San Francisco
16. Lage MJ, Platt GJ, Treglia M (2000) Inverting the classroom: a gateway to creating an inclusive learning environment. J Econ Educ 31:30–43
17. Bergmann J, Sams A (2012) Flip your classroom: reach every student in every class every day. Int Soc Technol Educ
18. London M, Smither JW (1999) Empowered self-development and continuous learning. Hum Resour Manage 38:3–15
19. Aspin DN, Chapman JD (2000) Lifelong learning: concepts and conceptions. Int J Lifelong Educ 19:2–19
20. Williamson MH (2018) Online exams: the need for best practices and overcoming challenges. J Publ Profess Sociol 10:2
21. Halbherr T, Reuter K, Schneider D, Schlienger C, Piendl T (2014) Making examinations more valid, meaningful and motivating: the online exams service at ETH Zurich. EUNIS J Higher Educ 1:14
22. Anderson LW et al (2001) A taxonomy for learning, teaching, and assessing: a revision of bloom's taxonomy of educational objectives. Longman, New York

Work Engineering for Sustainability: Required Education in Engineering

Sandra F. B. Gemma[1] ⓘ, Daniel Braatz[2](✉) ⓘ, Raoni Rocha[3] ⓘ,
and Flávia T. de Lima[4] ⓘ

[1] Universidade Estadual de Campinas, Limeira, SP, Brazil
gemma@unicamp.br
[2] Federal University of São Carlos, São Carlos, SP, Brazil
braatz@ufscar.br
[3] Federal University of Ouro Preto, Ouro Preto, MG, Brazil
[4] Mackenzie Presbyterian University, São Paulo, SP, Brazil

Abstract. Recent research shows that there is a significant lack of subjects related to the area we call 'work engineering' in undergraduate engineering courses in Brazil. The organization of work and the challenges posed by production models directly impact the health and safety of workers and indirectly affect society. It is argued that ergonomics, as a scientific discipline and professional practice, can serve as a means for engineering to analyze and understand the variability present in the occupational environment. The objective here is to discuss work from a critical perspective and its central role in society and in the constitution of health, aiming to position it as a protagonist in the training of engineers, as well as to debate the need to integrate different fields and knowledge proposed by 'engineering'. of work'. Finally, it discusses the importance of designing work situations in the context of engineering, highlighting how it can be a vector for transforming productive situations, enabling safe, healthy and efficient environments. In this sense, the expanded concept of 'work engineering' can contribute to social sustainability, laying the foundations for the construction of human relationships and more dignified work environments (SDG8, UN Agenda 2030).

Keywords: work · health · safety · ergonomics · design

1 Introduction

The term sustainability has gained different contours and meanings, being widely debated due to its importance in contemporary society. However, there is a certain ambivalence surrounding the term, sometimes adopted as a promising possibility for creating different modes of production and exchange, sometimes empty of meaning and "incapable of providing a conceptual basis for political actions of planetary dimensions that question demands" [1].

Despite the widely held notion, proposed by the UN in 1987, about sustainable development, the idea of meeting present needs without compromising the ability of

© The Author(s) 2025
H. Kohl et al. (Eds.): GCSM 2023, LNME, pp. 118–125, 2025.
https://doi.org/10.1007/978-3-031-77429-4_14

future generations to meet their own needs is quite complex, especially when considering the structural changes needed within a complex socioeconomic reality, aiming to balance the economic, social and environmental dimensions. Here, the focus is mainly on the social dimension of sustainability, specifically related to productive environments in terms of work sustainability, as partially addressed in SDG8 of the 2030 Agenda.

The debate here is limited to the sphere of work, especially in the context of labor engineering, which integrates knowledge related to labor sciences. Even when it comes to work directly linked to sustainable practices, working conditions do not necessarily favor workers. Social sustainability continues to be a challenge in terms of understanding the links between territory, work and the prevention of accidents and health problems. It is necessary to manage the different elements of work that influence working conditions if we want to think in terms of sustainable work because, despite work being identified as a fundamental need for individuals, it is often forgotten and neglected [2].

The greening of economic activities depends, in part, on the professional skills of workers, indicating that work needs to be properly considered if we want to encompass sustainability in a broader sense. After all, work is located at the economic, environmental and social interface, in constant tension and based on the integration of technical and political perspectives. Furthermore, the work represents the possibility of developing innovative techniques to guarantee the ecological transition and the economic performance of the system, also serving as a response to the challenges of integrating populations excluded from the world of work [2].

In Brazil, some research discusses the sustainability of work, such as in the agricultural sector, ecologically based production [3] and industrial and service work [4]. These studies indicate that even in companies committed to respecting social aspects, such as those proposed by ISO 26000 standards, there is little engagement in improving the work of those who contribute to the sustainability of the business.

The importance of work for the lives of individuals and for the constitution of subjectivity and health is well-known, as well as the contribution that work activity brings to the quality and productivity of organizations and to the development of culture and society, aiming, ultimately bring happiness [4]. It is argued that organizations need to promote respect and development of intelligence and creativity (as opposed to alienating work), promoting meaningful work for people [5]. Therefore, it is necessary to rethink some indicators to evaluate aspects related to organizational design and workers' mental health [6].

The knowledge proposed by work engineering can facilitate the design of sustainable work systems that contribute to health and efficiency, as it integrates a critical perspective on work, ergonomics, health and safety, as well as the design of work. Work. In this sense, the design, implementation and improvement of work systems are directly linked to engineering, as engineers were—and still are—responsible for significant advances in our society, being strongly linked to systems, works and products highly complex, sophisticated and technologically advanced, but in these projects we rarely observe the human being as a part or, even less, as a central element of intervention, as the oath proclaims [7].

It is understood that there is a deep-rooted problem in the training of engineers. The strictly technical and technological focus in the curriculum of undergraduate engineering courses, although fundamental, does not necessarily prioritize the human, social and environmental aspects of work, resulting in education primarily focused on the technical efficiency of production systems.

Thus, even Industrial Engineering, considered the most systemic of engineering due to its multidisciplinary knowledge of areas such as administration, economics and psychology, receives criticism regarding its curriculum and the lack of courses that promote more comprehensive and socially conscious training for these professionals [8].

While some courses address the topic of "working" with specific disciplines, others ignore it. In a recent survey [9] initially found 1,527 courses potentially related to the topic in 105 undergraduate engineering courses at 16 Brazilian universities (selected based on rankings and available places). After applying refinement criteria, 131 subjects were qualitatively analyzed, revealing a significant disparity between the courses examined. While the number of subjects that address the topic "Work" in Industrial Engineering courses was 3.65 per course, the corresponding indicator for Electrical Engineering and Civil Engineering was 0.67 and 0.80 subjects per course, respectively. On average, these courses have less than one subject covering aspects related to the world of work throughout their curriculum. It should be noted that even when the syllabus mentions the emergence of the topic, there are no guarantees that the content offered is aligned with the human and social perspective presented here [9].

In another study focused on Industrial Engineering programs in the 10 best public institutions in Brazil (international university rankings), it was shown that, on average, an Industrial Engineering course offers approximately four disciplines related to work sciences, such as Ergonomics, Occupational Health, Occupational Safety and Work Design. However, the majority of these courses are optional, highlighting the lack of prominent disciplines with critical, integrative and work sustainability perspectives [10].

In this scenario, the proposal presented here intends to dialogue with the "Engaged Engineering" movements, of national and international scope, which seek a new role for engineering. We ideally propose the development of extension projects, the questioning and reformulation of the engineering education system and the proposition of other initiatives, which allow us to intertwine with global issues, aligned with the UN objectives with a view to sustainable development.

The aim is to contribute to expanding the training of engineering students, aiming for future professionals to create appropriate solutions for the complex demands of the world of work, placing people at the center of discussions, taking into account various economic, social and cultural realities that permeate production contexts.

This initiative aims to contribute above all to changing the paradigm of engineering education, without however underestimating the great challenges that lie ahead, as engineering still faces ideological disputes that support reductionist practices [8]. Finally, it is also worth highlighting that the Work Engineering Initiative (IET) is fully aligned with a humanistic and innovative perspective, encompassing both theoretical reflections and concrete actions to improve engineering education and thus contribute to our society more broadly.

2 Work Engineering for Sustainable Work

The term Work Engineering is used in Engineering courses and is considered one of the areas of Production Engineering responsible for the "design, improvement, implementation and evaluation of tasks, work systems, products, environments and systems to make them compatible with the needs, skills and capabilities of people, aiming for better quality and productivity, preserving health and physical integrity" [11]. The Brazilian Association of Production Engineering also highlights that knowledge of Work Engineering can be used to understand interactions between humans and other elements of a system, further inferring that this field deals with the technology of the machine-environment-human-organization interface [11].

In practical terms, such content is often covered in engineering courses as something that enables future engineers to solve problems and challenges related to the world of work from a technical (and sometimes positivist) perspective. In this approach, it is assumed that engineers are capable of "optimizing" human work, defining an ideal way for actions to occur (strongly influenced by the ideas of Taylorism/Fordism that marked a paradigm shift in production models).

It is also common to see the term Work Engineering used erroneously as a synonym for Occupational Safety Engineering (OSE). This restricts it to a specific field of activity and application, as it is directly related to the professions of Occupational Safety Technicians and Engineers specializing in Occupational Safety (postgraduate level). There is no doubt that OSE has been and will continue to be an important area of knowledge and practice for professionals committed to improving working conditions and reducing workplace accidents and illnesses. The distinction is that Work Engineering is an integrative and broader field of knowledge that seeks to develop the competence to understand and transform real work, integrating different areas, work safety being one of them.

From this perspective, we propose an original approach to the term Work Engineering, with specific objectives of impacting the training of engineering professionals in different branches and specialties. This proposal integrates fundamental content intrinsic to human work, namely: Worker Health, Workplace Safety, Ergonomics and Work Design [12].

In this sense, we know that work in Brazil and around the world has been undergoing changes in recent decades, moving from industry to services, but maintaining the essence of timed work and rhythmic production. Although Brazilian legislation guarantees health and safety for workers, in practice, there is widespread legal flexibility in several sectors due to technological advances and the expansion of digital work, leading to unprecedented levels of precarious work.

As a result, new models of work organization emerge, directly impacting the physical and mental health of individuals. Some work contexts are historically emblematic in this discussion, such as meat processing plants and telemarketing, while others, more contemporary, are equally important, such as gig work (mediated by digital platforms and their applications) and informal work. In all these contexts, workers are subject to various physical and mental illnesses. Notably, mental illness has become the leading cause of work absenteeism in recent years [13, 14]. Thus, the idea of a "safe work environment" has evolved over time, as the absence of illnesses or the mere application of regulatory standards is not enough to guarantee the safety of the system and the people involved [15]. To achieve this, it is necessary to understand the work and real conditions of workers, both operational and managerial, and seek work organizations that are more compatible and aligned with the work carried out on the front line.

At this point, some of us may wonder: what does the engineer have to do with this? Aren't they the professionals who develop technical projects to solve problems? A possible answer to these questions certainly lies in the foundations of Work Engineering. From the perspective we have developed, this field of knowledge understands that any technical project, to make sense, must be supported by a social approach centered on its user, that is, on the worker himself. This is because the technical solution will only be successful if it is accompanied by solutions that promote the health of individuals and the sustainability of systems in an integral way. Therefore, the solution must be "sociotechnical". And to achieve this, the work of engineers, regardless of their specialization, must also be informed by Occupational Sciences, since human work crosses different fields of engineering.

Therefore, future engineers whose degree includes the contents of work engineering will be professionals who can work in processes and work environments with greater potential to promote more sustainable work, i.e. work that produces and generates wealth and at the same time is capable of preserving and even improving the health of individuals and the environment.

3 The Different Areas of Work Engineering

Work Engineering, as we propose it, is composed of content from different areas of knowledge related to occupational sciences, especially Occupational Health, Occupational Safety, Ergonomics and Work Design [12]. Before delving directly into topics such as Health and Safety, we consider it essential to reflect and level the understanding of what work is, how it was and is being organized, the meaning that this human activity has for individuals and the legal issues associated with it.

Therefore, the work must be the subject of reflection and study by engineers from all areas, so that they understand the impacts of the capitalist system, different forms of organization and the emergence of the phenomenon of precariousness in the lives of men and women in a globalized world. It is also crucial to understand the legal aspects related to occupational environments from a human rights perspective, as they should form the basis for a better understanding and development of concepts related to workers' health and safety at work.

A paradigm shift is urgently needed to move towards a sustainable agenda: engineering (and its professionals) needs to embrace the humanities and health sciences. The fragmentation of knowledge results in behaviors and practices that do not contribute to the search for integrated solutions and often make the work of different professionals alienated (and alienating). An engineer cannot fail to get involved in discussions involving the impacts of work on the environment and, above all, on the health conditions of individuals, simply because they do not have "training in the area". If this were reasonable, healthcare professionals could also claim that they do not know how to design or intervene directly in work (and its elements, such as machines and systems) and, therefore, could not do anything to prevent occupational illnesses and accidents.

Therefore, our proposal is that engineering courses incorporate content directly related to physical and mental health in the workplace into their curricula and training activities, making clear the relationship between engineering and this fundamental aspect of human life.

Furthermore, all engineering students must be aware that risks to the security of sociotechnical systems are created and managed by individuals, collectives and organizations. Just as we defend the expansion and recognition of engineers' responsibility for the health of workers, we maintain the same idea for the safety of all individuals in their work activities, in addition to the fundamental environmental issues that have been debated.

In this sense, we hope to free ourselves from a strictly normative paradigm, where the only concern is compliance with standards for the existence of safe working conditions, and where only technical and occupational safety engineers should be concerned with this. By providing engineering students with a broad understanding of work and the need for it to be sustainable in its multiple dimensions, they can be equipped with theories, methods, techniques and tools to understand and transform work effectively.

To this end, we propose that ergonomics, particularly the activity ergonomics approach, be presented in such a way that future engineers understand the importance of analyzing real work, the variability that exists in day-to-day productive situations, and recognize and consider the intelligence of workers in developing different strategies and regulations to deal with challenges in their work activities [16].

Therefore, we call for reflection on the typical engineering activity—design—to enable the sustainable development of work. Work design must be recognized and valued as an activity so that the knowledge acquired and awareness developed through previous content creates a fertile environment for future engineers to discover how they can design different elements of production systems (be it a tool, a physical arrangement, software interface, machine or any other artifact) in a way that considers the specificities of the work to be carried out and its social sustainability [9, 17].

4   Final Considerations

This text discusses the defense of Work Engineering as a training field that allows engineers to transform work situations in a sustainable way. To do this, it is first necessary to understand that Work Engineering integrates knowledge in worker health, work safety, ergonomics and design in work activities. It is essential to integrate this content into the curriculum of engineering courses in Brazil, providing any engineer with the ability to develop projects integrated into their professional practice. Ultimately, we see no other way to promote social sustainability at work, which means developing situations that align productive efficiency with the well-being of individuals and the environment involved.

To this end, some concrete actions have been developed such as: the publication and wide dissemination in Brazilian territory of a free book aimed at teaching Work Engineering in undergraduate courses, as well as extra material to be used in the classroom; technical forums and alignment meetings with deans and undergraduate coordinators of different Brazilian universities where there are engineering courses for the adoption of

Work Engineering content in undergraduate curricula; preparation of a training course for engineering teachers on the topic of Work Engineering; creation of an Advisory Council composed of engineering professors linked to Work Engineering themes with the function of improving our actions in both theoretical and practical fields; involvement of undergraduate students in the preparation of the previously mentioned book and in all Work Engineering actions; negotiations for the translation of the book in question into other languages.

Finally, continue to carry out research in the undergraduate context on the topic of Work Engineering with the aim of confirming or arguing against the proposals presented here.

The argument presented here serves to consider Work Engineering as a means of promoting sustainable work, creating real conditions to face the progressive wave of precarious work and life that has intensified in Brazil in recent years. The dynamics of the ultraliberal and financialized market, combined with the development of platform-mediated work and various political reforms in Brazil, have weakened labor rights and destabilized labor relations. We believe that only critical development through education can transform this harsh reality and bring to the fore the main element of wealth creation in our society: the work of men and women.

References

1. Silva Junior RD, Ferreira LC, Lewinsohn TM (2015) Between hybridisms and polysemies: towards a sociological analysis of sustainability. Environ Soc São Paulo 18(4):35–54
2. Boudra L (2016) Sustainability of work and adherence prevention: the case of the territorial dimension of waste in the activity of sorting domestic packaging. Diss. University of Lyon, France
3. Gemma SFB (2008) Complexity and agriculture: organization and ergonomic analysis of work in organic agriculture. Agricultural engineering thesis, State University of Campinas
4. Brunoro CM (2013) Work and sustainability: contributions from the ergonomics of activity and the psychodynamics of work. Thesis Polytechnic School, University of São Paulo, Brazil
5. Bolis I, Brunoro CM, Sznelwar L (2014) Mapping the relationships between work and sustainability and the opportunities for ergonomic action. Appl Ergonom 45:1225–1239
6. Brunoro CM, Bolis I, Sznelwar L (2015) Exploring work-related issues on corporate sustainability. Work 53(3):643–659
7. Braatz D, Rocha R, Gemma S (org) (2021) Work engineering: health, safety, ergonomics and design. Ex Libris Comunicação, São Paulo
8. Kleba J (2017) Engaged engineering—teaching and extension challenges. Revista Tecnologia e Sociedade 13(27):170–187
9. Paravizo E, Fonseca MLF, Lima, FT, Gemma SFB, Rocha R, Braatz D (2021) How are ergonomics and related courses distributed across engineering programs? An analysis of courses at Brazilian Universities. Lecture notes on networks and systems, 219 edn. Springer International Publishing, pp 567–574
10. Fonseca MLF, Paravizo E, Traldi F, Simões RR, Braatz D, Gemma SFB (2020) Analysis of the presence of occupational science disciplines in production engineering courses. In: XL national meeting of production engineering, Foz do Iguaçu, Paraná, Brazil
11. ABEPRO (2023) ABEPRO home page. Available in:https://abepro.org.br/interna.asp?c=362. Last accessed on 06/25/2023

12. Braatz D, Rocha R, Gemma S (2021) A new work engineering. In: Braatz D, Rocha R, Gemma S (org) Work engineering: health, safety, ergonomics and design. Ex Libris Comunicação, São Paulo
13. Lima MEA (2021) Mental health and work. In: Braatz D, Rocha R, Gemma S (org.) Work engineering: health, safety, ergonomics and design. Ex Libris Comunicação, São Paulo
14. Mendes R (2021) The relationship between health, work and illness. In: Braatz D, Rocha R, Gemma S (org.) Work engineering: health, safety, ergonomics and design. Ex Libris Comunicação, São Paulo
15. Rocha R, Vilela RAG (2021). For a culture of safety in organizations. In: Braatz D, Rocha R, Gemma S (org) Work engineering: health, safety, ergonomics and design. Ex Libris Comunicação, São Paulo
16. Gemma SFB, Abrahão RF, Traldi FL, Tereso MJA (2021) Ergonomic approach centered on real work. In: Braatz D, Rocha R, Gemma S (org) Work engineering: health, safety, ergonomics and design. Ex Libris Comunicação, São Paulo
17. Braatz D, Paravizo E (2021) Participatory work design: challenges and good practices. In: Braatz D, Rocha R, Gemma S (org) Work engineering: health, safety, ergonomics and design. Ex Libris Comunicação, São Paulo

Developing a Framework for Sustainable Education in a Production Engineering Study Program

Maxim Mintchev[1]($\boxtimes$), Valentin Eingartner[1], Bernd Muschard[1], and Nicole Oertwig[2]

[1] Technische Universität Berlin, Berlin, Germany
mintchev.krassimirov@tu-berlin.de
[2] Fraunhofer Institute for Production Systems and Design Technology, Berlin, Germany

Abstract. Given the current challenges faced by the world regarding climate change and the goals outlined by the United Nations in their Agenda 2030 for Sustainable Development, sustainability is increasingly gaining importance in education and industry. To be able to tackle the challenges of responsible consumption and production, societies need to prepare highly skilled and responsibly thinking engineers, who are fit for the market while also contributing actively to the transformative change towards sustainable economies and societies. The international master program Global Production Engineering (GPE) at Technische Universität Berlin aims to impart knowledge on production, engineering, and management with a major focus on sustainability. In this paper, policies of universities locally and worldwide are reviewed and lecturers of GPE are surveyed to determine which sustainability topics and approaches are integrated in their lectures and in the curriculum. From this, a framework for competence building for sustainable development in the economy is developed. The program is then reviewed and found to be teaching some of the core competencies outlined by the Education for Sustainable Development, albeit with a clear potential for improvement. To address this, corrective measures are proposed.

Keywords: education for sustainability framework · engineering education · competencies for sustainable development

1 Introduction

Fueled by an ever-increasing resource scarcity in many crucial areas for human prosperity and the threat of irreversible consequences brought about by climate change, the United Nations have adopted the resolution Agenda 2030 for Sustainable Development [1]. Among the 17 Sustainable Development Goals outlined in the resolution, importance is placed not only on changing the production and consumption patterns responsible for environmental degradation, but also on emphasizing the role of education as a lever for the transformative change towards more sustainable societies and economies. In its report [2], the German Advisory Council on Global Change points out that only an immensely transformation can ensure prosperity for around 9 billion people within the

H. Kohl et al. (Eds.): GCSM 2023, LNME, pp. 126–134, 2025.
https://doi.org/10.1007/978-3-031-77429-4_15

given global ecological limits in 2050. To achieve these goals, technological changes are not sufficient. Rather, a comprehensive transformative change is needed at all economic, institutional, and cultural levels. And, finally, better educated global citizens are needed who are aware of the impact of their actions and the changes that must be made. This is where education must play a crucial role.

The international master program Global Production Engineering (GPE) at the Technische Universität Berlin teaches a curriculum focusing on the fields of production, management, engineering, new-energy technologies, and intercultural communication, all with a strong emphasis on sustainable engineering practices [3]. As an international program, with students from over 70 nations worldwide, it offers a special opportunity to communicate the challenges of rational demanded sustainability and possible solutions, and to disseminate them to the world through the graduates as multipliers.

To fulfill this responsible role as an educational institution with respect to global challenges, the decision was made to assess the sustainability aspects included in the current curriculum of GPE. This assessment starts by analyzing the goals of the Education for Sustainable Development program and its road map for completion of the Sustainable Development Goals by 2030 [4]. After that, practices adopted by other universities nationally and internationally were reviewed to find best practices among them. To establish the current level of integration of sustainability contents into GPE's curriculum, a survey among its lecturers was carried out. With the knowledge gained from the policy review and the survey, a framework was developed with the objective of presenting a vision for the GPE study program. This vision includes relevant attributes and elements for the introduction of concepts and skills related to sustainability into a classic, purely industry-oriented study program. Lastly, the GPE study program was reviewed against the backdrop of the developed model, areas with required action are identified, and practical measures for improvement were derived.

2 Education for Sustainability

In recent years, different policies have been adopted with regards to sustainability in education, both from political entities and academic ones. In this section, a short review of prominent politics is presented.

2.1 Education for Sustainable Development (ESD for 2030)

ESD for 2030 is a framework adopted by UNESCO for the period between 2020 and 2030 [4] focuses on five priority action areas: supporting sustainable policy making, transforming educational environments, supporting competency development of teachers, mobilizing youth into action, and encouraging sustainable development on a local level. These five areas highlight the importance of education in reaching the 17 Sustainable Development Goals (SDGs) of the UN, not only by raising awareness about the SDGs, but also by promoting critical thinking which helps with multi-criterial problem optimization, such as it is common in sustainability problems [4].

2.2 University Strategies for Integration of Education for Sustainability

On a national level, several German universities have developed fleshed-out guidelines on sustainability in teaching. Some common attributes are a strong presence of study programs and courses in which sustainability plays a central role [5–9], with the presence of special programs to raise awareness on sustainability matters, such as special series of lectures [6], or Change Labs [8]. Providing advanced training on sustainability for university staff is another strategy adopted by some universities [9, 10]. Among the values pursued within these strategies, internationalization, inter- and transdisciplinarity, and empowerment for societal change stand out.

Internationally, the inclusion of sustainability-based courses and study programs is the way various universities choose to implement sustainable education in their portfolio [11–18]. Among those universities, the inclusion of special programs on sustainability such as Sustainable Labs [12], Living Labs [15, 16], or Massive Open Online Courses [18] are prevalent. Values sought to impart are critical thinking and interdisciplinarity [14, 15], as well as empowering students to become agents of change and responsible citizens [13].

Furthermore, examples of the integration of sustainability in single study programs can be found, providing further insight. In the Quality Management Engineering undergraduate program at Tongling University, a one-semester module called "ethics, involvement and sustainability" was introduced [19]. This module includes activities such as an industry-oriented field investigation, lectures on the social and environmental aspects of sustainability, where a two-way discussion is encouraged, and a concluding project. At Newcastle University, several case studies of engineering study programs with a strong component of sustainability are present [20]. Elements included in these programs are industry-based courses, capstone projects, and discussions of moral and ethical principles in the context of Corporate Social Responsibility. The programs are characterized by their increasing hands-on, interdisciplinary nature. For the sustainability integration in an Electronics Systems Design and Innovation major at the Norwegian University of Science and Technology, a pilot case was carried out, which focused on the perspective of academic staff involved [21]. It was found that for a successful integration of sustainability into a study program, academic staff need to be properly informed about sustainable development and motivated to integrate it into their lecture contents. Furthermore, they need to be inspired for it to become a bottom-up approach in which staff contributes proactively. Lastly, discussion and reflection among academic staff members on sustainable development support the adoption of sustainability into their modules.

3 Lecturer Survey on Sustainability

To understand the individual perspectives of the GPE lecturers, a survey about sustainability integration was carried out among them. 25 out of the 41 lecturers from 36 different modules answered the survey. Lecturers were asked about their perceived relevance of sustainability in their specific domain, their level of knowledge in this regard, their motivation to further integrate sustainability in their modules, and questions specifically about their module contents.

Relevance. A majority of GPE lecturers see the importance of sustainability in their domains, as 73% of them view sustainability as a relevant or very relevant matter in their field of expertise.

Level of Expertise. 76% of lecturers view themselves as capable or very capable of incorporating sustainability into their curricula.

Current Integration. 22 respondents, or 88%, already integrate sustainability topics into their modules. From those, 8 make it their main focus or use it as a common thread throughout lectures. For the other 14, it is integrated only as part of some topics or lectures.

Motivation. 12 lecturers saw the need of further integration of sustainability in their modules, while 13 did not. A similar outcome resulted when asking about how extensively they would be willing to modify their module contents for that purpose, with 48% willing to make major or extensive revisions, and 52% ready to make few or no revisions. When asked if they would be willing to participate in a working group for sustainability integration in GPE, 32% responded positively, 24% were against it, and a majority 44% responded "maybe". The contrast between these evenly distributed answers and the one-sidedness of the previous answers can be attributed to the fact that some lecturers feel that sustainability is sufficiently addressed in their lectures.

Module Content. When asked about sustainability topics addressed in their modules, only two lecturers do not provide any answer, while resource efficiency, circular economy, transformative economic change, and renewable energies are found among the most commonly stated lecture topics. Regarding critical discussion with students about pros, cons, and balances of transformative change towards more sustainable societies, 60% of the respondents engage often or very often in them, while the other 40% rarely or never do.

The answers to the survey show that perceived relevance and knowledge of sustainability are high among lecturers in their respective disciplines, and as such an overwhelming majority already integrate sustainability topics in their lectures. However, in many cases these topics are not deeply integrated into the curriculum but treated on occasion. Furthermore, a two-way critical discussion on sustainability matters with students often does not take place. This survey shows the current state of GPE regarding education for engineering sustainability. To derive improvement measures, a guiding principle for GPE has to be established, which is done by means of the development of a framework for the educations of responsible engineers of the future.

4 Framework for the Education of Responsible Engineers of the Future

According to the knowledge steps proposed in the knowledge oriented corporate management by North (Fig. 1), the path of transformation from information to competitive abilities can be seen as a sequence of levels, each adding to the current one. In this proposed model, the combination of information in combination with context, experience and expectations leads to knowledge, which in turn, combined with application and motivation, leads to competences that enable actions with the sequence finally leading to competitive capacity [22].

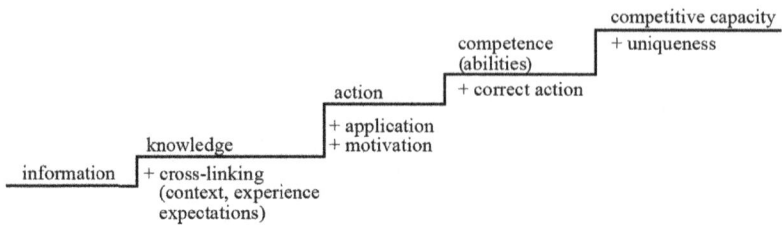

Fig. 1. The competence ladder [22]

Similar to corporate management, in which professionals and experts have to be further educated in order for corporations to stay competitive in the market, the framework can be modified to fit the education of engineering students. The goal however is not competitive capacity on its own, but to include all aspects of sustainability to finally have educated competitive engineers with the volition for sustainable value creation. Therefore, for the proposed model in Fig. 2 the steps are separated into two dimensions. To cover all sustainability aspects (economic, environmental, social) in the model, the overarching distribution of the aspects is separated: the professional dimension is focusing more on the economic dimension of sustainability, while the focus of the sustainability dimension in the model additionally is on the environmental and social aspects. While the theoretical separation is helpful for the development of the model, the aspects are always taught together.

The basis of the model in Fig. 2 is the knowledge level, which is divided into knowledge on the professional half and awareness about sustainability aspects. Knowledge is generated through lectures about state-of-the-art information, the best practice, while awareness includes both influences on the climate systems, the economic and social environment, and ways of mitigation as well as approaches for overcoming the challenges. Through the combination of knowledge and awareness with application, in this model represented by a hands-on approach, action is meant to be enabled. Besides lectures and exercises, the desired teaching forms here are real-life experiences close to the reality of industries combined with field trips, such as lab projects, conference visits, fare visits, and other field trips. The education in traditional engineering programs is mostly focusing on very specific professional subjects, which are rather self-contained without going beyond the boundaries of the respective subject areas. While this is necessary to obtain a good basic engineering education, it does not create the comprehensive view, which is needed. Students must therefore engage in interdisciplinary and transdisciplinary dialogue with other disciplines and with other international students. Teaching methods line interdisciplinary learning are summarized and proposed by Horbacauskiene [24].

The next level, professional competencies on the one hand and core competencies for the education for sustainable development proposed by the UNESCO on the other hand are reached through the correct application of the explicit knowledge. During this stage, the explicit knowledge is supplemented by implicit, or tacit knowledge. Like the other steps, competencies for an engineer, who is both educated well professionally and aware about their actions overlap on many occasions. According to the UNESCO framework for education for sustainable development, the core competencies include but are not limited to the competency of critical thinking and foresight, normative, strategic, and collaborative thinking, self-awareness competence, etc. [4].

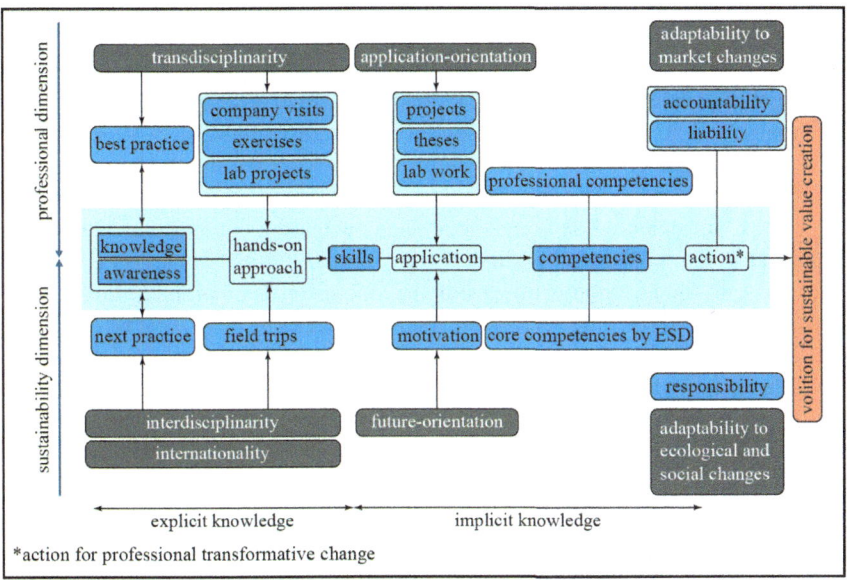

Fig. 2. Framework for the education of responsible engineers of the future

Combined with action for professional transformative change, these competencies finally enable the target state of engineers who are equipped with both the know-how and the motivation for sustainable value creation. To ensure correct action for professional transformative change, students must be taught adaptability to market changes, as well as ecological, social, and socio-economic changes, creating a sense of responsibility in the social and ecological dimension, and a sense of accountability and liability in the professional dimension.

5   Review of the Study Program with Regard to the Model

The curriculum is composed of five module groups, Manufacturing, Engineering, Management, Intercultural Communication, and Special Profile. In some limitations, students are allowed to choose their personal curriculum according to their interests and career plans. According to the study regulations of the TU Berlin, the modules consist of specific course forms, which are lectures, exercises, methodical exercises, integrated courses, seminars, colloquiums, lab exercises, projects, excursions, and courses [23]. Depending on the number of credit points, a module typically consists of two or three courses (e.g. lecture and exercise).

From the survey among the lecturers and by reviewing lecture materials, it is apparent, that almost every course has a connection to sustainability. Therefore, depending on their choices of modules, students of the GPE program graduate with some awareness about the challenges of the climate crisis. In some courses, approaches to face the challenges are taught; this is combined with field trips and extracurricular activities with a connection to sustainable value creation. This shows that the possibilities of teaching and learning methods are widely used but not to their full extend. In the curriculum, the dimension of professional competitive capability is mostly fulfilled, however a lot of courses don't yet teach skills and competencies concerning the ecological and social dimensions.

As a consequence to the current state, graduates are educated as somewhat responsible engineers, who are still lacking some of the core competencies by ESD, which must be given a greater consideration in most courses. In the international study program GPE, all students come from different cultures and therefore they show a social adaptability to the extent of daily interaction and navigating in unknown cultures, however adaptability to social changes must be expanded. Similar expansion is necessary for the ecological and economic dimension. In conclusion, training in the GPE program produces engineers, who are somewhat responsible, but don't have the absolute volition for sustainable value creation.

In order to bring the curriculum of GPE closer to the ideal, in the medium term, the curriculum must be refined to ensure a connection to sustainability and sustainable value creation is given in each of the modules. To this end, the GPE program management have to find lecturers who are willing to include the aforementioned concepts into their lecture and support them by doing so. Additionally, an analysis of topics in sustainable value creation that are not covered by the GPE curriculum should be done to determine if there is a need to expand the curriculum to include these topics.

In the first step, the model was developed based on the scientific principles of knowledge management and skills development with a particular focus on sustainable value creation. In the next step, individual experiences, suggestions and needs of various stakeholders such as lecturers, students and industry will be integrated into the model within the framework of working groups. In addition, there should be a constant comparison with the latest scientific and political recommendations.

References

1. UN General Assembly (2015) Transforming our world: the 2030 agenda for sustainable development
2. German Advisory Council on Global Change (2016) Humanity on the move: unlocking the transformative power of cities
3. Global Production Engineering. What is GPE? https://web.gpe.tu-berlin.de/. Last accessed 2023/07/19
4. UNESCO (2020) Education for sustainable development: a roadmap
5. RWTH Aachen. Nachhaltigkeitsbezug in Studiengängen. https://www.rwth-aachen.de/go/id/wkrdz. Last accessed 2023/06/13
6. Technische Universität Braunschweig, "Strategie". https://www.tu-braunschweig.de/nachhaltigkeit/strategie. Last accessed 2023/06/13
7. Leibniz Universität Hannover, "Nachhaltigkeit und Klimaschutz in der Lehre". https://www.sustainability.uni-hannover.de/de/lehre. Last accessed 2023/06/13

8. Universität Stuttgart, "Nachhaltigkeit". https://www.uni-stuttgart.de/universitaet/profil/nac hhaltigkeit/. Last accessed 2023/06/13
9. Leuphana Universität Lüneburg, "Nachhaltigkeit an der Leuphana". https://www.leuphana. de/universitaet/nachhaltig.html. Last accessed 2023/06/13
10. Technische Universität München, "Nachhaltigkeit". https://www.tum.de/ueber-die-tum/ ziele-und-werte/nachhaltigkeit. Last accessed 2023/06/13
11. University of California, Berkeley (2020) UC Berkeley sustainability plan
12. University of Toronto. Sustainable change programs. https://www.fs.utoronto.ca/sustainab ility/sustainable-change-programs/. Last accessed 2023/06/13
13. The University of British Columbia, UBC Sustainability. Teaching & applied learning. https:// sustain.ubc.ca/teaching-applied-learning. Last accessed 2023/06/13
14. UNSW Sydney. Learning & teaching|environmental sustainability. https://www.sustainab ility.unsw.edu.au/our-plan/learning-and-teaching. Last accessed 2023/06/13
15. The University of Sydney. Research and education. https://www.sydney.edu.au/about-us/vis ion-and-values/sustainability/research-and-education.html. Last accessed 2023/06/13
16. University of Pennsylvania, Penn Sustainability. Learning sustainability. https://sustainability. upenn.edu/initiatives/learning-sustainability. Last accessed 2023/06/13
17. University of Auckland. Studying sustainability. https://www.auckland.ac.nz/en/about-us/ about-the-university/the-university/sustainability-and-environment/studying-sustainability. html. Last accessed 2023/06/13
18. Lund University. Sustainability and education. https://www.lusem.lu.se/about/sustainability/ sustainability-in-our-education. Last accessed 2023/06/13
19. Qu Z, Huang W, Zhou Z (2020) Applying sustainability into engineering curriculum under the background of "new engineering education" (NEE). Int J Sustain High Educ 21(6):1169–1187
20. Haile S, Glassey J (2014) Teaching of sustainability: higher education (HE) case studies. In: Thomas K, Muga H (eds) Handbook of research on pedagogical innovations for sustainable development. IGI Global, Hershey PA, USA, pp 398–409
21. Verhulst E, Bolstad T, Reppe Lunde S, Randem Lunde H, Berntsen Henriksen R (2022) Integrating sustainability in an electronic engineering program: insights and experiences on academic staff involvement. In: SEFI 50th annual conference of the European society for engineering education. Universitat Politècnica de Catalunya, Barcelona, pp 1778–1785
22. Kubr M, Kubr M (2002) Management consulting a guide to the profession, 4th edn. International Labour Office, Geneva
23. Akademischer Senat der Technischen Universität Berlin: Ordnung zur Regelung des allgemeinen Studien- und Prüfungsverfahrens an der Technischen Universität Berlin (Allg. StuPO), §48, Berlin (2021). https://www.static.tu.berlin/fileadmin/www/10000000/Studie ngaenge/StuPOs/AllgStuPO_deu.pdf. Last accessed 2024/01/22
24. Horbacauskiene J (2019) Sustainable education methods. In: Leal Filho W (eds) Encyclopedia of sustainability in higher education. Springer, Cham. https://doi.org/10.1007/978-3-030-11352-0_113

Life Cycle Thinking

Identification of Lightweighting Potentials Towards More Sustainable Products via the Functional Lifecycle Energy Analysis (FLCEA)

Kristian König$^{(\boxtimes)}$ ⓘ, Janis Mathieu, and Michael Vielhaber

Institute of Product Engineering, Saarland University, Campus E2 9, 66123 Saarbrücken, Germany
koenig@lkt.uni-saarland.de

Abstract. In times of increasing material scarcity and still not widespread availability of renewable energy, it is inevitable to seize all opportunities to improve the resource efficiency of any product. In particular, this promotes lightweight design as a key technology to reduce both material and energy consumption across the entire product lifecycle. Therefore, systematically exploiting lightweighting potentials already on a functional level remains a major challenge requiring methods of product development to be applied in the early phase of each development cycle ensuring the implementation of the holistically best solution from a sustainability point of view. Thus, the proposed "functional lifecycle energy analysis" (FLCEA) provides a remedy to effectively implement lightweight design in products by identifying recommendations for action regarding future product generations, resulting in a holistic energy optimization. In this contribution, the methodology is presented in parallel to its implementation in the use case of conceptual designing a semi-mobile handling system. As a result of this study, it was not only possible to identify lightweighting potentials of the system-in-development, but also to analyze technical product functions requiring modifications in view of the circular economy to holistically improve the ecological sustainability of the entire system.

Keywords: Lightweight design · Design for sustainability · Functional design · Energy analysis · Lifecycle optimization

1 Introduction

Increasing trends in market dynamization are causing development times to become even ever shorter. Consequently, new digitized product development methods, methodologies and tools are needed taking into account current social and political dynamics like growing sustainability awareness on the one hand, while enabling lean and fast innovation processes for future product generations on the other hand. Facing the technological challenges, lightweight design is widely recognized as decisive not only for

H. Kohl et al. (Eds.): GCSM 2023, LNME, pp. 137–145, 2025.
https://doi.org/10.1007/978-3-031-77429-4_16

functionally, technically and economically outstanding solutions, but also for a reduced and targeted use of resources from an ecological sustainability viewpoint. In particular, an efficient use of material and energy in products can thus be achieved, whereby a holistic approach must always be retained for a meaningful evaluation of the usefulness of lightweight design. Indeed, highly sophisticated lightweight solutions can result in increased manufacturing and disposal efforts, contradicting, at least in part, the idea of circular economy. Therefore, development methodologies evaluating the trade-offs between lightweight design and other sustainability strategies to improve the ecological sustainability of a product while keeping development times and costs as low as possible are increasingly needed and unfortunately do not yet exist appropriately [9]. Therefore, the energy consumption is used within this contribution as the sustainability indicator since meaningful correlations can be derived in respect to other relevant variables such as CO_2 emissions or costs according to Ashby [3]. Besides, the functional design within product development is considered of exceptional importance for innovation generation because of its capability to architecturally rethink functionalities and interrelationships resulting in the highest possible potential for the optimization of existing product generations.

Hence, this contribution proposes the novel methodology "functional lifecycle energy analysis" (FLCEA) for the identification of lightweighting potential within functional design by treating the design task as a holistic energy optimization problem. For a better understanding of the methodology, it is illustrated closely at the example of a portal robot system throughout this contribution. First, the state of the art is outlined in Sect. 2 regarding existing deficits in research and promoting the FLCEA methodology, which is presented to the use case in Sect. 3. Finally, the results are summarized within Sect. 4 while also providing a short outlook on future work.

2 State of the Art

2.1 Product Development and Lightweight Design

Industrial product development is dominated by generation development, thus reference products are commonly available representing the starting point for optimization loops regarding future product generations [2]. Traditionally, this generation optimization is characterized by the interaction of methods firstly to analyze the current realization strategy regarding the product's deficits and secondly to synthesize new innovative solutions.

Examples of optimization goals for product generations include reduced costs, an improved functionality, or minimized sustainability impacts, for example, quantifiable in terms of a lower carbon footprint. As tool for the implementation of these optimization goals, lightweight design can suffice, capable of improving several impact dimensions simultaneously (e.g., weight, costs, and CO_2 emissions), if applied in an intelligent and appropriate way, always advanced by a holistic lifecycle view.

For this purpose, the SyProLei framework by Kaspar et al. [5] was recently published, which illustrates the multi-faceted nature of lightweight design including various disciplines of development ("product", "material", "production", "joining technology", "modeling, simulation & optimization") in a V-model based on the RFLP approach of the "model-based systems engineering" (MBSE) paradigm [7]. In addition to the targeted coupling of material, production and joining selection via multi-criteria decision making (MCDM) in between the component as well as the subsystem design, various individual lightweight design methods find application within the methodological framework enabling the mechanism of analyzing the current product generation and synthesizing it into an improved product to be appropriately realized for lightweight design. For instance, a methodological process for a creative lightweight and sustainability-oriented solution finding [8] is incorporated therein, which first requires the identification of weaknesses in the existing generation of the system-in-development embodying the highest optimization potential from a lightweight design perspective. Thus, in the next section, methods for the analysis of lightweight products will be discussed.

2.2   Methods for Identifying Lightweighting Potentials

The application of lightweight design in product development can be seen as a continuous activity along the entire product development process [9]. Accordingly, lightweight design methodologies can also be applied along this methodological process chain. At the functional design, methods to identify lightweight design potential intend for the analysis of functional design spaces. Therefore, the "function mass analysis" (FMA) proposed by Posner et al. [12] presents a common approach, which was further supplemented by the "extended target weighing approach" (ETWA) by Albers et al. [1] for the multidimensional analysis of the effects of the functional design in terms of mass, costs, and CO_2 emissions (can be summarized as "effort"). Their methodological approach leads to recommendations for action concerning a function to be improved in a successive product generation realized by lightweight design.

Regarding solution principles on the logical view of design Laufer's et al. "energy distribution analysis" (EDA) [10] offers a methodological framework to consider the translational and rotational state of motion as well as the mass distribution of individual components for the lightweighting optimization of products. This mass-related energy analysis gains in importance in the context of lifecycle engineering, a highly relevant factor for lightweight design investigations in future [9]. For this purpose, Laufer et al. merely emphasize the product's use phase, whereas the "life cycle energy optimisation" (LCEO) methodology of O'Reilly et al. [11] for the early integration of environmental considerations into vehicle development describes a more holistic approach. In this way, trade-offs between manufacturing, utilization and disposal are revealed on the basis of a comprehensive lifecycle energy analysis [4].

2.3   Deficits in Lightweight Design Methodology

However, the described methodologies still have shortcomings, especially in coupling a holistic sustainability assessment with the individual characteristics of lightweight design at the functional view of design.

For example, the ETWA triple-counts indicators that are interrelated in varying ways (a change in the weight of a function realization can affect costs and CO_2 emissions in both directions). Furthermore, it lacks a diversified analysis of the use phase, in particular, considering the translational and rotational behavior of components and their associated individual influences in sustainability effects. In contrast, the EDA addresses this issue from an energy perspective but lacks a holistic lifecycle view, notable in the incorporation of the product's begin—as well as end-of-life. Besides, a methodology for calculating the effects of component recirculation processes is missing in all the lightweight design methodologies and as of yet not available. Only the holistic approach of O'Reilly's et al. LCEO [11] seems to be promising, although a generalization for lightweight design and the view on the functional design for the ecology assessment is omitted.

The described deficits are tackled within this publication via introducing the "functional lifecycle energy analysis" (FLCEA) as a methodology for the holistic sustainability assessment of lightweight solutions with energy as ecological sustainability indicator. Thereby, the focus is placed on the functional product architecture since it offers the greatest possible leverage for system optimization. The methodology is descriptively presented in the following section incorporating the conceptual design of a portable robot system.

3 Methodology: FLCEA and Case Study

The implementation of the FLCEA is based on four consecutive steps, which are shown in Fig. 1 and presented in the following subsections on the example of a portal robot.

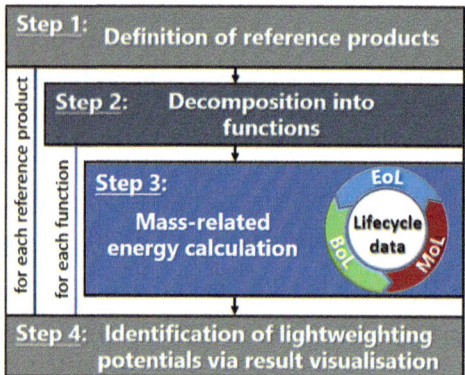

Fig. 1. Methodology of FLCEA implementation within functional design.

3.1 Definition of Reference Products

In the first step, it is initially necessary to define the product to be analyzed, which serves as reference for the optimization loop. For this purpose, either an already self-designed product or a competing product available on the market can be chosen. Multiple reference products are also conceivable. Thereby, all real components of the system must be identified, and the actual end-of-life strategy (retention option, RO) must be determined. In the example to be presented, the current product generation serves as the reference product for development within this contribution. The purpose of the robotic system is to transport tools between individual processing machines using a carriage and a 6-axis robot with a gripping unit.

3.2 Decomposition into Functions

In the second step the decoupling from real components to enable an abstract thinking in technical functions follows, finally opening up the solution space for the generation development. As shown in Fig. 2, the system-in-development is decomposed into its functional items as well as into its logical elements (parts and assemblies). At least on the highest level of detail, an allocation between each technical function and its corresponding logical element proceeds in the spirit of MBSE.

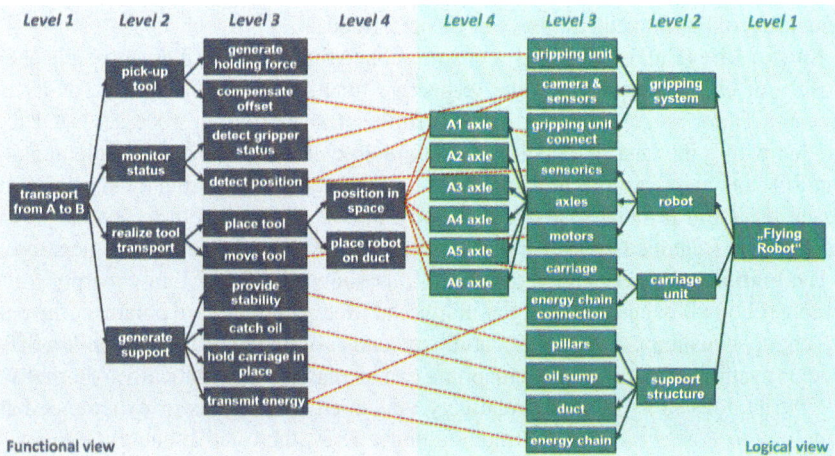

Fig. 2. Excerpt of the product architecture from a portal robot of the functional (green) and logical (blue) structure and allocation of functions to logical elements (red), adapted from [13].

3.3 Mass-Related Energy Calculation

The most sensitive step of the methodology is performed as a third step containing the energy calculation along the entire product lifecycle (PLC) for each function. It is assumed that each function is implemented by individual components, with functions overlapping components being proportionally prorated. To simplify this step, a division into three lifecycle stages as proposed by Kiritsis et al. [6] is foreseen, which are detailed in the following subsections.

Begin-of-Life (BoL). At the beginning of the PLC of each component in each function, the focus is on raw material extraction and production as well as the subsequent transportation of the product to its intended location of use. Thereby, the calculation methodology of Ashby [3] for the energy consumption is used, so that the system boundaries of the data acquisition are definable: starting with the material extraction, each of the first shaping and forming manufacturing processes are considered, followed by the calculation of the transportation and joining processes individually for each function. If single components have a shorter lifetime than the whole system, they need to be replaced during the usage causing the actual number of required components for the entire PLC of the system to be factored in the calculation.

Mid-of-Life (MoL). The energy consumption due to the motion of each component as well as to the dynamics of the entire system is subject of the analysis of the use phase from the lightweight design perspective. Here, the energy amount of each individual function must be determined separately, which can be realized most easily with a rigid multibody or finite element simulation model for either the full system or components separately. For this purpose, realistic use case scenarios based on the actual user behavior need to be determined and modeled accordingly. Alternatively, real measurements with a subsequent partial allocation of the energy amounts to the functions or only a rough manual approximation calculation can be performed.

End-of-Life (EoL). The energy consumption in the EoL describes a possible energy recovery of each component in the respective function as a consequence of its EoL treatment. Here, preparation, sorting and further transportation processes need to be included within the energy calculation methodology. Furthermore, the implementation of an RO can extend the lifetime of single components by integrating them into another or even the same product's lifecycle. This results in advantages in the overall energy consumption. Regained energy (e.g., by a thermal recovery) is included as negative.

Evaluation of Energy Calculation. If, as shown in Table 1 in excerpts for the portable robot, all balance data of the individual lifecycle stages are obtained, the entire PLC energy consumption can be accumulated and assigned to the functions. If a MBSE model is available, the calculative mapping can be realized automatically. Alternatively, the contributions must be made relatively via expert discussions or experience-based in analogy to Posner's et al. [12] methodology. Thus, the relative energy amounts for each lifecycle stage are apportioned to each function revealing their energetic contribution to the fulfillment of the overall product function and serving as the basis for the interpretation of results within the last step.

Table 1. Excerpt of the calculated energy consumptions across the three lifecycle stages for some parts of the portal robot.

Part/Assembly	Mass (kg)	E^{BoL} (GJ)	E^{MoL} (GJ)	EoL strategy	E^{EoL} (GJ)
Pillars	4140	66	0	Re-cyle	− 28
Carriage	675	105	150	Re-mine	0.004
A1 axle	440	35	15	Re-mine	0.002
Gripping unit	100	23	315	Re-manufacture	− 1.6
…	…	…	…	…	…

3.4 Identification of Lightweighting Potentials

Based on the relative energy contributions of each function within the individual lifecycle stages as well as accumulated over the entire PLC, weaknesses from a lightweight and sustainable design perspective can be identified in the graphical representation given in the form of a barplot as shown in Fig. 3 for the portal robot. Therein, the functions are plotted in increasing order of their importance to the overall system functionality (horizontal axis), and their associated relative energy expenditures are illustrated (vertical axis). Generally, a function should not consume more relative energy across the entire PLC (orange) than it is relatively important for the functional performance of the product (black line). Simultaneously, recommendations for action for optimization can be derived by observing the individual lifecycle stages (yellow for BoL; blue for MoL and green for EoL). As far as the MoL is significant for a function, the next optimization steps should be to reduce the product's environmental impact by an increasing effort in lightweight design of its actual implementation (e.g., "position in space"). In case of an energy-intensive BoL of a function, it is recommended to think about an improved retention treatment or the implementation of other materials or manufacturing processes (e.g., "hold carriage in place").

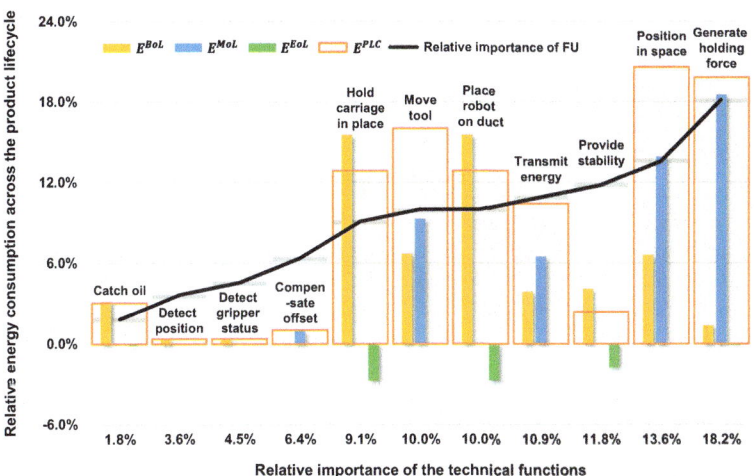

Fig. 3. Results of the FLCEA methodology for the portal robot.

4 Conclusion and Outlook

The presented FLCEA enables the lifecycle-spanning optimization of new product generations regarding their energy consumption on a functional system level. As shown in the robotics use case, the methodology is capable to reveal and visualize optimization potentials not only in lightweighting but also other in sustainability strategies (e.g., circular economy) at functional design.

A weak point of the FLCEA methodology is the high reliance on data for the energy calculation and the required expert knowledge for specifying the relative function importance. Subsequently, when applying the FLCEA, these aspects must always be questioned to prevent to the deliberate induction of wrong results. To validate the results of the FLCEA, we recommend applying more comprehensive methods such as the life cycle assessment in later detailing phases of development, once more knowledge about the product is available. Throughout the entire development process, it is essential to ensure the validity of the data and consider possible uncertainties. In addition, the systematic procedure for the allocation of functions to components has not yet been concretized in detail, which will be addressed in future work.

Acknowledgement. The authors thank the "Technologietransfer-Programm Leichtbau (TTP LB)" of the Federal Ministry for Economic Affairs and Climate (BMWK) for supporting their research within their project "SyProLei" (03LB2007H).

References

1. Albers A et al (2017) Extended target weighing approach—identification of lightweight design potential for new product generations. In: DS 87-4 Proceedings of the 21st International Conference on Engineering Design (ICED 17) Vol 4: Design Methods and Tools, Vancouver, Canada, 21–25.08.2017, pp 367–376
2. Albers A et al (2016) IPeM—integrated product engineering model in context of product generation engineering. Proc CIRP 50:100–105
3. Ashby MF (2021) Materials and the environment. Elsevier Inc, Philadelphia
4. Herrmann C et al (2018) Life cycle engineering of lightweight structures. CIRP Ann 67(2):651–672
5. Kaspar J et al (2022) (2022) SyProLei—a systematic product development process to exploit lightweight potentials while considering costs and CO_2 emissions. Proc CIRP 109:520–525
6. Kiritsis D et al (2008) How closed-loop PLM improves Knowledge Management over the complete product lifecycle and enables the factory of the future. IJPLM 3(1):54
7. Kleiner S, Kramer C (2013) Model based design with systems engineering based on RFLP using V6. In: Abramovici M, Stark R (eds) Smart product engineering. Springer, Berlin, Heidelberg, pp 93–102
8. König K et al (2023) Lightweight creativity methods for idea generation and evaluation in the conceptual phase of lightweight and sustainable design. Proc CIRP 119:1170–1175
9. König K et al (2023) Sustainable Systemic Lightweight Design (S2LWD)—guidance for the heuristic application of lightweight design within an integrated product development process. Proc CIRP 119:333–338
10. Laufer F et al (2018) Supporting engineers in lightweight design: the energy distribution analysis (EDA). In: DS 92: Proceedings of the DESIGN 2018 15th International Design Conference, pp 829–840
11. O'Reilly CJ et al (2016) Life cycle energy optimisation: a proposed methodology for integrating environmental considerations early in the vehicle engineering design process. J Clean Prod 135:750–759
12. Posner B et al (2013) Operationalisation of the value analysis for design for lightweight: The function mass analysis. In: DS 75-5: Proceedings of the 19th International Conference on Engineering Design (ICED13) Design for Harmonies, Seoul, Korea

13. Scholz J et al (2023) Lightweight design of a gripping system using a holistic systematic development process—a case study. Proc CIRP 187–192. Reprinted from Procedia CIRP, Vol 118, Johannes Scholz, Jerome Kaspar, Kristian König, Marco Friedmann, Michael Vielhaber, Jürgen Fleischer, Lightweight design of a gripping system using a holistic systematic development process—a case study, 187–192, 2024, with permission from Elsevier. Reprinted from The Lancet, Vol 118, Johannes Scholz, Jerome Kaspar, Kristian König, Marco Friedmann, Michael Vielhaber, Jürgen Fleischer, Lightweight design of a gripping system using a holistic systematic development process—a case study, 187–192, 2024, with permission from Elsevier

Sustainability Data Map: Framework for Data-Based Product Carbon Footprinting of Technical Products

Nick Schreiner[1]([✉]), Christian Kürpick[1], Arno Kühn[1], and Roman Dumitrescu[2]

[1] Fraunhofer Institute for Mechatronic Systems Design IEM, 33102 Paderborn, Germany
Nick.Schreiner@iem.fraunhofer.de
[2] Heinz Nixdorf Institute, University of Paderborn, 33102 Paderborn, Germany

Abstract. Growing awareness and political regulations increase the pressure on companies to become more sustainable. Manufacturers of technical products in particular face the challenge of decarbonization, as machinery and plants, for instance, have an impact on the emissions of almost all industries. However, information on the Product Carbon Footprint (PCF) is often not transparent for manufacturing companies. Digitalization has proven to be a key enabler, as up to 90% of manufacturers' emissions occur along the value chain, making it impossible to calculate the PCF without data. However, identifying the necessary data and the corresponding IT system is a major challenge, especially for technical products. The Sustainability Data Map supports the identification and structuring of relevant data and its sources for product carbon footprinting across all product life-cycle phases. In addition, the Sustainability Data Map serves as a workshop-based medium to communicate and integrate all relevant stakeholders in the value chain. It is based on an existing data map that has been extended to include the criteria required for PCF determination. As a result, the Sustainability Data Map enables manufacturing companies to create more transparency about the data needed to assess the PCF of technical products.

Keywords: Product carbon footprint · Data map · Technical products

1 Introduction

Nowadays, the megatrends of digitalization and sustainability are placing new demands on manufacturing companies [1–3]. As a result of digitalization, many companies have increasingly access to comprehensive data providing insights into each life cycle phase of the manufactured product [4, 5]. At the same time, growing stakeholder awareness and political regulation increase the pressure on companies to become more sustainable [6]. Especially manufacturers of technical products face the challenge of decarbonization, as technical products, for instance, machinery and plants, impact the emissions generated in almost all industries [7–9]. By calculating the Product Carbon Footprint (PCF), i.e., the carbon emissions along the entire life cycle of a product, companies can determine the baseline for their products and thus derive targeted measures to reduce emissions

© The Author(s) 2025
H. Kohl et al. (Eds.): GCSM 2023, LNME, pp. 146–154, 2025.
https://doi.org/10.1007/978-3-031-77429-4_17

[10]. However, information about the PCF and potentials for improvement are often not transparent to manufacturing companies [10]. Digitalization has emerged as a key enabler in this process, as it allows to collect and analyse vast amounts of data to evaluate the ecological sustainability of products [3, 11]. This is particularly important as up to 90% of manufacturers' emissions occur in the up- and downstream value chain [12], making it impossible to calculate the PCF without data. However, there are significant challenges in identifying and structuring the necessary data, particularly for technical products. Based on expert interviews and a literature review, a selection of relevant challenges has been identified and is described below:

Complexity of technical products: The effort required to determine a PCF depends significantly on the complexity of the product [10]. Especially for manufacturers of complex technical products, the large number of individual components, the variety of different materials and an often-multidimensional usage phase complicate the PCF calculation.

Lack of data transparency in the value chain: Generally, the calculation of the PCF relates to the entire life cycle of a product, from raw material extraction to production and disposal [10]. Selecting the appropriate scope of the PCF, such as cradle-to-gate, cradle-to-grave or cradle-to-cradle, depends on several factors, for instance, availability of data or customer requirements, and must be decided depending on the initial situation of the company [13]. In addition, activity data must be collected as part of the PCF, which is often the greatest difficulty for companies due to the lack of transparency in the value chain [14].

Missing integration of stakeholder from the value chain: Decarbonization is a cross-company mission [15]. The same applies to the determination of a PCF as a multi-stakeholder approach. Without the integration of relevant stakeholders, e.g., suppliers or customers, a PCF calculation is not feasible due to missing data access [10]. Thus, significant communication and integration efforts with numerous and diverse partners are required [15].

Due to the sustainability goals of companies, PCF will become a critical KPI in the purchase of materials, parts, and machines and thus PCFs need to provide a robust basis for decision making [16]. Therefore, we propose a method for specifying and structuring relevant data for a PCF across the value chain. The method should support in determining the PCF by providing a systematic approach to identify and structure necessary data for assessing the PCF of complex technical products. In addition, the method should outline all relevant information for each product life cycle phase and provide guidance on where to find the data and how to evaluate its reliability. Using the method, manufacturing companies should obtain a holistic overview of the required data and its storage location as well as of the relevant stakeholders in the value chain.

The structure of the paper is based on the research process for applied sciences according to Ulrich [17] because of its focus on problems with practical relevance. Based on the identification and detailing of the practical deficit (cf. Introduction), the literature is examined to determine if adequate solutions are provided. Subsequently, the newly proposed method is specified and validated using a real application example (cf. Results). The paper closes with a conclusion including a critical reflection and an outlook on further research needs (cf. Conclusion).

2 State of the Art

Existing publications from academia and practice offer a variety of modeling tools for visualizing, structuring, and contextualizing data. Selected approaches are examined in relation to the challenges mentioned from practice. The first modeling tool to consider is the Business Process Model and Notation (BPMN) [18], which enables inter-operation of business processes on a human level. This is achieved through providing business procedures in a graphical notation and a standardized communication of these procedures. Furthermore, Data Flow Diagrams (DFD) [19] are commonly used in software analysis, design, and business administration. They provide a visual representation of data flow and transformation within a system but do not depict the sequence of operations which makes them not a process or procedure modeling method. By combining a data model and an activity model the Structured Analysis and Design Technique (SADT) [20] can be used to visualize different aspects about data like generation and storage or the data flow between business processes. Although it does not provide a comprehensive view on business processes, because it focuses on data interfaces, it is still possible to visualize business processes together with the related interfaces in terms of input, output, and control data. The Unified Modeling Language (UML) [21] is used to model systems and consists of three major categories to describe the behavior and structure between objects and is mostly used in software modeling. It provides graphical notations like class diagrams, sequence diagrams or use case diagrams to depict various aspects of a system, such as structure, behavior, and interactions. Although it provides a variety of different diagrams and elements it is complex and requires time and effort to learn. In addition, the Data Map by Joppen et al. [22] is a modeling tool that addresses a variety of problems, for instance, time consuming processes and missing important context of information. By using objects derived from OMEGA notation [23] it gives a reasonable way to understand data to ensure business understanding. It also combines process- and data-oriented views with special focus on specific data and an easy identification of data generating and consuming objects and processes so that important knowledge is kept. Besides, the Process Mining Data Canvas provided by Brock et al. [24] focuses on making the emerging technology of process mining applicable in an industrial setting. Instead of modeling an entire life cycle or business process, the method identifies data processes and data silos, whereby keeping the required modeling time low. Finally, product specific carbon calculation standards and methods exist in the ISO standards 14040, 14044 and 14067 as well as the GHG protocol Product Standard, the Pathfinder Framework and sector specific guidelines [16]. While these standards and methods describe in detail what is required for a certifiable PCF, they provide less guidance on how this should be done. In particular, they do not provide assistance in identifying and structuring the relevant data for the PCF determination.

In Summary, there are numerous methods emphasizing the visualizing, structuring, and contextualizing of data. However, none of the approaches addresses the practical problem of identifying and structuring the necessary data and sources for PCF determination. The combination of the challenges in practice and the presented theory deficits form the research need for a framework for the data-based PCF of technical products.

3   Results

The result of this research work is a framework for data-based product carbon footprinting of technical products. The framework comprises the Sustainability Data Map and a procedure model. To develop the framework, existing approaches from literature for structuring data were tailored to the requirements arising from the PCF. As a result, the Sustainability Data Map enables a simple and intuitive documentation of all the necessary information relevant for a data-based PCF. Moreover, it serves as a communication tool during workshops with suppliers and customers. The elements of the Sustainability Data Map are described in the following Sect. 3.1. Besides, the framework is applied to a real case study from the machinery and plant engineering in Sect. 3.2.

3.1   Sustainability Data Map

The Sustainability Data Map is a semi-formal specification method, which is structurally based on the data map of Joppen et al. described in Sect. 2. In addition to a canvas, the Sustainability Data Map also accesses the necessary constructs as well as product life cycle phases, which are shown in Fig. 1.

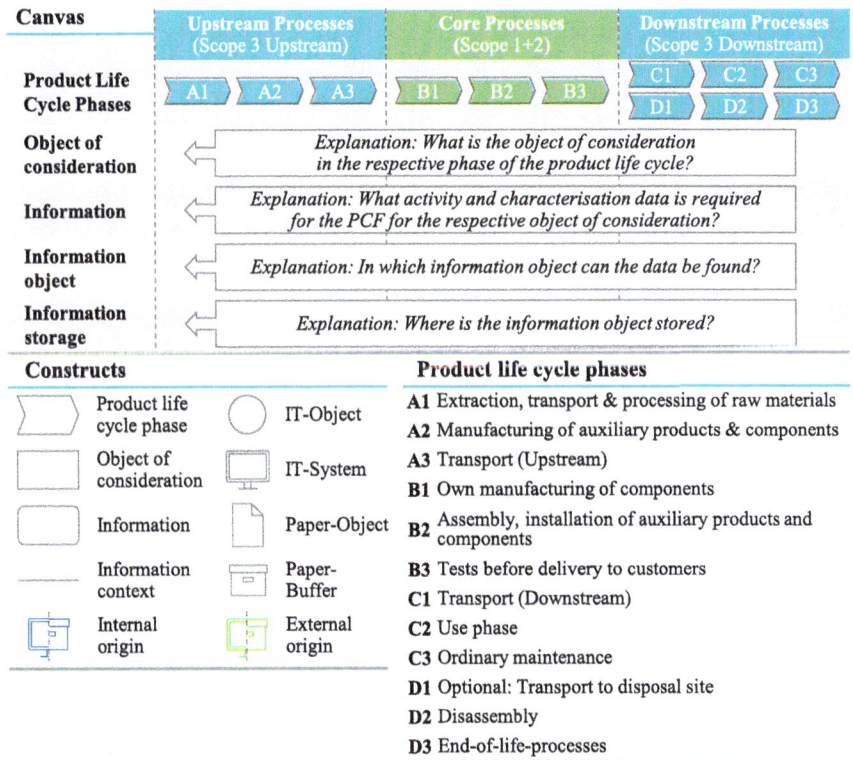

Fig. 1. Sustainability Data Map including canvas, constructs, and product life cycle phases

The Sustainability Data Map is structured along the relevant product life cycle phases (columns) that were aligned with the Greenhouse Gas Protocol's emission scopes (Scope 1, Scope 2, Scope 3 Upstream and Downstream). Thereby, *Scope 3 Upstream* reflects the company's upstream activities within the value chain especially to the suppliers. This is followed by *Scope 1* and *Scope 2*, which are essentially influenced by the company's internal value creation (production and product testing). The product life cycle concludes with the *Scope 3 Downstream* activities, which include the use and recycling of the product. The corresponding product life cycle phases of each scope mentioned above can be found in the product carbon footprinting literature [13]. According to the Greenhouse Gas Protocol, the manufacturing company of the product to be accounted for is always at the center of the map. The upstream and downstream activities in the value chain are adjusted depending on the product/company under consideration.

The information required in every product life cycle phase for a PCF is included in the Sustainability Data Map in additional four horizontal layers (rows). To determine the PCF of a technical product, the product must first be broken down into its constituent elements. The elements are included as objects of consideration in the first row of the map and represent subsets of the product (e.g., parts or components). To calculate the PCF of each object of consideration, it is necessary to consider additional information (second row) in terms of so-called activity and characterization data (e.g., energy amount and carbon emission factors). The specification of the required activity data depends on the product and its components, the respective product life cycle phase as well as the selected certification standard. The characterization data instead can be found in databases, e.g., Ecoinvent. To locate the information in the company, the corresponding information objects (e.g., BOMs) as well as their storage locations (e.g., IT systems) must be defined in the canvas in the third and fourth row. Constructs are provided for an easy and intuitive documentation of the sustainability information.

In addition to the constructs for the product life cycle phases, the object of consideration and its sustainability information, a distinction is made between IT objects and paper objects for either digital or analog information. Following this idea, a logical distinction is also made between the IT storage and paper buffer, that can be sourced from the company itself (internal source) or from customers or suppliers (external source). These constructs are colored accordingly. Finally, a line enables the contextualization of the respective information. An application example for the Sustainability Data Map including the procedure model is presented below.

3.2 Application Example

The Sustainability Data Map was applied in a project workshop on product carbon footprinting with a company from the machinery and plant engineering. The procedure model provides three essential steps: (1) selection of the PCF scope, (2) identification of the objects of consideration, (3) derivation of the relevant information for PCF calculation. The result of the procedure is a completed and documented Sustainability Data Map for further carbon footprint calculation. An example is shown in Fig. 2 and will be explained in the following.

Step 1—Selection of the PCF scope: As a first step, the scope of the product carbon footprint is determined. Depending on the PCF scope, a distinction between a *cradle-to-gate* or *cradle-to-grave* approach is made. Accordingly, the canvas is adjusted by omitting certain product life cycle phases (e.g., removing Scope 3 Downstream for the *cradle-to-gate* approach). During the project, a cradle-to-grave approach was chosen due to the application example, so the Sustainability Data Map was not adjusted.

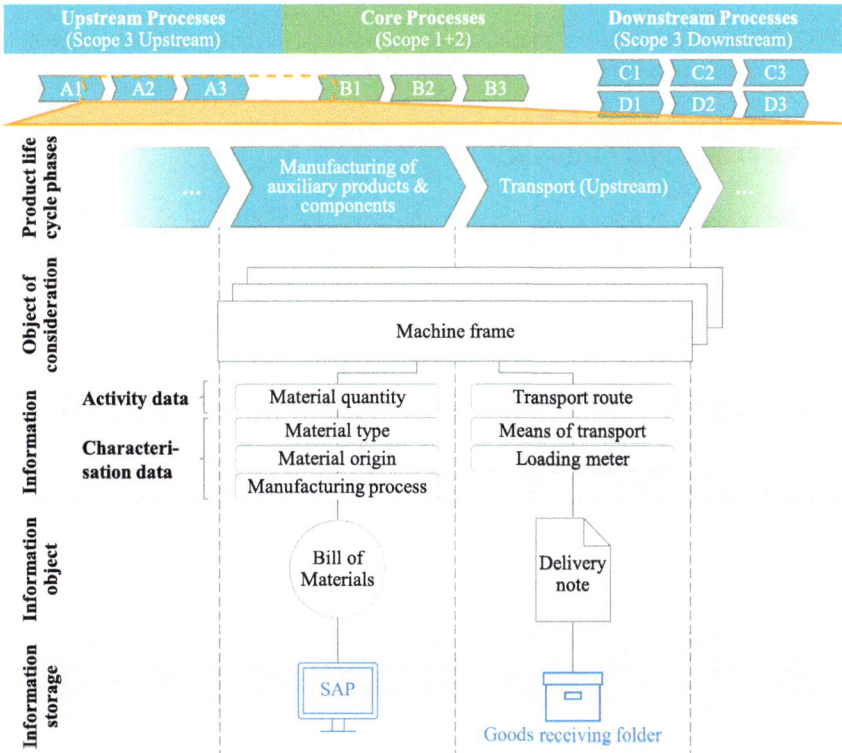

Fig. 2. Application example of the Sustainability Data Map

Step 2—Identification of the objects of consideration: In the second step, the corresponding objects of consideration are identified in each relevant product life cycle phase. The object of consideration represents an accounted subset of the product carbon footprint, which then needs to be detailed with the help of further information. In the application example shown in Fig. 2, the object of consideration "machine frame" has been identified in the product life cycle phase "manufacturing of auxiliary products and components" and "transport". The machine frame is a part of the entire product that needs to be accounted for the considered phase. The machine frame is represented with the help of the corresponding construct. Before the third step begins, all objects of consideration in all product life cycle phases are documented.

Step 3—Derivation of relevant information for PCF calculation: In the third step, the information about the corresponding activity and emission data, the required information object, and their storage locations are sequentially recorded for each consideration object. The relevant information objects with the needed activity and emission data are specified in the third row. In this course, the graphically intuitive constructs are applied as well. In the application example, the construct "IT object" indicates that the relevant data set for the activity and emission data can be found in the "bill of materials", which can be found in the SAP system. In contrast, the relevant information object to record the transport emissions of the machine frame, is located on the analog delivery note. This is not stored digitally but can be found in the goods receiving folder. Both the SAP system and the folder are internal storage locations, so that in this case no customer or supplier needs to be queried regarding the data.

4   Conclusion and Outlook

The paper provides a framework for the data-based determination of the PCF of technical products. The framework addresses the manufacturing industry's need for more sustainable business activities with an increasing digital maturity in the company. The core of the framework is the Sustainability Data Map and the corresponding procedure model. The Sustainability Data Map is a semi-formal specification method that identifies and structures the data required for carbon footprinting inside and outside the company under consideration. Thus, a canvas and constructs were developed to generate data transparency as a basis for the PCF determination. As the product life cycle phases extend beyond the company's own boundaries, the Sustainability Data Map is also a useful tool for workshops with suppliers and customers. The framework was applied with a company from the machinery and plant engineering sector and its suppliers and customers. The results show the practicality and applicability of the framework.

Despite its good applicability, the Sustainability Data Map represents only a part of the PCF determination of technical products, which needs to be embedded in a systematic overall approach. Furthermore, there is a need for research on the valid derivation of digitalization potentials in relation to data-based carbon footprinting to prepare companies for an automatically calculated PCF. In addition to the PCF, the Sustainability Data Map should in future be able to include other assessment dimensions of a life cycle assessment, such as water or toxicity. Finally, the Sustainability Data Map needs to be further validated as well as adapted to other sectors outside the manufacturing industry, for instance, the service sector, to support the decarbonization of all industrial sectors.

References

1. Fink H, Sauter R, Zirkelbach T (2021) CxO priorities study. Horváth, Stuttgar
2. Hallstedt S, Isaksson O, Rönnbäck AÖ (2020) The need for new product development capabilities from digitalization, sustainability, and servitization trends. Sustainability 12:10222
3. Diófási-Kovács O, Nagy J (2023) Beyond CO_2 emissions—the role of digitalization in multidimensional environmental performance measurement. Environ Sustain Indic 18:100252
4. Xin Y, Ojanen V (2017) The impact of digitalization on product lifecycle management: how to deal with it?

5. Seetharaman A, Patwa N, Saravanan AS, Vakhariya S (2018) Disrupting digital future of product lifecycle. Int J Manage Sustain 7:32–42
6. Kämmler-Burrak A, Kruck F (2023) Sustainability as a top priority on the corporate agenda. https://www.horvath-partners.com/en/media-center/featured-articles/sustainability-as-a-top-priority-on-the-corporate-agenda. Accessed 20 July 2023
7. Neuhold M (2021) Welchen Beitrag der Maschinenbau zur Dekarbonisierung leisten kann. Ernst & Young. https://www.ey.com/de_de/decarbonization/welchen-beitrag-der-maschinenbau-zur-dekarbonisierung-leisten-kann. Accessed 20 July 2023
8. Lorenz M, Lüers M, Ludwig M, Rees S, Rauen H, Zelinger M, Stiller R (2020) Grüne Technologien für grünes Geschäft. Boston Consulting Group (BCG)
9. VDMA e.V. Fachverband Nahrungsmittelmaschinen und Verpackungsmaschinen, McKinsey & Company (2022) Nachhaltigkeit—Chance für den Maschinen und Anlagenbau in Deutschland. VDMA e.V, McKinsey & Company
10. Hottenroth H, Joa B, Schmidt M (2013) Carbon Footprints für Produkte: Handbuch für die betriebliche Praxis kleiner und mittlerer Unternehmen. MV-Verlag
11. Sica D, Esposito B, Malandrino O, Supino S (2022) The role of digital technologies for the LCA empowerment towards circular economy goals: a scenario analysis for the agri-food system. Int J Life Cycle Assess 29:1486–1509
12. Greenhouse Gas Protocol, World Resources Institute (WRI), World Business Council for Sustainable Development (WBCSD) (2011) GHG Protocol Corporate Value Chain (Scope 3) and product life cycle standards. Greenhouse Gas Protocol
13. VDMA e.V. (2022) VDMA-Guideline "Berechnung des Product Carbon Foootprint im Maschinen- und Anlagenbau". VDMA e.V.
14. Together for Sustainability (2022) The product carbon footprint guideline for the chemical industry. Together for Sustainability
15. Hohlweck J (2023) SiGREEN making product carbon footprints actionable. Siemens, Munich
16. Catena-X (2022) CX-0029 product carbon footprint rulebook
17. Ulrich H (1984) Management. Bern & Stuttgart
18. Object Management Group (2012) Business Process Model and Notation (BPMN). Version 2.0, www.omg.org. Accessed 28 Nov 2012
19. Li QX, Chen Y-L (2009) Data flow diagram. Springer eBooks, Berlin, pp 85–97
20. Aktas AZ (1987) Structured analysis and design of information systems. Prentice Hall, Hoboken, NJ
21. Booch G, Rumbaugh J, Jacobson I (1999) Das UML-Benutzerhandbuch
22. Joppen R, Von Enzberg S, Kühn A, Dumitrescu R (2019) Data map—method for the specification of data flows within production. Proc CIRP 79:461–465
23. Fahrwinkel U (1995) Methode zur Modellierung und Analyse von Geschäftsprozessen zur Unterstützung des Business Process Reengineering. Heinz Nixdorf Institut, Paderborn
24. Brock J, Von Enzberg S, Kühn A, Dumitrescu R (2023) Process mining data canvas: a method to identify data and process knowledge for data collection and preparation in process mining projects. Proc CIRP 119:602–607

Life Cycle Assessment (LCA) of a Circular Saw Blade for Wood Processing

F. Schreiner[1]([✉]), M. Beller[1], J. Pohle[1,2], B. Thorenz[1], and F. Döpper[1,2]

[1] University of Bayreuth, Chair Manufacturing and Remanufacturing Technology, Universitätsstraße 30, 95447 Bayreuth, Germany
Florian.Schreiner@uni-bayreuth.de

[2] Fraunhofer IPA Project Group Process Innovation, Universitätsstraße 9, 95447 Bayreuth, Germany

Abstract. Climate change and its negative consequences are an inevitable topic that strongly affects all industrial sectors as well as private individuals. Therefore, it is necessary to assess and monitor the environmental impact of products, services, and processes. This paper presents a practical application of an LCA for a circular saw blade used in wood processing. By considering all stages of the circular saw blade's life cycle, including manufacturing, period of use, and end of life phase, a comprehensive assessment of its environmental impact is provided. The software SimaPro was utilized to conduct the LCA, which involved evaluating each individual element of the circular saw blade. As a result of this study, the environmental impact of the various components of a circular saw blade was quantified, enabling an assessment of its cumulative potential environmental impact. Consequently, recommendations for more sustainable manufacturing, use and disposal of circular saw blades in wood processing are derived from the findings, contributing to an overall improvement in their environmental impact.

Keywords: Life Cycle Assessment (LCA) · Climate change · SimaPro · Impact assessment · Circular saw blade

1 Introduction

Among the most pressing challenges we face today is the necessity to adjust our lifestyle to mitigate the impacts of climate change as effectively as possible. The shift towards sustainable materials, such as wood, presents a promising strategy to pursue this goal [1]. As such, the diversified use of wood in roles such as a raw material, a construction substance, and an energy source marks a significant stride towards sustainability [2]. This positive shift towards increased wood utilization causes a proportional rise in the demand for wood processing tools, thus amplifying their impact on the sustainability of the manufacturing process. In the realm of wood processing tools, the circular saw blade stands out due to its important role in modern woodworking. Since its inception, the circular saw blade has revolutionized woodworking, offering a blend of precision, efficiency, and versatility. This tool can be found in a variety of settings, from large-scale

H. Kohl et al. (Eds.): GCSM 2023, LNME, pp. 155–164, 2025.
https://doi.org/10.1007/978-3-031-77429-4_18

industrial operations to smaller domestic woodworking projects, making it one of the most widely used tools in the industry [3]. Given its prevalence and the multifaceted tasks it can perform, it becomes a focal point when discussing the environmental impact of woodworking tools.

However, while its benefits are numerous, the environmental implications throughout the life cycle of circular saw blades frequently go unnoticed. This oversight becomes problematic considering the volume of circular saw blades manufactured and discarded every year and the subsequent potential cumulative environmental impact.

This paper aims to provide a comprehensive LCA of a circular saw blade used in wood processing, delving into the environmental implications during its manufacturing, use, and end of life (EoL) phase. The goal is to guide stakeholders, from manufacturers to end-users, in making informed decisions about the production, waste management, and design of these blades. Ultimately, this study intends to steer the woodworking industry towards sustainability and align it with circular economy principles.

2   Goal and Scope

This paper examines the life cycle of a conventional circular saw blade used in sawmills. The cradle-to-grave approach is used, focusing on manufacturing, use, and EoL phase. The aim is to identify key areas for future improvements to enhance the product's sustainability [4]. Therefore, results of the LCA provide a foundation for comprehensive analysis, generating specific recommendations for the manufacturing process and circular saw blade design. A conventional circular saw blade, with detailed specifications of a diameter of 450 mm, a cutting width of 4.6 mm, a blade thickness of 3.2 mm and a bore diameter of 30 mm, serves as a reference. The conventional circular saw blade is made of 75Cr1 steel with 20 carbide teeth. The used data is reflecting current German science and technology practices. Primary data from industry-relevant companies supplements when secondary data is insufficient. The time frame of the study aligns with a circular saw blade's lifespan, conducted within a German facility using an 'average electricity mix Germany' dataset in the software SimaPro 9.5. Company-specific data was obtained from leading circular saw blade and corresponding machine manufacturers. The Ecoinvent database v.3 and the ReCiPe 2016 method are employed for analysis.

The system boundaries are summarized in Table 1. Letter-number combinations in the table represent life cycle modules as defined by DIN EN15804 and are grouped per life cycle stage. The production stage includes raw material supply, transport of materials to the manufacturer, and manufacturing. For this type of product, the construction stage (A4 and A5) can be translated to distribution of the circular saw blade to the customer and its installation, however this stage is not considered within system boundaries. For the use stage, the use of the circular saw blade, its maintenance (regrinding the saw teeth) and the associated energy use are included. The electric energy is needed to power the motor of the circular saw, converting electrical energy into mechanical energy to rotate the circular saw blade. The demolition, transport to waste, waste processing and final disposal are considered inside system boundaries. Benefits and loads beyond the system boundary (D) are excluded from the analysis.

Table 1. Life cycle modules according to DIN EN 15804

Production stage			Construction stage		Use stage							EoL stage				Optional information
Extraction and upstream production	Transport to factory	Manufacturing	Transport to site	Installation	Use	Maintenance	Repair	Replacement	Refurbishment	Operational energy use	Operational water use	De-construction demolition	Transport to waste process or disposal	Waste processing	Disposal	Potential net benefits from reuse, recycling and/or energy recovery beyond system boundary
A1	A2	A3	A4	A5	B1	B2	B3	B4	B5	B6	B7	C1	C2	C3	C4	D
X	X	X			X	X				X		X	X	X	X	

3   Life Cycle Inventory Analysis

The life cycle inventory (LCI) presents the basis for assessing the impact of a product's life cycle [5]. In the following, the life cycle of a conventional circular saw blade is visualized and explained step by step. The sequence of the individual processes is based on an analysis of the manufacturing steps of different manufacturers and information from literature [6–8]. The functional unit for this assessment is one meter of wood sawn by the circular saw blade. As this is the only way to ecologically evaluate and compare measures for extending the service life of circular saw blades in further research studies. For modelling, the cut-off system model was used.

3.1   Manufacturing Phase

The manufacturing of a circular saw blade's plate starts with CO_2 laser cutting which removes metal via vaporization [9]. Following laser cutting, the piece undergoes deburring via belt sandpapers, ensuring a smooth surface. Next is the tempering, a three-part process (heating, holding, and cooling at 450 °C). This step optimizes the blade's hardness and toughness [10]. The blade then undergoes milling and deburring called polishing at inner and outer rakers, preparing for carbide tip attachment.

As previous steps may induce deformations or stresses, tensioning and straightening using rolling circles is done, with two tension rings added to effectively distribute tension. Finally, manual straightening complements the clamping process. The sheet is clamped, checked for unevenness, and if necessary corrected using firm hammer blows [11]. Alongside machining the plate, the saw teeth are manufactured. Carbides are made using powder metallurgy. Materials are first mixed as powders, compacted using various presses to form a strong 'green compact', followed by sintering [10]. This heat treatment encourages atoms to form robust bonds, potentially reducing the volume by up to 50%. Sintering occurs in a blast furnace under a protective gas environment, at temperatures between 2/3 and 4/5 of the component's melting point. For tungsten carbide and cobalt, for example, this is roughly 1350–1500 °C [6]. Carbide is directly procured from manufacturers.

The blade and teeth are united via an automated induction brazing machine using Copper solder Cu49+. Balancing follows, with mass and weight imbalances rectified by flex-machining. Excess solder and flux residues are eliminated through sandblasting. To improve surface quality, the blade is polished using a corundum grinding wheel [10]. The saw teeth are ground with multi-axis CNC machines equipped with diamond grinding wheels and cooled with an emulsion or grinding oil during the process [7].

For circular saw blades used in sawmills, wipers are fitted with carbide tips, brazed at 800 °C. Post-brazing and flux residues are removed using a water-soda mixture. The blade undergoes quality and tension measurements via automated machines. Post-inspection, the finished product is labeled with a fiber laser, packaged, and shipped.

Table 2 depicts the individual processes used to model the manufacturing steps. The overview distinguishes between *Input to Technosphere* and *Output to Technosphere.*

Table 2. LCI for the manufacturing of a circular saw blade.

Part	Value	Unit	Ecoinvent processes used
Plate			
Input	5.43	kg	**Steel**, low-alloyed {GLO}	market for	Cut-off, U
	4.3085	kWh	**Electricity**, medium voltage {DE}	market for	Cut-off, U
	2.908	tkm	**Transport**, freight, lorry 16–32 metric ton, euro6 {RER}	market for transport, freight, lorry 16–32 metric ton, EURO6	Cut-off, U
Output	2.03	kg	**Scrap steel** {Europe without Switzerland}	market for	Cut-off, U
Saw teeth			
Input	0.0447	kg	**Tungsten carbide powder** {GLO}	market for	Cut-off, U
	0.0029	kg	**Cobalt** {GLO}	market for	Cut-off, U
	0.0028	kg	**Solvent**, organic {GLO}	market for solvent, organic	Cut-off, U
	0.0062	kg	**Nitrogen**, liquid {RER}	market for nitrogen, liquid	Cut-off, U
	0.0019	kg	**Paraffin** {RER}	market for paraffin	Cut-off, U
	0.5231	kWh	**Electricity**, medium voltage {DE}	market for	Cut-off, U
	0.0102	tkm	**Transport**, freight, lorry 16–32 metric ton, euro6 {RER}	market for transport, freight, lorry 16–32 metric ton, EURO6	Cut-off, U
Output	0.0062	kg	**Nitrogen**, atmospheric
	0.0019	kg	**Paraffins**
	0.0028	kg	**Spent solvent** mixture {Europe without Switzerland}	treatment of spent solvent mixture, hazardous waste incineration	Cut-off, U
Circular saw blade			
Input	0.0017	kg	**Brazing solder**, cadmium free {GLO}	market for	Cut-off, U
	0.0192	kg	**Flux**, for wave soldering {GLO}	market for	Cut-off, U
	0.282	kg	**Folding boxboard carton** {RER}	market for	Cut-off, U
	1.6172	kWh	**Electricity**, medium voltage {DE}	market for	Cut-off, U
	1.377	tkm	**Transport**, freight, lorry 16–32 metric ton, euro6 {RER}	market for transport, freight, lorry 16–32 metric ton, EURO6	Cut-off, U
Output	0.0047	kg	**Scrap steel** {Europe without Switzerland}	market for	Cut-off, U

3.2   Use Phase

The product's use phase includes sawing and regular regrinding the saw teeth. The circular saw blade's tool life determines tool change. It is the time from initial cut to wear limit. The lifespan is influenced by factors like wear, cutting speed, and material. Wear, mainly abrasive, is caused by hard mineral deposits in the wood [6]. The circular saw blade operates for eight hours per shift and can be used for up to four shifts, with active sawing taking up 68.2% and idle time 31.79% of the shift, there is no special cooling period as the machine cools between shifts. The average energy consumption during these shifts is 173.06 kWh, accounting for both active and idle phases of the blade.

Regrinding, a crucial step, sharpens the cutting edge of the saw teeth and reduces edge rounding, enhancing blade efficiency and longevity. Regrinding is carried out on the same machine as grinding, however, it is a quicker fine-tuning process targeting only the chest and back surfaces of the saw teeth.

The use phase is approximated by electricity and waste wood processes, cf. Table 3.

Table 3. LCI for the use phase of a circular saw blade.

Part	Value	Unit	Ecoinvent processes used
Sawing			
Input	173.06	kWh	**Electricity**, medium voltage {DE}I market for I Cut-off, U
Output	3174.0	kg	**Waste wood**, untreated {DE} I market for I Cut-off, U
Regrinding			
Input	3.8769	kWh	**Electricity**, medium voltage {DE}I market for I Cut-off, U

3.3 End of Life Phase

Sawmills typically do not disassemble circular saw blades. They are given to recycling companies as mixed scrap. The respective processes for dismantling and disposal are displayed in Table 4.

Table 4. LCI for the disposal of a circular saw blade.

Part	Value	Unit	Ecoinvent processes used
Dismantling			
Input	0.0333	kWh	**Electricity**, medium voltage {DE}I market for I Cut-off, U
Output	3.4429	kg	**Scrap steel** {Europe without Switzerland}I market for I Cut-off, U
Disposal			
Input	0.3448	tkm	**Transport**, freight, lorry 16–32 metric ton, euro6 {RER}I market for transport, freight, lorry 16–32 metric ton, EURO6 I Cut-off, U

4 Life Cycle Assessment and Results

To identify and evaluate the potential environmental impacts arising from the LCI, the inputs and outputs are assigned to impact categories. The assessment was carried out according to the ReCiPe 2016 method, which provides characterization factors that are representative for the global scale [12]. The endpoint impact assessment simplifies the interpretation of results and is useful for a first glimpse, cf. Figure 1. To allow a comparison of the impact categories, the diagram shows the impact in the unit [μPt]. μPt is the unit expressing the total environmental impact as a single score in which characterization, normalization, and weighting are combined and considered. The impact in the category human health is by far the largest, and 31% of its overall environmental impact originate from global warming potential. The impacts in the designated units for the individual categories are: Human health 2.40E−08 [DALY], Ecosystems 4.35E−11 [species.yr] and Resources 4.88E−04 [USD2013].

Impact in μPt

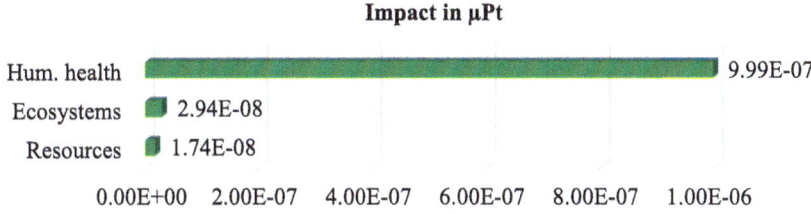

Fig. 1. Endpoint impact assessment of the lifecycle of a circular saw blade.

The midpoint impact assessment allows for a more detailed analysis of the aforementioned endpoint categories. In Sect. 4.4, the midpoint impact category *global warming* is further analyzed with the goal to identify the main drivers of climate change.

4.1 Manufacturing Phase

The global warming potential (GWP) of the manufacturing phase of one circular saw blade amounts to 7.39E−04 kg CO_2 eq per meter of sawn wood. 80% of the total volume is accounted for by the plate production, 11% by the saw teeth production and 9% by the final assembly of the circular saw blade. The evaluation of the manufacturing phase reveals that the steel used for the plate (64.55%) and the electricity consumed during its production (12.15%) are primary contributors to global warming. Moreover, when considering their weight, the impact of manufacturing the saw teeth is also notably significant. The saw teeth make up 3.6% of the total weight of a circular saw blade, while causing 11% of the GWP during the manufacturing phase. Besides the energy consumption during saw teeth production (1.47%), the highest contributors are the precious metals tungsten carbide powder (7.72%) and cobalt (1.40%). The main driver for the manufacturing of the circular saw blade is the consumed electricity (49.11%).

4.2 Use Phase

The GWP attributed to the use phase is equally shared between the processes for electricity and waste wood, each accounting for 3.67E−03 kg CO_2 eq per meter of sawn wood. Waste wood is generated solely during the active sawing phase and 98% of the GWP derived from energy consumption during the use phase is caused by the sawing process. 99% of the GWP during the use phase are generated from the sawing process whereas regrinding accounts for 1% of the GWP.

4.3 End of Life Phase

During EoL phase $4.99E{-}06$ kg CO_2 eq is generated. The transportation (disposal process) causes most of the GWP, amounting to $2.83E{-}06$ kg CO_2 per meter of sawn wood which is 56.71%. Another 29.26% can be accounted to the scrap steel and 13.87% to the energy consumption occurring during the dismantling process. It should be noted at this point that other system models, e.g., Allocation at the point of substitution, could lead to significantly different results, since the burdens for waste are shared between producers and subsequent users.

4.4 Overall

Figure 2 gives an overview of the GWP over the manufacturing, use and EoL phase. Therefore, the results of the individual life cycle stages, which were discussed in the previous sections, can be compared more easily. It can be noted that most of the GWP occurs during the use phase (89.81% due to sawing and 1% due to regrinding). 9.13% of the GWP originate from the manufacturing. Within this phase, the plate production is the biggest driver and contributes 7.31% to the total GWP. Lastly, during the EoL phase only 0.06% of the total GWP are produced.

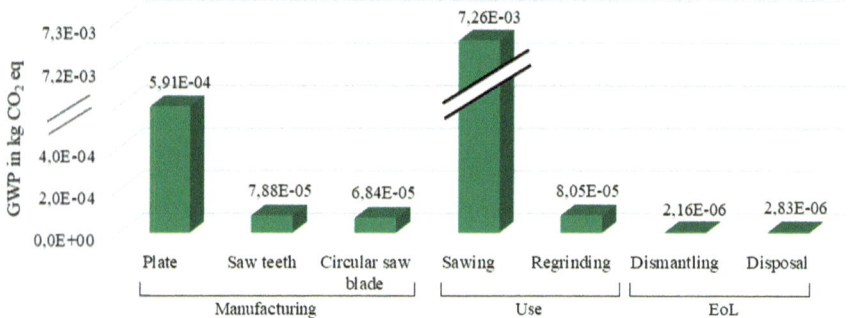

Fig. 2. GWP of the lifecycle of a circular saw blade per meter of sawn wood.

Based on the conducted hotspot analysis, several approaches can be derived which may prove useful for reducing the GWP of circular saw blades. One approach is to manufacture thinner circular saw blades, therefore conserving a substantial amount of woodware by reducing the width of the cutting kerf and moreover lowering energy consumption when sawing as less power is needed. Another approach is incorporating new saw tooth geometries and for example therefore enhancing the energy efficiency of the circular saw blade's use phase. These new geometries should aim to minimize cutting force without compromising the service life of the blade.

5 Conclusion and Outlook

The LCA of a circular saw blade presented in this paper provides valuable insights into the environmental impact, especially GWP, of this vital tool, encompassing all stages of its lifecycle, from manufacturing to use and EoL. The LCA, conducted with the software SimaPro, allowed an evaluation of the GWP of individual components of the circular saw blade. The greatest GWP was identified in the use phase, primarily involving electricity consumption and waste generation. In the use phase 90.81% of the overall GWP are produced, while during manufacturing 9.13% and EoL only 0.06%. Furthermore, this paper revealed the predominant influence of steel (64.61% of GWP in the manufacturing phase) and the total energy consumption during manufacturing (12.15% of GWP in manufacturing phase), therefore both being driving factors of the saw blade's environmental footprint.

The results of this hotspot analysis show the main drivers for the GWP of circular saw blades. Thus, indicating areas of improvement and providing a roadmap for sustainable practices in manufacturing, using and disposal of circular saw blades, which can significantly reduce their environmental footprint. This research contributes to reducing the woodworking industry's environmental impact and encourages further investigations into sustainable solutions.

References

1. Ramage MH et al (2017) The wood from the trees: The use of timber in construction. Renew Sustain Energy Rev 68:333–359
2. Sathre R, O'Connor J (2010) Meta-analysis of greenhouse gas displacement factors of wood product substitution. Environ Sci Policy 13(2):104–114
3. Nasir V, Cool J (2020) A review on wood machining: characterization, optimization, and monitoring of the sawing process. Wood Mat Sci Eng 15(1):1–16
4. Baumann H, Tillman AM (204) The hitch hiker's guide to LCA. Lund
5. Ciroth A, Arvidsson R (2021) Life cycle inventory analysis. In: Life cycle assessment: theory and practice. Springer, Switzerland, pp 73−97. https://doi.org/10.1007/978-3-030-62270-1_4
6. Ettelt B, Gittel H-J (2004) Sägen, Fräsen, Hobeln, Bohren: Die Spanung von Holz und ihre Werkzeuge, 3rd edn. DRW-Verlag, Leinfelden-Echterdingen
7. Gottlöber C (2014) Zerspanung von Holz und Holzwerkstoffen : Grundlagen—Systematik—Modellierung—Prozessgestaltung; mit 13 Tabellen. Fachbuchverlag Leipzig/Hanser, Leipzig
8. SWEDEX: Production saw blades. http://www.swedex.com/about-us/production/. Accessed 17 June 2022
9. Rüffler C, Gürs K (1971) Schneiden und Schweißen mit dem CO2-Laser. Materialwissenschaft Werkst 2(7):361–366
10. Schlegel J (2021) Die Welt des Stahls. Springer Fachmedien Wiesbaden, Wiesbaden
11. Fronius K (1989) Spaner, Kreissägen, Bandsägen. DRW-Verlag, Stuttgart. Arbeiten und Anlagen im Sägewerk/Karl Fronius, Band 2. ISBN 3-87181-332-X
12. Huijbregts MA, Steinmann ZJ, Elshout PM, Stam G, Verones F, Vieira M et al (2017) ReCiPe2016: a harmonised life cycle impact assessment method at midpoint and endpoint level. Int J Life Cycle Assess 22:138–147

A Systems-Based Framework for Product Circularity Assessment

Gisele Bortolaz Guedes[1(✉)], Junwon Ko[1], Fazleena Badurdeen[1], I. S. Jawahir[1], K. C. Morris[2], and Vincenzo Ferrero[2]

[1] Department of Mechanical and Aerospace Engineering, Institute for Sustainable Manufacturing (ISM), University of Kentucky, Lexington, KY, USA
gisele.guedes@uky.edu
[2] National Institute of Standards and Technology, Gaithersburg, MD, USA

Abstract. One of the key aspects of Circular Economy (CE), particularly when focusing on the product level, lies in its emphasis on designing products to facilitate the circulation of resources and maximize value throughout their entire lifecycle. To effectively develop such products an understanding of the characteristics of circular products (CP) is needed. Adopting a systems perspective that considers the total lifecycle and stakeholders involved and recognizing interdependencies among them is essential for identifying factors that characterize CPs. Existing methods for assessing product circularity do not identify all factors that characterize CPs and have limitations in the measurement methods proposed, leading to poor industry adoption. This research aims to develop a systems-based framework with clearly defined attributes, indicators, and metrics for product circularity assessment (PCA) to facilitate more effective design practices for CPs. A two-pronged approach, with industry engagement, is followed to address this gap: first, a systematic analysis of literature is conducted to identify key attributes and establish a clear foundation of what constitutes a CP; secondly, a comprehensive examination of indicators and metrics to evaluate the attributes is undertaken. This paper presents the preliminary systems-based framework for PCA with example attributes, indicators, and metrics, specifically for consumer electronic product circularity evaluation and directions for further research.

Keywords: Circular economy · Circular products · Product circularity

1 Introduction

The excessive and irresponsible use of resources for manufacturing products, and existing practices of their disposal at end-of-use, have resulted in significant adverse effects on the environment and society [1, 2]. To sustainably meet the needs of a growing global population striving for higher living standards, a shift from the linear economy's 'take-make-use-dispose' practices to a more Circular Economy (CE) that focuses on regenerative and restorative approaches through closed-loop material flow is necessary. For this transition to occur, CE implementation must be grounded in and driven by its application at the product level [3].

H. Kohl et al. (Eds.): GCSM 2023, LNME, pp. 165–173, 2025.
https://doi.org/10.1007/978-3-031-77429-4_19

CE principles must be incorporated into the design, manufacturing, and lifecycle management of products. Without a dedicated emphasis on integrating the CE principles during product design, i.e., to develop circular products (CPs), the benefits of circularity cannot be effectively derived across meso (e.g., industrial systems) or macro (e.g., cities, regions, or countries) levels. Product designers/engineers assume the role of change agents within CE [4] and need to be equipped with the necessary tools to make effective decisions to design CPs during product development process (PDP).

Sustainable manufacturing involves the application of 6Rs (Reduce, Reuse, Recycle, Recover, Redesign, and Remanufacture) [5] and total lifecycle consideration to ensure triple bottom line (TBL) benefits are optimized. By collectively implementing sustainable manufacturing practices at the product, process, and system levels additional benefits can be generated for all stakeholders, leading to more sustainable value creation [6]. CPs play a pivotal role in paving the path towards sustainability through increasing economic prosperity, closed-loop material flow, and reduced environmental impacts. As shown in Fig. 1, CPs can serve as one of the catalysts for the development of more sustainable products, ultimately leading to sustainable value creation for all stakeholders [6, 7].

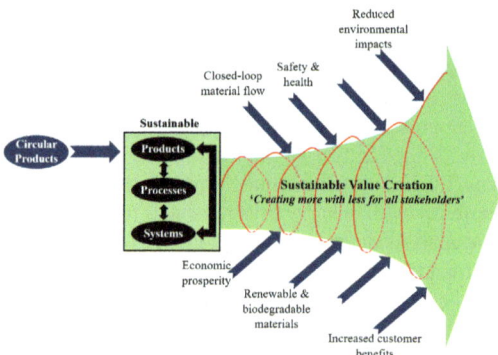

Fig. 1. Circular products as a path to sustainable value creation.

To effectively design such products, it is vital to have a concrete understanding of what constitutes a CP, and how to measure and assess product circularity. Numerous studies have proposed frameworks and metrics to assess product circularity. Despite extensive efforts, limitations persist [8]. Prior studies focusing on CPs and circular product design (CPD) have identified numerous and varied characteristics to describe CPs, but the consistency among them is lacking [9]. Limited awareness of benefits, challenges to operationalizing through CPD, misaligned regulations, and conflicting priorities may hinder widespread adoption of CPs. A clear understanding of key characteristics of CPs and a well-defined method for their assessment are still needed to operationalize the design of CPs. This paper presents a systems-based framework with clearly defined attributes, indicators, and metrics for product circularity assessment (PCA) to address this gap.

2 Overview of Methodology

A two-step approach is followed in this study to develop the PCA method (see Fig. 2). The first step focused on identifying attributes that characterize circular products. An attribute is defined as a constitutive characteristic that describes the desired features/properties of CPs [10]. The second step focused on a comprehensive examination of indicators and metrics to evaluate each of the selected attributes. Indicators provide valuable information about performance or describe the state of a specific attribute [11], while metrics offer more detailed information regarding the means of measuring these indicators [12]. The indicators and metrics relevant for evaluating circularity of a given product can highly vary from one industry to another. Therefore, to ensure industry relevance as well as for ease of verification and validation, consumer electronics products (CEP) are chosen as the use case and experts from the consumer electronics industry were engaged during both the steps.

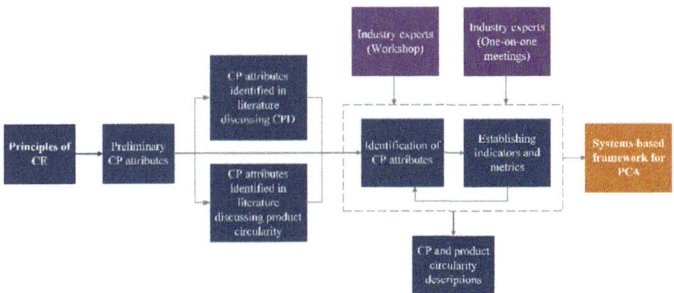

Fig. 2. Methodological process to design the systems-based framework for PCA.

Identification of attributes: Numerous definitions of the CE have been presented in the literature, often leading to conflicting viewpoints [3, 13, 14]. To develop a systems-based framework, efforts were initially dedicated to establishing a comprehensive understanding of the core tenets of the CE concept, delineating the scope to be considered for CPs. Then, a rigorous analysis of literature focused on CPs, CPD, and methods for evaluating product circularity. This process aimed to determine a list of CP attributes (as depicted in Fig. 2).

Establishing indicators and metrics: Upon determining the attributes, a comprehensive examination of indicators and metrics was undertaken. This process was also informed by reviews of scientific literature and industry reports. The metrics compiled are intended to serve as examples as they are likely to vary based on the industry and product being evaluated.

Industry stakeholders from the consumer electronics sector were actively engaged in the process first through workshops and then through one-on-one meetings to provide feedback (see Fig. 2). The entire process involved iterative refinement and continuous improvement of the descriptions for attributes, indicators, and metrics to ensure clarity and accuracy of the information conveyed for each of them.

3   Review of Circular Product Attributes

In authors' previous work [9], a comprehensive review of CP attributes identified in various literature sources was presented. Studies have pointed out the importance of designing products for multiple lifecycles, with opportunities for servitization and life-cycle extension and incorporating various value recovery strategies (e.g., recover, reuse, remanufacture, recycle), etc., to promote increased circularity. The significance of ensuring environmental and economic benefits through CPs [15], assessment over all lifecycle stages [16], incorporating strategies to enable material recirculation, utilization, and endurance in products and services [17], and many others have also been identified as relevant attributes. Space limitations do not permit a comprehensive discussion here; more details can be found in [9].

A comprehensive list of CP attributes, including those described above, were identified through a detailed literature review to develop the PCA framework. To address the variability and inconsistencies in the use of terms, the attributes were grouped based on their similarities in meaning. The results were then evaluated using a Pareto analysis, to prioritize and identify 20% of the attributes that can be considered as representative of describing the majority, or 80%, of characteristics of CPs. This analysis, shown in Fig. 3, highlights the attributes most frequently referenced in the literature as characteristic of CPs.

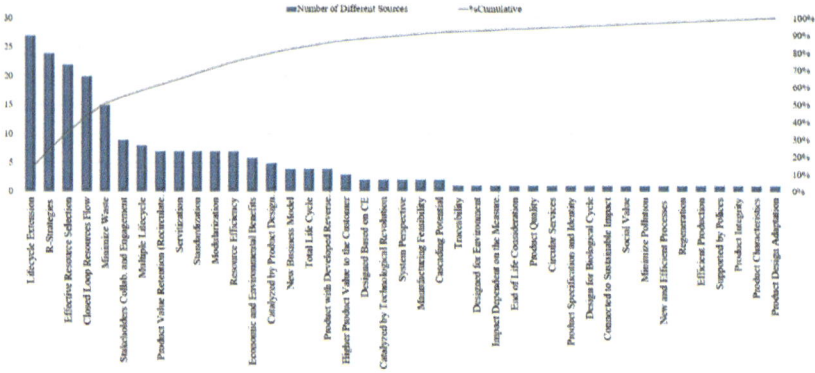

Fig. 3. Pareto analysis of CPs attributes.

4   Systems-Based Framework for PCA

The analysis presented in the previous section led to the definition of thirteen different attributes most frequently referenced in published literature as constituent characteristics of CPs. Following a closer examination of the criteria pertaining to each, these attributes can be classified into three distinct groups: (i) Drivers/Enablers, (ii) Outcomes, and (iii) Benefits and Implications. The drivers and enablers encapsulate the requirements to design, develop and deploy CPs. In other words, the drivers and enablers are capabilities necessary to offer CPs. The presence of such capabilities, however, does not automatically lead to CPs. The actual design and deployment of CPs, using forward and reverse supply chains and participation of all stakeholders, is what leads to the circular flow of resources necessary for transition to a CE.

Thus, the attributes belonging to the second group—Outcomes—denote the actual characteristics that are exhibited by the products designed, manufactured, used, and managed at end-of-life (EoL). When the drivers and enablers are utilized to deploy CPs, it can bring about numerous economic, environmental, and societal benefits. Designing and launching CPs is meaningless unless such benefits can be derived. However, inevitably trade-offs exist among them as well. These trade-offs will need to be considered and incorporated when different metrics are aggregated to convey the overall performance of a CP. The third category of attributes—Benefits and Implications—represent the potential consequences that arise from operationalizing CPs. In summary, the 'Drivers and Enablers' facilitate operationalizing the requirements to achieve the 'Outcome' of CPs which can bring about various 'Benefits and Implications'. From an industry perspective, the goal of developing CPs is to maximize the benefits and minimize any negative implications. The attributes that are drivers and enablers (D&E) will then become 'levers' that must be adjusted, to manage the benefits and implications (B&I), as desired.

As shown in Fig. 4, these attributes and the interdependent relationships among them offer a systemic framework to develop an approach for product circularity assessment (PCA) and for its industry adoption by identifying D&E to achieve desired B&Is.

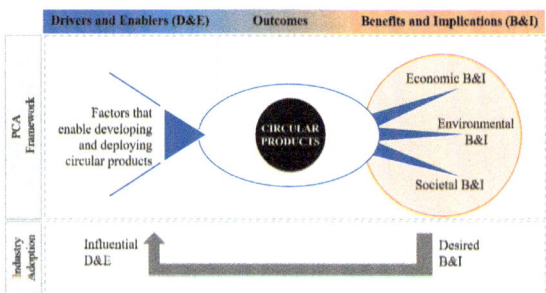

Fig. 4. Systems-based framework for PCA.

5   Attributes, Indicators and Metrics for PCA

For each key attribute identified from the Pareto analysis, different sources of literature were reviewed to determine the most suitable terminology that clearly conveys the constitutive characteristic, eliminating ambiguity. The identified attributes, along with a subset of literature sources (due to space limitations) recognizing their relevance, are organized into the three categories (Fig. 4 and presented in Table 1).

Three main attributes are considered as drivers and enablers: design/redesign to facilitate circular characteristics (e.g., modularity, easy maintenance, and repair), effective resource selection, and stakeholder consideration and engagement. In the outcomes category, CPs are defined by five main attributes: incorporate value recovery, extended product use-life, incorporate closed-loop resources flow, facilitate multiple lifecycles, and minimize waste. Five attributes are identified for the benefits and implications category: optimized resource efficiency, minimized harmful impacts, maximized economic value creation, and maximized economic value retention. Though implications to society were not a commonly identified attribute in literature, the economic and environmental benefits brought about by CPs can lead to advantages for society. Therefore, societal benefits and implications are included as an attribute in this PCA framework. These attributes were shared with consumer electronics industry stakeholders through a workshop for verification and validation. All the attributes were confirmed as important to evaluating product circularity.

Next, a comprehensive examination of scientific literature and industry reports was carried out to identify indicators for each attribute and define metrics to assess each indicator. Because indicators as well as metrics can be very industry- and even product-specific, the definition of indicators and metrics was conducted with a focus on CEPs. Often organizational size (e.g., small vs. large companies) as well as resources can limit data availability; thus, the work focused on defining example metrics for each indicator as a guide for industry stakeholders to subsequently adapt them depending on the specific use case and data availability.

Table 1. Key attributes of circular products*.

Drivers and enablers	Outcomes: circular products	Benefits and implications
Effective resources selection [18]	Incorporate value recovery strategies [20]	Optimized resources efficiency [23]
(Re) Design to facilitate circular characteristics [4]	Extended product use-life [21]	Minimized harmful environmental impacts [14]
Stakeholder consideration and engagement [19]	Incorporate closed-loop resources flow [22]	Maximized economic value creation [19]
	Facilitate multiple lifecycles [16]	Maximized economic value retention [17]
	Minimize waste [21]	Increased societal benefits

* Based on a comprehensive review supported by several other sources which are not shown due to lack of space

In total 29 indicators and 79 example metrics were developed. Figure 5 exemplifies one attribute with indicators and example metrics. The attribute 'effective resources selection' represents the emphasis on minimizing or eliminating virgin resources during CPD. Thus, alternative resources, such as recycled, upcycled, biodegradable, durable and safe (non-hazardous and non-toxic) materials, as well as recyclable and renewable resources, should be prioritized [18, 23]. Indicators such as material utilization, recovered material usage, and energy usage can be used to convey the extent to which resources are being selected effectively. They can be quantitatively measured using metrics such as the % of virgin material used, % of recycled materials, or the rate of non-renewable energy utilization.

The same approach was followed for the other attributes to identify indicators and metrics. The indicators are further verified through one-on-one meetings with experts from various consumer electronics companies to verify their relevance to evaluating the different attributes. The attributes are also being further verified during these meetings.

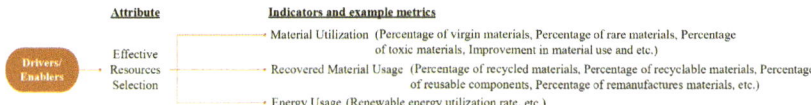

Fig. 5. Example of indicators and metrics for a circular product attribute.

6   Conclusions and Future Work

The successful operationalization of CE requires a comprehensive focus across the lifecycle—on product design, manufacturing, use, and EoL management—to establish seamless circular flow of resources. The cornerstone of this endeavor lies in the development of CPs which will be a catalyst for creating sustainable products, eventually enabling sustainable value creation for all stakeholders. One of the pre-requisites for designing CP is the ability to comprehensively evaluate products to assess the extent to which they satisfy circularity requirements. This paper presents a comprehensive approach founded upon a systems-based framework that integrates key attributes, indicators, and metrics for PCA. Industry experts from the consumer electronics sector have been continuously engaged in developing the method to ensure the usefulness and industry-relevance of the work. Future work will focus on implementing the systems-based framework for PCA to validate its effectiveness. Case studies will be conducted to assess the circularity of different products, starting with the consumer electronics sector. Further improvements will be made to facilitate a more effective design of CPs. Examining effective implementation by industry experts and striving for a practical PCA despite data limitations are imperative for tangible progress towards a CE and a sustainable future.

Acknowledgements. The work presented here is supported by industry partners from Amazon, Inc., Ryan Bradley and Ardeshir Raihanian Mashhadi, and a grant (No. 70NANB22H104) from the National Institute of Standards and Technology.

References

1. Duflou JR et al (2012) Towards energy and resource efficient manufacturing: a processes and systems approach. CIRP Ann 61(2):587–609
2. Haapala KR et al (2013) A review of engineering research in sustainable manufacturing. J Manuf Sci Eng 135(4):041013

3. Kirchherr J, Reike D, Hekkert M (2017) Conceptualizing the circular economy: an analysis of 114 definitions. Resour Conserv Recycl 127:221–232

4. Dokter G, Thuvander L, Rahe U (2021) How circular is current design practice? Investigating perspectives across industrial design and architecture in the transition towards a circular economy. Sustain Prod Consump 26:692–708

5. Jawahir IS, Dillon OW, Rouch KE, Joshi KJ, Venkatachalam A, Jaafar IH (2006) Total life-cycle considerations in product design for sustainability: a framework for comprehensive evaluation. In: Proceedings of the 10th international research/expert conference, Barcelona, Spain, vol 1, no 10

6. Badurdeen F, Jawahir IS (2017) Strategies for value creation through sustainable manufacturing. Proc Manuf 8:20–27

7. Bilge P, Badurdeen F, Seliger G, Jawahir IS (2014) Model-based approach for assessing value creation to enhance sustainability in manufacturing. Proc CIRP 17:106–111

8. Jerome A, Helander H, Ljunggren M, Janssen M (2022) Mapping and testing circular economy product-level indicators: a critical review. Resour Conserv Recycl 178:106080. https://doi.org/10.1016/j.resconrec.2021.106080

9. Ko J, Guedes GB, Badurdeen F, Jawahir IS (2024) TBD... A critical review of circular product attributes and their coverage in product circularity assessment tools. In: 2023 International Conference on Resource Sustainability (icRS 2023)

10. Oxford University Press (2023) The definitive record of the English language

11. Feng SC, Joung CB, Li G (2010) Development overview of sustainable manufacturing metrics. In: Proceedings of the 17th CIRP international conference on life cycle engineering, vol 6, p 12. PRC Hefei

12. Shuaib M, Seevers D, Zhang X, Badurdeen F, Rouch KE, Jawahir IS (2014) Product sustainability index (ProdSI) a metrics-based framework to evaluate the total life cycle sustainability of manufactured products. J Indus Ecol 18(4):491–507

13. Moraga G et al (2019) Circular economy indicators: what do they measure? Resour Conserv Recycl 146:452–461

14. Harris S, Martin M, Diener D (2021) Circularity for circularity's sake? Scoping review of assessment methods for environmental performance in the circular economy. Sustain Prod Consump 26:172–186

15. Saidani M, Kim H (2021) Design for product circularity: circular economy indicators with tools mapped along the engineering design process. In: International Design Engineering Technical Conferences and Computers and Information in Engineering Conference

16. Vimal KEK, Kandasamy J, Gite V (2021) A framework to assess circularity across product-life cycle stages—a case study. Proc CIRP 98:442–447

17. Boyer RH et al (2021) Three-dimensional product circularity. J Ind Ecol 25(4):824–833

18. Hildenbrand J, Lindahl E, Shahbazi S, Kurdve M (2021) Applying tools for end of use outlook in design for recirculation. Proc CIRP 100: 85–90. Based on: Ulrich KT, Eppinger SD (2012) Product design and development. McGraw-Hill/Irwin, NY

19. Hansen EG, Revellio F (2020) Circular value creation architectures: make, ally, buy, or Laissezfaire. J Ind Ecol 24(6):1250–1273

20. Bracquene E, Dewulf W, Duflou JR (2020) Measuring the performance of more circular complex product supply chains. Resour Conserv Recycl 154:104608

21. de Hollander MC, Bakker CA, Hultink EJ (2017) Product design in a circular economy: development of a typology of key concepts and terms. J Indus Ecol 21(3):517–525

22. Hapuwatte BM, Jawahir IS (2021) Closed-loop sustainable product design for circular economy. J Ind Ecol 25(6):1430–1446

23. Mestre A, Cooper T (2017) Circular product design. A multiple loops life cycle design approach for the circular economy. Des J 20(1):S1620–S1635

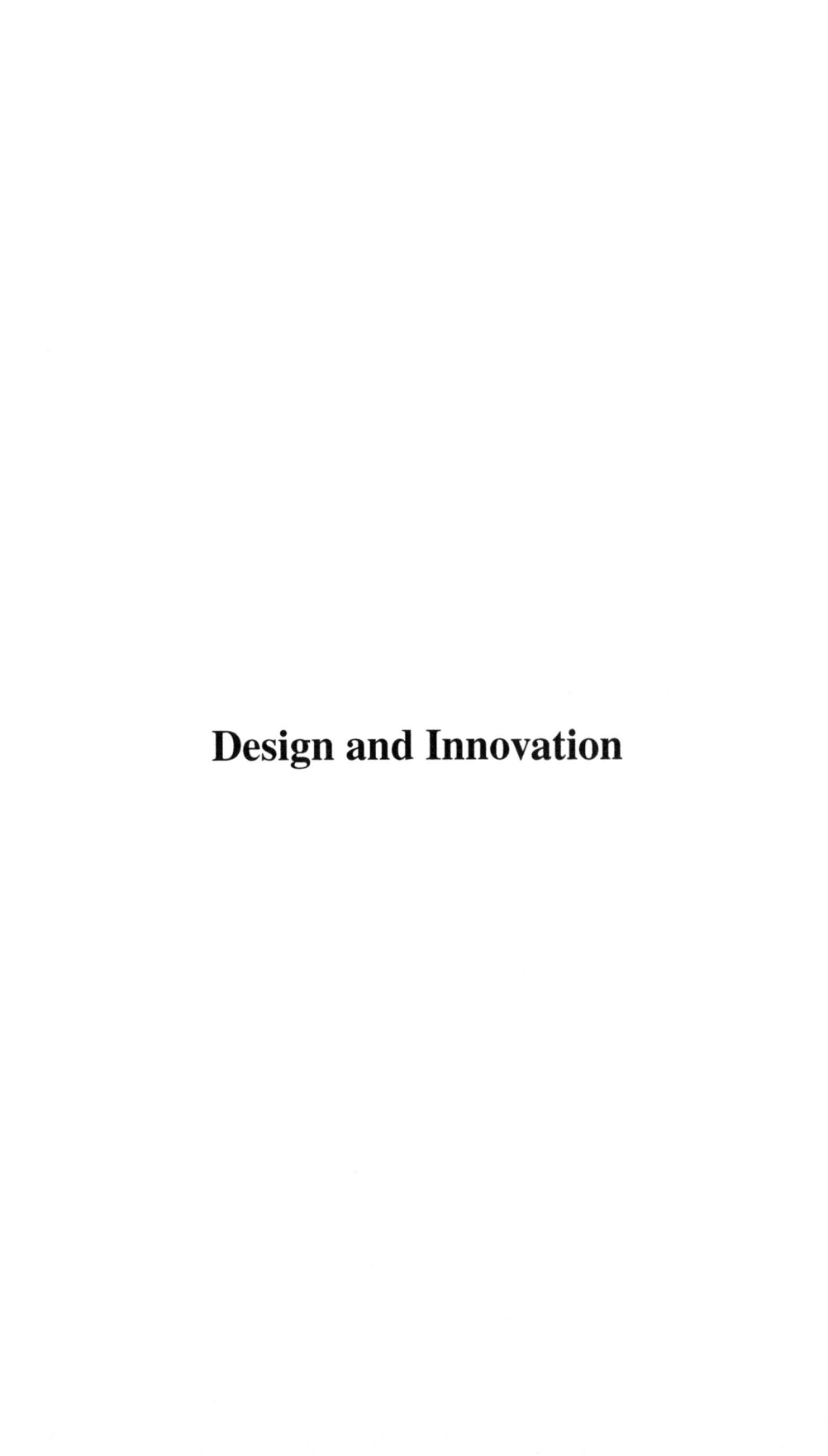

Design and Innovation

Technology Radar for a Sustainable Biological Transformation in the Manufacturing Industry

Janine Mügge[1]([⊠]), Magda Berkhahn[2], Tobias Knauf[1], Rainer Stark[2],
Lynn Faßbender[1], Annika Lange[1], Theresa Riedelsheimer[1], and Kai Lindow[1]

[1] Fraunhofer Institute for Production Systems and Design Technology IPK, Pascalstraße 8-9,
10587 Berlin, Germany
janine.muegge@ipk.fraunhofer.de
[2] Technische Universität Berlin, Pascalstraße 8-9, 10587 Berlin, Germany

Abstract. Bio-inspired, bio-integrated, bio-intelligent product and production systems are new research approaches aimed at achieving sustainability in manufacturing. For the industrial application of new technological solutions, it is crucial to convey their added value from not only an economic but also an environmental perspective. The evaluation of their technological readiness plays an important role in implementation. This paper presents a technology radar developed and evaluated within German research project BioFusion 4.0 involving 13 industry and research partners. The paper outlines the development approach and its application by means of three technological examples. The technologies are categorized and assessed using four significant attributes of an effective biological transformation: (1) the mode of action (bio-inspiration, bio-integration, and bio-interaction); (2) the transformative character and (3), the Technology—and (4) Sustainability Readiness Level. The technology radar comprises a step-by-step procedure to assess a technology by each criterion to enable different stakeholders, such as production planners or product developers to evaluate and identify suitable technologies according to their needs and business strategy. The application of the technology radar is demonstrated on the exemplary technologies of a digital twin with integrated life-cycle assessment, situation awareness monitor for networked production systems and bio-based 3D printing.

Keywords: Technology assessment · Biological transformation · Sustainability readiness level · Digital twin · 3-D printing · Holistic production systems

1 Introduction

Manufacturing companies are constantly changing to address global challenges such as climate change, increasing scarcity of natural resources, bottlenecks in supply chains and the resulting growing interest in sustainability in society [1]. To address these challenges from a manufacturing perspective, companies are exploring the Industry 4.0, Digital Transformation and Biological Transformation (BT) [2]. BT is defined as "a holistic approach to change industrial value creation towards sustainable optimized production systems, by an accelerating convergence of technical, digital, and biological systems

© The Author(s) 2025
H. Kohl et al. (Eds.): GCSM 2023, LNME, pp. 177–184, 2025.
https://doi.org/10.1007/978-3-031-77429-4_20

in the manufacturing environment" [3]. It is assumed that an increasing number of bio-inspired, bio-integrated, and bio-intelligent products and production systems will be developed in the future [3]. However, there is a lack of methodologies for assessing these technologies' sustainability in the context of sustainability. Therefore, the following research questions (RQ) are addressed: "How can bio-inspired, bio-integrated and bio-intelligent product and production systems be evaluated in the context of technology assessment? (RQ I)" and "How can their sustainability be evaluated?" (RQ II)".

As a main result, this paper presents a methodological framework to assess technologies regarding their Technology Readiness Level (TRL) and Sustainability Readiness Level (SRL) as well as its application with exemplary technologies.

2   State of the Art

In this section, the state of the art in BT manufacturing, the term technology as well as technology and sustainability assessment are examined.

2.1   Biological Transformation in Manufacturing

The convergence of bioinspired manufacturing and Industry 4.0 offers great opportunities for innovation in the manufacturing industry. Initially bionics involved transferring biological forms and functions to technical applications [4]. In the realm of industrial manufacturing the integration of production, information and communication technologies aims to create flexible, resource efficient and urban-compatible production systems [5]. BT in manufacturing is the process of optimizing industrial value creation towards sustainability with concepts and technologies inspired by nature, integrating biological aspects and operating in interaction between digital, and biological systems [3, 6, 7]. BT strives for a holistic change in society through two underlying currents of bionics and digitalization.

To provide a clear overview of effective principles that aid to identify and apply patterns in nature-based production implementation, the taxonomy of BT in manufacturing organizes principles hierarchically into three core modes[3]: *Bio Inspiration*, *Bio Integration* and *Bio Interaction*. Within these a total of 55 principles capture aspects such as self-awareness, closed loop recycling or 4D bioprinting [3].

2.2   Technology and Their Sustainability Assessment

Technology encompasses a wide range of artificial objects, human actions, and institutions involved in their creation and utilization [8]. Technologies can be classified into four categories based on their readiness level: embryonic technology, pacemaker technology, key technology or enabling technology [9]. Embryonic technologies are not yet relevant to industry, while pacemaker technologies have found their first relevant industrial applications. Key technologies are established in the market and enabling technologies are considered obsolete and possibly replaced soon [9]. To create a transparent information base about technologies and to identify opportunities and risks emerging in their development, early technology assessment is needed. Within the framework of early technology detection, there are three procedures: technology scanning, technology monitoring and technology scouting [9]. A widely used method for technology monitoring is the technology radar, which quantifies a technology's development status and perceived technical readiness [9, 10]. It involves compiling, analyzing, and clustering technologies and trends related to a particular subject [11].

Sustainability assessment (SA) ensures to align technologies with sustainability goals, evaluating environmental, social, and economic impacts throughout their life cycle [11]. Established frameworks, standards and specific sector-based guidelines exist to promote strategic planning, decision-making, and technology comparisons with SA results [12–15]. The Sustainability Readiness Level (SRL) concept simplifies technology's SA, guiding development, deployment, and improvement decisions. The SRL levels progress as follows [16–18]:

- Problem and requirement: Formulating sustainability targets and identifying potential solutions.
- Conceptual development: The technology's potential impacts are identified.
- Technology design: The technology undergoes iterative design to minimize negative impacts and maximize positive contributions to sustainability.
- Testing and validation: The technology's performance is tested, where sustainability claims are validated in experiments, simulations, pilot studies.
- Scale-up and deployment: The technology is implemented at larger scales, and its sustainability is monitored and assessed in real-world conditions.
- Continuous improvement: Feedback mechanisms to incorporate lessons learned and improve the technology's sustainability performance.

3   The Technology Radar for Biological Transformation

3.1   Methodological Approach

The methodical development of the technology radar is based on the method of early technology detection and monitoring. It can be divided into the four steps of identification, selection, assessment and dissemination [9, 19, 20].

Identification in the context of BT. In the first step, the search field is defined, which represents different trend clusters [19]. For a technology to serve the BT of the manufacturing industry it must function according to the principles of BT, evoke such a principle, or be able to implement it. Hence, the technology radar's search fields are based on the core and operational principles of the BT Taxonomy according to Berkhahn et al. [3]: *Bio Inspiration* (resilience, principles of circularity, self-x, functional morphology, biomimetic information modeling and processing), *Bio integration* (biosynthesis, biosubstitution, biodegradation and decomposition, bioenergetics, biotherapeutics, biomodification) and *Bio interaction* (biosensors, biological representation, biointelligent information processing, biohybrid actuation and biointelligent communication).

Selection of technology classes. The technology is chosen and classified according to its key innovation and its potential impact [20]. To this end, the framework of technology classes according to Schuh is adapted for biologically transformed value creation [9]. A technology will thus be classified as follows:

- *Enabling technology*: A technology that itself does not build upon principles of biological transformation but can be implemented to enable certain BT principles in a product or production system.
- *Technology to transform*: A technology that can be changed by introducing BT principles, whereby the product or production system is biologically transformed.
- *Transforming technology*: A technology that is built upon principles of BT and can be introduced into the production environment to bring about industry BT.

Assessment of technology and sustainability readiness. A core element of the technology radar is the assessment of technologies according to their TRL and SRL.

Technology Readiness Level. This is based on the well-established readiness levels, ranging from basic recording to successful operation on a 1–9 scale [10, 11]. For the TRL assessment detailed research of the technology is conducted, whereby the level of detail varies depending on the search field [19, 20].

Sustainability Readiness Level (SRL). The SRL assessment is designed in accordance with the recommendations in UNEP's guideline for Sustainability Assessment of Technologies on an operational level [11]. To determine appropriate criteria for the SRL, established frameworks of sustainability assessment were evaluated, and tested in expert workshops. As a result of this evaluation, six criteria are established for the SRL: (A) Reduction of resource consumption, (B) Reduction of carbon footprint, (C) Improvement in occupational safety, (D) Enhancement of employment relations (volume, diversity, inclusion), (E) Flexibility at the product and production level, (F) Regional supply chain integration. These criteria cover ecological, social, and economic aspects and are represented on a 10-point scale, akin to the TRL assessment's 9-point scale, to measure a technology's maturity in promoting specific sustainability aspects.

SRL 1: Not applicable.
SRL 2: Cause-effect relation identified for sustainability problem and technology.
SRL 3: Technology solution concept formulated to reduce the sustainability problem.
SRL 4: First laboratory trials of the technology solutions with relevant stakeholders.
SRL 5: Initial piloting the technology in a relevant environment to solve the sustainability problem.
SRL 6: Validation of the technology solution to the sustainability problem by applying the technology in a relevant environment as a first setup.
SRL 7: Prototype implementation in relevant environment to demonstrate the effectiveness of the technology solution to the sustainability problem.
SRL 8: Extended prototype testing in intended operation environment and, if necessary, refinement of technology solution.
SRL 9: Fully elaborated and tested technology deployment with implementation plan for solving the sustainability problem.
SRL 10: Proven technology in operation to solve the sustainability problem.

The rating "0: not applicable was introduced" was added, since the scale is universally suitable for technological developments, the introduced SRL criteria may not be fully addressable.

Dissemination: To enable a clear, graphical analysis and benchmarking, a visualization of the three technology radar with TRL (see Fig. 1) and SRL (see Fig. 2) was created.

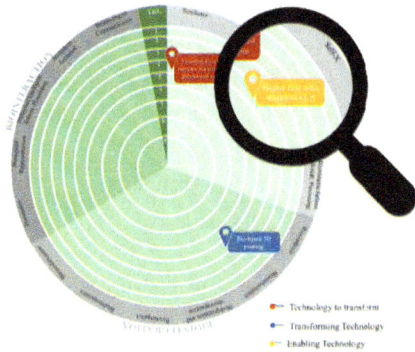

Fig. 1. Visualization of the technology radar for biological transformation (TRL).

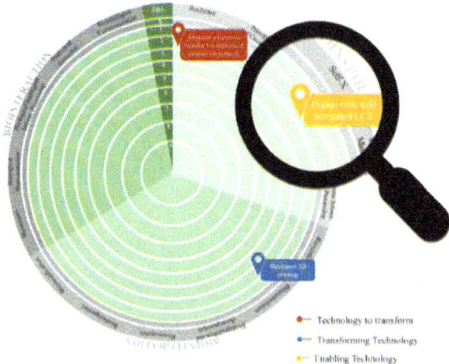

Fig. 2. Visualization of the technology radar for biological transformation (SRL).

3.2   Application of the Technology Radar for Biological Transformation

A step-by-step instruction for assessing technologies with the technology radar was developed, which guides participants through the presented steps of the technology radar assessment and gives supporting information and examples on all terms. The step-by-step guide includes the radar map visualization as displayed in Figs. 1 and 2.

3.3   Validation of the Technology Radar of Biological Transformation

To evaluate the applicability of the developed technology radar, a workshop with 10 participants from industry and research was conducted as part of the German research project BioFusion 4.0. Herein, an insight into the development status of three technologies is given in a comparable manner. The workshop groups assessed the technologies *digital twin with integrated* life-cycle assessment *(LCA) for products* [21], *bio-based 3D printing* [3] and a *situation awareness monitor for networked production systems* [22, 23], which are (further) developed within the research project. The overall results (assessment of BT core principle, BT principle subcategory, TRL and detail SRL) are presented in Table 1 and the visualization in Figs. 1 and 2.

Table 1. Exemplary application of the technology radar assessment.

Criteria	Digital twin with integrated LCA	Situation awareness monitor for networked production systems	Bio-based 3D printing
BT core principle	Bio inspiration	Bio inspiration	Bio integration
BT principle subcategory	Self X	Resilience	Biosubstitution
Technology class	Enabling technology	Technology to transform	Transforming technology
TRL	5	4	4
SRL average	2	2	2
Individual SRL categories			
SRL A	3	4	3
SRL B	3	4	3
SRL C	2	2	2
SRL D	1	1	Not applicable
SRL E	1	1	Not applicable
SRL F	3	2	5

It was concluded that situation awareness monitor for networked production systems and bio-based 3D printing are in the experimental stage, as is their development status to solve problems of resource and CO_2 efficiency. The digital twin with integrated LCA has been tested on an industrial scale (TRL 5). All evaluated technologies share that their contributions to the improvement of social sustainability and the flexibilization of product and production systems are in the early stage of cause-effect identification, but for bio-based 3D printing the cause-effect relation has not been identified. Bio-based 3D printing is more advanced in the development of a regional supply chain than the other two technologies and has already been validated in a relevant environment.

4   Conclusion

This technology radar provides current insights on the maturity of technology relevant to BT. It is utilized in the early stages of product development to assess the potential of technologies for enhancing the sustainability of product and production systems. This can be applied at regular intervals to highlight the technological progress of individual technologies. The assessment results can change with emerging innovation as they influence the TRL and SRL criteria. The technology radar follows multiple methodological approaches as recommended by VDI 3780 and UNEP's Guideline for SAT. The six SRL criteria align with the three pillars of sustainability and are understandable and assessable. However, the comparison of the different SRL-categories is a multicriteria-optimization problem and highly dependent on the prioritization and individual weighting of the environmental and social impact categories. It must also be acknowledged that with alternative SRL criteria a different result could be attained. In addition, it should be noted that a distortion of the SLR level assessment is possible when one or more SLS categories are not applicable to the technology under evaluation, as only the mean value is calculated. Consequently, sector and condition specific sets of criteria shall be established in the future. Assessment results are influenced by the knowledge and awareness of participating experts and a comprehensive assessment is ensured by an interdisciplinary team of experts for product development, sustainability, and biology. This paper introduced a method based on biomimetic product development, technology assessment, and sustainability assessment standards. Overall, the method of a technology radar for BT enables stakeholders like production planners and product developers to evaluate and identify suitable technologies for innovation and implementations.

Acknowledgments. This paper was written as part of the research and development project "Biological Transformation 4.0: Further development of Industry 4.0 by integrating biological principles (BioFusion 4.0)", which is funded by the German Federal Ministry of Education and Research (BMBF) and supervised by the Project Management Agency Karlsruhe (PTKA). The responsibility for the contents of this publication lies with the authors.

References

1. Neugebauer R, Ihlenfeldt S, Schließmann U, Hellmich A, Noack M (2019) A new generation of production with cyber-physical systems: enabling the biological transformation in manufacturing. J Mach Eng 19:5–15
2. Miehe R et al (2019) Enabling bidirectional real time interaction between biological and technical systems: Structural basics of a control oriented modeling of biology-technology-interfaces. Proc CIRP 81:63–68
3. Berkhahn M, Kremer G, Riedelsheimer T, Lindow K, Stark R (2022) Taxonomy for Biological Transformation principles in the manufacturing industry. In: Kohl H, Seliger G (eds) Proceedings of 18th Global Conference on Sustainable Manufacturing, Berlin
4. Steele JE (1960) Living prototypes. The key to new technology. In: Wright Air Development Division. Directorate of Advanced Systems Technology (ed) Bionics Symposium, Dayton, United States (1960)
5. Platform Industrie 4.0: Was ist Industrie 4.0?. https://www.plattform-i40.de/IP/Navigation/DE/Industrie40/WasIndustrie40/was-ist-industrie-40.html. Accessed 16 May 2022
6. Bauernhansl T, Brecher C, Drossel W (2019) Biointelligenz: eine neue Perspektive für nachhaltige industrielle Wertschöpfung. In: Ergebnisse der Voruntersuchung zur biologischen Transformation der industriellen Wertschöpfung (Biotrain)
7. Byrne G, Dimitrov D, Monostori L, Teti R, van Houten F, Wertheim R (2018) Biologicalisation: biological transformation in manufacturing. CIRP J Manuf Sci Technol 21:1–32
8. Verein Deutscher Ingenieure (2000) VDI 3780—Technology assessment—concepts and foundations
9. Schuh G (2011) Technologiemanagement. Handbuch Produktion und Management 2. VDI-Buch. Springer, Berlin, Heidelberg

10. DIN Deutsches Institut für Normung e. V. (2020) DIN EN 16603-11—Definition of the Technology Readiness Levels (TRLs) and their criteria of assessment (ISO 16290:2013, modified)
11. Stich V, Hicking J, Stroh M-F, Abbas M, Kremer S, Henke L (2021) Digitalisierung der Wirtschaft in Deutschland—Technologie—und Trendradar 2021. Studie im Rahmen des Projekts "Entwicklung und Messung der Digitalisierung der Wirtschaft am Standort Deutschland"
12. DIN ISO (2006) DIN EN ISO 14040:2006 Umweltmanagement—Ökobilanz—Grundsätze und Rahmenbedingungen. DIN-Norm, DIN EN ISO 14040:2009-11, Berlin
13. UNEP/SETAC Life Cycle Initiative (2009) Guidelines for social life cycle assessment of products. Lignes directrices pour l'analyse sociale du cycle de vie des produits. Canadian Electronic Library
14. Guinée J (2016) Life cycle sustainability assessment: what is it and what are its challenges? In: Clift R, Druckman A (eds) Taking stock of industrial ecology. Springer International Publishing, Cham, pp 45–68
15. Global Reporting Initiative (2023) GRI—Sector Program. https://www.globalreporting.org/standards/sector-program/. Accessed 28 July 2023
16. Holden NM (2022) A readiness level framework for sustainable circular bioeconomy. EFB Bioecon J 2:100031
17. Hallstedt S, Pigosso D (2017) Sustainability integration in a technology readiness assessment framework. In: Proceedings of the 21st International 2017, pp 229–238
18. BioInnovation (2019) BioInnovation Application Guide. BioInnovation
19. Schuh G, Zeller V, Stich V (2022) (eds) Digitalisierungs- und Informationsmanagement. In: Handbuch Produktion und Management, vol 9. Springer, Berlin, Heidelberg
20. Rohrbeck R, Heuer J, Arnold H (2006) The Technology Radar—an Instrument of Technology Intelligence and Innovation Strategy. In: 2006 IEEE International Conference on Management of Innovation and Technology. IEEE
21. Seegrün A, Mügge J, Riedelsheimer T, Lindow K (2023) Digital twins for sustainability in the context of biological transformation. In: Global Conference on Sustainable Manufacturing, pp 576–584. Springer, Cham (2023)
22. Lange A, Gering P, Agacayaklar IF, Knothe T, Busse D (2022) Bioinspirierte Erkennung von Abweichungen. Zeitschrift für wirtschaftlichen Fabrikbetrieb
23. Lange A, Knothe T, Kohl H, Seliger G (2022) Biological transformation: principles to enhance holistic production systems. Proc CIRP 110:293–298

Product-Production-CoDesign Thinking for Sustainable Manufacturing

Marvin Carl May[(⊠)], Louis Schäfer, Tobias Lachnit, and Gisela Lanza

wbk Institute of Technology, Karlsruhe Institute of Technology (KIT), Kaiserstr. 12, 76131 Karlsruhe, Germany
marvin.may@kit.edu

Abstract. Manufacturing needs to contribute towards a sustainable future for the sake of preserving and enriching humanity on planet earth. This goal is enshrined in the Sustainable Development Goals (SGD) set forth by the United Nations. SGD 9 aims at building a resilient, innovative and sustainable industrialization. SGD 12 ensures sustainable consumption and production patterns. Currently, manufacturing falls short of achieving these targets as product design and production engineering operate individually and sustainable practices are not focused. This industrial problem is reflected in the absence of holistic approaches that aim at sustainable production by providing applicable methods. To address this challenge, we propose Product-Production-CoDesign (PPCD) Thinking. With a clear focus on sustainability we delineate PPCD Thinking from Design Thinking and extend the notion towards manufacturing. It encompasses linear manufacturing (SGD 9) and circular production (SGD 12). Four case studies illustrate this software defined production enable PPCD Thinking and its customizability. In a nutshell, Product-Production-CoDesign Thinking, thus, can contribute to moving towards sustainable manufacturing and net zero.

Keywords: sustainability · design thinking · manufacturing

1 Introduction

Sustainable Manufacturing is key to achieve the sustainable development envisioned by the United Nations and hence contribute towards a sustainable future that preserves planet earth [23]. The main contribution of sustainable manufacturing lies in building a resilient infrastructure by promoting a sustainable, innovative industrialization enshrined in Sustainable Development Goal (SGD) 9 and enabling sustainable consumption and production patterns outlined in SGD 12 [23]. While SGD 9 fosters a sustainable industry through linear manufacturing, SGD 12 enhances this notion towards a circular economy. To achieving net zero solutions must encompass product design, business model, production and reverse logistics and remanufacturing. Thus, the future of manufacturing must regard both product development and production engineering in an integrative, holistic manner [2]. In an industrial symbiosis, research must empower a large scale

H. Kohl et al. (Eds.): GCSM 2023, LNME, pp. 185–193, 2025.
https://doi.org/10.1007/978-3-031-77429-4_21

industrial shift towards sustainable manufacturing and net zero. Agile approaches constitute such scalable and widely applicable methods [18]. While Design Thinking provides a holistic approach that encompasses product, service and business model design [6], a clear focus on manufacturing and sustainability is missing. However, this lack of coherently regarding product, production and sustainability hinders their application and diminishes their contribution to the SGDs. Novel approaches regard the entire product life-cycle at early stages, such as simultaneous engineering [16], software defined manufacturing [4] or Product Production-CoDesign (PPCD) [2]. To date they are hardly directly applicable to contribute to a transition to sustainable manufacturing. Thus, we introduce Product-Production-CoDesign Thinking as a Design Thinking process that unifies product development and product engineering approaches to achieve sustainable manufacturing and create holistic, net zero contributing solutions.

The paper is structured as follows. Section 2 introduced the fields of action that interplay in PPCD Thinking to enable sustainable manufacturing. In Sect. 3 the research scope and research questions are delineated. Then the general approach of extending Design Thinking towards PPCD is presented in Sect. 4. Enabling sustainable manufacturing through the presented approach is shown and discussed in Sect. 6. Section 7 concludes with a summary and outlook.

2 Fields of Action

2.1 Product-Production-CoDesign

Integrating product development and production engineering is a frequent scope of research. VDI2206 [19] describes the simultaneous development of product and production system and identifies the necessity to perform these concurrently to incorporate production system inflicted restrictions into the product development phase. In a similar vein, the product perspective is clearly illustrated in the integrated Product engineering Model (iPeM) [3] that extends into the starting of production and market opportunities. The initial approach, simultaneous engineering, dreamed of holistically integrating both product and production approach to simplify the complexity [2]. However, these approaches lacked a coherent integration of product generations [2] and production system generation and lifecycles. Hence, Product-Production-CoDesign (PPCD) was introduced in 2022 [2]. PPCD regards the timely paralleling of collaboratively developing, iterative planning and product creation within their systems [2]. This encompasses in particular the life-cycle of products and production systems, while integrating their development over product generations [17]. The latter includes the end of life decommissioning of products and production systems [17] and, hence, incorporates SGDs 9 and 12 for sustainable manufacturing. As sustainable manufacturing encompasses linear sustainability and circular production, PPCD serves as the stepping stone into holistically enabling sustainable manufacturing through both clearly describable approaches such as model based systems engineering [2] and innovative methods such as design thinking [18], that both have attracted research and industry alike.

2.2 Methods

Different methods and process models—such as the 6-3-5-method or Design Thinking—can be used to increase creativity in the solution finding process. The four phases of a creativity process are discovery, maturation, insight, and elaboration. These can be helpful to integrate creativity into problem solving activities. The whole procedure is accompanied by different emotions, such as fear and euphoria. [14] The necessity to use such methods to consciously promote and use creativity exists, as this allows to concentrate on one's own strengths. In general, creativity methods can be distinguished between intuition and discourse: While intuitive methods focus on the promotion of thought association, discursive methods involve the systematic search for solutions divided into individual logical steps. [5, 20] Examples of intuitive methods are brainstorming, -writing and TRIZ [13], while the morphological construction kit belongs to the discursive methods [8]. SPALTEN is a widely used and fractally structured method for general problem solving developed by [1]. The method is divided into seven different phases, whereby the whole solution process is included. [1]

While SPALTEN is a method to find solutions to problems in a structured way, other methods like *Design Thinking* (DT) are more focused on the implementation of creativity in solution processes. DT can help to accelerate the flow of ideas and, if necessary, to solve existing or emerging mental blocks, while being user-centered. [21] It always includes phases such as empathize, define, ideate, prototype, and test, each with minor adjustments in wording and content. There is a large body of literature on the topic of social innovation where [9] compiles a review of empirical research linking the current state of the art in applying DT in organizations.

2.3 Sustainable Manufacturing

In order to create a holistic environmentally sustainable product, close coordination between production and product systems is essential. Decisions made during the product design phase have a significant impact on the product's environmental footprint throughout its entire lifecycle, including manufacturing, usage, disassembly, reuse, remanufacturing, and recycling, which are largely predetermined [12]. Sustainable design is a vital element in this process, encompassing aspects beyond traditional ecodesign. This includes opportunities for design for Cradle to Cradle and Product-Service Systems (PSS) [7]. Several factors require attention and improvement, such as diagnosability, modularity, and the extension of product lifespan [22]. Achieving these goals necessitates not only proactive planning but also iterative improvement. Traceability plays a vital role, particularly in the success of circular economy practices, specifically in reverse logistics and their management [11]. The evaluation of data, optimization, and adaptation of design, as well as the identification of optimal routes for product and material reuse with minimal waste, are critical. User data collection is important for generating insights and integrating them into subsequent design processes and production improvements. The quantification of system design effectiveness can be facilitated through the utilization of Life Cycle Assessment (LCA) or Life Cycle Sustainability Assessment (LCSA), offering a way to assess its impact. [10] In conclusion, achieving a holistic environmentally sustainable product and manufacturing process necessitates close coordination between production and product systems.

3 Research Scope

The subject of the present research is the interface between product development and the associated production planning and control, as addressed by PPCD presented in Sect. 2.2. Both domains each address specific requirements on the methods and tools used: Activities in product design for example require creativity in finding solutions, whereas activities in production system planning have to deal with uncertain product characteristics. At the same time, digitalization offers new opportunities and possibilities through greater availability of information that can be used throughout the product development process. Here, approaches such as Design-for-X or other concepts are emerging. But the first fundamental question is which methods are suitable for integrated development and planning of product-production systems. For this reason, this article focuses on the following, first research question:

RQ1 Which method is suitable for application at the interface between product development and production system planning, and how can the concept of Product-Production-CoDesign (PPCD) be skillfully supplemented?

This article, thus, proposes a novel concept of Product-Production-CoDesign Thinking. In order to continue the motivation of the SDGs in terms of sustainable manufacturing of sustainable products, this article also investigates how the newly presented method can be applied in the context of sustainability. Thus, the second research question arises as follows:

RQ2 How can the developed methodology be applied to give greater consideration to sustainability in PPCD activities?

4 PPCD-Thinking Based on Design Thinking

To that end, PPCD-Thinking introduces a holistic Design Thinking Framework focusing enabling the interplay between product design and production planning. To address design thinking challenges and answer RQ1, the main stages Discover, Define, Ideate, Prototype, Test and Implement from design [18] are kept.

The examination of both production system and product is pivotal and illustrated in subsequent Fig. 1 delineating the two levels. After the described steps discover, define, and ideate have been completed, the feasibility on the product level can be assessed and refined through the utilization of virtual or physical prototypes. Further iterations loops, such as acceptability and comprehensibility of the solution, arise through the process of testing. The admissibility of user problems iterates back to the discovery process are also involved to discover new experiences and strategic topics. The examination of the production system levels presents the opportunity for additional iteration loops. Following the creation of the prototype, the implementability of the idea can be evaluated. Through testing, further loops can be enabled, including the iteration of data validity, the ability of the idea to address user concerns, and the assessment and enhancement of consistency with the production strategy. The coherent integration of product generations and production system generations distinguishes PPCD Thinking from other approaches such as life cycle management and simultaneous engineering.

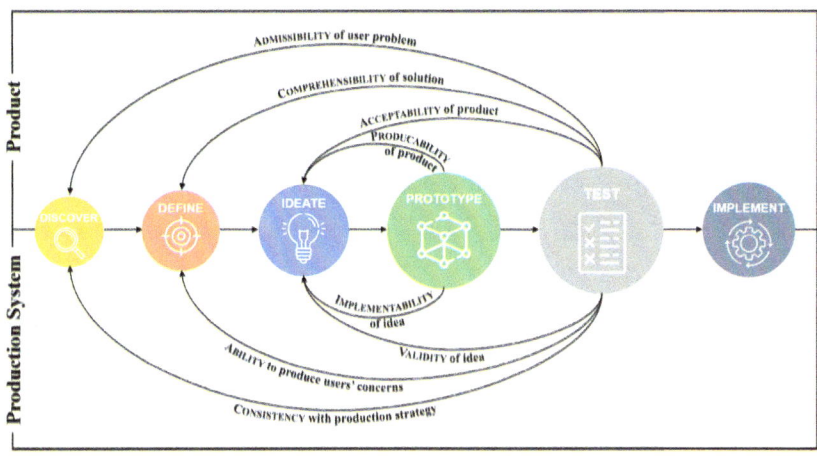

Fig. 1. Product-production-codesign thinking process model

5   Case Studies

The first case study, as highlighted in Fig. 2, highlights the admissability of the user problem. Model based Systems Engineering (MBSE) is a method commonly used in PPCD [17]. In this case MBSE based user requirement analysis structures and links identified user requirements before and during the prototype phase to improve subsequent product changes. The second case study makes use of a MBSE impact analysis, highlighting product producibility by mapping product features with production processes. In the third case study, consistency between production strategy and prototype are regarded. With a strategic fit analysis the effects of producing the prototype with required processes on the abilities and network footprint are assessed. Lastly, again on the production side, the implementability of a prototype production can be validated with a virtual prototype put into event discrete simulations on production system level and virtual commissioning on machine and system level.

6   PPCD-Thinking for Sustainable Manufacturing

A major contribution of PPCD is the extension and applicability to a circular production [2]. With PPCD-Thinking, this aspect should, hence, be in depth regarded as designing and engineering product and production systems [15] without regarding their end of life is still too common [17]. Thus, we couple the PPCD-Thinking process with the life cycle of products to address RQ2. As products are design in generations [3], the PPCD-Thinking process in linear production ends with the successful start of production. Independent of decommissioning being integrated into product and production engineering during this PPCD-Thinking, as soon as used products return to the manufacturer, the coherent PPCD need arises. Sustainable PPCD-Thinking comes into play as the next product generation shall incorporate learnings and potentially subsystems and components of previous product generations as shown in Fig. 3. This vastly increases complexity and

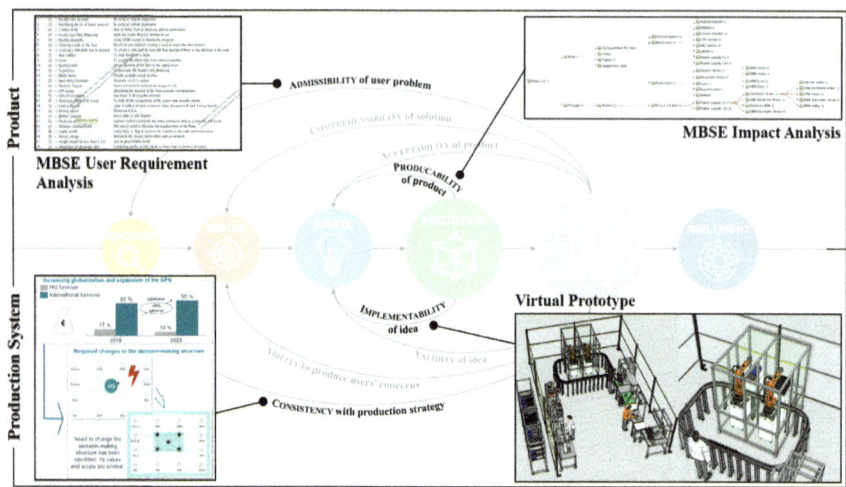

Fig. 2. Exemplary case studies highlighting four selected aspects of PPCD-Thinking: MBSE allows for understanding user req. And tracking these within complex systems and virtual prototypes were used to validate the implementability of generated ideas.

requires the integration of several PPCD-Thinking cycles. Figure 3 introduces major challenges to be solved during the PPCD-Thinking application.

7 Summary and Outlook

In a nutshell, PPDC-Thinking provides a novel design thinking approach highly customized for the realm of producing physical goods. Based on the PPCD approach, product design and production engineering are interlinked and relevant questions are addressed in the framework. The individual tools and solutions used in the framework can be taken from [17] and regularly extended. However, the framework aims at enabling sustainable manufacturing and circular production, so that a longer term validation will be necessary. As with any design thinking approach, educating engineers to properly apply the approach will be necessary during the application.

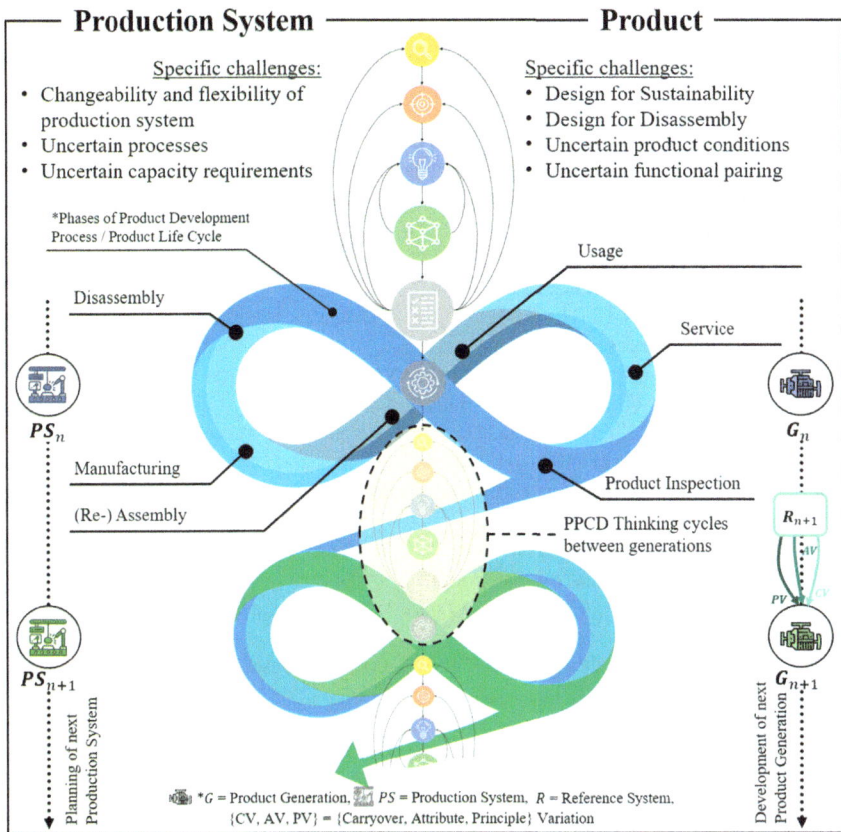

Fig. 3. PPCD-Thinking cycles (yellow) addressing specific circularity challenges within the development of new product and production system generations (vertical axis).

Acknowledgment. This research work was undertaken in the context of DIGIMAN4.0 project, a European Training Network supported by Horizon 2020 EU Framework Programme for Research and Innovation (Project ID: 814225). The authors gratefully acknowledge financial funding from the German Federal Ministry of Economic Affairs and Climate Action and the European Union (grant no. 13IK001ZF).

References

1. Albers A, Burkhardt N, Meboldt M (2005) Spalten problem solving methodology in the product development. In: ICED
2. Albers A, Lanza G, Klippert M, Schäfer L, Frey A, Hellweg F, MüllerWelt P, Schöck M, Krahe C, Nowoseltschenko K et al. (2022) Product-production co design: an approach on integrated product and production engineering across generations and life cycles. Proc CIRP 109:167–172
3. Albers A, Reiss N, Bursac N, Richter T (2016) Integrated product engineering model in context of product generation engineering. Proc CIRP 50:100–105

4. Behrendt S, Ungen M, Fisel J, Hung KC, May MC, Leberle U, Lanza G (2023) Improving production system flexibility and changeability through software defined manufacturing. In: Lecture notes in production engineering. pp 705–716
5. Breiing A, Flemming M (1993) Methoden zur Ideenfindung. Springer-Lehrbuch. In: Theorie und Methoden des Konstruierens. Springer, Berlin, Heidelberg
6. Brown T et al (2008) Design thinking. Harv Bus Rev 86(6):84
7. Ceschin F, Gaziulusoy I (2016) Design for sustainability: from product design to design for system innovations and transitions. Des Stud 47:118–163
8. Chulvi V, González-Cruz M, Mulet E (2013) Influence of the type of idea-generation method on the creativity of solutions. Res Eng Design 23:33–41
9. Elsbach K, Stigliani I (2018) Design thinking and organizational culture: a review and framework for future research. J Managem
10. Finkbeiner M, Schau M, Lehmann A, Traverso M (2010) Towards life cycle sustainability assessment. Sustainability 2(10):3309–3322
11. Gartner P, Benfer M, Kuhnle A, Lanza G (2021) Potentials of traceability systems—a cross-industry perspective. Proc CIRP 104:987–992
12. Herrmann C, Hauschild M, Gutowski T, Lifset R (2014) Life cycle engineering and sustainable manufacturing. J Ind Ecol 18(4):471–477
13. Koltze K, Souchkov V (2017) Systematische innovation: TRIZ-Anwendung in der Produkt- und Prozessentwicklung. Carl Hanser Verlag
14. Lubart T (2001) Models of the creative process: past, present and future. Creativity Res J 13:295–308
15. May MC, Kiefer L, Kuhnle A, Lanza G (2022) Ontology-based production simulation with ontologysim. Appl Sci 12(3):1608
16. May MC, Schmidt S, Kuhnle A, Stricker N, Lanza G (2020) Product generation module: automated production planning for optimized workload and increased efficiency in matrix production systems. Proc CIRP 96:45–50
17. May MC, Schäfer L, Frey A, Krahe C, Lanza G (2023) Towards product production-co design for the production of the future. Proc CIRP 119:944–949
18. Plattner H, Meinel C, Weinberg U (2009) Design-thinking. Springer
19. Plattner H, Meinel C, Weinberg U (2021) VDI2206. Beuth-Verlag
20. Pétervári J, Osman M, Bhattacharya J (2016) The role of intuition in the generation and evaluation stages of creativity. Front Psychol 7
21. Razzouk R, Shute V (2012) What is design thinking and why is it important? Rev Educ Res 82(3):330–348
22. Seliger G, Kim HJ, Kernbaum S, Zettl M (2008) Approaches to sustainable manufacturing. Int J Sustain Manuf 1(1–2):58–77
23. UN (2007) Indicators of sustainable development: Guidelines and methodologies—united nations department of economic and social affairs. 3rd edn New York

Research on the Development Mode and Path of Green Design of Industrial Products in China

Yizhi Song[1(✉)], Benxiao Yang[1], Xiaqing Liu[1], Jianhua Zhang[1], Frida Li[2], Xinyi Tong[2], and Jerome Feldman[2]

[1] Service-Oriented Manufacturing Institute (Hangzhou) CO., Ltd, Hangzhou, China
songyizhi@isom.org.cn
[2] Global Alliance of Innovators e.V., Berlin, Germany

Abstract. Green product design takes the coordinated development of humans and nature as the fundamental goal. The green design of industrial products plays the first lever function to start the sustainable development of industry and promotes the construction of green manufacturing system. As one of the world's largest consumer markets, China's green design of industrial products is of great significance to slowing down global warming and addressing global climate issues. By combing through China's nationwide green design initiatives, this paper interprets China's pre-planned green design concepts, analyzes the green design paths and development patterns of China's industrial products, summarizes achievements and lessons learned, and demonstrates future development.

Keywords: Manufacturing · Design · Sustainable development · Management

1 The Concept of Green Design of Industrial Products in China

From the perspective of ecological engineering, green design can be described as a design process that adopts reasonable technology and utilizes the self-organizing ability of the ecosystem to integrate the economy and society into the environment through practice [1].

In China Green design has been considered as a modern design concept and method based on the entire life cycle of a product and centered on the environmental resource attributes of the product [2]. The green design concept of Chinese industrial products emphasizes the comprehensive consideration of the impact of the whole life cycle on resources and environment at the design and development stage, including design and development, raw material selection, production, and recycling. During the design and development period, based on the support of the product life cycle database, the product life cycle analysis (LCA) should be conducted, the production processing should be optimized, and design for function, lightweighting, reliability and energy efficiency should be carried out. In the selection of raw materials period, through green supplier management, the green and environmentally friendly materials should be preferably selected by strengthening the control of toxic and hazardous substances. In the manufacturing process, the green manufacturing process and technology should be promoted, and the

H. Kohl et al. (Eds.): GCSM 2023, LNME, pp. 194–201, 2025.
https://doi.org/10.1007/978-3-031-77429-4_22

integration of design and manufacturing will be implemented. In the recycling process, minimal effect on the environment should be realized through adopting easy-to-recycle and dismantle design solutions, establishing a producer responsibility system, and setting up a recycling system. The green design concept in key industries in China is shown in Fig. 1 [3].

Fig. 1. Green design concept of industrial products in key industries in China

2 The Goal and Development Mode of Green Design for Industrial Products in China

Various countries around the world are pursuing a sustainable development direction, and some developed economies are planning or launching trade systems such as carbon border adjustment mechanisms. China recognizes that by accelerating the promotion of green design based on life cycle theory, it can help Chinese products effectively improve their competitiveness in the international market [4].

2.1 Objectives of Green Design for Chinese Industrial Products

In the "14th Five-Year Plan for Green Development of Industry" issued by the Chinese government, it is proposed to vigorously develop and promote new energy vehicles, promote alternative fuel vehicles such as methanol vehicles, to promote high-efficiency lighting, energy-saving air conditioners, energy-saving refrigerators, water-saving washing machines and other green and smart home appliances by using "old for new" and other means, to encourage the use of low VOC content of paints, cleaning agents, to accelerate the development of biomass, wood, gypsum, and other new building materials, and to increase the proportion of recycled materials consumption. The goal is to develop and promote 10,000 kinds of green products that meet the above requirements by 2025.

2.2   China's Industrial Product Green Design Development Model

Based on the government's policy cornerstone and guided by China's CO_2 peak and carbon neutrality goals, the overall blueprint of China's green product design has been planned. China Green Manufacturing Alliance and other public welfare organizations act as a linkage hub for green manufacturing resources, effectively organizing the construction of a green design architecture with the participation of industry, academia, and research. This architecture takes the green design product standard system as the benchmark, the industrial product green design demonstration enterprises as the main carrier, and the green design products as the output results. The Green Manufacturing Public Service Platform has created a new mode of green design on the Internet and formed a big pattern of green design ecological development for industrial products in China. China's industrial product green design development pattern is shown in Fig. 2.

Fig. 2.   China's industrial product green design development pattern

In terms of green design product standards, evaluation indicators are defined into four categories: resource attribute indicators, energy attribute indicators, environmental attribute indicators and quality attribute indicators. The "General Rules for Evaluation of Eco-designed Products", "Technical Specification for Evaluation of Green-designed Products Rare Earth Steel", "Technical Specification for Evaluation of Green-designed Products Small-power Electric Motors", "Eco-designed Product Evaluation Specification Part 1: Household Detergents", and "Technical Specification for Evaluation of Green-designed Products Fabric Products" have been formulated.

In terms of the green design demonstration enterprise, the selecting specifications mostly examine the enterprise's green development strategy, enterprise sustainable development management ability, green design innovation capability, design informatization software and database system, environmentally friendly production capacity and other ten major indicators. At the present stage, the green design demonstration enterprise norms in the industries of electronic and electrical appliances, textiles, machinery and equipment, automobiles and accessories, light industry and pharmaceuticals have been created around the dimensions of "green design + manufacturing" and "green design + service".

An important part of the green manufacturing public service platform is the green design database. It covers the basic database of ecological impacts of the whole life cycle of industrial products, the basic database of production such as green materials, green equipment and green process library, the database of material flow and energy flow in process. The online service platform also includes the green design case database and design service organization database and so on. The various databases online provide enterprises with personalized and customized "one-stop" services.

3   Analysis of Green Design Path for Industrial Products in China

China has adopted an all-round multi-level method to develop its green design architecture.

A. *Establishing green design policy system*

In 2013, the "Guiding Opinions on Carrying out Eco-design of Industrial Products", as an official document, clarified the concept of green design for the first time at the national level, and put forward the idea of implementing green design and its mission and goals. The "Green Development Plan for Industry (2016–2020)" and the "Green Manufacturing Project Implementation Guidelines (2016–2020)" put forward the construction tasks of the green design policy system in the 13th Five-Year Plan period (2016–2020) in an all-round and multi-level manner. Since the 14th Five-Year Plan, the "Outline of the 14th Five-Year Plan for National Economic and Social Development and Vision 2035" and the "14th Five-Year Plan for Green Development of Industry" have been issued, and the green design policy system has been formulated to encourage enterprises to carry out green design, organize green design innovation pilots, and build a green manufacturing support system. It has formulated comprehensive and three-dimensional measures to promote the construction of green design systems for industrial enterprises by encouraging enterprises to carry out green design, organizing green design innovation pilots, and building a green manufacturing support system.

B. *Formulating green design standards and evaluation norms*

A green design promotion mechanism for industrial products combining government guidance and market promotion has been initially established. A number of national, industrial and group standards have been formulated, released and gradually improved, such as the general rules for product evaluation, green design product labeling and certification and evaluation rules related to green design and green manufacturing. At the national level, the "General Rules for Eco-design Product Evaluation (GB/T32611)" has been released, which includes a series of standards and technical specifications for general evaluation methods, evaluation rules and labeling documents.

C. *Improving green design third-party evaluation indicators*

A third-party evaluation mechanism for green design products has been established. Over 100 third-party evaluation organizations are active with recommendation and evaluation. The third-party service organizations are promoted to innovate green manufacturing evaluation and service modes, to develop efficient tools, to conduct one-stop services such as consulting, testing, evaluation, certification, auditing, training, and provide overall solutions of design and manufacturing for customers in key industrial segments.

D. *Publishing green design-related directories*

Since 2017, China has begun to release green design demonstration enterprises, green design products, etc. The number of recommended lists has gradually increased, giving full play to the driving role of typical samples of green design, and the value of the continuous promotion of green design concepts has begun to emerge.

E. *Breakthroughs in key technologies for integrating green design and manufacturing*

The construction of green design platform has been identified as one of the core support directions in the national level. It is aimed at the management and evaluation of the whole life cycle of products. A green design information database for the whole life cycle of products will be established. The breakthroughs will be made in the selection of green raw materials, innovative design and application technologies, as well as key technologies for the integration of green design and manufacturing, so as the efficiency improvement of product development.

F. *Continuously promoting international exchanges and cooperation on green design*

By strengthening communication with the United Nations, the European Union and other international organizations, the sustainable green design exchanges and cooperation mechanisms has been established between China and Europe, China and Japan, and countries along the "Belt and Road".

4 Achievements of Green Design Products in China

As of 2022, 161 standards have been issued around the petrochemical industry, iron and steel industry, non-ferrous industry, building materials industry, machinery industry, light industry, textile industry, communication industry, and packaging industry, and this standard system is being continuously improved [5].

Six batches of green design product lists have been released, including a total of 3159 green design products, and the distribution of types of products is shown in Fig. 3.

China's green design product requirements are much higher than China's national standard requirements. For example, the air purifiers, their green design product evaluation standards are much higher than the corresponding national product standard. The energy-efficiency ratio (particulate matter) indicators are increased by 60%, the amount of particulate matter in clean air is required to be increased by more than 10%, and the rate of bacteria removal is required to be 98% improvement [6].

As of January 2022, China has selected and released a list of three batches of green design demonstration enterprises for industrial products, with a total of 245 green design

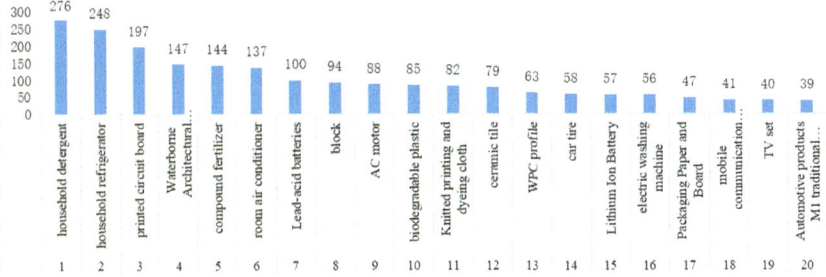

Fig. 3. Distribution of the top 20 green design product types by quantity

demonstration enterprises covering nine industries, including machinery and equipment, chemical industry, automobiles and accessories, building materials, textiles, and so on, and their distribution is shown in Fig. 4.

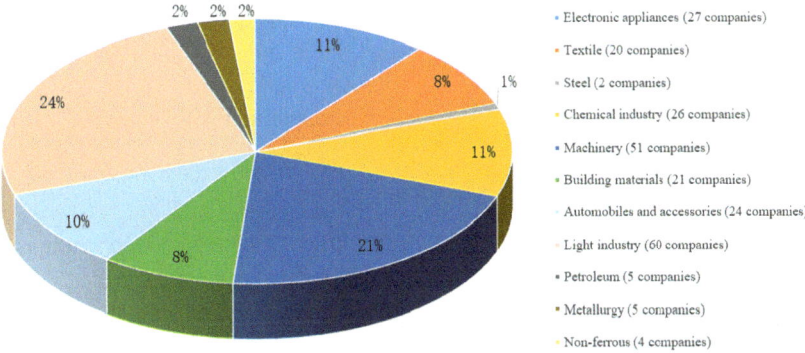

Fig. 4. Industry distribution of green design demonstration enterprises for industrial products

These enterprises are located in 30 provinces (autonomous regions and municipalities directly under the central government) across the country, with the top five interms of quantity being Zhejiang Province, Shandong Province, Guangdong Province, Henan Province and Beijing City [7–9].

5   Lessons Learned and Future Development

China attaches immense importance to green design and has established a relatively complete development system on a national scale, but the following problems still exist.

- Lack of organic system and guidance method for green design

The responsibility for recycling, dismantling and harmless treatment must be further defined. China's green design still needs to be worked out in these aspects. In addition, enterprises did not have the awareness of building green design databases in the past, and it is difficult to extract and apply data on operation status, energy consumption, including green features, like CO_2 footprint.

- The evaluation standard system to be completed

The mechanism of evaluating standards is still insufficient. At present, there are no more than 200 evaluation standards for green design products in China, which cannot meet the requirements for green design evaluation of all categories of industrial products.

- Less high-end green products exported from China

In the field of green products with high technological complexity, there is still a big gap between China and developed countries. In 2019, China produced 105 out of all 238 types of products with a comparative advantage, but only prevailed in 7 out of the top 50 types of green products in terms of technological complexity [10].

- Most green products in intermediate range

Compared with other major countries, China's exports of green products are more concentrated in intermediate products. On the one hand China's production can influence the production of green products in other parts of the world; on the other hand, it shows that China is still in the climbing stage of the global industry chain and has not yet mastered the manufacturing capacity of core advanced equipment.

It can be observed that China's green design development will be more active in the following aspects in the future.

- Cooperation and exchange with other countries

China will actively promote bilateral and multilateral enterprise cooperation in the field of industrial product design with other countries.

- Promotion of green product upgrading

Relying on China-EU cooperation, further improve the green product identification and the recognition of green products in the EU, and then more enter the EU green product market.

- Mutual recognition of green product through international cooperation

The standards and labels of China's green design product need to be recognized. China will conduct comparative analysis of domestic and foreign green product standards, promote green product standard certification and recognition and participate in the development of international standards and conformity assessment rules to improve the consistency of standards.

- Increase introduction of international green design intelligence

It will expand international green design talent exchange and cooperation.

References

1. Rui F (2014) Research on green design methods in product display design. Packaging Eng 18:123–126

2. Boyang L (2020) Development status and prospect of green design of industrial products. China's National Conditions and National Strength 2:36–38
3. Green Design of Industrial Products Demonstration Enterprises Experience Sharing No.3: Practical Experience of Green Design of Industrial Products in Electrical and Electronics Industry. Website of the Ministry of Industry and Information Technology of China. https://www.miit.gov.cn/jgsj/jns/qjsc/art/2020/art_7633669709724ba3baf93b0fd5b1016c.html
4. Liang C, Zhanrong H (2018) The internal logic, practical foundation and implementation path of innovation-led green development in the new era. Marxist Res 6:74–86, 160
5. List of green design product standards (updated in September 2022). Website of China's Ministry of Industry and Information Technology. https://wap.miit.gov.cn/jgsj/jns/gzdt/art/2022/art_dc8703ac6acf483484973b7a121f3321.html
6. China releases first batch of green design product evaluation group standards. People's Daily China Economic Weely. http://app.ceweekly.cn/?action=show&app=article&contentid=163524&controller=article
7. Notice of the General Office of the Ministry of Industry and Information Technology on the Announcement of the List of Demonstration Enterprises for Green Design of Industrial Products (First Batch). China Ministry of Industry and Information Technology website. https://www.gov.cn/xinwen/2019-11/25/content_5455456.htm
8. Notice of the General Office of the Ministry of Industry and Information Technology on the Announcement of the List of Green Design Demonstration Enterprises of Industrial Products (Second Batch). Website of the Ministry of Industry and Information Technology of China. http://jxj.beijing.gov.cn/ztzl/ywzt/jnhbfwpt/jnhbgzdt/202101/t20210104_2195783.html
9. Green Design Demonstration Enterprises of Industrial Products (Third Batch) List Announcement. Website of the Ministry of Industry and Information Technology of China. http://gxt.hunan.gov.cn/gxt/xxgk_71033/gsgg01/202109/t20210916_20593678.html
10. Special Report | Global Green Product Trade Characteristics and China's Export Opportunities. https://thinktank.phbs.pku.edu.cn/2021/zhuantibaogao_1022/44.html

Collaborative Development of Design Requirements for Designing Assistive Technology Devices for Early Infancy Children

Daniel Braatz$^{(\boxtimes)}$ ⓘ, Renato Luvizoto ⓘ, Andrea Fontes ⓘ, Mariana Santos ⓘ, Fabiane Lizarelli ⓘ, Gerusa Lourenço ⓘ, Luciana Agnelli ⓘ, and Rodrigo Martinez ⓘ

Universidade Federal de São Carlos, Rod. Washington Luiz, km 235, São Carlos, Brazil
braatz@ufscar.br

Abstract. This article presents the development of product requirements in the context of an assistive technology innovation project. It is emphasized that assistive technology plays a fundamental role in the development of sustainable products and environments, as it enables the inclusion and autonomy of people with disabilities, ensuring that all solutions are accessible and efficient. The initial stages of new product development are considered critical due to the need to define requirements that will guide the design process. The challenge in formulating these requirements lies in the contradiction between determining which criteria and factors should accompany the project at a stage where the understanding of the object and the problem is still limited. The methodological approach is primarily based on a descriptive method of the design process leading to the definition of requirements. The results are mainly divided into four topics: project description, description of the stakeholders and actors involved in the project, description of the process of defining design requirements, and finally, the presentation of the requirements developed in the project context. These results can contribute to two main ways: by sharing the defined requirements and by presenting a process of collective construction of design requirements.

Keywords: assistive technology · design · child

1 Introduction

Early childhood is a critical stage in human development during which rapid cognitive, emotional, physical, and social advancements occur. However, children in this age group who have physical, sensory, or cognitive disabilities face challenges that can impose significant barriers to their overall development. In this sense, Assistive Technology (AT) plays a pivotal role in providing support and opportunities for these children to explore their environment, learn, and interact with the world around them.

Creating AT devices for children in early childhood presents unique challenges. The understanding of the needs and capabilities of these children during the early stages of

H. Kohl et al. (Eds.): GCSM 2023, LNME, pp. 202–210, 2025.
https://doi.org/10.1007/978-3-031-77429-4_23

development is limited, which makes the process of defining design requirements complex. In this context, it is crucial to adopt a collaborative approach that involves healthcare experts, researchers, designers, and end users to ensure the creation of appropriate, efficient, and accessible solutions.

This article aims to present the development of design requirements through a collaborative approach in the context of an AT innovation project for children in early childhood. We explore the initial steps of the design process, highlighting the challenges faced in defining requirements and the importance of a collective approach to establishing those requirements.

The interconnection between AT and sustainability is a crucial aspect of this study. While traditionally sustainability has been primarily associated with environmental considerations, we are facing a new perspective: sustainability also encompasses social and economic aspects. In the context of early childhood assistance, AT is not only a means to improve the quality of life for children but also plays a role in building inclusive and equitable societies. This involves ensuring universal access to assistive technology devices, promoting active participation of families in the development process, and ensuring that these solutions are accessible and economically viable.

2 Literature Review

2.1 Concepts and Foundations of Assistive Technology

AT refers to an interdisciplinary field of knowledge that encompasses equipment, services, strategies, and practices specifically designed and implemented to mitigate the challenges faced by individuals with disabilities. Its primary goal is to enhance their functionality in various contexts, leading to increased social participation and improved quality of life [1, 2].

According to the World Health Organization (WHO), AT empowers individuals to lead healthier, more productive and independent lives by actively engaging in education, employment, and civic activities. Conversely, the lack of access to AT products exacerbates the negative impact of disability on individuals, their families, and society, as it restricts opportunities for participation and equal rights [3].

The effective utilization of assistive devices can significantly enhance various aspects of individuals' lives. For instance, hearing devices can improve language skills in children, while wheelchairs and appropriate computer resources can enhance educational participation and engagement in other contexts [3, 4]. However, access to AT often remains inadequate. Insufficient user-centered research and development, along with the absence of effective distribution systems and standards for quality, safety, and design, pose challenges to access the most suitable assistive devices for individuals [3].

2.2 Development of Design Requirements

The establishment of design requirements represents one of the primary steps within the scope of new product design. This process entails conducting a comprehensive assessment of existing market products, user needs, and other relevant sources to formulate a comprehensive list of requirements that will effectively guide the development stages, including initial concept generation, prototyping, testing, component refinement, and the final product iteration. By establishing these requirements, it becomes feasible to identify key priorities as the project progresses and the underlying concepts associated with the future device gradually unfold. Stuart Pugh's Product Design Specification (PDS) framework [5] offers one approach to delineating design requirements, providing a structured framework that aids in comprehending the critical aspects to be considered during the definition of the principal design requirements.

The initial phases of new product development are widely recognized as critical due to the imperative of defining requirements that will effectively steer the subsequent design process. To mitigate this inherent challenge, it is imperative to foster a collaborative approach that ensures the integration of user needs, preferences, and specific characteristics [6, 7]. This is particularly crucial in the domain of AT, where the design process necessitates a multidisciplinary perspective, with an essential minimum involvement of the health and engineering domains [8, 9].

Within the collaborative framework, it is vital to establish dedicated arenas for participatory engagement, employing diverse forms of representation [10]. Such representational approaches facilitate the sharing and evolution of projective activities, enabling individual perspectives to converge into group consensus and knowledge [11].

3 Methodology

The present study employed an exploratory qualitative research methodology to investigate the design requirements of AT devices. This methodological approach facilitated a comprehensive and systematic exploration of the subject matter. Data collection and analysis encompassed document analysis and recording of collective processes. These methodologies yielded detailed insights into the needs, experiences, and expectations of end users and other stakeholders engaged in the project.

3.1 Description of the Assistive Technology Innovation Project

The AT innovation project in which the present study is inserted aims to develop assistive devices for children in early childhood with physical, sensory or cognitive disabilities. This project involves a multidisciplinary team comprising AT researchers, engineering professionals, and designers, with active participation from individuals with disabilities and their families.

The adopted approach follows an iterative design process that centers on users' needs. From the initial stages of the project, extensive participation is emphasized to ensure the comprehensive consideration of the specific needs, preferences, and characteristics of children and their families throughout the entire development process.

The key actors involved in the project encompass:

1. Researchers and experts in AT: These individuals contribute their technical and theoretical knowledge of AT, identifying trends, and best practices. They collaborate in the analysis and evaluation of results, providing expertise in the field.

2. Engineering and computing professionals: This group comprises engineers and other professionals with expertise in developing technological solutions. Their involvement is essential in comprehending the specific requirements of children and offering valuable design insights.
3. Designers: Designers play a critical role in translating identified needs into practical design solutions. They leverage their expertise in areas such as ergonomics, usability, and aesthetics to create devices that are functional, accessible, and appealing to children.
4. Individuals with disabilities and their family members: The inclusion of individuals with disabilities and their families is vital as they provide firsthand perspectives in the project. Their active participation ensures that the solutions align with the actual needs of the end users.

3.2 Stages of Design Requirements Definition

The team collaboratively conducted the process of defining design requirements, involving all the aforementioned actors. The team divided the process into three steps, including: (1) Information gathering: The researchers conducted a literature review, examined patent databases, and consulted commercial suppliers to understand existing technologies and approaches in the field of AT. Additionally, the research collected information through presentations of previous projects, observations, and experiences with end-users and their families. (2) Needs and constraints analysis: Based on the gathered information, the team identified the primary needs and constraints pertaining to mobility devices for children in early childhood [12]. The analysis encompassed considerations of physical, cognitive, sensory, and emotional aspects. (3) Requirements definition: The team established design requirements based on the identified needs. These requirements encompassed technical features, functionalities, usability, safety, aesthetics, and other relevant aspects.

4 Results

4.1 Process of Design Requirements Definition

To address the specific objectives of the initial stages of the project, the methodological strategy employed consisted of regular and structurally organized meetings with the research team.

The project team comprises researchers from the fields of production engineering, architecture, computer science, occupational therapy, and physiotherapy, all of whom possess expertise in areas relevant to the development or prescription/development of AT devices. Over the course of ten collective meetings, the chosen format involved presentations on current challenges in the field of AT, trigger questions, and reflections on advancing knowledge production.

Each two-hour meeting provided dedicated spaces for dialogue among the participating team members. Records were kept in minutes and recordings as documentation of the process. The following topics were addressed throughout the ten meetings:

- Conceptualization and classification in AT.
- AT from the user's perspective.
- Insights on the product development process.
- Insights on software development.
- Available resources in the market and patents.
- Case studies covering children with disabilities and the use of AT devices.
- Previous projects conducted on the topic.

In all the meetings, the trigger questions guided the group in identifying current challenges in the development and accessibility of mobility aids targeted at the pediatric population, as well as the demands for technological innovation in the country based on the collective knowledge of the team. The compilation of these challenges and needs is presented as the results of the project's initial phase.

4.2 Presentation of Developed Requirements

The following 20 needs, identified and collectively discussed by the researchers, are presented below. Some pertain directly to devices, while others encompass aspects, such as fostering dialogue between health professionals, developers, and designers.

1. Devices developed for early childhood: There is a need for a wider range of devices designed to address the independent mobility challenges faced by young children. The growing international movement emphasizing early interventions for independent mobility, such as Go Baby Go [13], serves as an example.
2. Family participation: Encouraging the participation of users and their families in the implementation of AT is crucial to ensure that the devices effectively meet individual expectations and daily life demands.
3. Dialogue between health professionals, developers and designers: The choice of a particular device or ongoing monitoring of children utilizing previously prescribed options may necessitate adaptations or even changes to the products to align with the children's life contexts. Therefore, establishing channels of information exchange and fostering collaboration between the professionals involved is essential.
4. Time from prescription to care/dispensing: The financing structure via the Brazilian Unified Health System (SUS) for the provision of prescribed devices follows a regulated flow. Despite advancements in recent years, the waiting time for equipment to become available remains a costly process for families and professionals involved. Hence, reducing the time interval between prescription and equipment dispensation remains a pressing need.
5. Financially accessible, low-cost devices: Factors such as high-performance materials, limited production scale, importation of components or the devices themselves, among others, contribute to the high costs of effective AT devices. Therefore, there is a need to explore the development of devices that can be more affordable compared to those already commercialized.
6. Accessibility in terms of information: Locating available device options on the market is not a simple task, as it requires searching through national and international companies, gathering technical specifications, assembly instructions, target audiences, costs, and maintenance information. As a result, there is a demand

for easily accessible spaces and channels to disseminate this information, enabling professionals and families to access such information.

7. Ease of transport: The configuration of devices has a direct impact on their transportability in vehicles. For instance, rigid frame wheelchairs, motorized carts, and fixed walkers consume significant space and pose weight-related challenges.

8. Playful aesthetics: Initial research has revealed that AT devices primarily prioritize functionality over aesthetics. Given the target audience of this project and the broader objective of enhancing the overall attractiveness of device design, the need to pursue a playful aesthetic was identified.

9. Addressing stigmas associated with device use: In addition to a playful aesthetic, the developed technology devices should also counteract the existing stigma associated with such devices, ensuring they do not reinforce negative perceptions.

10. Professional training: The processes involved in the development and prescription of mobility devices require professionals to stay updated on existing technological innovations and evidence-based practices to facilitate informed decision-making based on the specific needs and demands of child users and their families.

11. Adjustability of devices: AT devices, in general, should incorporate adjustable features that cater to specific age ranges, movement envelopes, ranges of motion, and anthropometric measurements. Devices should offer adjustments and/or incorporate accessories that allow for versatility in use situations and accommodate various user profiles over an extended period.

12. Adaptability for different degrees of impairment: In addition to the aforementioned need, AT devices should be easily adjustable to cater to individuals with different levels of impairment, requiring minimal effort for adaptation.

13. Diversified range of device accessories: To address needs 11 and 12, it is essential for technology devices to offer a wide array of accessories that enable customization and expand the range of device use situations, among other purposes.

14. Integration among professionals, government/SUS, and families: Resolving the gaps between the primary actors involved in the prescription process, including users, prescribing professionals, relevant institutions, procurement processes, and responsible parties for purchasing, is crucial to address the real needs of users.

15. Selection of materials suitable for different environments: The materials chosen should allow to access various environments and adapt to different usage situations.

16. Ease of maintenance—cost, time, and effectiveness: The design and accompanying instructional materials should facilitate small adjustments, sanitization, and device safety enhancements, enabling users to perform maintenance activities with ease.

17. Usability-centered design for non-specialist users: Considering the various adjustments, device assembly, accessories, and usage situations, AT devices should be developed with the challenges of use in mind, catering to non-specialist users.

18. Design for "adaptations"—continuous design in use: Recognizing that individuals often make personal adaptations to devices based on their interactions, extending the range of usage situations or prolonging the device's lifespan, the design should be adaptable to accommodate unforeseen functionalities.

19. Provision of accessible spare parts (market availability and pricing): Ensuring the use of components readily available in the market or easily manufactured enables cost-effective and accessible replacement of components, facilitating both corrective and preventive maintenance.

20. Environmental impact: Incorporating a consideration for environmental sustainability involves minimizing equipment disposal by designing devices with future adaptations, adjustments, and parts replacements in mind. This way, devices can be repurposed for use by other individuals.

5 Discussion

5.1 Analysis of Results and Their Relevance

The engagement of stakeholders plays a pivotal role in the initial stages of the design process, particularly in defining the design requirements for AT devices targeted at children in early childhood. The active and collaborative involvement of AT experts, engineering professionals, designers, family members of individuals with disabilities, and, most importantly, the end users themselves is indispensable in comprehending and meeting the specific needs of the intended audience.

Therefore, collaborative engagement among these stakeholders is vital for accurately defining the design requirements for AT devices. This approach ensures that the resulting solutions are truly user-centered, accessible, effective, and capable of fostering the inclusion and autonomy of children with disabilities.

5.2 Potential Improvements and Future Directions

Although this study has made significant contributions to the definition of design requirements for AT devices targeted at children in early childhood, there are still numerous opportunities for further improvement and exploration of future directions.

One potential future action is to conduct additional studies to assess the effectiveness and usability of the developed devices based on the established requirements. Collecting feedback from end-users (and their families) and conducting pilot tests can offer valuable insights into the suitability of the proposed solutions, identifying potential adjustments and enhancements that may be necessary.

In summary, while this study has provided a solid foundation for defining design requirements for AT devices for children in early childhood, there are many opportunities for improvement and future advancements. Continued research efforts and ongoing engagement with stakeholders can drive innovation and further enhance the solutions offered, thereby contributing to the enhancement of the quality of life and inclusion of individuals with disabilities.

6 Conclusion

The needs assessment conducted to determine access to AT devices for children with physical disabilities yielded 20 requirements that were identified and discussed by the researchers. These requirements encompass various aspects, including the availability of specific devices for early childhood, the involvement of families in the development process, the communication between health professionals and developers, and the time interval between prescription and device delivery.

These requirements serve as a solid foundation for guiding the development of more efficient and accessible AT devices tailored to the needs of children with physical disabilities. Furthermore, they can serve as a reference for advances in related areas, addressing not only technical aspects, but also important social and economic considerations, within the perspective of sustainability. Notably, the collaborative and multidisciplinary nature of the development process, aimed at addressing the needs and limitations of all stakeholders involved, particularly prospective users, stands out as an essential aspect.

References

1. Cook AM, Polgar JM, Encarnação P (2020) Assistive technologies: principles and practice, 5th edn. Mosby
2. Brasil (2019) Estatuto da Pessoa com Deficiência—Lei no 13.146/2015. Brasil
3. World Health Organization (2022) Global report on assistive technology
4. Moen RD, Østensjø S (2023) Understanding the use and benefits of assistive devices among young children with cerebral palsy and their families in Norway: a cross-sectional population-based registry study. Disabil Rehabil Assist Technol. https://doi.org/10.1080/17483107.2023.2198563
5. Pugh S (1991) In: Total design: integrated methods for successful product engineering. Addison-Wesley Publishing Company
6. Binder T, Brandt E (2009) The design: lab as platform in participatory design research. 4:115–129. https://doi.org/10.1080/15710880802117113
7. Soares JMM, Fontes ARM, Ferrarini CF, Borrás MÁA (2020) Multicase study on product design in the area of assistive technology in Brazil. Disabil Rehabil Assist Technol 15:442–452. https://doi.org/10.1080/17483107.2019.1587019
8. Santos AVF, Silveira ZC (2020) AT-d8sign: methodology to support development of assistive devices focused on user-centered design and 3D technologies. J Braz Soc Mech Sci Eng 42:1–15
9. Borrás MÁA, Ferrarini CF, Marins PC et al (2023) 3D resources for visually impaired students. Global J Human-Soc Sci 23:9–21
10. Broberg O, Andersen V, Seim R (2011) Participatory ergonomics in design processes: the role of boundary objects. Appl Ergon 42:464–472
11. Luvizoto R, Fontes ARM, Torres I (2021) Técnicas de apoio ao projeto do trabalho. In: Engenharia do trabalho: saúde, segurança, ergonomia e projeto. Ex-Libris, pp 491–515
12. Soares JMM, Fontes ARM, Ferrarini CF, Borras MAA, Braatz D (2017) Assistive technology: revision of aspects related to the topic. Espacios 38:8–23
13. UD Mobility GoBabyGo! https://sites.udel.edu/gobabygo/. Accessed 19 Jul 2023

Efficient and Sustainable Production of Electrical Machines—Achieving a Higher slot Fill Factor Through an Innovative Forming Process Chain

M. Dix[1], M. Bach[1(✉)], V. Kräusel[1], and R. Wertheim[2]

[1] Fraunhofer Institute for Machine Tools and Forming Technology (IWU), Chemnitz, Germany
mirko.bach@iwu.fraunhofer.de
[2] Braude College, 2161002 Karmiel, Israel

Abstract. With the increasing electrification of mobility and the associated growing demand for electric machines, issues of efficient and resource-saving production and operation are becoming ever more central. Therefore, it is important to optimize the design of components, materials and the assembly of electric machines. An example, described in the paper, shows how to increase the slot fill factor of a stator. By optimising the wire content in the slots, electric machines can be made smaller or, at the same size, operate more powerfully resp. efficiently. Various applicable production methods have different drawbacks such as low productivity, high energy and resource consumption, and restrictions for geometric design. Here, forming technology offers remedy in the utilization of material with simple tools and its suitability for mass production. This paper shows the analysed process chain developed at Fraunhofer IWU for the forming-based production of coils with trapezoidal cross-sectional geometry. The investigation results provide a more sustainable structure and an efficient process chain.

Keywords: Forming · Electric machines · Coil · Sustainability · Process chain

1 Introduction

More than 43 % of the total electricity consumed worldwide is used for electric motor-driven systems [1]. In industry, the share of use for electric motor drives is as much as 69 % [2]. In the automotive sector, the overall efficiency of the internal combustion engine is comparatively less than 50 %, while it is around 65 % for an electric motor. Furthermore, the application of electric machines, above all thanks to electrification and the high number of safety standards and convenience equipment, has increased enormously, and it is to be expected that this will continue [3–5]. By 2030, electric cars will represent approximately 50 % of the overall automotive market [6] improving significantly environmental conditions. The continuously increasing demand for electric machines, including drives and generators and the latest increase in social and political requirements for production technology, such as sustainability, conservation of resources

H. Kohl et al. (Eds.): GCSM 2023, LNME, pp. 211–219, 2025.
https://doi.org/10.1007/978-3-031-77429-4_24

and climate protection, make it necessary to try new machine or component concepts and to improve manufacturing opportunities. Efficiency of electric machines, converting electrical into mechanical energy or the reverse direction, depends on design of single components and assembly. Developers steadily improve electric equipment to consume less energy and increase sustainability during usage and develop innovative technologies for increasing efficiency and reducing resources when manufactured.

2 Design and Manufacturing of Electric Machines

Figure 1, left is a split view of the main components of a typical electric motor. The two main components are the stator and the rotor. These two parts move in relation to each other by generating magnetic circuits. The rotor, supported by bearings to enable rotation normally has current carrying conductors, included coils or permanent magnets forced by the magnetic field of the stator. The stator contains electromagnets also made of wire windings around a ferromagnetic iron core to generate the magnetic poles. The stator core is made of metal laminates insulated one from each other to eliminate or decrease eddy currents. The geometry of the stator slots, the poles and the wire windings can significantly influence function, performance, cost and efficiency of electric machines and the structure of other components. The housing can be made of metals, mainly cast aluminium and partial also from plastic.

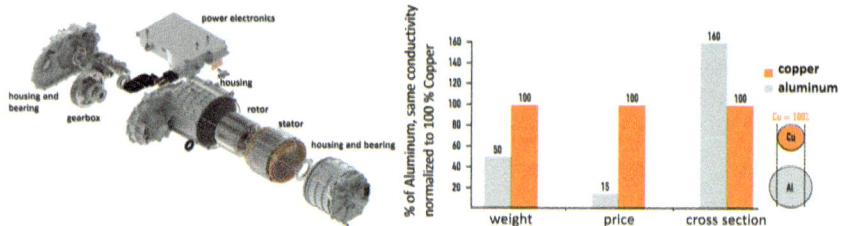

Fig. 1. a) components of an electric machine [7]; b): comparison of properties [8].

For producing electric machines with fewer resources and higher efficiency, the greatest potentials are expected from the customized use of either high quality materials (copper, rare earths [5], thinner stator sheets [9]) or winding design and procedure applied to the electric motor coils [10]. The winding design is distinguished between concentrated and distributed windings, with the concentrated type having greater benefits. This winding type is, due to its modest end turn geometry, very compact, and can be manufactured in more simple and cheaper processes [9].

Furthermore, the concentrated type has a less demand for conductor material and thereby caused lower ohmic losses. Copper and aluminium are the most common used conductor materials. Table 1 lists their electrical properties. In Fig. 1, right, the weight, the costs and the cross-section for the same conductance are compared. These properties have the largest influence on efficiency and sustainability. Assuming the same conductivity, using Aluminum requires a 60 % increase of cross-section and as a result much

Table 1. Electrical properties and electrical resistivity.

Material	Electrical conductivity γ in $\frac{MS}{m}$	Specific electrical resistivity ρ in $\frac{\Omega \cdot mm^2}{m}$
Copper-ETP	57	0,018
Aluminum 99.9	36	0,028

larger motor components and therefore more resources. The weight of copper is doubled, and the cost is much higher. However, for electric motors these disadvantages are compensated for by smaller sizes, less material including new manufacturing methods. An example of a forming process chain is developed and presented in the paper. Investigations of winding design aiming to reduce overall sizes, to minimize slot structure and the development of an efficient manufacturing chain are presented [11]. Further goals of this study are to form wires with complex shapes and to overcome production problems like risk hot spots or insufficient accuracy of winding geometry. Here an innovative manufacturing approach to tackle above problems and to compact coils with high power density is designed and scientifically investigated. Based on these strategies, the slot fill factor can be increased. This factor is the ratio between the actual filling of a slot with winding material and its available cross section.

3 Investigation of Winding Design

In classically manufactured electric coils made of round wire, the focus in adapting the winding process is on optimizing the wire layer pattern, and thus in increasing the slot fill factor. Slots are commonly filled by automated winding processes using round wire without controlled wire layering ("wild") or filled accurately to position (orthocyclically). However, under these conditions, fill factors of approximately 50 to 80 % are feasible [10, 12]. To achieve a significantly higher filling than with round wire, the cross section of the wire material must be adapted to the slot geometry. Many of the established manufacturing techniques, which are either too complex or energy consuming, are only suitable to a limited extent for wire cross section adaption for medium part quantities, and remain limited to niche applications [13].

Another disadvantage is that most of the techniques are unadaptable to be simply modified in response to changes in the available mounting space in electric machines. However, several forming techniques, such as mechanical compression inside the slot, offer a solution [14]. Figure 2 elucidates the several winding geometries in the slot areas – compacted round wire and rectangular wire. The position-adapted wire geometry is schematically shown and a main subject of this paper. The adaption of the geometry allows to increase the slot fill factor and potentially use of cheaper conduction materials, such as aluminium, or to reduce the whole electric machine's size. This saves additional resources and enables the use of more economical machine components.

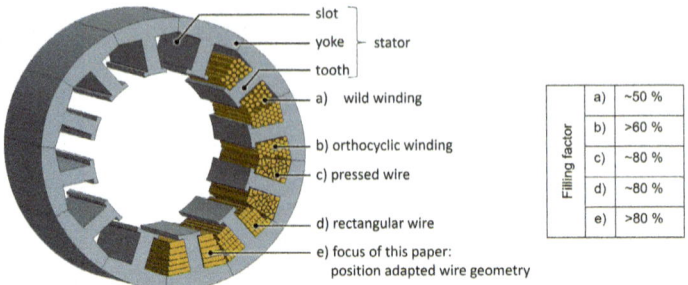

Fig. 2. Cross-section of the stator with selected coil winding geometries [11].

4   Process Chain for Manufacturing Position-Adapted Windings

When using round wire windings, a trapezoidal cross sectional slot geometry is selected in most cases [15]. Generally, when designing parallel flanked slots for a high slot fill factor, it is possible to use coils with rectangular cross-sectional wire. However, this slot design reduces the size of the cross-sectional tooth area, what leads to heterogeneously magnetic flux distribution and consequently only a limited strength can be amplified by the tooth. This diminishes the machine's efficiency [16]. In the investigations, essential steps for the manufacturing of a coil with trapezoidal cross section were identified, whose winding geometries are adaptable to their position in the slot with equal cross-sectional sizes. For the layout of electric machines, only the slot is considered in the section since the winding heads do not affect the magnetic operation. Based on this, local wire shaping or forming in the slot area or continuous forming of the whole semi-finished product will be referred to as cross section formation in the following. As a result, a sequence of differently shaped areas is obtained, as shown in Figure 3 a). The sections between the geometrically adapted areas have the initial semi-finished product geometry, such as the winding head. Otherwise, deviations from the nominal shape were to be expected during bending of adapted wire.

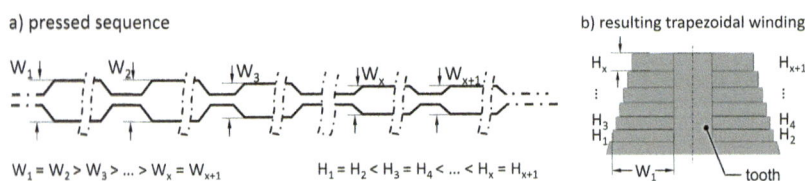

Fig. 3. a) Sequence with different shaped areas; b) Positioning of the different ares in the slot

The sections can be adapted by individual end turn design or subsequent forming. To generate a coil shape from the sequence subsequently, a bending step is required. The geometrically adapted sections $W1$ to $Wx+1$ are positioned in the slot at the corresponding spot according to the available width, as shown in principle in Figure 3 b). Independently of whether the profile cross section is adapted continuously or in the slot area, in addition to the formation of the winding assembly, a final calibration step is required to complete the nominal geometry.

5   Investigation and Implementation of the Forming Steps

First, in FE forming simulations, the material flow of sequential cross section formation, generation of a winding assembly, and calibration were analysed. As a result, for single process steps, challenges were identified, improvement proposals discussed, and forming tools and dies designed.

5.1   The Tested Die and Process Chain

For the cross-section formation step, a simple pressing tool was designed and applied (Figure 4). Upper and lower dies are guided together in a set to a specific pressing height by means of an adjustable impact plate in a press. The wire material starts flowing upon contact with the die surface, until the conductor's height and width defined for the winding dimension are reached.

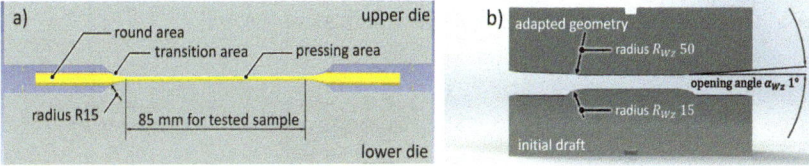

Fig. 4. a) Sequence with with different dimensions; b) Positioning of the sequential sections in the slot

At this point, in addition to widening, unintended material flow in axial direction was observed. During the investigation it was possible to identify an improved die geometry based on statistical design of experiments when modifying frictional coefficient, opening angle, and the radii of the die inserts, shown in Figure 4 b). As a result of these modifications, the expansion of the considered wire material was reduced down to approximately one third of the initial value. The deviations in length were considered to create a manufacturing model. This methodology was aimed to compensate for the conductor's axial expansion as a function of its pressed height via intentional roller feed control. Forces acting during the forming process cause the wire to expand and thus increase its surface area. Possible coatings of the wire are so far not able to withstand this strain and therefore crack. Consequently, uncoated wires are initially used for this process chain, which must then be coated with insulation after the forming steps.

The result of sequential pressing, a pressing sequence, was transferred into a modified standard tube bending machine for the follow-up process step (Figure 5 a). The bending process to generate a complete winding was implemented there. In this procedure, the researchers found it challenging to implement adequate feed travel to enable bending at just the right positions on the initially round wire. It was also necessary to make

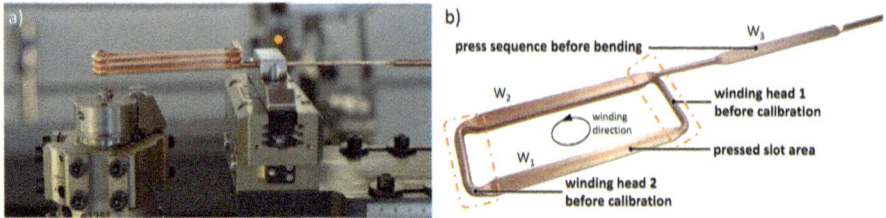

Fig. 5. a) pressing sequence during bending; b) intermediate state of a bent sequence [17]

occasional changes incorporating fastening, bending, and positioning tools and devices on the machine. The idea behind this was to enable reliable gripping/ handling and transport of wires with different cross section geometries. The result of a single bending process is illustrated in Figure 5 b). Due to the remaining round wire geometry in the winding head section, the modified pressing sections, which will be positioned in the slot of the electric machine later, do not yet contact one another. This circumstance would be disadvantageous for the slot's filling level. To avoid this, a final calibration step of the winding head is required. The calibration tool modified for pressing of the winding head has a cassette-like design. As shown in Figure 6 a), the tool consists of an upper and lower die plate, and several press plates positioned in between. The number of press plates is one less than that of the number of windings of the coil

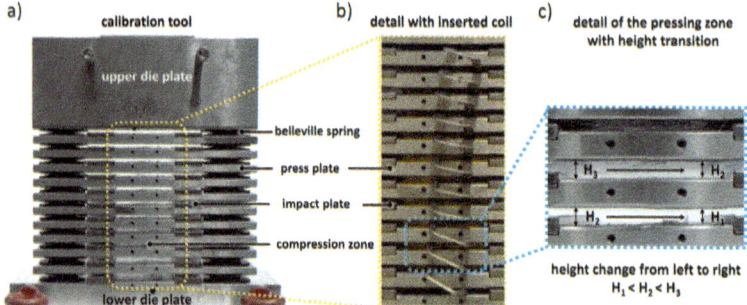

Fig. 6. a) Winding head pressing tool [17]; Detail: b) with aluminium coil, c) pressing zone

For the demonstrator coil, ten press plates were used. They are positioned atop one another, guided by four columns, and elastically kept at a distance from one another by belleville springs. The distance is necessary to provide sufficient space for the initially round wire in the region of the winding head when the coil is slid into the die before pressing. When formed with help of a punch coming from top with a force of about 2.000 kN (in case of copper), the tool is compressed until the impact plates contact each other and the correct height per winding is reached. In Figure 6 b), a section of the tool with an already completely pressed coil made of aluminium is shown. The two winding heads are pressed successively. It can be clearly seen that the winding head protruding from the press plates is used to change the winding level. The winding head located in the tool receives the final geometry with a height change from left to right, see c).

5.2   Accuracy of the Formed Windings

The completely formed demonstrator coil made of Al 99,9 with trapezoidal cross section after being removed from the calibration tool is illustrated in Figure 7. The view on the winding head 1 and 2 and the side view are visualised at the bottom (a). A Detail of the winding head 2 elucidates the transition in height from left to right implemented from the first to the second and from the second to the third winding. The shown heights H1, H2 and H3 have different values. The manufactured demonstrator coils deviate in two features from the designed final nominal geometry. One is the lateral flexure in the slot sections and the other one is the sidecut in the transition area from the slot to the winding head as predicted by FE simulation. Lateral flexure appears in the section of the coil situated in the die's outgoing radius during calibrating the end turn. The lateral flexure of each winding results from the required transition to the next winding layer. Each half coil winding that protruded without contact with the press plates must be led to the next tool layer. Consequently, bending of each single winding may be strongly influenced as a function of press plate thickness and coil length. Accordingly, thinner plates and longer coils lead to less bending. The area of the sidecut on the lowest coil winding before and after the calibration step of the demonstrator coil is shown in detail in Figure 7 b). The Sidecut results from axial material flow during the first step in the transition area and the follow-up calibration step in which a geometric deviation is resulting. The effect of reduction in wire lengthening during pressing with the modified die geometry positively affects the sidecut formation. Practical tests show that the wire width in the cross-sectional size in the sidecut area is on average almost 20 % larger than the wire width achieved with the die from the first design.

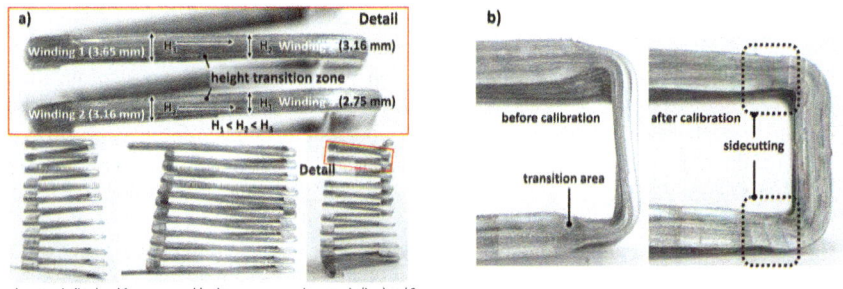

Fig. 7. a) Final demonstrator coil [17] b) left: coil before and after calibration [11]

6   Summary and Outlook

The development of the described process is an excellent example for improving efficiency, optimizing production, reducing resources and impact on environment. The paper describes the design and forming process of position-adapted coil windings for slots of a stator to increase the slot filling factor. Although the latest forming result has not yet met all the requirements for electromagnetic capability and geometric tolerances, it nevertheless shows high potential to manufacture coils with trapezoidal crosssectional geometry. The designed process chain demonstrates a way to implement a high slot fill factor thanks to position-adapted winding geometries with maximal material utilization. When using the technology described, a high potential for material and energy savings in the manufacturing and operation of electric machines can be expected. The challenges of the sidecut should be considered in further investigations to improve accuracy by optimizing the die and process, aiming also to eliminate the described effect. Further developments and experiments can decrease the number of process steps and cost for continuous improvement of sustainability.

References

1. Waide P, Brunner CU (2011) Energy-efficiency policy opportunities for electric motor-driven systems. Paris, Energy Efficiency Series. Accessed: 20 Oct 2021. [Online]. Available: https://www.oecd-ilibrary.org/content/paper/5kgg52gb9gjd-en
2. Binder A (2009) Technische optimierungspotentiale bei elektrischen antriebssystemen. In: Kosten- und Energiesparen durch effiziente elektrische Antriebe, Nürnberg, 2009. Accessed: 20 Nov 2020. [Online]. Available: https://www.ew.tu-darmstadt.de/media/ew/vortrge/090120_binder_optimierungspotentiale.pdf8
3. Bundesministerium für Umwelt, Naturschutz und nukleare Sicherheit (BMU), Ed., "Klimaschutz in Zahlen: Fakten, Trends und Impulse deutscher Klimapolitik," Ausgabe 2020, Referat Öffentlichkeitsarbeit, Online-Kommunikation, Social Media, Berlin 10034, May. 2020. Accessed: 20 May 2021. [Online]. Available: https://www.bmuv.de/service/publikationen-und-downloads/publikationen
4. Dobroschke A (2011) Flexible Automatisierungslösungen für die Fertigung wickeltechnischer Produkte. Dissertation, Technische Fakultät, Friedrich-Alexander-Universität Erlangen-Nürnberg, Erlangen
5. Schlick T, Hertel G, Hagemann B, Maiser E, Kramer M (2011) Zukunftsfeld Elektromobilität: Chancen und Herausforderungen für den deutschen Maschinen- und Anlagenbau. Roland Berger Strategy Consultants / VDMA, München
6. Wilker F. Elektrofahrzeuge werden ab 2030 voraussichtlich die Hälfte des weltweiten Automobilmarktes ausmachen. [Online]. Available: https://image-src.bcg.com/Images/20171106_BCG_Presse_Electric%20Car%20Tipping%20Point_tcm9-175823.pdf. Accessed 24 Feb 2021
7. Barrass P, Stover S, Fulton D (2020) Entwicklung und Optimierung elektrischer Antriebsstränge. MTZ - Motortechnische Zeitschrift 12:2020
8. Doduco Solutions GmbH, Vergleich zwischen Aluminium und Kupfer. [Online]. Available: https://www.alcunnect.de/Vergleich-Aluminium-und-Kupfer. Accessed 6 Jul 2021
9. Gerling D (2011) Trends und Herausforderungen bei zukünftigen E-Maschinen und Transformatoren. Fulda
10. Bickel B, Hübner M, Franke J (2014) Analyse des Optimierungspotenzials zur Erhöhung des Kupferfüllfaktors in elektrischen Maschinen. Ant J pp 16–21
11. Bach M, Dix M, Drossel W-G, Ihlenfeldt S (2022) Entwicklung und Charakterisierung einer Prozesskette zur umformtechnischen Herstellung elektromagnetischer Zahnspulen. Dissertation
12. Zerbe (2018) Innovative Wickeltechnologien für Statorspulen zur Erhöhung des Füllfaktors und Reduzierung der Beanspruchungen im Wickelprozess. Dissertation, TU Berlin
13. Groninger M et al. (2011) Casting production of coils for electrical machines. In: 1st International electric drives production conference EDPC. Nürnberg, pp 159–161

14. Rau E, Shendl A (2010) Stator for a polyphase electric machine and method for manufacturing same. US 2010/0295390 A1, USA 12/295,560

15. Müller G, Vogt K, Ponick B (2011) Berechnung elektrischer Maschinen, 6th ed. Weinheim: Wiley-VCH. [Online]. Available. http://swbplus.bsz-bw.de/bsz264767756cov.htm

16. Halwas M et al (2018) Entwicklung eines parallelen Technologieund Produktentwick-lungsprozesses/Development of a parallel technology and product development process using the example of winding design and manufacture as a part of the NeWwire funded project. wt, pp 301–306

17. Bach M, Psyk V, Linnemann M, Kräusel V, Bergmann M, Pohl N (2016) Towards the forming of concentrated windings with trapezoidal cross sections for increasing the slot filling factor. In: 11th International electric drives production conference EDPC, Erlangen, pp 35–39 Author F.: Article title. Journal 2(5):99–110

Technical Valuation

Environmental Impact Assessment
of Manufacturing of SiC/SiC Composites

Georgios Karadimas[(✉)], Yagmur Atescan Yuksek, and Konstantinos Salonitis

Sustainable Manufacturing Systems Centre, School of Aerospace, Transport, and
Manufacturing, Cranfield University, Cranfield MK43 0AL, UK
g.karadimas@cranfield.ac.uk

Abstract. SiC/SiC composites have attracted increasing attention in various
applications such as turbine blades, exhaust nozzles, and combustor chambers,
due to their exceptional mechanical and thermal properties. However, the environmental impact of these composites across their life cycle is an important aspect
that needs to be evaluated to support their responsible development and use. In this
study, a life cycle assessment of SiC/SiC woven laminate ceramic matrix composites to quantify their environmental impacts from cradle-to-gate was conducted.
Three different manufacturing methods to produce SiC/SiC woven laminates were
researched: chemical vapour infiltration (CVI), pyrolysis of a preceramic polymer
(PIP), and melt infiltration (MI). The Life Cycle Assessment approach was utilized to identify the effect outcomes for each process, analysing the raw material
extraction, raw material processing, and final product manufacturing phases to
develop the environmental impact assessment. The study's outcome showed that
CVI had the lowest average environmental impact between the two methods.

Keywords: SiC/SiC Ceramic Matrix Composites · Life Cycle Assessment ·
Environmental Impact

1 Introduction

Silicon Carbide/silicon Carbide (SiC/SiC) Ceramics Matrix Composites (CMCs) Have
Emerged as a Promising Class of Materials Due to Their Exceptional Mechanical Properties, High-Temperature Stability, and Corrosion Resistance [1]. These Materials Are
Composed of SiC Fibres Embedded in a SiC Matrix. SiC/SiC CMCs Have the Potential to
Replace Conventional Materials in Demanding Applications, Leading to Improved Performance Due to Their Lighter Weight and Thermal and Mechanical Properties [2]. However, a Comprehensive Understanding of the Environmental Implications Associated
with Producing SiC/SiC CMCs is Essential for Their Sustainable Deployment.

The application of SiC/SiC CMCs has been growing steadily in aerospace, nuclear,
and other high-temperature industries due to their exceptional thermal and mechanical
properties. However, despite their widespread use, there has been a notable scarcity of
LCA studies focusing specifically on SiC/SiC CMCs. This research aims to bridge this

© The Author(s) 2025
H. Kohl et al. (Eds.): GCSM 2023, LNME, pp. 223–231, 2025.
https://doi.org/10.1007/978-3-031-77429-4_25

gap by conducting a thorough evaluation of the environmental implications associated with the production of these materials.

This study aims to evaluate the environmental impacts of SiC/SiC composite manufacturing processes by comparing three possible manufacturing methods: Chemical Vapor Infiltration (CVI), Polymer Infiltration Pyrolysis (PIP), and Melt Infiltration (MI). The choice among these methods depends on factors such as ease of application, industry applicability, and end-product quality [3]. CVI is complex and technically demanding, suitable for high-performance SiC/SiC composites in aerospace, nuclear, and high-temperature applications, but may have limited use in small-scale or custom manufacturing [4]. PIP is more accessible and adaptable than CVI, commonly used in aerospace, automotive, and energy industries, producing composites with good properties. However, residual carbon can limit operating temperature and increase oxidation risk [6–8]. MI involves infiltrating a preheated fibre preform with a molten infiltrant material, commonly used in aerospace and high-performance applications for dense composites with excellent properties [9–11].

To assess their environmental impacts, Life Cycle Assessment (LCA) methodology is employed. LCA evaluates environmental impacts throughout a product's life cycle, considering energy consumption, materials, water usage, and waste generation [12–15]. It helps identify environmental hotspots, informs decision-making, and aids in product design, process optimisation, material selection, and waste management [15].

The outcomes of this study provide valuable insights into the environmental performance of SiC/SiC CMCs and facilitate informed decision-making for material selection, process optimisation, and sustainability improvements. The findings contribute to the growing body of knowledge on advanced composite materials and support the development of more sustainable and environmentally friendly manufacturing practices.

2 Methodology

The LCA approach employed in this study adheres to ISO 14040 and 14044 standards, which provide a systematic framework for analysing the environmental consequences of products throughout their life cycle. The approach involves four major stages: aim and scope definition, inventory analysis, impact assessment, and interpretation [16].

2.1 System Boundary

The system boundaries of the LCA study encompass the raw material extraction, material manufacture, and product manufacture stages. Figure 1 shows that SiC fiber and matrix are input materials for each method. Manufacturing of these input materials is encompassed in the LCA study and the manufacturing of composite materials using three different manufacturing methods is analysed separately. The functional unit is defined as 1 kg of SiC/SiC woven laminate, which serves as the reference unit for all impact assessments. The size and mechanical properties of the end product are not specified as these are out of the scope, and they do not impact the impact assessment results.

The environmental impact categories considered in the assessment include various midpoint indicators. The impact assessment methods employed are based on widely

Inputs   Processes   Outputs

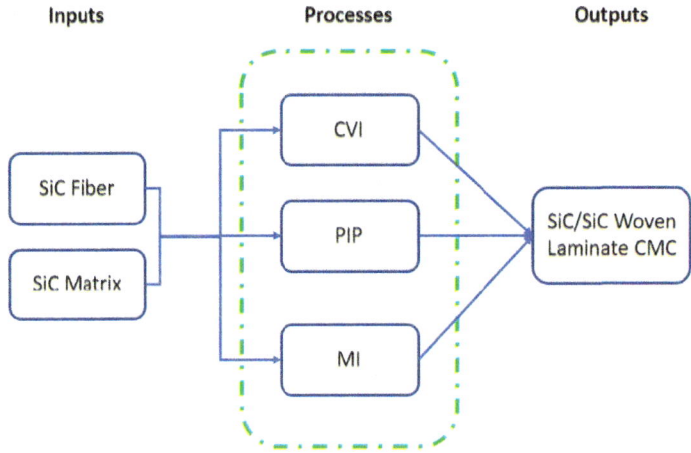

Fig. 1. System boundary for SiC/SiC woven laminate CMC manufacturing.

recognised characterization models and indicators, such as the ReCiPe method [17]. These methods enable the quantification and comparison of impacts across different environmental categories.

2.2 LCA Inventory

The data used in this study are sourced from up-to-date databases, scientific literature, industry reports, and manufacturers' specifications (20, 21). Specific assumptions are made in the result section to account for uncertainties and process hot spots, ensuring a comprehensive analysis. The inventory data include energy consumption, material inputs, emissions, waste generation, and transportation requirements at each life cycle stage.

The production processes for SiC/SiC CMC materials involve several key steps. These include the production of SiC base matrix material, the manufacturing of SiC fibres, and the weaving of SiC fibres into a laminate structure. Each process contributes to the production of SiC/SiC CMC materials and is crucial in determining their environmental impacts.

For the life cycle investigation, the different processes, materials, solvents, additives, and waste were accounted for 1kg of SiC/SiC composite material. More specifically, CVI process steps included the preparation of the fibre preform, the placement of the preform in a CVI reactor, the introduction of precursor gases, such as silicon-containing and carbon-containing compounds, chemical reactions on the surface of the preform to deposit SiC matrix material, repeated cycles of gas introduction and reaction to achieve the desired matrix thickness and the machining and finishing steps as shown in Fig. 2 [18, 19]. CVI does not typically involve the use of solvents. Waste gases, such as unused precursor gases, reaction by-products, and excess gases, may be generated [22].

With regards PIP, the first steps are the preparation of the fibre preform, impregnation of the preform with a preceramic polymer solution, the removal of the excess polymer solution from the preform, pyrolysis of the impregnated preform to convert the polymer

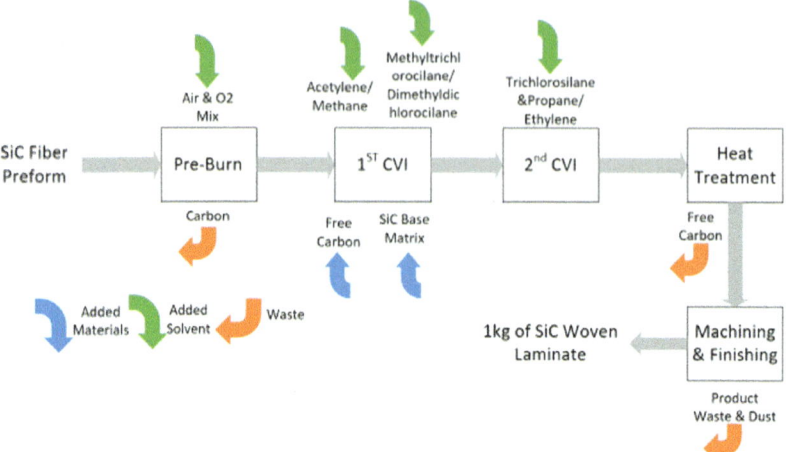

Fig. 2. SiC/SiC Woven Laminate CMC creation with CVI process.

into a SiC matrix, thermal treatment to achieve the desired matrix properties, and cooling and finishing steps as it can be seen in Fig. 3 [23, 24]. Common solvents used in PIP may include organic solvents, such as toluene or xylene, for dissolving the preceramic polymer. Waste solvents, excess polymer solution, and by-products of pyrolysis, such as gases and volatile organic compounds (VOCs), may be generated [25].

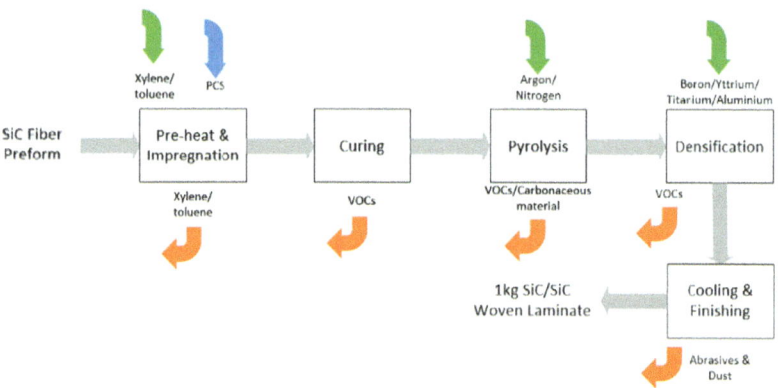

Fig. 3. SiC/SiC Woven Laminate CMC creation with PIP process.

MI process steps include the preparation of the fibre preform, preheating of the preform to a temperature suitable for infiltration, introduction of a molten infiltrant material, such as silicon or silicon alloy, infiltration of the molten material into the preforms void spaces, solidification and cooling of the infiltrated preform, removal of any excess infiltrant material and cooling and finishing steps as it can be seen in Fig. 4 [26, 27]. MI does not typically involve the use of solvents. Waste infiltrant material, excess infiltrant material, and any solidified or residual infiltrant may be generated as waste.

The life cycle inventory (LCI) results provide valuable insights into the energy consumption, material inputs, and emissions associated with the production of SiC/SiC CMC materials. Energy consumption data include the energy required for each process step, such as furnace heating, mixing, spinning, densification, and cooling. Material inputs encompass the quantities of raw materials, chemicals, and consumables used throughout production. Emissions data capture the release of greenhouse gases, air pollutants, and other substances resulting from the manufacturing processes.

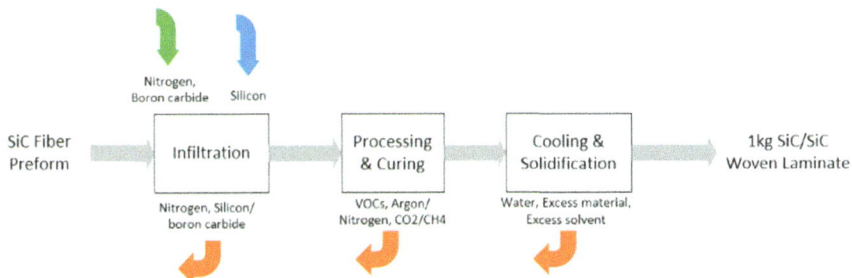

Fig. 4. SiC/SiC Woven Laminate CMC creation with M.I. process

In addition, the results allow for a comprehensive analysis of the environmental impacts associated with SiC/SiC CMC materials. This analysis encompasses various impact categories such as global warming potential, acidification potential, eutrophication potential, ozone depletion potential, and human toxicity. Additionally, resource depletion indicators can provide insights into the depletion of non-renewable resources throughout the life cycle.

3   Results

For each of the manufacturing processes illustrated in Figures, 2, 3, and 4, every step was researched separately in terms of operating temperature, time, energy consumption, solvent materials used, and occurring waste. This allows calculating and comparing the different impacts of each process regarding global warming, freshwater ecotoxicity, ionizing radiation, marine ecotoxicity, human carcinogenic, fossil resource scarcity, and landfill use. When comparing the environmental impacts between CVI, PIP, and MI processes, the highest values in these categories indicate that these processes have a relatively greater potential environmental impact than the others. The contributions of the base matrix, woven fabric, and other steps on the Endpoint Ecosystem impacts of each process are presented in Fig. 5 for each stage of the CVI, PIP, and MI processes.

In all three processes (CVI, PIP, MI), the SiC base matrix is the main contributor to the environmental impact, involving energy-intensive steps in its manufacturing. The creation of the SiC base matrix contributes significantly to the overall environmental impact of each process.

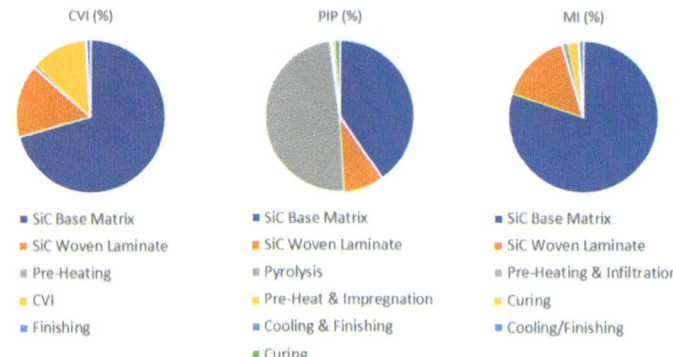

Fig. 5. Contribution of each manufacturing phase to the Environmental Impact

In the MI process, the SiC base matrix has a high impact due to the energy-intensive melt infiltration step. Similarly, in CVI, material selection and usage efficiency are crucial as the SiC base matrix significantly impacts the process.

The PIP process has a hotspot in the Pyrolysis stage, leading to a relatively higher environmental impact. However, the SiC Woven Laminate stage has a comparatively lower impact, which is a positive aspect.

Overall, reducing the environmental impact of the SiC base matrix remains a key focus in all processes. CVI shows a more evenly distributed impact, suggesting it may be a more environmentally friendly choice compared to PIP and MI.

According to Fig. 6 environmental impact results, it is evident that MI has the highest impact in terms of Ionizing Radiation. Whereas PIP has the highest impact in terms of Terrestrial Ecotoxicity, Marine Ecotoxicity, HCT, Freshwater Ecotoxicity, Land Use, and Fossil Resource Scarcity. Moreover, CVI has the highest Global Warming impact. From Fig. 6, it is concluded that PIP has the highest environmental impact of the compared manufacturing processes based on the compared indicators. Conversely, CVI consistently shows lower environmental impacts across most categories, making it a favorable option in terms of environmental sustainability. Overall, based on the provided data, CVI seems to be the most suitable process for the fabrication of SiC/SiC composites, considering its lower environmental impact in multiple categories.

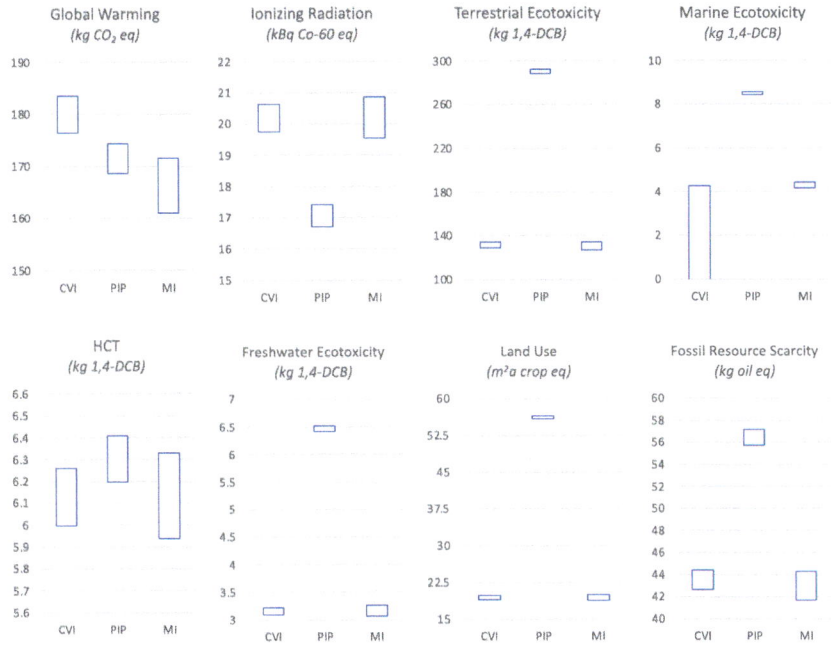

Fig. 6. Environmental Impact Assessment results.

4   Conclusion

In this study, the environmental impact of three different manufacturing processes of SiC/SiC CMCs was evaluated and compared. This study reveals the environmental profile of SiC/SiC CMC materials throughout the raw material extraction, the manufacturing processes, and the manufactured product. It summarizes the key findings related to, greenhouse gas emissions, resource depletion, and other environmental impact categories. The summary highlights the significant contributions of different life cycle stages to the overall environmental footprint of SiC/SiC CMC materials.

The study reveals distinct environmental impacts among the three manufacturing processes. CVI generally shows lower impacts than PIP and MI, suggesting potential environmental benefits. PIP has the highest environmental effect across most impact categories, while CVI has the lowest impact. MI falls between PIP and CVI in terms of overall environmental impact.

Acknowledgments. This project has received funding from the Clean Sky 2 Joint Undertaking (JU) under grant agreement No 886840.

References

1. Wang X, Gao X, Zhang Z, Cheng L, Ma H, Yang W (2021) Advances in modifications and high-temperature applications of silicon carbide ceramic matrix composites in aerospace: a focused review. J Eur Ceram Soc 41(9):4671–4688
2. An Q, Chen J, Ming W, Chen M (2021) Machining of SiC ceramic matrix composites: a review. Chin J Aeronaut 34(4):540–567
3. Amirthan G, Udayakumar A, Bhanu Prasad VV, Balasubramanian M (2009) Properties of Si/SiC ceramic composite subjected to chemical vapour infiltration. Ceram Int 35(7):2601–2607
4. Caputo AJ, Lackey WJ (1984) Fabrication of fiber-reinforced ceramic composites by chemical vapor infiltration. In: Proceedings of the 8th annual conference on composites and advanced ceramic materials: ceramic engineering and science proceedings, Smothers W (Ed.)
5. Delhaes P (2002) Chemical vapor deposition and infiltration processes of carbon materials. Carbon 40(5):641–657
6. Lee G, Fourcade J, Latta R, Solomon AA (2008) Polymer impregnation and pyrolysis process development for improving thermal conductivity of SiCp/SiC–PIP matrix fabrication. Fusion Eng Des 83(5–6):713–719
7. Santhosh U, Ahmad J, Easler T, Gowayed Y (2021) A polymer infiltration and pyrolysis (PIP) process model for ceramic matrix composites (CMCs). J Am Ceram Soc 104:6108–6130
8. He R, Ding G, Zhang K, Li Y, Fang D (2019) Fabrication of SiC ceramic architectures using stereolithography combined with precursor infiltration and pyrolysis. Ceram Int 45(11):14006–14014
9. Bernan JJ (2000) Interfacial characterization of a slurry-cast melt-infiltrated SiC/SiC ceramic-matrix composite. Acta Mater 48(18–19):4619–4628
10. Delpouve H, Camus G, Jouannigot S et al (2022) Relationship between both thickness and degree of crystallization of BN interphases and the mechanical behavior of SiC/SiC composites. J Mater Sci 57:17661–17677
11. Murthy PLN, Nemeth NN, Brewer DN, Mital S (2008) Probabilistic analysis of a SiC/SiC ceramic matrix composite turbine vane. Compos B Eng 39(4):694–703
12. Brusseau ML (2019) Chapter 32–Sustainable development and other solutions to pollution and global change, Editor(s): Mark L. Brusseau, Ian L. Pepper, Charles P. Gerba, Environmental and pollution science (Third Edition), Academic Press, 585–603
13. Widheden J, Ringström E (2007) Life cycle assessment, Editor(s): Johansson I, Somasundaran P, Handbook for cleaning/decontamination of surfaces, Elsevier Science B.V., 695–720
14. Farjana SH, Mahmud MAP, Huda N (2021) Introduction to life cycle assessment, Editor(s): Farjana SH, Parvez Mahmud MA, Huda N, Life cycle assessment for sustainable mining, Elsevier, pp 1–13
15. Huijbregts MAJ et al (2016) A harmonized life cycle impact assessment method at midpoint and endpoint level Report I: Characterization. ReCiPe RIVM Report 2016–0104
16. Pollini B, Rognoli V (2021) Early-stage material selection based on life cycle approach: tools, obstacles and opportunities for design. Sustain Product Consump 28:1130–1139
17. Luo H, Luo RY, Wang LY, Huang P, Cui GY, Song JQ (2021) Effects of fabrication processes on the properties of SiC/SiC composites. Ceram Int 47(16):22669–22676
18. Shimoda K, Hinoki T, Katoh Y, Kohyama A (2009) Development of the tailored SiC/SiC composites by the combined fabrication process of ICVI and NITE methods. J Nucl Mater 384(2):103–108
19. Nannetti CA, Ortona A, de Pinto DA, Riccardi B (2004) Manufacturing SiC-Fiber-Reinforced SiC matrix composites by improved CVI/Slurry Infiltration/Polymer impregnation and pyrolysis. J Am Ceram Soc 87:1205–1209

20. Karadimas G, Salonitis K (2023) Ceramic matrix composites for aero-engine applications—a review. Appl Sci Article No. 3017, 13 (5)
21. Coltelti M, Lazzeri A (2019) Chemical vapour infiltration on composites and their applications chemical vapour deposition (CVD). Adv Technol Appl 363
22. Lamon J (2005) Chemical vapor infiltrated SiC/SiC composites (CVI SiC/SiC). In: Bansal NP (eds) Handbook of ceramic composites. Springer, Boston, MA
23. Jin L, Zhang K, Xu T, Zeng T, Cheng S (2018) The fabrication and mechanical properties of SiC/SiC composites prepared by SLS combined with PIP. Ceram Int 44(17):20992–20999
24. Luo Z, Zhou X, Yu J (2014) Mechanical properties of SiC/SiC composites by PIP process with a new precursor at elevated temperature. Mater Sci Eng, A 607:155–161
25. Gu J, Lee SH, Kim D, Lee HS, Kim JS (2021) Improved thermal stability of SiCf/SiC ceramic matrix composites fabricated by PIP process. Process Appl Ceramics 15(2):164–169
26. Luthra KL, Corman GS (2001) Melt infiltrated (MI) SiC/SiC composites for gas turbine applications book editor(s): Krenkel W, Naslain R, Schneider H 12
27. Mital SK (2009) Modeling of melt-infiltrated SiC/SiC composite properties NASA/TM—2009–215806

Scenario-Based Life Cycle Assessment of an Automotive Wire Harness

Felix Funk[(✉)], Huong Giang Nguyen, and Jörg Franke

Friedrich-Alexander-Universität Erlangen-Nürnberg, 91054 Erlangen, Germany
felix.funk@faps.fau.de

Abstract. Wire harnesses are one of the the largest purchased part for many automotive Original Equipment Manufacturers (OEMs). They are highly complex assembly products, consisting of hundreds of wires, electrical connectors, electronic parts, and ancillary materials such as tape. Assembly still is a largely manual process, as the complexity of the wire handling and low level of standardization and digitization pose a challenge to automation efforts. Wire harnesses are expected to increase in complexity and size due to current developments in the automotive sector, namely electromobility, autonomous driving, and digitalization. As a result, the importance of these products to the overall environmental impact of vehicles is likely to increase, with copper playing a significant role. In this study, the environmental impacts of a rear door wire harness are quantified by performing a Life Cycle Assessment (LCA), considering the life cycle from cradle to gate and based on different supply chain scenarios. The impact of sourcing and transport is shown and potential for improvement is derived, especially with regard to reshoring and automation.

Keywords: LCA · reshoring · wire harness · supply chain · sustainability

1 Introduction

The quantification of the environmental impacts originating from industrial production is essential to ensure sustainable business practices. Governments are continuously introducing legislation obliging companies to provide information about their carbon footprint, especially in the automotive sector. This view might prove myopic as other negative environmental impacts are potentially reinforced as an effect of Pareto-optimization for low CO_2 emissions. Life cycle assessment of industrial products, especially in the automotive sector, is hence becoming increasingly relevant. Being the largest and most complex purchased part in the automotive industry, wire harnesses present a particular challenge. Their overall importance is likely to increase along with the continuous electrification of previously mechanical vehicle features, the introduction of novel comfort features, and the ever-impending outlook of autonomous vehicles [11].

A parallel development in recent years has been the growing desire for supply chain resilience. With global disruptions occurring from 2020 to 2021, companies have become more considerate of their procurement processes. Reshoring has been identified as a

© The Author(s) 2025
H. Kohl et al. (Eds.): GCSM 2023, LNME, pp. 232–240, 2025.
https://doi.org/10.1007/978-3-031-77429-4_26

potential strategy to increase supply chain resilience. The practice usually includes contracting suppliers in geographical proximity to the final assembly and rebuilding domestic production capacities to replace previously offshored ones. It has also frequently been purported that reshoring contributes to product sustainability due to reduced transport distances. However, such claims often lack formal confirmation. [4, 5].

In this paper, the environmental impacts resulting from wire harness production as well as potential improvements through reshoring are investigated by conducting a comparative attributional life cycle assessment of a rear door automotive wire harness for three different supply chains. The paper is structured as follows: Sect. 2 introduces the product and relevant research. Then, the LCA steps as defined by ISO 14040 [6] are followed with the goal and scope definition (Sect. 3) life cycle inventory (Sect. 4), the impact assessment (Sect. 5), and the interpretation (Sect. 6).

2   Background

2.1   Wire Harness Manufacturing

Wire harnesses connect all electric and electrical components within vehicles, providing power transfer and signal communication. They consist of wires, connectors, fixing components, protective components, tape, fuses, relay boxes, terminals, seals, and switching units. Wire harnesses are assembled separately by specialized suppliers and then mounted in the vehicle at the OEM. In the case of German car manufacturers, the so-called "Kundenspezifische Kabelsatz" (customer-specific wire harness, KSK) is still the predominating form, which means that each wire harness is individually manufactured for the vehicle configuration defined by the customer. KSK manufacturing is conducted in three steps: pre-cutting, pre-assembly, and final assembly. The plants of the wire harness suppliers are distributed across Africa, Asia, and Eastern Europe [12].

An intermediary product of the largely manual manufacturing process is the assembly modules. These can be manufactured separately and can later be joined to the main harness at select interface points. This study examines such a module, namely the rear door wire harness of a medium-sized vehicle by German manufacturer Opel. It provides sufficient complexity and already features a majority of the types of parts that would be found in a full harness.

2.2   Life Cycle Assessment of Automotive Wire Harnesses

There have been few LCA studies for automotive wire harnesses. Abouljalil and Amrani have written two papers on the gate-to-gate LCA for an instrument panel wire harness manufactured in Morocco. In their first paper [1], they investigate the impacts of manufacturing based on information provided by the undisclosed manufacturer. They rely on mass material balances and do not integrate transport distances or energy consumption, however, cuttings and other production waste is considered [1].

In their second paper [2], Abouljalil and Amrani consider the same system but apply different impact assessment methods to it, namely Recipe, Impact2002, and EcoIndicator99. They identify large differences between the three, especially in the impact category of human toxicity, which is due to the exclusion and weighting of several key metals (Arsenic, Nickel, and Manganese) across the methods. While a wire harness is considered, the paper's main contribution is of methodological nature [2].

Villanueva-Rey et al. [13] investigate an "innovative cable solution" and its environmental impact. They consider a full harness, but neglect all plastic and electronic components, purely focusing on the metals used for wiring. The authors find significant improvement in the environmental impacts throughout all phases by using a thinner and lighter CuSn0.3 FLMRY-A instead of 0.35 mm2 Cu FLRY-A, mostly due to weight reductions in the use phase.

Alonso et al. [3] also aim to identify optimization potential through design choices. To this end, they present three different designs and perform a materials-based LCA [3]. While LCA of wiring harnesses has been previously investigated, transport within the supply chain has, to the best of our knowledge, not been considered. Similarly, the environmental ramifications of reshoring for the wire harness industry have not been quantified.

3 Scope and Goal Definition

This study follows the methodology outlined in ISO 14040 [6] and ISO 14044 [7] and uses parts of the ILCD handbook [10] for reference. The target is to compare the effects of sourcing options and subsequent transport modalities for an automotive wire harness. Due to the focus on the supply chain, the usage and end-of-life stages are excluded, making it a cradle-to-gate study. Similarly, the emissions at the plants themselves are neglected, as the manufacturing process is largely manual and no data with sufficient resolution is available. The study uses an attributional model and can be categorized as situation C1 in the ILCD handbook. The functional unit is described qualitatively as "providing all energy and signal transmissions necessary in the rear door of a private consumer vehicle". The reference flow is the production of one rear door wire harness. Three sourcing scenarios are investigated: a global supply chain, a European supply chain, and a reshored supply chain in Germany (since Opel is the Original Equipment Manufacturer (OEM)).

4 Life Cycle Inventory

4.1 Material Mass Balance

For the inventory, a real harness is disassembled completely. Thermoplastic materials can be identified easily, as the material name is printed on most components. Other parts, such as tapes and terminals, are looked up in catalogues where material combinations are also listed. All parts are weighed to generate a material mass balance. Materials contributing less than 1 g are below the cutoff threshold. The cable mass is increased by 5% to account for clippings and stripping in the pre-cutting stages. For tape and filling materials, deviations may arise in the manual application. However, uncertainty is excluded, as only an individual harness is considered for which the balance is known. The inventory can be seen in Table 1.

Table 1. Material mass balance of a rear door wire harness

Material	Parts	Amount [g]
Copper and insulation	Wires	166.32
Brass	Terminals	10.92
EPDM	Rubber tubing	83.70
PA 6 GF 15	Tubing fixation	8.82
PA 66	Wire fixation	9.50
PA 66 GF 35	Plugs	13.46
PBT GF 15	Plugs	30.70
PET Fleece	Dampening tape	5.48
PVC	Bundling tape	6.89
Silicone	Adhesive mass	5.02

Material data is taken from EcoInvent 3.9, as it provides by far the most extensive list of flows per process [8]. EcoInvent provides a process for glass-fibre reinforced, injection moulded Nylon, but not for other processes. As the flows of that process are similar, it is recreated for other plastics based on the assumption that the differences in temperature throughout the process can be neglected.

4.2 Sourcing Scenarios

To quantify the impact of reshoring, three sourcing scenarios are developed, with wires being the only component that is always procured from the world market, for which there is a corresponding process in EcoInvent 3.9. For plug and furth component procurement as well as harness assembly, the location and the corresponding transport modes and distances are varied. All places are based on real plants of manufacturers providing the components.

In the global scenario, plugs are manufactured in Shanghai, China, and transported via an undefined type of lorry as well as sea freight to the assembly plant in Cape Town, South Africa. Other components are produced in Suzhou, China, and transported similarly. All components are produced using global (GLO) manufacturing processes defined in EcoInvent. The assembled harnesses are shipped from Cape Town to Rotterdam and then by EURO 5 lorry to the final assembly in Rüsselsdorf, Germany.

In the European scenario, plugs are manufactured in Brno, Czech Republic, and transported via EURO 5 lorry to the assembly plant in Timisoara, Romania. Other components are produced in Sparta, Greece, and transported similarly. All components are produced using European (RER) manufacturing processes defined in EcoInvent. The assembled harnesses are transported from Timisoara by EURO 5 lorry to the final assembly in Rüsselsdorf, Germany.

In the reshored scenario, plugs are manufactured in Dinkelsbühl, Germany, and transported via EURO 5 lorry to the assembly plant in Düsseldorf, Germany. Other components are produced in Hamburg, Germany, and transported similarly. All components are produced using European (RER) manufacturing processes defined in EcoInvent. The assembled harnesses are transported from Düsseldorf by EURO 5 lorry to the final assembly in Rüsselsdorf, Germany. Table 2 gives an overview of the sourcing scenarios and the transport distances (d).

Table 2. Sourcing scenarios, transport distances, and modes

Part	Parameter	Global	European	Reshored
Wires	Location	Global cable market		
Plugs	Location	Shanghai	Brno	Dinkelsbühl
	Lorry d. [km]	50	633	449
	Ship d. [km]	17131	0	0
Other parts	Location	Suzhou	Sparta	Hamburg
	Lorry d. [km]	50	1399	401
	Ship d. [km]	17131	0	0
Harness assembly	Location	Cape Town	Timisoara	Düsseldorf
	Lorry d. [km]	440	1290	214
	Ship d. [km]	12111	0	0
Final assembly	Location	Rüsselsheim, Germany		

5 Impact Assessment

The impact assessment is conducted in OpenLCA [9], a free software tool which integrates EcoInvent 3.9 as well as several impact assessment methods. Each harness sourcing scenario is modeled as a product system and all three are merged into a project. CML (after the Centrum voor Milieukunde at Leiden University) is chosen as the impact assessment method and all midpoint impact categories are computed, including eutrophication (EU), terrestrial ecotoxicity (TE), marine ecotoxicity (ME), freshwater ecotoxicity (FE), photochemical oxidant formation (PCOF), human toxicity (HT), elemental abiotic depletion (EAD), fossil fuel abiotic depletion (FAD), climate change (CC), ozone depletion (OD), and acidification (A). The absolute values for all three sourcing scenarios are shown in Table 3.

To better compare the different scenarios, the relations between them are considered. The highest value of each impact category is set as 100% and the ratio of the other two scenarios is computed. The results can be seen in Fig. 1.

Table 3. Environmental impacts by scenario and category

Impact category	Unit	Europe	Global	Reshored
EU	10^{-3} kg PO4-Eq	19.018	19.160	18.948
TE	10^{-3} kg 1,4-DCB-Eq	228.023	228.666	227.664
ME	kg 1,4-DCB-Eq	40311	40349	40271
FE	kg 1,4-DCB-Eq	34.342	34.329	34.321
PCOF	10^{-3} kg eth-Eq	2.784	2.856	2.759
HAT	kg 1,4-DCB-Eq	84.862	84.947	84.797
EAD	10^{-4} kg Sb-Eq	8.846	8.843	8.842
FAD	MJ	29.687	29.403	28.117
CC	kg CO_2-Eq	1.879	1.895	1.766
OD	10^{-5} kg CFC-11-Eq	0.035	0.035	0.035
A	10^{-4} kg SO_2-Eq	63.543	65.219	63.273

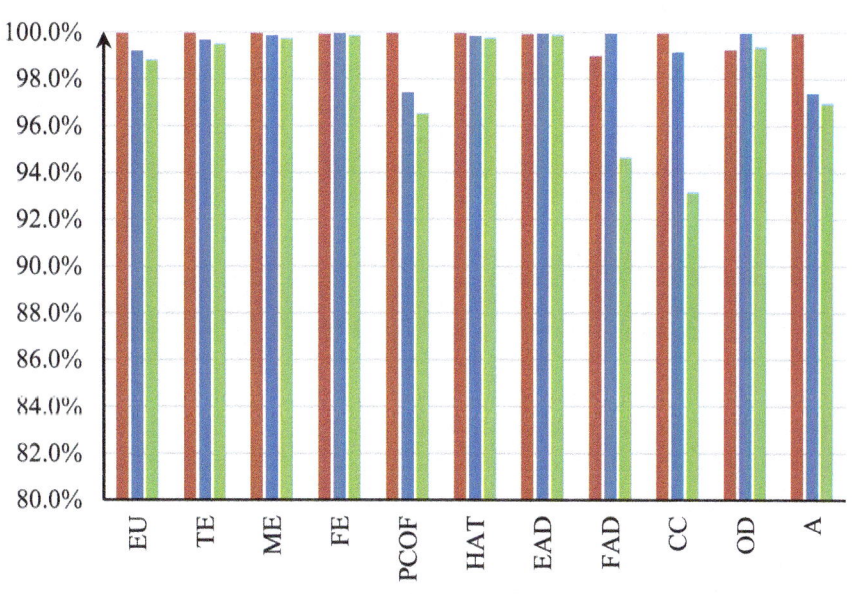

Fig. 1. Relative environmental impacts by category

6 Interpretation

6.1 Scenario Comparison

The magnitude of the differences between the sourcing scenarios varies between 0.04% and 6.78% across the impact categories. Ecotoxicity and elemental abiotic resource depletion are nearly unaffected, which can be explained by the fact that mining activities, which are the main contributors, barely change between the three scenarios. Eutrophication and ozone depletion also only show small differences of around 1%. While the global supply chain shows the highest impacts in most categories, the European scenario exceeds the global one in the case of fossil fuel depletion and ozone depletion. The change is the result of the significant increase in lorry transport, which is considerably more fuel intensive per ton-kilometer than sea freight by container ship. Larger differences are noticeable in the other categories. Photochemical oxidant formation shows an impact reduction of nearly 4% from the global to the reshored scenario, and 2.5% for the European scenario. Fossil fuel depletion is also significant, with a 5% reduction for the reshored, and a 1% reduction for the global case when compared to the European alternative. The largest difference can be observed for the climate change category, where the European case delivers a 0.8% improvement, whereas the reshored supply chain reduces emissions by nearly 7%. In terms of acidification, both the European and the reshored scenario show notable improvements of 2.6% and 3% respectively. Extrapolating the findings to a full wire harness is difficult, as the composition varies depending on the investigated submodule. Still, it is worthwhile mentioning that an expansion of this wire harness to the typical full harness weight of 20 to 30 kg results in the emission of 35.32 to 52.98 kg CO_2eq.

6.2 Completenes and Consistency

This study considered a module of a real wiring harness. Consistency was ensured by relying exclusively on EcoInvent and thus maintaining a high level of data quality and accuracy. The procedure itself, i.e. the disassembly of a real harness, ensured completeness as no part or material was neglected.

6.3 Discussion

The reshored supply chain is an ecological improvement across all categories. Especially in terms of climate change, significant gains can be made. Merely eliminating long-haul sea freight without considering the resulting road transport, however, is not a sensible solution. Whether the reduction in CO_2 emissions is sufficient to motivate reshoring efforts remains to be seen. As production is still largely manual, direct reshoring is unlikely, since labor wages are too high in the OEM countries. There have been consistent automation efforts in recent years, which could potentially tip the economic scales in favor of reshoring. A recalculation of the LCA would then be necessary to fully account for the changes in machinery, infrastructure, and peripherals. The increased impacts resulting from these resource investments may not exceed the reductions achieved through the transport adjustments. Additionally, the reshored case presented in this study can be seen as close to ideal, as plug and component manufacturing was reshored along with the assembly, with only cables being procured from a global market. Based on the current distribution of production sites, it is highly unlikely that a singular reshoring of the final assembly entails a positive environmental balance. Still, it might be beneficial with regard to supply chain resilience.

Further studies could extend this LCA by also considering peripherals, machinery, infrastructure, and tracking the interconnections between Tier 2 and Tier 3 suppliers. Furthermore, cable sourcing and packaging could be varied.

Acknowledgements. This Project is supported by the Federal Ministry for Economic Affairs and Climate Action (BMWK) on the basis of a decision by the German Bundestag.

References

1. Aboujlalil H, Amrani M (2020) Life cycle assessment of manufacturing processes of a cable harness in Morocco. J Env Sci Toxicol Food Technol 14
2. Aboujlalil H, Amrani M (2022) Life cycle assessment: comparative analysis of an electric wiring harness using different impact methods. Int J Sci Res 11
3. Alonso JC, Dose J, Fleischer G et al (2007) Electrical and electronic components in the automotive sector: economic and environmental assessment. Int J Life Cycle Assess 12:328–335. https://doi.org/10.1065/lca2006.08.263
4. Azmeh S, Nguyen H, Kuhn M (2022) Automation and industrialisation through global value chains: North Africa in the German automotive wiring harness industry. Struct Chang Econ Dyn 63:125–138. https://doi.org/10.1016/j.strueco.2022.09.006
5. Cosimato S, Vona R (2021) Digital innovation for the sustainability of reshoring strategies: a literature review. Sustainability 13:7601. https://doi.org/10.3390/su13147601
6. DIN EN ISO (2021) Environmental management–Life cycle assessment–Principles and framework (14040)
7. DIN EN ISO (2021) Environmental management - Life cycle assessment - Requirements and guidelines(14044)
8. Ecoinvent (2023) EcoInvent
9. GreenDelta GmbH (2023) OpenLCA
10. International reference life cycle data system (2010) ILCD Handbook. General guide for Life Cycle Assessment–Detailed guidance. https://eplca.jrc.ec.europa.eu/uploads/ILCD-Handbook-General-guide-for-LCA-DETAILED-GUIDANCE-12March2010-ISBN-fin-v1.0-EN.pdf. Accessed 15 Jun 2023
11. Kuhn M, Ngyuen HG (2019) The future of harness development and manufacturing. Results from an expert case study. https://www.researchgate.net/publication/336512177_The_future_of_harness_development_and_manufacturing_-_Results_from_an_expert_case_study
12. Nguyen HG, Kuhn M, Franke J (2021) Manufacturing automation for automotive wiring harnesses. Procedia CIRP 97:379–384. https://doi.org/10.1016/j.procir.2020.05.254
13. Villanueva-Rey P, Belo S, Quinteiro P et al (2018) Wiring in the automobile industry: life cycle assessment of an innovative cable solution. J Clean Prod 204:237–246. https://doi.org/10.1016/j.jclepro.2018.09.017

Feasibility Study and Economic Evaluation of Direct Contact Prelithiation of Lithium-Ion Batteries

Benedikt Stumper[✉] [iD], Henning Südfeld, Felix Diller, and Rüdiger Daub

TUM School of Engineering and Design, Department of Mechanical Engineering, Institute for Machine Tools and Industrial Management, Technical University of Munich, Boltzmannstr. 15, 85748 Garching, Germany
benedikt.stumper@iwb.tum.de

Abstract. The increasing need for energy storage across various sectors drives the growing demand for high-performance lithium-ion batteries. Prelithiation has emerged as a promising approach to enhance the capacity and cycle life of lithium-ion batteries. This study investigates the technical feasibility and economic viability of a scalable, roll-to-roll prelithiation process. Using an evaluation methodology adapted from literature, a comprehensive feasibility study and economic evaluation were conducted, considering the industrial-scale implementation of the process. The evaluation demonstrates that the direct contact prelithiation process utilizing lithium foil is technically feasible and suitable for industrial application. The economic viability of the process heavily relies on the lithium price, which constitutes a significant cost factor. The direct contact prelithiation process offers substantial benefits in terms of capacity and cycle life, resulting in enhanced cost-effectiveness of prelithiated lithium-ion batteries. The feasibility study and economic evaluation provide valuable insights implementing direct contact prelithiation as a viable strategy to improve lithium-ion battery performance. The findings contribute to advancing battery technology, manufacturing processes, and overall economic efficiency for diverse applications.

Keywords: Prelithiation · Lithium-ion battery · Battery production · Feasibility study · Economic evaluation

1 Introduction

Lithium-ion batteries (LIBs) have revolutionized the field of energy storage for mobile applications leading to an expected surge in production capacity in the coming years, primarily driven by the growing demand for electric vehicles [1, 2]. Improvements regarding energy density, cycle life, and costs are crucial to meet the increasing demand for LIB [3]. These improvements can be achieved through novel, high-capacity materials as well as innovative manufacturing concepts, such as the prelithiation of LIBs. Prelithiation is an approach in which the addition of extra lithium to the battery cell during its manufacturing can compensate for the capacity losses that occur during the

H. Kohl et al. (Eds.): GCSM 2023, LNME, pp. 241–249, 2025.
https://doi.org/10.1007/978-3-031-77429-4_27

formation and operation of the LIB, resulting in an increased energy density and cycle life [4, 5]. Silicon-containing anodes, in particular, lead to significant capacity losses due to the substantial volume changes of silicon particles during charging and discharging processes [4, 6]. Therefore, the utilization of prelithiation is essential for the deployment of silicon-based anodes.

The present publication investigates the technical feasibility of direct contact prelithiation for an industrial application and evaluating its economic viability. The assessment is carried out using a roll-to-roll process for applying metallic lithium foil to the anode for direct contact prelithiation [7]. An existing evaluation approach is expanded and customized to assess the direct contact prelithiation process. The adapted approach includes the evaluation aspects of technology, manufacturing, and economic feasibility, and employs a comprehensive assessment that incorporates both qualitative and quantitative evaluation factors.

2 Methodology

The present study utilizes a modification of the evaluation methodology introduced by Rummel [8], tailored to suit the analysis and evaluation requirements of the direct contact prelithiation process. To adapt the original product-based perspective of the methodology to a practical process-oriented viewpoint, the approach focused on three evaluation perspectives: technology, manufacturing, and economic viability. Figure 1 illustrates the adjusted framework. Each perspective can be assessed independently, with the resulting evaluations integrated into an overall assessment of the examined process.

The initial step involves evaluating the *technology perspective*, wherein the performance of the prelithiation process is assessed by comparing the achievable capacities and cycle lives of prelithiated lithium-ion batteries to those without prelithiation. This evaluation determines the technical maturity of the technology.

The subsequent phase of the study focuses on assessing the *manufacturing perspective*. Specific stages within battery production that are affected by the implementation of direct contact prelithiation are identified. Subsequently, the effects on these stages are described, and appropriate manufacturing methods are suggested to mitigate these effects. The result of this evaluation yields a clearly defined sequence of processes, which undergoes assessment to ascertain its maturity and feasibility battery production.

In the third perspective of the study, the assessment of *economic viability* is conducted. This evaluation involves the utilization of a cost model to quantify the associated costs. Specifically, the study compares the calculated capacity-specific costs for prelithiated and non-prelithiated cells as an initial analysis. This comparison considers the additional costs and capacity enhancements related to prelithiated cells. Subsequently, the capacity-specific costs are correlated with the achievable cycle life. The resulting parameters are compared between prelithiated and non-prelithiated cells. By analyzing these parameters, the economic efficiency of the two cell types can be determined.

Finally, a comprehensive qualitative conclusion is drawn based on the results from these three evaluation perspectives to assess the overall feasibility of the considered prelithiation process.

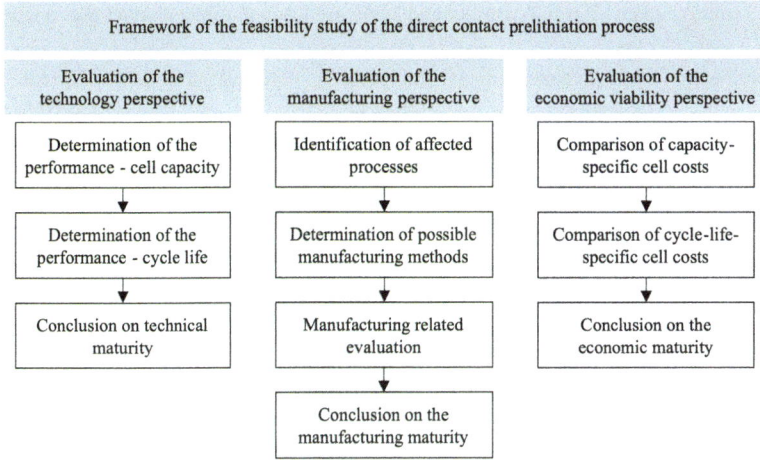

Fig. 1. Framework of the feasibility study (based on Rummel [8]).

3 Feasibility Study

The presented evaluation approach is applied to the direct contact prelithiation process according to [7]. For the necessary experimental data on the performance increase by direct contact prelithiation of lithium-ion batteries, results presented in [5] are used.

3.1 Technology Perspective

Experimentally gathered data from cell tests were utilized to assess the technological perspective. The investigations focused on single-layer pouch cells with NMC622 cathodes and silicon-graphite composite anodes. For precise compositions, dimensions of the components, and manufacturing steps see [5]. The prelithiated cells were prepared using two different prelithiation setups and compared with non-prelithiated reference cells. For the two prelithiation setups, varying amounts of lithium foil were introduced into the cells. For the first prelithiation setup (Compensation), lithium was introduced to compensate for initial capacity losses during the formation process. This amount corresponded to approx. 12% of the total anode capacity. In the second prelithiation setup (Reservoir), additional lithium was introduced to establish a lithium reservoir, compensating for further losses during operation and extending cycle life. This amount corresponded to approx. 28% of the total anode capacity of the reference cell. The related values for the initial capacity of the full cells and the cycle life achieved are shown in Table 1.

The consideration of the respective research results has shown that the technology of direct contact prelithiation with lithium metal foil enables an improvement of the considered cell characteristics. For the described cell setup, an increase in the initial discharge capacity of up to 9% and especially an extension of the cycle life of up to 151% can be achieved. The technological feasibility of the considered direct contact prelithiation is thus given.

Table 1. Improvement of initial capacity and cycle life through direct contact prelithiation [5].

Cell identifier	Initial capacity in mAh/g$_{AM}$	Increase to reference	Cycles at EoL	Increase to reference
Reference	143.12	–	49	–
Compensation	156.35	≈ 9%	68	≈ 39%
Reservoir	153.20	≈ 7%	123	≈ 151%

AM: active material; EoL: end of life (corresponds to dropping below 80% of the initial capacity)

3.2   Manufacturing Perspective

In assessing the manufacturing perspective, the initial stage involves identifying the segments of battery production that are affected by the investigated prelithiation process. In the examined process, the application of lithium foil takes place after the manufacturing of electrodes using a roll-to-roll approach. Consequently, the processes within electrode production remain unaltered. The affected production areas are primarily situated within cell assembly and finalization. The evaluation concentrates on the effort necessary for integrating direct contact prelithiation into battery production. This assessment is based on three specific criteria, namely the requirement for additional space, the increased complexity of processes and equipment, and the extended process and lead times. The necessary changes within the respective production areas are qualitatively evaluated based on literature and expert knowledge of the Institute for Machine Tools and Industrial Management of the Technical University of Munich. The affected processes and production areas which have been selected for evaluation as well as the qualitative evaluation are listed in Table 2. For the evaluation it is assumed that the lithium foil is applied after the anode calendering and is purchased in the required thickness, although the provision of lithium foils in the necessary thickness (< 20 μm) continues to be a challenge [9].

In the direct contact prelithiation process under consideration, the lithium foil is applied by an additional calender [7], which results in increased space requirement and process time. The additional use of carrier foils in the process also increases the complexity of the system compared to a standard calendering system. Nevertheless, there are no foreseeable hindrances from a plant and process perspective for implementing the application on an industrial scale. Separating electrodes by laser cutting is already an established process suitable for industrial applications [10]. The utilization of existing laser systems for cutting lithium has demonstrated promising potential for industrial-scale applications [11]. This suggests that existing plant technology can be utilized for electrode separation with certain required adjustments to accommodate the cutting of the multi-layer system (electrode, lithium foil, carrier foil). As a result, there is an increase in the complexity of the equipment for the separation process.

Removing of the carrier foil occurs within the stacking process, allowing for the utilization of existing single-sheet stacking systems. Moreover, there are already established concepts for handling multi-layer systems [12] and corresponding gripper technologies designed for lithium foils [13]. The adaption of the stacking process results in minimal

Table 2. Qualitative evaluation of the impact on the affected production areas.

Production area	Space requirement	Complexity of the process technology	Process time
Lithium application	↑	↑	↑
Separation	=	↑	=
Stacking	↑	↑	=
Formation	↑↑	=	↑↑
Atmosphere	↑	=	=

↑↑: strong increase; ↑: low increase; =: no change; ↓: low reduction; ↓↓: strong reduction

added complexity and space requirements. The process equipment required for monitoring prelithiation does not impose any additional demands, allowing the utilization of the same equipment employed in the formation process. However, due to the extended process duration, which increases from approximately 48 to around 90 h or even more, there is a substantial impact on the overall process time and on the space requirements due to the need for additional formation devices. To process lithium foil, it is recommended to maintain a dew point of at least −45 °C [7]. Considering the existing procedure of carrying out cell assembly within a dry room featuring a dew point of up to −60 °C [3], it is viable to incorporate the application of lithium foil within the same dry room. This integration would result in a marginal increase in space of the dry room due to the additional calendering system for lithium foil application.

The evaluation of the manufacturing perspective of direct contact prelithiation indicates the presence of limitations that are linked to external factors. The analysis demonstrates that technical remedies have been identified for all the recognized challenges; however, they have not undergone extensive testing on an industrial scale. Since there are existing approaches to address the challenges, the technical feasibility of implementing the prelithiation process is potentially given.

3.3 Economic Viability Perspective

For the economic evaluation, the BatPaC model [14] is used to calculate the costs of prelithiated and non-prelithiated LIBs. The model follows a comprehensive calculation approach and uses up-to-date calculation parameters given in the model. The Excel-based calculation tool was extended by a prelithiation dashboard, which summarizes the input and output values relevant to prelithiation. This allows adjustments to the cell properties and the affected processes, as well as considering the amount of lithium for prelithiation. The calculation is based on a battery factory scenario with an annual production volume of 50 GWh. The cost rate used for the lithium foil is 400 \$/kg [2], for the cathode material (NMC622 cathode) 25 \$/kg, and for the anode material (silicon-graphite composite anode with 10 wt% silicon) 15 \$/kg. The other values and cost factors were not changed. The necessary adjustments to the production areas can be referenced from Table 3.

The adjustments include additional calendering systems for applying the lithium foil, which are integrated into the dry room, making it larger and more expensive. It

Table 3. Adjusted values for the production areas influenced by the prelithiation process.

Identifier	Lithium application	Stacking	Formation	Dry room
Workload in h/year	138,000 (100%)	700,000 (0%)	1,050,000 (87.5%)	8,000 (0%)
Investment costs in m$	50 (100%)	255 (50%)	1,556 (87.5%)	203 (6.9%)
Required space in m²	4,200 (100%)	n/a	206,250 (87.5%)	65,070 (6.9%)

The percentages in brackets indicate the increase compared to the baseline factory

is also assumed that the increased complexity of the stacking process will increase the investment costs of the equipment. For the considered process, the prelithiation is monitored within the cell test systems of the formation, which significantly increases the process time and the number of formation systems. This increases space requirements, investments costs and workload for the formation step. As there is no increase in space requirements and process time for the separation process, no modified considerations are required in the calculation. The remaining values within the calculation tool remain unchanged. The calculated cell costs and their comparison are shown in Fig. 2.

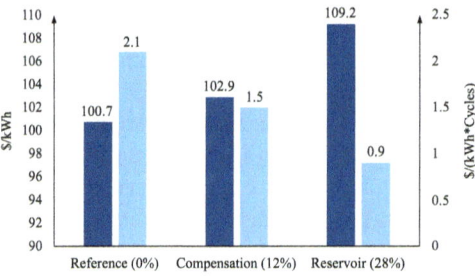

Fig. 2. Comparison of capacity-specific (dark blue) and cycle-life-specific (light blue) cell costs of the prelithiated (Compensation and Reservoir) and non-prelithiated (Reference) cells.

The analysis shows that the cell costs increase due to prelithiation, which is mainly attributed to the additional lithium required. Table 1 demonstrates that surpassing the compensation amount of lithium does not result in a proportional increase in capacity. Thus, higher levels of prelithiation lead to elevated cell costs without providing additional capacity benefits. However, when considering the connection between cell costs and cycle life, prelithiated cells prove to be more economically favorable than reference cells due to their extended cycle life. Figure 3 shows that material costs exert a significant impact on the overall cost increase, constituting approximately 80% of the total cell costs in this analysis. The increased process costs primarily arise from the additional requirement for formation capacities, contributing to nearly 90% of the increased process costs. The economic viability can thus be evaluated as established. Despite the current higher cell costs associated with capacity, there exists significant potential for improvement, as these costs are predominantly driven by material expenses and are currently inflated by a high lithium price. In terms of cycle-life-specific cell costs, prelithiated cells already demonstrate an economic advantage over non-prelithiated cells.

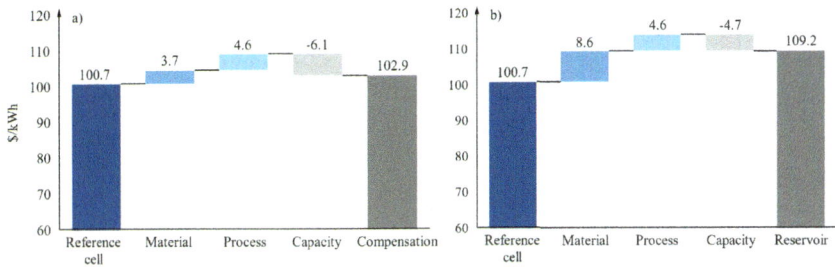

Fig. 3. Cost increase from reference cell to the respective prelithiated cell for the different prelithiation setups a) Compensation and b) Reservoir and the cost factors considered.

4   Overall Feasibility of the Direct Contact Prelithiation Process

From the combination of the individual evaluation perspectives for the investigated process and the considered experimental cell data, the overall feasibility of direct contact prelithiation is given. The findings from the technology perspective evaluation demonstrate that prelithiation leads to significantly improved cell capacity and cycle life, indicating a high level of technical maturity. Regarding the manufacturing perspective, it is evident that solutions exist for the required process-related changes, although they are not yet established for large-scale production. The implementation of direct contact prelithiation has a notable impact on battery manufacturing, which for now, hinders the maturity of the manufacturing perspective. When evaluating the economic viability of a prelithiation process, relying solely on the resulting capacity-specific cell costs is insufficient. To conduct a thorough evaluation of economic feasibility, it is crucial to include cycle-life-specific cell costs. By taking these costs into account, the economic efficiency of the process can be accurately determined and is given for the investigated process. The technical feasibility and economic viability of the investigated process for direct contact prelithiation are given under the considered conditions.

5   Conclusion and Outlook

This publication has presented a comprehensive feasibility study and economic evaluation of a direct contact prelithiation process for lithium-ion batteries with research-grade silicon-graphite composite anodes. The feasibility study demonstrated that direct contact prelithiation could be scaled up to an industrial scale using existing technical approaches. Moreover, the economic evaluation of direct-contact prelithiation is promising when cycle-life-specific costs are included in addition to capacity-specific costs. Due to the large share of material costs, future price developments for lithium foils have an immense impact on the widespread implementation of the process. Due to the increased cell costs, direct contact prelithiation is currently more suitable for special applications where the focus is on increasing battery performance and not primarily on cost reduction. Further research regarding the ideal prelithiation time and the shortening of the formation times through prelithiation also have significant leverage on costs, which may allow for better cost efficiency. An additional limitation arises from the current restricted availability of ultra-thin lithium foils, necessitating further investigation into the process of reducing the thickness of lithium foils.

References

1. Mauler L, Duffner F, Leker J (2021) Economies of scale in battery cell manufacturing: the impact of material and process innovations 286, p 116499
2. Schmuch R, Wagner R, Hörpel G, Placke T et al (2018) Performance and cost of materials for lithium-based rechargeable automotive batteries 3, p 267
3. Kwade A, Haselrieder W, Leithoff R, Modlinger A et al (2018) Current status and challenges for automotive battery production technologies 3, p 290
4. Chevrier VL, Liu L, Wohl R, Chandrasoma A et al (2018) Design and testing of prelithiated full cells with high silicon content 165, pp A1129–A1136
5. Stumper B, Mayr A, Mosler K, Kriegler J et al (2023) Investigation of the direct contact prelithiation of silicon-graphite composite anodes for lithium-ion batteries 170, p 60518
6. Goodenough JB, Park K-S (2013) The Li-ion rechargeable battery: a perspective. J Am Chem Soc 135:1167
7. Stumper B, Mayr A, Reinhart G (2020) Application of thin lithium foil for direct contact prelithiation of anodes within lithium-ion battery. Production 93:156
8. Rummel S (2014) Eine bewertungsbasierte Vorgehensweise zur Tauglichkeitsprüfung von Technologiekonzepten in der Technologieentwicklung. Universität Stuttgart
9. Stumper B, Dhom J, Schlosser L, Schreiner D et al (2022) Modeling of the lithium calendering process for direct contact prelithiation of lithium-ion batteries 107, p 984
10. Kriegler J, Binzer M, Zaeh MF (2021) Process strategies for laser cutting of electrodes in lithium-ion battery production 33, p 12006
11. Kriegler J, Duy Nguyen TM, Tomcic L, Hille L et al (2022) Processing of lithium metal for the production of post-lithium-ion batteries using a pulsed nanosecond fiber laser 15, p 100305
12. Konwitschny F, Schnell J, Reinhart G (2019) Handling cell components in the production of multi-layered large format all-solid-state batteries with lithium anode 81, p 1236
13. Fröhlich A, Gresens D, Vervoort B, Dröder K (2020) Design and evaluation of a material-adapted handling system for all-solid-state lithium-ion battery. Production 93:143
14. Knehr K, Kubal J, Nelson P, Ahmed S (2022) Battery Performance and Cost Modeling for Electric-Drive Vehicles (A Manual for BatPaC v5.0)

Life Cycle and Cost Assessment of CO₂ Assisted Hard Turning of AISI 52100 in Comparison to Conventional Cooling Techniques

Iñigo Llanos[1](✉), Iker Urresti[1], David Bilbatua[2], and Oier Zelaieta[1]

[1] Ideko–Basque Research and Technology Alliance (BRTA), Arriaga Industrialdea 2, 20870 Elgoibar, Spain
`illanos@ideko.es`
[2] Danobat S. Co op., Arriaga Industrialdea 21, 20870 Elgoibar, Spain

Abstract. The use of cryogenic CO_2 as a cutting fluid is a promising sustainable manufacturing technology capable of substituting oil-based emulsions during hard turning, showing a significant potential for both productivity improvement and environmental impact reduction. While previous works have focused on the assessment of the technical impact of the cryogenic CO_2 on the process and quality performance, the present paper faces the evaluation of both the environmental and economic impact of this manufacturing technique. By the application of Life Cycle Assessment (LCA) methodology and Return-on-Investment (ROI) calculations for different manufacturing scenarios, the environmental and economic performance of the cryogenic CO_2 cooling technique is evaluated in comparison to conventional cooling. Results indicate that the application of the cryogenic cooling can be successful depending on the manufacturing scenario. The performed analysis shows the possibility to identify the cases where the use of cryogenic CO_2 would outperform conventional cooling strategies, proving a valuable methodology for minimizing the environmental impact and ensuring the cost effectiveness during the design phase of industrial turn-key projects for hard-turning operations.

Keywords: First keyword · Second keyword · Third keyword

1 Introduction

The global manufacturing industry pushes towards higher competitiveness and sustainability requirements, increasing the demands for lower environmental impact solutions while increasing their flexibility and productivity. In this context, hard turning emerges as an advantageous process over costly and cutting fluid intensive grinding processes for the production of high performance parts [1, 2]. However, challenges concerning the hard turning process hinder its wide industrial implementation. The high thermal loads on the tools produce rapid tool wear of costly pCBN cutting tools [3]. In addition, cutting tool deterioration induce a detrimental surface integrity on the parts, decreasing drastically hard turning process reliability.

H. Kohl et al. (Eds.): GCSM 2023, LNME, pp. 250–258, 2025.
https://doi.org/10.1007/978-3-031-77429-4_28

Therefore, removing the heat from the cutting zone is essential to improve tool and process performance. Traditionally, mineral oil emulsions with high environmental impact are used for this purpose, which are also hazardous to health [4, 5]. Hence, traditional metal working fluid replacement is essential for environmentally friendly manufacturing process development, where cryogenic cooling techniques emerge as promising alternatives.

In this sense, multiple works reported advantages of implementing cryogenic cooling during hard turning. In [6], a remarkable tool life increase (up to 370%) of pCBN tools was reported when using cryogenic cooling in comparison to conventional and dry hard turning. Significant improvement of machined surface integrity by cryogenic cooling is also reported in bibliography [7]. Although cryogenic cooling presents a promising technology, still few commercial delivery systems are available in the market (e.g. Fusion Coolant Systems, BeCold® and ICEFLY®). In [8] a novel system enabling the stable and controlled LCO_2 delivery was presented, together with the benefits regarding part quality and tool wear in comparison to conventional flooding and blown air.

Also, the implementation of cryogenic cooling system in industrial machining installations is very limited. While multiple works reported specific process performance improvement due to cryogenic cooling [4–10], the actual environmental and economic benefits in industrial turn-key solutions is scarcely addressed in literature [11]. Moreover, the impact from cutting tools is usually not considered, even if the cooling techniques analysed induce significantly improvements in tool life [5, 10, 11].

In this paper, the environmental impact and cost evaluation for hard turning turn-key solutions is evaluated through the application of Life Cycle Assessment (LCA) and Return-on-Investment (ROI) calculations for different manufacturing scenarios. Such calculations are performed on data provided by [8] for hard turning of AISI 52100 with pCBN tools using different cutting fluids. Results indicate that LCO_2-assisted hard turning can improve the cost and environmental performance of solutions based on conventional cooling depending on the manufacturing scenario.

2 Life Cycle Assessment

In this study, a gate-to-gate LCA to compare the effect of using LCO_2 in comparison to blown air was performed. This way, only the impact from the consumables and energy consumption during the hard turning process are considered. The impact from building the turning machines and the different components is not taken into account. Neither is considered the material of the parts, as it is the same in both cases analysed.

The impact categories of Global Warming Potential [kg CO_2 eq], Acidification potential [kg SO_2 eq], Eutrophication potential [kg P eq], Ozone depletion potential [kg CFC-11 eq], net water consumption [m3] and smog potential [kg NMVOC eq] are considered as they are the most commonly used ones for LCAs of metals [13]. Next, the identification and quantification of the different process input flows for the selected indicators is performed.

2.1 Life Cycle Inventory

The evaluation of the impact from the cutting tools considered both the cemented carbide body and the pCBN tips. Based on the data provided by [13] for the manufacturing of WC-Co, the impact of the 16.02 gr of WC-Co per cutting tool considered in the present study was calculated. The impact from the pCBN tool tips was calculated in a two-step manner. For the generation of CBN granulates, [14] identified the high energy consumption during the CBN synthesis as the main environmental impact generator, providing the value of 2.19 kWh per CBN gram. The CBN granulates are processes by High-Pressure High-Temperature (HPHT) sintering to produce pCBN discs and then, cut by WEDM to generate the final tip. While no data was found available for this process on pCBN, the analogous process for PCD was analysed by [15], obtaining the value of 2.5 kWh per tool tip. Thus, considering 0.08 gr of pCBN for a tool tip, 2.675 kWh energy consumption was calculated for the manufacturing of each tool tip.

To evaluate the impact from the LCO_2, the CO_2 capture process must be considered [10]. In the last decade, the Direct Air Capture (DAC) technologies have evolved greatly as a mean to act against climate change by removing CO_2 from the atmosphere or CO_2 emissions. Thanks to this, the International Energy Agency (IEA) estimates that Liquid DAC technologies can capture and compress 1 ton CO_2 using 1.3 GJ (361 kWh) energy [16]. Also, the power consumption for the LCO_2 delivery is considered in this study, measuring 1.37 kW during the cutting tests when providing a stable 0.5 l/min LCO_2 mass flow. The use of LCO_2 is compared to blown air (7 bar compressed air), generated by a standard compressor with 22 kW power and a 3800 l/min flow capacity. Also, the power consumption by the turning machine during the hard turning process was considered in the present study. The cutting power was measured during cutting tests through the TRACE function of the Siemens Sinumeric 840D CNC, yielding average values of 575 W for the use of blown air and 580 W for the case of using LCO_2. These power values were used to calculate the electrical energy consumption during the machining processes based on the process duration.

To quantify the impact of the electrical energy consumption, the environmental impact from different electricity generation technologies per generated kWh in the Europe (EU28) region was considered [17]. Then, the electricity generation technology mix was taken from European Council data based on the Eurostat Dataset [18]. Table 1 shows the quantification of the environmental impact indicators for each of the different process input flows considered.

It is noteworthy the differences in the values from Table 1 in comparison to the ones reported in bibliography [10], mainly for power generation and LCO_2. The increase of the share for renewable power sources and the improvements in CO_2 capture and storage techniques during the last decade has led to significant reductions in the environmental impact for these process input flows, yielding the values reported here.

3   Return on Investment Calculations

The cost evaluation for using LCO_2 in comparison to blown air is performed considering a Turn-key solution. As so, the costs related to the equipment, tooling, hourly labour and electricity consumption were considered for the evaluation of the final part cost.

Table 1. Summary of the Life Cycle Inventory.

Impact indicator	1 kWh	Tool tip	1 kg LCO_2	1l/min air × min
Global warming potential [kg CO_2 eq]	2.48 e-1	7.30e-1	8.94e-2	4.59e-5
Acidification [kg SO_2 eq]	3.56e-1	3.15e-3	1.28e-4	6.59e-8
Freshwater eutrophication [Kg Peq]	8.07e-5	4.15e-4	2.91e-5	1.49e-8
Water depletion [m3]	5.77e-2	1.55e-1	2.08e-2	1.07e-5
Ozone depletion [kg CFC-11 eq]	1.72e-8	5.87e-8	6.19e-9	3.18e-12
Photochemical oxidant formation [kg NMVOC eq]	2.70e-4	1.36e-3	9.76e-5	5.00e-8

Defining a manufacturing scenario, yearly production and a part cost, yearly return-on-investment (ROI) values are obtained for an installation working 3 shifts, with an Overall Equipment Effectiveness (OEE) value of 85% and an equipment amortization timeline of 8 years. Besides the cutting time, the non-productive stages are considered, using a value of 1 min for the part handling and 0.5 min for tool change time. Next, the cost assessment for the different items considered in the calculations is detailed.

3.1 Cost Assessment

The cost of the pCBN tools was roughly 80€, with 4 usable cutting edges. The average hourly labour costs for the EU (€30.5) in 2022 [19] was considered in the calculations. Concerning electricity prices, a cost of 0.2104€ per KWh was used, the EU average price for non-household consumers in the second half of 2022 [20]. The cost used for 1 L LCO_2 was 1.5 €, the average of the cost when performing the tests in [8].

In the case of the equipment, an estimated cost of 500,000 € for a precision hard turning machine was employed. The price for the LCO_2 delivery system from [8] was roughly 40,000 €. In the case of the blown air, the cost of the compressor was not considered as it is a standard equipment that would not be directly related to the hard turning installation. These costs can be seen in a summarized manner in Table 2.

Table 2. Costs considered in the ROI calculations.

Item	Cost - €
Hard Turning machine	500,000
LCO_2 equipment	40,000
Tool edge	20
LCO_2 litre	1.5
Labour hour	30.5
1 kWh energy	0.2104

4 Experimental Analysis

The LCA and ROI calculations were performed for two test-cases based on actual turn-key solutions. Figure 1 shows the geometry for each test-case in mm, which was simplified to avoid proprietary issues. The considered test-cases are similar parts with different dimensions, where the finishing of the outer diameter must be performed achieving a mean surface roughness (Ra) of 0.2 μm. The first test-case is a small part manufactured in a high-throughput scenario with low part values, while the second one represents a bigger part with lower batch sizes and high added value. The values taken for these scenario conditions in the present study for both parts can be seen in Table 3.

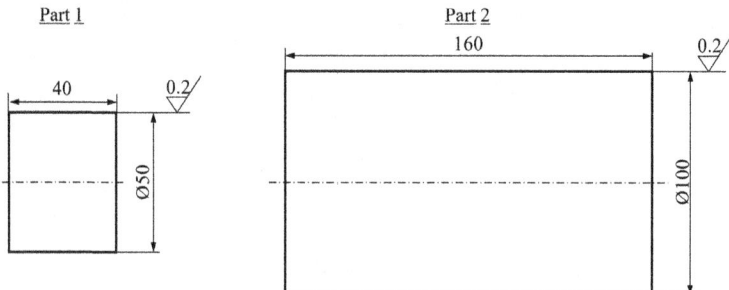

Fig. 1. Geometry of the parts analysed.

Table 3. Manufacturing conditions for both test-cases.

Condition	Part 1	Part 2
Yearly production - Parts	100,000	20,000
Part cost - €	5	35

4.1 Process Data

The analyses on the present paper are founded in the process, tool life and part quality data provided in [8] for hard turning of AISI 52100 with different cutting fluids (Table 4). While in [8] results for LCO_2, blown air and oil emulsions are analysed, only the use of LCO_2 and blown air yielded mean roughness values bellow 0.2 μm. Thus, only these cutting fluids are considered here.

5 Results

5.1 Test-Case 1

Figure 2 shows the part cost and ROI calculations for test-case 1. The part cost shows the contribution from each of the items considered. While the use of LCO_2 reduces the cost associated to the cutting tool, the higher cost from the LCO_2 yields a higher overall

Table 4. Process and tool life data from [8].

Cutting fluid	Vc – m/min	fv – mm/v	ap - mm	Tool Life
LCO$_2$	200	0.05	0.1	7000
Blown Air	200	0.05	0.1	4000

part cost. Also, the cost for the LCO_2 delivery equipment generates lower ROI when using LCO_2 in the turn-key installation. Figure 3 shows the results for the LCA impact indicators, with the contribution to each indicator from the cutting tools, cutting power and the cutting fluid. Notice the exponents used for each indicator to scale the figure. Analogously to the results obtained for the costs, the use of LCO_2 reduces the impact from the cutting tool usage, but the contribution of the LCO_2 is much higher.

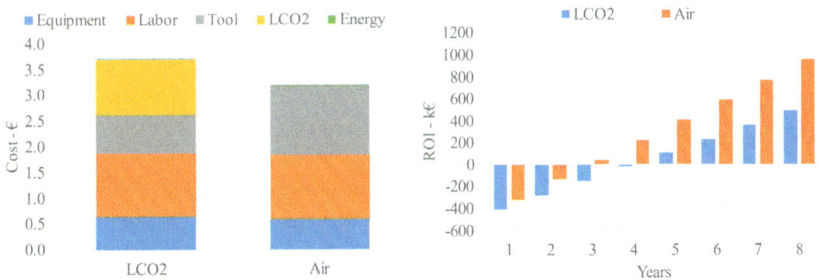

Fig. 2. Part cost and ROI calculations for test-case 1.

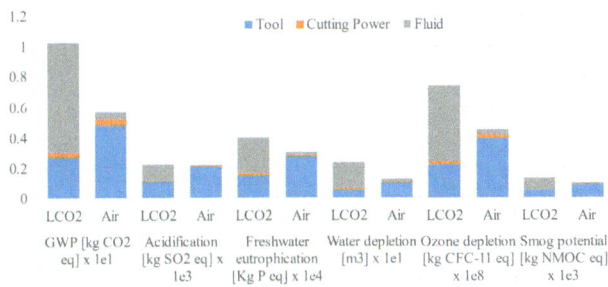

Fig. 3. Environmental impact indicator values for each manufactured part from test-case 1.

5.2 Test-Case 2

Figure 4 shows the cost and ROI results for test-case 2. The higher part dimension generates higher tool consumption and thus, higher cost associated to this item. This way, when using LCO_2 the overall part cost is reduced. This lower part cost yields higher ROI values for the use of LCO_2 as a cutting fluid.

Similar results are obtained for the environmental impact indictors on this test-case (Fig. 5). Except for water depletion, the impact from LCO_2 usage is lower than the reduction of the impact from tool usage when using LCO_2 during the hard turning process in comparison to blown air.

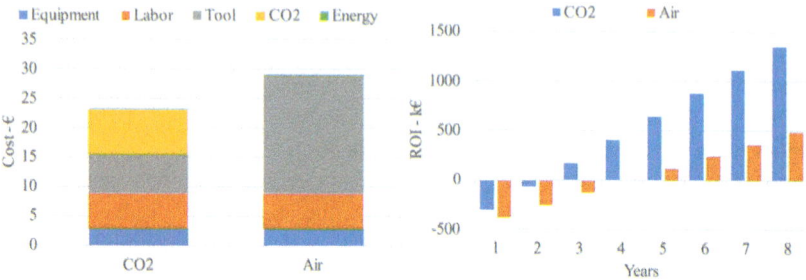

Fig. 4. Part cost and ROI calculations for test-case 2.

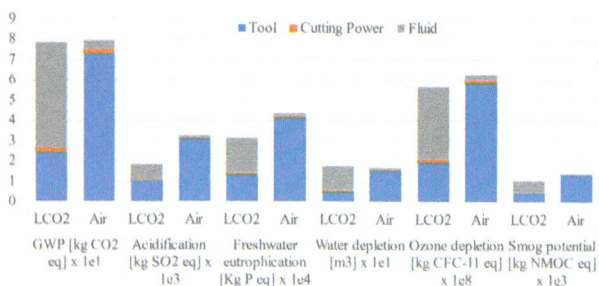

Fig. 5. Environmental impact indicator values for each manufactured part from test-case 2.

6   Conclusions

Next conclusions can be drawn from the work presented above:

- Thanks to the application of LCA and ROI calculations, it is possible to provide measurable results for the evaluation of replacing conventional cutting fluids by LCO_2 in industrial turn-key installations.
- While usually considered as a sustainable technology in literature, the use of cryogenic cutting fluids can generate higher environmental impact than conventional techniques depending on the application.
- The capability of a cutting fluid to outperform another one regarding productivity, cost and environmental impact will depend on the part shape and the manufacturing scenario to be applied.
- The lower environmental impact and process cost reduction during cryogenic assisted hard turning are related to increased tool life produced by the use of LCO_2 as a cutting fluid.

Acknowledgements. Authors are grateful to the Spanish government for the financial support provided through CRYOMACH (INNO-20182038), a Project funded under the SMART EUREKA CLUSTER on Advanced Manufacturing programme.

References

1. König W, Berktold A, Koch KF (1993) Turning versus grinding–a comparison of surface integrity aspects and attainable accuracies. CIRP Ann 42(1):33–43
2. Klocke F, Brinksmeier E, Weinert K (2005) Capability profile of hard cutting and grinding processes. CIRP Ann 54(2):22–45
3. M'Saoubi R, Johansson MP, Andersson JM (2013) Wear mechanisms of PVD-coated PCBN cutting tools. Wear 302(1–2):1219–1229
4. Jawahir IS (2016) Cryogenic manufacturing processes. CIRP Ann 65(2):713–736
5. Khanna N, Shah P, López de Lacalle LN, Rodríguez A, Pereira O (2021) In pursuit of sustainable cutting fluid strategy for machining Ti-6Al-4V. Sustain MaterTechnol 29:e00301
6. Biček M, Dumont F, Courbon C, Pušavec F, Rech J, Kopač J (2012) Cryogenic machining as an alternative turning process of normalized and hardened AISI 52100 bearing steel. J Mater Process Technol 212(12):2609–2618
7. Umbrello D, Micari F, Jawahir IS (2012) The effects of cryogenic cooling on surface integrity in hard machining: a comparison with dry machining. CIRP Ann 61(1):103–106
8. Llanos I, Urresti I, Bilbatua D, Zelaieta O (2023) Cryogenic CO_2 assisted hard turning of AISI 52100 with robust CO_2 delivery. J Manuf Processes 98:254–264
9. Shah P, Bhat P, Khanna N (2021) Life cycle assessment of drilling Inconel 718 using cryogenic cutting fluids while considering sustainability parameters. Sustainable Energy Technol Assess 43:100950
10. Pereira O, Rodríguez A, Fernández-Abia AI, Barreiro J, López de Lacalle LN (2016) Cryogenic and minimum quantity lubrication for an eco-efficiency turning of AISI 304. J Cleaner Prod 139:440–449
11. Pusavec F, Kramar D, Krajnik P, Kopac J (2010) Transitioning to sustainable production–part II: evaluation of sustainable machining technologies. J Cleaner Prod 18(12):1211–1221
12. LNCS homepage. http://www.springer.com/lncs. Accessed 21 Nov 2016
13. Furberg A, Arvidsson R, Molander S (2019) Environmental life cycle assessment of cement-ed Carbide (WC-Co) production. J Cleaner Prod 209:1126–1138
14. Winter M, Ibbotson S, Kara S, Herrmann C (2015) Life cycle assessment of cubic boron nitride grinding wheels. J Cleaner Prod 107:707–721
15. Furberg A, Fransson K, Zackrisson M, Larsson M, Arvidsson R (2020) Environmental and resource aspects of substituting cemented carbide with polycrystalline diamond: the case of machining tools. J Cleaner Prod 277:123577
16. IEA homepage. https://www.iea.org/energy-system/carbon-capture-utilisation-and-storage/direct-air-capture. Accessed 2023/10/02
17. United nations economic commission for Europe (2022) Carbon neutrality in the UNECE region: Integrated life-cycle assessment of electricity sources. United Nations, Geneva
18. European council homepage. https://www.consilium.europa.eu/en/infographics/how-is-eu-electricity-produced-and-sold/. Accessed 2023/10/02
19. Eurostat statistics explained, hourly labour costs. https://ec.europa.eu/eurostat/statistics-exp lained/index.php?title=Hourly_labour_costs. Accessed 18 July 2023
20. Eurostat statistics explained, electricity price statistics. https://ec.europa.eu/eurostat/statis tics-explained/index.php?title=Electricity_price_statistics#Electricity_prices_for_non-hou sehold_consumers. Accessed 2023/10/02

Social Valuation

The Role of Behavioural and Environmental Economics in Sustainable Manufacturing

Rashmeet Kaur$^{(\boxtimes)}$, John Patsavellas, and Konstantinos Salonitis

SMSC, Cranfield University, Bedford MK430AL, UK
rashmeet.kaur@cranfield.ac.uk

Abstract. Sustainable manufacturing is a rapidly growing field that primarily seeks to reduce the environmental impact of manufacturing processes. Although the three-lens approach of social, environmental, and economic aspects remain the primary focus in any sustainability study, the domains of behavioral, and environmental economics along with smart data technologies have not been used in a unified approach. Through a review of the state of the art, this paper establishes the individual cases for each one of these domains and underscores the research interest in their combinatorial application and possible complementary efficacy for advancing the development of sustainable manufacturing strategies. A research agenda involving comparative testing and the development of pertinent policies and interventions for sustainable manufacturing is proposed for the integration of behavioral economics and environmental economics, within the context of sustainable manufacturing.

Keywords: Sustainable Manufacturing · Behavioral Economics · Environmental Economics

1 Introduction

As the global manufacturing sector grapples with the challenge of advancing sustainable production, it has increasingly become desirable to identify suitable transformational approaches and tools that can comprehensively address all the dimensions of sustainability and facilitate a seamless transition towards a more sustainable future. One such transformational approach could be the deployment of behavioral economics (BE), which combines insights from psychology and economics to comprehend, predict, and influence the behavior of individuals and organizations towards more sustainable practices and outcomes. Similarly, environmental economics (EE) could offer a transformational approach through the systematic assessment of the economic ramifications of the environmental impact of contemporary manufacturing practices along with the costs and benefits of any improvement strategies aimed at reducing carbon emissions. This is depicted in Fig. 1.

© The Author(s) 2025
H. Kohl et al. (Eds.): GCSM 2023, LNME, pp. 261–268, 2025.
https://doi.org/10.1007/978-3-031-77429-4_29

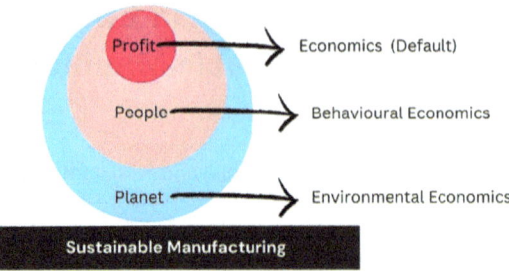

Fig. 1. Three Lens view of Sustainable Manufacturing

2 Behavioral Economics and Sustainable Manufacturing

Over the last few decades, the genre of BE has been developed and deployed through credible research methods to identify empirical laws that can accurately describe and influence human behavior when the neoclassical economics assumptions of linear cause-and-effect factors in human behavior around customer satisfaction and producer profit prove inadequate or invalid [1, 2]. It has also been used to investigate the implications of deviations from neoclassical human behavior assumptions for the functioning of broader economic systems, institutions, and public policy [3, 4]. It seeks to provide empirical evidence about the structure and content of the economics utility function, or any construct that may replace it, in a valid behavioral theory to enhance the accuracy of predictions about human economic behavior [5, 6].

The keyword combination phrase of "Behavioral economics" was used to search for relevant literature on three large academic databases: Scopus, Web of Science, and IEEE Xplore, with an open time filter to capture the maximum number of works. Initial title-filtering to identify and exclude duplicates was carried out both horizontally and vertically, i.e., within each database and across all three. Further screening of papers was done by title and abstract content, to classify them into industry-sector relevance groupings. A systematic literature analysis of this domain of knowledge as it relates specifically to the manufacturing sector (using the PRISMA statement [7]) yielded 16 research papers as shown in Fig. 2.

Fig. 2. PRISMA statement summary for BE

The taxonomy of research endeavors across the spectrum of human activity, as depicted in Fig. 3, aptly demonstrates a substantial body of scholarly work in the realm of BE. There is a category labelled as agnostic in Fig. 3, which covers a broad range of concepts, theories and tools that constitute the BE toolkit and are not about the implementation of BE toolkit in different sectors.

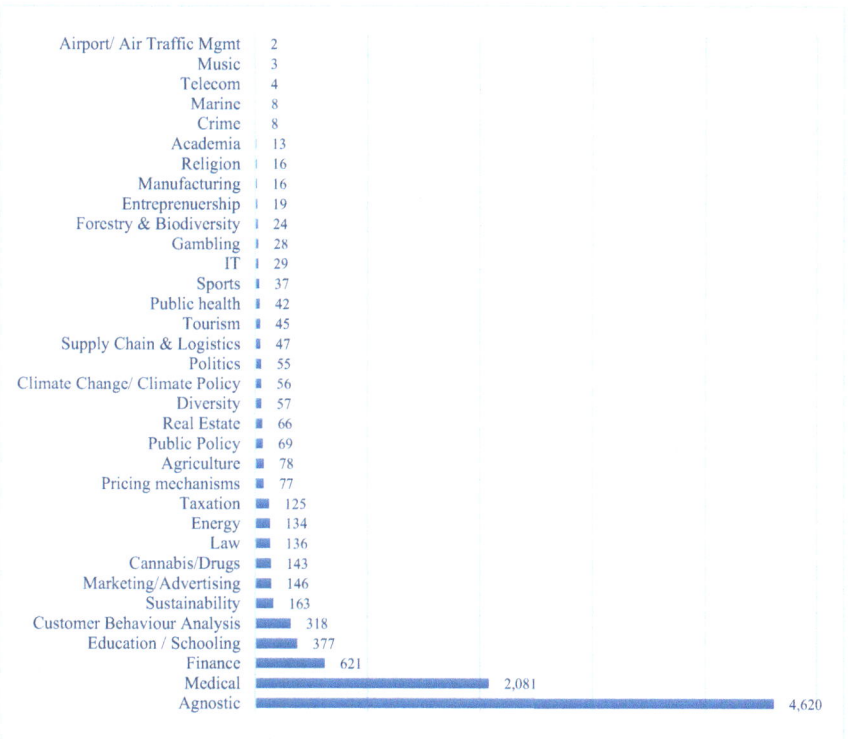

Fig. 3. Industry wise use of BE

Among the various sectors represented, the medical and healthcare domains emerge as the primary areas of focus, showcasing the highest prevalence of implementation and utilization of BE principles. Following suit, the field of finance exhibits a noteworthy presence in terms of adoption and practical application of behavioural economics concepts. However, full text analysis of the 16 papers referencing manufacturing in their titles revealed that only 1 paper demonstrates using applying a BE theory for optimisation of their predictive maintenance optimisation [8]. The themes represented in the remaining 15 papers are focused on human-robot collaboration and the impact of technology on human decision-making processes along with behaviourally informed financial decision making and pricing strategies. Evidently, the findings suggest a dearth of research work utilizing BE in the various aspects of manufacturing. Notably, substantial research gaps exist in important domains such as end-to-end supply chain management, integrated manufacturing systems, as well as the emerging realm of digital manufacturing, among others that could warrant further investigation. Only the sub-set of prospect theory which suggests that people tend to be risk-averse when it comes to potential gains but risk-seeking when facing potential losses has been used in the context of predictive maintenance for risk assessment [8] whilst other popular BE theories such as rational choice theory, libertarian paternalism, and choice architecture, have yet to be examined in the context of manufacturing. The absence of research on the application of these theories presents a notable gap in knowledge that warrants attention, specifically within the context of sustainable manufacturing systems.

3   Environmental Economics and Sustainable Manufacturing

Environmental economists use a variety of methods to study the environment, including cost-benefit analysis, environmental impact assessment, and natural resource economics [9, 10]. EE has grown rapidly in recent years due in part to the increasing awareness of environmental problems, such as climate change, pollution, and resource depletion [10, 11]. Developing policies and solutions that can help to address these problems is the perpetual target of economists in this field [12, 13]. An additional systematic literature review was conducted to identify the combined use of EE and BE specifically in manufacturing. The outcomes of the PRISMA literature analysis of the entire literature pool with no time filter using the results of keyword for search "environmental economics" are summarized in Fig. 4.

Fig. 4. PRISMA statement summary for EE

Figure 5 shows the top-20 themes from a detailed thematic analysis of the diverse areas of research focus within EE. A total of 197 papers were identified to be directly or indirectly associated with manufacturing. However, a detailed analysis of these papers clearly reveals that there is no research work combining BE and EE in the context of manufacturing or sustainable manufacturing.

While the combination of BE and EE has been explored in relation to sustainability from a consumer perspective, based on this analysis there is an absence of work with regards to human decision-making models that relate to adopting sustainable practices in manufacturing organizations. This offers an opportunity to examine the potential of BE in combination with environmental costing, sustainable performance measurement and economic analysis for better outcomes in the drive for sustainable transformation of manufacturing.

4   Discussion

The combination of BE and EE within the realm of sustainable manufacturing refers to the converging interaction of these two disciplines towards advancing sustainable outcomes in manufacturing organizations. It acknowledges that addressing sustainability challenges in manufacturing necessitates an understanding of both human behavior and the environmental impacts of industrial activities. BE focuses on analyzing the

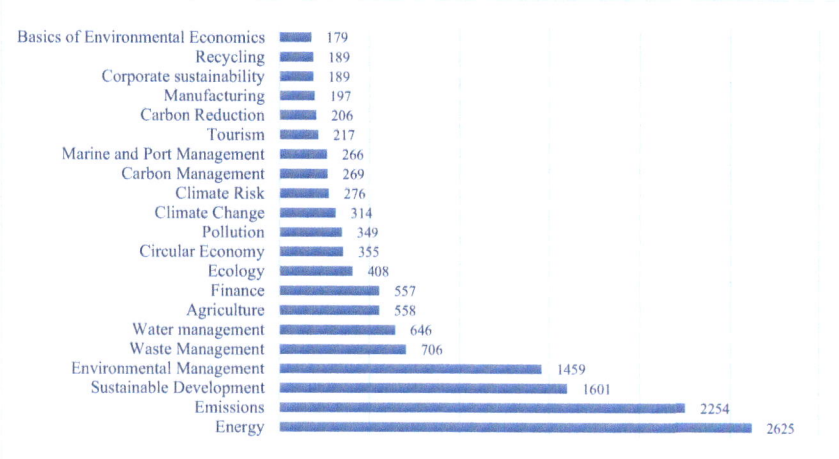

Fig. 5. Top 20 themes in EE literature

decision-making and actions of individuals and organizations within economic contexts, acknowledging that human behavior is not always rational or solely driven by economic considerations. In the context of sustainable manufacturing, the theories of behavioral economics can aid in identifying and comprehending the factors that influence decision-making processes related to environmental sustainability. This may encompass studying how individuals or organizations perceive and respond to environmental risks, how they prioritize sustainability in their decision-making, and which behavioral interventions can be employed to foster and even "nudge" manufacturing professional towards more sustainable practices. The main tools, concepts, and theories from the BE body of knowledge that could be critically examined for this purpose are rational choice theory, anchoring, prospect theory, loss aversion, status quo bias, mental accounting, framing, choice architecture, reciprocity, nudging and libertarian paternalism [3, 14–16]. Given the use of EE for investigating the economic consequences of environmental policies, regulations, and resource utilization and its utility in quantifying the costs and benefits associated with environmental degradation and sustainability measures, its application in the manufacturing sector is of legitimate interest. Most importantly, EE can help with the setting of net zero goals in the manufacturing sector and providing useful key performance indicators that are could be used to track sustainable transformation. To assess the current state of sustainable manufacturing transformation, it is important to examine the progress made towards established net zero goals and targets. Key performance indicators from environmental economics literature can provide insights into how the sector is performing against defined sustainability objectives. However, achieving these targets also requires comprehending behavioral factors. Previous research shows BE has been effectively utilized in sectors like healthcare and finance to positively influence behaviors. Identifying examples where behavioral techniques have been successfully deployed in manufacturing could illuminate possibilities for wider adoption across the sector. Analyzing which specific components of sustainable manufacturing strategies are most amenable to behavioral interventions could further guide integration. Overcoming barriers to sustainability transformation relies on aligning economic incentives and behavioral motivations. EE can quantify the costs and benefits of various policy options, while behavioral analysis provides understanding of how to effectively motivate change. Harnessing insights from both disciplines could facilitate more impactful and scalable progress towards sustainable manufacturing.

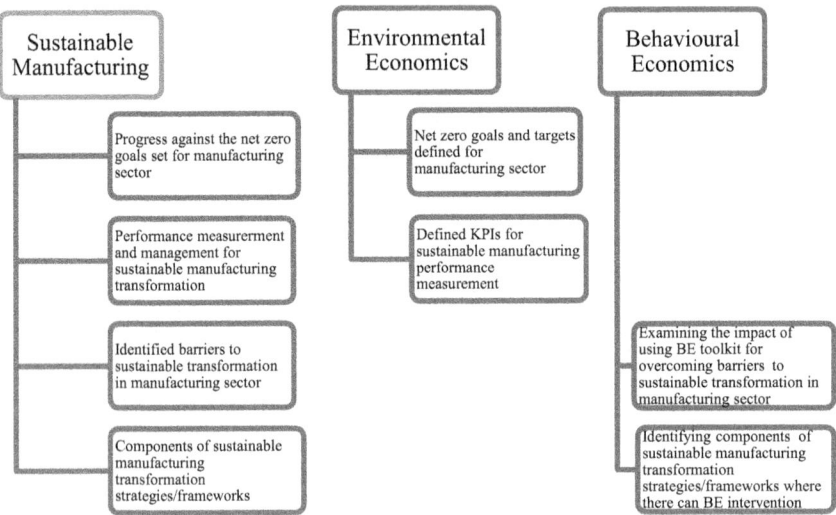

Fig. 6. Proposed research direction

In summary, a research direction as shown in Fig. 6, focused on integrating BE and EE has potential to advance sustainable manufacturing. Key priorities include cataloguing examples of BE deployment in manufacturing much like the concepts of established research like behavioural safety or behavioural operation management, determining high-impact integration points, and co-developing solutions that synthesize economic and behavioural insights. The overall research questions that could be covered in the future research agenda are listed below in Table 1. More empirical research blending theory and practice will be essential to realizing the benefits of this multidisciplinary approach.

By identifying the most suitable BE tools that could influence behavioural change and decision making and integrating them with key sustainable performance measurement parameters from EE analysis, the combinatorial impact on manufacturing sustainability may be critically investigated. Consequently, it is necessary to assess and measure the outcomes of such combination through a suitable performance management system and establish a guiding framework for facilitating the transition towards more advanced sustainable manufacturing practices.

Table 1. Research questions for interdisciplinary future research agenda

Sustainable Manufacturing	Environmental Economics	Behavioural Economics
What is the progress of sustainable manufacturing transformation in set net zero goals and targets?	What specific net zero goals and targets have been set for manufacturing sector?	Which industries or sectors have utilised BE tools and how may have these enhanced their performance?
How is the performance of sustainable manufacturing transformation measured currently against the defined targets?	What are the key performance indicators defined in EE literature specifically for manufacturing and sustainable manufacturing transformation?	Are there any use cases of BE being deployed within manufacturing?
What barriers is the manufacturing sector facing in sustainable transformation?		Can BE toolkit be harnessed to facilitate a more widespread adoption of sustainable manufacturing?
What strategies or frameworks are most frequently used for sustainable manufacturing transformation?		What specific areas or components of sustainable manufacturing strategies can BE techniques be introduced more effectively in?

5 Conclusion

A systematic review of the state of the art reveals a lack of existing research on the individual and combined contributions of BE and EE in sustainable manufacturing. The proposed research agenda provides potential directions for future studies to investigate the integration of insights from these two domains to advance sustainable manufacturing practices. Bridging theory with practice, the converging perspective of BE and EE could allow for a more comprehensive understanding of human decision-making, economic incentives, and environmental outcomes. This could pave the way for a novel, approach towards the green and clean transformation of the manufacturing sector. Empirical research will be key to assessing the promise of such a cross-disciplinary approach.

References

1. Katona G (2023) Psychological economics. 1975. Accessed 22 Sep 2023. [Online]. Available. https://psycnet.apa.org/record/1975-27395-000
2. Krstić M, Pavlović N (2020b) Behavioral economics. In: Advances in logistics, operations, and management science book series, pp 281–298. https://doi.org/10.4018/978-1-7998-4601-7.ch015
3. Krstić M, Pavlović N (2020) Behavioral economics: new dimension in understanding the real economic behavior. In: Handbook of research on sustainable supply chain management for the global economy, pp 281–298. IGI Global
4. Schwartz H (2002) Herbert Simon and behavioral economics. J Socio-Eco 31(3):181–189. https://doi.org/10.1016/s1053-5357(02)00161-0

5. Angner E (2016b) A course in behavioral economics. Red Globe Press
6. Manski CF (2000) Economic analysis of social interactions. J Eco Perspect 14(3):115–136. https://doi.org/10.1257/jep.14.3.115
7. Sarkis-Onofre R, Catalá-López F, Aromataris E, Lockwood C (2021b) How to properly use the PRISMA statement. Systematic Rev 10(1). https://doi.org/10.1186/s13643-021-01671-z
8. Louhichi R, Pelletan J, Sallak M (2022) Application of prospect theory in the context of predictive maintenance optimization based on risk assessment. Appl Sci 12(22):11748. https://doi.org/10.3390/app122211748
9. Tietenberg T, Lewis L (2018b) Environmental and natural resource economics. In: Routledge eBooks. https://doi.org/10.4324/9781315208343
10. Hanley N, Shogren JF, White B (1997b) Environmental economics in theory and practice. https://doi.org/10.1007/978-1-349-24851-3
11. Turner RK, Pearce DW, Bateman I (1993b) Environmental economics: an elementary introduction. Johns Hopkins University Press
12. Pearce D (2002) An intellectual history of environmental economics. Annu Rev Energy Env 27(1):57–81. https://doi.org/10.1146/annurev.energy.27.122001.083429
13. Kahneman D, Tversky A (1979) Prospect theory: an analysis of decision under risk. Econometrica 47(2):263. https://doi.org/10.2307/1914185
14. Thaler RH (1999) Mental accounting matters. J Behav Decis Mak 12(3):183–206. https://doi.org/10.1002/(sici)1099-0771(199909)12:3
15. Thaler RH, Sunstein CR (2012) Nudge: the final edition. Penguin UK
16. Leonard TC (2008b) Richard H. Thaler, Cass R. Sunstein, Nudge: Improving decisions about health, wealth, and happiness. Constitutional Political Eco 19(4):356–360. https://doi.org/10.1007/s10602-008-9056-2

Work Throughout the Industrial Revolutions and the Impacts of Industry 4.0 on Workers

Mariana Lazari Kawashima[(✉)], Daniel Braatz, and Fabiane Letícia Lizarelli

Federal University of São Carlos, São Carlos, Brazil
marianalazkawa@gmail.com

Abstract. The concepts of work found in literature distinguish themselves and approach the subject from different perspectives. Aiming to expand and update the reflection on work and its implications for workers, this article presents the main characteristics of human work throughout history from the perspective of industrial revolutions and the potential impacts that the Industry 4.0 paradigm has on it. As critical components of the social sustainability pillar, work and workers must be a focal point in the face of technological changes with potential impacts. To achieve this, impressions of professionals from academic and industrial institutions in Brazil and Germany were collected through self-administered questionnaires and subjected to content analysis. The process highlights impacts ranging from physical aspects of work environments to psychological effects on workers. Ultimately, the article concludes by offering a thoughtful reflection on future guidelines for human work in the context of the new industrial model.

Keywords: Work · Occupational health and safety · Working conditions · Social sustainability

1 Introduction

Scientific and academic literature offers a broad spectrum of work definitions, each viewing the subject from unique angles. This divergence in conceptualization stems from the idea that definitions are culturally shaped entities, developed over time and influenced by individual interpretations of the concept. These shifts in our understanding of work are intricately tied to the progression of human knowledge, changes in societal structures, and shifts in production methods and relationships [1].

Among the existing definitions, some portray work from a perspective restricted to the activities performed, conceiving it as a set of coordinated activities supported by effort, aimed at a goal [2]. However, this view can be expanded to encompass the social and human aspects, understanding work as the performance of an activity, its conformation, or its impacts on the psychological and social life of the worker [3]. This inclination towards the social and human nature of work is also seen in Marxist conceptions and supported by various authors. Marx argued that the expenditure of human force is merely the material foundation of work, while its essence is social [4]. In his conception, work

© The Author(s) 2025
H. Kohl et al. (Eds.): GCSM 2023, LNME, pp. 269–276, 2025.
https://doi.org/10.1007/978-3-031-77429-4_30

consists of a condition of human existence, a means through which humans distinguish themselves from their animal nature [5].

Bringing out from the discussed perspectives and aiming to expand and update the reflection on work and its implications for workers, this article presents the main characteristics of human work throughout history from the perspective of industrial revolutions and the potential impacts that the Industry 4.0 paradigm has on it. As critical components of the social sustainability pillar, work and workers must be a focal point in the face of technological changes with potential impacts in order to ensure rights, improve working conditions, foster continuous skill development, and create fair working environments that value workers and guarantee an equitable distribution of the benefits of technological progress.

2   Work Throughout the Industrial Revolutions

The Industrial Revolutions consisted of processes characterized by unprecedented technological transformations with impacts on the economic and social systems [6]. Although commonly described in terms of attributes such as the predominance of energy resources, transportation modes, and technologies with high economic impacts, these periods also generated social effects and changes in working conditions.

Pre-industrial manufacturing was carried out by artisans in home-based units, and work, which was almost entirely manual, had little division [7]. The production was characterized by technological knowledge as "craft specialization". Those with this specialization had control over the production processes, which meant they could determine, for example, the sequencing of operations and the workday length. This scenario applied to individual artisans and those working in workshops [8].

With the onset of the First Industrial Revolution, characterized by the invention of the steam engine, production was shifted from private homes to central factories [9]. Manufacturing, once associated with domestic life, was replaced by a system within large factories, where work became mechanized and segmented into specialized tasks [10]. This new production model resulted, among other things, in losing control over the production process by workers and in relationships of buying and selling labor power [11]. The daily work schedule was 12 or 13 h, carried out in dangerous and unhealthy conditions, with high risks of occupational accidents [12].

The Second Industrial Revolution, which began in the 19th century with the invention of the combustion engine, was marked by rapid industrialization and the use of petroleum and electricity as power sources for production [13]. The mechanization of industries was accompanied by the simplification of operations, with the implementation of automated and repetitive tasks that unskilled workers could perform. This work organization followed the Scientific Management, or Taylorism [14].

This model has faced criticism from specific authors who argue that the aims of work rationalization ultimately establish control mechanisms over workers and dominate their behavior [2]. Another argument is that the social division of labor entails increased productivity and the appropriation of knowledge by factory owners [15]. It is also reported that worker specialization prevents individuals from showing initiative and originality, makes work tedious, and reduces feelings of accomplishment [7].

Technological and scientific developments characterized the Third Industrial Revolution by introducing informatics, microelectronics, and robotics. Simultaneously, new management and production paradigms gradually replaced Taylorism with Toyotism [16]. This period was also marked by impacts on labor rights, which were deregulated and flexible to provide the necessary tools for the productive system to adapt to the new phase [5]. The flexibilization of work has been heavily criticized as many workers are forced to submit to informal work, wage reductions, and the loss of labor rights [16].

The unceasing stream of innovations persists with the advent of the Fourth Industrial Revolution [17], also known as Industry 4.0, whose main differentiating factor is the introduction and integration of technologies in the industry, such as autonomous robots, cybersecurity, augmented reality, Internet of Things, additive manufacturing, big data, and analytics [9, 18, 19]. This model significantly modifies products and production systems and has consequences for future employment by creating new business models [20]. Similar to the previous revolution, there is a fear that individuals may be forced to leave skilled jobs due to the introduction of automated and robotic processes, leading to increased voluntary and involuntary turnover and levels of unemployment [21].

The qualifications required for professionals in Industry 4.0 involve using digital technologies, understanding supply chains and customer relationships, and adapting to changes in career requirements [18]. At the organizational level, professionals tend to act altruistically with evident demonstrations of going "beyond", to become less dispensable. They may also adopt extreme attitudes such as working 24 h a day, 7 days a week [21]. Industry 4.0 can also bring some benefits, such as flexibility, improved coordination, and a better work-life balance. However, there is concern that these advantages may be accompanied by deregulation of work relationships, pressure for constant availability, and new possibilities for evaluating digitally employee performance [18]. Industry 4.0 technologies such as, Big Data and Artificial Intelligence, enable new dimensions of automation, from lightweight robots in the industrial sector, humanoid robots or chatbots used by banks and insurance companies in customer service to predictive maintenance in large technical facilities [22]. Technologies such as Big Data and robots can, respectively, identify patterns and provide dexterity to tasks that were previously manual, and, probably, this can change the nature of work in all sectors and professions [23]. The authors, using a predictive model, predict that most workers in transportation/logistics occupations, bulk of office and administrative support, and labor in production occupations, are at risk [23].

In summary, the theoretical framework used in this research allows for an expansion of the schemes and images that illustrate the Industrial Revolutions and their characteristics. This broader view, which includes the impacts on work, is presented in Fig. 1. In addition to the theoretical framework presented and to better understand the potential impacts of the ongoing revolution on human work, this research gathered perceptions from professionals in Brazil and Germany, the country of origin of the term Industry 4.0, on the subject.

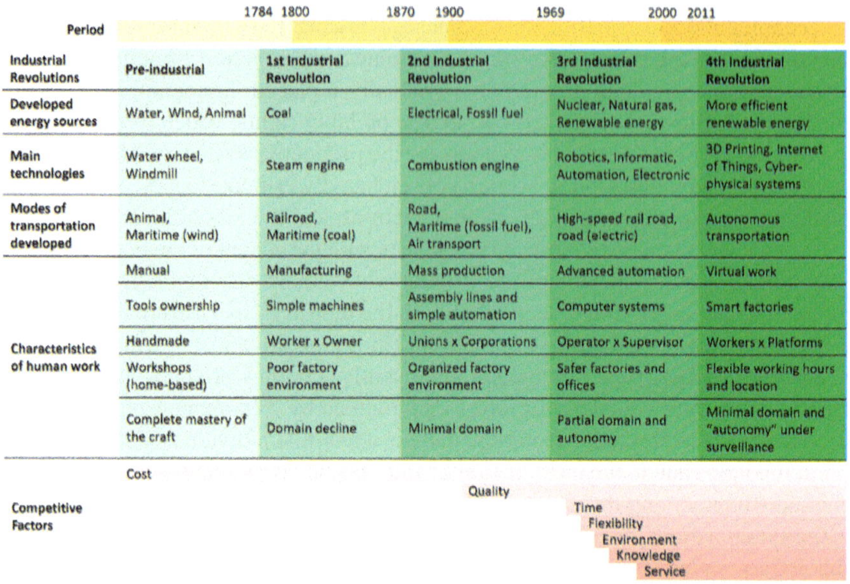

Period		1784 1800	1870 1900	1969	2000 2011
Industrial Revolutions	Pre-industrial	1st Industrial Revolution	2nd Industrial Revolution	3rd Industrial Revolution	4th Industrial Revolution
Developed energy sources	Water, Wind, Animal	Coal	Electrical, Fossil fuel	Nuclear, Natural gas, Renewable energy	More efficient renewable energy
Main technologies	Water wheel, Windmill	Steam engine	Combustion engine	Robotics, Informatic, Automation, Electronic	3D Printing, Internet of Things, Cyber-physical systems
Modes of transportation developed	Animal, Maritime (wind)	Railroad, Maritime (coal)	Road, Maritime (fossil fuel), Air transport	High-speed rail road, road (electric)	Autonomous transportation
Characteristics of human work	Manual	Manufacturing	Mass production	Advanced automation	Virtual work
	Tools ownership	Simple machines	Assembly lines and simple automation	Computer systems	Smart factories
	Handmade	Worker x Owner	Unions x Corporations	Operator x Supervisor	Workers x Platforms
	Workshops (home-based)	Poor factory environment	Organized factory environment	Safer factories and offices	Flexible working hours and location
	Complete mastery of the craft	Domain decline	Minimal domain	Partial domain and autonomy	Minimal domain and "autonomy" under surveillance
Competitive Factors	Cost		Quality	Time Flexibility Environment Knowledge Service	

Fig. 1. Modifications and developments in the Industrial Revolutions. Prepared by the authors based on the theoretical framework [7, 24, 25].

3   Research Method

This article is based on survey-type research conducted through a self-administered questionnaire, which allowed the identification of participants' sociodemographic characteristics and the collection of their perceptions on the topic through an open-ended question [26]: "In your opinion, what will be the main impacts of Industry 4.0 on people's work?". The sampling method employed was non-probabilistic convenience sampling [27].

The target respondents are professionals from academic and business institutions in Brazil and Germany. Therefore, the questionnaire was prepared in Portuguese and English using the Google Forms tool and made available through email, WhatsApp, Facebook Messenger, and LinkedIn. 78 responses were obtained, three disregarded during data processing due to their incompatibility with the target respondents.

Most participants (59%) work in the Industrial sector, with engineers and analysts of various specialties being the most common job positions. Academic professionals, represented by professors and researchers, account for 41% of the data. Regarding their affiliation with institutions in Brazil and Germany, 75% of participants reported having a professional relationship only with Brazilian corporations, 9% exclusively with German ones, and 16% with both.

The content analysis of the open-ended question was conducted in three steps: 1. Understanding and summarizing the main concepts from the 75 responses; 2. Counting the frequency of the concepts; and finally, 3. Grouping similar concepts into categories, which were progressively developed during the analysis [28].

The first step allowed synthesizing the content into 55 distinct concepts, subsequently grouped into 12 categories. For example, the responses "Reduced need for manual labor" and "High likelihood of operational tasks being replaced by machine work" were assimilated into the concept of "Reduction of operational job positions", which was included in the category "Volume of job positions". Other concepts such as "Increase in tactical and strategic job positions" and "Rise in unemployment", extracted from other responses, were also included in this category.

4 Results and Discussion

The stage of quantifying the recurrence of each concept revealed that only one respondent mentioned 51% of them, while 4 (7%) gathered 10 or more occurrences. By grouping the concepts into broader sets that would provide a more essential understanding of the responses, the 12 categories displayed in Table 1 were obtained.

Table 1. Content analysis of the main impacts of Industry 4.0 on human work. Elaborated by the authors (2020)

Categories	Occurrences	% Occurrences
Competencies	45	29%
Volume of job positions	23	15%
Professional occupations	17	11%
Technology benefits	17	11%
Form and content of work	14	9%
Flexibilization	10	6%
Work environment	9	6%
Training	8	5%
Psychological effects	6	4%
Competitiveness	2	1%
Teams	2	1%
Others	3	2%
Total	**156**	**100%**

The category Competencies encompasses the most mentioned concepts by the participants. They mentioned impacts on the development of functional, behavioral, social, and general competencies (41 occurrences). Additionally, three respondents highlighted the importance of forming a broader understanding of the production flow.

Participants expressed concerns regarding the rise in unemployment (9 occurrences) in the Volume of Job Positions category. They also indicated a possible reduction in operational job positions (7 occurrences) and an increase in tactical and strategic positions (3 occurrences). Notably, the opinion of these participants aligns with some authors who mention that job positions involving simpler and repetitive tasks are more susceptible to replacement than those involving more complex knowledge [13, 17].

Professional Occupations retains 17 occurrences and encompasses new professions, occupations' extinction, replacement, and development. There was also a mention of the increase in some activities at the expense of others. "Creation of new professions" and "Extinction or significant reduction of some professions" are among the responses included in this category.

The concepts included in Technology Benefits depict improvements in the technological field that generate impacts on workers' daily lives. For example, the enhancement in data processing, which includes responses such as "Agility in data compilation", and the increase in information in task execution reflect more efficient ways of performing an activity and affect workers' day-to-day routines.

In the category Form and Content of Work changes in the content and execution methods of work were indicated. Additionally, human-machine interaction was referenced in 3 responses, understood as beneficial in one occurrence ("More collaborative human-machine relationship") and unspecified in the others (e.g., "Increased bilateral interaction with machines and equipment"). As for Flexibilization, participants pointed out the flexibility of the workplace, the working hours, the tasks performed, and the labor legislation. The "Blending of work and personal life" was also mentioned, stemming from increased connectivity among individuals.

Regarding the concepts related to the Work Environment, more excellent safety, ergonomics, and work quality were mentioned as positive impacts of Industry 4.0. The other points involved physical changes in workstations and the environment, but they were not accompanied by expressions indicating benefits or drawbacks to the workers. Additionally, participants indicated increased control and monitoring of work activities.

The need for continuous training and increased professional development were mentioned in the Training category. Given that Industry 4.0 encompasses a technological landscape and technologies are continually advancing, it is reasonable to infer that professionals must consistently invest in updating their knowledge and skills. Moreover, there was a reference to a high number of trainees within organizations due to the need for constant training.

The Psychological Effects on workers were also noted in line with authors who propose mental stress as a field of action for implementing the Fourth Industrial Revolution [29]. Some participants cited a higher psychological burden, loss of focus with simultaneous activities, and increased pressure at work. Furthermore, resistance to change and workers' need for mental preparation were pointed out. Lastly, one participant portrayed a lower psychological burden for "professionals operating machines and equipment".

In the Competitiveness category, two participants indicated increased concurrence and competitiveness. In the first case, "Concurrence and accelerated pressure from companies for implementation" was mentioned, while the second case did not specify the agent generating the increase in competitiveness. In Teams, the reduction in team size and the maintenance of "collectively integrated work" were quoted.

Lastly, the concepts in the "Other" category, which, although mentioned only once, are relevant for bringing different points such as: "Significant investment in knowledge management", "Increase in wage disparity between jobs with and without technical skills for Industry 4.0", and "No significant impact".

5 Conclusions and Final Remarks

The impacts of Industry 4.0 include creating, extinction, and replacement of professional occupations, augmented safety, ergonomics, and work quality, physical changes in workstations and the work environment, and greater control and monitoring of activities. Additionally, some participants highlighted the need for intensified and continuous training, leading to a growing number of trainees within organizations.

Psychological effects on workers include increased psychological burden, loss of focus with simultaneous activities, and elevated work pressure. The Competitiveness category denotes higher concurrencies in the workplace and competitiveness, related to companies pressure. Lastly, the reduction in team size and the maintenance of collectively integrated work were mentioned.

As observed, the impacts of Industry 4.0 on work extend across different areas, encompassing physical modifications of the environment and psychological effects on workers. Therefore, it is of utmost importance to incorporate in studies and implementation projects in industries topics that highlight the influences of technologies from the new industrial model on work, ensuring that they bring benefits and more efficient, sustainable, and safe production systems, rather than becoming an end in themselves.

A critical reflection indicates that with the advent of Industry 4.0 and the dissemination of the use of technologies will significantly change work, new competencies will have to be developed by employees, while there is also the prospect of reducing tasks and disappearing certain types of jobs, while others will emerge. Therefore, there is a need for future research to monitor changes that have already occurred at work, observing impacts of specific technologies, such as Big Data and Artificial Intelligence.

References

1. Borges LO (1999) Conceptions of work: a content analysis study of two national circulation journals. J Contemporary Adminis 3(3):81–107
2. Lhuilier D (2013) Work. Psychol Soc 25(3):483–492
3. Araújo JNG (2007) Work, organization, and institutions. In: Jacó-Vilela AM, Sato L (eds) Dialogues in social psychology. Evangraf, Porto Alegre, pp 397–411
4. Monteiro AQ (2010) Work, information, and value: the process of infoexploitation. In: Souza JS, Araújo R (eds) Work, education, and sociability. Massoni, Maringá, pp 67–86
5. Antunes R (2005) Goodbye to work?: essay on the metamorphoses and centrality of the work world. 10th edn. Cortez, São Paulo
6. Castells M (2010) The rise of the network society: the information age: economy, society and culture, 2nd edn. Wiley-Blackwell, Chichester
7. Kwasnicka EL (2004) In: Introduction to administration. 6th edn. Atlas, São Paulo
8. Bruland K (1993) The transformations of work in European industrialization. In: Mathias P, Davis JA (eds) The first industrial revolutions, vol 1. Dom Quixote, Lisbon, pp 215–232
9. Drath R, Horch A (2014) Industry 4.0: hit or hype? IEEE Indus Electron Magazine 8(2):56–58
10. Sparta M, Lassance MC (2003) Vocational guidance and transformations in the world of work. Brazilian J Vocat Guidance (4)2:13–19
11. Matos E, Pires D (2006) Administrative theories and work organization: from Taylor to the present day, influences in the health sector and nursing. Text and Context—Nursing 15(3):508–514

12. Chiavenato I (2011) Introduction to general theory of administration, 8th edn. Elsevier, Rio de Janeiro
13. Min X, Jeanne MD, Suk HK (2018) The fourth industrial revolution: opportunities and challenges. Int J Finan Res 9(2):90–95
14. Fleury ACC, Vargas N (1983) Conceptual aspects. In: Fleury ACC, Vargas N (eds) Work organization: an interdisciplinary approach: seven Brazilian cases for study. vol 1. Atlas, São Paulo, pp 17–37
15. Decca ES (1982) In: Everything is history: the birth of factories. 1st edn. Brasiliense, São Paulo
16. Prieb S (2007) The working class facing the third industrial revolution. In: V Brazilian Congress of Marxist Studies. Cemarx, Campinas, pp 1–6
17. Postelnicu C, Câlea S (2019) The fourth industrial revolution. global risks, local challenges for employment. Montenegrin J Econ 15(2):195–206
18. Heine I, Schmitt R (2019) In: Humans and production: the future of work. pp 1–40
19. Cruz RJM, Tonin LA (2022) Systematic review of the literature on digital twin: a discussion of contributions and a framework proposal. Managem Prod 29
20. Ślusarczyk B (2018) Industry 4.0: are we ready? Polish J Managem Stud 17(1):232–248
21. Coldwell DAL (2019) Negative influences of the 4th industrial revolution on the workplace: towards a theoretical model of entropic citizen behavior in toxic organizations. Int J Environ Res Public Health 16(15):1–13
22. Digitalisation of Working Worlds. pp 2267. https://digitalisierung-der-arbeitswelten.de/home
23. Frey CB, Osborne MA (2017) The future of employment: how susceptible are jobs to computerisation? Technol Forecast Soc Chang 114:254–280
24. Sube S, Biswajit M, Manoj KT (2019) Framework and modeling of inclusive manufacturing system. Int J Comput Integr Manuf 32(2):105–123
25. Prisecaru P (2016) Challenges of the fourth industrial revolution. Knowledge Horizons—Econ 8(1):57–62
26. Hair J, Page M, Brunsveld N (2019) Essentials of business research methods. Routledge
27. Haenssgen MJ (2019) Sampling methods. In: Interdisciplinary qualitative research in global development: a concise guide. Emerald Publishing Limited, pp 53–61
28. Kleinheksel AJ, Rockich-Winston N, Tawfik H, Wyatt TR (2020) Demystifying content analysis. Am J Pharm Educ 84(1):7113
29. Dombrowsky U, Wagner T (2014) Mental strain as field of action in the 4th industrial revolution. Proc CIRP 17:100–105

Quantifying the Carbon Footprint of Events: A Life Cycle Assessment-Based Framework for Evaluating Impact of Location and Timing

Yagmur Atescan-Yuksek[1]([⊠]), Sanjooram Paddea[1], Sharon Jackson[2], Mark Jolly[1], and Konstantinos Salonitis[1]

[1] Sustainable Manufacturing Systems Centre, School of Aerospace, Transport and Manufacturing, Cranfield University, Bedford MK43 0AL, UK
`yagmur.atescanyuksek@cranfield.ac.uk`
[2] European Sustainability Academy, 73008 Drapanos, Greece

Abstract. This research proposes a Life Cycle Assessment-based framework to quantify the carbon footprint of events, considering the event's location and timing. The framework aims to standardise environmental impact calculations through inventory analysis. To validate it, a comparative analysis on conducting an event in different locations and time periods, while maintaining similar scale and nature is conducted. The assessment includes emissions from attendee transport, accommodation, food and drink, and venue. Additionally, it considers emission reductions resulting from attendees not using their personal household resources. This accounts for the actual additional emissions released into the atmosphere as a consequence of the event. The results highlight variations in emissions across different consumption categories based on the selected location and timing. By providing this information, the LCA-based framework provides valuable guidance for event organizers and policymakers to assess event environmental impacts and promote sustainability.

Keywords: Carbon footprint · Life cycle assessment · Events · Quantitative approach

1 Introduction

Over the past few decades, the global events industry has experienced remarkable and rapid expansion, leading to significant tourism and socio-economic development advancements. However, this substantial growth has pressured global climate change mitigation goals [1]. Consequently, there has been increasing recognition of the urgent need to address the environmental impacts and carbon emissions associated with scientific and social events. This recognition has resulted in a notable surge in demand for sustainable event management practices.

The COVID-19 pandemic has profoundly impacted event planning and execution, forcing event organisers to shift to online platforms. It has been demonstrated that transitioning from in-person to virtual conferencing can substantially reduce carbon footprint

H. Kohl et al. (Eds.): GCSM 2023, LNME, pp. 277–285, 2025.
https://doi.org/10.1007/978-3-031-77429-4_31

by approximately 94% and decrease energy use by around 90% [2]. However, the quick return to in-person and hybrid events following the easing of COVID-19 restrictions has highlighted the limitations and drawbacks of entirely virtual events from an efficiency and socio-economic perspective [3]. This has underscored the importance of balancing environmental considerations and participant experience in event planning and execution.

In the pursuit of sustainable event management, it is crucial to identify and accurately assess the environmental impacts of events. Several quantitative methods have been proposed to conduct environmental assessments of events to achieve this. For instance, Collins and Cooper utilised the ecological footprint method to examine the environmental impact associated with the 2012 Hay Literary Festival [4]. Their analysis focused on attendees' transport, food and drink consumption, and energy use in overnight accommodation. The scientific community has also increasingly used the life cycle assessment (LCA) method to assess event organisations' multi-dimensional impact [5]. Toniolo et al. conducted an LCA study to quantify the impact associated with an exhibition [6]. Their analysis considered the organisation, assembly, utilisation, and dismantling stages of the exhibition area for 120 exhibits over 3 days. Similarly, Edwards et al. conducted an LCA to measure the carbon footprint and other impact indicators of two Homecoming events at the University of Arizona [7]. Additionally, Toscani et al. provided a uniform life cycle model to assess the environmental impacts of any event by adapting a multi-case implementation and a multi-actor perspective [8]. These quantitative studies have provided valuable insights into the environmental impacts of various events, enabling event organizers to make informed decisions and implement sustainable practices in event planning and execution. However, a standardized approach for examining the impact of event's location and timing on the environmental impacts is missing.

This research aims to develop an LCA-based framework for quantifying the carbon footprint of events, considering the event's location and timing. The methodology of this framework involves a comparative analysis of conducting an event in different locations and time periods, while maintaining similar scale and nature. The carbon footprint is assessed by considering various consumption categories such as transportation and accommodation, which can lead to different emissions based on the selected location and timing. To validate and demonstrate the applicability of the framework, a case study is performed, focusing on the organization of a workshop in the context of TransFIRE hub (https://transfire-hub.org/) either in Greece or in the United Kingdom (UK). The findings of this study will offer valuable insights to event organizers, policymakers, and sustainability practitioners.

2   LCA-Based Framework for Sustainable Event Management

The LCA-based framework, as illustrated in Fig. 1, adheres to the four essential stages of LCA: goal and scope definition, life cycle inventory (LCI), life cycle impact assessment (LCIA), and interpretation. However, it incorporates two additional stages, identification, and analysis, before the LCI stage. This extended framework employs a comparative analysis methodology to determine the optimum time and location to conduct the events with minimum impacts to the environment. To conduct this comparison, the first step is the identification of the potential event locations and timing options. The selection of options should be based on various factors, such as weather conditions, the average distances of potential attendees to the event location candidates, required organisational arrangements, availability of local suppliers, distances from possible suppliers, and the nature of the event (indoor/outdoor, event size, duration, etc.). Following the identification stage, the analysis stage aims to carefully assess the potential locations and timing for hosting the event. During this analysis, the framework defines the consumption categories that may vary depending on the selected location and timing. These consumption categories include transportation, accommodation, venue, organisational requirements, food and drink, and waste management.

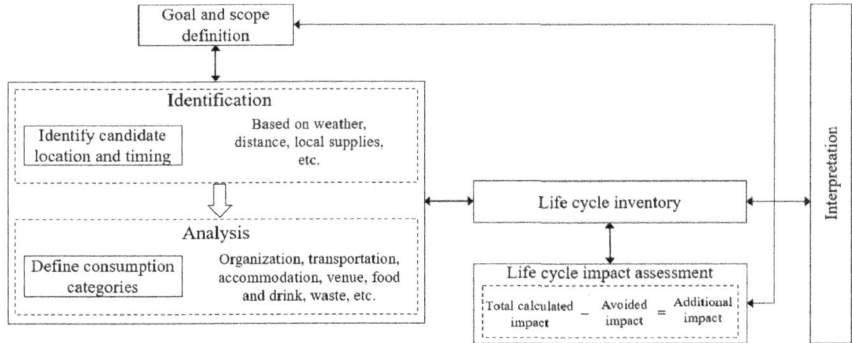

Fig. 1. The LCA-based framework for sustainable event management.

Following the identification of location and timing options, and the analysis of related consumption categories, the subsequent stages in the framework are the LCI and LCIA stages of the LCA. The LCI stage functions as a comprehensive database, encompassing all relevant information regarding resource usage and emissions associated with each selected consumption category. The data gathered during this stage forms the foundation for the subsequent LCIA stage, during which the environmental implications associated with each location and timing option are assessed and quantified. Once the environmental impact of each category is determined, the additional impact of the event is calculated by subtracting the attendees' total impact at their home location during the event from the total impact of the event. In the interpretation stage, the results are analysed to compare the selected event locations and timing in terms of their environmental impact. By conducting this comprehensive analysis, the framework empowers event organizers to make informed decisions regarding the most sustainable location and timing for the events.

3   Case Study: Organization of Workshop

This section demonstrates the applicability and usefulness of the sustainability assessment framework through a case study. For the case event, a workshop held in Chania, Greece in April 2023 was selected. This workshop lasted for four whole days with 21 attendees and took place in European Sustainability Academy (ESA), which operates wholly off-grid and is powered by a photovoltaic (PV) system. A comparison is made between the CO_2eq emission results of the workshop event held in Greece and the emissions that would be generated if this workshop were held in Lake District (UK), with the same scale and nature.

The analysis stage involves defining the consumption categories that may vary between each event location. Since the events were scheduled for the same time of the year but held in different locations, a comparison of the consumption categories; transportation, accommodation, food and drink, and venue was conducted for both event venues. Table 1 presents a summary of the scope of these consumption categories calculated for the Greece event, while estimations were used for the UK event. For the Greece event, primary data were collected from the event organizer, and government data and reports were utilized. However, for the UK event, the consumption patterns were estimated within a range, covering both low and high consumption scenarios. Details of these estimations are provided in Table 1 and further explained for each consumption category in the following paragraphs. Temperature differences between the two locations were taken into consideration as they can significantly impact energy usage for heating and cooling, thus affecting emissions from accommodation and venue operations.

The transport consumption category includes return journeys of all attendees to event accommodation and journeys made during the event. In the case of the workshop held in Greece, attendees travelled by train from their homes to the airport, flew from London to Chania, and used taxis between the airport and the accommodation. The actual flight distances between London and Chania were determined from https://www.flightradar24. com/ using the flight numbers of attendees. In the case of the workshop held in the UK, in the low consumption scenario, all attendees used rail transport, and taxis were utilized between rail station and accommodation. No transportation was required for event days, assuming that the event venue and accommodation were at the same location. In the high consumption scenario, all attendees used their own cars, and attendees from Greece rented a car after their flight to London. As most journeys made within the UK are by rail and road, domestic air travel was not considered in the calculations. In the high consumption scenario, it was also assumed that the event venue's location differed from that of the accommodation, and for this assumption, distances and transportation types were defined same as for the Greece event. Additionally, since the common fuel type in the UK is petrol, it was used for vehicles, as opposed the diesel fuel used in the Greece event.

The accommodation consumption category encompasses the total energy consumption caused by attendees in the hotel. For the Greece event, a Four-Star hotel served as the accommodation, while a hotel of the same class was selected for the UK event. The number of room-nights was determined based on the total number of attendees and their length of stay. Energy consumption per guest-night for a four-star hotel was defined from [9], covering heating, cooling, lighting, domestic hot water (DHW) and process equipment. For the Greece event, energy consumption for lighting, DHW and process equipment was included. Electricity was used as the source for lighting and process equipment while a combination of electricity and solar power was utilized for DWH due to the presence of solar panels for water heating at the hotel in Greece. In the UK event's low consumption scenario, energy consumptions for lighting, DHW and process equipment were included, with heating omitted because of the possible warm weather in April. However, in the high consumption scenario, heating was also considered. For the UK event, electricity was used for lighting and process equipment, while natural gas was employed for DHW and heating.

The impacts from the consumption of food and drinks were calculated based on the quantity of food and drinks consumed and the sourcing of their ingredients. For the drinks based on menu information provided by the organizer, along with portion sizes for an average person. The food and drink items served at the Greece event were prepared using locally grown ingredients within a 10 km distance to the event venue. Consequently, only the impacts associated with agriculture, processing, and use were considered, while transportation and packaging impacts were excluded from the assessment. The energy embedded in agriculture, processing, and the use of food and drink items at the Greece event was determined using data from the Joint Research Centre Science and Policy Report [10]. In the low consumption scenario of the UK event, the same type of food as in the Greece event was served, with the same quantity. However, considering the distinct nature of the food served in the UK, meals in the high consumption scenario were defined as premium buffet-style for lunch, and pub-style for dinner. Specific meals and quantities were determined based on example menus. For both scenarios of the UK event, impacts from transportation and packaging of ingredients were also taken into consideration in addition to the impacts from agriculture, processing, and use. This comprehensive approach was necessary due to the global sourcing of ingredients for the food served in the UK. The embedded CO_2eq emissions resulting from agriculture, processing, use and packaging of food and drink items for the UK event were determined using data from [11]. The transportation impact was calculated based on the countries of origin for food ingredient imported into the UK and the distances between these countries and the UK.

The venue's impact was calculated based on the total energy consumed during the event. For the event held in Greece, the venue operates entirely off-grid and is powered by a PV system, resulting in no associated emissions. The total energy used during the event was determined from PV analytics and used for the UK event, encompassing energy consumption for lighting, laptop usage, kitchen usage, router, security video camera system, and end of day cleaning. For the UK event, in the high consumption scenario, heating energy consumption was additionally considered. However, the low consumption scenario used the same amount of energy as in the Greece event.

The LCA of all consumption categories was performed utilising SimaPro 9.2 software. The background data for each category were retrieved from the ecoinvent 3 database. As for the LCIA process, the ReCiPe 2016 methodology (H) has been chosen.

The total CO_2eq emission results for both the Greece event and the low and high consumption scenarios of the UK event were determined using LCA. To calculate the additional CO_2eq emissions resulting from the attendance at the event, the total CO_2eq emissions of attendees at their home location during the event time were subtracted from the total CO_2eq emission results. The calculation of attendees' total emissions at their home locations during the event time was based on data from the Department for Environment, Food & Rural Affairs, UK Carbon Footprint results for 2019 [12]. The

Table 1. Scope of the consumption categories for Greece and UK events.

Consumption categories	*Chania, Greece*	
Transport	Return distances travelled from the home location to the event accommodation Return distances travelled from the accommodation to the event venue. Two 8-seater vans and one taxi car with diesel fuel	
Accommodation	Energy consumptions for lighting, DHW and process equipment	
Food and drink	Quantity and type of food and drink items served during the event Locally grown ingredients within 10 km distance to event venue, no transportation, no packaging	
Event venue	Total energy consumed during the event	
	Lake District, United Kingdom	
	Low consumption	High consumption
Transport	All attendees used public transportation. Train and taxi from train station to accommodation were utilized. No domestic flights were used No event day transportation	All attendees used their own car. No domestic flights were used Event days transportation: two 8-seater vans and one taxi car were used with petrol, as this is the common fuel type used in the UK
Accommodation	Energy consumptions for lighting, DHW and process equipment	Energy consumptions for lighting, DHW, process equipment and heating
Food and drink	Quantity and type of food and drinks were determined same as for the Greece event Transportation and packaging impact of food ingredients were also included	Quantity and type of food and drinks were determined in accordance with what is typically served in the UK Transportation and packaging impact of food ingredients were also included
Event venue	Same total energy with Greece event was used	Heating energy consumption was also considered

calculated values in kg CO_2eq were 423 for transportation, 254 for accommodation and 88 for food and drinks.

Figure 2 shows the additional CO_2eq emission results for each consumption category. Results for the Greece event are represented with a single line, as the actual consumptions data were collected from event organizer. UK event results on the other hand, are presented in a range covering both low and high consumption scenarios. Among the consumption categories, the transport category has the highest impact, measured in the scale of thousands, while the venue has the smallest impact, measured in the scale of tens. In the transport category, the Greece event resulted in higher additional emissions. The primary reason for this difference was that most of the workshop attendees travelled via international flights from the UK to Greece, whereas they used trains or cars for transportation on the UK event scenarios. Emissions for accommodation, food and drink, and the venue were lower for the Greece event. This was because heating was unnecessary in the accommodation, food and drink items were prepared with locally grown ingredients, and the venue operated off-grid with all power generated from solar energy, resulting in no emissions.

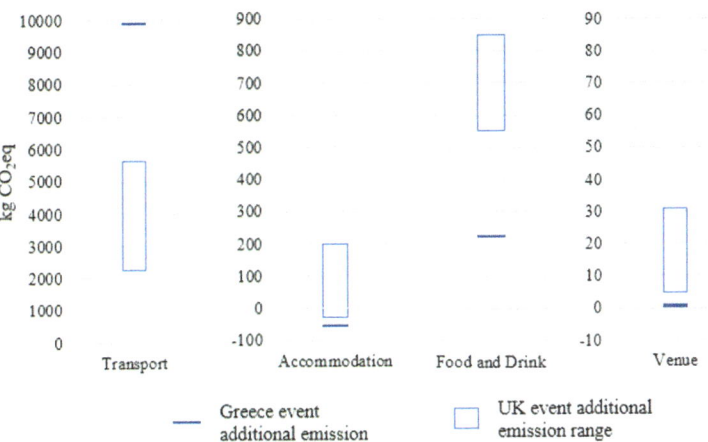

Fig. 2. Additional CO_2eq emission results for consumption categories: transport, accommodation, food and drink, and venue.

4 Conclusion

In this paper, an LCA-based framework has been presented, aimed at quantifying the environmental impact of events, taking into account the event's location and timing. The methodology involved conducting a comparative analysis of hosting an event in various locations and time periods, while ensuring a consistent scale and nature. By considering various consumption categories, such as transportation and accommodation, which can result in varying emissions based on the selected location and timing, valuable insights were provided. To validate the applicability and usefulness of the framework, a case study was conducted, focusing on a workshop that took place in Chania, Greece. The carbon emission results from this workshop were then compared to the emissions that would be generated if the same workshop were held in the Lake District, UK, with a similar scale and nature. It has been demonstrated through this research that location and timing play a significant role in determining the carbon footprint of events. By analysing factors such as weather conditions, attendees' travel distances, and organizational arrangements, the framework enables event organizers to make informed decisions to minimize the environmental impact of their events.

In conclusion, research emphasizes the importance of proactive measures in selecting event locations and implementing sustainable practices to minimize the environmental impact of events. In future work, other environmental factors, such as water usage, waste generation, and ecosystem impacts, could be incorporated to provide a more comprehensive sustainability assessment for events. Furthermore, the framework should be applied to a wider range of event types and case studies to validate its applicability.

References

1. Holmes K, Hughes M, Mair J, Carlsen J (2015) Events and sustainability, 1st edn. Routledge, London
2. Tao Y, Steckel D, Klemeš JJ, You F (2021) Trend towards virtual and hybrid conferences may be an effective climate change mitigation strategy. Nat Commun 12. https://doi.org/10.1038/s41467-021-27251-2
3. Moss VA, Adcock M, Hotan AW, Kobayashi R, Rees GA, Siégel C, Tremblay CD, Trenham CE (2021) Forging a path to a better normal for conferences and collaboration. Nat Astron 5
4. Collins A, Cooper C (2017) Measuring and managing the environmental impact of festivals: the contribution of the ecological footprint. J Sustain Tour 25. https://doi.org/10.1080/09669582.2016.1189922
5. Arzoumanidis I, Walker AM, Petti L, Raggi A (2021) Life cycle-based sustainability and circularity indicators for the tourism industry: a literature review. Sustainability (Switzerland) 13
6. Toniolo S, Mazzi A, Fedele A, Aguiari F, Scipioni A (2017) Life cycle assessment to support the quantification of the environmental impacts of an event. Environ Impact Assess Rev 63. https://doi.org/10.1016/j.eiar.2016.07.007
7. Edwards L, Knight J, Handler R, Abraham J, Blowers P (2016) The methodology and results of using life cycle assessment to measure and reduce the greenhouse gas emissions footprint of "Major Events" at the University of Arizona. Int J Life Cycle Assess 21. https://doi.org/10.1007/s11367-016-1038-4
8. Cavallin Toscani A, Macchion L, Stoppato A, Vinelli A (2022) How to assess events' environmental impacts: a uniform life cycle approach. J Sustain Tour 30. https://doi.org/10.1080/09669582.2021.1874397
9. Charalambous A, Kastanias P, Koutsokoumnis N, Ikkos A, Kyriakides M, Mylonas S, Schneider M, Weir J, Barckhausen A (2020) Assessment of cyprus and greece hotels structural characteristics, energy and GHG emissions performance indicators
10. Monforti-Ferrario F, Dallemand J-F, Pinedo Pascua I, Motola V, Banja M, Scarlat N, Medarac H, Castellazzi L, Labanca N, Bertoldi P, Pennington D, Goralczyk M, Schau EM, Saouter E, Sala S, Notarnicola B, Tassielli G, Renzulli P (2015) JRC science and policy report: energy use in the EU food sector: state of play and opportunities for improvement. Italy
11. Poore J, Nemecek T (2018) Reducing food's environmental impacts through producers and consumers. Science (1979) 360. https://doi.org/10.1126/science.aaq0216
12. Department for Environment, Food & Rural Affairs. UK carbon footprint results. https://www.gov.uk/government/statistics/uks-carbon-footprint

Social Inclusion Challenges of Low-Income People Through Entrepreneurship in Industry 4.0

Celia Hanako Kano[1][(✉)], Marly Monteiro de Carvalho[1],
Roberta de Castro Souza Piao[1], and Jairo Cardoso de Oliveira[2]

[1] Polytechnic School, University of Sao Paulo, Sao Paulo, Brazil
`{chkano,marlymc,robertacsouza}@usp.br`
[2] PPGA, Nove de Julho University, Sao Paulo, Brazil
`jairo.oliveira@uni9.edu.br`

Abstract. Industry 4.0 raises concerns about the inclusion of low-income people. Researchers approach entrepreneurship as a path to social inclusion. In this context, this paper aims to investigate how Industry 4.0 engages low-income people under the perspective of entrepreneurship. This paper is based on a systematic literature review using bibliometrics and content analysis. The paper combines RStudio, VOSviewer and NVivo software analysis for conducting a systematic literature review. The paper also conducted a content analysis. This study understands that entrepreneurship has different challenges in Industry 4.0 depending on the social group. To better understand these challenges, low-income people were divided into three social groups: low-literacy and poor people, rural communities and women. The results indicated that the main challenges to engage low-income people on Industry 4.0 are digital inclusion, digital financial inclusion, adoption of digital platforms, and others.

Keywords: Entrepreneurship · Low-income people · Industry 4.0

1 Introduction

In 2011, the German government introduced Industry 4.0 developing intelligent production, internet of things, cloud technology and big data (Oztemel & Gursev, 2020). Industry 4.0 modified business models and triggered social changes (Kotarba, 2018). This paper will focus on low-income people and it will be based on the "bottom of the pyramid" concept. This concept refers to the bottom-tier of the world income pyramid, living in a situation of poverty (Gold et al., 2013; Sharma & Jaiswal, 2018; Prahalad & Hart, 2002) with annual incomes of no more than $3000 (Du et al., 2021).

Digitalization seems to fail especially with low-income people. As an example, industrial robots, automated vehicles and intelligent machines will replace humans in inventory tracking, quality control and other activities, eliminating low to medium-skilled jobs (Ghobakhloo, 2020). Another example is related to the changes in social

© The Author(s) 2025
H. Kohl et al. (Eds.): GCSM 2023, LNME, pp. 286–294, 2025.
https://doi.org/10.1007/978-3-031-77429-4_32

and labor relations. There are indications that fixed-term labor contracts, stable employment and well-developed social protection will be replaced by more flexible part-time work, short-term labor contracts, lower social protection, reduced free time and labor income less stable (Azoeva et al., 2020). This scenario opens space to a discussion about entrepreneurship as a social inclusion strategy to overcome the challenges imposed by Industry 4.0.

Although "Industry 4.0", "bottom of the pyramid" and "entrepreneurship" are highly studied individually, as far as we are concerned the integrated analysis of these topics are still open and may bring a contribution to practitioners and researchers in operations management. This paper seeks to answer the research question: "what are the challenges of entrepreneurship to low-income people inclusion in Industry 4.0?".

2 Research Method

This paper is based on a systematic literature review. The search protocol to collect data is presented in Table 1 and was divided into two parts: sampling process and data analysis. Table 2 summarizes criteria adopted during screening stage.

Table 1. The search protocol to collect data from the literature.

Sampling process	Search string	(digital* or "industr* 4*") AND (bop OR "base of the pyramid" OR "bottom of the pyramid" OR "base of pyramid" OR "bottom of pyramid" OR "low income" OR poverty OR inequalit*) AND (entrepreneur*)	
	Search in	Article title, Abstract, Keywords	Topic (title, abstract, author keywords, and Keywords Plus)
	Database	Scopus - August 25, 2022	Web of Science - August 25, 2022
	Inclusion criteria	(i) English documents and (ii) peer-reviewed journal articles (Article or Review)	(i) English documents and (ii) peer-reviewed journal articles (Article or Early Access or Review Article)
	First result	92 articles	95 articles
	Merging result	127 articles = 92 + 95 − 60 removed with Rstudio	
		122 articles = 127 − 5 removed manually	
Data analysis	Screening stage	63 articles were removed	
	Final sample	59 articles	

3   Bibliometric Analysis

The final sample has 59 papers published from 2010 to 2022 in 53 scientific and academic sources. The international journal with the largest number of publications (4 papers) in the sample was Technological Forecasting and Social Change.

The thematic map was generated by RStudio (Fig. 1). From this analysis, important issues were identified after full reading process. The covid-19 pandemic, digital divide and digital inclusion were motor themes and anticipating digital learning to low-literacy people was an important issue to understand and explored in Sect. 4.1. Gender inequality in entrepreneurship was identified as a motor and niche theme and explored in Sect. 4.3. In niche themes, we identified digital economy, digital entrepreneurship, digital innovation and digital technologies as relevant constructs to be explored. From niches and basic themes, financial inclusion was identified, anticipating it is a relevant issue to support low-income entrepreneurs advance in an Industry 4.0 context.

Table 2. Screening criteria.

Included	Not included
1) BOP as entrepreneurs—Examples: – Poor / Low-income – Low digital literacy – Rural – Women – BOP as Sellers in platforms **2) BOP as entrepreneurs and clients** Example: SMEs and fintechs/banks **3) Digital Divide/Inequality studies focused on entrepreneurship**	**4) BOP as client of entrepreneurs**—Examples: – People with disability – Rural population and telemedicine healthcare service / health startups – Low-income people and ICT services – Social enterprises / entrepreneurs / frugal innovation – BOP as clients in platforms **5) Digital Divide/Inequality studies <u>NOT</u> focused on entrepreneurship**

Another bibliometric analysis generated was co-ocurrence analysis by VOSviewer. Figure 2 shows the network and 7 clusters. From this network analysis, we found some themes are not directly connected to entrepreneurship such as cluster 5, cluster 6 and cluster 1. This analysis pointed these keywords are relevant constructs to be explored by are not the main focused to be studied in the full reading step. In addition, six keywords were highlighted as emerging themes: digital technology, covid-19, platform, poverty alleviation, entrepreneurial opportunity and developing country.

4   Discussion

After bibliometric analysis, papers were inserted in NVivo and the automatic coding oriented the full-reading process. During the content analysis, the paper understood entrepreneurship has different challenges in Industry 4.0 depending on the social group. To better understand these challenges, low-income people were divided in three social groups: low-literacy and poor people, rural communities and women.

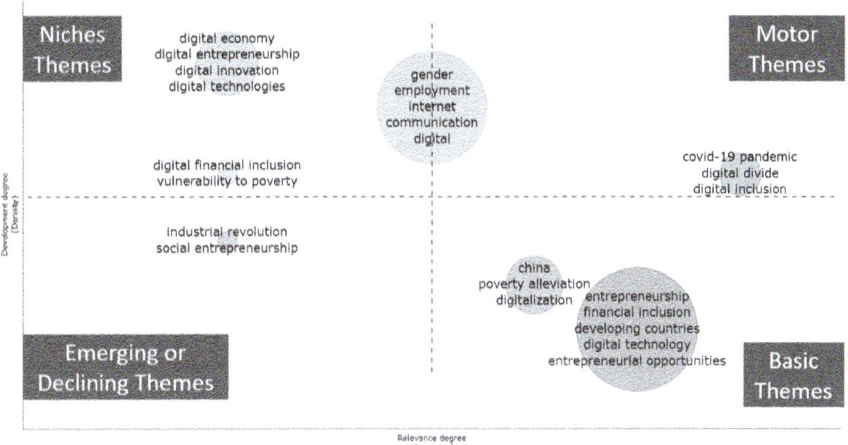

Fig. 1. Thematic map by Rstudio (Bibliometrix).

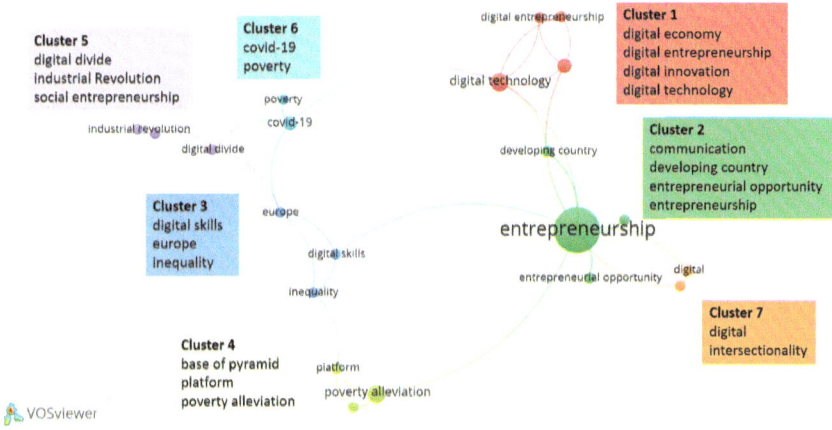

Fig. 2. Co-ocurrence analysis, network visualization by VOSviewer.

4.1 Low-Literacy and Poor People

Digital divide literature and digital gap for disadvantaged people such as low-literacy poor people and homeless has been widely discussed. A study explained the learning challenge is to help low-income entrepreneurs advance through technology and digital literacy (Neumeyer et al., 2020). Digital inclusion of these population even in developed economies is not easy and some initiatives fail to address this social issue. A study case in United States proposed a social media workforce education to train up to 500,000 marginalized and low-literacy residents for jobs in the information and knowledge economy (Wiig, 2016). However, the author explained this initiative did not meet expectations and pointed out a workforce education initiative needed to create and provide job opportunities to participants.

In developing countries, learning digital inclusion is also challenging. In Bangladesh, telecenter initiatives providing affordable information and communication technology (ICT) services to the poor is necessary (Rashid, 2017). Another case study in Taiwan shows low-income people and homeless wanderers need not only financial aids to obtain ICT hardware, software and training, but also patience and compassion to enable their using the Internet to move up social layers (Huang & Cox, 2016).

Due to costs and other advantages, poor entrepreneurs in Colombia manage web presence with social media such as Facebook and Instagram (Garcés et al., 2020). However, their use is based on the intuition and experience of the entrepreneurs, rather than on strategic planning (Garcés et al., 2020).

Associated to digital technologies adoption, technology maintenance is another issue. In United States and Ghana, the cost of constantly upgrading to new technologies is too high relative to income and so entrepreneurs often sought help to repair devices or to learn to use them with newer systems or programs (Avle et al., 2019).

Several studies pointed out digital financial inclusion is connected to promotion of entrepreneurship. In China, the digital finance has created a better business environment, which enables migrant to obtain more employment opportunities (Li et al., 2022). Another study also in China showed digital financial inclusion reduced the vulnerability to poverty for rural households with low income, a low level of financial development and low human capital (Chen et al., 2022). In Kenya, Tanzania and Uganda, women reinforced this perspective showing the significant positive association between mobile money, digital savings and access to digital credit and the promotion of entrepreneurship (Koomson & Martey, 2023). In Bangladesh, bKash, a mobile money service acceptance among marginalized micro-entrepreneurs, also demonstrated the benefits of digital financial inclusion (Rahman et al., 2017).

Other macro recommendations on supporting low-income inclusion through entrepreneurship in Industry 4.0 are cited in the sample: (1) a study in Africa claims that entrepreneurial and digital ecosystems framework is the key to creating a robust digital economy (Mafimisebi and Ogunsade, 2022) and (2) "impact sourcing" by using microwork centers instead of large for-profit vendors (Gino & Staats, 2012).

4.2 Rural Communities

In developing countries, technology and information asymmetry between rural and urban regions is common, the market is characterized with many microbusinesses operating in the informal market, government failure and corruption are expected, and for these reasons, entrepreneurial is challenging (He, 2019).

The development of digital technology can help entrepreneurs, since faster and easier information acquisition helps rural communities to identify new opportunities and innovate (Fahmi & Savira, 2023). However, a study with coffee entrepreneurs and craft producers in Indonesia shows the identification of market opportunities and new business ideas vary depending on socio-cultural values and collectivism influencing the individual valuation and ability to adopt digital technology into rural business (Fahmi & Savira, 2023). In this way, since not all people benefit from digitalization, higher disparity may occur. The authors suggest that it might not be relevant to force the use of digital technology in rural enterprises, instead, it would be more appropriate to nurture social and environmental conditions that enable knowledge sharing about how digital technology can be valuable to each individual (Fahmi & Savira, 2023).

Considering collectivism, there are no boundaries between family and professional life in the rural villages of emerging economies. Family and community are important for businesses in rural emerging markets (Soluk et al., 2021). By surveying more than 1000 microentrepreneurs in rural India, a study found that both the families and communities of rural entrepreneurs have a positive and significant effect on entrepreneurship when digital technologies are used. This happens because the adoption of digital technologies improves the interaction with family and community members and communication of feedbacks, best practices sharing, discussion of new ideas, emotion support and financial resources support.

The adoption of digital platforms has divergent opinions. According to a fieldwork focused on vegetable supply chains in Indonesia, e-commerce enables value co-creation at the bottom of the pyramid and a "social justice logic", because (1) digital technologies created online 'consumption communities' where information and educational interactions supported online food purchases and innovation in the supply chain 'pull' strategies and (2) the short chains of e-commerce enable better partnerships that accommodate fair-trade and faster payment for farmers (Utami et al., 2021). On the other hand, in East China, digital platforms created opportunities for peasants and marginalized urban youth to achieve social mobility (Zhang, 2021). However, it also shapes a new regime of value to rural population with the disappearance of boundaries between manual and intellectual labor, copying and innovation, and individualized entrepreneurship and networked collective production (Zhang, 2021).

4.3 Women

Some studies are focused on understanding if and how digital businesses help reduce inequality between female and male entrepreneurships. In the sample appears study cases focused on women digital entrepreneurs in developed countries such as UK (Dy et al., 2017), United States (Duffy & Pruchniewska, 2017), Australia (Heizmann & Liu, 2022), Spain (Mora-Rodríguez et al., 2020). These studies explain some digital entrepreneurship's advantages: (1) easier to launch a business idea on to the digital market than in the traditionally market, (2) flexible work–life balance, (3) socialization is diminished since they can easily access contacts with digital social networks. However, these studies also highlight the privileges depending on social positions of gender, race and class status.

In the other hand, studies in developing countries or rural areas explore several limitations to women entrepreneurs. The presence of strong market failure in developing countries acts against women's likelihood to pursue a successful business (Mazhar et al., 2022). A 7-year field study in 20 rural villages in India showed that through an ICT intervention disseminate useful information that promotes entrepreneurial activity among women and provide information that can aid in increased savings, wealth and profits (Venkatesh et al., 2017). The study suggested Internet centers in rural locations are useful and staffing such centers with women may facilitate communication profits (Venkatesh et al., 2017). Another survey study with 450 rural women in Pakistan suggested: (1) the local government might devise policies for more inclusive participation of rural women in accessing information, learning and family enterprise loans, (2) local internet service providers can significantly contribute through removing gender-digital-divide where improved digital literacy would open-up new business opportunities, (3) empowering rural-based organizations through soft and hard incentives can play a more inclusive role in nurturing digital and financial abilities and (4) optimizing sites and selling platforms could be an effective tool for removing local marketing barriers (Mazhar et al., 2022).

5 Conclusion

The paper contributes with an initial overview of the challenges of entrepreneurship to low-income people inclusion in Industry 4.0, focusing on three groups: low-literacy and poor people, rural communities and women. It was a first exercise to understand the particularities of the challenges of entrepreneurship in Industry 4.0. However, it is important to reinforce that these social classifications can be combined in reality, just as the challenges can be combined.

Another relevant finding that the study demonstrates is that the challenges of including low-income people in Industry 4.0 can be similar between developed and developing countries. Digital inclusion learning has been studied in countries such as the United States (Wiig, 2016) and Bangladesh (Rashid, 2017). Another example is the challenges faced by women in Spain (Mora-Rodríguez et al., 2020) and Pakistan (Mazhar et al., 2022). Such studies can generate complement insights.

Like other systematic literature reviews, this study has limitations. Considering the research method, this article has limitations due to the keywords used. Furthermore, this article did not adopt the snowballing technique, as this article was limited to carrying out an initial and broad mapping of the challenges of including low-income people through entrepreneurship in Industry 4.0. As a suggestion for future studies, this study suggests an in-depth study of social groups and their specific challenges.

References

Avle S, Hui J, Lindtner S, Dillahunt T (2019) Additional labors of the entrepreneurial self. Proc ACM Human-Comput Interact 3:1–24

Azoeva OV, Yu Mikhalevich L, Ostapenko VA, Shim GA (2020) The labor nature changes and its regulation challenge caused by global digitalization of business. Int J Organ Leadersh 9(4):170–183

Chen S, Liang M, Yang W (2022) Does digital financial inclusion reduce China's rural household vulnerability to poverty: an empirical analysis from the perspective of household entrepreneurship. SAGE Open 12(2):21582440221102424

Du HS, Xu J, Li Z, Liu Y, Chu SKW (2021) Bibliometric mapping on sustainable development at the base-of-the-pyramid. J Clean Prod 281:125290

Duffy BE, Pruchniewska U (2017) Gender and self-enterprise in the social media age: a digital double bind. Inf Commun Soc 20(6):843–859

Dy AM, Marlow S, Martin L (2017) A Web of opportunity or the same old story

Fahmi FZ, Savira M (2023) Digitalization and rural entrepreneurial attitude in Indonesia: a capability approach. J Enterp Commun People Places Glob Econ 17(2):454–478

Garcés LPA, García Nieto MT, Romero González GC (2020) Digital communication in micro and small business the case of the cultural sector in the Colombian department of Bolivar. Rev Comun SEECI 52:149–169

Ghobakhloo M (2020) Industry 4.0, digitization, and opportunities for sustainability. J Clean Prod 252:119869

Gino F, Staats BR (2012) The microwork solution. Harv Bus Rev 90(12):92

Gold S, Hahn R, Seuring S (2013) Sustainable supply chain management in "Base of the Pyramid" food projects - a path to triple bottom line approaches for multinationals? Int Bus Rev 22(5):784–799

He X (2019) Digital entrepreneurship solution to rural poverty: theory, practice and policy implications. J Dev Entrep 24(01):1950004

Heizmann H, Liu H (2022) "Bloody wonder woman!" Identity performances of elite women entrepreneurs on Instagram. Hum Relat 75(3):411–440

Huang SC, Cox JL (2016) Establishing a social entrepreneurial system to bridge the digital divide for the poor: a case study for Taiwan. Univ Ac Inf Soc 15:219–236

Koomson I, Martey E, Etwire PM (2023) Mobile money and entrepreneurship in East Africa: the mediating roles of digital savings and access to digital credit. Inf Technol People 36(3):996–1019

Kotarba M (2018) Digital transformation of business models. Found Manage 10(1):123–142

Li Y, Tan J, Wu B, Yu J (2022) Does digital finance promote entrepreneurship of migrant? Evidence from China. Appl Econ Lett 29(19):1829–1832

Mafimisebi OP, Ogunsade AI (2022) Unlocking a continent of opportunity: entrepreneurship and digital ecosystems for value creation in Africa. FIIB Bus Rev 11(1):11–22

Mazhar S, Sher A, Abbas A, Ghafoor A, Lin G (2022) Empowering Shepreneurs to achieve the sustainable development goals: exploring the impact of interest-free start-up credit, skill development and ICTs use on entrepreneurial drive. Sustain Dev 30(5):1235–1251

Mora-Rodríguez C, Verdú-Jover AJ, Gómez-Gras JM (2020) Analyzing opportunities for eliminating inequality in female digital entrepreneurship in Spain. Eurasian Stud Bus Econ 331–340

Neumeyer X, Santos SC, Morris MH (2020) Overcoming barriers to technology adoption when fostering entrepreneurship among the poor: the role of technology and digital literacy. IEEE Trans Eng Manage 68(6):1605–1618

Oztemel E, Gursev S (2020) Literature review of industry 4.0 and related technologies. J Intell Manuf 31:127–182

Praharad CK (2022) The fortune at the bottom of the pyramid. Strategy+Bus 26:54–67

Rahman SA, Taghizadeh SK, Ramayah T, Alam MMD (2017) Technology acceptance among micro-entrepreneurs in marginalized social strata: the case of social innovation in Bangladesh. Technol Forecast Soc Chang 118:236–245

Rashid AT (2017) Inclusive capitalism and development: case studies of telecenters fostering inclusion through ICTs in Bangladesh. Inform Tech Int Dev 13:14

Sharma G, Jaiswal AK (2018) Unsustainability of sustainability: cognitive frames and tensions in bottom of the pyramid projects. J Bus Ethics 148:291–307

Soluk J, Kammerlander N, Darwin S (2021) Digital entrepreneurship in developing countries: The role of institutional voids. Tech Forecast Soc Change 170:120876

Utami HN, Alamanos E, Kuznesof S (2021) 'A social justice logic': how digital commerce enables value co-creation at the bottom of the pyramid. J Mark Manage 37(9–10):816–855

Venkatesh V, Shaw JD, Sykes TA, Wamba SF, Macharia M (2017) Networks, technology, and entrepreneurship: a field quasi-experiment among women in rural India. Acad Manag J 60(5):1709–1740

Wiig A (2016) The empty rhetoric of the smart city: from digital inclusion to economic promotion in Philadelphia. Urban Geogr 37(4):535–553

Zhang L (2021) Platformizing family production: the contradictions of rural digital labor in China. Econ Labour Relat Rev 32(3):341–359

Digital Product Passports for Light Electric Vehicles
A Tool for Reducing Environmental Impacts, Meeting Regulatory Requirements and Implementing a Circular Economy

Nora Boßung[1,2] and Semih Severengiz[1(✉)]

[1] Sustainable Technologies Laboratory, Bochum University of Applied Sciences, Am Hochschulcampus 1, 44801 Bochum, Germany
semih.severengiz@hs-bochum.de

[2] Universidad de Congreso, Av. Colón 90, M5500 GEN Mendoza, Argentina

Abstract. Rapid population growth and urbanization pose significant challenges to transportation systems and vehicle technology. Expansion of the existing routes alone is inadequate to tackle the problem. To address environmental and health concerns, sustainable structures are necessary. This has led to a growing demand for Light Electric Vehicles (LEVs), which include electrically powered "micro-mobility." The Digital Product Passport (DPP), first researched in electric vehicle batteries, has potential for broader sustainable usage. This concept digitizes product life cycles, improves supply chain transparency, and aids in material recycling in line with green and digital transformation objectives. This study addresses the absence of complete DPP discussions in the electric mobility field. It combines academic insights with feedback from LEV value chain stakeholders, addressing diverse information needs and opportunities. In assessing the DPP's launch, it centers on criteria such as lessening environmental impact, adhering to regulations, and establishing a circular economy. The outcome is a comprehensive DPP framework for LEVs that outlines crucial details for sustainable and circular value chains. This versatile concept has numerous applications, sparking debates about the potential and implementation of DPPs, especially with regards to current product policies.

Keywords: Digital product passport · DPP · Light electric vehicle · LEV · Electromobility · Micromobility · Circular economy

1 Introduction

The European Union strives for climate neutrality by 2050 and has introduced the "Fit for 55 Package" to cut greenhouse gas emissions by at least 55% by 2030 compared to 1990 levels [1]. Road transport, responsible for around a fifth of EU emissions, has seen a 21% rise in CO_2 emissions from 1990 to 2021 due to increased traffic, higher energy consumption, and limited electric adoption [2]. These factors offset the efficiency gains achieved in transportation, impeding emission reductions. Electromobility is vital for transport decarbonization, yet the entire value chain's sustainability challenges must be addressed, particularly for materials like cobalt, nickel, and lithium for batteries.

© The Author(s) 2025
H. Kohl et al. (Eds.): GCSM 2023, LNME, pp. 295–305, 2025.
https://doi.org/10.1007/978-3-031-77429-4_33

The Circular Economy (CE) promotes shared, leased, repaired, and recycled materials and products, extending their life cycle and reducing waste [3]. Despite growing circular economy efforts, challenges such as transparency and data sharing remain. Digital Product Passports (DPPs) offer solutions, backed by the European Commission's Circular Economy Action Plan. While DPPs have been studied, particularly for electric vehicle batteries, a comprehensive analysis of requirements across the electromobility value chain is needed. This study delves into DPPs information needs and potential in sustainable electric vehicle management. A literature review identifies general DPP information requirements. A questionnaire targets LEV value chain stakeholders, uncovering e-mobility specifics and discussing potentials and challenges.

1.1 State of Research

The literature review serves as a basis for the description of the state of the art. Based on the literature selection, the most important existing approaches for the description of product-related data are explained, which serve as a basis for the analysis of the information needs and the development of the questionnaires. As it is crucial for understanding the use of DPPs, the technical implementation is also discussed.

Literature Research. Since the research topic of DPPs is still relatively new and current in relation to electromobility and other research areas, a semi-systematic literature review is conducted, focusing on qualitative information, topic areas and theoretical perspectives, rather than using quantitative data. The semi-systematic literature review refers to published knowledge in the areas of digital product passport/material passport, circular economy, electric mobility sector and LEVs. Since there is little literature on the combination of the three overarching topics, the first step is to analyze the general state of knowledge on DPPs. The literature used for the analysis is identified using ten different search terms and combinations listed in Table 1. The search is conducted using both English and German language terms in order to obtain a larger amount of scientific literature.

Depending on the search term and the database, the number of search hits varies greatly. The next step was to define the inclusion criteria. These are based on the title or abstract that explicitly refers to the search term and relates to one of the areas of circular economy, value chains, product/battery/material passport or the (electric) automotive industry. The literature source was included if it covered general requirements for DPPs. For all 16 searches, the first 50 results were then subjected to a title check. After removing duplicates and checking the abstract, a total of 28 publications remained suitable for analysis. The texts of the 28 results were fully examined, focusing on the different passport approaches, information requirements, use cases and technical implementation. Three additional publications that were not relevant for the analysis were discarded. Thus, three German and 22 English literature sources published in the period 2017–2022 constitute the result of the literature search.

Existing passport approaches for describing product-related data. Legislation is being enacted worldwide to support the development of the CE. However, the standardized and digitized product data required for CE remains limited. According to literature comparison, DPPs are increasingly being acknowledged as an essential tool to achieve

Table 1. First results of the literature search, by search term and database.

Search terms and combinations	Google Scholar	ScienceDirect
"Battery passport" OR "Batteriepass"	153	21
"Circular economy" AND "digital Passport"	94	15
"Circular economy" AND "digital Passport" AND "Automotive"	24	4
"Digital Product Passport" OR "Digitaler Produktpass"	173	9
"Kreislaufwirtschaft" AND "Produktpass"	53	–
"Light electric vehicles" AND "passport"	12	–
"Material Passport" OR "Material Pass"	2900	4907
"Vehicle passport" OR "Fahrzeugpass"	46	1

sustainability and circular systems. Despite the absence of extensive DPP research, initial prototypes are currently being developed, particularly in the production of electric vehicle batteries (EVB) due to the projected growth of EVB, inadequate recycling methods, and sustainable development issues.

Table 2. Comparison of three digital passport sets for the battery sector.

Passport	Description and purpose	Ref
(Digital) Battery passport	Collect and use individual battery data more efficiently. Enable planning activities to make more informed decisions. Facilitate battery recycling. Ensure the sustainability and safety of all batteries in the EU	[4]
Battery identity global passport	Track (global) battery life cycle data to improve recycling (automated sorting and more efficient separation of EVB) and reuse. Contribute to the development of environmentally sustainable recycling technologies	[5]
Battery pass	Document data that fully describes the sustainability and responsibility of the supply chain. Develop cross-industry content and technical standards for a Battery Passport and demonstrate them in a pilot project using a software prototype to demonstrate and verify the basic concept and functionalities	[6]

Table 2 demonstrates the key passport approaches employed in the battery industry. Most research discusses these approaches conceptually, but often lacks practical use cases and comprehensive views. As a result, these approaches only capture a fraction of the capabilities of DPPs, such as involving all the stakeholders in the life cycle. DPPs are either widely applied or focused on sectors such as construction and shipbuilding. Battery studies discuss the limited potential of DPPs in electromobility. This paper bridges the research gap by using the LEV example to outline DPP information needs, criteria, and possibilities.

Requirements for the implementation of DPPs. The existing literature predominantly explores the conceptual aspects of DPP applications and highlights their general benefits. While the discussion of technical implementation has been limited, comprehending DPPs and their design hinges on it. Figure 1 illustrates both the technical and content-related aspects of DPP implementation, outlining the requirements for the theoretical DPP concept.

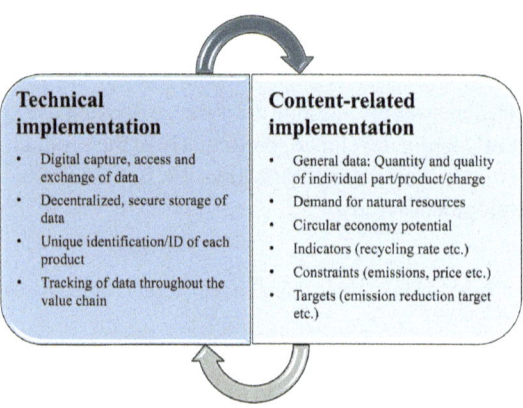

Fig. 1. Requirements for a technical and content-related implementation of the DPP.

Regarding technical implementation, machine-readable DPP information can be stored on servers or in the cloud. Data ownership and access rights for diverse user groups are established features. Additionally, the information should be accessible through a distinct identifier, such as a number or QR code on the product, packaging, or accompanying documents. This facilitates end-to-end data tracing, aligned with literature analysis and regulations like the Sustainable Product Initiative, a cornerstone of the CEAP.

In terms of content, aside from general material information, three core themes should be addressed: recycling potential, diverse indicators (environmental, etc.), and constraints/targets from regulatory frameworks. This paper formulates requirements for a holistic DPP approach for LEVs, catering to stakeholders' information gaps, promoting CE, adhering to regulations, and curbing environmental impacts. These criteria stem from a range of electromobility-related backgrounds, including evolving regulations, the imperative for CE, suboptimal battery recycling, and shifting environmental concerns.

1.2 Theoretical Framework

As a starting point for the rest of the work, the basics, and connections necessary for a better understanding are explained.

Electrification of transport. Amidst the pressing global challenges of climate change, the discourse on novel mobility and propulsion concepts is gaining prominence. The conventional internal combustion engine is increasingly criticized for its adverse environmental effects, including emissions and CO_2 output. This has propelled a demand for transportation alternatives, with electromobility emerging as a promising solution for the energy transition.

However, the production of lithium-ion batteries (LIBs) is complex due to critical raw materials often associated with human rights concerns and substandard working conditions during mining. Furthermore, recycling of used batteries remains underdeveloped, intensifying raw material demand during the ongoing shift to electromobility. The EU anticipates a nearly 18-fold rise in lithium demand by 2030 (close to 60-fold by 2050), along with a 15-fold cobalt demand increase by 2030 [7]. While electric vehicles exhibit lower life cycle CO_2 emissions, particularly when powered by renewable sources, addressing battery production and disposal challenges is vital to enhance electric mobility's efficiency and environmental benefits. The transition to electric vehicles shifts potential negative impacts from usage to production. Focused research and innovation in these realms are pivotal for sustainable and resilient mobility.

Micromobility. The UN predicts a global population increase to 9.7 billion by 2050, with an uneven regional distribution [8]. Urbanization is set to reach approximately 70%, driving megacity growth [9]. This presents fresh challenges for transportation and vehicle technology. Merely expanding roads won't suffice. Societal shifts towards environmental and health awareness are spurring demand for sustainable mobility, including sharing concepts and micro-mobility. Lightweight, electric small vehicles are gaining traction, especially for short urban trips. The market for micromobility, particularly in stationless sharing models, is poised for robust growth in the US, Europe, and China [10]. This transformation holds the potential to permanently reshape the mobility landscape.

Synergies between electromobility and circular economy. Electromobility introduces shifts in mobility and environmental impacts, influencing material cycles. The circular economy strives to elongate product life cycles through strategies like design, sharing, and recycling, encompassing all stages. This strategy has gained prominence in politics, industry, and business, and holds significance in electric mobility. Circular economy measures can enhance the ecology and economy of electric vehicles, offering benefits like cost savings, job stability, and economic resilience. The interplay between electric vehicle batteries and the circular economy is crucial for productivity, longevity, and effective recycling of battery materials. The circular economy serves as a vital component in transitioning from traditional engines to electromobility. While regulations for the circular economy exist, a cohesive framework is absent. A 2020 Capgemini study reveals that circular economy implementation in the automotive sector remains nascent, with only 32% of companies contributing via their supply chains. A functional electric vehicle recycling system is pivotal for sustainable electromobility. Yet, end-of-life electric vehicle reuse and recycling remain underutilized due to economic inviability and inadequate quality assessment, resulting in value losses. Battery recycling is limited and costly. Enhanced transparency is needed to facilitate efficient recycling by providing accurate information on end-of-life product quality. However, the current linear economy lacks this transparency due to information loss along the supply chain, partly due to the absence of standardized protocols and technologies for data transfer. Promoting the circular economy in electromobility and addressing data gaps is essential, and digital technologies can aid this transition. The development of the DPP concept must intricately incorporate circular aspects.

2 Methodology

This research adopts a mixed methods approach, involving a semi-systematic literature review for theoretical insights and a questionnaire series to gather industry perspectives. It initiates by identifying pertinent stakeholders and their roles, followed by a comprehensive analysis of general DPP information requirements. Building on these findings and anticipated questionnaire outcomes, five hypotheses are formulated, which will be validated or invalidated through the questionnaire responses. Moreover, the study explores prospects and challenges in advancing and implementing DPPs within the electromobility sector. Specific LEV-related DPP information needs are inferred from responses and integrated with general requirements to form a comprehensive LEV DPP framework. The provided table addresses the research question: "What information should a DPP for LEVs encompass to enhance data exchange across the value chain, facilitate optimized LEV recycling, diminish environmental impacts, and meet regulations?" The methodology, hypotheses, and surveyed stakeholder groups are summarized in Fig. 2.

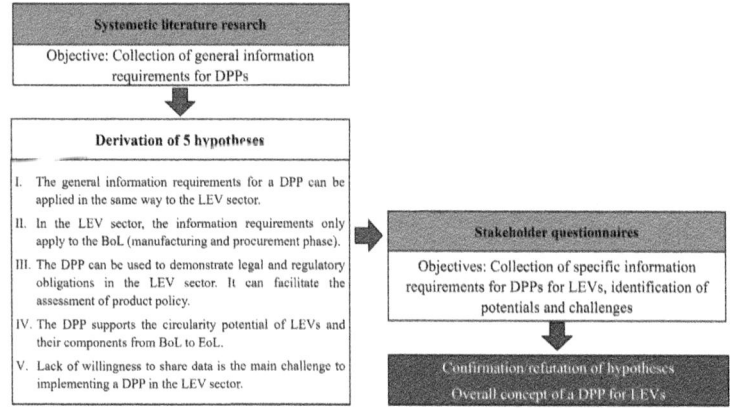

Fig. 2. Methodology.

3 Results

The DPP concept demonstrates its practical relevance by contrasting the information flow in a linear economy without DPP with that in a circular economy with DPP. In this context, the actors involved in DPP play distinct roles. It becomes evident that a comprehensive and high-quality requirement for information exists throughout the entire product life cycle.

3.1 Information Requirements

General information requirements. To identify general information requirements, a comparison of publications and regulations was conducted, forming the DPP concept [12]. These needs were categorized into four life cycle phases: product development, use, repair, and disposal/recycling. Regulatory literature reveals more comprehensive and specific requirements compared to the defined literature, covering physical flows, chemical properties, manufacturing processes, and environmental impacts. Phase allocation of information highlights significant needs in the initial product development, procurement, and manufacturing phase, with suppliers and manufacturers as key sources. Holistic value chain analysis points to various focuses, including actor responsibilities, product transparency, sustainability services, carbon footprint, social standards, and battery status across life cycle phases.

Specific information requirements. Although the general information requests provided valuable insights, specific information gaps remain concerning LEV production, usage, and recycling. Addressing this gap, the developed questionnaires effectively targeted these areas.

Table 3. Excerpt from the questionnaire: For which of the following possibilities does the use of a DPP make the most sense for your company?

Possibilities	Questionnaire							
	Research & development					Producer	Recycler	Sharing service provider
	1	2	3	4	5	6	7	8
Stakeholder collaboration	6	1	4	7	1	4	3	4
Preparation of life cycle assessments	2	1	5	5	3	5	7	5
Communication within the value chain	4	1	2	1	1	3	6	1
Sustainability reporting	2	1	7	2	2	6	5	2
Proof of legal and regulatory obligations	3	7	6	3	1	2	4	2
Decision tool: design, procurement, manufacturing, purchasing, and financing decisions	5	1	3	4	1	1	1	6
Circulation of materials	1	1	1	6	1	7	2	6

Diverse DPP user groups assign varying importance to comprehensive product information, leading to distinct information needs. Meeting sustainability and circular economy criteria, including intricate battery details such as cell chemistry, requires complex data. Collaboration of experts across the LEV life cycle stages during DPP implementation is crucial to ensure transparent and complete information flow across the value chain. The DPP's potential applications vary, reflecting its evolving role in the LEV industry. Nonetheless, optimizing logistics, improving supply chain communication, and facilitating data-driven decision-making emerge as key functions. A summary of questionnaire responses is presented in Table 3, with items ranked in descending order of importance. The varied prioritization highlights diverse stakeholder groups and DPP intentions.

3.2   Potentials of the DPP in the LEV Area

The potential analysis of DPPs has revealed their versatile applications, especially evident through questionnaire responses from different stakeholders. Each actor in the LEV life cycle seeks data for distinct purposes, illustrating diverse utilization intentions. This interaction reinforces the chosen assessment areas. The identified objectives and possibilities were amalgamated into the overall concept, concluding the second section. Figure 3 encapsulates major findings from stakeholder response analysis, illustrating potential across all three evaluation criteria.

Fig. 3. Potentials of the DPP in the LEV area.

Numerous opportunities identified center on supporting the circular economy (CE). The DPP can assist the KrWG by providing information for closed-loop LEV management.

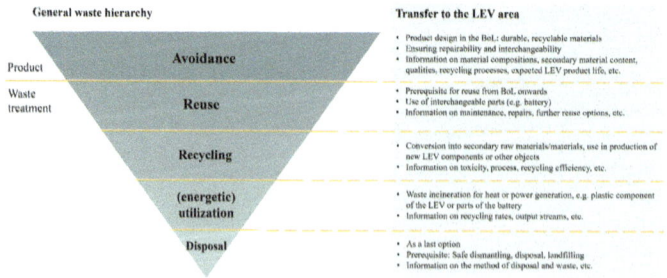

Fig. 4. General waste hierarchy and transfer to the LEV area (own illustration based on [11]).

Illustrated in Fig. 4, the waste hierarchy (§6 KrWG) is applied to LEVs, depicting actions and information needed to achieve each hierarchy level. Avoidance, with the highest CE potential, emphasizes value retention. The DPP's general information and usage data can evidence the KrWG's hierarchy level attainment at a given time. Furthermore, leveraging the DPP for precise LEV component condition classification can avert value losses. Each CE strategy's implementation relies on dependable, up-to-date data, inferred from specific information needs identified by stakeholders. Thus, for the DPP to enhance resource efficiency, it must evolve into a robust digital information system. Reliability and clarity of information are pivotal for recyclates to meet primary material standards, particularly for performance and product safety.

4 Discussion

Confirming Hypothesis 1, all questionnaires echoed previously identified general information requirements from literature and regulatory analysis. These requirements are transferable to the LEV sector. Hypothesis 2 is disproven by survey findings. While literature often portrays manufacturers as pivotal DPP sources due to essential BoL phase data, the survey emphasizes the importance of information on the use and EoL phases. Such data is crucial for ongoing LEV sustainability improvements and identification of recycling potential. Though current regulations hint at or mandate DPP implementation, Hypothesis 3 remains unconfirmed. Questionnaire responses indicate di verse perspectives on DPP potential within the LEV sector, particularly as evidence of legal and regulatory obligations. DPPs aren't proof of these obligations but can aid compliance by collecting data and facilitating external audits. Hypothesis 4 is firmly confirmed. Existing research and passport approach descriptions establish this as a key DPP development factor. Certain passports, like the C2C passport, primarily target CE support. Hence, DPPs are foundational for global CE strategies, enhancing LEV circularity through life cycle monitoring, driving sustainable development and production. Partially supported by interviewees' expertise, Hypothesis 5 highlights data sharing as a significant challenge but not the sole one. Legal regulations and standardization absence are equally significant hurdles, as shown in Table 4.

Table 4. Excerpt from the questionnaire: What do you consider to be the biggest challenges in implementing and using a DPP?

Challenge	Questionnaire								
	Research & development						Producer	Recycler	Sharing service provider
	1	2	3	4	5	6	7	8	9
Threats to data security		x		x			x		x
Manipulation of the data		x	x	x			x		x
Costs		x		x		x			x
Technical hurdles		x						x	x
Expenditure of time		x	x			x			x
Unwillingness to share data	x	x	x		x	x	x	x	x
Lack of documentation of data within the value chain		x				x		x	x
Less flexibility		x							
Competitive disadvantages		x			x				
Lack of legal requirements and standardization	x	x	x	x		x		x	

5   Conclusion

The DPP concept for LEV [12] is rooted in literature research, analysis, and questionnaire discussions, constituting the study's central finding. This concept outlines essential information for establishing sustainable, circular LEV value chains, specifying stakeholder-specific data usage. Its goal is enhancing LEV actors' and external stakeholders' comprehension. The information covers composition, production and usage environmental impacts, and end-of-life cycle recycling solutions. Thus, the concept fully addresses the research question on DPP-required LEV information. A well-crafted DPP holds short- and long-term advantages, enabling forthcoming concrete use cases. Starting in 2026, the EU battery regulation mandates selective DPPs for electric scooter batteries exceeding 2 kWh. If tightened further, DPPs might encompass smaller battery LEVs. While currently limited to battery components, this study suggests extending DPPs to entire vehicles is viable. Content will be consistently updated and expanded. Given evolving policy requirements, additional EU DPP directives are anticipated, impacting LEV supply chain due diligence and product directives in the future.

Acknowledgements. We thank Mr. Víctor Duplanic[2] for useful discussions which led to the generation of new ideas. We extend our gratitude to the participating companies for their valuable contributions by completing the questionnaire, which significantly enhanced the research findings.

References

1. Statistisches Bundesamt, https://www.destatis.de/Europa/DE/Thema/GreenDeal/inhalt.html. Last accessed 28 Sep 2022
2. Statistisches Bundesamt, https://www.destatis.de/Europa/DE/Thema/Umwelt-Energie/CO2_Strassenverkehr.html. Last accessed 30 Jul 2023
3. European Parliament, https://t1p.de/p6i7a. Last accessed 30 Jul 2023
4. Berger K, Schöggl J-P, Baumgartner RJ (2022) Digital battery passports to enable circular and sustainable value chains: conceptualization and use cases. J Clean Prod
5. Bai Y, Muralidharan N, Sun Y-K, Passerini S, Stanley Whittingham M, Belharouak I (2020) Energy and environmental aspects in recycling lithium-ion batteries: concept of battery identity global passport
6. Acatech. https://en.acatech.de/project/battery-pass-made-with-germany-implementation-of-a-new-generation-of-digital-product-handling/. Last accessed 05 Aug 2023
7. European Commission Directorate General for Environment. https://data.europa.eu/doi/10.2779/430546. Last accessed 07 Feb 2023
8. United Nations, https://www.un.org/development/desa/pd/sites/www.un.org.development.desa.pd/files/wpp2022_summary_of_results.pdf, last accessed 2022/10/27
9. The World Bank. https://www.worldbank.org/en/topic/urbandevelopment/overview#1. Last accded 31 Oct 2022
10. Heineke K, Kloss B, Scurtu D, Weig F. https://t1p.de/71doi. Last accessed 01 Nov 2022
11. Glöser-Chahoud S, Huster S, Rosenberg S, Schultmann F (2022) Rücklaufmengen und Verwertungswege von Altbatterien aus Elektromobilen in Deutschland
12. The concept of a DPP for LEVs (supplementary material). https://hsbochumde-my.sharepoint.com/:b:/g/personal/semih_severengiz_hs-bochum_de/EcgzScXJA8FJrDSQYEvr67UB_2h08b8RvQVQF3CGs8IvUQ?e=li4mQW. Last accessed 07 Mar 2024

Remanufacturing

Challenges of Automatic Optical Inspection of Used Turbine Blades with Convolutional Neural Networks

J. Lehr[1]([✉]), C. Briese[1], S. Mönchinger[1], O. Kroeger[1], and J. Krüger[2]

[1] Fraunhofer IPK, Pascalstr. 8-9, 10587 Berlin, Germany
jan.lehr@ipk.fraunhofer.de
[2] TU Berlin, Strasse Des 17. Juni 135, 10623 Berlin, Germany

Abstract. This paper presents an automatic optical inspection task for used turbine blades. The defects arising are very small and occur very rarely. The paper analyzes to what extent state of the art deep learning methods of image processing help to solve the inspection tasks. A total of 34 different turbine blades were acquired image-wise for this work. For the localization and classification of the defects, detection methods such as YOLOv7 were used on the one hand, and segmentation methods such as Mask R CNN and QueryInst on the other. Despite a very small amount of data, the methods can be trained to learn the defects and recognize unseen defects. A maximum mAP 0.5 of 60.9% was achieved. Even though the inspection task was challenging in terms of defect characteristics and the number of training data was low, reliable models could be created. The accuracy is not sufficient for full automation, but it can initially generate useful suggestions for the workers and focus attention on critical areas.

Keywords: MRO · automated optical inspection · machine learning · supervised learning · object detection · object segmentation

1 Introduction

MRO is a process that refers to the Maintenance, Repair and Overhaul of plant, equipment and machinery. The general MRO process can be divided into different phases: Planning, Execution, Review and Documentation. The present work investigates to what extent machine learning methods and especially supervised learning can support the review phase.

Industrial gas turbines generate electrical energy by converting thermal energy into mechanical energy to drive a generator. The main elements of the turbine are the gas compressor, the combustor and the expansion turbine. Both the compressor and the expansion turbine consist of multiple stages, each consisting of numerous turbine blades. A typical turbine blade is usually made of a nickel-based alloy such as Inconel or Nimonic and consists essentially of three parts: the blade, the platform, and the base with its dovetail geometry. The concave part of the blade is called the suction side and the convex pressure side.

© The Author(s) 2025
H. Kohl et al. (Eds.): GCSM 2023, LNME, pp. 309–316, 2025.
https://doi.org/10.1007/978-3-031-77429-4_34

Classification to determine the limits of defects that occur (cracks, notches) is used with the aim of estimating the remaining service life. In the entire process chain, a distinction is made between three states and the process is thus divided into three phases. These are:

1. Goods receipt: Bucket is operationally stressed and (residually) coated. Not in focus here.
2. Repair process (VT): Blade is operationally stressed and decoated. In the VT visual inspection, defects are searched for in the visible wavelength range and without aids.
3. Goods issue: Blade is repaired and coated. The inspection takes place immediately before shipment to the customer. It is in the nature of this inspection step that only very few defects are present.

Table 1 represents which defect types are to be detected in which repair phase. In phase 2 defect are present in the base material and in phase 3 defects are present in the coating. Detection via the presented approach is conducted in phases 2 and 3. AI-based optical inspection is much needed in many industries, as manual inspection of products requires a lot of documentation. Some defects are easy to find, while others are difficult to find, requiring several years of training and experience. Employees should not waste time looking for easy defects and documenting them, as this is an inefficient use of their skills. Implementing AI-based inspection systems can help increase productivity and reduce costs by automating the search for easy defects and shifting employee focus to the more difficult cases.

Table 1. Defect types in the different repair phases

Defect type	Repair process	Goods issue
Spallation		X
Material volumetric loss (MAVL)	X	
Linear indicator (FLIN)	X	
Non-linear indicator (FISO)	X	
Double drilling	X	
Discoloration		X
Recessed trailing edge		X

2 State of the Art

For products and goods with high investment costs and long service life, aftersales services are among the biggest profit drivers for companies, but are often ignored in terms of innovation and research [3]. MRO processes, as an after-sales service can be optimized in terms of cost, time and quality through intelligent data and knowledge management [4]. The inspection of the components at the incoming goods is a crucial process for the subsequent repair process. Incorrect inspections can lead to additional work within the MRO process. Aust et al. [5] showed in a sample of 2600 observations of a manual visual inspection process for aircraft turbine blades that the error rates are not insignificant. Only 60% of the participants were able to obtain acceptable results in the study. This is where AI-based assistance systems can provide support.

Aust et al. compared humans, 2D machine learning (ML) and 3D detection for the inspection process in a study [6] and showed that humans had a better performance compared to the AI available on the market but had disadvantages in the area of objectivity and reproducibility. The 3D method showed high performance in detection but only under high investment costs, long process times and high sensitivity to environmental influences. In addition, the acceptance of the technologies is also considered and it is shown that especially the 2D ML systems often have to struggle with a low acceptance due to no transparency regarding the decision and unclear position of responsibility. This work shows on the one hand that ML in the 2D area can support humans but not replace them. On the other hand that in terms of performance and acceptance of 2D ML systems there is still a high need for development in terms of commercial solutions.

Studies [7] show the freed-up productivity from employees and the improved MRO process also increase sustainability. Improved and complete repair processes increase both, component life and cycles. The modern methods also increase the quality of work, through less assembly work and more opportunity to productively create added value. Increasing efficiency of the repair process also contributes to a higher share of repaired blades.

3   Methods Used

3.1   Yolov7

With the publication of "You Only Look Once" (YOLO), the field of object detection was reframed as a regression problem for spatially separated bounding boxes and their associated class probabilities. The authors are using a singular neural network in order to predict the bounding boxes and class probabilities [8]. YOLOv7 by Wang et al. with its architecture derived from YOLOv4 [9], integrate several other works which lead to a breakthrough in other computer vision areas. Compound model scaling is one of the proposed methods added by the authors in order to balance between high accuracy and inference speed. In comparison to other model scaling methods, compound model scaling allows the model to maintain its properties it had at initial design and an optimal structure. Furthermore the authors introduce a new architecture called Extended Efficient Layer Aggregation Networks (E-ELAN) based on ELAN [11]. More specifically E-ELAN changes the architecture in computational block, ultimately leading to the learning of more diverse features. [10].

3.2   Mask R CNN

Mask R-CNN is built on the Faster R-CNN model [12] and extends it to generate precise pixel masks for the detected objects. In a first step, a feature map of the input image is created using a CNN, on which region proposals are generated using a trained region proposal network. In the second step, bounding boxes and object classes are predicted for each region proposal. The region proposals are transformed to a fixed size using RoIAlign and passed to the mask header consisting of a Fully Convolutional Network and a pixel-by-pixel segmentation is generated [13].

3.3 QueryInst

QueryInst is a multi-level end-to-end instance segmentation system based on Sparse Regional-CNNs [14] and introduces a new perspective by treating instances to be detected as learnable "queries". Here, a "query" is shared between detection and segmentation using dynamic convolution and parallel supervised multilevel learning. The model consists of a query-based object detector and six dynamic mask heads for pixel-by-pixel segmentation and is controlled by parallel supervision [15].

4 Concept

The following Section describes the acquisition of the image-based data set, the challenges of labelling the data, several pre-processing steps, and the training and testing of the machine learning methods.

4.1 Data Set

For the two applications of inspection after decoating and outgoing goods, 20 blades are available in each case. These are captured by the five cameras once on the pressure side and once on the suction side. This results in a data set of $20 \times 5 \times 2 = 200$ images per application. One image set consisting of ten images per blade is shown in Fig. 1. With a resolution of 4000×6000 pixels, the images are on the one hand too large to be processed by state of the art network architectures.

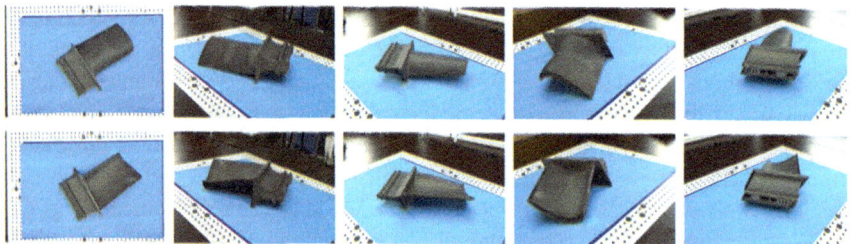

Fig. 1. Top row: Five images of the suction side are shown. Bottom row: The pressure side is shown. The blue background has a length of 42 cm and a width of 30 cm.

On the other hand, the defects on the blades are also very small, so a simple resize would cause many defects to disappear. The images are divided into patches. There are three different patch sizes: 256 px, 512 px, 1024 px. In addition, patches that d not

Fig. 2. The large 4000 × 6000 pixel images are divided into small patches. Only patches where the blade can be seen and where a defect is annotated are included in the data set.

Table 2. The number of patches used for training and testing shown here.

Use case		256 px	512 px	1024 px
VT	Train	525	393	346
	Test	138	105	85
Goods issue	Train	2185	1713	782
	Test	576	429	258

represent a blade are not used further. For supervised learning, only patches on which a defect is visible are used. Figure 2 illustrates this. The remaining patches are distributed for training and testing. Table 2 shows the distribution of patches.

The number of patches is not linear with respect to the patch area, since some defects are very small and far from each other. Finally, all defects are annotated pixel-wise to create a ground truth for the supervised learning methods. From the pixel-wise annotation, bounding boxes are then computed, which are required by the YoloNet.

4.2 Methods and Metrics

The training images are then used for the training. The test images are held out. Only after the convergence of the training, the test data are presented to the methods. The results are presented with two different metrics. On the one hand, the accuracy is used, which indicates how many patches could be correctly assigned to the OK or NOK state. On the other hand, the localization of the defect should also be evaluated. For this purpose, the intersection over union (IoU) is used as a basis. It describes the overlap of the predicted object area with the real object area. For several IoUs, a so-called Average Precision (AP) results. For each defect category, the AP is the average of all IoUs for each image. If the mean is now calculated for all defect types, the result is the mean Average Precision (mAP). In this work, only detections that have at least an IoU of 50% are considered. The metric is now called mAP 0.5.

5 Experiments and Results

5.1 Decoated

For the decoated blade state, the defect types MAVL and FISO are investigated. Table 3 gives a general overview of how exactly the defects are localized. This is indicated using the mAP 0.5. Table 4 presents an analysis of which defects are well recognized by the QueryInst method and which are not. The accuracy is used for this purpose.

Table 3. For the QueryInst and Mask R CNN methods, the mAP 0.5 is shown for different patch sizes.

Method	256 px (%)	512 px (%)	1024 px (%)
QueryInst	51.8	**52.5**	42.4
Mask R CNN	49.1	43.9	15.8

Table 4. The confusion matrix for QueryInst with a patch size of 512 px.

Ground truth		MAVL (%)	FISO (%)	Background (%)
Prediction	MAVL	**62**	–	59
	FISO	–	61	41
	Background	38	39	–

5.2 Outgoing Goods

An analogous analysis is performed for the outgoing goods, but only the Yolov7 was used here. Table 5 shows the mAP 0.5 for different patch sizes. The detection rates for each defect type (doublets hole, discoloration, spallation, and scalloped leading edge) are shown in Table 6.

Table 5. For the Yolov7, the mAP 0.5 for different patch sizes is shown.

Method	256 px (%)	512 px (%)	1024 px (%)
Yolov7	56	**48.4**	42

6   Conclusion

The localization of the defects is similarly high for the patch sizes 256 and 512. The ratio of defect size to the respective patch size is not critical for defect detection here. Only when the patch size is increased to 1,024 px the localization rate in terms of the mAP decreases significantly. The defects simply become too small in relation to the patch size, so that resulting gradients no longer differ significantly from other patterns that are no defects. Basically, the values with a mAP 0.5 greater than 50% are very robust values. As an evaluation baseline, the COCO data set should be mentioned on which state of the art methods achieve just over 60% mAP 0.5. The correct assignment of the defect classes can be improved with 62% and 61%, respectively. A more detailed analysis shows that QueryInst recognized many defects correctly, but did not mark them as defects during annotation. This once again shows the difficulty of the inspection problem, since a large part of them are marginal patterns for which it is not clear whether they are defects or not. Even though the inspection task was challenging in terms of defect characteristics and the number of training data was low, reliable models could be created. The accuracy is not sufficient for full automation, but it can initially generate useful suggestions for the workers and focus attention on critical areas.

Table 6. The confusion matrix for QueryInst with a patch size of 512 px.

Ground truth		Drilling (%)	Discolor (%)	Spallation (%)	RTE (%)	Background (%)
Prediction	Double drilling	61	–	–	–	55
	Discoloration	–	81	–	–	44
	Spallation	–	–	**100**	–	–
	Recessed trailing edge	–	–	–	29	1%
	Background	39	19	–	71	–

In the future, it will be investigated whether anomaly detection methods provide better results. These methods are trained exclusively with defect-free images. At run time, any deviation from this good state should then be detected. Even if the turbines themselves all look very different due to their life cycle and there is no really uniform good state, anomaly detection could work at patch level. There are enough patches that do not have a defect. The potential of anomaly detection will be exploited in future work.

Acknowledgements. This work was co-financed by the European Union (EFRE).

References

1. Vieira DR, Loures PL (2016) Maintenance, repair and overhaul (MRO) fundamentals and strategies: an aeronautical industry overview. Int J Comput Appl 135(12):21–29. Published by Foundation of Computer Science (FCS), NY, USA

2. Han J-C (2004) Recent studies in turbine blade cooling. Int J Rotating Mach 10(6):443–457. PII: H24B20JU4K69LEB6. ISSN: 1023-621X. https://doi.org/10.1080/10236210490503978

3. Uhlmann E, Bilz M, Baumgarten J (2013) MRO—challenge and chance for sustainable enterprises. Procedia CIRP 11:239–244. ISSN 2212-8271. https://doi.org/10.1016/j.procir.2013.07.036

4. Bierer A, Götze U, Köhler S, Lindner R (2016) Control and evaluation concept for smart MRO approaches. Procedia CIRP 40:699–704. ISSN: 2212-8271. https://doi.org/10.1016/j.procir.2016.01.157

5. Aust J, Pons D (2022) Assessment of aircraft engine blade inspection performance using attribute agreement analysis. Safety 8:23. https://doi.org/10.3390/safety8020023

6. Aust J, Pons D (2022) Comparative analysis of human operators and advanced technologies in the visual inspection of aero engine blades. Appl Sci 12:2250. https://doi.org/10.3390/app12042250

7. Pelt M, Stamoulis K, Apostolidis A (2019) Data analytics case studies in the maintenance, repair and overhaul (MRO) industry. MATEC Web Conf 304. https://doi.org/10.1051/matecconf/201930404005

8. Redmon J, Farhadi A (2018) YOLOv3: an incremental improvement. arXiv

9. Bochkovskiy A, Wang C, Liao HM (2020) YOLOv4: optimal speed and accuracy of object detection. arXiv:abs/2004.10934

10. Wang C-Y, Bochkovskiy A, Liao H-YM (2022) YOLOv7: trainable bag-of-freebies sets new state-of-the-art for real-time object detectors. arXiv:2207.02696

11. Wang C-Y et al (2022) Designing network design strategies through gradient path analysis. J Inf Sci Eng 39:975–995

12. Ren S, He K, Girshick R, Sun J (2017) Faster R-CNN: towards real-time object detection with region proposal networks. IEEE Trans Pattern Anal Mach Intell 39(6):1137–1149. https://doi.org/10.1109/TPAMI.2016.2577031

13. He K, Gkioxari G, Doll´ar P, Girshick R (2017) Mask R-CNN. In: 2017 IEEE International conference on computer vision (ICCV). Venice, Italy, pp 2980–2988. https://doi.org/10.1109/ICCV.2017.322

14. Sun P et al (2021) Sparse R-CNN: end-to-end object detection with learnable proposals. In: 2021 IEEE/CVF conference on computer vision and pattern recognition (CVPR). Nashville, TN, USA, pp 14449–14458. https://doi.org/10.1109/CVPR46437.2021.01422

15. Fang Y et al (2021) Instances as queries. In: 2021 IEEE/CVF International conference on computer vision (ICCV). Montreal, QC, Canada, pp 6890–6899. https://doi.org/10.1109/ICCV48922.2021.00683

Developing a Novel Eco-design Approach for Disassembly Based on Fuzzy Sustainable QFD, Customer Segmentation and Circularity

Hermès Tang[(✉)] and Samira Keivanpour

Mathematics and Industrial Engineering Department, Polytechnique Montréal, Montréal, Canada
`hermes.tang@polymtl.ca`

Abstract. Design for Disassembly (DfD) is a challenging concept that facilitates the disassembly of products for refurbishing and reusing their components. In the context of circular economy, DfD minimizes value loss at the end of product's life and remanufacture costs and maximizes environmental benefits. Therefore, DfD considers technical, environmental financial and social factors, but they are rarely integrated. Today, many studies state that the use of Quality Function Deployment (QFD) approach as a decision support tool helps to make choice by promoting one criterion over one another. However, a systematic approach should also consider uncertainties associated with DfD such as technical features, the recovered parts, the disassembly process, and the optimal disassembly sequence due to the product complexity. The current paper analyzes and compares different QFD approaches in the literature review and then provides a new Fuzzy Sustainable QFD (FS-QFD) methodology, which integrates the three pillars of sustainability. Finally, it shows the effectiveness of the suggested approach through a numerical example.

Keywords: Sustainable QFD · Design for disassembly · Fuzzy numbers · Decision support tool

1 Introduction

At the dawn of the 21st century, despite a more environmentally aware and inclusive society, there are still goals to reach in order to make the world more sustainable [1]. While designers were looking to optimize the technical performance of products, they are now interested in design for remanufacturing (DfRem). Indeed, the notion of remanufacturing is in line with circular economy which is an emerging concept that also takes into account social and environmental criteria [2]. Thus, the present paper will focus on disassembly which is an essential stage in the remanufacturing process. It enables the reconsideration of product components at their End-of-Life (EOL), thus extending its lifespan [3]. As far as designers are concerned about social and environmental stakes, they need decision support tools in order to make choices and improve the disassembly operation. Although the traditional QFD methodology as an eco-design tool is popular, the increasing complexity of products has led designers to rethink the approach. Unlike DfRem, a fuzzy

H. Kohl et al. (Eds.): GCSM 2023, LNME, pp. 317–325, 2025.
https://doi.org/10.1007/978-3-031-77429-4_35

approach for DfD is not widely used. The current paper intends to propose a FS-QFD approach for DfD based on the uncertainties related to customers' expectations. The paper extends the existing FS-QFD approach by incorporating customer segmentations and circularity criteria, which are key factors for eco-design. Customer segmentations can help to identify the different needs and expectations of different customer groups, while circularity features can help to reduce the environmental impact and increase the resource efficiency of the product. In addition, it focuses on the design for disassembly of laptops, which is a specific and complex product that requires a tailored eco-design tool. The paper is conducted in several sections. The first one is the literature review, followed by a presentation of the proposed method. Then, a numerical example illustrates the model. Finally, the conclusion and remarks are given in the last section.

2 Literature Review

The aim of this literature review is to understand the various issues involved in the disassembly process and its parameters in the context of the circular economy. It also aims to identify and understand some of the QFD approaches used in the literature to propose a novel eco-design approach.

2.1 Design for Disassembly and Parameters Related to Circular Economy

Disassembly is one of the essential steps in the remanufacturing process because parts and components have to be evaluated, cleaned and refurbished before being reused and reassembled [3]. While the preeminent philosophy of the circular economy is essentially to extend the products lifespan, den Hollander et al. [4] discussed the stage of product design and the environmental benefits of remanufacturing. According to Charter and Gray [5], remanufacturing also offers others benefits, both economic and social. Indeed, the process of remanufacturing including the disassembly's stage creates employment, new skills and leads to innovation and creativity. Thus, remanufacturing covers the three pillars of sustainability [6]. Therefore, disassembly must be carefully considered in the design stage of the product. While DfD improves the ease of remanufacturing, Soh et al. [7] underlined the importance of scrutinizing the disassembly sequence and reminded that the three main challenges in DfD are methodology, technology and human factors. Some authors have proposed different methodologies which optimize the disassembly sequence for saving time and cost, however operators may find it difficult to access fasteners or parts [3, 8]. There are also others factors that affect DfD particularly materials, architecture and fasteners and connectors. Thanks to materials studies, DfD improves the environmental impact of products [9]. Finally, Crowther [10] discussed that DfD can be based on three-dimensional hierarchic level: resources, lifecycle stage and strategies like conserving, reusing, recycling, protecting or quality. Crowther believed that DfD can reduce the number of materials and components, avoid toxic and hazardous materials, and design a simple product that is easy to dismantle and interchange with other products.

2.2 Quality Function Deployment as an Eco-design Tool

The trend of an eco-design tool is reducing environmental impact and costs by comparing these indicators at every stage of product life cycle. However, the modern concept of circular economy leads designers to rethink the relevance of the eco-design tool because the concept is more centered on the three pillars of sustainability, with particular attention paid to the EOL phase [11]. Ultimately, some new methodologies that integrate the concept are developed in the literature. Soh and Wong [12] introduced an advanced methodology which explains a new approach of eco-design in the context of circular economy. In the literature, the first eco-design tools developed were checklists or tables for identifying hot spots [13]. A survey summarized many eco-design methods and tools related with circular economy, one of them is Quality Function Deployment which will be the main point of this paper [14]. By using a House of Quality (HoQ), the designer can favor technical features based on the results provided by the classical QFD approach. In this way, it allows better products design [15].

2.3 Fuzzy Sustainable QFD

However, as products and their objectives become more complex, novel approaches must be constantly proposed. Some papers have shown other alternative QFD approaches. Indeed, Siwiec et al. [16] proposed a QFD-CE method based on the goals of the circular economy and customers' needs. The method modifies the HoQ by adding a floor granted to the circular economy. Romli et al. [17] suggested a QFD approach whose procedure is divided into four successive stages. They justified the environmental selected parameters affiliated to the manufacturing process. Shi et al. [18] developed a fuzzy QFD approach for improving an engine's remanufacturing process. Fuzzy information was used to correlate customers' needs with remanufacturing process solution alternatives. Yang et al. [19] submitted a fuzzy QFD approach for DfRem and bolstered their model with an automotive case study. The authors focused on environmental issues and costs associated with remanufacturing activities. As Gun and Sang [20] discussed in their construction case study, a fuzzy QFD approach allows the manager to conduct the work properly and in line with the objectives set.

2.4 Synthesis of Literature Review

To sum up, the present literature review emphasizes the parameters linking the concept of DfD and circular economy and highlights different framework of fuzzy QFD approaches. Table 1 provides the differences or similarities between the numerous studies on DfD or DfRem using QFD approach. As a result, gaps in the literature are identified. Despite the amount of work around fuzzy QFD, DfRem is more covered than DfD, that is why the paper focuses on DfD. Also, it shows the lack of consideration for the uncertainties associated with customers' expectations. Only the weights provided by customers' expectations are given by fuzzy numbers, and yet the panel of customers surveyed is very heterogeneous. It would be interesting to categorize and weight the different customers since they have different requirements [21].

Table 1. Parameters' comparison between DfD or DfRem case studies

Reference	Case study / Numerical example	Main parameters used
Shi et al. (2023) [18]	Remanufacturing an engine for heavy-duty trucks – Fuzzy QFD and fuzzy FMEA approach	Remanufacture process Failure mode
Yang et al. (2013) [19]	Remanufacturing in automotive industry – Traditional HoQ with deviation of parameters weights	Remanufacturability Environmental impact and costs Operators' workload
Keivanpour (2023) [22]	Disassembling parts in industry 4.0 – Factors' importance is represented by triangular fuzzy numbers	Ease of disassembly Environmental impact and costs Operators' workload
Zhang et al. (2019) [23]	Remanufacturing an engine's crankshaft – Traditional QFD with fuzzy weighting	Remanufacturability Failure mode
Zhang et al. (2015) [24]	The use of spindle in a remanufacturing process – Results from fuzzy linear regression complete the HoQ	Failure mode Process characteristics
Yüksel (2010) [25]	Remanufacturing in automotive industry – Traditional QFD approach	Remanufacturability
The present paper	Disassembling a laptop – FS-QFD approach	Disassembly, environmental and social impact

3   Methodology

In this section, we propose a FS-QFD approach, which is an improvement over the conventional QFD method. The FS-QFD approach differs from the classical QFD in diverse ways. Unlike the traditional QFD, which only considers the functional requirements of products, this approach considers the environmental and social impacts of products in the design process. The methodology uses fuzzy numbers to represent the customers' needs, which are often vague and uncertain. The fuzzy numbers quantify the degree of satisfaction of customers with different product attributes. The novel approach also includes customers segments in the QFD structure allowing us to identify the different needs and preferences of various customers groups and tailor the product design accordingly. The structure of our methodology is shown in Fig. 1.

The FS-QFD approach has the following steps:

- Step 1: Identify and prioritize customers' needs.

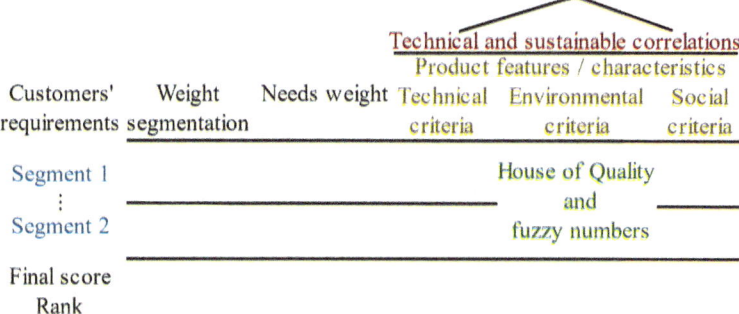

Fig. 1. FS-QFD approach for DfD

- Step 2: Assign the weights for customers' requirements are calculated by averaging all values between 1 and 5, and then translated by a triplet of numbers: (1, 2, 3) is for weak level, (4, 5, 6) is for medium, (6, 7, 8) is for strong and (7, 8, 9) for very strong. In this stage, customers' needs are also segmented according to categories of people.
- Step 3: Identify and prioritize technical features, and environmental and social criteria.
- Step 4: Construct the HoQ matrix and fill it with fuzzy triangular numbers that translating the mutual influence between customers' needs and technical features, and sustainability criteria. Then, scores are calculated by multiplication of the elements in the relationship matrix and weights. The variable k represents the segment of customers' expectations, n the number of customers' needs for each segment, A_k represents the matrix of needs' weight, B_k fuzzy triangular numbers and C weights of each segment of customers' needs. Equation (1) shows the score $\tilde{S}_k$ of each column and Equation (1) reveals the final score S.

$$A_k = \begin{pmatrix} a_{11} & a_{12} & a_{13} \\ \vdots & \vdots & \vdots \\ a_{n1} & a_{n2} & a_{n3} \end{pmatrix}_k , B_k = \begin{pmatrix} b_{11} & b_{12} & b_{13} \\ \vdots & \vdots & \vdots \\ b_{n1} & b_{n2} & b_{n3} \end{pmatrix}_k \text{ and } C = (c_k), \forall k, \forall n$$

$$\tilde{S}_k = \sum_{i=1}^{n} \sum_{j=1}^{3} \frac{(a_{ij_k} . b_{ij_k})}{3}, \forall k, \forall n \tag{1}$$

$$S = \sum_{k=1}^{m} c_k . \tilde{S}_k, \forall m \tag{2}$$

- Step 5: Evaluate final scores and rank in order of importance.

4 An Illustrative Example

As far as FS-QFD approach can be applied to any products in the context of manufacturing, the aim of the following case study is to demonstrate the methodology of circular design for the electronics industry, focusing on laptops as an example. The first methodology's step is to elicit and segment customers' needs including circular parameters. The second step is to segment and weight customers' expectations according to the target users. This step recognizes that different customers may have different product needs and priorities. For example, a student may value low cost and durable product more than a professional who may value high performance and large storage space more. The third step is to identify the sustainability criteria that affect the disassembly operation of laptops. The criteria are divided into three categories: technical, environmental, and social. Based on the guidelines and recommendations from the literature, technical criteria that affect DfD include weight, shape, size, material, structure, number of components, and connection and fasteners. Other technical criteria are comprised such as software and electronic hardware. Environmental criteria are composed of energy use including gas, petroleum, electricity, renewable energies, and coal [26], carbon footprint, raw materials consumption, and waste generation. Social criteria include operators' workload, stress and fatigue, and risk of injury. Then, after constructing the HoQ, the final scores are calculated to rank the most influential parameters. The details and results are shown in Fig. 2. The main finding if that Material is the most crucial factor to consider for circular design of laptops, followed by Energy use and Waste.

Customers' needs	Segment weight	Needs weight	Weight	Shape	Size	Material	Structure	Numbers of components	Connection and fasteners	Software	Electronic hardware	Energy use	Carbon footprint	Material usage	Waste	Operators' workload	Stress and fatigue	Injury risk
			Technical features									**Environmental criteria**				**Social criteria**		
High performance		(1,2,3)				(1,2,3)				(7,8,9)	(6,7,8)	(4,5,6)						
Long battery life	0,30	(4,5,6)				(4,5,6)	(1,2,3)	(1,2,3)		(4,5,6)	(6,7,8)	(4,5,6)	(6,7,8)					
Large storage space		(1,2,3)								(4,5,6)	(6,7,8)							
Light weight		(4,5,6)	(7,8,9)	(6,7,8)	(6,7,8)	(4,5,6)	(4,5,6)	(4,5,6)	(4,5,6)									
Score Segment 1			40,6	35,6	35,6	56	36,3	36,3	25,6	53	65	10,6	0	0	0	0	0	0
Affordable price		(7,8,9)	(1,2,3)			(6,7,8)	(6,7,8)	(1,2,3)		(6,7,8)	(6,7,8)							
Durable and rugged	0,30	(6,7,8)	(1,2,3)			(6,7,8)	(4,5,6)	(4,5,6)	(4,5,6)		(4,5,6)							
Security features		(4,5,6)										(1,2,3)	(4,5,6)					
Score Segment 2			31,3	0	0	106,3	40,6	40,6	40,6	67,3	118	0	0	0	0	0	0	0
Less energy		(6,7,8)										(1,2,3)	(7,8,9)					
Less carbon footprint		(6,7,8)				(1,2,3)								(7,8,9)				
Less wastes		(6,7,8)				(1,2,3)								(6,7,8)	(7,8,9)			
Less material usage	0,40	(4,5,6)				(6,7,8)								(7,8,9)				
Ease of disassembly		(4,5,6)	(1,2,3)	(4,5,6)	(1,2,3)	(1,2,3)	(7,8,9)	(7,8,9)	(7,8,9)								(4,5,6)	(4,5,6)
Recyclable materials		(7,8,9)				(6,7,8)		(1,2,3)				(1,2,3)	(1,2,3)	(4,5,6)	(4,5,6)			
Safety standards		(4,5,6)													(1,?,3)		(4,3,6)	(4,3,6)
Workers' condition		(1,2,3)											(1,2,3)	(4,5,6)		(7,8,9)	(6,7,8)	(6,7,8)
Score Segment 3			10,0	13,6	10,8	132,3	40,6	57,3	40,6	0	14,6	73,3	73,3	131	108	16,6	66	66
Score			25,81	20,92	14,92	101,61	39,31	45,99	36,10	36,09	60,74	32,50	29,32	52,40	43,20	6,64	26,40	26,40
Rank			12	13	14	1	6	4	7	8	2	9	10	3	5	15	11	11

Fig. 2. A FS-QFD approach applied to a laptop DfD case study

5   Conclusion

Traditional QFD approach is a design tool that allows to associate customers' requirement with technical criteria and that solves conflicting design requirements. The paper presents an alternative approach that adds social and environmental features in response to a more circular product policy. Also, due to uncertainties and vague information, fuzzy numbers are included in the decision support tool to bring more reflection about solutions. As customers' categories are heterogenous, weighting customers' expectations seems decisive. An example from the electronics field is provided for illustrating the

FS-QFD methodology. However, there are some limitations of the case study. The sustainability criteria and circular parameters may not be representative of all the aspects of circular design for laptops. Indeed, it would be relevant to scrutinize other environmental criteria such as water consumption, eco-toxicity, greenhouse gas or acidification, in future research. The customers' needs may depend on the context, culture and personal preferences of the customers and its segmentation and weighting may not reflect the actual market demand or users' behaviors. The final score may be sensitive to the choice of scoring method, data sources and assumptions. Thus, this model could be upgraded in several ways. Firstly, using interval-valued fuzzy numbers instead of triangular fuzzy numbers would gain control more accurately and flexibility on uncertainties and variabilities of relationships between customers' expectations and product features. Also, to improve the segments' weight, it could be interesting to use an Analytic Hierarchy Process to arrange customers' requirements as well. So, the results obtained from feedback or surveys would allow us to determine more precisely needs to be well weighted. Also, a sensitivity analysis should be performed to complete the AHP. With an One-At-a-Time (OAT) methodology changing one parameter at a time, it enables to compare different solution and evaluate, if certain weights generate strong imbalance of decisions [27].

Acknowledgements. The authors gratefully acknowledge the financial support from the Natural Sciences and Engineering Research Council of Canada (NSERC) and PIED of Polytechnique Montreal.

References

1. United Nations (2015) Transforming our world: the 2030 agenda for sustainable development. Department of Economic and Social Affairs. https://sdgs.un.org/2030agenda. Accessed 17 Jul 2023
2. Khan S, Haleem A, Fatma N (2022) Effective adoption of remanufacturing practices: a step towards circular economy. Jnl Remanufactur 12:167–185. https://doi.org/10.1007/s13243-021-00109-y
3. Soh SL, Ong SK, Nee AYC (2015) Application of design for disassembly from remanufacturing perspective. Procedia CIRP 26:577–582. https://doi.org/10.1016/j.procir.2014.07.028
4. den Hollander M, Bakker CA, Hultink E (2017) Product design in a circular economy: development of a typology of key concepts and terms: key concepts and terms for circular product design. J Ind Ecol 21. https://doi.org/10.1111/jiec.12610
5. Charter M, Gray C (2008) Remanufacturing and product design. Int J Prod Dev 6. https://doi.org/10.1504/IJPD.2008.020406
6. Hunkeler D, Rebitzer G (2005) The future of life cycle assessment. Int J Life Cycle Assess 10:305–308. https://doi.org/10.1065/lca2005.09.001
7. Soh SL, Ong SK, Nee AYC (2014) Design for disassembly for remanufacturing: methodology and technology. Procedia CIRP 15:407–412. https://doi.org/10.1016/j.procir.2014.06.053
8. Matthieu G, François P, Tchangani A (2012) Optimising end-of-life system dismantling strategy. Int J Prod Res 50:3738–3754. https://doi.org/10.1080/00207543.2011.588263
9. Bogue R (2007) Design for disassembly: a critical twenty-first century discipline. Assem Autom 27:285–289. https://doi.org/10.1108/01445150710827069
10. Crowther P (2005) Design for disassembly - themes and principles

11. Bourgeois I, Queirós A, Oliveira J, Rodrigues H, Vicente R, Ferreira VM (2022) Development of an eco-design tool for a circular approach to building renovation projects. Sustainability 14:8969. https://doi.org/10.3390/su14148969

12. Soh KL, Wong WP (2021) Circular economy transition: exploiting innovative eco-design capabilities and customer involvement. J Clean Prod 320:128858. https://doi.org/10.1016/j.jclepro.2021.128858

13. Szendiuch I, Schischke K (2006) Eco-design - new part of technological integration. In: 2006 29th International Spring seminar on electronics technology, pp 521–526

14. Royo M, Chulvi V, Mulet E, Ruiz-Pastor L (2023) Analysis of parameters about useful life extension in 70 tools and methods related to eco-design and circular economy. J Ind Ecol 27:562–586. https://doi.org/10.1111/jiec.13378

15. Delgado-Hernandez D, Bampton K, Aspinwall E (2007) Quality function deployment in construction. Constr Manag Econ 25:597–609. https://doi.org/10.1080/01446190601139917

16. Siwiec D, Pacana A, Gazda A (2023) A new QFD-CE method for considering the concept of sustainable development and circular economy. Energies 16. https://doi.org/10.3390/en16052474

17. Romli A, Prickett P, Setchi R, Soe S (2015) Integrated eco-design decision-making for sustainable product development. Int J Prod Res 53:549–571. https://doi.org/10.1080/00207543.2014.958593

18. Shi J, Ren M, Shu F, Xu H, Cui J (2023) A remanufacturing process optimization method based on integrated fuzzy QFD and FMEA. Jnl Remanufactur 13:121–136. https://doi.org/10.1007/s13243-022-00123-8

19. Yang S, Ong SK, Nee AYC (2013) Design for remanufacturing - a fuzzy-QFD approach. In: Nee AYC, Song B, Ong S-K (eds) Re-engineering manufacturing for sustainability. Springer, Singapore, pp 655–661

20. Lee GH, Park SH (2021) Fuzzy QFD-based prioritization of work activities of construction for safety. ICIC Express Lett Part B Appl 12:1–8. https://doi.org/10.24507/icicelb.12.01.1

21. Zhou J, Zhai L, Pantelous AA (2020) Market segmentation using high-dimensional sparse consumers data. Expert Syst Appl 145:113136. https://doi.org/10.1016/j.eswa.2019.113136

22. Keivanpour S (2023) A fuzzy sustainable quality function deployment approach to design for disassembly with industry 4.0 technologies enablers. In: Kohl H, Seliger G, Dietrich F (eds) Manufacturing driving circular economy. Springer International Publishing, Cham, pp 772–780

23. Zhang X, Zhang S, Zhang L, Xue J, Sa R, Liu H (2019) Identification of product's design characteristics for remanufacturing using failure modes feedback and quality function deployment. J Clean Prod 239:117967. https://doi.org/10.1016/j.jclepro.2019.117967

24. Zhang X, Zhang H, Jiang Z, Wang Y (2015) An integrated model for remanufacturing process route decision. Int J Comput Integr Manuf 28:451–459. https://doi.org/10.1080/0951192X.2014.880804

25. Yüksel H (2010) Design of automobile engines for remanufacture with quality function deployment. Int J Sustain Eng 3:170–180. https://doi.org/10.1080/19397038.2010.486046

26. Suliga M, Wartacz R, Kostrzewa J, Hawryluk M (2023) Assessment of the possibility of reducing energy consumption and environmental pollution in the steel wire manufacturing process. Materials 16:1940. https://doi.org/10.3390/ma16051940

27. Librantz AFH, Dos Santos FCR, Dias CG, Da Cunha ACA, Costa I, De Mesquita SM (2016) AHP modelling and sensitivity analysis for evaluating the criticality of software programs. In: Nääs I et al (eds) Advances in production management systems. Initiatives for a sustainable world. Springer International Publishing, Cham, pp 248–255

Enabling Aircraft Recycling Through Information Sharing and Digital Assistance Systems

Dennis Keiser[1(✉)], Birte Pupkes[1], Thorsten Otto[2], Matthias Reiß[2], Matthias Poggensee[2], Sonja Rehsöft[2], Antje Terno[2], Rafael Mortensen Ernits[2], and Michael Freitag[1,3]

[1] BIBA – Bremer Institut für Produktion und Logistik GmbH at the University of Bremen, Hochschulring 20, 28359 Bremen, Germany
ked@biba.uni-bremen.de

[2] Airbus Operations GmbH, Kreetslag 10, 21129 Hamburg, Germany

[3] Faculty of Production Engineering, University of Bremen, Badgasteiner Straße, 28359 Bremen, Germany

Abstract. Due to the environmental targets of the aviation industry, opportunities for optimization are being explored along the entire life cycle of commercial aircraft. In this context, the recycling of aircraft is increasingly the focus in the aviation industry. Previous research work has therefore examined the overarching recycling process. However, current approaches to improve and increase the aircraft recycling are not sufficient to achieve the defined goals. Based on this motivation, this paper presents first the general challenges of aircraft recycling. Subsequently, the paper shows a conceptual framework whose focus is on the lifecycle phases as well as on its stakeholders and its data. Furthermore, a first conceptual data sharing architecture for the implementation of the approach is introduced. The operationalization in the form of a user interface for a digital assistance system which makes the data available to a recycler in a structured way concludes this paper.

Keywords: End-of-life · Aviation industry · Recycling · Data sharing

1 Introduction

The aviation industry is facing a variety of challenges striving to meet self-imposed sustainability goals [1]. In addition to the development of completely new aircraft concepts and the incremental optimization of existing aircraft concepts, the end-of-life (EOL) phase of aircraft is of great interest [2, 3]. In the context of aircraft recycling, the subject of current research is the investigation of economic efficiency [3] as well as the modeling of the environmental impact of various EOL strategies [4]. Furthermore, approaches for process-oriented optimization of planning processes have been investigated [5, 6]. However, the opportunities for improved stakeholder collaboration throughout the lifecycle to collect relevant data for aircraft recycling and make it available to aircraft recyclers

H. Kohl et al. (Eds.): GCSM 2023, LNME, pp. 326–334, 2025.
https://doi.org/10.1007/978-3-031-77429-4_36

at the end of the lifecycle has not been investigated yet. This motivates the contribution and represents an overarching approach for collecting and sharing recycling-relevant data along the life cycle of commercial aircraft. The aim is to integrate the data into the recycling process by means of a digital assistance system, which are already widely used in other industries and process such as assembly [7]. This is of particular importance for components and assemblies with a low resale value, as such components are currently not being recycled. This paper is structured as follows to introduce the developed concept: In Sect. 2, the basics of aircraft recycling and digital assistance systems are presented. Section 3 shows current challenges of information sharing and aircraft recycling in particular. In Sect. 4, the developed concept, a general framework for information sharing in the aviation industry is presented and a user interface is introduced. This paper concludes with a critical discussion and the further research possibilities.

2   Aircraft End-of-Life

A typical commercial aircraft has a lifetime of 25 years and is then no longer used in flight operations. It is either initially stored in specially designated aircraft graveyards or sold to specialized recycling companies. Regarding the forecast of worldwide retirement of commercial aircraft, various simulations and models can be found in the literature [2, 3]. The projection by the German Aerospace Center used a Monte Carlo simulation and shows a continuous increase in aircraft to be recycled [3]. In addition to the forecasts, the state of the art in end-of-life treatment of aircraft is described in various scientific publications using process diagrams [6, 8]. Most publications refer to the 'Process for Advanced Management of End-of-Life of Aircraft' project from 2005 [9]. In collaboration with Airbus, a French recycling company dismantled and recycled an entire Airbus A300. The goal of the project was to demonstrate that 85% of the total weight of the aircraft could be recycled [9]. As part of the project, an overall process for aircraft end-of-life procedures was developed and presented: After an aircraft has been decommissioned by the airline, it is usually first put into storage by specialized companies. A decision is then made as to whether the aircraft should be resold. This may be the case, for example, if the aircraft is to be converted into a freighter. If the aircraft is not resold, the next step is decommissioning. During this step, hazardous materials and substances in particular are disposed of. After all hazardous materials have been removed, the planning of the recycling process begins, which includes a profitability calculation. In the last step of the process, the components and parts are disassembled. Finally, the components are reprocessed according to the four possible end-of-life options: re-use, recycling, recovery and disposal [6].

3   Challenges

3.1   Aircraft Recycling

Based on the current state of the art of aircraft end-of-life, seven overarching challenges for aircraft recycling were identified. These are listed in Table 1. The first challenge is the high proportion of manual activities for recycling (1). In conjunction with the high technical complexity (2) of commercial aircraft and the supply chain (3), the economic recycling (4) of non-high-value components has so far only been possible to a limited extent. This can lead to a situation where the theoretically achievable recycling rate is not reached and thus opportunities for reducing the environmental impact are not used. Furthermore, there is currently a lack of material and component tracking (5) to enable reliable profitability calculations and to make the complexity of recycling controllable through the targeted provision of information. Future aircraft types will also have an increased proportion of composite materials (6) and thus pose further challenges for aircraft recycling. Finally, the last challenge results from the lack of information. Due to the patchy data situation, research and practice currently continue to lack consistent decision-making and modeling models (7) for optimizing recycling processes.

These challenges specifically apply to the aircraft cabin, as there is a high material mix (e.g. seats, galleys). Furthermore, the cabin components are not high-value parts and are therefore rarely optimally recycled. Given the relatively short lifecycles of cabin components, there is significant potential for improving recycling rates in this area. However, the challenges are also valid for other low-value parts such as landing and wheel components.

3.2 Information Sharing

Due to the cooperative character of the approach presented here, the provision of data and information with the aim of improving recycling and re-use can be referred to information sharing on a conceptual level. This describes the cross-company exchange of information to achieve specific goals such as the optimization of processes or the increase in transparency [15]. The increasing horizontal and vertical integration of information systems enables targeted information sharing across company boundaries, but this is only one necessary prerequisite [16]. Rather, other dimensions need to be considered in the successful design and implementation of information sharing. Against this background, a systematic analysis of the challenges related to information sharing for establishing a circular economy was investigated by [17]. The authors divide the challenges into three overlapping dimensions [17]: First, the already initiated technological integration. In addition, high costs and a lack of standards are often responsible for the fact that technological integration cannot be implemented by various stakeholders along the product lifecycle for the purpose of information exchange. Second, challenges at the administration level are mentioned. This includes e.g. a lack of incentives for stakeholders or the low quality of information. The third dimension describes the general management level. According to [17] here one challenge is the lack of knowledge regarding the benefits of information sharing.

4 Concept Development of a Digital Assistance System Supporting Aircraft Recycling

Based on the research on the state of the art and the derived challenges of information sharing as well as aircraft recycling, a concept for an digital assistance system is presented in this section. The basis for developing the concept is the following hypothesis for the aviation industry and recycling:

Table 1. Identified challenges for aircraft recycling.

No.	Challenge	Description	Reference
(1)	High proportion of manual activities	Aircraft recycling has so far been characterized by manual processes and is therefore very labor-intensive. Automation has not yet taken place	[3]
(2)	Product complexity	Aircraft are high-tech products of high technical complexity and disassembly is therefore a challenge. In addition, an individualization of the aircraft is taking place	[10–12]
(3)	Complexity of the supply chains	The entire aircraft industry consists of complex supply chains and therefore it is difficult to track information	[10, 11]
(4)	Economic efficiency	The economic recycling of non-high-value parts is only partially possible due to the high manual labor share	[13]
(5)	Lack of material and component tracking	Up to now, there have been no solutions to identify materials directly and to have components and data available for recycling	[14]
(6)	Increasing share of composites	The proportion of composite materials is currently increasing in the aviation industry. The separation and recycling of such materials has so far only been partially solved technically	[3]
(7)	Lack of decision and modeling models	Due to the relatively young research field of aircraft recycling, no sufficient decision and modeling models exist so far. These are based on data from stakeholders and processes	[2]

H1. To implement a technology-based strategy for recycling, a common and open framework is needed.

H2. To achieve a broader adoption, a user friendly and context-based interface for recyclers is necessary.

4.1 Conceptual Framework

Based on the needs formulated in Sect. 1 and the actual process and current challenges presented in Sects. 2 and 3, the solution-oriented conceptual design is presented in this section. As presented in Sect. 3, one of the core challenges, besides economic issues, is the lack of information and traceability of commercial aircraft components. Here, the classical life cycle consists of four phases: First, the design and production of parts, followed secondly by final assembly, and then the aircraft is handed over to an airline, followed by the operational phase. During the operating phase of about 25 years, maintenance and servicing are carried out.

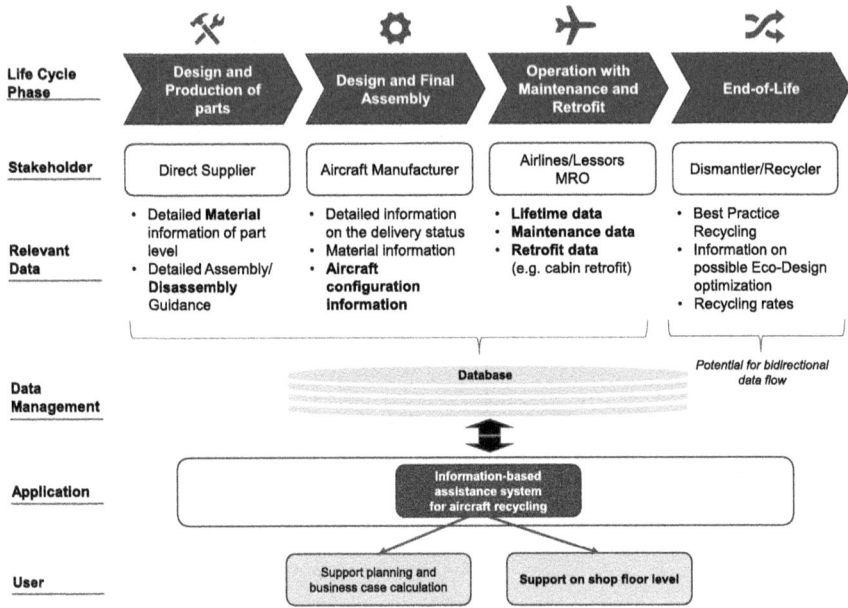

Fig. 1. Conceptual framework for digital assistance system based on life cycle stages, stakeholder and data.

In addition, so-called retrofits are carried out for the cabin. This involves replacing parts or even the entire interior. Reasons for a retrofit include classic signs of wear and tear, rebranding by airlines or an adjusted strategic orientation of the airline. As shown in Fig. 1, the phases can be assigned to stakeholders involved and to data generated for recycling or from recycling. Suppliers at the beginning of the life cycle have two types of relevant data for recycling in particular. First, suppliers have detailed material information of components and assemblies, which is of high importance for recycling. In addition, detailed assembly and disassembly instructions are known, which are particularly necessary for non-destructive and sorted disassembly. The aircraft manufacturers know the delivery condition of the aircraft with the selected configuration as well as further material information. From the time of delivery, the airlines, lessors and MROs (Maintenance, Repair and Operation service provider) are responsible for operating the aircraft and maintaining its airworthiness. Here, data such as operating hours of the individual assemblies are recorded. This is relevant for a possible re-use of assemblies. This also includes maintenance and repair data. In addition, there is information on retrofits at the aircraft level.

The data collected in this way over the entire life cycle must be made available to the recycler/dismantler in a structured form by means of a database. The application level enables two possible usage scenarios for the recycler for the data as shown in the conceptual framework in Fig. 1. First, the data can be used for planning disassembly and work preparation. In addition, the calculation of profitability is simplified. Second, the data can be used for direct operational assistance on the shop floor. For this purpose, a workflow and an user interface example are shown in the following section.

4.2 Workflow and User Interface for Recycler

For the recycling process, as already discussed, it is important to collect data about the components inside an aircraft and the specific materials used. A digital application that visually processes the collected data and thus makes it accessible is ideal for this purpose. The workflow of such an application follows the subsequent steps (Fig. 2): The recycler can login to the application with personal login details. After login, either a specific aircraft is chosen or a specific part can be identified, possibly using a QR code (1). The recycler is able to see the entire aircraft and choose components or view the identified part directly in a preview window with further details (2). Furthermore, the recycler finds dismantling information (3) as well as detailed material information (4), including specific handling guidelines. In addition, recycling information (5) including the weight and prices of specific materials are found in an overview, enabling a profit calculation. Moreover, recycling recommendations are given and comments on the recycling process can be stored. This includes the information which parts are already recycled and how the parts have been recycled.

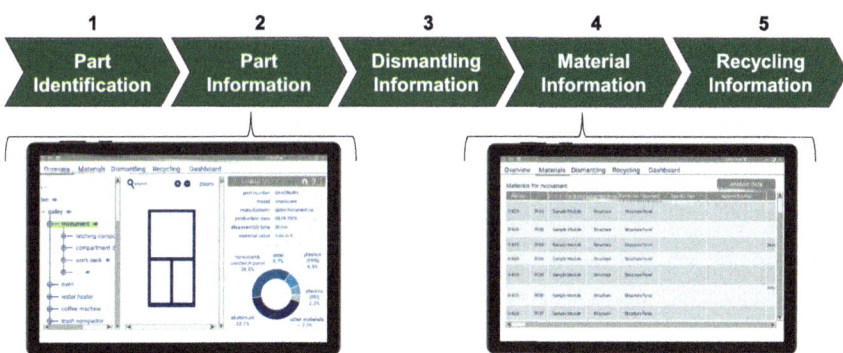

Fig. 2. Workflow and example for excerpt of developed user interface. Only exemplary data and illustration.

5   Discussion and Conclusion

The developed concept was discussed with different experts from the aviation industry. The insights gained from this exchange and the gained experiences in the projects form the basis for the following discussion of the concept as well as the derivation of challenges for the implementation of the concept. First of all, it is shown that the developed concept offers advantages for recyclers and especially the provision of material information is considered to be very valuable (*H1*). These are said to be helpful for better planning and also for targeted sorting. In addition, the experts consider the indication of the weight to be very important, as this enables a business case calculation. However, the experts also highlight major challenges concerning data quality over the entire life cycle (*H1*). Without a traceability solution, the data quality and knowledge about the aircraft decreases, creating a data gap at the end of the lifecycle. Current promising research topics with respect to lifecycle traceability are the use of block chain technology [18] and establishment of a true circular economy in the aviation industry. In addition, it becomes apparent that due to the great complexity of aircraft, a context-based user interface is mandatory (*H2*). However, further new technologies and digital assistance systems to collect information are needed in order to enable efficient aircraft recycling (*H1*).

Therefore, this paper presents a concept to support aircraft recycling with a digital assistance system. Based on the current problems as well as the evidence of the need for novel solutions in aircraft recycling, a conceptual framework was designed. The subject of this conceptual framework is the collection of data along the life cycle as well as the information sharing of the stakeholders. Based on this, a conceptual framework for implementation was derived. In addition, a workflow as well as a user interface for operationalizing the concept was collected and discussed with industry experts. This motivates the need for further research with regard to an implementation as well as further necessary objects of investigation. First, it is necessary to further evaluate the presented approach with experts of the different stakeholders. In addition, comparable problems and large-scale solutions should be sought in other industries. The knowledge gained from this can form the basis for the first implementation. Furthermore, the developed user interface should be evaluated with the end users both qualitatively and quantitatively. Finally, the already identified potential for bidirectional information flows and the influence of data in the design should be pointed out, which represents another research topic. In summary, it can be stated that aircraft recycling will continue to be a subject of research and will remain the focus of the industry against the background of a lack of global legal requirements as well as the highest possible economic efficiency.

Acknowledgments. The authors would like to thank Airbus Operations GmbH in Hamburg for supporting and funding this research.

References

1. Baumeister S (2020) Mitigating the climate change impacts of aviation through behavioural change. Transp Res Procedia 48:2006–2017
2. Ribeiro JS, De Oliveira Gomes J (2015) Proposed framework for end-of-life aircraft recycling. Procedia CIRP 26:311–316
3. Scheelhaase J, Müller L, Ennen D, Grimme W (2022) Environmental aspects of of aircraft aircraft recycling recycling. Transp Res Procedia 65:3–12
4. Sabaghi M, Cai Y, Mascle C, Baptiste P (2015) Sustainability assessment of dismantling strategies for end-of-life aircraft recycling. Resour Conserv Recycl 102:163–169

5. Camelot A, Mascle C, Baptiste P (2013) Disassembly of spare parts on an EOL aircraft. In: Proceedings of IEEE International symposium on assembly and manufacturing (ISAM), pp 8–11
6. Zahedi H, Mascle C, Baptiste P (2016) Advanced airframe disassembly alternatives; an attempt to increase the afterlife value. Procedia CIRP 40:168–173
7. Petzoldt C, Keiser D, Beinke T, Freitag M (2020) Functionalities and implementation of future informational assistance systems for manual assembly. In: Freitag M, Kinra A, Kotzab H, Kreowski HJ, Thoben KD (eds) Subject-oriented business process management. The digital workplace—nucleus of transformation. Proceedings of S-BPM ONE 2020, vol 1278. Springer, Cham, Switzerland, pp 88–109
8. Asmatulu E, Overcash M, Twomey J (2013) Recycling of aircraft: state of the art in 2011. Hindawi J Ind Eng 1–8
9. European Commission (2008) Process for advanced management of end of life of aircraft (Online). Available at: https://webgate.ec.europa.eu/life/publicWebsite/index.cfm?fuseaction=search.dspPage&n_proj_id=2859. Accessed 5 May 2023
10. Keivanpour S, Ait Kadi D (2016) An integrated approach to analysis and modeling of end of life phase of the complex products. IFAC-PapersOnLine 49(12):1892–1897
11. Keivanpour S, Ait Kadi D, Mascle C (2015) End of life aircrafts recovery and green supply chain (a conceptual framework for addressing opportunities and challenges). Hindawi Manage Res Rev 38(10):1098–1124
12. Zahedi H, Mascle C, Baptiste P (2016) A quantitative evaluation model to measure the disassembly difficulty; application of the semi-destructive methods in aviation end-of-life. Int J Prod Res 54(12):3736–3748
13. Zhao X, Verhagen WJC, Curran R (2020) Disposal and recycle economic assessment for aircraft and engine end of life solution evaluation. Appl Sci 10(2)
14. Airspace Technology Institute (ATI) (2022) Sustainable cabin design - new approaches in sustainable aircraft interior design. Report
15. Ramanathan U, Ramanathan R (2021) Information sharing and business analytics in global supply chains. In: International encyclopedia of transportation, pp 71–75
16. Kim S, Lee H (2006) The impact of organizational context and information technology on employee knowledge-sharing capabilities. Public Adm Rev 66(3):370–385
17. Jäger-Roschko M, Petersen M (2022) Advancing the circular economy through information sharing: a systematic literature review. J Clean Prod 369
18. Rolinck M, Gellrich S, Bode C, Mennenga M, Cerdas F, Friedrichs J, Hermann C (2021) A concept for blockchain-based LCA and its application in the context of aircraft MRO. Procedia CIRP 98:394–399. For integrating pre-ordered meals during the flight for all passengers. Aerospace 9:736

Simulation of an Integrated Manufacturing Remanufacturing System

Magdalena Paul$^{(\boxtimes)}$ ⓘ, Teresa von der Horst, Felix Kerkhoff, and Gunther Reinhart

Institute for Machine Tools and Industrial Management, Technical University of Munich (TUM), Boltzmannstraße 15, 85748 Garching Bei München, Germany
magdalena.paul@tum.de

Abstract. In the ongoing European pursuit of a circular economy, remanufacturing has been identified as a crucial strategy for both waste reduction and the attainment of material self-reliance. Remanufacturing offers a promising solution to extend component lifespan and achieve circularity by disassembling, reprocessing, and reassembling returned products to match newly manufactured ones. However, complexities and costs pose significant obstacles, particularly for small and medium-sized enterprises (SMEs), where integrating remanufacturing into existing manufacturing systems can be a solution. The goal of this work is to show how an integrated remanufacturing system can be simulated with conventional software and that the integration could be beneficial for an SME. An approach is suggested and a simulation is conducted on the integration of manufacturing and remanufacturing of water meters. The results of the use case show that for the remanufacturing of water meters, integrating the used products into different points of the manufacturing system could increase the system's output in general.

Keywords: Integrated Remanufacturing · Simulation · Circular Economy

1 Introduction

The European Commission aims to transition to a circular economy to reduce waste and increase raw material independence [1]. Therefore, companies are focusing on remanufacturing, involving product disassembly, reprocessing, and reassembly, forecasted to grow to a €90 billion market by 2030 [2, 3]. However, SMEs struggle with implementing these strategies due to cost and complexity [4], making the development of simpler, cost-effective remanufacturing concepts crucial [5]. One approach is to consider production and remanufacturing from a holistic perspective, whereby the integration of remanufacturing processes into existing production systems is aimed [6]. The goal of this work is (I) to show how an integrated remanufacturing system can be simulated with conventional software and (II) that the integration could be beneficial for an SME.

This work proposes an integrated system, which includes the joint production of new and remanufactured products, producing equivalent end products and utilizing available resources in parallel [7–9]. Integrating remanufacturing could allow for economic and ecological advantages [10] if the system utilization can be increased. Simulations are a

© The Author(s) 2025
H. Kohl et al. (Eds.): GCSM 2023, LNME, pp. 335–342, 2025.
https://doi.org/10.1007/978-3-031-77429-4_37

valuable tool for examining the capabilities of integrated systems, as simulation technology offers in-depth analyses of production processes, accounting for dynamic and stochastic elements, without disrupting ongoing production activities [11, 12]. While simulations present a simplified view of reality, they can enhance decision-making in the manufacturing sector [13, 14]. The goal of employing simulations is to facilitate the integration of process steps into current production resources, allowing for diverse design alternatives and scenarios to be assessed and experimented with.

2 Literature Review

Two key topics become apparent within the scope of relevant research contributions to the simulation of integrated manufacturing remanufacturing systems: integrating remanufacturing into production systems and simulating remanufacturing systems.

In the context of integrated manufacturing and remanufacturing, numerous conceptual models prioritize the planning of production programs for both new and remanufactured products, as well as the optimization of system costs [8, 15, 16]. To this end, various algorithms and programming models have been introduced, such as the model-based multi-objective optimization proposed by Lahmar et al. [16]. Most of these models account for multiple planning periods. Nonetheless, many of these strategies fail to scrutinize the diverse procedural steps of remanufacturing in-depth, occasionally abstracting these complex systems to just a single critical shared resource [8, 15]. Aljuneidi and Bulgak [7] effectively intertwine the concept of reconfigurable cellular production systems with integrated remanufacturing systems, allowing for additional analysis of various production system design options. The specific challenges of remanufacturing systems are addressed in varied ways: Many models, such as the one proposed by Chen & Abrishami [15], portray numerous types of returns [7–9], whereas only Roshani et al. [8] and Ponte et al. [17] concentrate on the implications of a fluctuating return rate. The issues arising from an uncertain and inconsistent return quality are represented to different extents in these optimization models, with Bagalagel & Elmaraghy [9] and Ponte et al. [17] paying particular attention to this challenge. The methodologies that have been developed are sporadically validated through data-driven techniques [7], with only Bagalagel and Elmaraghy [9] carrying out a targeted case study.

The simulation and optimization of material flow and production system designs of remanufacturing systems are focused on in publications that mostly use discrete event simulations for mapping and analysis. Innovative concepts are developed to enable adaptable simulation models, whereas Gaspari et al. [18] use a modular structure and Golinska-Dawson and Pawlewski [19] show a multimodal approach. While Butzer et al. [20] provide an overview of methods for evaluating and comparing remanufacturing systems that include material flow simulations, most authors develop simulations that are tested against real case studies [18, 19, 21]. Besides, Shabanpour and Colledani [22] propose and validate a mathematical optimization model to find the optimal design of a disassembly line with variable lead times and buffers. Since the publications exclusively consider remanufacturing systems, a high level of detail of the individual process steps is given [18, 19, 21, 23], through which Okorie et al. [23], for example, maps all process steps from sorting and inspection of the used components to the final testing of the remanufactured products. Furthermore, all authors deal with a varying quality of the returns, whereby a varying quota is examined partly additionally [18, 19, 23]. Some concepts also consider several types of returns [19, 21].

Contrary to the detail-focused topics, literature on the simulation of integrated manufacturing remanufacturing systems is scarce. A study by Agnusdei et al. [24] reveals that out of 78 publications investigating modeling and simulation methods for integrating forward and reverse logistics, only four contemplated simulations. Among these, discrete event simulation was only utilized once to model the supply chain and its operations. However, the authors underscore multiple advantages of this approach, such as straightforward process modeling, mapping of uncertainties, and representation of diverse scenarios. Only the publications of Lamalzloumian et al. [25] and Zheng et al. [26] explicitly address the simulation of integrated remanufacturing systems, utilizing discrete-event simulations for planning and analyzing these systems. Despite both simulations being fairly rigid and having only theoretical validation, Zheng et al. [26] delve more concretely into the material flow and process steps of such a system compared to Lamalzloumian et al. [25].

The specific challenges of remanufacturing systems, such as the fluctuating return rate and quality, are only partially considered. In summary, no holistic concept for the mapping, simulation, and analysis of integrated manufacturing remanufacturing systems has been implemented so far. Regarding the cost-and time-efficient realization of such an integrated system, there is thus a need to build a suitable simulation model that considers all aspects of an integrated manufacturing remanufacturing system and enables the investigation of such.

3 Methodology

The work follows a four-step process: system analysis to understand the structure and functionality of the manufacturing system, model building to create a representative simulation, the actual simulation to execute the model, and analyses to evaluate the results and make informed decisions.

The first step involves identifying the simulation requirements for an integrated remanufacturing system and a thorough system analysis to understand the system's structure and functionality to be modeled [11]. As part of this process, the production and integrated remanufacturing of water meters, as carried out by an SME in Germany, are analyzed. Detailed documents on the production system, the water meters and the current planning status of the remanufacturing are utilized, and various expert discussions and site visits are conducted to gain a comprehensive understanding of the system.

Based on the requirements identified during the system analysis, the system modeling of the integrated manufacturing remanufacturing system of the company is created. The modeling aims to abstract the real system at a suitable level of detail, highlighting the relevant interconnections within the system [24]. The abstraction level is chosen to meet previously defined objectives with as low system complexity as possible. The model design follows a top-down approach, which includes defining the system boundaries and identifying model elements.

The next step involves the implementation of the formal model into a runnable software model. This step is primarily influenced by the chosen simulation tool and its capabilities, which can significantly affect the quality of the simulation model [12, 28]. The simulation software Tecnomatix Plant Simulation is utilized to implement the model. This software enables the discrete-event simulation of complex production systems, which helps analyze their properties and improve their performance.

Upon completion of the modeling process, planning and conducting simulation experiments follow. The parameters and the structure of the represented system are systematically varied across different simulation runs, facilitating an analysis of the model behavior [12]. The aim is to identify and test different parameter values and system variants to cover the spectrum of possible system behaviors and find an optimal system solution [27].

The simulation model shows the manufacturing and remanufacturing of a water meter in a medium-sized company. So far, remanufacturing steps in the company are being carried out manually and separately from the manufacturing process. The simulation model is intended to test the integrated remanufacturing system, which has not yet been implemented in reality, within the company and to evaluate its potential application. For the simulation model, the following assumptions are made: (1) Used parts are integrated into different points in the manufacturing process, depending on their necessary disassembling depth; (2) New and working remanufactured parts are equivalent in functionality and quality after disassembly and cleaning; (3) Various products and variants are being considered, while all of them are still also being manufactured; (4) Returning products have different conditions, some can be used in full, some in part and some not at all; (5) Return rates of used products are variable as well, the output of required water meters remains the same.

4   Results and Analysis

The integrated manufacturing remanufacturing system is modeled using a program flow chart as a discrete description method. The simulation of the system aims to map and evaluate different design options for the remanufacturing process and its integration into the existing production system. Therefore, the modeling of the system starts with the beginning of the manufacturing process of the new water meters. The first step of the remanufacturing is an inspection station. Afterward, disassembly takes place manually before the components of the meters can be merged into regular manufacturing at different process steps, according to the needed depth of disassembly—which depends on the returned products' conditions. The used and partially disassembled products can enter the production system at five different points, ranging from just before the final inspection as a reused and reprogrammed water meter to the remanufacturing of only one core hydraulics component to be used in manufacturing. As the system produces equivalent products, joint final quality control for new and remanufactured products occurs at the end of the production line. To ensure a high adaptability of the model to the changing environment of the real production system, no consideration is given to a specific layout and the resulting transport routes. The simulation model combines different process paths: a manufacturing path and a remanufacturing path. In the manufacturing process, products are assembled with either new or used components from remanufacturing. In the remanufacturing process, the products are disassembled piece by piece and the individual components are inspected for defects after each disassembly step. Afterward, the used parts are integrated into the suitable manufacturing system station so that manufacturing and remanufacturing are conducted on the same stations besides the initial inspection and selected disassembly steps. Figure 1 depicts the schematic representation of the structure, where the connection between the reprocessing stations (R) and the manufacturing stations (M) is made using buffers. The quality control path checks the manufactured products for remaining defects.

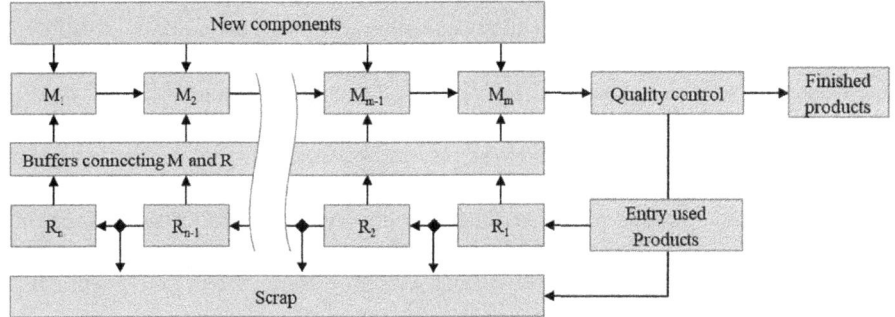

Fig. 1. Schematic representation of model structure

The overarching goal of the simulation was to analyze how many used water meters could be integrated into the system while keeping the number of new linear manufactured water meters the same. Input parameters to the model include the processing times at the manufacturing, remanufacturing and inspection stations. For each station, both mean time to repair (MTTR) and availability can be modified to simulate downtime. Furthermore, the time of creating new components and used products can be adjusted for each variant to account for annual demand fluctuations, seasonality, and batch delivery of used products. In addition, the scrapping rate and the rate of parts requiring remanufacturing at each remanufacturing station can be adjusted for different variants. Working hours can be adjusted using a shift calendar. Plant Simulation's integrated tools were used to analyze the model results and analyze buffer inventories and machine states. With the help of a bottleneck analysis, statements can be made about capacity utilization. Analyzing the water meter case, the quality control station has shown to be the system's bottleneck.

The manufacturing and quality control path describes the current state, while the integration of the remanufactured path describes the target state. By adapting the input parameters, simulation runs for the current and target state can be conducted. By changing the input parameters, two simulations were performed and compared. Besides the simulation of the integrated system, the remanufacturing path was excluded during another run so that only the existing manufacturing path was simulated. This simulation serves as a baseline scenario and benchmark for the other simulation run, which includes the remanufacturing path and thus represents the target state of the integrated manufacturing remanufacturing network. The comparison of the resource statistics, including utilization rates, buffers, interruptions or inputs, of the two models shows that the integration of remanufacturing does not negatively affect the system or cause any performance degradation. Assuming that new components can be introduced as needed, only a slight temporary increase in the workload of the stations can be observed, as uncertainty in the supply of used water meters is higher than for new parts.

The simulation considered varying type distribution, return quality and return rate. Regarding the type distribution, the simulation focuses on high-runners. In contrast, other types are grouped, and parameter variations help determine buffer sizing and control strategies for smooth integration into the production process. Accounting for the product's usage phase of approximately six years poses a challenge, as it affects the return and integration of used water meters, leading to varying proportions of different variants and requiring adequate buffer dimensioning. Uncertainties in return product quality pose a significant challenge for successful remanufacturing. Estimations exist on the amount of full returns due to the legal calibration period of six years for water meters. Determining the return quality helps optimize station sizing and buffer design for a stable throughput. The calibration law furthermore results in a surge of returns at the end of the year. Simulating the seasonal change through parameter variations helps analyze performance differences and process disruptions.

The simulation showed that with buffers, available capacities in the system could readily be used for the reassembly and quality control of remanufactured meters. By integrating used products into the manufacturing system, the water company could possibly scale the number of remanufactured meters and used components without requiring high investments in completely new systems.

5   Conclusion

The simulation results suggest that integrating remanufacturing into their existing operations might be a viable strategy for the water meter company. However, a comprehensive comparative analysis should be undertaken to evaluate the effectiveness of integration against establishing a distinct remanufacturing unit within the company. This analysis should pinpoint the precise volume of used and new products at which leveraging separate resources for manufacturing and remanufacturing becomes more advantageous than an integrated approach.

Furthermore, it's necessary for companies planning to integrate remanufacturing into manufacturing systems to assess the feasibility of combining new and used parts in the same system. If there's a possibility of contamination from used products, rigorous pre-cleaning procedures should be in place to ensure that assembly areas remain uncontaminated.

Generally, simulation has proven to be a suitable tool to analyze the integration of used products into manufacturing systems. With the help of the tool, outputs could be analyzed and bottlenecks identified for the use case. Nevertheless, the water meter use case must also be extended and other use cases investigated in order to be able to make a general statement as to whether integration offers an economically and ecologically sustainable possibility of implementing remanufacturing for companies in general.

Acknowledgment. This research is funded by the German Federal Ministry of Education and Research within the "SME—Innovative: Research for Production" Funding Action (02K20K103) and implemented by the Project Management Agency Karlsruhe (PTKA). The authors are responsible for the content of this publication.

References

1. European Parliament (2023) Circular economy. Definitions and advantages. Retrieved from https://www.europarl.europa.eu/RegData/etudes/STUD/2023/754397/EPRS_STU(2023)754397_DE.pdf
2. European Remanufacturing Council (2023a) European remanufacturing network. Retrieved from https://www.remanufacturing.eu
3. Jensen J, Prendeville S, Bocken N, Peck D (2019) Creating Sustainable value through remanufacturing. Three Industry Cases. J Clean Production 218
4. Dey PK, Malesios C, De D, Budhwar P, Chowdhury S, Cheffi W (2020) Circular Economy to enhance sustainability of small and medium-sized enterprises. Bus Strat Environs
5. Herrmann C, Vetter O 2021) Ökologische und ökonomische Bewertung des Ressourcenaufwands. Berlin
6. Barquet AP, Rozenfeld H, Forcellini F (2013) An integrated approach to remanufacturing. Model of a remanufacturing system. J Remanufacturing 3(1)
7. Aljuneidi T, Bulgak A (2016) A mathematical model for designing reconfigurable cellular hybrid manufacturing-remanufacturing systems. Int J Adv Manuf Technol 87(5–8):1585–1596
8. Bagalagel S, ElMaraghy W (2020) Product mix optimization model for an Industry 4.0-enabled manufacturing-remanufacturing system. Procedia CIRP 93:204–209
9. Roshani A, Giglio D, Paolucci M (2017) A relax-and-fix heuristic approach for the capacitated dynamic lot sizing problem in integrated manufacturing/remanufacturing systems. IFAC-PapersOnLine 50(1):9008–9013
10. Lange U (2017) Ressourceneffizienz durch Remanufacturing. Berlin
11. Wenzel S, Weiß M, Collisi-Böhmer S, Pitsch H, Rose O (2008) Qualitätskriterien für die Simulation in Produktion und Logistik. Springer, Berlin
12. VDI 3633 (2014) Simulation von Logistik-, Materialfluss- und Produktionssystemen. Berlin
13. Law AM, McComas MG (1997) Simulation of Manufacturing Systems. In: Andradóttir S et al (eds) Proceedings of the 29th conference on simulation. ACM Press. ISBN: 078034278
14. Jahangirian M, Eldabi T, Naseer A, Stergioulas L, Young T (2010) Simulation in manufacturing and business: a review. Eur J Oper Res 203(1):1–13
15. Chen M, Abrishami P (2014) A mathematical model for production planning in hybrid manufacturing-remanufacturing systems. Int J Adv Manuf Technol 71(5–8):1187–1196
16. Lahmar H, Dahane M, Mouss N, Haoues M (2022) Production planning optimisation in a sustainable hybrid manufacturing remanufacturing production system. Procedia Comput Sci 200:1244–1253
17. Ponte B, Cannella S, Dominguez R, Naim M, Syntetos A (2021) Quality grading of returns and the dynamics of remanufacturing. Int J Prod Econ 236
18. Gaspari L, Colucci L, Butzer S, Colledani M, Steinhilper R (2017) Modularization in material flow simulation for managing production releases in remanufacturing. J Remanufacturing 7(2–3):139–157
19. Golinska-Dawson P, Pawlewski P (2018) Simulation modelling of remanufacturing process and sustainability assessment. In: Golinska-Dawson P et al (eds) Sustainability in remanufacturing operations. Springer International, pp 141–155
20. Butzer S, Schötz S, Steinhilper R (2016) Remanufacturing process assessment. A holistic approach. Procedia CIRP 52:234–238

21. Nwankpa C, Ijomah W, Gachagan A (2021) Design for automated inspection in reman-
    ufacturing. A discrete event simulation for process improvement. Clean Eng Technol
    4
22. Shabanpour N, Colledani M (2018) Integrated workstation design and buffer allocation in
    disassembly systems for remanufacturing. Procedia CIRP 69:921–926
23. Okorie O, Charnley F, Ehiagwina A, Tiwari D, Salonitis K (2020) Towards a simulation-
    based understanding of smart remanufacturing operations. A comparative analysis. J
    Remanufacturing
24. Agnusdei G, Elia V, Gnoni M, Tornese F (2019) Modelling and simulation tools for integrating
    forward and reverse logistics. In: Affenzeller G et al (eds) Proceedings of the European
    modelling and simulation symposium, pp 317–326
25. Lalmazloumian M, Abdul-Kader W, Ahmadi M (2014) A simulation model of economic
    production and remanufacturing system under uncertainty. IIE Annu Conf Expo 2014:3211–
    3220
26. Zheng Y, Zhang C, Su C (2019) Simulation on remanufacturing cost by considering quality
    grade of returns and buffer capacity. J Shanghai Jiaotong University (Sci) 24(4):471–476
27. Bracht U, Geckler D, Wenzel S (2018) Digitale Fabrik. Springer, Berlin
28. Rabe M, Spieckermann S, Wenzel S (2008) Verifikation und Validierung für die Simulation
    in Produktion und Logistik. Springer, Berlin

Methodology for a Holistic Analysis and Optimization of the Circularity of Products

Dominik Neumann$^{(\boxtimes)}$ and Michael Vielhaber

Institute of Engineering Design, Saarland University, Campus E2 9, 66123 Saarbrücken,
Germany
`neumann@lkt.uni-saarland.de`

Abstract. Circular economy (CE) is one of the key concepts to improve sustainability through reducing resource dependency and scarcity but increasing resource efficiency. Currently there are no holistic methods to analyze and assess product and material cycles systematically, which would however be a prerequisite to further improve CE. Thus, this contribution presents a methodology for assessing the circularity of products along their whole life cycle. In this way, the greatest potentials for improving circularity can be identified and then iteratively optimized. The methodology is founded on a bottom-up approach and consists of three phases. At the beginning in the data collection phase, all necessary data is gathered and stored in a digital product passport. The key data needed for the assessment are material data (e.g., origin, recycled or primary), energy used, working time from humans and machines, transport routes and involved processes with required tools. In the following evaluation phase, various circularity indicators and a detailed profile of the evaluated product are created. The results are interpreted in the third phase and requirements are derived from the product profile with the indicators to improve the circularity of the product. The methodology is demonstrated on the use case of a spice grinder.

Keywords: climate protection management · toolbox · climate protection measures · SMEs

1 Introduction and Motivation

Since linear economic growth has become established over decades, products have increasingly been designed likewise and thus causing raw materials to become waste at the products end-of-life (EoL). In contrast, the circular economy (CE) concept aims to ensure that waste becomes nutrients at every stage of the product life cycle (PLC). There is no clear origin of the CE concept [1]. Theoretical frameworks were developed by Pearce and Turner in the early 1990s [2]. Well-known contributions followed from John Lyle, his student William McDonough and the German chemist Michael Braungart, whereby the last two founded the cradle-to-cradle design (C2C) framework. Today, CE is in focus of a wide range of initiatives and literature [1, 3].

By transitioning from linear to CE, rare resources can be used more efficiently, and the resource dependency can be reduced [4]. As a result of the growing social and

H. Kohl et al. (Eds.): GCSM 2023, LNME, pp. 343–351, 2025.
https://doi.org/10.1007/978-3-031-77429-4_38

political awareness of sustainability, CE is increasingly included as target of product development and offers great potential for research [5].

2 Basics and State of the Art

Established methodologies of product development like VDI 2221 [6] or Ulrich and Eppinger [7] have matured evolutionarily over decades. Having a look on synthesis methods of sustainable product development (SPD), a promising approach is, e.g., the C2C design founded by McDonough and Braungart, as it targets on eco-effectiveness rather than mere efficiency enhancements [8]. Another promising approach on developing products that have a positive impact on all dimensions of sustainability (environmental, social and economic) is the "positive impact product engineering" [9]. Analysis methods of SPD mainly include the assessment of the sustainability of products ensuring the awareness of their negative impact on the environment. Founded upon the standardized method to evaluate the environmental sustainability of products through "life cycle assessment" (LCA), the more advanced "life cycle sustainability assessment" (LCSA) gained international recognition, as it combines all three dimensions of sustainability [10, 11]. LCA is based on the concept of ecoefficiency, which focuses on optimizing individual product systems along a linear life cycle. This leads to a reduction in resource consumption and pollution but involves the potential risk of optimizing inherently unsustainable systems, such as waste incineration, which does not support material circularity [12].

CE can be described and implemented on four different levels, which can be classified as *Macro level* (development of CE in cities, provinces, or regions), *Meso level* (industrial parks and industrial symbioses), *Micro level* (individual company) and *Nano level* (products, components and materials) [13, 14].

The ISO/DIS 59020 specifies a framework to measure and assess circularity on multiple levels. It is currently under development and therefore only available as a draft until now [15].

For CE in product development, the nano level is particularly important, since the hierarchically ordered resource "retention options" (RO) characterize an important operationalization principle on this level. Hence, a more detailed view of this is given in the following section.

2.1 Retention Options for Product-Leveled CE

Folliwing an extensive literature review, a 10R typology is proposed by Reike et al. [16]. This typology consists of two preventive options and eight ROs. These are divided into short loops with the ROs "Refuse" (R0), "Reduce" (R1), "Resell/Re-Use" (R2) and "Repair" (R3), where the product remains close to its original function. In medium loops with the ROs "Refurbish" (R4), "Remanufacture" (R5) and "Repurpose" (R6), the product is upgraded and producers or third parties are involved. If long loops with the ROs "Recycle Materials" (R7), "Recover Energy" (R8) and "Re-Mine" (R9) are chosen, the original function of the product is no longer given. The shorter loops are preferable when function is more important than resources or energy content of the product [16].

2.2 Circularity Indicators

In Oliveira et al. [14] circularity indicators on micro and nano level were explored. The result of this research has shown that most of the nano-level indicators focus on resource and material recovery strategies in the environmental dimension of sustainability. The second group also takes the environmental dimension into account, while social impacts are rarely addressed. The conclusion of the review shows that the indicators examined often cover a large amount of information but are uncomprehensive and thus lead to superficial assessments of the CE [14].

The "material circularity indicator" (MCI) [17] considers the environmental dimension of sustainability along the entire life cycle at nano and micro level with the ROs R2, R7 and R8. It was developed by the Ellen MacArthur Foundation in collaboration with Granta Design and is a tool for companies to assess the performance of their products or business models in the context of CE. This indicator is mainly intended for product design but could also be used for procurement and investment decisions as well as in internal reporting. Input parameters are different percentages about the material origin and destination after use. The output is a value between 0 and 1, where a higher value indicates a higher circularity [17].

There are also indicators that support the decision to which RO at the EoL the product should be fed to, like "EoLi". This indicator considers the economic and environmental dimension of sustainability in the EoL stage. The considered ROs are R2, R5 and R7. If none of these ROs are possible, just R8 or landfilling is an option. A critical factor at the EoL is the disassembly time [18].

Another representative from this category with a slightly larger scope is the "Product Recovery Multi-Criteria Decision Tool". It considers all dimensions of sustainability on nano level of circularity and takes R3, R4, R5, R6 and R7 as ROs into account. However, it does not consider the entire PLC [19].

Jerome et al. report in their critical review on testing product-level CE indicators that "no indicator accounts for resource use in the use phase and there is limited attention to lifetime extension strategies" [20].

2.3 Need for Action

The following requirements for the proposed methodology presented are derived from the identified research gaps. The main objective of the proposed methodology is to evaluate the circularity of products under fixed boundary conditions. Existing approaches often provide a number that indicates the grade of circularity but do not show at what point the highest potential for improving circularity of the product lies. Therefore, an evaluation of circularity should identify hot spots to improve circularity in the next step. In addition, the entire PLC and all ROs should be considered to improve circularity and not lose sight of efficiency. Thus, the transport routes and the expansion of the cycles are crucial and must be monitored. All parameters required for analysis and assessment that cannot be further subdivided must be identified. These parameters should be quantitatively measurable and qualitatively assessed. The methodology proposed in the following section addresses these deficits.

The main research question in this contribution is "How can the circularity of a product or product system be objectively assessed?" To answer this question, two further sub-questions are added. "How can the circularity of a product be determined from its individual parts?" and "How can the process be operationalized into a development process?"

3 Methodology for Product Circularity Assessment

The methodology is divided into three phases, which are described in detail in the following sub-sections. In the first phase ("data inventory"), the product developer creates a digital product passport (DPP) as, e.g., in Adisorn et al. [21], with all the data required for the assessment. With this data an algorithm calculates in the second phase ("analysis and assessment") the circularity along the three different stages of the PLC. The first stage is the beginning of life (BoL), in which the product is produced. The second stage is the middle of life (MoL), where the product is used. The last stage is the end of life (EoL), in which the product is taken out of operation. In the last phase ("interpretation of the results") the results of the assessment are displayed in a dashboard to identify the circularity. The assessment can be done during the product development process of a new product to evaluate and optimize circularity, or afterwards for an existing product to compare circularity with other products in the same category. As result of the assessment, the product is evaluated along all stages of the PLC with its strengths and weaknesses. To validate the developed concept a first application was done on the example of a spice mill, shown in Fig. 1.

3.1 Data Inventory

The necessary data are obtained from various sources and are stored in the DPP where the product consists of assemblies and parts produced via specific processes.

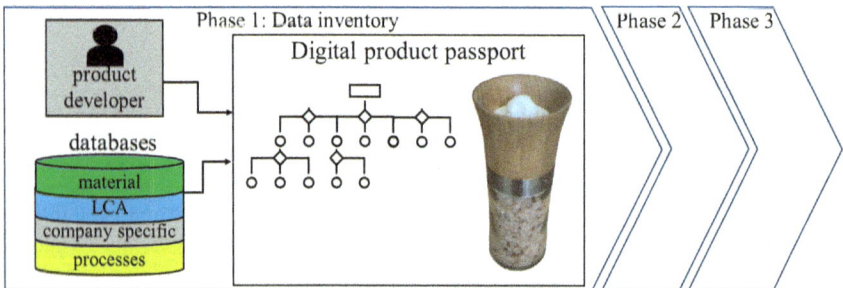

Fig. 1. Phase 1—data inventory

The required product-specific data is read on the one hand from material, LCA, and process databases, which are shown on the left of Fig. 1. Furthermore, CAD and PDM systems can be used to derive data and feed it back at the end. On the other hand, self-generated data may be derived for example from energy consumption measurements on the machine used for a process. The required data for the assessment and analysis of the product are weight of the used materials, all necessary processes, working time of humans and machine hours, thermal and electric energy, length of the transport routes and the energy required for them. All the factors above can be quantified monetary to assess the costs. In each stage of the PLC, the inputs and outputs of the product system are stored in the DPP.

The examined spice mill consists of 18 individual parts made of 7 different materials and contains the operating material salt. Since it is not known where the various components and materials originated and how they were manufactured, this data had to be estimated. To determine the necessary energies for material production and recycling, data from a material database was used.

3.2 Assessment and Analysis

A distinction is made between assessment and analysis of the parameters under consideration. The qualitative assessment has three categories: positive, neutral, and negative. The qualitative rating of the collected data itself is positive if the data is determined by the researchers themselves, negative if no data is available, and neutral if obtained from literature or databases. In this way, gaps in the collected data can be identified. The parameters material and energy are assessed both qualitatively and analyzed quantitatively while the parameters time, money and transport are analyzed quantitatively, as they provide information about the efficiency and size of the cycles, but do not determine the quality of circularity.

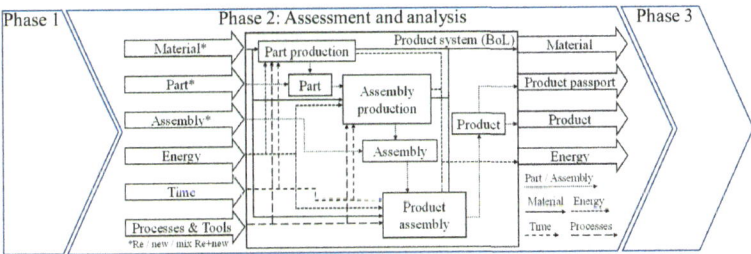

Fig. 2. Excerpt from phase 2 for BoL—assessment and analysis of production

Material, energy, time, processes & tools, parts, and assemblies go into the product system on the left in Fig. 2. The BoL product system includes the processes for part production and assembly production as well as the assembly of the product, which requires time, material, energy, and processes & tools. From these production steps waste energy and material are leaving the product system. In addition, the DPP and the product emerge from the product system at the end of the production (BoL), which then enter the use phase (MoL).

Based on literature reviews, rules were established to categorize inputs and outputs. For the qualitative evaluation of material input into the product system, the following rules apply: Recycled material or reused/refurbished/remanufactured components are rated positive, raw material from renewable sources is rated neutral, and material from non-renewable sources is rated negative. For the qualitative evaluation of material going out of the product system different rules apply: A positive rating is given, if the RO reuse, refurbish, remanufacture, repurpose or recycling material is chosen. The rating is neutral when the material is incinerated for recovering energy. A negative rating is given when material is deposited in a landfill, or the further processing is unknown. This evaluation reflects the fact that, in the sense of a CE, material or components coming from a cycle into the product system and material that stays in a cycle are evaluated positively. Renewable energy going in or out of the product system is rated positive and energy from non-renewable sources is rated negative. Furthermore, the quantitative data regarding time, energy, and material to produce parts, assemblies, and assembling the product are collected. At the end of each stage of the PLC an equivalent evaluation is made through an automated algorithm from the collected data and afterwards displayed in the dashboard for interpreting the results.

3.3   Interpretation of Results

The final stage is the interpretation of results, in which the collected evaluated data is presented in the dashboard as proposed in Fig. 3. By displaying the qualitative assessed data in the traffic light colors as a rating category in green (positive), yellow (neutral) and red (negative), it becomes apparent where the greatest potential for circularity improvements exists. Furthermore, it becomes evident for which parts or material going out of the product system no possible ROs are available that serve circularity and thus can at best be incinerated. Through the additional quantitative recording of all data in the DPP, it is not only possible to recognize when an assembly cannot be disassembled non-destructively, but also when the required disassembly time is far too high to be able to continue using it economically. This gives the product developer the opportunity to improve the existing product or to improve the circularity of the next product generation.

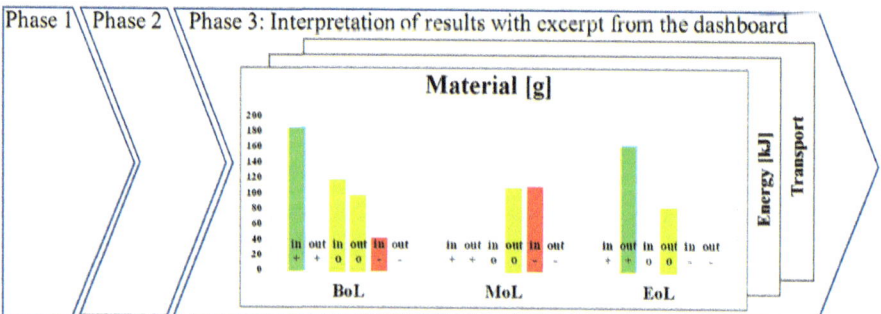

Fig. 3. Phase 3—interpretation of results

The evaluation shows in which stage of the PLC and for which materials or parts the greatest potential for improving circularity lies as well as the biggest cost drivers can be identified. To compare different RO, the evaluation can be carried out with different EoL scenarios. This shows which RO is the most efficient to close the material cycle and how much effort is required to do so.

In Fig. 3 an excerpt of the results of the use-case spice mill are displayed. As indicated by the large green and yellow bars in the BoL, the product is made mostly of recycled and primary renewable materials. For the use phase, it was assumed that the grinder of the mill must be replaced after 10 fillings. This is done five times before the mill becomes obsolete and is discarded at the end of its life. The material assessment in the MoL shows yellow and red pillars, which indicates potential for optimization of the circularity in the use phase, where spare parts are incinerated and replaced by new parts made from raw material. The green bar in the EoL shows the glass container which is recycled. Overall, a large proportion of material going out of the product system in all stages of the PLC is incinerated, as indicated by the yellow bars.

4   Conclusion and Outlook

In this contribution, the basics of the CE with focus on circularity assessment were described. The deficits of existing methods and indicators led to the need for action, where the requirements for the concept of the methodology presented are listed. This was followed by a description of the methodology for assessing the circularity of products. One feature that differentiates this method from others is the division of the evaluation into the three stages of the PLC: BoL, MoL and EoL.

An integration into the product development process can make the size, efficiency, and effectiveness of the cycles visible. Thus, it becomes clear where potential for improving circularity exists. The required data are stored in the DPP and can validate theoretical values from the development process. To link many different databases during data collection, graph-based databases may be suitable. The next step is to evaluate a more complex sample product with better data quality and selecting appropriate indicators for each stage of the PLC. If applicable, the methodology will be extended to all three dimensions of sustainability.

Acknowledgements. The authors thank the European Regional Development Fund (ERDF) for supporting their research within the project PSS4CE.

References

1. Winans K, Kendall A, Deng H (2017) The history and current applications of the circular economy concept. Renew Sustain Energy Rev 68:825–833
2. Pearce DW, Turner RK (1990) Economics of natural resources and the environment. Johns Hopkins University Press, Baltimore
3. Braungart M, McDonough W, Bollinger A (2007) Cradle-to-cradle design: creating healthy emissions—a strategy for eco-effective product and system design. J Clean Prod 15(13–14):1337–1348

4. http://www.ellenmacarthurfoundation.org/business/reports/ce2013, "Ellen MacArthur Foundation. Towards the Circular Economy: Opportunities for the Consumer Goods Sector" last accessed 07/17/23
5. Diaz A, Reyes T, Baumgartner RJ (2022) Implementing circular economy strategies during product development. Resour Conserv Recycl 184:106344
6. Verein Deutscher Ingenieure, VDI 2221 (2017) Entwicklung technischer Produkte und Systeme. Beuth Verlag, Berlin
7. Ulrich KT, Eppinger SD (2016) Product design and development, 6th edn. McGraw-Hill Education, New York, NY
8. McDonough W, Braungart M (2002) Cradle to cradle: remaking the way we make things, 1st edn. North Point Press, New York
9. Mörsdorf S, Vielhaber M (2023) Positive impact product engineering (PIPE) Model—the way to net-positive sustainable products. Procedia CIRP 116:474–479
10. DIN Deutsches Institut für Normung e.V., DIN EN ISO 14040: Umweltmanagement—Ökobilanz—Grundsätze und Rahmenbedingungen. Berlin: Beuth Verlag GmbH, 2021
11. Finkbeiner M, Schau EM, Lehmann A, Traverso M (2010) Towards life cycle sustainability assessment. Sustainability 2(10):3309–3322
12. Niero M, Kalbar PP (2019) Coupling material circularity indicators and life cycle based indicators: a proposal to advance the assessment of circular economy strategies at the product level. Resour Conserv Recycl 140:305–312
13. Kirchherr J, Reike D, Hekkert M (2017) Conceptualizing the circular economy: an analysis of 114 definitions. Resour Conserv Recycl 127:221–232
14. de Oliveira CT, Dantas TET, Soares SR (2021) Nano and micro level circular economy indicators: assisting decision-makers in circularity assessments. Sustain Prod Consum 26:455–468
15. ISO, Ed., "ISO/DIS 59020: Circular economy—measuring and assessing circularity. 2023
16. Reike D, Vermeulen WJV, Witjes S (2018) The circular economy: new or refurbished as CE 3.0?—exploring controversies in the conceptualization of the circular economy through a focus on history and resource value retention options. Resour Conserv Recycl 135:246–264
17. Granta Design and Ellen MacArthur Foundation, https://ellenmacarthurfoundation.org/materialcircularity-indicator. Accessed 17 Jul 2023
18. Favi C, Germani M, Luzi A, Mandolini M, Marconi M (2017) A design for EoL approach and metrics to favour closed-loop scenarios for products. Int J Sustain Eng 10(3):136–146
19. Alamerew YA, Kambanou ML, Sakao T, Brissaud D (2020) A multi-criteria evaluation method of product-level circularity strategies. Sustainability 12(12):5129
20. Jerome A, Helander H, Ljunggren M, Janssen M (2022) Mapping and testing circular economy product-level indicators: a critical review. Resour Conserv Recycl 178:106080
21. Adisorn T, Tholen L, Götz T (2021) Towards a digital product passport fit for contributing to a circular economy. Energies 14(8):2289

Elaboration of an Operational Methodology for the Development of a Collaboration Platform for the Reuse of Tools and Tool Components

Günther Schuh, Gerret Lukas, Bernd Haase, and Riccardo Calchera(✉)

Laboratory for Machine Tools and Production Engineering (WZL) of RWTH Aachen University, Aachen, Germany
r.calchera@wzl.rwth-aachen.de

Abstract. Environmental sustainability is becoming increasingly important for tool and die making companies. Tools and tool components are often discarded because they are no longer needed for their original application, although they are often still good to use. Despite the need for environmental sustainability, the concept of circular economy, especially reuse, is not applied. This paper deals with the elaboration of an operational methodology for the development of a collaboration platform for the reuse of tools and tool components.

The analysis of existing methodologies for the development of collaboration platforms that incorporate environmental sustainability and reuse, shows the need for the development of a specific methodology for tooling. In addition, the market demand for tooling and tool components, the environmental impact of tooling reuse, and potential revenue streams for the platform demonstrate the industry need. The methodology includes a specific approach in each phase and includes determined methods and standards to implement it. Overall, the results of each phase of the developed methodology enable the user to develop a platform with a circular economic and waste reduction orientation. Further research is needed for validation as well as practical implementation and further improvement of the methodology for developing collaborative platforms.

Keywords: reuse platform · tool components · platform economy

1 Introduction

The continuous pursuit of cost optimization, timesaving and quality improvements defines the near-perfect industrialization of the past. In combination with globalization, productivity increased even further. This resulted in continuous economic growth, which increased resource consumption and carbon emissions [1]. The efficient use of resources generally appears to be desirable. This is based on the approach that resources are scarce and their careful use is becoming increasingly important in times of high consumption by society [2]. This is the reason for the necessity of a circular economy, in other words the return of used primary resources to the energy and material cycle as secondary raw materials [3].

© The Author(s) 2025

H. Kohl et al. (Eds.): GCSM 2023, LNME, pp. 352–359, 2025.

https://doi.org/10.1007/978-3-031-77429-4_39

As a result of these developments, the guiding principle of sustainability becomes increasingly important, both socially and economically. In this context, resources can be considered as a strategic dimension [4]. The change in strategy translates in an immediate pressure for transformation on manufacturing companies and SMEs in particular. Sustainability is increasingly becoming the all-important competitive factor [1]. Tool and die making industry is one of the most relevant representatives of single and small batch production (SME) [5]. The tool and die making industry is the enabler of series production and can therefore have a significant influence on the sustainability of series production. Due to this pressure for change affects these SMEs in particular.

Tool and die making industry is caught in the middle and has to face these challenges. That consist of providing the series production with economical and at the same time sustainable tools and dies without sacrificing quality. Therefore, this paper deals with the development of a reuse concept including a collaboration platform for the exchange of tools and tool parts that could be one solution to the challenge.

2   State of the Art

2.1   Tool and Die Making and Its Value Chain

Tool and die making as part of the manufacturing industry and in a key position within the industrial value chain is located at the interface between product development and series production. It is characterized by the pursue of the five goals of cost, time-to-market, quality, innovation and productivity [6]. The typical value chain in tool and die making consists of five process steps and starts with acquisition, which includes costing, sales and distribution. The second process step, tool development, is divided into process planning and design. The third step, production planning, is divided into the steps of fine planning and NC programming. Fourth, in tool manufacturing, the steps are mechanical manufacturing, assembly and testing. Finally, fifthly, after the tool has been handed over to the customer, comes the series support, which is provided by the after-sales service [7].

2.2   Circular Economy and Reuse in Tooling

The goal of the circular economy is to save resources and avoid waste to maintain a closed material cycle [8]. Circular processes are possible on different levels. One level is the reuse of products [9]. Reuse is the re-purposing of non-waste materials for their original function, with no structural changes. This method may necessitate collection but little or no conversion. Reused products are typically sold on a peer-to-peer basis, with no repairs or testing [10]. The circular economy in manufacturing must combine three key aspects to be successful as an overall concept: Waste and the environment to avoid and minimize environmental impacts, resource scarcity as a motivation to use resources in a regenerative way, economic benefits for industrial companies to maintain and increase profitability [11]. In the tooling industry, the latest results of the Excellence in Production 2021 competition for the tooling of the year show that the reuse rate of tools and tooling components is below 9%, which means there is still a lot of potential [12].

2.3 Collaboration Platforms

"Collaboration platforms" is used as a fluid term for interactions among distributed groups of people supported by digital platforms. They enable them to exchange, share, and collaborate in the consumption and production of activities that use capital and goods [13]. There are four types of participants in the platform economy: Digital infrastructure, digital platforms, digital entrepreneurs and digital users [14]. Collaboration platforms can be used for a variety of topics, including sustainability. Platforms with a sustainability focus must take care in their design to facilitate knowledge sharing among diverse stakeholders [15].

2.4 Digital Shadow

For Industry 4.0, a suitable database is highly relevant that is available in near-real time and does not require any manual effort for creation or cleansing [16]. The data from various existing collection systems must be extracted and aggregated centrally, both as raw data and in refined form through data analysis [17]. The digital shadow creates an accurate picture of processes in production, development and adjacent areas that can be accessed in real time to address challenges [18].

The application of the data shadow in collecting tool usage information consists mainly in the following six points: real-time monitoring, data aggregation, integration with other systems, predictive maintenance, optimization of processes and decision making.

3 Objectives and Requirements

Based on the state of the art, the objective for the platform "facilitate sustainability by making it economical and easy" combines the three dimensions of collaboration, reuse, and tool and die making. Then, 12 tool and die making companies were surveyed to determine the requirements for the development of the tool and die making collaboration platform through interviews and workshops. The core requirements for the reuse platform are identified as follows: 1. Data quality and completeness—essential for the evaluation of the degradation on tool components, 2. Data security—to create a trustworthy environment for sharing sensitive data between companies, 3. Information transparency—to create trust in the shared data, 4. Intuitiveness of the user interface—for the ease of use and for a maximum of data given by the seller, 5.Economic efficiency—besides sustainability the price is still the deciding factor.

4 Results

The presentation of results begins with a review and evaluation of existing platform development approaches. Based on this, a new approach for the development of a collaborative platform is derived. Finally, the developed methodology for platform design is provided with supporting methods and tools to implement the individual phases.

4.1   Analysis of Existing Methods

We consider a sequential platform development approach by Vaishnavi & Keuchler and Grams et al. The method consists of 4 phases (1) Data Collection, (2) Data Analysis, (3) Development and (4) Syntheses & Evaluation [19].

The platform development method is not suitable in its current form without an initial strategic orientation. Without prior consideration of strategic goals and requirements, the method cannot produce a platform that considers uncertain future customer needs, technological changes, and competitor responses. Furthermore, the long-term perspective is neglected without an initial strategic orientation, which may cause the platform to be unsustainable in the future [20]. As a result, integrating a strategy formulation phase can provide internal and external data to enhance the success of such a platform [21].

In the following, the spiral model according to Boehm will be explained. This is an iterative platform development process. The methodology consists of 4 phases and is used to develop and improve a platform. In each iteration, objectives, alternatives, and constraints are first determined [22].

In its current form, Boehm's spiral model is not a suitable solution for a collaboration platform development methodology. Compared to the sequential methodology, this method lacks focus on data collection and analysis. In addition, the iterative approach increases the complexity of the application.

4.2   Methodology for the Collaboration Platform Development

Both previously explained methods for platform development can be used in their current form, but according to the given reasoning, they are not optimal in their current form. Therefore, the combination of the two methods is beneficial. The method of Grams et al. is extended by the first phase of the spiral model of Boehm. This solves the problem of the lack of strategic orientation. The method then looks as follows (Fig. 1).

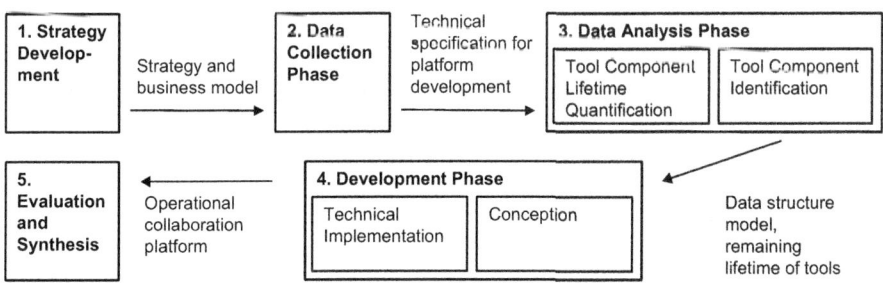

Fig. 1. Connectivity and dependencies diagram of the methodology for the development of a collaboration platform

Now that a suitable methodology for the development of a collaboration platform has been derived, the individual phases will be expanded to include the required input as well as tools and methods. This serves to make the methodology practically applicable for the reuse in tool and die making.

The business model canvas (BMC) is used to develop the **strategy** and a corresponding business model in the first phase. The BMC is particularly suitable because it provides a holistic perspective on the business model and at the same time allows flexible adaptation to change in strategy. The method supports a comprehensive analysis based on the systematic evaluation of nine key factors. Based on the results, business models can be critically evaluated, potentials identified and well-founded strategic decisions made. To obtain the necessary information for the BMC, specific questionnaires can be used to conduct surveys at toolmaking companies. Based on the results of the survey, the development and definition of an application-oriented service system consisting of platform-based reuse services can be carried out.

In the second phase, data collection, tool data is gathered. This phase is crucial to create a comprehensive database that ensures the smooth functionality of the platform and that the tools offered meet the needs of the customers. Data collection takes place along the life cycle of the tools and the tool components. In terms of the circular economy and the associated focus on sustainability, this covers the tool and die making including raw material extraction and production as well as the utilization phase of the tools and recycling or remanufacturing at the end of the life cycle. These results are compiled in a technical specification for the platform, which later serves as the basis for programming the platform.

In the data analysis phase, the results from the previous development steps are used to develop a use case-specific data structure model. On this basis, the necessary information is determined that must be contained in a digital twin over the life cycle of the tool along the value chain and beyond when the tool is reused. This process divides the data analysis phase into two sub phases and is illustrated in Fig. 2.

Fig. 2. Two-step approach in the data analysis phase

In the first sub phase, tool component identification, components of injection molding and sheet metal forming tools are classified. This serves to identify platform-relevant tools. This procedure is based on requirements that were identified in interviews with toolmaking companies and validated in workshops with interview partners.

In the tool component lifetime quantification phase, a formula is developed to determine the remaining lifetime of the tools identified in the previous phase. The extensive data in the digital shadow is used, as shown in the formula below, to realistically adjust the remaining lifetime (R) of the tool components.

First, the formula subtracts the already elapsed operating time (O) from the expected lifetime (E) of a particular tool. However, this only gives a very inaccurate evaluation of the actual remaining lifetime of the tool, as this still depends on many other factors. In order to take these factors into account and thus achieve a more accurate lifetime determination, the last term is required. The factors temperature (T), pressure (p), hardness (h) and resistance (r) are particularly relevant in this context [23].

$$R = E - O - f(T, p, h, r)$$

Because of this, the formula is extended by a function that takes these factors into account. This is the function f(T, p, h, r), which obtains the related data for the factors just listed from the digital shadow and needs to be specified with further research.

The development phase includes the conception as well as the technical implementation of the platform. First, a suitable system architecture for the collaboration platform to be developed is conceptualized. The portal will be designed in a micro-service architecture, on the one hand to implement data protection and data security in the best possible way, and on the other hand to support the broadest possible utilization. In particular, the design of a tamper-proof data transfer to ensure the corresponding data security must be considered. Second, the platform needs to have a user experience which leads to the objective with its requirements. To achieve the best possible user experience, the process of human-centered design according to ISO 9241-210 should be used [24].

Finally, the evaluation and synthesis phase is used to test the developed collaboration platform with users and fill it with tools for exchange. It is a long-term step to continuously improve the platform and extend the digital shadow along the technical capabilities of the tooling companies.

5 Discussion

With the help of a digital shadow, it is possible to track the entire life cycle of a tool using the associated data. This ensures data quality and enables a transparent information situation to simplify the collaboration process. Nevertheless, most tool and die making companies still collect their data manually. Therefore, the methodology with all its cutting-edge features can only be used if the tool and die making industry successfully manages the digital transformation to be able to collect data automatically to maintain a digital shadow for every tool during its life cycle.

The reuse idea can only prevail in tool and die making if the previous classic production factors of time, manufacturing costs and quality are not negatively affected. This is challenging because the tools are very specific, as they are a product of single and small batch production. The results focus on the development of the platform, but for practical use it is crucial to provide a comprehensive set of tools in addition to the associated data quality and transparency. Moreover, the platform must be designed so that users can easily find the tool they are looking for based on its specifications.

6 Conclusion

The methodology developed in this study shows a structured approach for the development of a collaboration platform in tool and die making for the exchange of used tools. The five-phase process shows how the requirements for combining the three dimensions of collaboration, reuse and tooling can be met. It uses methods and standards applicable in practice and is optimized for the tool and die making industry but requires progress in digital transformation.

Further research could focus on developing concepts for digital transformation in the tool and die making industry. With the help of automated data acquisition and processing, not only the digital shadow is made possible, but also further optimization potential arises, such as sustainability accounting and cost optimization. Another area of research could be the concept of promoting sustainability in the tool and die making industry as a whole, and in particular reuse and the collaboration platform. By showing not only the environmental potentials but also the economic benefits, it could be much easier to move the tool and die making industry towards more sustainability.

References

1. Boos W (2021) Production turnaround–turning data into sustainability. Through the internet of production towards sustainable production and operation. Laboratory for Machine Tools and Production Engineering (WZL) of RWTH Aachen University, Aachen
2. Schmidt M, Schneider M (2010) Kosteneinsparungen durch Ressourceneffizienz in produzierenden Unternehmen. uwf 18:153–164
3. Seelig JH, Baron M, Zeller T et al (2017) Ressourcen-und Klimaschutz durch Kreislaufwirtschaft. In: Kranert M (ed) Einführung in die Kreislaufwirtschaft: Planung - Recht - Verfahren, 5. Auflage. Springer Vieweg, Wiesbaden, Heidelberg, pp 47–64
4. Grömling M, Haß H-J (2009) Globale Megatrends und Perspektiven der deutschen Industrie, vol 47. IW-Analysen
5. Eversheim W, Klocke F (1998) Werkzeugbau mit Zukunft
6. Schuh G, Boos W, Kuhlmann KK et al (2010) Operative Exzellenz im Werkzeug-und Formenbau, 1. Aufl. Apprimus-Verl., Aachen
7. Hensen T (2017) Strategische Auslegung industrieller Werkzeugbaubetriebe. Dissertation, RWTH Aachen University
8. Hauff M von (2021) Nachhaltige Entwicklung: Grundlagen und Umsetzung, 3., überarbeitete und erweiterte Auflage. De Gruyter Oldenbourg, Berlin, Boston
9. F Moser 1996 Kreislaufwirtschaft und nachhaltige Entwicklung H Brauer Eds Handbuch des Umweltschutzes und der Umweltschutztechnik Springer Berlin Heidelberg, Berlin, Heidelberg 1059 1153
10. E Suzanne N Absi V Borodin 2020 Towards circular economy in production planning: challenges and opportunities Eur J Oper Res 287 168 190
11. M Lieder A Rashid 2016 Towards circular economy implementation: a comprehensive review in context of manufacturing industry J Clean Prod 115 36 51
12. Fraunhofer Institute for Production Technology IPT (2021) Competition excellence in production (EiP)
13. M Fuster Morell R Espelt M Renau Cano 2020 Sustainable platform economy: connections with the sustainable development goals Sustainability 12 7640
14. ZJ Acs AK Song L Szerb 2021 The evolution of the global digital platform economy: 1971–2021 Small Bus Econ 57 1629 1659
15. I Hellemans AJ Porter D Diriker 2022 Harnessing digitalization for sustainable development: understanding how interactions on sustainability-oriented digital platforms manage tensions and paradoxes Bus Strat Env 31 668 683
16. T Bauernhansl S Hartleif T Felix 2018 The Digital Shadow of production–a concept for the effective and efficient information supply in dynamic industrial environments Procedia CIRP 72 69 74
17. Change request im Produktionsbetrieb (2017) Apprimus Verlag Aachen, Germany
18. Bauernhansl T, Krüger J, Reinhart G et al (2016) WGP-Standpunkt Industrie 4.0
19. S Grams F Schwade J Mosen 2021 A method for developing and applying metrics profiles for the benefits management of enterprise collaboration platforms Procedia Comput Sci 181 553 561

20. MV Martin K Ishii 2002 Design for variety: developing standardized and modularized product platform architectures Res Eng Design 13 213 235

21. H Dadfar JJ Dahlgaard S Brege 2013 Linkage between organisational innovation capability, product platform development and performance: the case of pharmaceutical small and medium enterprises in Iran Total Qual Manag Bus Excell 24 819 834

22. BW Boehm 1988 A spiral model of software development and enhancement Computer 21 61 72

23. K Lange L Cser M Geiger 1992 Tool life and tool quality in bulk metal forming CIRP Ann 41 667 675

24. DIN EN ISO 9241-210:2020-03, Ergonomie der Mensch-System-Interaktion_- Teil_210: Menschzentrierte Gestaltung interaktiver Systeme (ISO_9241-210:2019); Deutsche Fassung EN_ISO_9241-210:2019

Repair and Maintenance

Boosting Resource Efficiency and Circular Economy of the Manufacturing Companies in Finland

Katri Salminen[✉], Silja Kostia, Mika Ijas, Jere Siivonen, Erkki Kiviniemi, and Markus Aho

Tampere University of Applied Sciences, Kuntokatu 3, 33520 Tampere, Finland
katri.salminen@tuni.fi

Abstract. Increased resource efficiency and circular economy among manufacturing industry could help to turn sustainability into economic value while simultaneously supporting policy objectives set to decrease CO_2 emissions. Overall, this will require changes in organizational strategies, fast uptake of new technologies and skilled workforce. The paper investigates the status of the Finnish manufacturing industry in the uptake of both practices in product design and production, development and future needs in resource efficiency and circular economy in terms of research, skills and investments. A total of 38 Finnish manufacturing companies (18 small and medium size enterprises) answered the survey. The results show that overall Finnish industry considers that the uptake of resource efficiency and circular economy is in at a good level. However, respondents were able to identify several enablers and barriers that could further facilitate the transition towards sustainable manufacturing. Finally, the paper investigates collaboration needs of the manufacturing companies with research organisations and universities in several different sustainable manufacturing technologies such as material efficiency. In the discussion part a collaboration model between a Finnish university and regional manufacturing industry companies is represented and regional testbeds supporting twin transition will be discussed in detail.

Keywords: Business strategies · Industry 5.0 · Circular Economy

1 Introduction

Despite growing pressure by consumers and policy makers, manufacturing industry is still creating enormous amounts of waste and pollution with its products causing up to 30% of the CO_2 emissions [1]. The rise of themes such as circular economy has a potential to change the current trajectory of the industry permanently [2]. However, there are emerging signs that suggest companies are not able to uptake and develop sustainable manufacturing technologies [3]. Several studies show that even basic digital technologies such as Industrial Internet of Things essential to collect and analyze product information are currently not used in manufacturing companies in Europe [4]. This can be expected to become a significant problem for businesses once new regulations to

© The Author(s) 2025
H. Kohl et al. (Eds.): GCSM 2023, LNME, pp. 363–371, 2025.
https://doi.org/10.1007/978-3-031-77429-4_40

improve resource efficiency and circular economy such as Digital Product Passport will be mandatory [5]. Unfortunately, from industry's point of view the biggest drivers to uptake green technologies are often political (i.e., measures are taken when mandatory) instead of actively creating value from leading the twin transformation [6].

To gain in-depth understanding about the manufacturing industry's ability to make a greener shift in their business, it will be imperative to find out current state-of-the art and enablers and barriers for different central measures that can enable sustainable manufacturing practices (e.g., circular economy). Large Enterprises (LEs, 250+ employees) and Small and Medium Size Enterprises (SMEs, less than 250 employees) have different resources to investigate towards sustainable manufacturing technologies. This should be considered when collecting information and analyzing it [e.g., 7 and 8].

The current study collected information about promoting resource efficiency and circular economy for industry. A total of 38 Finnish manufacturing companies (18 SMEs) answered the survey where the responders were asked first to estimate how well the resource efficiency principles and circular economy are taken into account in their current business practices and what is the significance of both [see 9, 10 for the materials submitted to the respondents]. Then, the respondents were asked to estimate the green technologies they would focus within the next two years, decision making process behind the selection, barriers, and potential topics for collaboration with universities. The discussion provides recommendations for policy making and university's role in creating stronger regional partnerships and integration for sustainability.

1.1 Background Literature

Industry 5.0 (I5.0) was launched by the European Commission as a policy guideline to highlight the upcoming manufacturing technologies that would be human-centric and environmentally sustainable [e.g., 11]. I5.0 builds upon industry digitalization [12]. Uptaking circular and resource-efficient manufacturing technologies and practices in industry require resources such as the availability of skilled employees, the adoption of lean methods and ability to collaborate with universities to try the technologies before investing to them [e.g., 13]. There is a large variation between companies regarding their maturity to uptake sustainable manufacturing principles [14]. A large number of previous studies indicate that industry digitalization has a positive impact on manufacturing industry's sustainability, e.g., in performance measurement and management [12, 15]. Examples of identified barriers for the uptake of sustainable manufacturing in industry include but are not limited to poor understanding of the concept [16], outdated technologies [17], and company culture [18]. Singh et al. [19] is a recent review of organizational barriers on the topic. Success factors, on the other hand, include but are not limited to aligning the organizational strategy along the objectives of I4.0/I5.0, commitment of the management, employees (e.g., skills), digitalization of both organization and supply chain, changes to management practices and clear business models [e.g., 20]. A critical factor hindering the implementation of any novel manufacturing practices is company size. A number of papers have shown that LEs are able to invest in the uptake and development of any given emerging technologies that can reduce the environmental effects of the production [21, 22]. However, SMEs are in a different position. The

SME's drivers for sustainable manufacturing have remained external such as social pressure and policies [6]. Further, resources, unclear return-of-investment and management level inability to understand the drivers behind digital and/or sustainable manufacturing hinder the progress [e.g., 23–27]. SMEs struggle with production planning and business model transformation related to circular economy [28, 29]. From university's perspective it is possible to support industry in better understanding of circular economy and sustainable development by offering testbeds for tryouts before investigating to the new technologies, skills and pilot studies [e.g., 30]. It can be expected that measuring of the supply chain emissions, the use of material resources and testing end-of-life properties of a product will be increased by the forthcoming Digital Product Passport initiative [5].

2   Methods

2.1   Data Collection

A total of 38 respondents from Finnish manufacturing companies (18 SMEs, mostly metal industry or service providers) participated the study. The data collection consisted of an online survey on manufacturing companies. The respondents had a possibility to give anonymous answers. The respondents were informed about data management procedure, privacy and other ethical aspects of the survey as well as the expected publication schedule. The survey questions were selected from previous literature and SITRA's policy documents [e.g., 9, 10]. The respondents were asked to familiarize themselves with resource efficiency and circular economy from SITRA's documents.

2.2   Survey Questions

The responses were given in a Likert Scale from 1 (low) to 5 (high) with potential for a verbal explanation. The questions were categorized under six sections:

Section 1: Resource efficiency. Introduction: How well are things related to resource efficiency implemented in your company? (Likert 1–5, 1 not at all, 5 very well). Questions to be replied using the Likert scale: (1) Resource efficiency guides the company's planning and production; (2) The company has the know-how to develop resource efficiency in planning and production; (3) The company has clear development goals to improve resource efficiency. (4) Personnel have the opportunity to bring forward ideas to improve resource efficiency; and (5) The company has a process for advancing resource efficiency ideas. After answering the five questions, the following question was presented: What is your own view on resource efficiency as a future competitive factor for the company?

Section 2: Circular economy. Introduction: How well are things related to the circular economy implemented in your company? (Likert 1–5, 1 not at all, 5 very much). Questions to be replied using the Likert scale: (1) The principles of the circular economy guide the company's planning and production; (2) The company has the know-how to develop circular economy business in design and production; (3) The company has clear development goals for the development of circular economy business; (4) The personnel have the opportunity to bring forward ideas that promote the circular economy; and (5) The company has a process for advancing circular economy ideas. After answering the five questions, the following question was presented: What is your own view of the importance of development activities promoting the circular economy as a competitive factor in the company's future?

Section 3: Identified opportunities for sustainable manufacturing practices in the next two years. Introduction: Which of the following sustainable development opportunities will your company utilize in its own operations in the next two years? (Likert 1–5, 1 not at all, 5 very much). Questions to be replied using the Likert scale: (1) Material and resource efficiency; (2) Circular economy; (3) Product or Service Lifecycle Management (PLM); (4) Responsibility; (5) Operational opportunities for sustainable development; (6) Lifelong learning; (7) Renewable energies; (8) Energy saving; and (9) The pursuit of carbon neutrality.

Section 4: Decision making behind the selected sustainable manufacturing technologies. Introduction: On what basis does your company choose the technologies to be developed? Options: We are ready to adopt sustainable development operating models and technologies more widely (check the box if your answer is yes): Because customers and partners consider it important; Because we hope to get cost savings from it; Because there is a lot of funding available for promoting sustainable development; The owners consider the themes of sustainable development ideologically important, e.g. preventing climate change is important; Company profiling and branding, e.g. individual products and better quality; and Advantage in recruitment and attractiveness.

Section 5: Largest barriers hindering the uptake of sustainable manufacturing. Introduction: What are the biggest obstacles to making the leap to sustainable development? Options: The price of environmental technologies; Earning logic in the area of sustainable development; Lack of knowledge, not knowing what to do in one's industry; Skills gap; It is difficult to reconcile sustainable development with other goals; Topics related to sustainable development are not communicated clearly enough to operational level; Difficulty choosing which technologies to focus on; Conflicts between goals and practical needs; and Missing networks and partner.

Section 6: RDIE environments. Introduction: What equipment, environments and services related to RDIE environments and services of sustainable development does your company need and/or is it willing to cooperate with in the future (also e.g., RDIE cooperation related to the company's own technologies and environments)? (RDIE = Research, Development, Innovation, Education). Options: Material and resource efficiency; Circular economy; Product or service life cycle management (PLM); Responsibility; Business opportunities for sustainable development; Lifelong learning; Renewable energies; Energy saving; and The pursuit of carbon neutrality.

3   Results

The following section represents comparison between LEs and SMEs for the sections. A t-test assuming unequal variances was used for Sections 1–2 while responses to Sections 3–6 are presented as percentages (MD > 10% bolded). The verbal replies obtained from Sections 1 and 2 are handled in the discussion to highlight the results (Tables 1 and 2).

Section 3: (1) **Material and resource efficiency** (SME 44%, LE 74%); (2) **Circular economy** (SME 17%, LE 37%); (3) **Product or Service Lifecycle Management**

Table 1. Section 1: Question number (see methods), means for SMEs and LEs, T-stat and p-value (significant differences are highlighted with *)

Question	SME	LE	T-stat	p-value
1	3.4	3.9	2.4	0.01^{*}
2	3.6	3.9	1.03	0.15
3	3.5	4	1.5	0.07
4	3.8	2.5	1.1	0.1
5	3.4	3.8	1.6	0.05^{*}

Table 2. Section 2: Comparison between SMEs and LEs.

Question	SME	LE	T-stat	p-value
1	3.1	3.3	0.8	0.2
2	2.9	3.5	2.0	0.02^{*}
3	3.3	3.5	0.7	0.24
4	3.3	3.4	0.2	0.4
5	3.1	3.4	1.1	0.14

(PLM) (SME 39%, LE 53%); (4) Responsibility (SME 50%, LE 58%); (5) **Operational opportunities for sustainable development** (SME 33%, LE 53%); (6) **Lifelong learning** (SME 50%, LE 63%); (7) Renewable energies (SME 33%, LE 42%); (8) **Energy saving** (SME 39%, LE 47%); and (9) The pursuit of carbon neutrality (SME 33%, LE 32%). Section 4: Because customers and partners consider it important (SME 61%, LE 63%); **Because we hope to get cost savings from it** (SME 61%, LE 47%); Because there is a lot of funding available for promoting sustainable development (SME 0%, LE 0%); The owners consider the themes of sustainable development ideologically important, e.g. preventing climate change is important (SME 33%, LE 32%); **Company profiling and branding, e.g. individual products and better quality** (SME 33%, LE 52%); and **Advantage in recruitment and attractiveness** (SME 16%, LE 0%). Section 5: **The price of environmental technologies** (SME 44%, LE 32%); Earning logic in the area of sustainable development (SME 22%, LE 21%); Lack of knowledge, not knowing what to do in one's industry (SME 5%, LE 0%); **Skills gap** (SME 22%, LE 0%); **It is difficult to reconcile sustainable development with other goals** (SME 44%, LE 16%); Topics related to sustainable development are not communicated clearly enough to operational level (SME 6%, LE 5%); Difficulty choosing which technologies to focus on (SME 0%, LE 16%); Conflicts between goals and practical needs (SME 33%, LE 32%); and Missing networks and partners (SME 11%, LE 5%). Section 6: Material and resource efficiency (SME 39%, LE 42%); **Circular economy** (SME 11%, LE 21%); Product or service life cycle management (PLM) (SME 22%, LE 26%); Responsibility (SME 22%, LE 16%); Business opportunities for sustainable development (SME 11%,

LE 16%); Lifelong learning (SME 17%, LE 16%); **Renewable energies** (SME 0%, LE 16%); **Energy saving** (SME 44%, LE 21%); and The pursuit of carbon neutrality (SME 11%, LE 11%).

4 Discussion

4.1 General Discussion

As expected [e.g., 21], LEs were ahead of SMEs in resource efficiency and circular economy, especially regarding company's strategic priorities such as know-how and planning. Overall, the results support the line of thinking that SMEs focus on narrow product segments and sales, while LEs are beginning to consider sustainable manufacturing as an integrated part of business strategy and values. This is supported by the Section 3, where LEs were able to identify more sustainable manufacturing technologies they will focus on the near future than SMEs and Section 4 where SMEs were focusing on cost savings while LEs understood the value of the sustainable manufacturing for product design, profiling and branding.

The verbal responses backed up the survey data. The SME respondents argued how resource efficiency and circular economy are important to cut expenses or improving the profit. Only one SME respondent highlighted the competitive advantage for circular economy. On the other hand, the respondents from the LEs focused on developmental topics and organizational development and saw clear business value for circular economy. Section 5 aimed at identifying barriers in the sustainable manufacturing uptake. Close to 50% of the SMEs found pricing, skills gap and/or strategic barriers. While previous studies [e.g., 24, 27] support partially the findings, the current results provide clear indications for policymakers to support SMEs (see Sect. 4.2).

Finally, Section 6 focused on university-industry collaboration within selected topics. Overall, LEs were more favorable for collaboration than SMEs in potential RDI topics such as circular economy or renewable energies. However, SMEs were intrigued by the idea to collaborate within energy saving topics which in line with the other sections' notions about financial motives being the main driver for SMEs to uptake sustainable manufacturing technologies.

4.2 Recommendations

Policy making: Roughly a third of the LEs and SMEs find the sustainable manufacturing theory conflicting with the practice of the business. In order to improve the uptake of the sustainable manufacturing principles, policymakers and researchers should focus on finding the current state-of-the-art of the industry and adapt measures, key performance indicators and financial and RDI support that ensure that the sustainable manufacturing goals can be realistically achieved within the next decade. The current data further supports the previously identified troubles of SMEs to integrate sustainability to the manufacturing business and this needs to be addressed in the regional, national and international policy initiatives. First steps into developing strategic research roadmaps for manufacturing industry in Finland already take place [1]. Collaboration model: There is a great need to understand the differences between the SME and LE needs when developing collaboration models. LEs have a good understanding about the specific technologies they want to develop together with the universities while SMEs do not. LEs are more innovation-driven and SMEs more practical. In Tampere region, FieldLab [31] has targeted the industry's twin transition offering a set of sustainable technologies such as remanufacturing and data for testing. FieldLab has an extensive portfolio for university-industry collaboration and its own service model for industry. Test before invest services include but are not limited to: (1) Emerging technology introductions such as DeepDives, workshops and seminars for specific topics in the field of emerging technologies in Industry 4.0; (2) Knowledge and technology transfer: Technology pilots support companies to make business investment decisions on their technology roadmap for the future; and (3) Testing and experimentation services and use case demonstrations: Service provides generic demos, small-scale company-specific demos and tangible examples, showcasing how new technologies can be implemented in the use-cases to support companies in developmental activities. This service model is targeted to suit the needs of SMEs and LEs as it is scalable, adaptable, and flexible.

References

1. Heilala J, Paasi J, Lanz M, Aho M, Salminen K, Kutvonen A (2022) Sustainable Industry X Research Ecosystem (SIRE): strategic research and innovation agenda SRIA 2022: project deliverable December 2022
2. Bjørnbet MM, Skaar C, Fet AM, Schulte KØ (2021) Circular economy in manufacturing companies: a review of case study literature. J Clean Prod 294
3. Gbededo MA, Liyanage K, Garza-Reyes JA (2018) Towards a life cycle sustainability analysis: a systematic review of approaches to sustainable manufacturing. J Clean Prod 184:1002–1015
4. Pacchini APT, Lucato WC, Facchini F, Mummolo G (2019) The degree of readiness for the implementation of Industry 4.0. Comput Ind 113:103125
5. European Commission: https://hadea.ec.europa.eu/calls-proposals/digital-product-passport_en, 21st of June, 2023
6. Harikannan N, Vinodh S, Gurumurthy A (2020) Sustainable industry 4.0-an exploratory study for uncovering the drivers for integration. J Model Manag
7. Calabrese A, Levialdi Ghiron N, Tiburzi L (2021) 'Evolutions' and 'revolutions' in manufacturers' implementation of industry 4.0: a literature review, a multiple case study, and a conceptual framework. Prod Plan Control 32(3):213–227
8. Zheng T, Ardolino M, Bacchetti A, Perona M, Zanardini M (2019) The impacts of Industry 4.0: a descriptive survey in the Italian manufacturing sector. J Manuf Technol Manag
9. SITRA: https://www.sitra.fi/en/topics/a-circular-economy/, 21st of June, 2023
10. SITRA: https://www.sitra.fi/en/dictionary/resource-wisdom/, 21st of June, 2023
11. European Commission: Industry of the Future—a transformative vision for Europe: governing systemic transformations towards a sustainable industry, 10th of Jan, 2023
12. Beltrami M, Orzes G, Sarkis J, Sartor M (2021) Industry 4.0 and sustainability: towards conceptualization and theory. J Clean Prod 312:127733
13. Hariyani D, Mishra S (2022) Organizational enablers for sustainable manufacturing and industrial ecology. Clean Eng Technol 6:100375
14. Ahmad S, Wong KY, Rajoo S (2019) Sustainability indicators for manufacturing sectors: a literature survey and maturity analysis from the triple-bottom line perspective. J Manuf Technol Manag 30(2):312–334

15. Ejsmont K, Gladysz B, Kluczek A (2020) Impact of industry 4.0 on sustainability-bibliometric literature review. Sustainability 12(14):5650
16. Ching NT, Ghobakhloo M, Iranmanesh M, Maroufkhani P, Asadi S (2022) Industry 4.0 applications for sustainable manufacturing: a systematic literature review and a roadmap to sustainable development. J Clean Prod 334:130133
17. Alayón CL, Säfsten K, Johansson G (2022) Barriers and enablers for the adoption of sustainable manufacturing by manufacturing SMEs. Sustainability 14(4):2364
18. Singh RK et al (2022) Integration of green and lean practices for sustainable business management. Bus Strateg Environ 31(1):353–370
19. Hariyani D, Mishra S, Sharma MK, Hariyani P (2022) Organizational barriers to the sustainable manufacturing system: a literature review. Environmental Challenges
20. Salminen K, Siivonen J, Hillman L, Rainio T, Ukonaho M, Ijas M, Aho M (2023) Sustainable digital transformation of manufacturing industry: needs for competences and services related to industry 5.0 technologies. In: 2023 Portland international conference on management of engineering and technology (PICMET). IEEE, pp 1–9
21. Rafael LD, Jaione GE, Cristina L, Ibon SL (2020) An industry 4.0 maturity model for machine tool companies. Technol Forecast Soc Chang 159
22. Müller JM, Buliga O, Voigt KI (2021) The role of absorptive capacity and innovation strategy in the design of industry 4.0 business models-a comparison between SMEs and large enterprises. Eur Manag J 39(3):333–343
23. Pech M, Vrchota J (2020) Classification of small-and medium-sized enterprises based on the level of industry 4.0 implementation. Appl Sci 10(15):5150
24. Da Silva VL, Kovaleski JL, Pagani RN, Silva JDM, Corsi A (2020) Implementation of Industry 4.0 concept in companies: empirical evidences. Int J Comput Integr Manuf 33(4):325–342
25. Moktadir MA, Ali SM, Kusi-Sarpong S, Shaikh MAA (2018) Assessing challenges for implementing Industry 4.0: implications for process safety and environmental protection. Process Saf Environ Prot 117:730–741
26. Rossini M, Costa F, Tortorella GL, Portioli-Staudacher A (2019) The interrelation between industry 4.0 and lean production: an empirical study on European manufacturers. Int J Adv Manuf Technol 102(9):3963–3976
27. Marcon É, Soliman M, Gerstlberger W, Frank AG (2021) Sociotechnical factors and industry 4.0: an integrative perspective for the adoption of smart manufacturing technologies. J Manuf Technol Manag
28. Cantú A, Aguiñaga E, Scheel C (2021) Learning from failure and success: the challenges for circular economy implementation in SMEs in an emerging economy. Sustainability 13(3):1529
29. Pedone G, Beregi R, Kis KB, Colledani M (2021) Enabling cross-sectorial, circular economy transition in SME via digital platform integrated operational services. Procedia Manufacturing 54:70–75
30. Siivonen J, Pöysäri S, Hakamäki AM, Lanz M, Salminen K, Ijas M., Nieminen H et al (2022) Reconfigurable pilot lines enabling industry digitalization: an approach for transforming industry and academia needs to requirements specifications. Procedia CIRP 107:1226–1231
31. FieldLab. https://sites.tuni.fi/fieldlab-en/, 21st of June, 2023

Development and Implementation of Reliability Centered Maintenance Framework in Large Organisations

Mesuli Percival Mhlungu[1], Ilesanmi Daniyan[2(✉)] ⓘ, Boitumelo Ramatsetse[3], and Thembeka Behrens[1]

[1] Department of Industrial Engineering, Tshwane University of Technology, Pretoria 0001, South Africa
[2] Department of Mechatronics Engineering, Bells University of Technology, P. M. B. 1015, Ota, Nigeria
iadaniyan@bellsuniversity.edu.ng
[3] Department of Mechanical and Mechatronics Engineering, Stellenbosch University, Stellenbosch, South Africa

Abstract. Large organisations across the world constantly seek to ensure the optimal operation of their processes to remain competitive against their industry peers. This study seeks to help the organisation improve maintenance performance by developing a framework that can be adopted by the organisation in implementing Reliability Centered Maintenance (RCM). This study develops a model to help companies to adopt RCM and to do that, a company were selected to test the model. A questionnaire was sent to a selected population to gather the views of people regarding maintenance within the selected organisation. This is achieved by first investigating the current maintenance regime in Company Y via a survey. Historical data relating to maintenance were obtained and analysed to highlight problem areas. Plant documents relating to maintenance policies were sought for the understanding of the status quo. The results obtained indicate that from 2016 to 2021, the organisation did not meet the maintenance target of 95% and there are major maintenance issues, which have affected the equipment availability and organisation's productivity. Thus, a guideline for the implementation of RCM was developed for use. The guideline will be useful for organisations in their quest to achieve RCM for optimal productivity.

Keywords: Maintenance · Optimal operation · Productivity · Reliability Centered Maintenance

1 Introduction

Until recently, maintenance activities have been regarded as unavoidable by the various management functions in an organisation [1]. The business has placed an emphasis on the improvement of maintenance. With this, a need to implement sound maintenance strategies has arisen. The notion of maintenance strategy has been evolving slowly over

H. Kohl et al. (Eds.): GCSM 2023, LNME, pp. 372–381, 2025.
https://doi.org/10.1007/978-3-031-77429-4_41

the years. The development of new maintenance strategies has been slow, mainly due to the relative lack of importance afforded to maintenance in the various industries, with a greater focus falling on production. However, in recent times, there has been a renewed focus on maintenance, mainly due to the pressures on operating margins and the need to continuously reduce downtimes [2]. As businesses compete to gain or maintain their share of the market, they have come under pressure to be effective and efficient. The selected organisation to test the developed RCM is Company Y. The company is an excellent coal export terminal situated in KwaZulu-Natal, South Africa. The company has an initial capacity of 12 million tons per annum (mt/a). The company provides South Africa's Coal Exporting Parties (CEPs) with an excellent logistics service that facilitates the export of coal. Handling and effectively moving coal is a vital activity in the energy value chain. The coal handling facility is faced with a record low availability of machines, which presents a bottleneck in the value chain. The recorded availability of machines for the past six years, dating from 2016 are as follows: 79.9% for 2016, 82.9% for 2017, 81.22% for 2018, 89.00% for 2019, 90% for 2020 and 90.00% for 2022. These figures indicate that there is an underlying problem to be addressed for the facility to be able to achieve its desired 96% machine availability. This facility has different business units with different maintenance teams and each business unit adopts its own maintenance strategy, as it deems necessary. Company Y is faced with several plant failures, particularly in the yard machines. The number of failures experienced in the yard machines in the past 6 years has increased the maintenance costs in the organisation. The organisation has spent a major portion of its budget in an attempt to rectify these recurring issues either through the utilisation of internal resources or, in some cases, outsourcing the maintenance and engineering specialists. This study sought to improve the existing maintenance approaches by providing RCM framework to minimise the frequency of machine failures in Company Y and other industries that can adopt the developed framework.

The challenge for managers is that the maintenance approaches within these different business units need to be standardised in compliance with the company's overall way of performing maintenance management. At present, the larger business units implement the RCM approach in their operations. These business units exert more influence on the entire organisation due to their significant contribution to the bottom line. Decisions regarding maintenance approaches, technologies, and tactics can be easily made by larger business units and cascaded down to smaller business units and supporting workshops for implementation. The problem is whether the same maintenance approach be used for all business units, irrespective of the production method used in the coal handling facility.

Thus, this study contributes significantly to business growth and development by providing an insight into the possible way to implement RCM. It also proposes a RCM framework that can help organisations attain achieve RCM in a time and cost effective way.

Many organisations meet with failures on a regular basis, due to lack of a proactive strategy such as RCM. United Airlines was ordered by the United States Department of Defense in 1974 to investigate the methods utilised in the civil aviation business to build aircraft maintenance programmes. This investigation gave birth to a report titled "Reliability Centred Maintenance" in 1978 [11]. The investigators set out to track hundreds of electrical, mechanical, and structural aircraft parts to which they concluded that, as these parts aged, a specific failure pattern emerged [12]. RCM has since been applied with sizable success for more than twenty years; first in the aircraft industry, and later within the military forces, the nuclear power industry, the offshore oil and gas industry, and many other different industries [13]. Experiences from these industries show significant reductions in preventative maintenance (PM) costs while maintaining, or even improving, the availability of the system [14]. RCM is a proactive maintenance method that is regarded to be one of the most potent asset management strategies. It has one of the most diversified uses of any technology on the planet [3]. Prabhakar and Raj [4] defines RCM as a structured and logical process for developing or optimising the maintenance requirements of a physical resource. This is to realise 'inherent reliability', where 'inherent reliability' is the level of reliability that can be achieved with an effective maintenance programme. It constitutes the optimum mix of reactive, time- or interval-based, condition-based, and proactive maintenance practices [5].

The high costs associated with preventative maintenance, including frequent over-hauls or replacements of aircraft components, spurred a special investigation into the effectiveness and efficiency of these maintenance activities [6, 7]. Existing studies have indicated that it is more time and cost effective to implement proactive or predictive maintenance using some digital technologies such as sensors, artificial intelligence and machine learning approaches [8–10]. Marinho et al. [15] developed a total productive maintenance framework that can assist can help organisations increase their productivity due to less equipment stoppages while the use of Lean tools for the evaluation of equipment malfunctions to improve the overall equipment effectiveness has been highlighted [16].

Reliability centered maintenance (RCM) is a decision-making tool [17]. For example, the act of performing the RCM decision-making process promotes better co-operation among all those involved in the process [18]. The major goal of RCM is to cut maintenance costs by focusing on the most vital operations of the systems and avoiding or eliminating maintenance tasks that are not necessary. It is based on the notion that the equipment's inherent reliability is a product of its design and construction excellence [19]. It is very useful when developing an effective preventative maintenance [20]. Breakdowns will be less frequent in a well-maintained system. Less frequent breakdowns imply higher plant availability and reliability. Breakdowns are not just a test of dependability. Taking a system out of operation for preventative maintenance on a regular basis makes it less available. Thus, RCM can reduce the frequency of failures by implementing maintenance procedures that reduce the likelihood of component failure; thus, enhancing its dependability [21]. In order to attain a high degree of safety and dependability, the RCM guidelines are designed to assist in identifying failure modes and associated root causes. The guidelines include: preservation of system function, identification of failure modes, prioritization of failure modes, and selection of effective task to control identified failure modes [21, 22]. High levels of system reliability and plant availability are intended to be improved or maintained by the RCM [21, 22].

The literature review indicated that many organisations are aware of RCM, however, but its implementation framework has not been sufficiently highlighted by the existing literature. The novelty of this study lies in the fact that it provides a practical guideline for RCM implementation to assist organisations to achieve reduction in equipment breakdown.

The succeeding sections present the methodology and results discussion as well as the presentation of the proposed RCM framework. The study ends with conclusion and recommendations based on the findings.

2 Methodology

This study employed the mixed (both qualitative and quantitative) method using the questionnaire as the survey instrument. This method is well-suited for this study as it allows the collection of information via a survey in a form of a questionnaire and the collection of statistical data recorded in the company database to investigate a complex issue such as maintenance [23–25]. Amongst others, the questionnaire probed the frequency of equipment breakdown, maintenance availability and the number of unplanned maintenance. The statistical data represent historical maintenance performance of the organisation while the survey seeks to understand the perceptions of the personnel in relation to the maintenance culture within the organisation. The analysis of the data gathered was carried out with the aid of a histogram which depicts the organisation's plant major breakdowns as well as the maintenance availability and the target between 2016 and 2021.

For this study, two divisions were selected, that is, maintenance and operations because these divisions are responsible for maintaining the plant machines and operating them. The maintenance division has 64 employees, which includes managers, engineers, artisans, planners, and assistants. The operations division has 39 permanent employees comprising managers, shift leaders and operators. A target population for this study was 103. The sample size for this study was mathematically determined to be 40 using Eq. 1.

$$n = \frac{NZ^2 \times P(1-P)}{d^2 \times (N-1)Z^2 \times P(1-P)} \tag{1}$$

where: n is the sample size, N is the target population, Z is the confidence level, P is the expected proportion and d is the precision.

The level of confidence for this study is 90% and the expected proportion is 38% with a margin of error of 10%. This then means that out of the 103 employees from both maintenance and operations divisions, only 40 were selected to participate in this study. The method of sampling used in this study is random sampling. This method was chosen because it presents simplicity and eliminates any potential biasness. This reason is validated by Sharma [26] who explained that "random sampling" makes it easy to put together the sample, which is one of its key features. Additionally, because each member has an equal chance of being chosen, it is seen as a fair method of picking a sample from a particular population." The organisation system of Company Y keeps track of all production delays that occurs throughout the production time. Based on daily targets established for the plant, the management uses these delays to assess how well the plant is performing. The Key Performance Indicator (KPI) reports are also examined to learn more about the performance of the plant. These reports provide understanding of how the previous shift performed and the equipment faults experienced from the daily reports. Based on the conclusions reached after analysing the production delays that happened, recommendations with action plans were generated. For this study, permission to access and use the historical information of the company was sought and granted. Data relating to maintenance performance and the machine availability rate over the years were garnered and analysed. Data relating to the performance of the plant against the set targets for the same period were also collected, analysed, and quantified. The Computerised Maintenance Management System (CMMS) of the organisation served as the primary source of secondary data about the performance of the plant's machinery. When a request for corrective maintenance following equipment failure is made, the maintenance function generates the data in the form of a notification. This information was chosen over others because of its dependability and suitability for the compilation of management reports. The accuracy and completeness of the data are also carefully examined before the maintenance function uses it to take the necessary steps to monitor the functioning of plant equipment.

3 Results and Discussion

This section discusses the survey results and the guidelines for the implementation of RCM.

3.1 Survey Results

Figure 1 shows major breakdown recorded at Company Y between 2016 and 2021. The results obtained indicated that that the number of breakdowns in Company Y increased between 2016 and 2018 but later reduced and stabilized between 2019 and 2021 owing to the formation of a maintenance task group in 2017 to address this issue. A major breakdown is defined as the asset failure that has a Mean Time to Repair (MTTR) of more than 2 h [27]. In the context of this organisation, a major breakdown is classified as the one in which the MTTR is over 12 h and the production cost is over R300, 000.00. Stacker reclaimers (SR) continue to be one of the most serious problems plaguing the organisation. According to the results obtained from the survey, one of the reasons for the low production performance (low export volumes) has been poor maintenance performance, either through low availability of the machines or poor reliability of the machines.

Figure 2 presents the maintenance availability in the organisation from 2016 to 2021. It is evident from Fig. 2 that the organisation has struggled to meet world-class manufacturing (WCM) standards. According to Idoniboyeobu and Ojeleye [28], the WCM recommends a minimum availability of 90%. Figure 2 also indicates that between

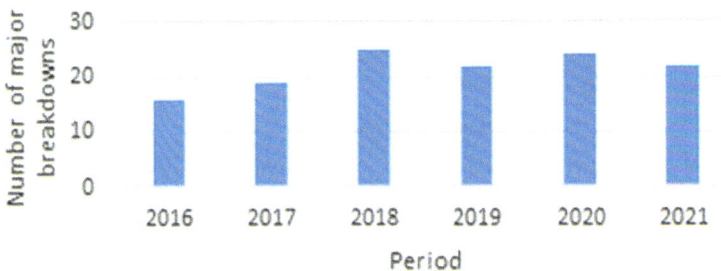

Fig. 1. Plant major breakdowns.

2016 and 2018, the organisation did not meet this standard. In the years 2019–2021, this threshold was met but it is not considered good enough since this achievement is still below the 95% target set by the organisation. Moreover, a 90% achievement for three successive years shows no improvement as it places the organisation on the borderline and risks a performance dip in the coming years.

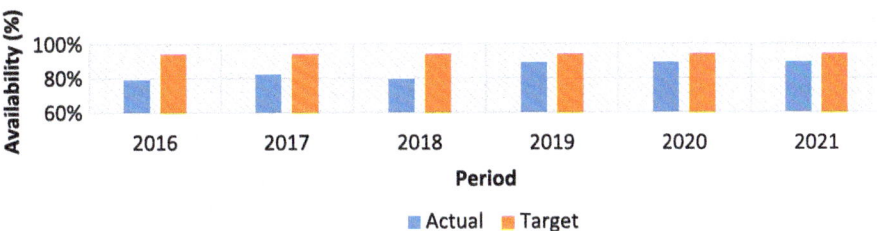

Fig. 2. Organisation's maintenance availability vs target 2016–2021.

Table 1 presents the plant unplanned stoppages (the total minutes lost due to unplanned stoppages as well as the per maintenance section within the organisation. The sections are classified as mechanical, electrical/instrument/automation and services. The unit of measure for machine stoppages is minutes/10000t. The terminal uses this unit of measure to determine machine downtime per unit.

Drawing from Table 1, from 2016 to 2021, most unplanned plant stoppages largely occurred due to mechanical breakdowns. The results obtained from the survey indicated that there are major maintenance issues with the organisation, which have affected the equipment availability and organisation's productivity.

3.2   RCM Maintenance Framework

For the successful implementation of RCM, each piece of equipment must be treated individually and data related to it must be examined in order to gain a better understanding of the process. A blanket approach must be avoided to ensure that the strategy for each machine is the most appropriate one. Figure 3 illustrates how the framework can be used to adopt and implement RCM in the plant. The first step is to select the machine and establish the scope of maintenance. Here, decision has to be made on the selection

Table 1. Organisation's unplanned stoppages 2016–2021.

Plant stoppages per section between 2016–2021						
Year	Mechanical		Electrical/Instrument/Automation		Services	
	Total minutes	Per unit	Total minutes	Per unit	Total minutes	Per unit
2016	18508	12.75	3683	2.54	3894	0.27
2017	19115	12.75	3092	2.06	3359	0.22
2018	17262	11.62	3270	2.20	855	0.06
2019	20348	13.02	3512	2.50	984	0.07
2020	20711	13.19	2784	2.01	4581	0.31
2021	20400	13.09	3097	2.06	1727	0.92

criteria and the data component. Thereafter, there is a need to organize the criteria to aid the findings, data analysis and maintenance. The criticality of a component is determined in terms of how it impact the production process. Furthermore, Asset Criteria Ranking (ACR) should consider the operation, uses, safety, quality and environment of the component or machine. The higher the ACR, the more important the machine is and the more important it is to track its failure symptoms for prompt maintenance. Next to this is the analysis of the failure modes and effect including its associated risks and the development of maintenance schedules. The Risk Priority Number (RPN) can be used to determine the level of risk in any event of machine break down considering its severity, likelihood and detectability. Machine with high RPN and Critical Failure Number (CFN) can have their spare parts stocked in quantities on site while the ones with medium RPN and CFN can be registered on the system for quick ordering. On the other hand, the parts of machines with low RPN and CFN can be ordered on site. Next to this phase is the monitoring of the machine's performance and components through data analysis to measure its reliability. Once failure modes are detected, they easily be remediated.

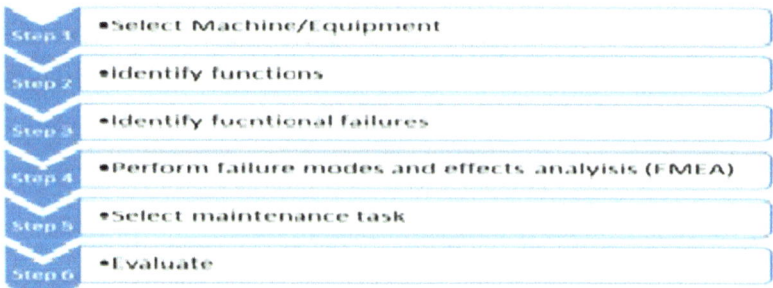

Step 1   •Select Machine/Equipment

Step 2   •Identify functions

Step 3   •Identify fucntional failures

Step 4   •Perform failure modes and effects analyisis (FMEA)

Step 5   •Select maintenance task

Step 6   •Evaluate

Fig. 3. Guidelines for RCM implementation.

# 4	Conclusions

The research investigated the current maintenance strategy that the organisation uses, and it was found that there is currently no comprehensive maintenance strategy that informs all the maintenance decisions. According to the survey results obtained, one of the reasons for the low production performance of the Company Y selected is poor maintenance performance, either through low availability of the machines or poor reliability of the machines. The results obtained indicate that from 2016 to 2021, the organisation did not meet the maintenance target of 95% and there due to major maintenance issues, which have affected the equipment availability and organisation's productivity. It was also found that the organisation is aware of the risks presented by this RCM approach and has been very slow in mitigating them. The positive finding was that the organisation is aware of the need of an RCM based strategy and a few individuals are trying to adopt this. However, individuals cannot apply it only; it needs to be an attitude of everyone within the organisation. Furthermore, this study developed a guideline for RCM implementation. The RCM framework can be used for the optimization of maintenance programmes by establishing a safe threshold for the upkeep of machines or equipment. By matching an asset with the chosen maintenance technique, an organisation can achieve a cost-effective maintenance programme. Thus, the developed RCM framework can help organisations attain achieve RCM in a time and cost-effective way. Hence, it is recommended for implementation. Future works can consider the validation of the proposed framework as well as the integration of the RCM with Total Preventive Maintenance (TPM).

References

1. Eti M, Ogaji S, Robert S (2007) Integrating reliability, availability, maintainability and supportability with risk analysis for improved operation of the Afam thermal power-station. Appl Energy 84(2):202–221
2. Dekker R (1996) Applications of maintenance optimization. Erasmus University, Rotterdam
3. CENESCO (2022) 7 benefits of reliability-centered maintenance (RCM). [Online] Available at: https://cenosco.com/blog-post/7-benefits-of-reliability-centered-maintenance-rcm/. Accessed 05 04 2022
4. Prabhakar D, Raj DJ (2014) CBM, TPM, RCM and A-RCM—a qualitative comparison of maintenance management strategies. Int J Manag Bus Stud 4(3):49–56
5. NASA (2008) Reliability-centred maintenance guide for facilities and collateral equipment, Washington D.C: Pearson Education
6. Cotaina M, Carretero J, Garcia F (2020) Study of existing reliability centered maintenance (RCM) approaches used in different industries. Technical Report, Universidad Politécnica de Madrid
7. Mungani D, Visser J (2013) Maintenance approaches for different production methods article. S Afr J Ind Eng 24(3):1–16
8. Daniyan IA, Mpofu K, Adeodu AO (2020) Development of a diagnostic and prognostic tool for predictive maintenance in the railcar industry. Procedia CIRP 90:109–114
9. Daniyan IA, Mpofu K, Muvunzi R, Uchegbu ID (2022) Implementation of artificial intelligence for maintenance operation in the rail industry. Procedia CIRP 109:449–453

10. Daniyan IA, Mpofu K, Oyesola M, Ramatsetse BI, Adeodu AO (2020) Artificial intelligence for predictive maintenance in the railcar learning factories. Procedia Manufacturing 45:13–18
11. Barros C, Masala M (2020) Reliability centered maintenance (RCM)—implementation and benefits. Available [Online] at https://manwinwin.com/wp-content/uploads/2021/04/Reliab ility-Centered-Maintenance-RCM-Implementation-and-Benefits.pdf. Accessed 26 Nov 2021
12. Takele T (2009) Maintenance program development and Import/Export of Aircraft in USA. Malardalen University Sweden
13. Zeinalnezhad M, Chofren G, Goni A (2020) Critical success factors of the reliability-centred maintenance implementation in the oil and gas industry. Symmetry 11(1):1–14
14. Naik D, Soni KP (2016) Research review on reliability centred maintenance. Int J Innov Res Sci Eng Technol 5(6):9605–9607
15. Pinto GFL, Silva FJG, Campilho RDSG, Casais RB, Fernandes AJ, Baptista A (2019) Continuous improvement in maintenance: a case study in the automotive industry involving Lean tools. Procedia Manufacturing 38:1582–1591
16. Marinho P, Pimentel D, Casais R, Silva FJG, Sa JG, Ferreira LP (2021) Selecting the best tools and framework to evaluate equipment malfunctions and improve the OEE in the cork industry. Int J Ind Eng Manag 12(4):286–298
17. Ebrahim R, Parvari A (2015) The impact of RCM on reducing the maintenance costs of construction machinery machinery compared with the maintenance method. J Eng Constr Manag 1(1):15–25
18. Mammo M, Haileluel M (2016) Equipment maintenance policies from two different perspectives. Ethopian E-J Res Innov Foresight 8(1):38–58
19. Khasanah S, Pwnirewod A, Richmad B (2021) The Reliability-Centered Maintenance (RCM) effect on plant availability and downtime loss in the process industry. IOP Conf Ser Mater Sci Eng 1(1):15–52
20. Chopra A (2021) Applications and barriers of reliability centered maintenance (RCM) in various industries: a review. Ind Eng J XIV(1):15–24
21. Kim TW, Singh B (1996) Guidlines for implementation of RCM on safety system. Yusong: Korea Atomic Energy Institute
22. Available at: https://www.geeksforgeeks.org/how-to-implement-rcm/. Accessed 04 06 2022
23. Akinbowale OE, Klingelhöfer HE, Zerihun MF (2023) The assessment of the impact of cyberfraud in the South African banking industry. J Financial Crime 1–15
24. Akinbowale OE, Klingelhöfer HE, Zerihun MF (2023) Application of forensic accounting techniques in the South African banking industry for the purpose of fraud risk mitigation. Cogent Econ Finance 11(2153412):1–21
25. Akinbowale OE, Klingelhöfer HE, Zerihun MF (2022) Analytical hierarchy process decision model and Pareto analysis for mitigating cybercrime in the financial sector. J Financ Crime 29(3):884–1008
26. Sharma G (2017) Pros and cons of different sampling techniques. Int J Appl Res 3(7):749–752
27. Njomane L, Telukdarie A (2017) Corrosion management: a case study on South African oil and gas company. In: Proceedings of the international conference on industrial engineering and operations management Paris, France, pp 581–595
28. Idoniboyeobu D, Ojeleye J (2011) Maintenance and management of thermal power stations case study: the Egbin thermal power station Nigeria. NJEM 12(2):29–39

Development of a Maintenance Framework for Addressing Power Outage in South Africa

Thembeka Behrens[1], Ilesanmi Daniyan[2(✉)] [iD], Felix Ale[3], and Sesan Peter Ayodeji[4]

[1] Tshwane University of Technology, Pretoria 0001, South Africa
[2] Department of Mechatronics Engineering, Bells University of Technology, P. M. B. 1015, Ota, Nigeria
iadaniyan@bellsuniversity.edu.ng
[3] National Space Research and Development Agency, Abuja, Nigeria
[4] Federal University of Technology, Akure, Nigeria

Abstract. In recent times, South Africa has been experiencing power outages, which has significant effect on businesses and economic wellbeing of the nation. To address this challenge, this study develops a maintenance framework that power-generating industries in South Africa can implement. Primary data was collected from company X that is a major subsidiary of the power generating industry in South Africa to probe the root causes of power outages in recent times. The analysis of the data obtained was carried out using Pareto chart. The results obtained identified delay in maintenance operations as one of the possible causes of power outages. The rework job was observed to be tardy by 50 days. This length of days is critical enough to cause power outage. To address this, a framework was developed to assist in effective scheduling of maintenance activities. Hence, this work provides an insight into the possible causes of power outages in South Africa and how they can be mitigated to avoid load shedding.

Keywords: Maintenance framework · Load shedding · Power outage · Pareto chart

1 Introduction

Company X is one of subsidiaries of Eskom, which provides a distinct service in respect of power generation, machine maintenance, and overall refurbishment of components in South Africa. The machines used at Company X are subjected to unforeseen and frequent breakdowns, which often lead to their unavailability for refurbishment processes, thus leading to load shedding. Mehmeti et al. [1] asserted that, with increased equipment complexity, professional access to maintenance plays a key role in being competitive in the market and the increase in productivity. Maintenance outages within Eskom are often not completed within the planned time. These outages are often scheduled in consideration of the available capacity within the National Power Grid. Eskom schedules outages that are essentially for maintenance with a scope of work that is required to be carried out based on the cycle of a set number of operating hours that the power plant

H. Kohl et al. (Eds.): GCSM 2023, LNME, pp. 382–390, 2025.
https://doi.org/10.1007/978-3-031-77429-4_42

turbine has completed. Eskom generates power for consumers within South Africa and neighbouring countries in Africa. Eskom as a power utility faces challenges in meeting the demand in the supply of electricity. The heart of the problem is the unavailability of the required skills to conduct periodic maintenance work on mandatory outages thus leading to delay in the completion of maintenance activities [2]. Existing works have indicated that predictive maintenance is often more cost and time effective and ensures equipment availability compared to the periodic maintenance strategy [3–5] currently implemented by Eskom and its subsidiaries. The Planned Capability Loss Factor (PCLF) governing planned maintenance has been above 10%, since 2014 due to a loss in critical Eskom skills [6]. This results in continuous load shedding for extended periods where customers are negatively affected. In view of these challenges, this work has identified the maintenance issues as one of the root causes of power outages. It has also developed a framework aimed at addressing the maintenance issues affecting outage execution.

Eskom influences the development of South Africa through six main areas: economic development, job creation, skills growth, local communities, environmental effect and the improvement of the country's growth through the provision of electricity, which has become a challenge in recent time [7–9]. Thus, this study contributes significantly to South Africa's growth and development by an insight into the possible causes of power outages in South Africa and how they can be mitigated. Power outages and load shedding in South Africa is an emerging problem and investigation of the root causes has not been sufficiently reported by the existing literature. The succeeding sections present the literature review followed by the methodology and results and discussion. The study ends with conclusion and recommendations based on the findings.

2 Literature Review

Power outages in South Africa have far-reaching effects on the socio-economic and financial developments [8–10]. An understanding of the impact of power outages on the economy and businesses in South Africa is vital [9]. Furthermore, Goldberg [11] conducted a study on the impact caused by power outages on the economy by looking at the South African retailers. The results highlighted that R13.72 billion was lost in revenue for retailers during the first six months of 2015. Furthermore, Memane *et al.* [12] acknowledged that power outages lead to a decrease in revenue and customer dissatisfaction and proposed a framework for the reduction load shedding. Recently, Onaolapo et al. [13] carried out a study that looked into forecasting power outages in KwaZulu-Natal using artificial neural networks (ANNs) and Trend Projection (TP) methods. The positive outcome of the results is that the results derived from ANN are vigorous and acceptable for electricity outage projections. It is important for power outages to be executed within the allocated period of maintenance; hence, ensuring that various industries within South Africa are not affected. The delays in outages do not only affect generation, transmission and distribution but they affect the mining, manufacturing, clothing and many more sectors. Not only does the load-shedding saga affect activities within the country, it has a significant negative impact on the economy.

Electric power transmission and distribution planning cannot be undertaken without a clear idea of the economic worth of a continuous electric power supply [14]. Continuous production of goods and provision of services can only be achieved where there is continuous electricity. Enhanced industrial productivity leads to increased sales, which creates additional systemic values.

Shuai et al. [15] indicated that the economic impact of power outages can be divided into the direct and indirect impact. Meanwhile, the direct economic impact of power outages can be further divided into two namely: the effects of blackouts on the power industry and the economic impact on power users. Shuai et al. [15] explains that the main task of power enterprises is to provide the power users with reliable and high-quality electricity at reasonable prices. The author further emphasised that power outages would cause huge economic loss to the power industry, including a reduction in sales revenue due to a decline in electricity sales and troubleshooting costs. The key highlight of this study is that there is limited research on the indirect economic impact of electricity failure.

Shuai et al. [15] identified the following factors that affect the economic loss because of power outages. The economic loss owing to a power outage is mainly affected by factors such as interruption frequency, interruption duration, user types, blackout time, time in advance notice of power outages, among others. South Africa is now constrained by an electricity supply-side shortage that could have severe implications for future economic growth [11]. The impact of power outages decreases revenue in the business sectors [11]. Goldberg [11] further stated that the energy sector of the economy remains an integral part of infrastructural development that will in effect set the foundation for broad based sustainable long-term economic growth and development. Therefore, as most sectors rely on consistent power supply to function and be productive, the lack of power supply and power outages will cause increasing severe economic losses for the users. Having established the impact of power outage on the South African economy, this study employed a structured questionnaire to further gain an understanding of the probable root causes of power outage in order to develop a framework aimed at tackling it.

Thus, the proposed framework presented in this study can be followed in identifying the critical path of the equipment, scope of maintenance, availability of spare parts, risk and severity to facilitate condition-based monitoring.

3 Methodology

Historical data of rework was collected from Company X business management system for the period January 2021 to June 2022. This is to identify the root causes of delay in maintenance activities that leads to incessant load shedding. In the context of this study, a rework is considered as any equipment that fails in less than or within 60 days after maintenance operation has been carried out. This research has adopted a mixed (qualitative and quantitative) research approach, through the implementation of questionnaire surveys. The choice of the mixed approach stems from the fact it is suitable for probing the root cause of complex issues [16, 17], such as delays in maintenance activities and load shedding. The questionnaire surveys were distributed to a sample of employees within Company X. Overall the turbo generator service departments comprise of five major departments namely; mechanical, bearing & OSM, quality, non-destructive service (NDT), and generator service departments having 133 employees. These departments consist of engineers, senior supervisors, supervisors, senior technicians, technicians and artisans who completed the research questionnaire. The questionnaire probes the number of reworks experiences in these departments as well as the root causes. The population size in this case is 133, and the confidence level is 95%, as this is the true population size based on the number of employees within these departments. The confidence interval is ± 2% based on the 95% confidence level. The sample size is calculated according to Eq. 1 [18].

$$n = \frac{N \frac{z^2 p(1-p)}{e^2}}{\frac{z^2 p(1-p)}{e^2} + N - 1} \tag{1}$$

where: n is the sample size, N is the population size, z is the confidence level (in percent, such as 96%), p is the sample proportion (in percent, such as 50%), e is the confidence interval (margin of error) (in percent, such as 2% = 0.02). The criteria for accuracy in quantitative research outcomes was met by a sufficient sample size. The computation from Eq. 1 demonstrates a sample size of 125.796, where a population size is 133 within Company X. The results however will be more accurate based on the larger the sample size. The data analyses determined important factors that contribute to delay in maintenance activities within the organisation. To determine the reasons for rework, the types of rework, the cost of rework, and the impact the rework exerts on the organiatisonal skills shortage, the data were sorted and statistically analysed. The Pareto Chart enabled the prioritization of the necessary actions and make well-informed decisions to reach the intended goal [19, 20]. Following that, the Pareto tool was then used to determine the root causes of delay within the various departments as it is well recognised that it is a good tool for illustrating the relative importance of issues for this research. The Pareto tool was used to evaluate which area that needs to be improved and given more attention based on the number of times that rework was detected.

4   Results and Discussion

This section presents the results obtained from the survey and the proposed maintenance prediction completion model.

4.1 Survey Results

The information obtained from the survey indicates that lack of skills in the organisation, is partly responsible for the need for rework, which resulted in delay in maintenance activities with resulting significant rework costs. The outage programmes in Table 1 shows the impact of rework on outages when compared with the programme based on the original scope of work that was planned to be executed during the outages. The impact of the rework will often exert a direct bearing on the critical path depending on the scope of work. Table 1 shows the impact that rework has had on the programmes by noting the increase in duration of the programmes. The rework job was observed to be tardy by 50 days. This length of days is critical enough to cause power outage.

Table 1. Outage programme reworks.

Project description	Original duration (days)	New duration (days)	Tardiness (days)
BFPT rotor	15	24	9
LP 1 rotor	44	47	3
IP casing	50	60	10
Generator rotor	37	49	12
Generator stator	29	35	6

The information gathered from the survey reveals the number and important aspects of rework, the root causes and the departments involved within the organization. This is depicted using Pareto charts in Figs. 1 and 2. The results show that the majority of the rework occurs in the mechanical section due to poor workmanship. This lends credence to the fact that the skills shortage is a concern in the organisation and most of the rework can be traced to it. Shah and Burke [21] describe skill as the capacity to carry out a productive task at a specific level of competence. The level of competence required to carry out refurbishment is a challenge that is reflected through the rework shown in Fig. 1 and 2.

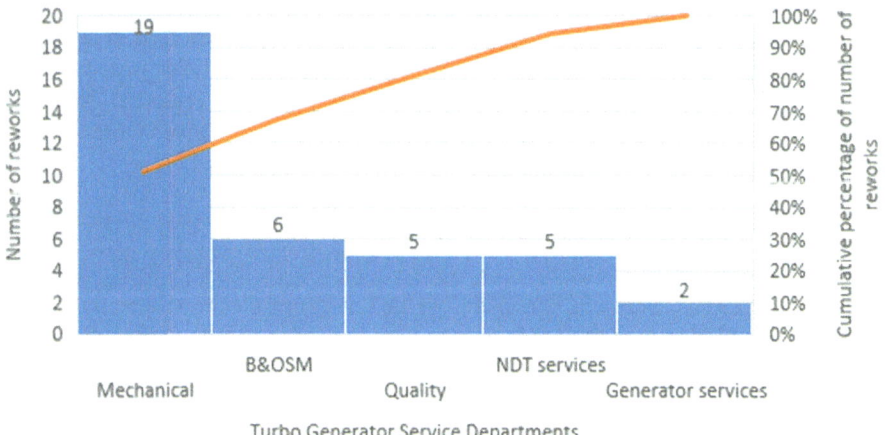

Fig. 1. Pareto chart for the reworks.

Some of the causes of rework that have been identified within these departments include poor performance, the human factor/discipline, procedure or process, and lack of training. The findings agree with some existing works which identified specific variables that cause rework, for example, inadequate communication, bad site management, incorrect supervision, and inspection, were first brought up decades ago but are still relevant today in explaining why rework occurs [22, 23]. According to Bosworth et al. [24], training lead times may make it necessary to wait a while before a skill shortage is resolved, even though market indications are recognised from both the supply and demand sides.

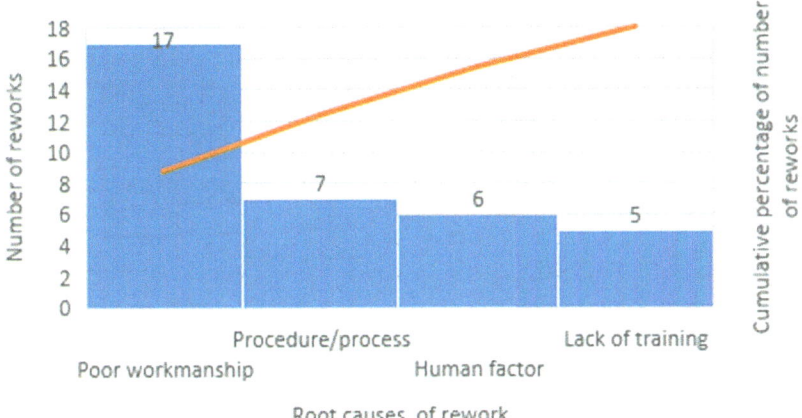

Fig. 2. Pareto chart for the root causes.

4.2   Maintenance Completion Prediction Model

It is noted that maintenance refurbishment of the components within the organisation are carried out without following all the best practices that could help the department to perform better. There is continuous pressure to execute the work in a rush without considering the challenges that the department may have. These challenges were revealed from the outcome of the survey. In line with best maintenance practices, Fig. 3 shows the prediction model that can be implemented for refurbishment scope for projects that are executed to avoid tardiness. Four major factors link to the proposed model. First is the critical path analysis, which identifies the machine r equipment that needed to be monitored continuously. This is because any breakdown along the equipment's critical path could affect power generation. Secondly, the scope should be interrogated with scenario planning, readiness, delivery date and critical path analysis. Most projects within the organisation are undertaken without interrogating issues such as the availability of spares; some jobs require specialised spares replacements, which are often taken for granted. Technical competence of the nature of skills required carrying out activities such as de-blading and re-blading should be interrogated before the job begins. Third, the refurbishment scope readiness highlights the scope of the maintenance to be carried out and lastly the scenario planning will enable condition-based maintenance. This will also aid the decision-making process whether there is a need for maintenance or not based on the real time asset condition. The prediction model should form part of the mandatory steps of the execution process, which will identify the need for specialised resources should there be any incidents.

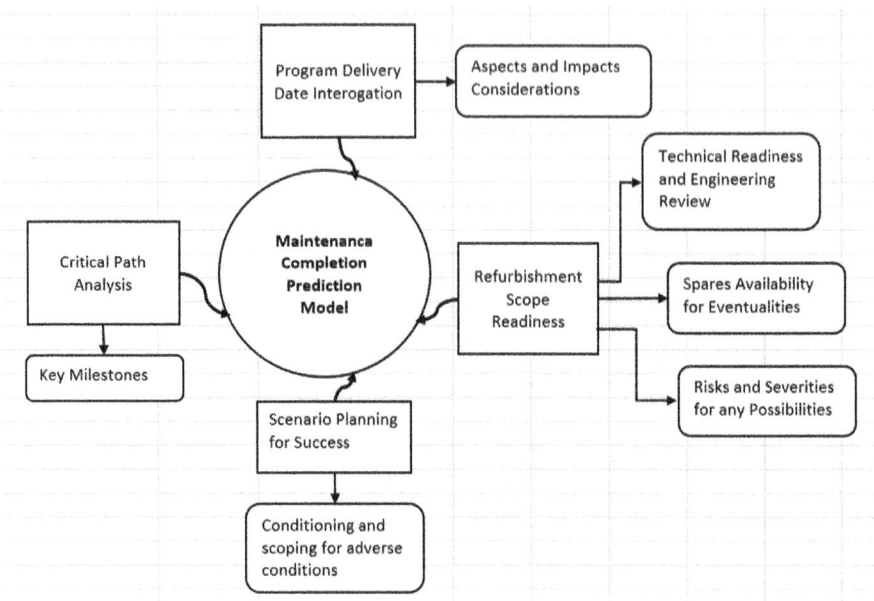

Fig. 3. Maintenance completion prediction model.

5   Conclusions

The unavailability of the power station units for the generation of power continues to lead the country into load shedding. At the heart of these challenges lies the issue of the lack of the required skills to execute the maintenance outage work. The results obtained from the survey indicate that lack of skills in the organisation, is partly responsible for the need for rework, which resulted in delay in maintenance activities with resulting significant rework costs. The results also show that the majority of the rework occurs in the mechanical section due poor workmanship. Some other causes of rework that have been identified within these departments include poor performance, the human factor or discipline, procedure or process and lack of training. The rework job was observed to be tardy by 50 days. This length of days is critical enough to cause power outage. The developed maintenance completion prediction model is recommended for implementation to address tardiness during maintenance operation. Future work can consider the investigation of other root causes leading to load shedding in South Africa and the development of framework for their mitigation.

References

1. Mehmeti X, Mehmeti B, Sejdiu R (2018) The equipment maintenance management in manufacturing enterprises. IFAC-PapersOnLine 51(30):800–802
2. MYBROADBAND (2020) The real reason for load-shedding—and it's not what Eskom is telling you. Available at: https://mybroadband.co.za/news/energy/338506-the-real-reason-for-load-shedding-and-its-not-what-eskom-is-telling-you.html. Accessed 24 July 2020
3. Daniyan IA, Mpofu K, Adeodu AO (2020) Development of a diagnostic and prognostic tool for predictive maintenance in the railcar industry. Procedia CIRP 90:109–114
4. Daniyan IA, Mpofu K, Muvunzi R, Uchegbu ID (2022) Implementation of artificial intelligence for maintenance operation in the rail industry. Procedia CIRP 109:449–453
5. Daniyan IA, Mpofu K, Oyesola M, Ramatsetse BI, Adeodu AO (2020) Artificial intelligence for predictive maintenance in the railcar learning factories. Procedia Manuf 45:13–18
6. Eskon (2020) Eskom weekly system status report. Available at http://www.eskom.za/Wha twereng/Supplystatus/Pages/AdequacyReports2018.aspx?Paged=TRUE&p_SortBehavior= 0&p_Created=202004008. Accessed 23 July 2020
7. Rambe P, Modise D (2016) Power distribution at Eskom: putting self-leadership, locus of control and job performance of engineers in context. Af J Bus Econom Res 11(1):45–92
8. Nkosi NP, Dikgang J (2018) Pricing electricity blackouts among South African households. J Commod Mark 11:37–47
9. Dumisa B (2023) The catastrophic impact of load shedding. [Online] Available at https://www.iol.co.za/news/politics/opinion/the-catastrophic-impact-of-load-shedding-will-the-eco nomy-recover-0d086bb3-148b-45aa-b525-c92f2f214f95. Accessed 5 Apr 2023
10. Lawlor P (2023) SA's load shedding constraint and its impact on different sectors [Online] Available at https://www.investec.com/en_za/focus/economy/sa-s-load-shedding-how-the-sectors-are-being-affected.html. Accessed 5 Apr 2023
11. Goldberg A (2016) The economic impact of load shedding: the case of South African retailers', MBA Thesis, (November). Gordon Institute of Business Science, University of Pretoria, South Africa. Available at: https://repository.up.ac.za. Accessed 23 July 2020
12. Memane NP, Munda JL, Popoola OM, Hamam Y (2019) An Improved Load shedding tech- nique for optimal location and profitability for contingency conditions. In: Southern African universities power engineering conference/robotics and mechatronics/pattern recognition association of South Africa (SAUPEC/RobMech/PRASA), pp 241–246
13. Onaolapo K, Carpanen RP, Dorrell DG, Ojo EE (2021) Forecasting electricity outage in KwaZulu-Natal, South Africa using trend projection and artificial neural networks techniques. IEEE PES/IAS Power Africa 1–5
14. Küfeoğlu S (2015) Economic impacts of electric power outages and evaluation of cus- tomer interruption costs. Doctoral Dissertation, Department of Electrical Engineering and Automation, Aalto University, Finland, pp 1–64. https://doi.org/10.1109/ISGTEurope.2014. 7028868
15. Shuai M, Chengzhi W, Shiwen Y, Hao G, Jufang Y, Hui H (2018) Review on economic loss assessment of power outages. Procedia Comput Sci 130:1158–1163
16. Akinbowale OE, Klingelhöfer HE, Zerihun MF (2023) The assessment of the impact of cyberfraud in the South African banking industry. J Financial Crime 1–15
17. Akinbowale OE, Klingelhöfer HE, Zerihun MF (2023) Application of forensic accounting techniques in the South African banking industry for the purpose of fraud risk mitigation. Cogent Econ Finance 11(2153412):1–21
18. CHIPETA C (2020) Sample Size Calculator. Available at: https://conjointly.com/blog/sam ple-size-calculator/. Accessed 01 Aug 2022

19. Brown J, Mellott S (2016) The Janet A. Brown healthcare quality handbook: a professional resource and study guide, 30th edn. JB Quality Solutions
20. Akinbowale OE, Klingelhöfer HE, Zerihun MF (2022) Analytical hierarchy process decision model and Pareto analysis for mitigating cybercrime in the financial sector. J Financ Crime 29(3):884–1008
21. Shah C, Burke G (2003) Changing skill requirements in the Australian labour force in a knowledge economy.Working paper no. 48, 2003. ACER-Monash University Centre for the Economics of Education and Training, Melbourne
22. Yap JBH, Low PL, Wang C (2017) Rework in Malaysian building construction: impacts, causes and potential solutions. J Eng Des Technol 15(5):591–618
23. Yap JBH, Chong JR, Skitmore RM, Lee WP (2020) Rework causation that undermines safety performance during production in construction. ASCE J Constr Eng Manag 146(9):04020106
24. Ye G, Jin Z, Xia B, Skitmore RM (2015) Analyzing the causes for rework in construction projects in China. ASCE J Constr Eng Manag 31(6):04014097
25. Bosworth D, Dutton P, Lewis J (1992) Skill shortages: causes and consequences. Aldershot, Avery, 1–9

Approach for Structured Repairability Assessment for Automated Repair Processes

Hannah Lickert[1]([✉]), Tobias Lachnit[2], and Franz Dietrich[1]

[1] Technische Universität Berlin, Institute of Machine Tools and Factory Management, Chair of Handling and Assembly Technology Research, Pascalstraße 8-9, 10587 Berlin, Germany
h.lickert@tu-berlin.de

[2] Wbk Institute of Technology, Karlsruhe Institute of Technology (KIT), Kaiserstraße 12, 76131 Karlsruhe, Germany

Abstract. Repair is a crucial process to recover resources and reduce waste within a circular economy. Automating manual and laborious repair processes has the potential to establish greater economization and efficiency. While research on assessing product repairability exists, there is currently limited research specifically focused on the repairability regarding automated processes. Therefore, this paper investigates the existing knowledge on automation, manual and automated repair processes from theory as well as practical applications, to identify benchmarks for reparability assessment for automated processes. Based on these requirements, solutions, processes, and operating resources are identified. A standardized and structured approach for evaluating the repairability of products for automated repair is proposed. This research also highlights the economic and ecological advantages, as well as the sustainability challenges and potentials of using automated repair processes. Overall, this research contributes to the development of a more sustainable and efficient CE through the advancement of automated repair processes.

Keywords: Repair · Repairability Assessment · Automated repair

1 Introduction

The transition towards circular economy is vital in view of the climate crisis. Currently, the global economy operates only at a circularity level of merely 7.2%, which has also declined nearly 2% over the past five years [1]. While the circular economy framework primarily emphasizes remanufacturing and recycling, the vital role of repair often remains overlooked. Repair, however, constitutes an indispensable component of the circular transformation, offering both economic and ecological advantages [2]. Among the various options to recover value, repair stands out as a process with low cost and labor intensity as well as energy consumption. However, as compared to remanufacturing it involves reduced warranty, reliability, and performance of the product [3, 4].

But the political focus on product reparability is growing. Corresponding measures address the right to repair, assess repairability and facilitating reuse of electronic and

© The Author(s) 2025
H. Kohl et al. (Eds.): GCSM 2023, LNME, pp. 391–398, 2025.
https://doi.org/10.1007/978-3-031-77429-4_43

electrical waste, such as the the circular economy action plan of the European Commission, the DIN EN 45554 on the assessment of the ability to repair, reuse and upgrade energy-related products and the EU directive on waste electrical and electronic equipment [3, 5, 6]. These regulations highlight the necessity for the effective recovery of electronic and electrical waste to achieve long-term climate objectives.

Regardless of the necessity and feasibility, product repairability depends on the monetary investment required for the repair process, as it competes with product replacement [7]. This is a challenge establishing repair, especially for low-priced products and in countries where labor costs and spare part expenses are high. The integration of automation technologies offers opportunities to reduce manual labor costs and increase the efficiency of the repair process. Nevertheless, economies of scale must be incorporated for automation. Considering the mass production of electronic and electrical products and assuming a product service system or similar business models, the repair of similar products can be centralized in large quantities. Therefore, it becomes essential to explore how repairs can be efficiently performed automated for large quantities, enabling environmentally friendly circular economy practices.

Chapter 2 covers the barriers and limitations for repair in general and the potential for automated repair as well as its current implementation in practice. To create a basis for assessing whether products and processes are suitable for automated repair, a structured approach is developed in Chap. 3. Concluding in Chap. 4, that even though automated repair holds great potential, technical solutions to make automation more versatile are still being developed, which will allow further possibilities in automated repair.

2 Potentials of Automated Product Repair

Although repair is possible and significant efforts have been made by policymakers, manufacturers, and repair initiatives, there are barriers and limitations that prevent repairs from being offered commercially on a large scale. To overcome these barriers the automation of repair processes offers a potential. In order to understand what is intended by automated repair and how this can reduce barriers, the term repair and the existing automation solutions must first be examined. The potential for automating repairs is then derived on the basis of the identified barriers to repair.

The intension of repair is defined by Thierry et al. as: "to return used products to working order" [4]. Meaning, that the"quality of repaired products is generally less than the quality of new products" [4]. General process steps of repair are inspection, disassembly, repair, reassembly, and functional test. These steps can be repeated iteratively until the function of the product is restored, as shown in Fig. 1.

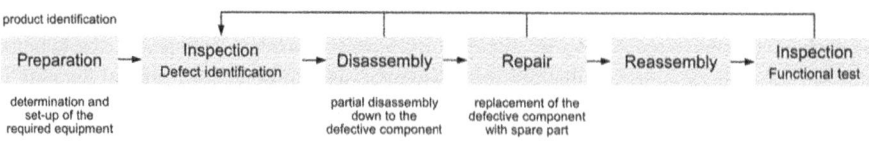

Fig. 1. General process steps of repair

Currently, only repairs by replacement are considered in the examination, meaning repair at the product level including the replacement of defective components and parts with spare parts. This reduces the complexity of the overall process and focuses on the automation potential of (dis)assembly and handling operations.

The term automated repair is often used in connection with automated maintenance or refurbishment and can be clearly distinguished from automated program repair in the software field, yet no uniform term usage was found in literature. Using the definition of automation and repair, automated repair is specified in this paper as: Automated product repair is the process by which machines or computer-controlled systems perform partially or fully automated repairs on products to restore their functionality.

Automated repair is being selectively applied in some industries on material, component, and product level. At the material level, the repair of turbine blades by automated laser powder build up welding is an example, or the repair of rotor blades as maintenance of wind turbines [8]. At the component level, the automated repair of electronic components, e.g. circuit boards, is widely applied and already scientifically elaborated in the mid-1990s [9]. At the product level, some applications have been developed for the automotive industry, such as automated tire changing using image recognition and machine learning [10].

2.1 Barriers and Limitations of Repair

A list of factors was compiled that hinder users from engaging in repairs, which are categorized into the general aspects regarding economic and ecological sustainability, product, and process related aspects. The most relevant are explained as follows.

General. The repair must be economically and ecologically sustainable as well as safe and reliable. From an economic point of view, the cost of repair exceeds the customer's profit margin or willingness to invest. In addition, due to an improperly performed repair and bad experiences, some users associate repaired products negatively, while repair is generally considered to be more resource-efficient than other recovery processes. This does not apply in particular and must therefore be individually assessed and verified that the repair is also ecologically beneficial [5, 11]. Maintenance and repair for high-quality products and especially those for commercial use are established, as these products are often designed for longer periods of use, whereas for low-quality products, the effort and cost of repair are not viable. Automation can create economic efficiency for low-priced products in repair through economies of scale.

Product. A significant reason for reduced reparability is the limited availability of spare parts, documentation, and repair information. Also, the product design is crucial, as it often leads to unfeasible or complex repair processes [11, 12]. These aspects become especially important for automation, which raises additional requirements for the design and information set of the product. Furthermore, the product is decisive for the design of repair offers. In addition to repairability, the value of the product and the available quantity play an important role. Obtaining enough defective products is obvious for automation, but knowledge and equipment must also be gained in preparation for manual assembly.

Process. Manual repairs are associated with a high effort of time either a specialized repairer with product-specific knowledge is required, which is often difficult to obtain, or the private repairer must acquire this knowledge, which is limited by his willingness, skills and the effort involved [11]. In addition, quality variance can arise in manual repair depending on the skills of the repairer [13]. These factors can be addressed by automation, but the difficulty in scheduling repair processes presents a challenge. Also, automated resources must achieve a level of flexibility and adaptability that allows for consistent processing across a variety of different repair cases [14].

In summary, several barriers to repair can be identified, including the time required for repairs, the costs associated with manual labor, the need for reliable and consistent quality of repairs. The automation of repair processes enables the economic repair of products in large quantities while maintaining consistent quality [13]. Automated repair is already being applied on the material, component, and product level. However, there is a lack of research establishing general principles, regarding requirements, processes, and development of automated repair applications.

This work aims to give an overview for requirements and an approach to assess the feasibility of automated repair applications on a product-level, to provide fundamentals. Tools exist for assessing the reparability of products and processes for repair are being studied. But product and process required for repair need to be considered coherently, to evaluate whether they can be implemented, automated and thus applied on a large scale. Since circular processes are integrated into closed-loop systems, general criteria are included in the consideration. Thus, for the structured approach, requirements are first specified in the three areas of product, process including equipment and general aspects, derived from literature and the above-mentioned barriers. Proceeding from this, a workflow is created to apply these requirements in a structured manner.

3   Structured Approach to Assess Automated Repair Processes

The developed approach consists of two elements: a list of requirements regarding criteria about the product, process and equipment and general factors related to automated repair; and a workflow as a kind of guide for the sequence and purpose of using the requirements list.

3.1   List of Requirements

The list of requirements was compiled based on literature and derived from the barriers described. The list does not claim to be exhaustive, but since it is quite extensive, only an overview is presented here, and the complete list is assessable via the platform DepositOnce[1] of the TU Berlin.

Product. These requirements address product features concerning general repairability, condition, and automated repair. For example, that the product:

– has documentation, which includes repair instructions and information.

---

[1] https://depositonce.tu-berlin.de—The file will be published with DOI for the final publication on this platform. For submission, the file was attached to the paper.

- has no contamination or deformation, which affects the process.
- has visible and automated separable connecting elements.

Process and equipment. These requirements relate to general versatility of equipment including software, actuators, end effectors and sensors for process execution, data processing, detection, control, and inspection. for example, that the equipment offers:

- low-effort and reusable programming of the system.
- universal fixture and handling of different product variants.
- automatic recognition of product information, connecting elements and defects.

General. These requirements include economic and ecological criteria. Additional organizational and market factors are included. For example, that:

- the expected number of units and the value added is sufficient.
- the availability of spare parts is given.
- the automated repair is beneficial compared to manual processes, new purchase, or other recovery methods.

3.2 Workflow

In order to assess the feasibility of an automated repair process, several steps are necessary. These steps are divided into product, process, and general factors. Through iteration loops, emerging obstacles can be overcome by the manual execution of a process step, measures to increase sustainability or the redesign of the product. In the following this procedure is illustrated by Fig. 2 and described in detail below.

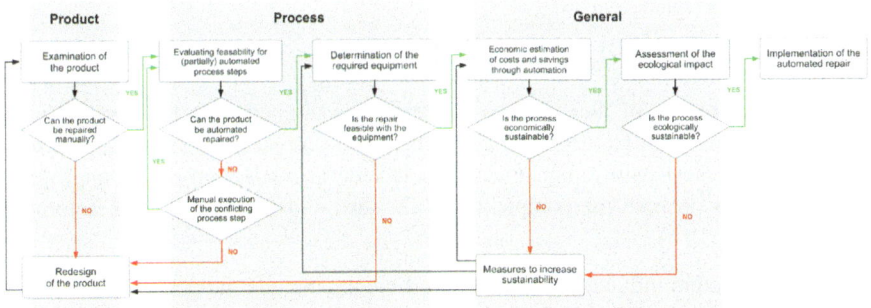

Fig. 2. Workflow to assess the feasibility for automated repair.

Product. During the examination of the product the manual repairability is evaluated. This requires a definition of all repair cases and verification of the availability of appropriate spare parts. It is important to identify and analyze potential product-related obstacles that could affect repair. By examining manual repairability, potential deficiencies in the product can be identified early and appropriate measures can be applied. If it becomes apparent during the evaluation process that manual repair is not feasible, a redesign of the product is required.

Process. For evaluating the feasibility of automating process steps, it is necessary to expand and breakdown the manual repair processes. Subsequently, each individual step of the repair process must be examined regarding its automatability, whereby the following Table 1 can serve as a guide. If it is not possible for certain process steps, manual execution must be considered as an alternative.

Table 1. Assessment of the option to automize the process steps.

No.	Process step	Substep	Automation option
1	Preparation	Product identification Determination of the required operating resources Set up with required equipment	
2	Inspection	Incoming inspection (deformation/contamination) Defect identification	
3	Disassembly	Defect specific derivation of disassembly steps Disassembly step 1 to n Intermediate storage of components Check for complete execution	
4	Repair	Replacement of the defective unit Deposit of the defective unit	
5	Reassembly	Reassembly planning Picking up the components from intermediate storage Resassembly step 1 to n Check for complete execution	
6	Inspection	Functional test	

The determination of the required equipment is aligned with the product to be repaired. Suitable end effectors, such as grippers and screwdriving systems, which are necessary for a successful repair, must be selected. The selection of the appropriate systems depends on both the complexity of the task and the scope of the automated processing.

General. The factors influencing economic efficiency include the costs resulting from processing time for machines and equipment, for manual labor, and for setting up the automated repair. Since the setup costs are dependent on the number of pieces, a quantity estimation should be conducted in advance. Depending on the required accuracy, process times can be estimated by an expert employee, a simulation, or a calculation.

The assessment of the ecological impact is only partially quantifiable. The repair process is usually ecologically sustainable since most of the value is retained, and a completely new product has usually a higher ecological impact. This is also shown by Bracquené's study on the example of laptops [15]. But the process can also entail ecological disadvantages, especially in the case of a failed repair or energy-intensive products. It is necessary to consider the effort of repair in relation to various alternatives, such as buying a new product as well as the expected extension of the product's service life. In terms of automation per se, based on Wang's study, repairs performed by robots compared to human labor is expected to lead to lower environmental impacts [16]. One method to quantify the ecological impact is to perform a life cycle assessment.

The implementation of the automated repair can be realized, presumed that all relevant criteria are met. For this purpose, the corresponding operating resources must be procured, and the system must be set up. After the functionality of the repair process has been successfully confirmed by tests, the serial repair can begin.

4 Conclusion and Outlook

Automated repair holds great potential for ecologically and economically sustainable product repair in large quantities. By automating monotonous and repetitive manual repair tasks, future-oriented, safe, and ergonomic work environments ensuring consistent quality are enabled. Compared to remanufacturing, automated repair offers advantages in terms of efficiency, energy consumption, and cost-effectiveness. It complements existing methods and enhances product circularity. The benefits could drive a greater realization of circular services. To advance automated repair, this study outlines basic requirements for finding new solutions. By integrating repair, automation, and additional criteria into a comprehensive list, the necessary preconditions for automated repair were detected. This facilitates evaluation and measures for users who want to automate product repairs, considering criteria for automated during repair planning and development.

Several obstacles hinder the widespread realization of automated repair. However, integration within a centralized circular economy, alongside remanufacturing, could enhance its role in sustainable value chains. Manual repair still possesses advantages in flexibility, simplicity, and setup requirements. Nonetheless, trends in automation technology, rising wages, and increasing product repairability favor the application of automated repair. Despite its potential, the complexity of automated repair compared to manual repair remains a challenge. Handling the variability of used products and enabling easy programming of different processes are unresolved issues. Intelligent fault detection and functional control are essential for fully automated processes. Further research is needed to advance automated repair and address these complexities.

References

1. https://www.circularity-gap.world/2023
2. Knäble D, de Quevedo Puente E, Pérez-Cornejo C, Baumgärtler T (2022) The impact of the circular economy on sustainable development: a European panel data approach. Sustain Prod Consum, Jg 34:233–243
3. Gharfalkar M, Ali Z, Hillier G (2016) Clarifying the disagreements on various reuse options: repair, recondition, refurbish and remanufacture. Waste Manag Res, ISWA 34(10):995–1005
4. Thierry M, Salomon M, Van Nunen J, Van Wassenhove L (1995) Strategic issues in product recovery management. Long Range Planning 28(3):120
5. Rudolf S et al (2022) Extending the life cycle of EEE—findings from a repair study in Germany: repair challenges and recommendations for action. Sustainability 14(5):2993
6. Šajn N, European parliamentary research service: right to repair. https://www.europarl.eur opa.eu/RegData/etudes/BRIE/2022/698869/EPRS_BRI(2022)698869_EN.pdf

7. Cooper T (2004) Inadequate life? Evidence of consumer attitudes to product obsolescence. J Consum Policy 2004(27):421–449
8. Bergmann A, Grosser H, Graf B, Uhlmann E, Rethmeier M, Stark R (2013) Additive Prozesskette zur Instandsetzung von Bauteilen. LTJ 10(2):31–35
9. Leicht T (1995) Automatische Reparatur elektronischer Baugruppen. Dissertation, Fraunhofer Institut für Produktionstechnik und Automatisierung (IPA), Stuttgart
10. Donlon M (2023) Robots are primed to replace auto mechanics (or are they?). [Online]. Verfügbar unter: https://electronics360.globalspec.com/article/18552/robots-are-primed-to-replace-auto-mechanics-or-are-they (Zugriff am: 13. Februar 2023)
11. Terzioğlu N (2021) Repair motivation and barriers model: investigating user perspectives related to product repair towards a circular economy. J Clean Prod 289:125644
12. Sabbaghi M, Cade W, Behdad S, Bisantz AM (2017) The current status of the consumer electronics repair industry in the U.S.: a survey-based study. Resour Conserv Recycl 116:137–151
13. Uhlmann E, Bilz M, Baumgarten J (2013) MRO—challenge and chance for sustainable enterprises. Procedia CIRP 11:239–244
14. Uhlmann E, Heitmüller F, Manthei M, Reinkober S (2013) Applicability of industrial robots for machining and repair processes. Procedia CIRP 11:234–238
15. Bracquené E, Peeters J, Duflou J, Dewulf W (2021) Sustainability assessment of product lifetime extension through increased repair and reuse. In: Proceedings, 3rd PLATE conference, Berlin, Germany, 18–20 September 2019
16. Wang J, Wang W, Liu Y, Wu H (2023) Can industrial robots reduce carbon emissions? Based on the perspective of energy rebound effect and labor factor flow in China. Technol Soc 72:102208

Factory Planning and Production Management

Sustainable Manufacturing for SMEs: An Agile Readiness Model of Decarbonization Through Theory and Practice

Xiaohui Tang[1], Shun Yang[2(✉)], Yanwen Qian[3], and Sebastian Thiede[2]

[1] Karlsruhe Institute of Technology (KIT), Hertzstrasse 16, 76187 Karlsruhe, Germany
[2] University of Twente, Drienerlolaan 5, 7522 NB Enschede, The Netherlands
s.yang-1@utwente.nl
[3] Siemens Ltd., China, Suyuan Avenue 19, Nanjing 211100, China

Abstract. Decarbonization is a critical area as companies work to meet the environmental mandates associated with their environmental, social, and governance (ESG) commitments. It provides significant potentials for the sustainable manufacturing of Small and medium-sized enterprises (SMEs) to match the demands from the downstream value chain. However, SMEs find themselves not ready to take the first step due to a lack of sufficient professional resources. This paper presents a literature review and practical investigation to find out the root causes of decarbonization issues with regard to SMEs. As novel countermeasure, an agile readiness model with consideration of production and environment engineering is developed which allows collectively identifying the current status and targeted performance. Finally, the proposed approach is validated through a case study with industrial partners (in developed countries like Germany and The Netherlands, and emerging countries like China).

Keywords: Sustainable Manufacturing Systems · Decarbonization · Readiness Model · Assessment Tool

1 Introduction

Compared with the last decades, much more attention from the public and stakeholders is spent on Environment, Society and Governance (ESG) of the enterprises. That means the entrepreneur should also take care of legal compliance, industry waste pollution, Green House Gas (GHG) emission [1]. Many large-scale companies have responded very rapidly. For instance, Siemens AG, Schneider Electric and Robert Bosch GmbH have joined in Science Based Targets Initiative (SBTi) since 2019 [2].

However, as the backbone of economies, the small and medium-sized enterprises (SMEs) face more difficulties to start sustainability related initiatives in contrast to the large-scale enterprises. On the one hand, there are huge amounts of literatures highlighting this issue [3, 4]. On the other hand, five major obstacles have been identified according to the practical investigation.

H. Kohl et al. (Eds.): GCSM 2023, LNME, pp. 401–409, 2025.
https://doi.org/10.1007/978-3-031-77429-4_44

First of all, the knowledge basis of decarbonization is insufficient at SMEs. Secondly, the standard regulation and corporation cultures still don't include the elements of decarbonization. It leads to lack of integration of decarbonization with business. Thirdly, it is still lack of a clear implementation strategy for transforming the decarbonization. Fourthly, specific talents of decarbonization in SMEs are too deficient to build up a team. Lastly, there are quite fewer practical methods and tools as solution, which can support the SMEs to identify the status and find the improvement direction.

Despite the challenges, SMEs are eager to find out their own way to implement decarbonization and sustainable development. In this context, a simple and fast assessment tool needs to be developed to guide SMEs to find their own way. This paper will introduce a novel approach for developing a simple and fast readiness model.

2   State of the Art

To elaborate the readiness model, the existing readiness or assessment model and decarbonization indicators need to be discussed. First of all, the general assessment models are investigated. Secondly, the environmental assessment including Life Cycle Assessment (LCA) as the key area is further analyzed as well.

In terms of general assessment models, Monostori & Kádár et al. established a CPS maturity model which has been divided into five levels. Particular reflections in practice are meanwhile mentioned for every single level in this model [5]. Schuh and Anderl et al. generated the acatech Industry 4.0 Maturity Index [6]. Most authors pay attention to the implementation of Industry 4.0 technologies on the benefits of value-added chain [7], paradigms such as the proper integration of employees [8], and the design of the system infrastructure [9]. However, there is still lack of an agile readiness model for decarbonization in SMEs.

In terms of environmental assessment, there are many existed common tools for evaluating the environmental consequences, such as life cycle assessment, risk assessment, cost–benefit analysis (CBA), environmental impact assessment [10]. All those models are quantitative model, which need large database and lot of resource to input during the investigation and data collection stages. A briefly comparison of the ad-vantages and disadvantages listed in Fig. 1.

As one of the popular assessments for environmental accounting and management methodology, Life Cycle Assessment (LCA) aims to support decision making in early design phases to enable a higher cost effectiveness and increase the degree of sustainability [11]. As a calculation model, LCA is very good in systematization, quantification, standardization, universality. However, the large case studies and big database are very important [12]. Environmental Product Declaration (EPD) methodology is based on the LCA tool as following ISO series 14040 [13]. There would be a lot of pressure to spend much resource in the early stage.

Although a conceptual framework and different scenarios were presented for the improvement of energy and resource efficiency in manufacturing [14, 15], the organizational factors are not covered next to the energy. In fact, an intelligent waste management system was introduced to extend the improvement of decarbonization [16], how-ever, the other environment factors were not included. Furthermore, a holistic sustain-ability

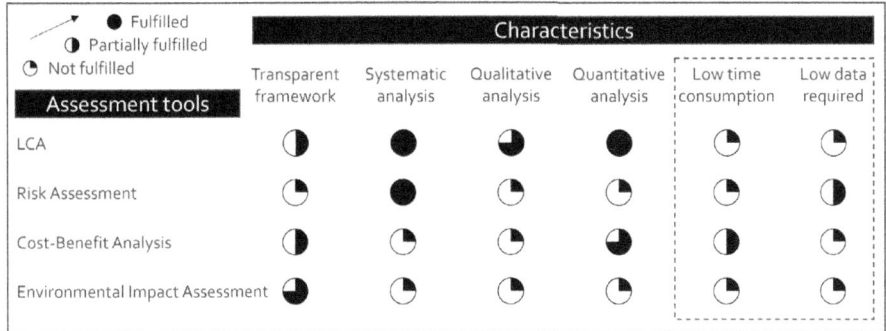

Fig. 1. Comparison of the different assessment tools

assessment tool for manufacturing SMEs was developed [17], however, it is difficult to visualize the desired status of future. Moreover, the deeper analysis was contacted to understand potentials for a decarbonization via LCA [18]. Unfortunately, it required amount of data, which leads to the hug time consumption. Despite the variety [19], it is still not sufficient research on how to create the model to rapidly assess the status of decarbonization and transfer to the applicable level for SMEs.

3 Methodology

In this paper, a three steps method is proposed for designing an agile readiness model (ARM), respectively definition of goal and scope, modeling of structure, visualization of assessment results.

3.1 Definition of Goal and Scope

Goal of the study is to provide the qualitative overview of decarbonization status from different perspectives. The ARM is evaluating the current status, linking the desired status and visualizing the gaps between both statuses. It helps to align actors and organizations to instigate the decarbonization breakthroughs to meet a collective challenge. By considering the scope of ARM, the manufacturing industry has been focused since it is significantly supporting to fight global warming.

3.2 Modeling of Structure

In this paper, the environmental product declaration (EPD) has been taken as reference since it combines the whole value chain and it is well recognized in the industry. The overview procedure of structural modeling is introduced in Fig. 2. A starting point is the identification of applied fields. By considering pragmatical perspective, the ARM only consists of two major part, basics information and assessment dimensions.

In terms of basics information, the contact information, company related questions (i.e. location, domain, size, etc.), and product related questions (i.e. variety, complexity, batch size and product value) have been generated. It brings the value to make the benchmarking according to the specific typology.

Assessment dimensions consists of four aspects, namely dimension products, organization, processes, and EPD. Each dimension has its own application fields, which consists of different elements. To evaluate each element, the criteria have been defined into 5 different legends. Take dimension Products as example, this dimension consists of three application fields, namely product design, product usage, product recycling. The product design indicates the extent to which a company uses digital technologies and data to develop a circular product design. One of elements for product design is to indicate the extent to which the products incorporate digital technologies that enable to continuously generate data on product operation and condition to implement circular economy solutions. To evaluate the element, the five criteria have been identified. The legend can be determined according to the fulfillment of criteria. As-is is to indicate the current status while To-be is meaning the desired status.

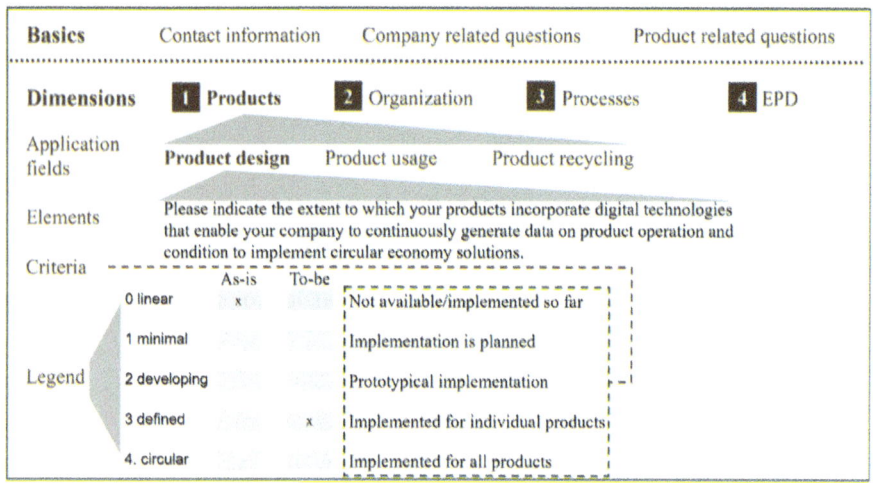

Fig. 2. Overview procedure for modeling of structure (an exemplar)

3.3 Visualization of Assessment Results

By considering the rapid principle, it requires the simplified visualization of assessment results. Therefore, a rapid decarbonization assessment tool (RDAT) has been generated through the MS excel. A brief report will be summarized via radar diagrams.

For the calculation of the decarbonization legend, it is determined as followed.

$$V_{Dx} = \sum_{i=1}^{n} W_i \times V_{DxAFi} \tag{1}$$

$$V_{DxAFi} = \sum_{j=1}^{m} a_j \times V_{DxAFiEj} \tag{2}$$

$$\sum_{i=1}^{n} W_i = 1 \text{ and } \sum_{j=1}^{m} \alpha_j = 1 \tag{3}$$

VDx represents the overall value of assessment result of dimension x. $VDxAFi$ means the assessment value of application field i. The value of n equals to the number of application fields of specific dimension. $VDxAFiEj$ is identifying the assessment value of element j in the specific application field. The value m equals to the total number of elements in this application field. Wi is the weight of application field i and αj is the weight of element. The scores have been collected. Then the four ranges from linear to circular has been defined according to the legend of assessment tool (see Table 1).

Table 1. Range of assessment tool legend

Assessment tool legend	Range	Description of range and legend
0 (include)-1 (exclude)	1	Above linear, but below minimal
1 (include)-2 (exclude)	2	Above minimal, but below developing
2 (include)-3 (exclude)	3	Above developing, but below defined
3 (include)-above	4	Above defined (towards circular)

4 Validation

To serve the SMEs to have an effective assessment of their performance in decarbonization, five companies from 3 countries are invited to attend the assessment (Table 2).

The proposed approach has been successfully conducted in five SMEs (see Appendix 2) through on-site analysis and online interviews. The assessment results of Company H has been presented in Fig. 3 while the rest results can be found in the Appendix 3.

Company H appears positive result in the dimension of EPD and wants to develop the performance in the fields of Organization, Processes and EPD in the future. Mean-while, Company H has performed the high scores in energy consumption of EPD di-mension and manufacturing of processes dimension, but still low score in product re-cycling, treatment of solid waste, and treatment of wastewater.

The comprehensive comparison has been conducted among these five companies. The overall assessment results have been transferred into four ranges, through which the comparison among five SMEs has been generated as below (Table 2).

From above chart, there are mainly three findings derived from the comparison. Firstly, all five SMEs currently perform low scores in EPD dimension. Meanwhile, all five SMEs' management express high motivation and expectation on the topic EPD. It seems that there is huge deviation between the As-is and To-be situation (Table 3).

Secondly, there are four SMEs perform low score in Product dimension, only one SME achieve range 2. Since SMEs play the role of manufacturing among whole value

Table 2. Overview of company typology

Name	Location	Industry & technologies	# of employee	Customer	Production site distribution
H	China, Asia Pacific	Metal stamping	350	Global	Regional
T	China, Asia Pacific	Machining	170	Global	International
U	China, Asia Pacific	Metal parts	250	Global	(Inter) continental
P	Netherland, West Europe	Assembly	200	International	International
I	Germany, West Europe	Wastewater treatment	250	Global	(Inter) continental

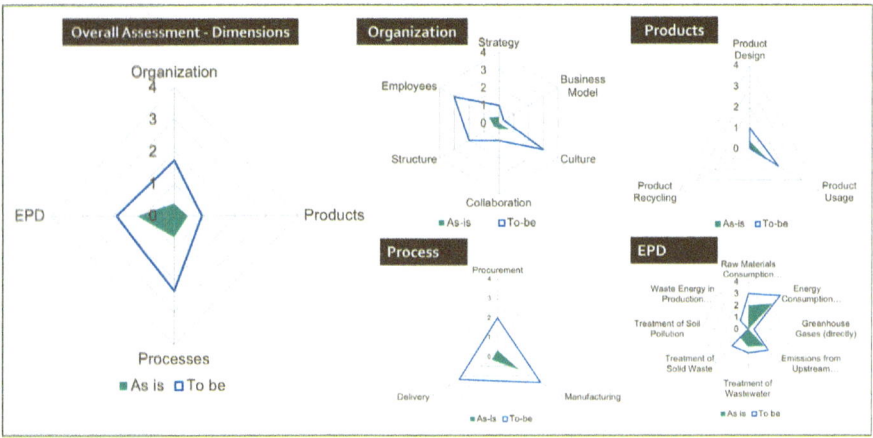

Fig. 3. The assessment result of Company H.

Table 3. The comparison of assessment results

	Organization	Products	Processes	EPD
Company H	Range 1	Range 1	Range 1	Range 2
Company T	Range 1	Range 1	Range 2	Range 1
Company U	Range 2	Range 1	Range 3	Range 2
Company P	Range 1	Range 1	Range 1	Range 1
Company I	Range 2	Range 2	Range 2	Range 3

chain, the business model of Original Equipment Manufacturer (OEM) will strongly affect SME's performance in Products. For Company I, normally it provides whole solutions within products, the business model of Original Design Manufacturer (ODM) leads SMEs to pay more attention on Products.

Thirdly, Company U & Company I have obtained higher scores compared with other 3 SMEs. According to the general information of Company U & Company I, their production sites are (inter) continental and their products are globally distributed to the customers. Therefore, there are high requirements in procurement, delivery, and manufacturing are expected from the global customers (some of main customers are world-leading companies & large-scale companies).

5   Discussion

According to the results of the validations, the row materials flow and the energy consumption are the most important issues from scope 1 and scope 2 according to greenhouse gas Emission Protocol. ARM can be highly valuable for decision making, particularly in management board meetings for their sustainability strategy plan. The companies spend two to three hours to conduct the readiness model via workshops. By comparing with typical assessment tool such as LCA, it is much less time consuming. By focusing on the critical factors, decision-makers can quickly grasp the key areas that require attention.

There are also improvement potentials of the proposed method. For instance, by as-signing weights or scores to different questions, decision-makers can prioritize their importance based on the specific goals and context. Additionally, it would be beneficial to establish a clear link between the legend's five levels and the sub-dimensions of the assessment tool. Moreover, the "Efforts—Benefits" model can be applied within ARM to evaluate the feasibility and return on investment (ROI) of different initiatives.

6   Conclusion

As one of significant steps of sustainability, the decarbonization shows its importance for industry to safeguard their future. This paper proposed a rapid assessment model which enables the SMEs to quickly assess the current state, identify gaps, and provide the possibility to chart a roadmap towards to desired outcomes. The value of this systematic and structured methodology has been demonstrated through the validation in developed countries and emerging countries. SMEs are able to overcome resource constraints and enhance their decision-making capabilities in the area of decarbonization by applying assessment tool. As the next step, the roadmap will be further developed and analyzed, which can support SMEs to reduce the gap between the current state and the desired future state.

References

1. Gelles D (2023) How environmentally conscious investing became a target of conservatives. The New York Times. Retrieved, 2

2. SBTi Homepage, https://sciencebasedtargets.org/companies-taking-action/case-studies. Accessed 10 July 2023
3. Siegel R, Antony J, Garza-Reyes JA, Cherrafi A, Lameijer B (2019) Integrated green lean approach and sustainability for SMEs: from literature review to a conceptual frame-work. J Clean Prod 240:118205
4. Isensee C, Teuteberg F, Griese KM, Topi C (2020) The relationship between organizational culture, sustainability, and digitalization in SMEs: a systematic review. J Clean Prod 275:122944
5. Monostori L, Kádár B, Bauernhansl T, Kondoh S, Kumara S, Reinhart G, Ueda K (2016) Cyber-physical systems in manufacturing. CIRP Annals 65(2):621–641
6. Schuh G, Anderl R, Gausemeier J, Ten Hompel M, Wahlster W (eds) Industrie 4.0 maturity index. Herbert Utz Verlag
7. Bauernhansl T (2017) Die vierte industrielle Revolution–Der Weg in ein wertschaffendes Pro-duktionsparadigma. Handbuch Industrie 4.0 Bd. 4: Allgemeine Grundlagen, 1–31
8. Deuse J, Weisner K, Hengstebeck A, Busch F (2015) Gestaltung von produktionssystemen im kontext von industrie 4.0. Zukunft der Arbeit in Industrie 4.0, pp 99–109
9. Wang S, Wan J, Li D, Zhang C (2016) Implementing smart factory of industrie 4.0: an outlook. Int J Distrib Sens Netw 12(1):3159805
10. Goodfellow HD, Wang Y (eds) Industrial ventilation design guidebook: volume 2: engineering design and applications. Academic press
11. Ingrao C, Messineo A, Beltramo R, Yigitcanlar T, Ioppolo G (2018) Investigating life cycle assessment applications for energy efficiency and environmental performance. J Clean Prod 201:556–569
12. Herrmann IT, Hauschild MZ, Sohn MD, McKone TE (2014) Confronting uncertainty in life cycle assessment used for decision support: developing and proposing a taxonomy for LCA studies. J Ind Ecol 18(3):366–379
13. Minkov N, Schneider L, Lehmann A, Finkbeiner M (2015) Type III environmental declaration programmes and harmonization of product category rules: status quo and practical challenges. J Clean Prod 94:235–246
14. Thiede S, Posselt G, Herrmann C (2013) SME appropriate concept for continuously improving the energy and resource efficiency in manufacturing companies. CIRP J Manuf Sci Technol 6(3):204–211
15. Loftus PJ, Cohen AM, Long JC, Jenkins JD (2015) A critical review of global decar-bonization scenarios: what do they tell us about feasibility? Wiley Interdiscip Rev Clim Chang 6(1):93–112
16. Aivaliotis P, Anagiannis I, Nikolakis N, Alexopoulos K, Makris S (2021) Intelligent waste management system for metalwork-copper industry. Procedia CIRP 104:1571–1576
17. Chen D, Thiede S, Schudeleit T, Herrmann C (2014) A holistic and rapid sustainability assessment tool for manufacturing SMEs. CIRP Ann 63(1):437–440
18. Gebler M, Cerdas JF, Thiede S, Herrmann C (2020) Life cycle assessment of an automotive factory: Identifying challenges for the decarbonization of automotive production–a case study. J Clean Prod 270:122330
19. Madanchi N, Thiede S, Sohdi M, Herrmann C (2019) Development of a sustainability assessment tool for manufacturing companies. Eco-Factories Future 41–68

Planning Multiproduct Assembly Lines Through a Continuous-Time Model Accounting for Carbon Footprint

Nélida B. Camussi[1] and Diego C. Cafaro[1,2]($\boxtimes$) (ID)

[1] INTEC (UNL-CONICET), Güemes, 3450, 3000 Santa Fe, Argentina
dcafaro@fiq.unl.edu.ar

[2] Facultad de Ing. Química, Centro Interuniversitario de Investigaciones en Gestión Analítica de Procesos (GAP), ITBA-UNL, Santiago del Estero, 2829, 3000 Santa Fe, Argentina

Abstract. This work presents an efficient mathematical formulation for the optimal planning of multiproduct assembly lines. The aim is to establish a cyclic production agenda that minimizes the sum of inventory holding and transition costs per unit time. Besides, carbon footprint of assembled products is tracked along the line while optimally determining task times and operating modes. Batch sizing and sequencing in synchronous assembly lines have been typically addressed through discrete-time approaches that imply significant computational burden. In contrast, our novel representation resembles continuous-time pipeline scheduling models that permit to obtain optimal solutions in reasonable times. Results show the capabilities of the optimization approach to solve large instances of the problem, also demonstrating how task times and operating modes can be handled to reduce carbon footprint with no loss of productivity.

Keywords: Assembly lines · Sequencing · Batch sizing · Optimization

1 Introduction

Assembly lines are extremely effective flow-line production systems that consist of a number of workstations arranged along a unidirectional conveyor. In multiproduct assembly lines, campaigns or lots of different products are launched down, and every item in a lot moves from station to station such that in each of them, pieces of work or tasks are performed to manufacture the product. Every task usually requires material and/or energy consumption, which contribute to carbon (CO_2) emissions. There are usually alternative modes to perform these tasks (equipment, energy source, duration) yielding different carbon emissions per unit of product (i.e., carbon intensities). This has motivated the need to achieve systematic ways of measuring the carbon footprint of the products along their value chain [1]. Manufacturing tasks increase the carbon footprint of the products along the assembly line according to how they are performed. Operating modes affect not only costs and yield rates but also the magnitude of the carbon emissions, often being greener at the expense of longer times. In modern value chains it is becoming critical to plan operations sustainably and efficiently, meeting demand under carbon footprint specifications, at minimum total cost.

© The Author(s) 2025
H. Kohl et al. (Eds.): GCSM 2023, LNME, pp. 410–418, 2025.
https://doi.org/10.1007/978-3-031-77429-4_45

In turn, multiproduct pipelines are the most reliable and efficient mode of transportation of fluid products over land. A multiproduct pipeline moves different products (typically oil derivatives) over long distances in batches, which are pumped one after the other into the same duct, usually with no physical separation between them. Although the aim of pipelines and assembly lines is totally different, the operational planning of both systems share common features that deserve a deeper study. The first similarity is that both involve very expensive capital investment, from which a rapid return based on massive usage and high performance is expected. Moreover, predictable batch tracking in pipelines avoids environmental issues by strict control on volume balances, while synchronous assembly lines (simultaneously moving individual items between adjacent stations) reach similar standards by keeping the workforce balanced and work-in-process at low values.

Similar to pipeline scheduling models in which several batches of oil products are conveniently arranged to flow through the system one after the other, the movement of different products along the assembly line needs to be optimally planned (see Fig. 1). In a multiproduct assembly line, each product may have its own cycle time (time between two subsequent movements) which makes the problem challenging. Recent works have addressed the optimal planning of multiproduct assembly lines, which seek for the optimal sizing of lots and the most suitable sequence of products to be processed along the line. At the same time, new developments in process planning and production scheduling prove to be important for environmental performance [2].

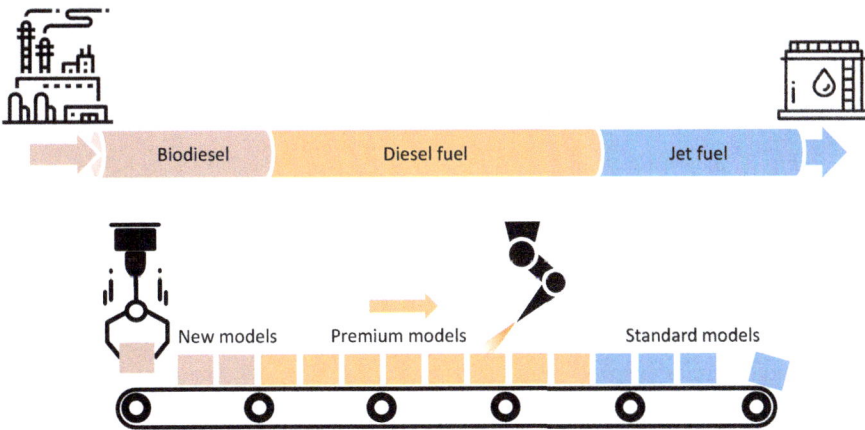

Fig. 1. A simple example illustrating the analogy of multiproduct pipelines and assembly lines

2 Problem Definition

In unidirectional pipelines carrying multiple products, different batches with different volumes move from one extreme to the other, as illustrated on Fig. 1. It is a simple, efficient, safe and economical way for the massive transportation of fluids. After the installation of the pipeline, the sequencing, timing and sizing of batch injections is critical, so as to timely meet demands while satisfying pipeline operational constraints. Many authors have proposed efficient mathematical models with those purposes [3, 4]. A schematic representation of this problem is shown on Fig. 1. Three batches, move from the beginning to the end of the pipeline following a certain chronological order or sequence (jet fuel, diesel fuel, biodiesel). By the incompressibility assumption, if the diesel fuel is denser and more viscous than the other products flowing through the pipeline, it controls the speed of the flow stream.

On the other hand, an assembly line is a production system focused on partitioning the whole work into smaller parts which are assigned to different workstations so as to maximize the throughput and the use of resources (workers, tools, equipment, etc.). In addition to balancing multiproduct assembly lines problems [5], research has been directed towards the organization of the production agenda in order to minimize inventory holding and changeover costs at the same time [6]. Figure 2 shows a paced assembly line with several workstations processing different products where different items synchronically move from one station to the next downwards. Analogously to pipelines, the slowest product controls the speed of the assembly line.

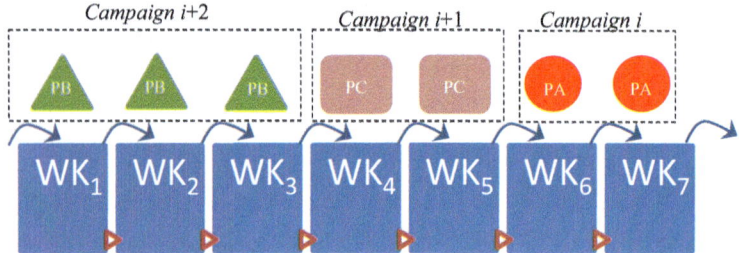

Fig. 2. Schematic outline of production items moving along an assembly line

In this case, instead of seeking for the length, duration and sequence of product batches sent through a pipeline, the aim is to find the optimal number of items comprised by a production campaign which must be processed along the assembly line so as to minimize, simultaneously, inventory holding and changeover costs. Furthermore, average carbon emissions per unit of product should be kept within admissible ranges.

3   Main Decision Variables and Constraints

3.1   Time Control

Every balanced assembly line has a cycle time, denoted as ct, which represents the time spent by the line for processing a new item of final product. It is clear that the cycle time is inversely proportional to the production rate p, i.e., $ct=1/p$. The cycle time stands for the rhythm at which the line moves synchronously. If the line processes several models $j \in J$ with different cycle times ct_j, the simultaneous presence of two or more different products on the workstations makes the cycle time become equal to the maximum value among all of them, which limits the speed of the line. Figure 3 illustrates such condition, with three products (A, B, C) being simultaneously processed by the line. If cycle times satisfy the inequalities $ct_B > ct_C > ct_A$, then the input of at least one piece of model C (indicating the beginning of a new campaign) imposes the reduction of the line speed even though A has a faster cycle time. Similarly, the input of at least one piece of model B imposes the reduction of the production rate even though there are faster products along the line. Finally, and due to environmental constraints, cycle times for different products may need to be optimally managed by adopting "greener" (slower) or "grayer" (faster) processing modes $m \in MD$ during different runs. As a result, the cycle times in our optimization model are given by the parameter $ct_{j,m}$, indicating the time to process a single unit of j when the line runs in mode m.

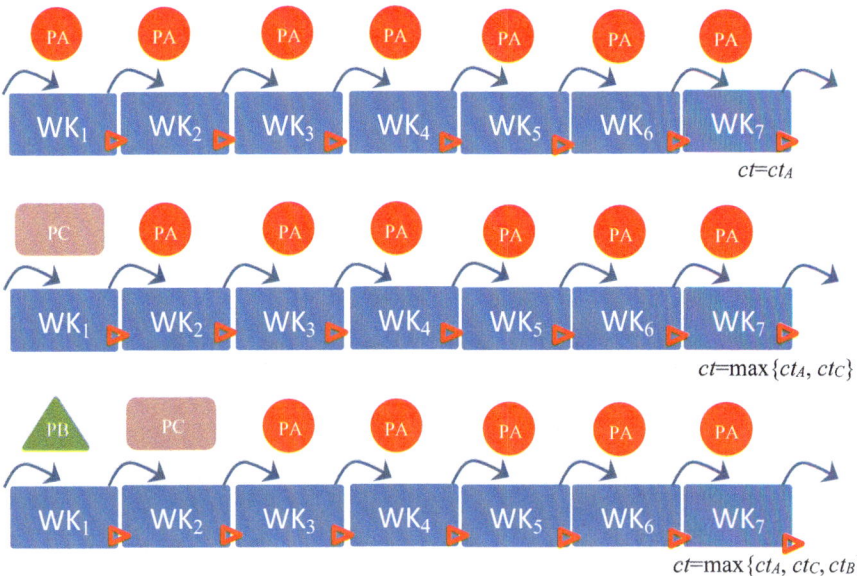

Fig. 3. Illustrative example comparing different cycle times, controlled by the slowest product

It is clear that each time a piece of any product enters the line in workstation 1, a new item of (the same or other) final product leaves the last workstation and is ready for fulfilling demand. There is a synchronic displacement downwards that makes each piece move from workstation k to workstation $k + 1$. If the precedent idea is now thought as a decision variable involving a group of products of the same kind instead of a single element, the notion of campaign is introduced. When a campaign i' displaces another group of items called campaign i (already in transit and exiting the line) we can use the variable $DP_{i,i',j}$ to account for the number of elements in the campaign i leaving the assembly line during the input of campaign i'. Through a simple mass balance equation, the total number of elements from previous campaigns i exiting the line while i' is entering should be equal to $N_{i'}$ and is imposed by Eq. (1). Note that $N_{i'}$ represents the number of elements in campaign i'.

$$N_{i'} = \sum_{i \leq i'} \sum_{j:products} DP_{i,i',j} \quad \forall i' \in Runs \tag{1}$$

The elapsed time between the initial and final elements of the campaign i' is governed by $N_{i'}$ times the maximum cycle time of the products laying into the assembly line, according to the selected operating mode m, that is $ct_{j,m}$. This allows to calculate the start and end times, $S_{i'}$ and $C_{i'}$, respectively, expressed mathematically by Eq. (2). In that constraint, variables $x_{i,i'}$, $y_{i'',j}$ and $w_{i'',i',m}$ are binary variables which take value one when campaign i' makes some items in the previous campaign i be finished at the other extreme of the line ($x_{i,i'} = 1$); when product j is associated with campaign i'', moving between i and i' ($y_{i'',j} = 1$); and when campaign i'' runs in mode m during i' ($w_{i'',i',m} = 1$), respectively. MT is a large enough constant, in time units. Figure 4 shows how three items of PB in the new run i' displace the three items of PA in the last three workstations of the assembly line, belonging to campaign i.

$$C_{i'} \geq S_{i'} + N_{i'} \, ct_{j,m} - MT\big(3 - y_{i'',j} - x_{i,i'} - w_{i'',i',m}\big),$$
$$\forall i'' \in Runs : i \leq i'' \leq i', \forall j \in Prod, m \in Modes \tag{2}$$

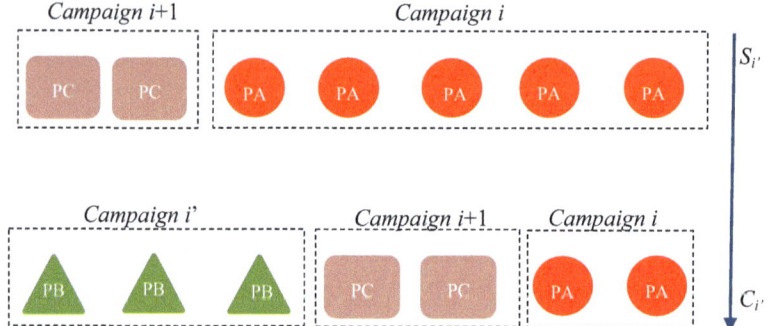

Fig. 4. Assembly line arrangement before and after the input of run i' with product PB

3.2 Inventory Control

We now focus on the calculation of the inventory level of the final product j after the input of production run i', $Inv_{i',j}$, which basically consists of the stock of j previous to i' plus the production of product j during campaign i', minus the consumption of j in the elapsed time interval (given by the demand rate r_j), as it is expressed by Eq. (3).

$$Inv_{i',j} = Inv_{i'-1,j} + \sum_{i'':1 \leq i'' \leq i'} DP_{i'',i',j} - r_j\big(C_{i'} - C_{i'-1}\big), \quad \forall i' \in Runs, j \in Prod \tag{3}$$

3.3 Carbon Footprint

The carbon footprint from the manufacturing of a campaign i during run i' can be obtained from the operating mode selected for processing the products in i that are in transit along the line. If $W_{i,i'}$ is the number of elements of run i into the assembly line at the end of run i' then the carbon footprint of tasks performed during run i' depends on the selected mode m, as shown in the block of Eq. (4). Note that the parameter $cf_{j,m}$ is the carbon footprint (total emissions) in kg of CO_2 per unit of product j, per station and cycle, when the assembly line runs in mode m for product j. MF is a large enough constant, in carbon emissions units.

$$CF_{i,i'} \geq N_{i'} \, W_{i,i'} \, cf_{j,m} - MF\left(2 - y_{i,j} - w_{i,i',m} + x_{i,i'}\right)$$

$$CF_{i,i'} \geq [N_{i'} \, W_{i,i'-1} - D_{i,i'}(D_{i,i'} + 1)/2] \, cf_{j,m}$$
$$- MF\left(3 - y_{i,j} - w_{i,i',m} - x_{i,i'}\right), \; i' > 1$$

$$CF_{i',i'} \geq (N_{i'}(N_{i'} + 1)/2) \, cf_{j,m} - MF\left(2 - y_{i',j} - w_{i',i',m}\right)$$

$$\forall i, i' \in Runs : i < i', j \in \text{Prod}, m \in Modes \qquad (4)$$

Finally, the total carbon emissions from a production cycle are computed as in the left-hand-side of Eq. (5). In one of its simplest versions, carbon footprint control is imposed as a maximum carbon intensity (emissions per unit of product), in an aggregate form. More specifically, total emissions are evenly distributed among all product units finished in a production cycle, and the maximum carbon intensity target is imposed as in the right-hand-side of constraint (5).

$$\sum_{i,i':i \leq i'} CF_{i,i'} \leq \sum_{i'} N_{i'} \, \text{maxci} \qquad (5)$$

3.4 Objective Function

We seek to optimize the total cost per unit time, derived from the average inventory holding and changeover costs, as expressed in Eq. (6). In the first term, the inventory holding cost for product j per unit time is given by ic_j while the other term is related with the transition cost incurred between two consecutive campaigns processing different products ($TCost_i$), as captured by Eq. (7). Parameter $tc_{j,j'}$ is the transition cost when model j in a production run $i'-1$ changes to another model j' in the next campaign i'. T is the length of the whole production cycle and is defined by Eq. (8).

$$z = \left[\sum_{j \in P} \sum_{i:i>1} ic_j \frac{\left(Inv_{i,j} + Inv_{i-1,j}\right)}{2} \left(C_i - C_{i-1}\right) + \sum_i TCost_i\right] \Big/ T \qquad (6)$$

$$TCost_{i'} \geq tc_{j,j'}\left(y_{i',j'} + y_{i'-1,j} - 1\right) \quad, \forall j, j' \in \text{Prod} : j \neq j' \qquad (7)$$

$$T \geq C_i \quad \forall i \in Runs \qquad (8)$$

4 Case Study

In order to evaluate the performance of the proposed mathematical model, a motivation example from a truck trailers manufacturing industry in Argentina is addressed in this section. The aim is to find the optimal sizing and sequencing of a synchronous assembly line with five workstations that processes four products: P_1, P_2, P_3 and P_4. The sequence of products found by the MINLP solver comprises 6 successive campaigns i_1, i_2, i_3, i_4, i_5 and i_6 processing items of P_1, P_3, P_4, P_2, P_2, and P_1, respectively, along a total production cycle of 13.33 h. The lot sizes of each campaign are 1, 2, 4, 2, 2 and 5, respectively. On Table 1 we report the different cycle times and operating modes that can be adopted for each campaign. Figure 5 shows how the different production runs progress through the assembly line.

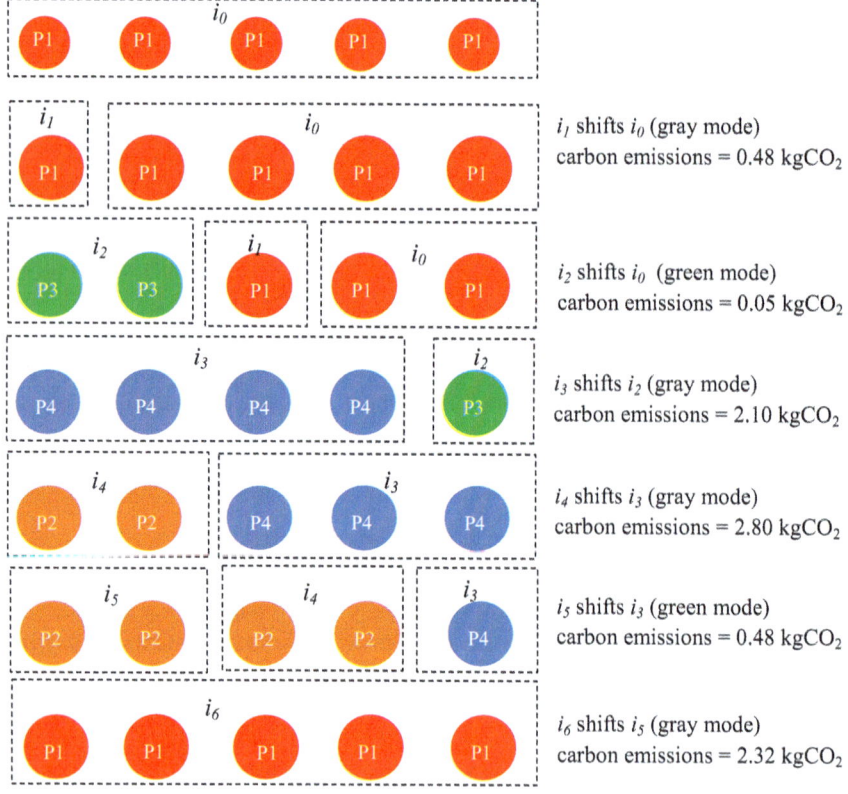

Fig. 5. Evolution of production runs along the assembly line, with illustrative carbon emissions.

Two out of the six runs are accomplished in green mode, with the illustrative carbon emissions given at the right of each line of Fig. 5. For confidentiality reasons, actual emissions are not disclosed. Note that a maximum of 1 kg of CO_2 (average) per unit of product is admitted as carbon footprint, thus making the model to favor production on green mode when other products of slower cycle times are in transit along the line.

Table 1. Cycle times (in hours) for each product and operating mode

	P_1	P_2	P_3	P_4
"Green"	0.888	0.833	0.833	0.750
"Gray"	0.800	0.750	0.750	0.675

5 Conclusions

As previous authors have emphasized [2], performance improvements can be obtained along with energy consumption and emissions reduction by properly managing batch sizes and product sequences. In that sense, we have developed a novel mixed-integer non-linear programming (MINLP) formulation for the optimal sequencing and sizing of runs in multiproduct, synchronous assembly lines. By means of a continuous representation in both time and spatial domains, which has been adapted from pipeline scheduling optimization models, the problem complexity can be tackled in reasonable computational times. Furthermore, the optimization model has been extended to assess the impacts of operating modes ("greener" or "grayer" modes, emitting different amounts of CO_2). Operating modes affect not only costs and yield rates but also the magnitude of the carbon emissions, often being greener at the expense of slower cycle times. This fact is especially relevant in assembly lines which process multiple products. Planning operations sustainably and efficiently, meeting demand under carbon footprint specifications at minimum total cost is a worthwhile effort. In fact, optimally planning operations to solve the tradeoff between emissions and costs is a key goal of modern production systems. Future research will focus on solving larger instances of the problem, with more workstations and a wider set of products for which continuous time models can be even more competitive in comparison to discrete counterparts.

References

1. Jeswiet J, Nava P (2009) Applying CES to assembly and comparing carbon footprints. Int J Sustain Eng 2:232–240
2. Shuterland JW, Skerlos SJ, Haapala KR, Cooper D, Zhao F, Huang A (2020) Industrial sustainability: reviewing the past and envisioning the future. J Manuf Sci Eng 142:110806
3. Rejowski R, Pinto JM (2003) Scheduling of a multiproduct pipeline system. Comput Chem Eng 27:1229–1246
4. Cafaro DC, Cerda J (2008) Efficient tool for the scheduling of multiproduct pipelines and terminal operations. Ind Eng Chem Res 47:9941–9956
5. Scholl A (1999) Balancing and sequencing of assembly lines. 2nd ed. (1999). Physica-Verlag, Heidelberg, Germany
6. Camussi NB, Cerdá J, Cafaro DC (2021) Mathematical formulations for the optimal sequencing and lot sizing in multiproduct synchrpnous assembly lines. Comput Ind Eng 152:107006

System Boundaries, Data Sources and Assessment Methods in the Ecological Evaluation of Complex Assembly Products

Felix Funk[(✉)] and Jörg Franke

Friedrich-Alexander-Universität Erlangen-Nürnberg, 91054 Erlangen, Germany
`felix.funk@faps.fau.de`

Abstract. Life Cycle Assessment (LCA) has become the most popular method for assessing the environmental impact of products. It is internationally standardized by ISO 14040 and ISO 14044, which outline the general methodology, steps, and issues to be considered. While the standards ensure a consistent general understanding of the method, it does not provide explicit recommendations on the decisions to be made when conducting an LCA. These decisions include, in particular, the definition of system boundaries, the prioritization of data sources, and the selection of appropriate assessment methods. For some industries, particularly raw materials and process goods, the gap has mostly been filled by industry standards. The vast number of manufactured goods that are the result of complex assemblies originating from multi-tiered supply chains still lack such industry standards. This work addresses the issue by conducting a thorough literature review on the subject. As a result of the literature review, methodological gaps are identified and quasi-standards are derived from previous studies, thus providing general guidance for future LCA of complex assembly products.

1 Introduction

The generic process for Life Cycle Assessment (LCA) defined in ISO 14044 [1] provides a general understanding of conforming studies. However, it remains vague in its demands and recommendations. The gap has been addressed for several industries, such as agricultural produce, raw materials, and chemical as well as semi-finished products. For more complex technical products, however, database entries, transparent public case studies, as well as detailed methodological guidelines are missing. Previous literature reviews focus on specific products or product groups and aim to compare product impacts instead of scrutinizing the methodology. Still, the variation in outcomes caused by procedural differences is frequently noted, e.g. for photovoltaic systems [2] and fuel cells [3]. More general studies are scarce but come to similar conclusions. Frequently cited papers attest large methodological divergence [4] and call for "harmonized " [5]. The large divergence between studies is further supported by continuous standardization efforts, e.g. by the EU through Product Environmental Footprints. The target of this study is to form a better understanding of the quasi-standard that has developed in research by analyzing the state of the art. First, some of the most prominent standards are examined

H. Kohl et al. (Eds.): GCSM 2023, LNME, pp. 419–428, 2025.
https://doi.org/10.1007/978-3-031-77429-4_46

to identify the most common and impactful decisions to be made during LCA. Then, an analysis of scientific LCA studies with regard to these decisions is conducted. The results are discussed and recommendations are derived. Finally, the limitations of the study and potential for future research is outlined.

2 Norms and Standards

The most commonly cited basis for LCA studies are ISO 14040 and ISO 14044. While ISO 14040 defines the basic principles of the methodology, ISO 14044 specifies further requirements and aims to be a guideline for practitioners. However, it remains rather vague. ISO 14044 generally presents readers with lists of LCA aspects that need to be described in studies, without giving clear recommendations regarding their specific implementation. Consequently, studies claiming accordance with ISO 14044 cannot generally be assumed to fulfill a certain quality standard.

A more detailed methodology is presented in the "General guide for Life Cycle Assessment", which is part of the ILCD Handbook published by the European Union's European Platform on LCA [6]. It adds a separate layer of detail to the ISO specifications. However, it still leaves room for interpretation. Provision 7.4.2.4 e.g. demands for "all relevant inputs and outputs" to be collected—whether this includes materials, services, emissions, or wastes, is left to the practitioner [6].

To derive a quasi-standard of LCA studies for complex industrial products, a list of necessary high-impact decisions is derived from the ILCD handbook and the answers (henceforth "decisions") given in current scientific LCA studies are analyzed (Table 1).

3 Literature Review

3.1 Literature Sampling

The goal of the literature review is to generate a sufficiently representative sample of high-quality publications, which conduct an LCA for parts or products. These parts or products should present a higher level of complexity than semi-finished products, be produced (or intended to be produced) at an industrial scale, and involve some level of assembly. To ensure high quality and prioritize influential works, only peer-reviewed journal publications with at least 5 citations are considered. Scopus is used as the sole database, as it covers some of the most relevant journals on the topic. The final search string is: *"TITLE("Life Cycle Assessment") AND TITLE-ABS-KEY(Industry OR Assembly OR Automotive OR Electronics) AND PUBYEAR > 2017 AND DOCTYPE(ar)*. It was applied to the Scopus database on June 15 2023 and all subsequent findings are in reference to that state of the database. The initial query results are filtered further based on titles and abstracts. Titles and abstracts indicating non-matching products are excluded. The review is enhanced by subjecting contributions to the 2022 Global Conference on Sustainable Manufacturing to a similar process. The individual papers are filtered using the same criteria for title and abstract. Of the original 1019 publications found through the Scopus query, 34 remain relevant. Of the original 121 papers of GCSM 2022, 2 remain relevant.

Table 1. High-impact questions to be answered in the literature

LCA step	Topic	Question
Goal and scope definition	Motivation	Why was the study conducted?
	Type	Which stages were examined and how?
	Background	What is the main expertise of the authors?
	Model	Is the study attributional or consequential?
	Reference flow	How is the reference flow quantified?
	Cutoff criteria	How are the cutoff criteria defined?
Life cycle inventory	Processes	Which processes remain after cutoff?
	Data sources	Which data sources are used in the study?
	Databases	Which standard databases are employed?
	Software tools	Which software tools are used?
Life cycle impact assessment	Uncertainty	Is uncertainty statistically considered?
	Applied method	Which assessment method was chosen?
	Impacts	Which impact categories are analyzed?
Interpretation	Completion	Did the authors conduct a completion check?
	Consistency	Did the authors conduct a consistency check?
	Sensitivity	Did the authors conduct a sensitivity check?

To put the literature review into context, some bibliometric results are presented in the following. Despite the small sample size, several reputable journals are featured. Five journals are represented by multiple papers in the final sample, while eleven journals provide one contribution each. The Journal of Cleaner Production (10 papers) and the International Journal of Life Cycle Assessment (5) rank highest, which can be interpreted to demonstrate the suitability of the procedure. The growing importance is supported by the continuous increase from 142 publications in 2018 to 239 in 2022.

3.2   High-Impact Decisions in Literature

The overall results of the review are presented in Table 2. The most prominent decisions with regard to each topic are accompanied by their prevalence in the overall sample. The sum per topic may exceed 100%, as some decisions are not mutually exclusive.

In the sampled literature, comparative LCA between different products is the most prominent, with pure accounting being second. Parameter optimization of production processes is rather uncommon. It is worth noting that nearly all studies use attributional models, despite the authors rarely explicitly stating it. Consequential studies are scarce. Cutoff criteria are also usually not formally specified and only quantified in two cases. It is noticeable that most papers use the terms functional unit and reference flow interchangeably. Raw materials and energy use are deemed the most relevant impact contributors, being considered by at least three-quarters of studies. Finished part data is

Table 2. Decisions made in the literature sample, sorted by incidence [7–42]

Topic	Decision	[%]
Motivation	Comparison	56
	Accounting	33
	Optimization	14
Type	**Cradle to grave**	**64**
	Cradle to gate	**31**
	Gate to cradle	**3**
	Gate to gate	**6**
Background	Researchers	94
	LCA experts	14
	Industry	8
	Other	6
Model	**Imp. attributional**	**64**
	Exp. attributional	**31**
	Exp. consequ	**3**
	Imp. consequ	**3**
Cutoff criteria	Not specified	92
	Argumentative	6
	Weight	3
	Value	3
Reference flow	**SI unit**	**70**
	Product/process	**39**
Processes	Raw materials	92
	Energy use	75
	Part/product data	67
	End of life	53
	Usage	50
	Waste	42
	Aux. material use	39
	Gas emissions	36
	SC[a] transport	33
	Distribution	17
	Prod. Peripherals	11

(*continued*)

Table 2. (*continued*)

Topic	Decision	[%]
	Infrastructure	3
	Sup. Processes	3
Data bases	**EcoInvent**	**75**
	GABI	**22**
	SimaPro	**11**
	Other	**6**
	None specified	**6**
	Agribalyse	**3**
	ILCD	**3**
	WSA[b]	**3**
Data sources	Public sources	94
	By authors	50
	Simulation	28
	Private sources	22
Uncertainty	**Not Considered**	**69**
	Considered	**31**
Software tools	SimaPro	39
	GABI	28
	Not specified	25
	OpenLCA	8
Applied method	**ReCiPe**	**44**
	CML	**22**
	None specified	**14**
	JRC-based	**11**
	IMPACT 2002+	**8**
	ILCD 2011	**6**
	USEtox	**6**
	Other	**6**
	UBP 2006	**3**
	Riskpoll	**3**
	EDP	**3**
	AEM[3]	**3**

(*continued*)

Table 2. (*continued*)

Topic	Decision	[%]
	TRACI	**3**
	CRAES	**3**
Impacts	GWP	97
	RD	81
	HT	81
	EU	75
	A	67
	PCOF	58
	ET	56
	OD	56
	PM	47
	IR	28
	LO	17
	EQ	11
Completion	**Not checked**	**94**
	Checked	**6**
Consistency	Not checked	94
	Checked	6
Sensitivity	**Checked**	**64**
	Not checked	**36**

[a] Supply chain, [b] World Steel Association, [c] Accumulated Exceedance Model

also used where it is available. Only 50% of papers employ data empirically recorded by the authors, with reliance on external sources or simulation being considerably more present. EcoInvent is the most popular database among researchers, while industry-specific databases are unpopular. In terms of software tools, GABI and SimaPro account for 67%, with OpenLCA only having a minuscule share. Despite all software tools providing the automation potential, uncertainty, e.g. through Monte-Carlo-simulations, is not considered in most studies. Instead, common impact assessment methodologies are applied to a single inventory. Of the methodologies, ReCiPe and CML are the most prominent. Due to the interdependency between some methodologies, a clear differentiation is not always possible. Overall, the variety is quite large, with niche methods also finding occasional use. This variety, however, necessitates grouping impact categories to some degree. Greenhouse warming potential (GWP), resource depletion (RD), and eutrophication (EU) are the most commonly analyzed impacts, with ecosystem quality (EQ) and land occupation only being considered in a small number of individual studies. Despite being recommended by the ILCD, completion and consistency checks are uncommon, or at least not explicitly discussed in the sampled literature. Conversely, sensitivity checks or analysis are the standard and explicitly provided in nearly every paper.

4   Quasi-Standard and Deficits

The review has shown that there is an unofficial consensus for life cycle assessment in current manufacturing science. This quasi-standard is defined by the several attributes:

- Comparative: Several products are compared
- Attributional: Only the immediate processes and effects are considered
- Life-cycle-holistic: The entire life-cycle is investigated (cradle-to-grave)
- Contribution-myopic: Only raw materials and energy are considered
- Database-dependent: Secondary data from EcoInvent is heavily featured
- Midpoint-selective: Global warming, eutrophication, and acidification are evaluated.

This, in principle, is a solid basis and helps to extend the public knowledge base. The proliferation of public databases and previous research create a basic level of comparability. Still, several improvements to the process can be made. Considering the manufacturing research background of most authors, it is regrettable that independently captured empirical data is only used in half of the studies. The capacity and competence for detailed measurements should be readily available in academia. Similarly, vast parts of the production process are frequently categorically excluded from examination, such as auxiliary materials used during production, emissions during production, transport between sites or to the end customer, and infrastructure. The latter is especially relevant, as experience has shown that for many companies, peripheral processes such as lighting, heating, and intralogistics account for a higher energy use per product than the machinery itself. Again, these values are readily available and not essential to the competitiveness of companies. Especially for SMEs, they pose easy opportunities to assess and improve their environmental impact. A proper investigation of the production infrastructure is hence recommended for future LCAs. Significant improvements can be made with regard to the definition of functional unit and reference flow. Clear distinctions are often lacking in the literature, which complicates the comparison of different studies. The same issue arises as a result of vague definitions of cutoff criteria. More clarity in both regards would greatly benefit the quality of future studies. Finally, it is worth noting that consequential studies are greatly underrepresented. While attributional studies are of course valid and valuable contributions, a further examination of the consequences of technological development also presents an interesting challenge. This is especially true for some currently prominent topics, such as e-mobility and battery technology. Any significant innovation in these areas is likely to find large adaption and hence have an immense impact on global supply chains and subsequently the environment.

5   Limitation and Conclusion

The present study has shown the current standard in scientific LCA based on a literature review. The identified papers were examined in depth, and conventions as well as deficiencies were shown. Still, the methodology has several limitations. The search string only targets select industries and the overall sample size could be increased. Furthermore, some publications don't clearly state their methodology, necessitating contextual interpretation. Based on the findings, recommendations for the improvement of scientific LCAs were made. Future research may include the development of a more generalized scientific LCA approach as well as the development of a concise LCA labelling-system describing a study's most significant features to simplify comparisons.

Acknowledgements. This work has been funded by the Federal Ministry for Economic Affairs and Climate Action based on a decision by the German Bundestag.

References

1. DIN EN ISO (2021) Environmental management—Life cycle assessment—Requirements and guidelines(14044)
2. Muteri V, Cellura M, Curto D et al (2020) Review on life cycle assessment of solar photovoltaic panels. Energies 13:252
3. Abdelkareem MA, Elsaid K, Wilberforce T et al (2021) Environmental aspects of fuel cells: a review. Sci Total Environ 752:141803
4. Zamagni A, Guinée J, Heijungs R et al (2012) Lights and shadows in consequential LCA. Int J Life Cycle Assess 17:904–918
5. Curran MA (2013) Life Cycle Assessment: a review of the methodology and its application to sustainability. Curr Opin Chem Eng 2:273–277
6. International Reference Life Cycle Data System (2010) ILCD handbook. general guide for life cycle assessment—detailed guidance. https://eplca.jrc.ec.europa.eu/uploads/ILCD-Handbook-General-guide-for-LCA-DETAILED-GUIDANCE-12March2010-ISBN-fin-v1.0-EN.pdf. Accessed 15 Jun 2023
7. Andersson Ö, Börjesson P (2021) The greenhouse gas emissions of an electrified vehicle combined with renewable fuels: life cycle assessment and policy implications. Appl Energy 289:116621
8. Auer J, Meincke A (2018) Comparative life cycle assessment of electric motors with different efficiency classes: a deep dive into the trade-offs between the life cycle stages in ecodesign context. Int J Life Cycle Assess 23:1590–1608
9. Bay C, Nagengast N, Schmidt H-W et al (2023) Environmental assessment of recycled petroleum and bio based additively manufactured parts via LCA. In: Kohl H, Seliger G, Dietrich F (eds) Manufacturing driving circular economy. Springer International Publishing, Cham, pp 669–677
10. Booto GK, Aamodt Espegren K, Hancke R (2021) Comparative life cycle assessment of heavy-duty drivetrains: a Norwegian study case. Transp Res Part D: Transp Environ 95:102836
11. Bushi L, Skszek T, Reaburn T (2019) New ultralight automotive door life cycle assessment. Int J Life Cycle Assess 24:310–323
12. Calado EA, Leite M, Silva A (2019) Integrating life cycle assessment (LCA) and life cycle costing (LCC) in the early phases of aircraft structural design: an elevator case study. Int J Life Cycle Assess 24:2091–2110
13. Cecchel S, Chindamo D, Collotta M et al (2018) Lightweighting in light commercial vehicles: cradle-to-grave life cycle assessment of a safety-relevant component. Int J Life Cycle Assess 23:2043–2054
14. Cossutta M, Vretenar V, Centeno TA et al (2020) A comparative life cycle assessment of graphene and activated carbon in a supercapacitor application. J Clean Prod 242:118468

15. Degen F, Schütte M (2022) Life cycle assessment of the energy consumption and GHG emissions of state-of-the-art automotive battery cell production. J Clean Prod 330:129798
16. Iturrondobeitia M, Akizu-Gardoki O, Minguez R et al (2021) Environmental impact analysis of aprotic li–O$_2$ batteries based on life cycle assessment. ACS Sustainable Chem Eng 9:7139–7153
17. Iturrondobeitia M, Akizu-Gardoki O, Amondarain O et al (2022) Environmental impacts of aqueous zinc ion batteries based on life cycle assessment. Adv Sustain Syst 6:2100308
18. Jia X, Zhou C, Tang Y et al (2021) Life cycle assessment on PERC solar modules. Sol Energy Mater Sol Cells 227:111112
19. Kallitsis E, Korre A, Kelsall G et al (2020) Environmental life cycle assessment of the production in China of lithium-ion batteries with nickel-cobalt-manganese cathodes utilising novel electrode chemistries. J Clean Prod 254:120067
20. Kokare S, Asif FMA, Mårtensson G et al (2022) A comparative life cycle assessment of stretchable and rigid electronics: a case study of cardiac monitoring devices. Int J Environ Sci Technol (Tehran) 19:3087–3102
21. La Souza LP, de Lora EES, Palacio JCE et al (2018) Comparative environmental life cycle assessment of conventional vehicles with different fuel options, plug-in hybrid and electric vehicles for a sustainable transportation system in Brazil. J Clean Product 203:444–468
22. Lopes Silva DA, de Oliveira JA, Padovezi Filleti RA et al (2018) Life cycle assessment in automotive sector: a case study for engine valves towards cleaner production. J Clean Prod 184:286–300
23. MA (2019) An empirical study on life cycle assessment of double-glazed aluminium-clad timber windows. IJBPA 37:547–564
24. M. Lunardi M, Alvarez-Gaitan JP, Chang NL et al (2018) Life cycle assessment on PERC solar modules. Solar Energy Mater Solar Cells 187:154–159
25. Nordelöf A, Romare M, Tivander J (2019) Life cycle assessment of city buses powered by electricity, hydrogenated vegetable oil or diesel. Transp Res Part D: Transp Environ 75:211–222
26. Ozkan E, Elginoz N, Germirli Babuna F (2018) Life cycle assessment of a printed circuit board manufacturing plant in Turkey. Environ Sci Pollut Res Int 25:26801–26808
27. Palazzo J, Geyer R (2019) Consequential life cycle assessment of automotive material substitution: replacing steel with aluminum in production of north American vehicles. Environ Impact Assess Rev 75:47–58
28. Park J, Hengevoss D, Wittkopf S (2019) Industrial data-based life cycle assessment of architecturally integrated glass-glass photovoltaics. Buildings 9:8
29. Roy P, Miah MD, Zafar MT (2019) Environmental impacts of bicycle production in Bangladesh: a cradle-to-grave life cycle assessment approach. SN Appl Sci 1
30. Salvador R, Chapieski GA, Oshiro LI et al (2023) Screening life cycle assessment of thermoacoustic panels from agricultural byproducts. In: Kohl H, Seliger G, Dietrich F (eds) Manufacturing driving circular economy. Springer International Publishing, Cham, pp 678–684
31. Schenker V, Oberschelp C, Pfister S (2022) Regionalized life cycle assessment of present and future lithium production for Li-ion batteries. Resour Conserv Recycl 187:106611
32. Shanbag A, Manjare S (2020) Life cycle assessment of tyre manufacturing process. J sustain dev energy water environ syst 8:22–34
33. Tuan DD, Wei C (2019) Cradle-to-gate life cycle assessment of ships: a case study of Panamax bulk carrier. Proceed Insti Mech Eng Part M: J Eng Maritime Environ 233:670–683
34. Velandia Vargas JE, Falco DG, Da Silva Walter AC et al (2019) Life cycle assessment of electric vehicles and buses in Brazil: effects of local manufacturing, mass reduction, and energy consumption evolution. Int J Life Cycle Assess 24:1878–1897

35. Villanueva-Rey P, Belo S, Quinteiro P et al (2018) Wiring in the automobile industry: life cycle assessment of an innovative cable solution. J Clean Prod 204:237–246
36. Wang Q, Liu W, Yuan X et al (2018) Environmental impact analysis and process optimization of batteries based on life cycle assessment. J Clean Prod 174:1262–1273
37. Wang F, Deng Y, Yuan C (2020) Life cycle assessment of lithium oxygen battery for electric vehicles. J Clean Prod 264:121339
38. Xiao L, Liu W, Guo Q et al (2018) Comparative life cycle assessment of manufactured and remanufactured loading machines in China. Resour Conserv Recycl 131:225–234
39. Xie M, Ruan J, Bai W et al (2018) Pollutant payback time and environmental impact of Chinese multi-crystalline photovoltaic production based on life cycle assessment. J Clean Prod 184:648–659
40. Xu L, Zhang S, Yang M et al (2018) Environmental effects of China's solar photovoltaic industry during 2011–2016: a life cycle assessment approach. J Clean Prod 170:310–329
41. Yang GG et al (2019) Comparative life cycle assessment of mobile power banks with lithium-ion battery and lithium-ion polymer battery. Sustainability 11:5148
42. Zhao S, You F (2019) Comparative life-cycle assessment of li-ion batteries through process-based and integrated hybrid approaches. ACS Sustain Chem Eng 7:5082–5094

Towards Real-Time Condition Monitoring of Electroplating Plants

M. Lindner[1]([✉]) [iD], R. Duckstein[2] [iD], M. Mennenga[1] [iD], and C. Herrmann[1,2] [iD]

[1] Chair of Sustainable Manufacturing and Life Cycle Engineering, Institute of Machine Tools and Production Technology, Technische Universität Braunschweig, Langer Kamp 19B, 38106 Braunschweig, Germany
marija.lindner@tu-braunschweig.de

[2] Fraunhofer Institute for Surface Engineering and Thin Films IST, Riedenkamp 2, 38108 Braunschweig, Germany

Abstract. For securing a high-quality plating process, one of the main challenges in electroplating is the dosing of electrolytes. The optimal dosing and the related demand for resources do not only influence the coating quality but can also have a high environmental and economic relevance. Currently, the condition monitoring approaches related to dosing are mostly model-based and are seldom real-time capable. Therefore, a concept for real-time data-based condition monitoring of electrolytes is proposed. The paper discusses the challenges of pre-processing and modeling time series data with machine learning algorithms and quality requirements and availability of data within the electroplating process. Moreover, the usage of neural networks for condition monitoring of time series data is presented and discussed in a case study with a focus on anomaly detection. With this example, the applicability of a data-based approach for dynamic prediction of electrolyte chemicals is presented and evaluated.

Keywords: Condition monitoring · Data mining · ANN · Electroplating · Electrolytes · Chemicals · Real-time

1 Introduction

1.1 Electroplating Process

Electroplating is a surface technology for applying metallic coatings to protect components from corrosion and wear or to attain specific surface characteristics. The process is usually carried out with aqueous electrolytes. These are metal salt solutions in which the salt of the metal to be deposited is present in the form of ions. With the aid of an external voltage source, the metal ions are discharged at the workpiece surface, resulting in metal deposition [1]. Electroplated surface finishes are used to significantly extend the service life of components, equipment, and systems by applying thin metallic layers, thus making a decisive contribution to reducing waste and saving resources [2].

The process consists of several individual process steps, which can be grouped into pre-treatment, coating, and post-treatment processes. Figure 1 schematically illustrates

© The Author(s) 2025
H. Kohl et al. (Eds.): GCSM 2023, LNME, pp. 429–436, 2025.
https://doi.org/10.1007/978-3-031-77429-4_47

such a process chain using the example of galvanic zinc-nickel plating. Galvanic zinc-nickel coatings offer reliable corrosion protection of steel [3–5] and are used, for example, in the automotive, aviation and aerospace industries. The quality of galvanic applied coatings depends on several parameters, such as the applied current density, temperature, bath movement and electrolyte composition [6].

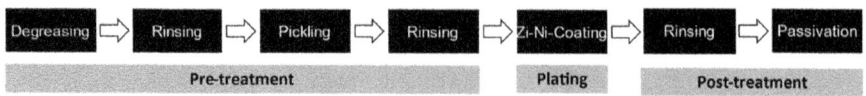

Fig. 1. Zi-Ni plating process chain including pre- and post-treatment steps.

In addition to metal salts, electroplating electrolytes also contain other additives which have the task of optimizing the deposition process, while serving to improve the quality of the metal layer to be deposited. If the pH value is used as a general distinguishing feature, these electrolytes can be classified as acidic (pH < 7), neutral (pH ≈ 7), or alkaline (pH > 7). For example, in the case of the alkaline zinc-nickel electrolyte, a high content of complexing agents is added. These are used to deposit metals in a controlled manner and to prevent precipitation as insoluble metal hydroxides in the higher pH range. In addition to complexing agents, other inorganic and organic additives are added to electrolytes to influence their conductivity and mode of operation (e.g. leveling, wettability, spreadability) as well as the resulting layer properties (e.g. hardness, ductility, wear resistance) [1].

1.2 Challenges of Real-Time Monitoring of Chemical Concentrations

One of the main challenges is the dosing of chemicals and electrolytes. Not only does the optimal dosing affect the coating quality, but it also holds significant environmental and economic implications related to the demand for resources [7, 8]. Process solutions become increasingly ineffective when the operating concentrations of chemicals fall below a certain set point. Therefore, the operating concentrations are kept above the set point to avoid serious deposition and coating defects as well as to extend the service life of the electrolyte. In practice, overdosing occurs due to increased set point concentration specifications by the electrolyte supplier to reliably guarantee product quality. Additionally, operating personnel tend to overdose with the aim of saving work steps. However, with increasing concentrations, considerable volumes of chemicals are carried over into subsequent process steps via the transport of the components to be coated, the wetting of their surface, or due to scooping effects caused by the component geometry. As a result, the chemicals in the electrolyte are consumed unused while a lot of rinsing water is needed for cleaning the components. This leads to lower resource efficiency and an increased effort required for reprocessing and disposal of the rinsing baths. Therefore, it is of significant importance that the electrolyte concentration is monitored and only dosed if necessary. Many of the basic chemicals found in the electrolyte, such as metals, conductive salts, acids and bases can be analyzed by titration in the company laboratory.

Furthermore, an X-ray fluorescence spectrometer is capable of detecting metal concentrations. The analysis becomes more challenging when it comes to organic additives.

Typically, this knowledge is limited to the electrolyte supplier alone, leading to the practice of taking electrolyte samples and sending them for analysis. However, this process often extends beyond a week, causing delays in maintenance.

State-of-the-art system control of an electroplating plant performs the dosing of chemicals based on a specific number of ampere-hours [9]. The real-time chemical concentrations are not considered, since the sampling from the tanks occurs at longer intervals (daily, weekly, etc.) depending on the analyte. Therefore, process transparency is not given, and the tank's condition is often estimated by an experienced employee. In addition, the pre-processing and matching of data from various sources, e.g., stationary data from a laboratory with time-series data from the manufacturing execution system (MES) is characterized by a high level of complexity and time effort, which sets limitations to monitoring in real-time. However, cyber-physical production systems (CPPS) and the underlying methods and technologies offer a systematical approach to the implementation of condition monitoring in production processes [10]. Since it consists of four elements (physical system, data acquisition, cyber system, and control/decision support) it incorporates different interfaces for communication between machines, MES, humans, and products.

1.3   Data-Based Modelling of Time Series Data

Deep learning (DL) algorithms have been used for decision support in numerous applications [11–13]. Different architectures of artificial neural networks (ANN) such as feedforward, convolutional or recurrent neuronal networks are often applied for anomaly detection, product quality prediction, image segmentation, object recognition and similar. For processing time series data, suitable architectures are presented in Table 1, which differ in prediction performance and computational complexity. In comparison to recurrent neural networks (RNN), an autoencoder shows better performance in forecasting, but requires more time for computing. For considering parameters from previous process steps, the long-short-term-memory (LSTM) is more suitable due to its ability to save information within a long-term memory [14]. Various architectures can be employed for modeling, taking into account factors such as data quality, optimal processing time, and the specific application use case [15].

So far, model-based approaches have been developed to estimate chemical concentrations for planning and controlling electrolyte and rinsing cascades [16–19]. However, they have rarely been implemented in a real-time process, since the electrolyte behavior can be estimated only between single measurements in static snapshots [16]. In contrast, data-driven approaches offer a viable solution to this challenge, as they have demonstrated successful implementation in various domains [12, 20]. In the context of electroplating, a novel visual data-based inspection and control method for nickel tank composition has been proposed in [11]. However, a holistic monitoring approach for the entire process chain is required to achieve process transparency and enable more sustainable production. Addressing the resulting questions, a concept for real-time monitoring of the plating process chain is presented.

Table 1. Comparison of DL architectures for time series prediction [15]

Deep learning model	Performance (prediction accuracy etc.)	Computational complexity
Autoencoder (AE)	High	High
Long-short-term-memory (LSTM)	Medium	Medium
Gated recurrent unit (GRU)	Medium	Medium
Bidirectional LSTM	Medium	High
Recurrent neural network (RNN)	Low	Medium

2   CPPS Framework for Data-driven Monitoring of a Plating Plant

The implementation of an automated control system in a plating plant requires proven reliable and robust models. The validation of data-based models can become time-consuming, as the models need to be iteratively optimized to encompass all crucial use cases and perspectives necessary for process control. Conducting a trial-and-error experimental data acquisition on an industrial plant would not be economically feasible, requiring the consideration of alternative approaches. However, data-driven approaches offer the opportunity for real-time monitoring and control of a plant, gaining process and process chain transparency as well as reducing resource consumption. Based on these reflections, a CPPS-based framework for data-driven monitoring and control of a plating plant has been developed.

The framework consists of four main elements in general: physical system (I), data acquisition (II), cyber system (III) and control/decision support (IV). These elements are organized in two data-acquisition control loops: one is connected to a test environment (inner ring) and the second is connected to the industrial environment (outer ring) (Fig. 2).

Within the first loop, the test environment (I a) serves as a lower-scaled replica of the industrial process in which different experiments are conducted with varied parameters such as temperature, amperage and different dosing scenarios. Therefore, product, electrochemical and physical process data (II) are gathered, establishing a foundational data set for modeling. In the cyber system of the first loop (III a), as a part of feature engineering, the data undergoes pre-processing and regression models are developed for variables whose measured values are acquired time-delayed. Hence, the comprehensive modeling of chemical concentrations and parameters takes into account the entire process chain with the following objectives: (a) Forecasting of chemical concentrations of a specified future time frame aiming to achieve real-time data-driven control which considers the plating schedule and incoming plating jobs; (b) Optimization of chemical dosing and process parameters settings, due to narrowing the upper/lower range of their values; (c) Virtual modeling of defined quality gates throughout the process chain to identify critical parameters in various pre-processing steps that significantly influence plating quality. Ultimately, these models are implemented within a test environment, enabling decision support (IV a) for process adaptation and electrolyte analytics.

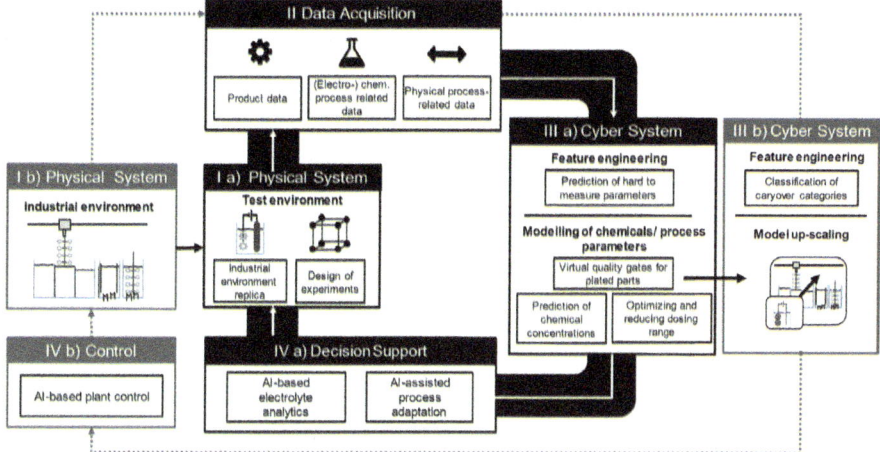

Fig. 2. Framework for a data-based modelling and control of a plating plant

Simultaneously, within the outer ring, the data is acquired from the second physical system (I b), the industrial environment. Compared to the test environment, the industrial plating plant operates at a higher plating rate, resulting in measurable carryover across the plating tanks. Accordingly, within the cyber system's feature engineering (III b), the plating parts are categorized into low or high carryover index based on their shape [21]. This categorization supplements existing models and enables predictive control of the plant, considering incoming plating parts. Due to widely differing environment conditions, the data acquired from both physical systems is compared in order to identify model up-scaling transferability from the test to the industrial plant. Finally, after the validation of the models at the industrial plant, a real-time AI-based control (IV b) is implemented and deployed. The proposed concept has yet to be experimentally applied in a chemical dosing use case.

3 Case Study: Anomaly Detection of Peripheral Devices of an Electroplating Plant Using Autoencoder

A comparable case study has been concluded in accordance to the presented CPPS framework, demonstrating the applicability of an online monitoring concept. In this example, the exhaust system of an electroplating plant represents the physical system (I). The health monitoring of this peripheral system plays an important role in the plating process chain, since a device failure can lead to a blockage of the whole plating line due to safety requirements.

Within the data acquisition (II), a multi-sensor device containing vibration, temperature and acoustic emission sensors is attached to the exhaust system and monitored together with the data from a manufacturing execution system such as pressure, electrical current, conductance, various spatial temperature measurements, volume flows, flow rates within the exhaust system, etc. resulting to 28 features (Fig. 3). The data was gathered continuously over 9 months with a sampling interval of 1 s. The focus of the cyber system (III) persists of a statistical analysis of the daily, weekly, and monthly average of raw data and real-time anomaly detection within the system. To identify anomalies, an autoencoder architecture has been applied. The model training was concluded on a normalized dataset of 2 months with a train/test split of 80/20. The autoencoder configuration contained 28 input layers which were gradually reduced within 7 layers to 15 features and reconstructed using a ReLu (Rectified Linear Unit) activation function. Within the regression model, a mean square error (MSE) loss function and Adam optimizer were applied and the model was trained during 15 epochs with a batch size of 100. Since the autoencoder can be used for regression tasks, the model is applied to new data to predict 28 feature values. A high prediction error (MSE) is defined as an anomaly: when the new data displays a different behavior in comparison to the training data set which was defined as "normal". Thus, the anomaly (MSE), resp. The model output, is defined as a mean of the squares of prediction errors $|X_i - X'_i|^2 0$ with X being the actual and X' the predicted value.

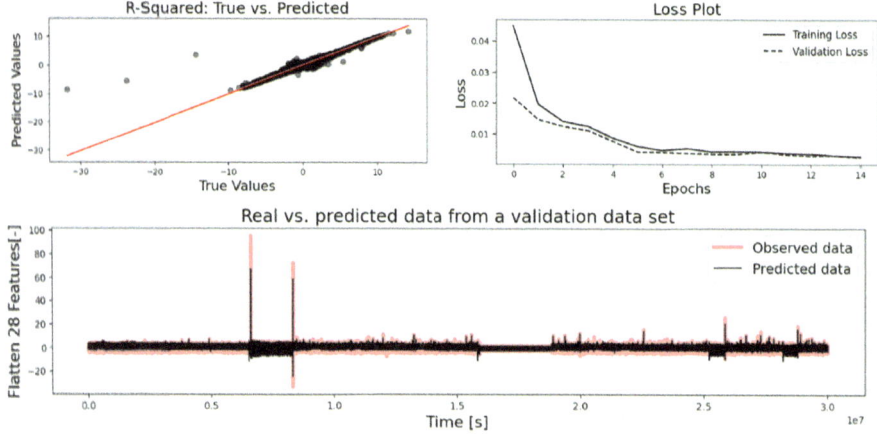

Fig. 3. Model accuracy on the training data and model validation

The results show a high model accuracy on training data with $R^2 = 0.996$, which is also evident in the low loss curves from training and validation (Fig. 3). Further, the model was applied on a data set from the following seven months, validating the accuracy over a longer period. An event log was used for matching the abnormal events to identified anomalies. Thus, different events were assigned to occurred device issues, their maintenance and their repair, proving the feasibility of the model. E.g., the identified anomalies (two peaks in Fig. 3) occurred during a dosing pump issue and the maintenance two days later. Additionally, for decision support (IV) during monitoring, data has been visualized in a local dashboard.

Accordingly, real-time data is accessed from a database and the model is applied directly to it, visualizing raw data, weekly and monthly mean, the calculated MSE, as well as historical MSE from the past 3 months.

4 Discussion

As shown above, deep learning techniques can enable real-time condition monitoring of processes. The overall system can be monitored with an autoencoder algorithm which can detect anomalies or predict different chemical concentration values. For dynamic monitoring of a plating job throughout the process chain, Long-Short-Term-Memory can be applied. However, to enable real-time monitoring, requirements have to be fulfilled, such as sufficient digitalization of a plant and processes, as well as data availability and high data quality. Furthermore, developed models have to be proven stable and robust, yet flexible and adjustable in the case of a process change. Finally, in addition to deep learning techniques, to enable dynamic process control, reinforcement learning approaches should be considered.

5 Conclusion and Outlook

A concept for data-based monitoring of an electroplating process has been presented in this paper with the aim of optimizing resource consumption by smart chemical dosing and control throughout the process chain. This approach has been embedded within a CPPS framework with a focus on the cyber system. To display the applicability of ANN in this research question, a case study with comparable research scope has been presented. This approach offers a high potential for increasing resource efficiency within the process chain. Not only can dosing of different chemicals be reduced, but also by categorizing parts based on their geometry from low to high carryover volume, within a quick reaction time, the rinsing cascade water flow can be adjusted, reducing the overall water consumption. Future work includes implementing the cyber system by modeling of chemicals and process parameters as described in the CPPS framework.

Acknowledgment. The authors thank German Ministry of Education and Research for funding the project KI-InGatec (01IS222014D). The project KI-InGatec further develops concepts and approaches initiated with the project SmARtPlaS (02K18D115).

References

1. Kanani N (2020) Galvanotechnik. Grundlagen, Verfahren und Praxis einer Schlüsseltechnologie. Hanser, München
2. Leiden A et al (2020) Transferring life cycle engineering to surface engineering. Procedia CIRP 90:557–562
3. Conde A, Arenas MA, de Damborenea JJ (2011) Electrodeposition of Zn–Ni coatings as Cd replacement for corrosion protection of high strength steel. Corros Sci 53:1489–1497
4. Feng Z et al (2016) Corrosion mechanism of nanocrystalline Zn–Ni alloys obtained from a new DMH-based bath as a replacement for Zn and Cd coatings. RSC Adv 6:64726–64740
5. Narasimhamurthy V, Shivashankarappa LH (2020) Physico-chemical properties of Zn-Ni alloy deposits from an acid sulphate bath containing ethanolamines. J Adv Electrochem 6:184–187
6. Yli-Pentti A (2014) Electroplating and electroless plating. In: Comprehensive materials processing. Elsevier, pp 277–306

7. Kaufmann T, Niemietz P, Bergs T (2023) Leveraging peripheral systems data in the design of data-driven services to increase resource efficiency. In: Liewald M, Verl A, Bauernhansl T, Möhring H-C (eds) Production at the leading edge of technology. Springer International Publishing, Cham, pp 799–809
8. Kölle S, Schmid K, Mock C (2019) Elektrolytführung neu gedacht. WOMag
9. Leiden AT Integrated planning and operation of plating process chains
10. Monostori L (2014) Cyber-physical production systems: roots, expectations and R&D Challenges. Procedia CIRP 17:9–13
11. Katirci R, Danaci KI (2023) The optimization of nickel electroplating process parameters with artificial intelligence methods. J Appl Electrochem
12. Farias JLCB, de Bessa WM (2022) Intelligent control with artificial neural networks for automated insulin delivery systems. Bioengineering (Basel, Switzerland) 9
13. Gellrich S et al (2021) Deep transfer learning for improved product quality prediction: a case study of aluminum gravity die casting. Procedia CIRP 104:912–917
14. Pacella M, Papadia G (2021) Evaluation of deep learning with long short-term memory networks for time series forecasting in supply chain management. Procedia CIRP 99:604–609
15. Ahmed SF et al (2023) Deep learning modelling techniques: current progress, applications, advantages, and challenges. Artif Intell Rev
16. Leiden A et al (2020) Model-based analysis, control and dosing of electroplating electrolytes. Int J Adv Manuf Technol 111:1751–1766
17. Granados GE, Lacroix L, Medjaher K (2020) Condition monitoring and prediction of solution quality during a copper electroplating process. J Intell Manuf 31:285–300
18. Luo KQ, Huang YL (1997) Intelligent decision support for waste minimization in electroplating plants. Eng Appl Artif Intell 10:321–333
19. Gong JP, Luo KQ, Huang YL (1997) Dynamic modeling & simulation for environmentally benign cleaning & rinsing. Plat Surf Finish 84:63–70
20. Alshehri AK, Ricardez-Sandoval LA, Elkamel A (2010) Designing and testing a chemical demulsifier dosage controller in a crude oil desalting plant: an artificial intelligence-based network approach. Chem Eng Technol n/a-n/a
21. Leiden A, Herrmann C, Thiede S (2021) Cyber-physical production system approach for energy and resource efficient planning and operation of plating process chains. J Clean Prod 280:125160

Planning and Control of Value Creation Networks in Timber Construction

Sebastian Orozco[1], Nicole Oertwig[2(✉)], Annika Feldhaus[2], Valentin Eingarnter[1], Maxim Mintchev[1], and Holger Kohl[1,2]

[1] Technische Universität Berlin, Berlin, Germany
[2] Fraunhofer Institute for Production Systems and Design Technology, Berlin, Germany
`nicole.oertwig@ipk.fraunhofer.de`

Abstract. Almost 40 percent of the global carbon emissions are caused by the construction and building sector. Additionally, the production of traditional building materials like concrete and steel requires a large amount of energy. The demand for residential buildings is increasing none the less and can potentially be met by utilizing more sustainable materials like timber in construction. However, timber construction is still connected to a rather manual value creation chain and analog information transfer. To make the process more time- and energy efficient, this paper proposes a framework for the value creation system of urban timber construction from the forest to the city. The three-leveled framework, based on the principles of Industry 4.0, connects the urban development planning with the vertical networking of digital technologies and the horizontal networking in the value chain. The framework is implemented to a dashboard using Microsoft Power BI to demonstrate the close relations of the layers.

Keywords: Value creation network · Timber construction · Digital transformation

1 Introduction

The potential of networking the value creation systems in the field of timber construction can be demonstrated in many places. The greatest possible effect can be achieved not only through the cooperation of timber construction companies themselves, but rather through a vertical as well as horizontal networking of the entire value creation system (timber) construction. The diversity of innovation needs, the span of the system as well as existing traditional structures in the construction industry itself require a gradual development and expansion of the value creation system for sustainable construction. This paper uses an integrated framework and a prototypical implementation of a planning and control system to demonstrate the possibilities and added value of vertical and horizontal networking of the stakeholders involved from forest to city.

H. Kohl et al. (Eds.): GCSM 2023, LNME, pp. 437–444, 2025.
https://doi.org/10.1007/978-3-031-77429-4_48

1.1 Value Creation Networks in Prefabricated Timber Construction

A value creation network (VCN) refers to a network of relationships among participants in an enterprise system that generates both tangible and intangible value [1]. In the context of prefabricated timber construction, VCNs exhibit unique characteristics that are specific to the industry. These networks involve the integration of multiple stakeholders, including architects, engineers, manufacturers, suppliers, and contractors, into a cohesive system. In Germany, the timber construction industry has a rich heritage and is predominantly comprised of small and medium-sized timber construction companies [2]. The presence of numerous participants in timber construction reflects the decentralized nature of the industry and poses challenges for collaboration which can only be met by networking all stakeholders within the forest to city VCN.

In timber construction, the exchange of information spans the entire life cycle of a building, encompassing activities such as foresting, lumber production, transport, serial prefabrication, building assembly, maintenance, disassembly, and recycling. As a result, there is a significant demand for effective communication and coordination, which is further increased by the intended circularity for buildings and the establishment of material cycles. The resulting complexity is nowadays addressed by Building Information Modeling (BIM) as a holistic approach to building design, execution, and management using unified data availability [3].

Additionally, knowledge management poses another hurdle in the advancement of VCNs as it often takes place only within small and established networks, mostly in an informal manner. The knowledge generated rarely finds its way out of these closed networks, limiting the potential for cooperative learning and the establishment of VCNs [4]. Accordingly, the network mentality in Germany is often weak, which is due to the heavy workload of companies and the lack of suitable network structures that align with their goals and foster the necessary trust and experience for cooperation [5].

As a result, there is currently no consistent institutionalized format for cooperation and networking along the entire value chain, whether in conventional construction or specifically in the timber construction industry [5]. This also leads to the trend that the planning of buildings is often entirely shifted to prefabrication companies, thus further increasing complexity of planning activities in VCNs.

1.2 Management Challenges in Construction Processes

The construction sector in Berlin is currently experiencing a significant boom, especially due to the growing demand for ecological and sustainable construction methods, leading to an increase in timber construction. This growth is further supported by the resolution on "Sustainability in construction: Berlin builds with wood," which was approved by the House of Representatives in 2019 [6]. However, despite the upswing in timber construction, the industry faces management challenges that hinder its full potential. One of the challenges is the difficulty in meeting the surging demand, especially for large-scale projects such as new schools and kindergartens. Regional companies are struggling to meet the high demand for construction contracts, and as a result, high-volume projects, including multi-story constructions, are primarily undertaken by non-regional and conventional companies [5]. This competition restricts the participation of local timber construction companies and hampers further networking within the industry.

As the coordination with multiple stakeholders throughout the construction process is complicated, delays in planning are caused due to restricted information accessibility. Regulatory requirements in Germany, which strictly separate planning and execution processes, further complicate matters by limiting the involvement of construction companies in the planning activities of buildings [7]. This becomes especially crucial considering the customized nature of architectural plans and the limitations of prefabrication execution, which necessitates a certain degree of standardization for cost-effective implementation of construction projects.

Additionally, there is still limited experience in sustainable and circular construction methods and materials, like wood and other renewable resources. While modern timber construction offers highly advanced solutions in prefabrication, many companies also lack the necessary experience, technical capabilities, and capacities. However it is crucial to integrate timber construction into comprehensive production systems that facilitate serial prefabrication within regional value networks [5]. Thus, designing production systems according to lean construction principles can lead to a minimization of time, labor and materials and to improved value creation in prefabrication [8].

To address the challenges mentioned above, it is crucial to foster digital networking within the timber construction industry and support VCNs as it holds the potential to optimize value creation processes and enhance competitiveness in the timber construction sector. In this context the following section shows a Framework of Planning and Control of Timber Construction for integrating such VCNs.

2   Framework of Planning and Control of Timber Construction

The challenges mentioned above can be tackled by using the principles and technologies of Industry 4.0. To create a digitally integrated VCN in the production industry, three aspects are highly relevant: Horizontal integration, digital continuity of engineering and vertical integration of digitalization [9].

However, in order to achieve complete digital networking of the value network for urban timber construction, the systemic framework conditions of the construction industry must be considered. A key feature is that the production of a building is highly influenced by legal regulations and sets of rules. The construction sector is highly dependent on local and regional developments, e.g., regional planning policies so that initial decisions are already made on the design of cities and its buildings. Furthermore, specific products can be typed and released as product groups, while buildings and their construction process are always released as individual projects [10]. Even if such type approvals are being sought, especially in timber construction. The actors in the value creation system each have a specific function in the construction process (e.g., planning, building construction, civil engineering, installation), which keeps the effort for communicating changes particularly high. Therefore, the proposed framework is divided into several levels, each focusing on a different relation to the value creation. Figure 1 shows the three levels and their connections in the Urban Timber Construction System.

The *from forest to city level*, considering the flow of information for the urban planning, represents the first stage of continuous engineering for the entire VCN. Information on demand for residential and commercial buildings can be aligned with data on existing

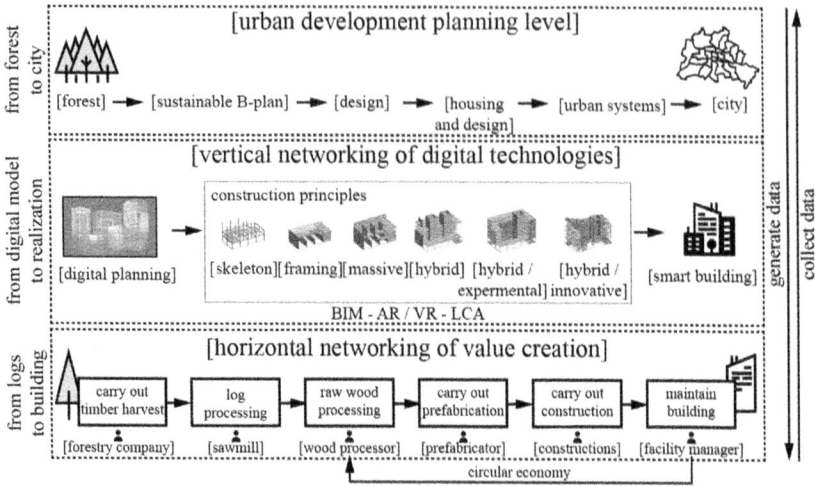

Fig. 1. The three levels of the VCN "Urban Timber Construction".

regional forest resources, which can be harvested sustainably. The information generated provides the basis for decisions on the building plans, their design and building principles, and so forth. The *level of vertical networking* of digital technologies refers to the integration from the digital models coming from several systems, e.g. digital forest models (availability of wood), urban planning (design and requirements for urban development and buildings) or the realization and the use phase of timber buildings. Using BIM as a platform, digital planning and control for the processes of production and construction can be realized as well as for the use phase and maintenance of buildings [11]. Through the additional integration of Life Cycle Assessment (LCA), the ecological impact of planned projects can be determined in the early planning stages. To be able to use the data and information generated throughout the steps during the value creation, the *horizontal networking* refers to connecting all stakeholders along the value chain. The development of open standards, preferably using BIM as a platform, enables unified data usage and exchange and thus the continuous communication from logs to the building and simplifies knowledge generation and management. This adds to the interoperability of the VCN of timber construction and creates synergies, making the planning process, as well as the production- and construction process more efficient.

The given framework has been applied within an integrated management model that encompasses vertical path through components from the forest through digital planning, including construction principles, to prefabrication. The integrated management tool is presented in the following sections. However, further components are to be integrated.

3 Integrated Management Tool for a Digital Value Creation Network in Timber Construction

Initially developed as part of the Bauhütte 4.0 initiative, the Value Chain Digital Model (VCDM) provides access to the 'Forest-to-City' value chain digital twin. Based on a Business Intelligence (BI) Platform, this analytic tool is hosted online and accessible through an interface of five interactive dashboards which encompass four different analysis scales: Overview (conceptual), Forestry (regional), Manufacture (local), Building (architectural), and City (urban). The initial overview landing page provides users with a conceptual context, describing the different phases of the value chain according to the ISO 14001 LCA standard [12]. This serves to clarify which processes and activities are involved in the 'Forest-to-City' transformation, along with their average percentual contribution to either carbon emissions or storage. As the user sets different values for driving parameters, such as wood processing efficacy or transportation mode, specific **Key performance Indicators** (KPIs) are updated in real time. A visual breakdown from total biomass to laminated products yield is displayed in the dashboard as well as other relevant information, such as dominant tree species, carbon net flux and environmental protection regime if applicable.

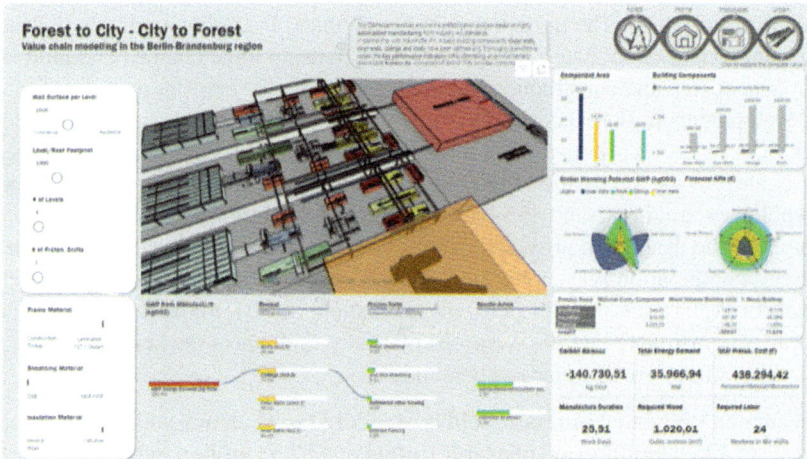

Fig. 2. Manufacturing Dashboard—"Urban Timber Construction".

The next step along the process is the manufacturing dashboard (Fig. 2), which revolves around the prefabrication process based on a highly **automatized manufacturing scenario** configured after Industry 4.0 principles. Four basic building components (**Outer walls, inner walls, ceilings, and roofs**) have been defined as initial prefabrication elements, and thoroughly quantified to obtain the following KPIs: (1) Prefabrication carbon balance, (2) Total energy demand, (3) Estimated prefabrication cost, (4) Manufacture duration, (5) Required wood volume, (6) Required human resources. Simultaneously, the user can adjust several design parameters according to specific project requirements, which combined with financial, material, and time indicators, describe an environmentally responsible **Building Kit**, composed of approximately 70% wooden components. Further along the value chain, both the architectural and urban scale are accessible for exploration. The contents of the dashboard´s main window are scale-dependent, allowing users to select specific building parts (architectural dashboard), or specific buildings within the study case urban area (urban dashboard). This provides access to their material composition breakdown and their contribution in terms of carbon emissions according to four building typologies with increasing percentages of wood components (15%-50%-70%-80%). For these types of analysis, detailed BIM models were created, and validated embedded carbon values were consistently referenced from current German literature. Specifically, the *Ökobaudat Sustainable Construction Information Database V.2022* [13] and interview-based research conducted with local design firms were the main input for database creation. The following section provides a Use Case, demonstrating the application of such an analysis for an exemplary project in Berlin.

4 Use Case

The developments around the Schumacher Quartier (SQ) project in the neighborhood of Berlin Tegel serve as a case study for the analyses of the forest-to-city VCN, forming the scope of this study. The SQ is a model development in Berlin, aiming to use timber as its main construction material to provide new housing units planned and built using sustainable design and construction methods. Therefore, it serves as a model for future city developments in European cites [14].

The Urban Carbon Storage dashboard describes the effect of combined prototypes in the context of the SQ. It allows the user to combine determined percentages of building typologies ranging from "15%" (present day) to ambitious "80% +" wood-based prototypes. The dashboard provides a series of KPI´s that instantly adjust according to the user´s combination presenting the achievable carbon balance. In time this type of hypothesis exploration method is meant to facilitate decision-making processes on a political-administrative level. Some of the main key-findings of the first implementation of this BI platform are summarized hereunder:

- The region´s potential **biomass yield is of 255.152 million m3**, which could produce a total of **125.02 million m3 of usable timber** and **87.52 million m3 of CLT/Glulam.**
- This amount of wood can **store up to 116.27 million tons of CO_2** over a minimum of **70 years** when used for new building construction.
- An approximate amount of **5.25 million housing units** could be built using this material as its main input (circa. 70%), with an estimated **market value** of **1.564 trillion euro**.
- Currently existing pioneer building projects only incorporate 8–13% of wooden components which renders a carbon emission of approximately **1218.86 tons of CO_2** into the atmosphere per every 100 inhabitants.
- Highly ambitious experimental building prototypes use up to 73.11% wooden components and have a **carbon storage capacity of 120.87 tons CO_2** per every 100 inhabitants.

5 Discussion and Conclusion

This study suggests the forests of Brandenburg to be processed into sustainable architecture for housing (initially) in Berlin through the development of a regional industry branch in the form of a networked value chain. Regarding this, the study found that there currently are 1.12 million hectares of forest in the region with an average Net Carbon Flux of -159.93 yearly tons per hectare. 1.07 million hectares of this surface corresponds to the Pinus Sylvestris species, or common pine wood.

Two of the main benefits of such approach are: (1) The progressive regeneration of the forest biotope, based on a more ecologically healthy, multi-species agroforestry [15]; and (2) The capacity enhancement for large-scale carbon storage within the urban tissue in the city of Berlin [16]. As a testbed scenario for this principle, the new residential development in Schumacher Quartier has the capacity to store 39586.92 tons of CO_2 when using a mix of 80% experimental wood-predominant building prototypes and 20% of at least carbon neutral building typologies, in combination with locally sourced wood. Additionally, 57450 tons of emitted carbon from concrete based materials in traditional building typologies could be saved through the substitution of concrete elements with wood components in the new experimental building prototypes.

Technical limitations of the current VCDM online platform must be acknowledged and improved during the development of the DiKieHo project. Specifically, those regarding the GIS surveying interface of the forestry panel (currently restricted to the simultaneous display of 30.000 polygons), and the exploration of "live" BIM models in the architectural panel, instead of the current static-image representation. Finally, the idea of implementation for urban planning, needs to detach from the SQ as a pilot exemplary project, and instead be able to explore any building (or group) within the urban footprint of Berlin. The involvement of local actors (along the different stages of the VC) and public administration is crucial to profit from a data-based production network such as the VCDM. A constructive debate, where all the actors have easy access to technically reliable data, and parametric scenario exploration could open the door to a better informed, more inclusive, and faster structural change towards a more sustainable built environment. Similar implementations of this analytical workflow can be applied to other German or European regions in order to discover new bio-based materials reservoirs and simulate the contribution of new forest-to-city VCs across the continent.

By means of the given integrated framework and management tool a constructive debate can be supported, where all actors have easy access to technically reliable data, and parametric scenario exploration to open the door a better informed, more inclusive, and faster structural change towards a more sustainable built environment.

Acknowledgements. We thank the Federal Ministry of Food, Agriculture and Consumer Protection, Berlin, Germany, as well as the Federal Agency of Renewable Resources (FNR) for funding of the research project DiKieHo (Funding Code 2221HV082B).

References

1. Kong E (2013) The future of knowledge: increasing prosperity through value networks: Verna Allee. Knowl Process Mgmt 10(5):137–138. https://doi.org/10.1002/kpm.169

2. proHolz Austria: Ausbildung Holzbau (2020) Zuschnitt Zeitschrift über Holz als Werkstoff und Werke in Holz 20(78):1–28
3. Baumgärtel L, Schönbach R, Hartung R, Ruwoldt A, Klemt-Albert K (2020) BIM-basierte Kollaboration. Bautechnik 97(12):817–825. https://doi.org/10.1002/bate.202000098
4. von Krutschenbach M (2016) Analyse des Informations- und Wissensflusses bei Forstunternehmen: Dargestellt am Beispiel von Forstunternehmen innerhalb des Verbandes der Agrargewerblichen Wirtschaft e.V.
5. Kohl H, Bunschoten R, Oertwig N, Kulick C, Moritz KM, Blackburn P (2023) Studie zur Stärkung der Holzbauwirtschaft in der Metropolregion Berlin-Brandenburg. Fraunhofer-Institut für Produktionsanlagen und Konstruktionstechnik IPK, Berlin
6. Abgeordnetenhaus Berlin (2019) Nachhaltigkeit auf dem Bau: Berlin baut mit Holz: Drucksachen 18/1471 und 18/1726 – Schlussbericht
7. Hauptverband der Deutschen Bauindustrie e. V (2018) Bauen statt streiten.: Partnerschaftsmodelle am Bau - kooperaiv, effizient, digital
8. Erik Eriksson P (2010) Improving construction supply chain collaboration and performance: a lean construction pilot project. Supply Chain Manag Int J 15(5):394–403. https://doi.org/10.1108/13598541011068323
9. Kagermann H, Wahlster W, Helbig J (2013) Umsetzungsempfehlungen für das Zukunftsprojekt Industrie 4.0: Abschlussbericht des Arbeitskreises Industrie 4.0. München
10. Coppola M (2012) Industrielle Beziehungen zwischen Konflikt und Stabilität.: Eine qualitative Studie über den Arbeitskonflikt um den Landesmantelvertrag im Bauhauptgewerbe 2007/2008. Dissertation. Freiburg
11. Borrmann A (2015) Building information modeling: Technologische Grundlagen und industrielle Praxis. Springer Vieweg, Wiesbaden
12. ISO/TC 207/SC 5 Life cycle assessment: ISO 14040:2006 (2014). https://www.iso.org/standard/37456.html?browse=tc. Accessed 25 Jul 2023
13. Bundesministerium für Wohnen (2023) Stadtentwicklung und Bauwesen: ÖKOBAUDAT Sustainable Construction Information Portal. https://www.oekobaudat.de/no_cache/datenbank/suche.html. Accessed 25 Jul 2023
14. Tegel Projekt GmbH: Schumacher Quartier (2022). https://schumacher-quartier.de/. Accessed 26 Jul 2023
15. Zerbe S, Brande A (2003) Woodland degradation and regeneration in Central Europe during the last 1,000 years—a case study in NE Germany. Phyto 33(4):683–700. https://doi.org/10.1127/0340-269x/2003/0033-0683
16. van der Lugt P, Harsta A (2020) Tomorrow's timber: towards the next building revolution. The Netherlands: MaterialDistrict

Technologies and Concepts for E-Waste Treatment in Urban Secondary Raw Material Factories

Kolja Meyer$^{(\boxtimes)}$ ⬤, Severin J. Görgens ⬤, Christoph Persch ⬤, Klaus Dröder ⬤, and Christoph Herrmann ⬤

Junior Research Group Urban Flows and Production, Technische Universität Braunschweig, Langer Kamp 19 b, 38106 Braunschweig, Germany
`kolja.meyer@tu-braunschweig.de`

Abstract. Supply chains for cities are currently challenged due to geopolitical conflicts and general resource scarcity of substantial raw materials. One currently discussed strategy to reduce the dependency of external supply is a relocalization of production facilities to cities enabling local secondary raw material extraction integrated in Circular Economy value creation chains. Urban regions offer the potential for the installation of circular and secondary raw material factories because of the concentration of waste- and byproducts, which can reduce transport efforts. The installation of urban factories is however highly dependent on the economic feasibility as well as the acceptance of the local environment. One critical source for the retrieval of critical raw materials are e-wastes. The current retrieval of these critical raw materials from e-wastes is connected to high spatial requirements and production volumes. These characteristics would hamper an installation in urban areas. Current literature provides different approaches with the intention to adapt e-waste treatment to these special conditions. To assess which of these are suitable for utilization in urban environments, this paper provides an analysis of their possibility to generate an *urban fit*. To do so, criteria containing ecologic, social, and economic, technological indicators are defined and an assessment of the suitability is conducted.

Keywords: Urban manufacturing · secondary raw materials · reverse production technology · urban metabolism · urban mining

1 Introduction

Urban areas are often economically strong and therefore also characterized by a high demand for products resulting in higher greenhouse gas (GHG) emissions and a higher share in other pollutants in comparison to other settlement forms. With an expected increase in urbanization, an increase of material consumption as well as waste materials is anticipated [1]. Especially e-wastes are a challenge in this regard. The challenge will increase as return rates are low in regions with a high per capita generation of e-waste (Europe, North and South America, Oceania) and, in addition, emerging countries

© The Author(s) 2025
H. Kohl et al. (Eds.): GCSM 2023, LNME, pp. 445–453, 2025.
https://doi.org/10.1007/978-3-031-77429-4_49

such as Asia and Africa are expected to generate a higher proportion of e-waste in the future, which may lead to a strong effect considering they also have similar or even lower treatment infrastructure [2]. Some primary raw materials for the production of electronic components are becoming scarce and critical raw materials (CRM) supplied by singular regions are in focus [3], with the potential of further stresses caused by geopolitical conflicts [4] or national disasters. The increasing demand and the resulting challenges call for action in regard to Circular Economy (CE) approaches to retrieve materials [5]. Low local recycle rates, even in regions with high return rates, lead to exports of waste material [6]. Governing bodies, especially in Europe and Asia have become aware of this problem and adapt their policies [2] adding pressure to manufacturers. Based on the current urban related CE challenges, it is discussed whether and to which extend CE could be performed in a more local level by urban factories [7] utilizing the close proximity between waste generation, waste disposal and recycling. Additionally, it is discussed whether the materials should be converted to new end-products instantly [8] or secondary raw materials (sRMs) in order to supply for smaller production facilities scattered around the city. Meyer et al. [9] introduce a factory type, the Urban Secondary Raw Material Factory (USeRMaFa), which retrieves sRMs supplying for makers and producers around the city, thus allowing for a higher efficiency and reducing the overall waste output of cities or even enable the substitution of waste exports by purified material exports. For the establishment and operation of urban factories and USeRMaFas in particular, the embedding in the city must be taken into account. Especially other urban functions such as living and recreation should be supported rather than inhibited. Furthermore, value adding must take place also with respect to competition with other high yielding urban uses [10]. The retrieval processes for use in USeRMaFas must therefore overcome the tension between economic viability and *urban fit*. In this paper, we use the term *urban fit* to describe the potential of a factory to operate in urban environment. E.g., e-wastes containing printed circuit boards (PCB) can bear a high economic potential due to the used materials such as precious metals or rare earths. PCBs are however complex, integrated products and most production chains for recycling arc currently operating at large, non-urban centered scales [11]. Thus, the installation of USeRMaFas for e-wastes is hampered and it is unclear whether production approaches addressing this waste type with the intention of an *urban fit* exist. From the field of tension between the need to apply urban sRM retrieval to e-wastes and the possible stresses between producers and their urban environment, the following research questions and sub-questions are derived:

RQ1 "Which production processes are available to transform e-waste flows to sRMs under the constraints of urban areas?"

"With the following sub-questions:"

RQ2 "Which degree of vertical integration can be realized by USeRMaFas to minimize land use?"

RQ3 "Which criteria exist to evaluate the *urban fit*"?

To answer the research questions, this paper follows the following structure: Sect. 2.1 introduces the position of urban sRM production in supply chains for PCB as an essential e-waste component and introduces different concepts for a sRM harvesting while

Sect. 2.2 introduces different aspects for *urban fit* derived from literature sources. In Sect. 3, the defined criteria are applied to the introduced approaches and it is discussed if, and under which conditions they could be installed for urban sRM production.

2 E-Waste Recycling and Its Role in Urban Supply Chains

2.1 Technologies for the Retrieval of Raw Materials from PCB

The recovery of sRM from PCBs can be performed utilizing different approaches. Tesfaye et al. [12] and Cui et al. [13] give an overview over possible steps for the retrieval of metallic parts and especially the copper and gold components. In general, a mechanical pre-treatment which consists of the steps of dismantling, shredding, and physical processing is performed. Subsequently, the materials are metallurgically processed with three general options: Pyrometallurgy, hydrometallurgy, and biometallurgy. Pyrometallurgic treatment is usually performed in several steps where the material is treated in molten metal baths with supercharged air, which oxidizes the non-precious metals. The copper matte is then upgraded to anodes in converters and electrochemically treated to retrieve pure materials [14]. The second option, hydrometallurgy, is understood as the extraction of a soluble constituent from a solid. Cyanide, Thiasulfate or Halide leaching ("Aqua Regia") are utilized in different types of this procedure [13]. Another option is biometallurgy, which includes bioleaching and biosorption. Here, micro- or biologic organisms are used to retrieve precious or heavy metals [13]. Subsequent to these procedures usually an electrochemical purification is performed [12]. Figure 1a shows the different process stages, which also define the degree of vertical integration for the generation of secondary precious metals from PCB. While the described processes are not traditionally performed with the goal of urban application, several researchers describe approaches for the local extraction of precious metals. To determine the urban fit of such processes, the following sources are considered.

- Zeng et al. [15] describe an integrated mobile recycling plant which covers the mechanical process steps dismantling, crushing, and multi-level separation including electrostatic and magnetic process steps integrated in two shipping containers.
- De Michelis et al. [16] introduce a mobile PCB recycling plant with the same dimensions, which consists of the reverse production steps of dismantling, shredding and physical processing followed by hydrometallurgical material recovery.
- Copani et al. [17] introduce an approach using integrative technologies for PCB recycling in SMEs. Here, an automated disassembly adapting to different products and product variants is followed by hydrometallurgic, HF-free retrieval of rare metals.
- Torihara et al. [18] perform pulverization of PCBs in remotely controlled factories using sieving technologies to separate between the different materials.

Figure 1b shows the discussed approaches based on the described process steps of pre-processing, retrieval, and post-processing. Here, it can be seen that the different authors address different depths of added value. The left-hand side indicates the literature source, while the bars describe which process steps are fulfilled in the individual approaches and the color describes whether pyro-, hydro-, or biometallurgy or mechanical treatments are performed to retrieve sRMs. [15, 18] focus on disassembly and mechanical processing, while [16, 17] also include metallurgy. The approaches utilizing metallurgy use hydrometallurgy. This may result from the high spatial demand and production volume necessary for pyrometallurgy [13] which usually cannot be provided by urban environments. The final electro-metallurgy is not discussed in any of the publications. Thus, the answer to RQ2 is, that it is possible to provide a sRM which still needs additional refinement as well as downstream processes like commissioning.

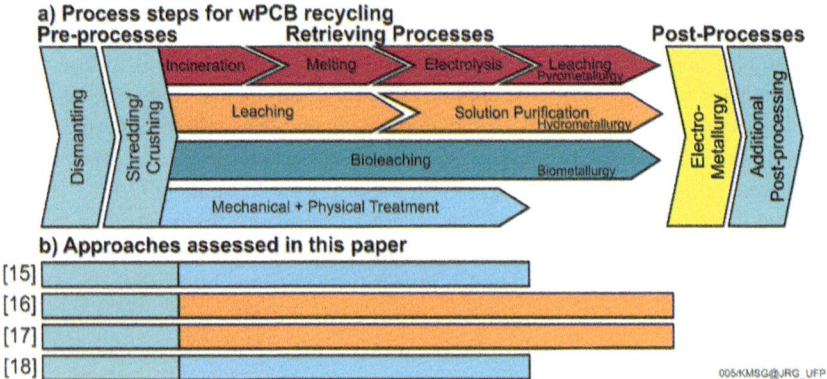

Fig. 1. Process steps covered in the different approaches

2.2   Criteria for the Implementation into Urban Areas

The implementation of factories into urban areas bears several challenges due to the dense surrounding population and rapid urban growth. Literature therefore differentiates between *unintended* and *intended* urban factories [10]. Unintended urban factories exist due to urban spread around existing factories, intended factories are built with the intention to operate in urban environments. Both provide opportunities for symbiotic integration, leading to greater acceptance within their surroundings. In contrast, unintended factories require more adjustments [19]. For such an integration, Lentes et al. [20] propose to install ultra-efficient factories which adapt to the urban surrounding by operating as sustainable as possible. Tsui et al. [7] state that the spatial impact of factories in urban environments should be as small as possible. Meyer et al. [21] provide guidelines for the installation of urban factories in urban environments and focus on the production system design. Not all of these criteria can apply in the specific case of local e-waste recycling and have to be put into context. By combination of the mentioned sources, the following aspects should be considered. While some of these aspects, such as the necessity to low emissions have a higher prioritization, a clear order cannot be provided because some of the aspects go hand in hand and an urban-suitable production should comply to all aspects equally.

- **Small spatial footprint.** According to [7], other urban functions are less negatively impacted when installing small to medium sized production units.

- **Low Emissions/emission-free production**. According to [20, 21], urban producers should emit as few as possible emissions regarding particles, noise, vibration, water/air/soil contamination. Especially in regard to harmful materials such as e-waste, the emissions have to comply to high standards.
- **Automated handling** [21]. Ensures worker safety especially in regard to hazardous products. This could also enable ergonomic production, which is a criterium from [20].
- **Adaptability and flexibility** [21]. An urban production unit should be able for flexible and even adaptable production to react to high variant diversity. In regard to e-waste, it would be required to adapt to different kinds of input products to be resilient against changes in products.
- **Waste-less production**. According to [20], the production processes should not produce garbage and utilize every input component to increase efficiency on a material level.
- **High efficiency of the process chain**. In order to operate economically and ecologically optimal, the process chains need to operate as efficient as possible [20]. An increase in process efficiency could enable an economical viable operation to overcome the loss of scale in production.
- **Ability to integrate into local energy networks** (smart grids) and potential for urban symbiotic integration. According to [22], the efficiency of an urban factory could be increased, if it would be part of urban symbiotic energy networks.

3 Multi Criterial Assessment of Technologies for USeRMaFa

To assess the urban fit of the described approaches, the individual criteria from chapter 2.2 are evaluated qualitatively. The information provided in the references differs in regard to the level of detail in description because of the different scope addressed. The assessment is therefore performed under consideration of the information from the references as well as of the general process knowledge. Table 1 shows the urban fit in regard to the defined categories. Here, a fully filled circle means that the processes provide full accordance to urban areas while an empty circle indicates disagreement to known standards. A half filled circle indicates compliance to a certain degree but additional measurements have to be taken into account, for example to comply with legal standards. The quarter- and three-quarter filled circles are used when gradients between these states occur.

In regard to the criterion **spatial footprint**, [15, 16] state, that their process chains are placed in two 45-foot shipping containers, which is a small size in comparison to most buildings and traditional production units. Copani et al. [17] do not mention the process chain sizes, but the described machines are designed for small to medium production scales. Hence, at least an integration into the existing building structure could be performed [18] do not provide information about their process chain size. In regard to the **local emissions**, Veldhuizen and Sippel [14] state that local air and water pollution standards are met, while especially container 2 containing the grinding processes emits noises with ca. 100 dB. Those noise levels can be expected for the similar processes performed in [18]. [16, 17] perform a combination of mechanical processing and hydrometallurgy, thus a similar noise level for the mechanical steps can be expected, hydrometallurgy may additionally involve the risk of gaseous emissions as well as low-frequency noise due to stirring.. All of the mentioned approaches aim for a higher **flexibility and adaptability** than traditional process chains, where the hydrometallurgic treatments (detailed information in [16]) require a certain batch size to operate. They are therefore only able to adapt to some amounts of material, ideally continuous, inputs. [17] integrate an automated recognition for disassembly to adapt to a higher number of possible PCB variants. The described characteristics also apply for **grid integrability**: The mechanical processes can be ramped up and down depending on the availability of electric energy, while hydrometallurgic treatments have to be carried out in batches. With regard to **automated handling** hydrometallurgic treatment has a high potential because of defined process stages. [18] describe a tele-operation, which does not ensure a full automation but a reduction or avoidance of physical labor, while [15] still contain manual disassembly processes. **Waste-less manufacturing**: While all processes aim to reduce existing waste, the hydrometallurgic processes deal with hazardous materials and byproducts must be treated properly. Due to the lack of information for a fair comparison, the criterium **process chain efficiency** is left blank. A comparison with defined boundary conditions and standardized measurements would be necessary to provide suitable information. The introduced concepts differ in the level of production depth and therefore the resulting type and quality of retrieved materials. In this regard, it can be seen, that with an improved mechanical treatment, the necessary metallurgy can be reduced. If metallurgy is applied, the concepts utilize hydrometallurgy [16, 17], the current large-scale standard of pyrometallurgy is not pursued, due to the high financial and spatial invest, as well as the inflexibility regarding input volumes.

Table 1. Harvey ball matrix assessing the categories of *urban fit*

Reference	[15]	[16]	[16]	[18]
Low spatial footprint	●	●	◕	-
Low local emissions	◑	◔	◔	◑
Adaptability/flexibility	◕	◑	◑	◕
Automated handling	◑	●	•	◕
Waste-less production	◕	◑	◑	◕
Process chain efficiency	–	–	–	–
Grid integrability	●	◑	◑	●

The input PCB can be very various; therefore, a high adaptability of disassembly processes would be beneficial for the subsequent steps in the process chains, indicated by the comparably higher flexibility of [15, 18]. The earlier separation is performed in the process chain, the more possibilities for retrieval exist. Hence, the utilization of automated production lines utilizing flexible and adaptable production processes has a high potential and should be further investigated. By integrating automated type detection and combining these with a high degree of non-metallurgic separation, potentially dangerous chemicals can be avoided in cities while maintaining high quality and flexibility.

4 Conclusion and Outlook

This paper discusses the installation of USeRMaFas to retrieve precious metals from PCB. They aim for an increased retrieval rate by treating the materials in close proximity to the products end-of life location. To enable production in urban environments, several criteria have to be fulfilled because of the tension field between production and other urban functions. In this paper, a set of criteria which assesses the *urban fit* of production chains is thus developed. Approaches, which are setup to locally recycle PCB are evaluated using these criteria. The research shows, that while some criteria are met, no current approach enables a full urban fit. Mechanical treatments should be preferred, because they offer higher flexibility and adaptability for different scenarios. Further research should also be performed in regard to the definition of different scenarios. Different cities have different outputs and possibilities for sRM retrieval. Therefore, it would be viable for some cities to apply the physical processes and export pre-purified materials, the subsequent metallurgy could be performed in larger centers. For other cities, it could be viable to have integrated sRM retrieval with full production depth, if the threshold for pyrometallurgy or hydrometallurgy is met. For both scenarios, automated disassembly would be beneficiary to reduce the cost or in the best case, create economic opportunities - one of the main benefits of urban sRM production [9]. It should also be considered, that the precious metals are only one part of PCBs – the other, possibly lower-yielding parts can have negative ecologic influences too. Legislation must provide appropriate frameworks for this. Producers could enable economic viability of ecologically necessary processes through the highly value-added processes addressed in this paper. Another aspect for further research could be whether the aspects discussed for urban fit might be beneficial for production in rural areas too.

References

1. United Nations. The weight of cities resource requirements of future urbanization. https://wedocs.unep.org/handle/20.500.11822/31624. Last assessed 30 June 2023
2. Murthy V, Ramakrishna S (2022) A review on global e-waste management: urban mining towards a sustainable future and circular economy. Sustainability (2022)
3. Martins F, Castro H (2020) Raw material depletion and scenario assessment in European Union—A circular economy approach. Energy Rep (2020)
4. Kalantzakos S (2020) The race for critical minerals in an era of geopolitical realignments. Int Spect (2020)
5. Abdelbasir S, Hassan S, Kamel A, El-Nasr R (2018) Status of electronic waste recycling techniques: a review. Environ Sci Pollut Res Int. https://doi.org/10.1007/s11356-018-2136-6
6. Balde C, van den Brink S, Forti V, van der Schalk A, Hopstoken F (2020) The Dutch WEEE Flows 2020: What happened between 2010 and 2018? United Nations University (UNU), United Nations Institute for Training and Research (UNITAR)
7. Tsui T, Peck D, Geldermans B, van Timmeren A (2021) The role of urban manufacturing for a circular economy in cities. Sustainability (2021)
8. Nascimento D, Alencastro V, Quelhas O, Caiado R, Garza-Reyes J, Rocha-Lona L, Tortorella G (2019) Exploring Industry 4.0 technologies to enable circular economy practices in a manufacturing context. JMTM. https://doi.org/10.1108/JMTM-03-2018-0071

9. Meyer K, Görgens S, Juraschek M, Herrmann C (2023) Increasing resilience of material supply by decentral urban factories and secondary raw materials. Front Manuf Technol. https://doi.org/10.3389/fmtec.2023.1106965

10. Herrmann C, Juraschek M, Burggräf P, Kara S (2020) Urban production: state of the art and future trends for urban factories. CIRP Annals

11. Kaya M (2016) Recovery of metals and nonmetals from electronic waste by physical and chemical recycling processes. Waste Manage (New York, N.Y.)

12. Tesfaye F, Lindberg D, Hamuyuni J, Taskinen P, Hupa L (2017) Improving urban mining practices for optimal recovery of resources from e-waste. Minerals Eng (2017)

13. Cui J, Zhang L (2008) Metallurgical recovery of metals from electronic waste: a review. J Hazardous Mater. https://doi.org/10.1016/j.jhazmat.2008.02.001

14. Veldhuizen H, Sippel B (1994) Mining discarded electronics. Ind Environ (Switzerland) 17:7–11

15. Zeng X, Song Q, Li J, Yuan W, Duan H, Liu L (2015) Solving e-waste problem using an integrated mobile recycling plant. J Clean Prod. https://doi.org/10.1016/j.jclepro.2014.10.026

16. de Michelis I, Kopacek B (2018) HydroWEEE project. In: Waste electrical and electronic equipment recycling. Elsevier, pp 357–383. https://doi.org/10.1016/B978-0-08-102057-9.00013-5

17. Copani G, Colledani M, Brusaferri A, Pievatolo A, Amendola E, Avella M, Fabrizio M (2019) Integrated technological solutions for zero waste recycling of printed circuit boards. In: Tolio T, Copani G, Terkaj W (eds) Factories of the future. Springer, Cham, pp 149–169. https://doi.org/10.1007/978-3-319-94358-9

18. Torihara K, Kitajima T, Mishima N (2015) Design of a proper recycling process for small-sized e-waste. Procedia CIRP (2015)

19. Herrmann C, Blume S, Kurle D, Schmidt C, Thiede S (2015) The positive impact factory–transition from eco-efficiency to eco–effectiveness strategies in manufacturing. Procedia CIRP. https://doi.org/10.1016/j.procir.2015.02.066

20. Lentes J, Mandel J, Schliessmann U, Blach R, Hertwig M, Kuhlmann T (2017) Competitive and sustainable manufacturing by means of ultra-efficient factories in urban surroundings. Int J Prod Res. https://doi.org/10.1080/00207543.2016.1189106

21. Meyer K, Aschersleben F, Reichler A, Mennenga M, Dröder K, Herrmann C (2023) Dezentrale Fabriken als Chance für die Produktion in der Stadt? In: Gärtner S, Meyer K (eds) Die Produktive Stadt. Springer, Berlin, Heidelberg, pp 99–115. https://doi.org/10.1007/978-3-662-66771-2_6

22. Hertwig M, Lentes J (2023) Smart eco-factory—Aspects for next generation facilities supporting sustainable and high-tech production. In: Huang C, Dekkers R et al (eds) Intelligent and transformative production in pandemic times. Lecture notes in production engineering. Springer, Cham, pp 661–670. https://doi.org/10.1007/978-3-031-18641-7_61

Sovereign Services for Machine Tool Components for Resource-Efficient Machining Processes

Oliver Kohn[(✉)], Viktor Berchtenbreiter, and Matthias Weigold

Institute of Production Management, Technology and Machine Tools (PTW),
Technical University of Darmstadt, Otto-Berndt-Str. 2, 64287 Darmstadt, Hessen,
Germany
o.kohn@ptw.tu-darmstadt.de
https://www.ptw.tu-darmstadt.de/

Abstract. The increasing use of modern edge computing solutions in production is creating new opportunities for collecting and processing internal machine data. This data provides information on the load on a machine tool during operation. Using the technical expertise of component manufacturers, specific loads on individual components can be derived. This knowledge provides valuable insights for machine and component manufacturers to analyze the usage behavior of machine operators and use it for sustainable product development. Additionally, this information can be used to implement new business models, creating financial incentives for sustainable usage through load-oriented payment models. As a result, new potentials for users in efficient design and optimization of manufacturing processes arise. Data and service ecosystems play a crucial role in this context by establishing the technical framework that enables sovereign data exchange and the deployment of expert knowledge in the form of digital services. Therefore, this paper demonstrates a concept and the potential for sustainable machining based on digital services for two central components of a machine tool, the spindle and the feed axis.

Keywords: Machine tool, Sustainable machining, Digital manufacturing system

1 Introduction

The inherent flexibility of machine tools enables the manufacturing of products for diverse applications. Particularly, machining centers enable the manufacturing of a vast range of customized and complex products, which lead to varying input parameters of the machining process due to distinct product requirements. This wide solution spectrum of possible input variables, such as material, cutting tool, or process parameters, results in different loads being applied to the machine tool. The consequence is varying stress on individual components of

© The Author(s) 2025
H. Kohl et al. (Eds.): GCSM 2023, LNME, pp. 454–462, 2025.
https://doi.org/10.1007/978-3-031-77429-4_50

the machine tool, which has an impact on the wear and failure performance of the components. Thus, the information about the applied stress on the components is an important database to analyze the impact of the input variables and optimize the cutting process. In recent years, the development of modern edge computing solutions has significantly enhanced the ability to gather and process information directly from production processes. But despite the increasing availability of data from machine tools, the potential has not been fully utilized yet. Therefore, in the context of the emerging data and service ecosystems, the paper presents a new approach of sovereign services for machine tool components, which analyze the stress of the individual components and thus helps to implement measures avoiding service life-reducing loads for resource-efficient machining. The concept is demonstrated using a 3-axis machining center, where sovereign services, including the expert knowledge of the component supplier to determine the stress for the main assemblies of the feed axis and the spindle, are offered. At the component level, the load on critical components such as ball screws and spindle bearings are considered. Using representative machining processes, the resulting stress on the components is evaluated to show the potential for the machine user to optimize the input variables of the machining processes. Additionally the component suppliers want to use this information to improve their product development. By establishing a sovereign data exchange this approach aims to address the interests of the involved parties such as machine users, machine manufacturers, and component suppliers.

2   State of the Art

2.1   Load and Wear Behaviour of Machine Tools

The machine tool industry can rely on many years of experience in the design of machine tools and their components so that a broad knowledge of expected loads and wear behavior has already been accumulated [1,2]. An established method to determine an adequate design of components to avoid fatigue failures of components is described in standards like ISO 3408-5 for ball screws or in ISO 281 for bearings [3,4]. The calculation of the service life L_{10} is based on the dynamic load rating C and the equivalent load F_m calculated from an estimated duty cycle:

$$L_{10} = \left(\frac{C}{F_m} \right)^3 \cdot 10^6 \tag{1}$$

$$F_m = \sqrt[3]{\frac{\int |F(t)|^3 \cdot |n(t)| dt}{\int |n(t)| dt}} \tag{2}$$

Instead of an estimated duty cycle, this method enables the prognosis of the service life based on the historical data directly from the machine. The required forces $F(t)$ and speeds $n(t)$ during the operation time t can be derived from the sensor data available from the CNC [5]. In the next step, the load integral from the actual machine, shown in Eq. 2, can be used to indicate the current condition of the component [6].

However, it should be noted that further investigations show that a prediction of the calculated service life for ball screws can lead to inaccurate results due to life-limiting factors that are not taken into account such as short stroke movements, shocks or strong vibrations caused by impact stress for example [7]. In regard to the high requirements of spindle bearings due to high operating loads and speeds, kinematic parameters such as the spin-roll-ratio, ball advance and retardation or the maximum contact pressure are considered in addition to the theoretical service life. The simulation models used during the design process describe the relation between operating conditions (process forces, rotational speed, tool length) and the bearing load, but are not further being used during subsequent life cycle phases of the machine to identify service life-reducing operating conditions. However, the availability of this information enhances transparency regarding component stress and enables the optimization of machining processes to extend the lifespan of these components. Therefore this paper presents a concept of how this existing know-how can be offered as a service in an ecosystem without putting the corporate intellectual property in danger using the example of a machine tool.

2.2   Data and Service Ecosystems

In 2020 the EU-Commission published its data strategy with the goal to establish a single market for data according to European values and rules [8]. This market consists of several data spaces in different domains, with one of them being the manufacturing sector. Since 2020 the Gaia-X Initiative (GX) as well as others, like the Data Spaces Business Alliance, are working on realising this strategy. In its terms, several data spaces and ecosystems with a focus on supply chains and manufacturing have emerged (e.g. Smart Connected Supplier Network[1], Catena-X[2], GEN-X[3]). Data spaces allow sharing of knowledge in a sovereign way [9]. The sovereignty is based on so-called verifiable credentials (VCs). In order to be legally binding these VCs are signed by the respective legal or natural person who issued the VC. A holder of such a VC, e.g. a person who wants to consume an algorithm to analyse a component's condition, is able to identify itself by presenting this VC [10].

There are three different kinds of participants: Federators, Providers, and Consumers. In this paper, only the Consumers and Providers will be considered. The roles are defined in the GX Architecture Document [11]. Providers are operating resources and offer them as GX-Services. Resources can be physical or virtual. An example of physical resources are machine tools or data centres. Virtual resources are for example databases or measurements. Consumers consume GX-Services and are using the specified resources.

---

[1] www.smart-connected.nl.

[2] www.catena-x.net.

[3] www.docs.genx.minimal-gaia-x.eu.

Every GX-Service that providers like to offer in an ecosystem needs a verifiable service description. Verifiable descriptions in a GX context consist of VCs. The mandatory content of VCs is defined in the Trust Framework [12]. Due to the signing of the VCs and self-sovereign identities based on trust anchors in accordance with the eIDAS Regulation [13], contracts established in this ecosystem are legally binding. The contents of services or data sets become retraceable.

3   Concept for Edge-Services for Components of Machine Tools

Regarding the implementation of sovereign services for the components, the concept of the Asset Administration Shell (AAS) already offers standardized functionalities regarding the distribution of services or implementing data models. As shown in Fig. 1 the AAS is instantiated from the AAS Type of the component supplier and consists of instances that are running on both the execution environment of the component supplier on the one hand and the end user's on the other hand. The AAS instances of the end user and component supplier might communicate via the Industry 4.0 language [14] in the future which is defined in the VDI/VDE 2193-1 and -2. The service instances require access to

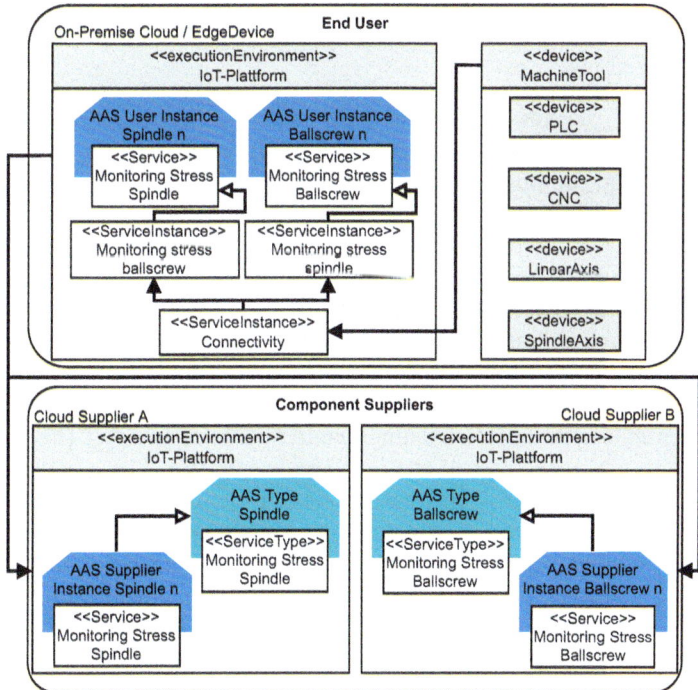

Fig. 1. Interaction between component supplier and end user

the machine data and can be executed on an edge device close to the machine or on cloud resources on premise or at a data center. The storage of the results of the monitoring services, in this case the stress of the components, are referenced by the AAS.

The services themselves are provided via AAS as well. Since the suppliers design and manufacture the components they possess the expertise to design stress monitoring services. All assets of the same type will use the same service. Therefore the services are part of the Type AAS. As soon as an instance of an asset is created, the service becomes part of the respective Instance AAS. When the end user buys the component an own AAS Instance is created. It references the supplier's AAS Instance and is able to call on the service which is then instantiated in the end user's execution environment.

The results of the stress monitoring services are of interest for the suppliers as well. Based on the values, condition monitoring or the proposal of further services are possible. It might even support the development of new components. What the results may be used for, is regulated in ecosystems by smart contracts. For these, it is necessary for framework conditions to be defined by each company as to what they want to make their data available for. In the future, this information could be the content of a separate submodel of the AAS and then be exported to a verifiable service description (see Sect. 2.2).

4 Use Case: Stress-Based Operator Model for Machine Tools

In the following, the new potential of the introduced concept will be demonstrated for a 3-axis machining centre using component-services to monitor the stress of individual components to implement a new stress-based operator model. For combining multiple services of different providers as a service composition it is required to ensure interoperability and portability. Therefore, the first step is to collect the required data from the devices. The presented approach in Fig. 2 is based on the available data from the CNC, but can be extended with additional sensors if necessary to improve the accuracy. In the data preprocessing process forces are derived from the available CNC data providing the input for the services to monitor the stress of the spindle bearings and the ball screws as in [5]. In the presented scenario, the machine manufacturer aggregates the component loads and determines the rate of the payment model based on a contracted cost model of the machine as in [15]. Besides offering knowledge in form of a service to monitor the stress, it is also possible for the component supplier to propose new payment models, which can be aggregated by the machine manufacturer.

Based on the presented approach in Fig. 2 an implementation of the load integral is demonstrated for a milling process using an end mill with a diameter of 16 mm and evaluated for two different materials. The toolpath of the cutting process and the resulting process forces as well as the feed rates are illustrated in Fig. 3. The time fraction q on the x-axis indicates the percentage for a particular load regarding the entire duration of the load spectrum. Based on the

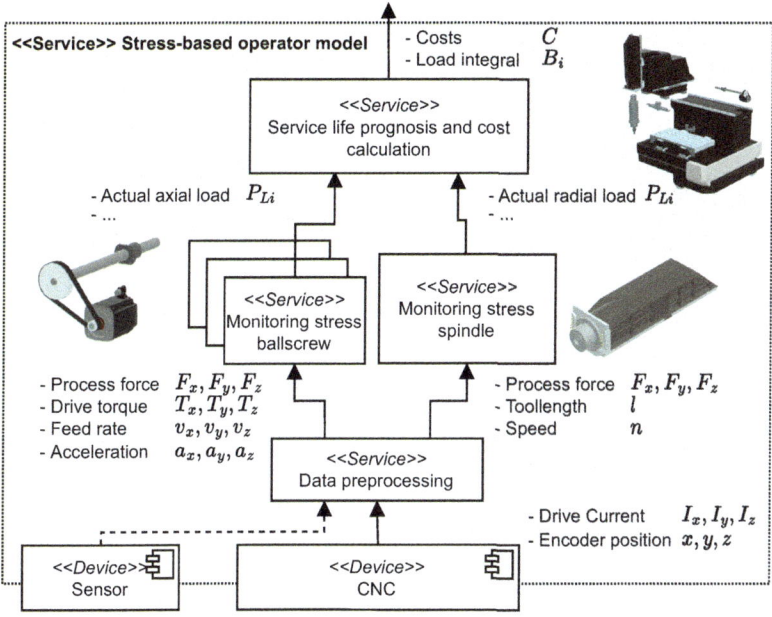

Fig. 2. Composition of component services as a stress-based operator model

applied loads the equivalent speed n_m and equivalent load F_m is calculated. The figure shows that the processing of the workpiece from steel (42CrMo4) results in significantly higher process forces than the aluminum-based (AlCuMgPb) workpiece. For this process the feed rates and spindle speed are higher for processing the aluminum-based workpiece leading to a reduced machining time by about 50 % for the workpieces.

For the evaluation of the individual components the load integral is shown in Fig. 4. Since the machining operations are performed in the x-y plane, the load integral of the z-axis is lower. For the ball screws, the acceleration during positioning movements causes high loads and results in the increase of the illustrated load integral for the machining process. Therefore, the difference between the load integrals for the different materials is small, despite the different machining duration. On the contrary, the load integral for the spindle bearing is significantly higher, despite the lower process loads. This is primarily caused by the higher cutting speed when machining aluminum. However, additional factors such as tool length or spindle displacements due to rotational speed may also lead to increased loads on the spindle bearing. However the considered machining processes do not show critical operating conditions, but make the stress on the individual components more transparent.

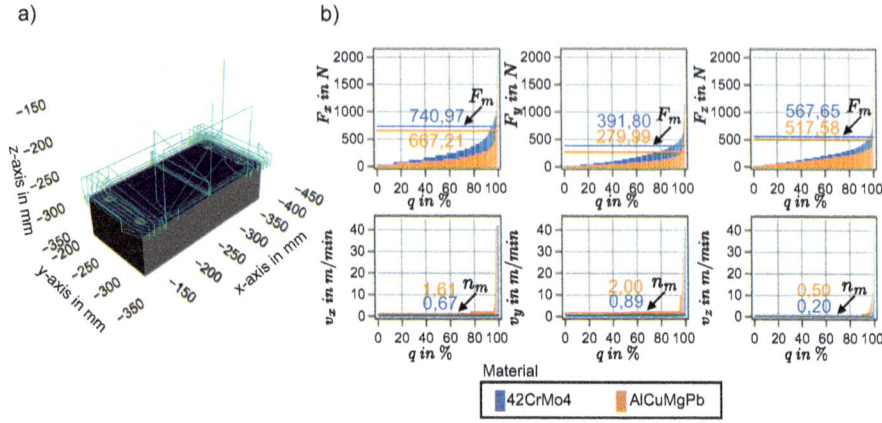

Fig. 3. Comparison of the process-related loads for workpieces with different materials; a) Workpiece toolpath; b) Duty cycles

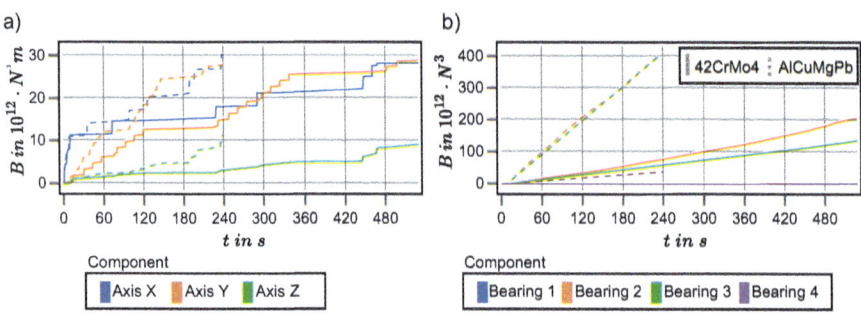

Fig. 4. Resulting load integrals of the individual components for a) ballscrews and b) spindle bearings

5   Conclusion and Future Research

By establishing an open service and data ecosystem, component suppliers will be able to offer digital services, to measure the loads on individual components or monitor their condition. In this context, sovereign data exchange between companies and the use of standards to ensure interoperability are important drivers for a successful implementation. The concept of the AAS helps integrating all required information or references for individual components in a structured and standardized data model. This creates a database to complement existing approaches for predicting the service life with data-driven approaches. By correlating the loads during operation with the service life, life reducing operating conditions can be identified and thus reduced by optimizing the process to ensure a resource-efficient operation of the machine. Furthermore, this is an essential step for creating new opportunities to implement innovative data-driven business models. The presented concept shows the potential of an open service and data ecosystem for production on the basis of a concrete use case for a machine tool.

Acknowledgements. The research is financed with funding provided by the German Federal Ministry for Economic Affairs and Climate Action (BMWK) within the "Edge Datenwirtschaft" program under the project ESCOM. It is implemented by the DLR Project Management Agency. The author is responsible for the content of this publication.

References

1. Abele E, Altintas Y, Brecher C (2010) Machine tool spindle units. CIRP Ann 59(2):781–802
2. Altintas Y, Verl A, Brecher C, Uriarte L, Pritschow G (2011) Machine tool feed drives. CIRP Ann 60(2):779–796
3. ISO 3408-5 (2006) Ball screws—Part 5: static and dynamic axial load ratings and operational life
4. ISO 281 (2007) Rolling bearings—Dynamic load ratings and rating life
5. Altintas Y, Aslan D (2017) Integration of virtual and on-line machining process control and monitoring. CIRP Ann 66(1):349–352
6. Huf A (2012) Kumulative Lastermittlung aus Antriebsdaten zur Bewertung des Zustands von Werkzeugmaschinenkomponenten. Jost-Jetter Verlag, Heimsheim
7. Yagmur T (2014) Analyse. Verbesserung und Beschreibung des Verschleißverhaltens von Kugelgewindetrieben für Werkzeugmaschinen. Apprimus Verlag, Aachen
8. European Commission (2020) Shaping Europe's digital future. https://commission.europa.eu/system/files/2021-01/communication-shaping-europes-digital-future-feb2020_en_3.pdf
9. Steinbuss S, Ottradovetz K, Langkau J, Punter M (2020) IDSA rule book. Zenodo, International Data Spaces Association, https://internationaldataspaces.org/wp-content/uploads/dlm_uploads/IDSA-White-Paper-IDSA-Rule-Book.pdf
10. Maier B, Pohlmann N (2022) Gaia-X secure and trustworthy ecosystems with Self Sovereign Identity: developing a decentralised, user-centric, and secure cloud ecosystem. https://www.gxfs.eu/ssi-whitepaper/
11. Gaia-x architecture document (2022). https://docs.gaia-x.eu/technical-committee/architecture-document/latest/
12. Gaia-X trust framework (2022). https://docs.gaia-x.eu/policy-rules-committee/trust-framework/latest/
13. eIDAS regulation (2023). European Commission. https://digital-strategy.ec.europa.eu/en/policies/eidas-regulation
14. Vialkowitsch J et al (2018) I4.0-Sprache - Vokabular, Nachrichtenstruktur und semantische Interaktionsprotokolle der I4.0-Sprache, Bundesministerium für Wirtschaft und Energie. https://www.plattform-i40.de/IP/Redaktion/DE/Downloads/Publikation/hm-2018-sprache.pdf
15. Kohn O et al (2022) Development of a stress factor as an indicator for stress-based payment models for machine tools. In: Production at the leading edge of technology, lecture notes in production engineering. Springer, pp 239–247

Optimal Design of Pipeline Networks for Inter-plant Water and Energy Integration

Renzo O. Piccoli[2,3]($\boxtimes$), Diego J. Trucco[1], Demian J. Presser[1,2], and Diego C. Cafaro[1,2]

[1] GAP (ITBA-UNL), Facultad de Ingeniería Química, Universidad Nacional del Litoral, Santiago del Estero 2829, 3000 Santa Fe, Argentina
[2] INTEC (UNL-CONICET), Güemes 3450, 3000 Santa Fe, Argentina
rpiccoli@frsf.utn.edu.ar
[3] Facultad Reg. Santa Fe, Univ. Tecnológica Nacional, Lavaise 610, 3000 Santa Fe, Argentina

Abstract. The concentration of specific industries in relatively small areas exists in many countries. This characteristic incentivizes cooperation and facilitates the optimal use of common resources. That is the case of the cluster of crop and chemical manufacturing industries in the center of Argentina, next to the Parana River. All of these plants are closely linked to the use of water streams in associated processes. The objective of this work is to minimize the overall costs of water and energy consumption through the optimal design of an inter-plant pipeline network. This is achieved by considering the daily availability and demand of hot water at each node over time, while ensuring that all pipelines are capable of reversing flows, if needed, at the expense of specific equipment and operating costs. The optimization model developed in this work can be categorized as a MILP formulation that considers fluid dynamics variables, temperatures, energy and flow in order to properly determine the pipeline diameters. The construction and operation of water pipelines in concentrated industrial regions can leverage positive impacts on the sustainable use of natural resources. In contrast to previous contributions, explicit consideration of flow reversals offers benefits that can significantly reduce investment costs.

Keywords: Energy · Water · Interplant · Optimization · Pipeline Network

1 Introduction

Manufacturing industries from chemical and crop sectors make heavy use of hot and cold water in their production processes. Additionally, and in particular in Argentina, most of these industries are clustered together in regions near large rivers. From these facts, establishing a shared network of pipelines to optimally supply water and energy in an integrated fashion is being emphasized, with the aim of reducing costs and environmental impacts. This problem begins by considering the specific demands and working schedules of nearby industries. The waste water stream from one of the plants can be transported to another industry that makes use of it, and this flow may revert over time.

© The Author(s) 2025
H. Kohl et al. (Eds.): GCSM 2023, LNME, pp. 463–471, 2025.
https://doi.org/10.1007/978-3-031-77429-4_51

Water/energy sources, such as thermal power plants and rivers are also linked with this network. Pipelines are highly efficient means for the transportation of liquid and gas products across extensive areas. However, investment costs in pipeline networks are significant, and are only justified if they operate at high utilization levels over long periods of time. That is the case of water transportation networks that predominantly rely on pipelines. As water/energy production and demand patterns for most industries vary with the hour of the day, and with the day of the year, building efficient pipeline networks has become increasingly challenging.

Budak Duhbaci et al. [1] present a thorough review on recent mathematical programming methods for water and energy minimization in industrial processes. Merket and Castro [2] introduce a mixed-integer linear programming (MILP) model to enhance the operational flexibility of a combined heat and power (CHP) plant that is connected to a district heating load through a pipeline. They seek for the optimal schedule of water slugs to be sent through pipeline, which offers significant benefits by utilizing the thermal inertia of the pipeline as energy storage. Zhang et al. [3] present an optimization framework for the integrated design of intra- and inter-plant water pipeline networks in the steel industry. The optimal design of the water network is built from a superstructure of alternatives, using a mixed-integer nonlinear programming (MINLP) formulation. The model is validated with an illustrative problem from a typical steel park. The aim is to identify individual water systems for each industry, and inter-plant connections are established to describe potential configurations of the water network. The integrated solution permits a reduction of 22% in freshwater consumption and a 21% reduction in the total annual cost when compared to decentralized water usage. Similarly, Aguilar-Oropeza et al. [4] propose a superstructure-based mathematical model that addresses the recycle and reuse of water within an eco-industrial park. Water flows serve as optimization variables, with the ultimate goal being the synthesis of a network that enables the fulfillment of water requirements at the lowest total cost. Caballero & Ravagnani [5] propose an MINLP model to minimize the total cost of a water distribution network. They consider predefined pipe diameters and different flow directions for every segment. Finally, Cafaro & Grossmann [6] present a general optimization framework for the design of water pipeline networks in the shale gas industry, accounting for flow reversals. However, capital and operational expenditures of flow reversals are not included in the formulation, from which the actual value of reversals is not clear. Changing from direct to reverse mode requires time- and cost-consuming tasks, which should be properly monetized in the objective function.

Similar to previous studies, this work aims to establish an integrated network of water pipelines among industries sharing a common area. The primary objective is to mitigate the environmental impact and resource consumption while fulfilling the individual demands of each company. In contrast to previous studies, however, there are two novel features addressed in this paper: (a) pipeline sizing, and (b) flow reversals. Based on well-established fluid-dynamic correlations, we seek to optimally determine pipeline diameters, which typically follow economy-of-scale cost functions. In turn, we also account for flow reversals in pipeline segments with the aim of utilizing the existing transportation infrastructure to make the water flow in both directions. We propose a MILP model that is capable of managing flow reversals in any segment through-out the specified time horizon. By means of this generalized framework, we are able to assess the performance of the pipeline network under different conditions and scenarios. This framework is capable of accommodating flow reversals in any segment of the pipeline throughout the specified time horizon. By employing this generalized framework, we can explore and evaluate the performance and feasibility of the pipeline network under different flow conditions, considering various operational scenarios efficiently. The implementation of flow reversals in pipelines can be particularly beneficial in achieving cost-effective network designs under time varying demand patterns. Needless to say, including this feature into the optimization model is challenging, as will be shown later in this work.

The remaining of the paper is organized as follows: We formally define the problem in Sect. 2 and introduce the model assumptions in Sect. 3. The optimization model is presented in Sect. 4, and a case study from the South of the Santa Fe, Argentina is addressed in Sect. 5. Conclusions are drawn in Sect. 6.

2 Problem Definition

In general terms, the design of a water pipeline network involves three primary decisions: First, how to connect the nodes, i.e., determining the pipeline layout. Second, selecting the pipeline diameter for each segment. Third, identifying flow directions and rates along the pipeline segments during the time horizon. For planning purposes, the time domain is usually divided into discrete intervals [7]. Our particular water/energy integration problem can be stated as follows:

Given are: (a) the location of water sources and manufacturing plants, (b) the water/energy production and demand curves for a representative day at every node, and (c) alternative pipeline diameters for potential interconnections. The aim is to determine: (1) whether or not to construct a unidirectional/bidirectional pipeline segment between a pair of nodes, (2) the diameter of such a pipeline segment, and (3) the flow direction, flowrate and temperature of the water moving along a segment, at every period of the time horizon. The optimal plan should minimize the net present cost (accounting for capital and operational expenditures), while fulfilling water/energy demands in every node, at every time period. Figure 1 illustrates a simple example with three factories and a source node (a power plant) potentially supplying hot water to meet energy demands. Possible pipeline interconnections are also depicted in Fig. 1.

3 Model Assumptions

To develop the mathematical programming model presented in upcoming sections, the following assumptions are made:

1. Geographical locations of the water source and factories are known data.
2. Energy demands from each industry are given for a discrete set of time periods, for a single representative day. In our case study, three periods per day are considered, corresponding to three eight-hour work shifts.

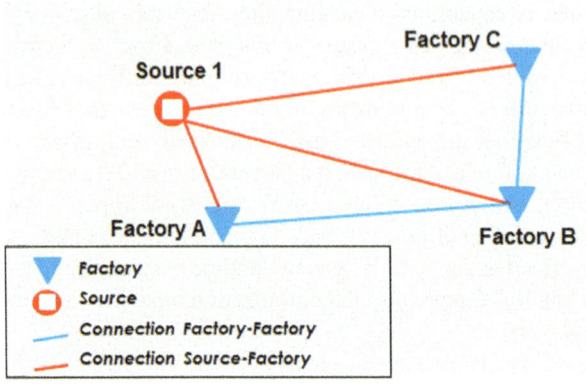

Fig. 1. Nodes and possible connections

3. It is assumed that there are no heat losses in the pipeline segments.
4. The temperature range of water flows is discretized. Constant density and heat capacity is assumed for the water at any temperature.
5. All connections between nodes are free to let the flow run in one direction or the other at every period. Moreover, no more than a single pipeline can be built between a pair of nodes.
6. A simplified form of the Hazen-Williams correlation is adopted to size the pipeline diameters, based on a maximum linear velocity of the flow.
7. Each water source has a maximum capacity for supplying thermal energy, associated with the temperature and maximum flowrate (available waste water) at each period of time.
8. A demand profile for each factory is given beforehand, and includes heat needs in each period and temperature range for inlet/outlet flows.

4   Mathematical Formulation

An MILP formulation is presented with the aim of obtaining the optimal network of inter-plant pipelines to integrate water/energy flows across an industrial area. The elements $j \in J$ are the nodes that require hot water for production, and $i \in I$ are the sources that offer their excesses of hot water. The set $t \in T$ represents the work shifts used for time discretization. The nonnegative variable $FS_{i,j,t}$ indicates the volume of water transferred from source i to node j in time interval t while the variable $FF_{j,j',t}$ represents the amount sent from node j to node j' at work shift t. V_j is the subset of nodes that can be connected to j. Equation (1) represents the mass balance for node j over time period t. The total amount transferred into the node must either be sent to another node j' or discarded, which is indicated by the positive variable $FO_{j,t}$.

$$\sum_{j' \in V_j} FF_{j,j',t} + FO_{j,t} = \sum_{i} FS_{i,j,t} + \sum_{j' \in V_j} FF_{j',j,t} \quad \forall j, t \tag{1}$$

Equation (2) restricts outlet flows to the production capacity of the source given by $cap_{i,t}$.

$$\sum_j FS_{i,j,t} \leq cap_{i,t} \quad \forall i,t \tag{2}$$

Analogously, the variable $QS_{i,j,t}$ measures the heat delivered to node j from source i during period t, and $QF_{j,j',t}$ represents the heat transfer between nodes j and j'. Given that $qd_{j,t}$ is a parameter indicating the expected heat required by node j from external sources at time interval t, Eq. (3) ensures the fulfillment of that demand.

$$qd_{j,t} \leq \sum_{j' \in V_j} QF_{j',j,t} + \sum_i QS_{i,j,t} \quad \forall j,t \tag{3}$$

We also introduce the set $d \in D$ to account for the available pipeline diameters, with the parameter $pipecap_d$ fixing the water transportation capacity of a pipeline of diameter d. The binary variables $XS_{i,j,d}$ and $XF_{j,j',d}$ represent the decision of installing a new pipeline of diameter d between source i and node j, and between j and j', respectively. Equations (4) and (5) entail the installation of a pipeline segment of a certain diameter and its associated transportation capacity. Similarly, Eqs. (6) and (7) impose that the heat exchange can only occur if a pipeline between the nodes is installed. The scalars *maxheat* serve as upper bounds for the heat transfer.

$$FS_{i,j,t} \leq \sum_d pipecap_d \cdot S_{i,j,d} \quad \forall i,j,t \tag{4}$$

$$FF_{j,j',t} \leq \sum_d pipecap_d \cdot \left(XF_{j,j',d} + XF_{j',j,d} \right) \quad \forall j,j' \in V_j, t \tag{5}$$

$$QS_{i,j,t} \leq maxheat_{i,t} \cdot \sum_d XS_{i,j,d} \quad \forall i,j,t \tag{6}$$

$$QF_{j,j',t} \leq maxheat_{j,j'} \cdot \sum_d XF_{j,j',d} + XF_{j',j,d} \quad \forall j,j' \in V_j, t \tag{7}$$

Furthermore, Eqs. (8) and (9) state that the installation of parallel pipelines is avoided.

$$\sum_d XS_{i,j,d} \leq 1 \quad \forall i,j \tag{8}$$

$$\sum_d XF_{j,j',d} + XF_{j',j,d} \leq 1 \quad \forall j,j' \in V_j \tag{9}$$

To model the energy balances in linear terms, we define the set $r \in R$ including discrete temperature levels that can be selected for heat transfer. The parameter $temp_r$ represents the reference temperature associated with level r. Additionally, we introduce the 0–1 variables $YS_{i,j,t,r}$ and $YF_{j,j',t,r}$ to indicate that the temperature level r is selected for water/heat transfer during work shift t, between source i and node j, or between j and j', respectively. Likewise, the binary variable $YO_{j,t,r}$ equals one if the water emerging from node j during time t is at the temperature level r. Equations (10) to (12) represent energy balances between the source i and the demand node j. Parameters cp and ρ refer to the specific heat capacity and density of water, respectively.

$$QS_{i,j,t} = \sum_{r,r'|_{r>r'}} QSR_{i,j,t,r,r'} \forall i,j,t \tag{10}$$

$$QSR_{i,j,t,r,r'} \leq cp \cdot \rho \cdot FS_{i,j,t} \cdot (temp_r - temp_{r'}) \forall i,j,t,r,r' \tag{11}$$

$$\sum_{r'|_{r'<r}} QSR_{i,j,t,r,r'} \leq maxheat_{i,t} \cdot YS_{i,j,t,r} \quad \forall i,j,t,r \tag{12}$$

Equations (13) to (15) describe the heat transfer between nodes j and j' in a similar manner. In turn, Eq. (16) ensures that if the water emerging from node j is at temperature level r, the nodes j' that receive water from node j must also receive it at temperature r.

$$QF_{j,j',t} = \sum_{r,r'|_{r>r'}} QFR_{j,j',t,r,r'} \forall j,j' \in V_j,t \tag{13}$$

$$QFR_{j,j',t,r,r'} \leq cp \cdot \rho \cdot FF_{j,j',t} \cdot (temp_r - temp_{r'}) \quad \forall j,j' \in V_j,r,r',t \tag{14}$$

$$\sum_{r'|_{r'<r}} QFR_{j,j',t,r,r'} \leq maxheat_{j,j'} \cdot YF_{j,j',t,r} \quad \forall j,j' \in V_j,t,r \tag{15}$$

$$YO_{j,t,r} \geq YF_{j,j',t,r} \quad \forall j,j' \in V_j,t,r, \sum_r YO_{j,t,r} \leq 1 \forall j,t \tag{16}$$

Moreover, only a temperature level r is allowed to exit node j in a time period. To incorporate the flow reversal feature in this model we introduce the 0–1 variable $XD_{j,j',t}$ that equals one if the flow direction over period t is from j to j'. The fact that opposite directions cannot be operating simultaneously is imposed by Eq. (17).

$$XD_{j,j',t} + XD_{j',j,t} \leq 1 \forall j,j' \in V_j,t \tag{17}$$

Equations (18)-(19) state that water/heat transfer is only allowed in the selected direction.

$$FF_{j,j',t} \leq maxflow_{j,j'} \cdot XD_{j,j',t} \quad \forall j,j' \in V_j,t \tag{18}$$

$$QF_{j,j',t} \leq maxheat_{j,j'} \cdot XD_{j,j',t} \quad \forall j,j' \in V_j,t \tag{19}$$

Equation (20) detects a change in the flow direction over a pipeline segment at a specific point in time. $FRev_{j,j',t}$ equals one if the flow direction changes from (j',j) to (j,j') at the initial time of period t. Finally, the construction of bidirectional pipelines incurs in an additional cost (*revcost*), whenever variable $XR_{j,j'}$ is equal to one.

$$XR_{j,j'} \geq FRev_{j,j',t} \geq XD_{j,j',t} - XD_{j',j,t-1} - XD_{j',j,T}|_{t=1} \forall j,j' \in V_j,t \tag{20}$$

Lastly, the model aims to minimize the net present cost *NPC* of the infrastructure to collect and distribute flows over the specified time horizon. The objective function is defined by Eq. (21), where $dist_{i,j}$ and $dist_{j,j'}$ represent distances and $pipecost_d$ denotes the investment cost of a unidirectional pipeline of diameter d per unit length.

$$Min\,NPC = \begin{bmatrix} \sum_{i,j,d} pcost_d \cdot dist_{i,j} \cdot XS_{i,j,d} \\ + \sum_{j,j' \in V_j, d} pcost_d \cdot dist_{j,j'} \cdot XF_{j,j',d} \\ + \sum_{j,j' \in V_j} revcost \cdot XR_{j,j'} \end{bmatrix} \qquad (21)$$

Summarizing, the MILP seeks to minimize Eq. (21), subject to Eqs. (1) to (20).

5   Case Study and Results

An illustrative case study of realistic dimensions is proposed as a means of validating the optimization model. The example consists of eight factories with their respective demands over a given time horizon, distributed across 53 km². The possible connections between them can be seen in Fig. 2A. There are two sources of water/energy in the area which are two thermoelectric power plants. The time horizon is discretized into three periods per day, corresponding to eight-hours work shifts. Three alternative pipeline diameters can be used to build the network: 6, 10 and 12 inches. The cost of the pipelines is set at USD 55,000, USD 67,500, and USD 80,000 per kilometer of length for each respective diameter. The cost for the installation of equipment and accessories to add the flow reversal feature in a pipeline segment is set at 30,000 USD. For simplicity, capital and operating cost for reversals are independent of the segment length and diameter. The temperature range goes from 15 °C to 95 °C and is divided into eight slots of 10 °C each. The case study is implemented on GAMS 36.1 and solved to global optimality (0% gap) using GUROBI 10.0.1, on an Intel Core i5-8265U CPU, 8 GB RAM, 6 threads. For illustrative purposes the models have been solved accounting for different water availabilities at the sources.

Figure 2B, C compare solutions with and without flow reversals, respectively, considering a maximum availability of 200 m³/h of hot water at the sources. The CPU time required to solve the model with no flow reversals is 25 s, resulting in an NPC of 15.18 MMUSD. The same model, allowing flow reversals, yields a total cost of 14.21 MMUSD in 75 s. A bidirectional pipeline has been suggested for one of the segments of the resulting network, as shown in Fig. 2B. The NPC of the network with flow reversals reduces in 0.97 MMUSD. Savings are a direct consequence of the better use of available pipelines, particularly the segment connecting factories C and D. Figure 3 describes the optimal results for different water availability levels, within the same case study. It is interesting to note that when the overall amount of water supplies from the two sources is lower than 200 m³/h, there is no feasible solution with unidirectional pipelines. In other words, there is no chance to supply plants with external heat sources unless bidirectional segments are used. The importance of flow reversals is smaller as the total amount of available water at the sources increases.

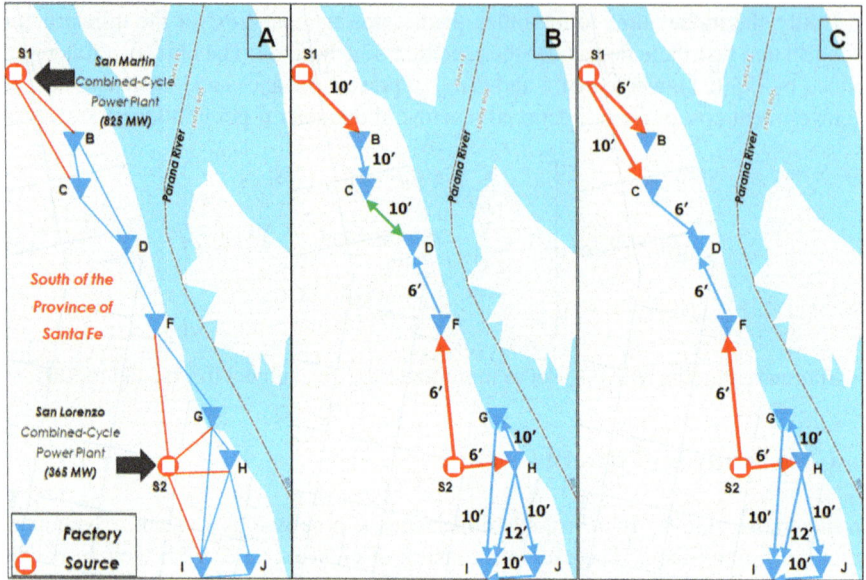

Fig. 2. Case study and optimal solutions with and without flow reversals

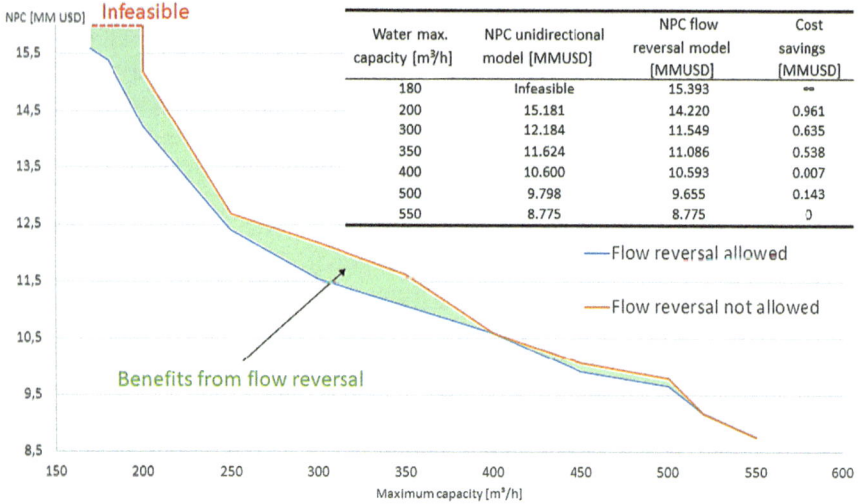

Water max. capacity [m³/h]	NPC unidirectional model [MMUSD]	NPC flow reversal model [MMUSD]	Cost savings [MMUSD]
180	Infeasible	15.393	∞
200	15.181	14.220	0.961
300	12.184	11.549	0.635
350	11.624	11.086	0.538
400	10.600	10.593	0.007
500	9.798	9.655	0.143
550	8.775	8.775	0

Fig. 3. Comparison of NPC from both models for different scenarios

6 Conclusions

This work has presented a novel formulation to promote water streams integration between manufacturing plants as a source of energy in their operations. The optimization model has been tested on a real-world case study involving chemical and crop industries located along the Parana River in Argentina. Results show the benefits of flow reversals in designing water pipeline networks. This feature proves to enhance pipeline capacity utilization over time, resulting in reduced diameters and installation costs. By comparing the NPC of the networks with and without flow reversals for different scenarios we conclude that bidirectional pipelines increase their importance as the water availability from energy sources is more constrained. In fact, under very restricted conditions, unidirectional designs are directly infeasible. Flow reversals have the potential to foster industry symbiosis in resource and energy usage, thus minimizing environmental impacts. In future research, the model's application is expected to expand to other types of inter-industry networks, further encouraging resource-sharing. This development can pave the way for a more sustainable industrial landscape, where the reuse of water resources for energy and heat leads to increased efficiency and environmental conservation.

References

1. Budak Duhbacı T, Özel S, Bulkan S (2021) Water and energy minimization in industrial processes through mathematical prog.: a literature review. J Clean Prod 284:124752
2. Merkert L, Castro PM (2020) Optimal scheduling of a district heat system with a comb. heat and power plant considering pipeline dynamics. Ind Eng Chem Res 59:5969–5984
3. Zhang K, Zhao Y, Cao H, Wen H (2018) Multi-scale water network optim. considering simult. intra- and inter-plant integration in steel industry. J Clean Prod 176:663–675
4. Aguilar-Oropeza G, Rubio-Castro E, Ponce-Ortega JM (2019) Involving acceptability in the opt. synthesis of water nets. in eco-indus. Parks. Ind Eng Chem Res 58:2268–2279
5. Caballero JA, Ravagnani MA (2019) Water distribution networks optimization considering unknown flow directions and pipe diameters. Comput Chem Eng 127:41–48
6. Cafaro DC, Grossmann IE (2020) Optimal design of water pipeline networks for the development of shale gas resources. AIChE J 67:e17058
7. Saldanha-da-Gama F (2018) Comments on: extensive facility location problems on networks: an updated review. TOP 26:229–232

Empirical Investigation into the Drivers of Green Manufacturing Technologies

Ganesh Prasad Shukla⬤ and Pranav Gupte(⊠) ⬤

Department of Industrial and Production Engineering, Guru Ghasidas Vishwavidyalaya (A Central University), Bilaspur, Chhattisgarh 495009, India
guptepranav01@gmail.com

Abstract. Global manufacturing sectors are progressively using state-of-the-art green technologies (GTs) to alleviate the detrimental effects of production on the environment. Adoption of GTs is motivated by a number of variables referred to as drivers. Although a variety of drivers and GTs are described in the literature, few models demonstrate the intricate relationships that occur between them. Although there are evidences that drivers might be distinctive for every particular GT in a given business, there is a dearth of empirical study on how and to what extent the drivers are influencing the adoption and implementation of GTs. To address these gaps, this study first categorizes drivers and GTs based on existing literature. Subsequently, a theoretical framework is developed grounded in natural resource-based view (NRBV), establishing the link between adoption of specific GTs and their potential drivers through the development of enabling key resources. A case study of an Indian steel manufacturing company is conducted to empirically test the framework by analyzing various implemented green initiatives. Observed compliances and deviations to the framework are examined. Additionally, this research offers valuable insights to manufacturing managers for GT selection and recommends avenues for future research in this domain.

Keywords: Green technology · Natural resource based view · Drivers · Key resources · Competencies · Case study

1 Introduction

Due to the growing consciousness for the negative consequences that manufacturing activities have on people and natural environment, implementation of green manufacturing (GM) is becoming a pressing necessity for manufacturing firms [1]. Specifically, all the GM-related equipments, methods, practices, programmes, and policies that assist manufacturing firms, either technically or non-technically, in integrating ecological sustainability into their products, procedures, and workflows are collectively referred to as "green technologies" (GTs) [2]. The adoption of GTs in a manufacturing firm is influenced by a variety of factors having a positive effect on their execution, are often referred to as drivers. While implementing GTs, it is essential to understand the elements that will influence a firm's adoption of such practices. Study on drivers aids in understanding the elements that assists to an organization's current level of sustainability, which is necessary for identifying hurdles and potential for GT adoption [3].

Various scholars such as Shukla et al. [3], Plooy et al. [4], and Shankar et al. [5] have widely categorized the drivers of GM, but have primarily focused on assessing the drivers within the context of specific countries. Although Karurkar et al. [6] and Shang et al. [7] highlights the impact of stakeholders on GM in the automobile sector and the advantages of stringent regulations in the electronics industry, respectively, earlier research emphasizes sustainable initiatives at the business, sector, or commercial levels. Research on GTs is progressed by Potrich et al. [8] and Shukla and Adil [9] using maturity stage models. Studies by Shukla and Adil [10] and Aboelmaged [11] add green performance measures, sustainable manufacturing capabilities, and competitive advantages. Despite extensive literature on drivers and GTs, empirical research establishing the relationship between adoptions of GT in accordance with the driver is scarce. There is also a lack of research on how and to what extent the drivers are impacting the implementation of GTs in manufacturing organizations. Therefore, the aim of this study is to construct a theoretical framework to address the research question of "how and to what extent the drivers are influencing the adoption of a specific GT in manufacturing firm". A qualitative case study technique is employed to contextualize and validate the significance of driver in GT selection.

This paper is organized as follows. The theoretical framework is presented in Sect. 2. Section 3 of this paper discusses the techniques used for selecting the case firm, collecting the data and observations from the case analysis. In Sect. 4, the paper ends with its findings and conclusions.

2 Theoretical Framework

2.1 Green Technologies

Hart's seminal study [12] on the Natural Resource-Based View (NRBV) posits that "strategic orientation and the attainment of competitive advantage should be grounded in competencies that intrinsically facilitate the pursuit of sustainable economic initiatives." Within the NRBV framework, GTs serve as instrumental tools to augment methods and resources required for cultivating essential competencies, specifically Pollution Prevention, Product Stewardship, and Sustainable Development. This study, aligning with Shukla and Adil [10], categorizes GTs into Pollution Control (PC), Pollution Prevention (PP), and Management System (MS). PC encompasses capital investments, including equipment and techniques, to oversee or mitigate pollutants or manufacturing by-products. Conversely, PP involves methods minimizing or eliminating pollutant generation from the manufacturing origin. MS includes structural investments beyond physical equipment, establishing a systematic and proactive environmental management approach. This ensures sustainability efforts become ingrained in the organizational culture, fostering holistic commitment to environmental responsibility and long-term sustainability practices across the entire business operation [10].

PC, PP and MS are further categorized as End-of-pipe technologies (PC-GT1), PP-Resource conserving technologies (PP-GT2), PP-Product/Process Innovation technologies (PP-GT3), MS-Environmental Infrastructural Support (MS-GT4). Remarkably, the GTs related to PC and PP, namely PC-GT1, PP-GT2, and PP-GT3, actively reduce the negative environmental impacts of manufacturing processes. On the other hand, GTs associated with MS, specifically MS-GT4, provide the infrastructure and procedures necessary to achieve the environmental goals rather than immediately responding to lower environmental harm [10]. The NRBV theory's strategic capabilities are mapped onto the GT categories to give the aforementioned classifications a solid theoretical foundation, which lacks in the work of Shukla and Adil [10]. Table 1 shows the GT descriptions together with their NRBV mapping.

Table 1. GTs and mapping with NRBV theory

GT categories	Description	NRBV capability [12]
End-of-pipe technologies (PC-GT1)	The application of techniques or machinery at the last step of production, either before or after, to reduce or get rid of impurities without changing the final product or process [2]	Reaction to legal requirements needed to obtain a license to conduct business. It doesn't appear to be guided by any strategic capabilities of NRBV
PP- Resource Conserving technologies (PP-GT2)	Strategies that include actions, initiatives, and methods to reduce pollutants, emissions, and waste of raw materials at the origin of production. Emphasizing continual improvement, involving recycling, reusing, and minimizing hazardous material usage [2]	Pollution Prevention (Ecological initiatives aimed at improving the maintenance and resource efficiency at in-house process levels)
PP-Product/Process Innovation technologies (PP-GT3)	A strategic approach that modifies operational procedures and product designs to reduce harmful environmental consequences over the course of the product life cycle. These technologies signify a fundamental shift towards environmental sustainability [10]	Product Stewardship (Minimize the influence of a product's lifespan on the ecosystem along the full value chain, which calls for stakeholder involvement)
MS-Environmental Infrastructural Support (MS-GT4)	Investments in infrastructure, indicative of a company's commitment to environmental policy and strategies for achieving environmental goals. It includes key performance indicators, division creation, ecological management systems, and monitoring of sustainable innovation [10]	Sustainable Development (Becoming environmental pioneers by combining strategies with operations of elements of world network and interacting with a spectrum of stakeholders)

2.2   Characterization of GTs for Their Drivers, Key Resources and Competency Outcomes

Competencies, intricate combinations of institutional resources and operational procedures strategically implemented through GTs [12], play a pivotal role in achieving environmental stability. GTs facilitate the development of competencies that prevent effluent discharge, reduces resource usage in internal production activities, lower product life cycle expenses, and yield positive environmental outcomes [2]. These competencies serve as foundational elements supporting and enabling GM activities. NRBV asserts that future competency foundations, constrained by ecosystem limitations, will be established by emerging GTs supported by strategic resources that acts as "hindrances to replication" due to their tacit, socially complex, and unique nature [2, 12]. Tacit resources, developed through labor-intensive initiatives and expertise growth, may be challenging to document or quantify. Socially complex resources rely on diverse team efforts, emphasizing collaboration and collective expertise [2]. Managerial involvement underscores the significance of innovative capabilities and continual enhancement of unique, firm-specific resources essential for success in manufacturing [2]. The adoption of specific GTs is driven by efficient utilization of strategic resources, encompassing regulatory compliance, cost-effective processes, market advantage, and leadership aspirations.

The theoretical framework in Table 2, derived from the above discourse, delineates three critical attributes namely drivers of GTs, enabling key resources, and realized competency/adoption of GTs. This framework elucidates how drivers influence the selection of a specific GT. Firms are influenced by numerous factors serving as driving forces for GT adoption. In response, firms seek various strategic resources pivotal for successful GT implementation, fostering diverse competencies. In other words, the decision to adopt a GT, driven by various factors, prompts the acquisition of distinct resources, forming the foundation for GT implementation and contributing to varying competency levels. The continuum of GTs from PC-GT1 to MS-GT4 represents the ascending competency levels guided by underlying drivers in manufacturing businesses. This prioritization aligns GTs with organizational goals, addressing immediate needs and long-term sustainability aspirations. Table 2 synthesizes and details each GT category, encompassing drivers, enabling key resources, and realized competencies through a literature review.

3   A Case Study

3.1   Case Selection and Data Collection

The case study centers on an Indian steel manufacturing company chosen for its extensive use of raw materials, presenting an opportunity for environment friendly manufacturing techniques. The selected company, referred to as ABC, was chosen based on (1) convenient researcher access, (2) availability of environmental reports, and (3) evidence of implemented environmental initiatives. To ensure ethical standards, official permission from ABC and approval from the academic institution's ethics board were obtained. A comprehensive case study protocol covered research background, company details, and specifics of implemented green initiatives (GIs). A structured questionnaire, aligned with the theoretical framework, was designed. Specifically, the questionnaire included,

Table 2. Theoretical Framework on "how the drivers influence the adoption of GTs"

Drivers for GTs	Key resources	GT adoption	Realized competency
Regulations, criteria, and permit to run business [10]	Implementation of investment-intensive methods [2]	End-of-pipe technologies (PC-GT1)	Towards the completion of production process or last process phases, contaminants must be collected, treated or recycled [2]
Minimize the price of goods, reduce risk and regulatory costs, establish benchmarks for usage of materials [10]	Staff participation, continual growth, and a focus on individuals and skills (tacit/causally ambiguous) [12]	PP-resource conserving technologies (PP-GT2)	Reduce pollutants, contaminants, and particles from internal production processes, source mitigation/conserve resources [10, 12]
Outpace competitors by differentiating and producing items that are reliable and environment friendly [10, 12]	Collaborative efforts of multidisciplinary groups, incorporation of "life cycle approach" and "shareholder viewpoint" (Socially complex) [12]	PP-product/process innovation technologies (PP-GT3)	Reduces the cost of product's lifespan, improves business's reputation, sustainability in its external endeavors, and draws in new clients concerned about the environment [2, 10]
Effective collaboration, involvement of stakeholders, national acknowledgment for sustainability contributions [10]	Participation from high leadership, restructuring in-house proficiency, and business abilities (rare, firm-specific) [12]	MS-Environmental Infrastructural Support (MS-GT4)	Supports in minimizing the overall negative impact on ecosystem, tracking sustainability outcomes, recognition for sustainability initiatives in the global market [10, 12]

why a GI was implemented? (i.e. what are the drivers behind the adoption of this GI?), how was it implemented? (i.e. what resources are to be accumulated towards installation of this GI?), to what extent (i.e. what are the competencies developed after installation of the GI?). Examination of ABC's environmental reports and discussions with the Head of EHS guided the selection of three significant GIs for data collection. Triangulation involving interviews, documents, and cross-verification enhanced research validity. Furthermore, the internal validity was assured by drawing on explanations obtained from the existing body of literature. Data from interviews, plant visits, and documents were meticulously consolidated for analysis.

3.2 Case Analysis

GI-1 Installation of Effluent Treatment Plant for Water Recovery Water is a vital resource for ABC, used in crucial operations like the fly ash brick factory, lime and dolo plants, coal washery, steel smelting shop, and process boilers. To address its water footprint, ABC deployed Effluent Treatment Plants (ETP) targeting pollutant elimination. The main driver for ETP is to comply with zero liquid discharge (ZLD) policy of Central Pollution Control Board, defining GI-1 as an "End-of-pipe" technology (PC-GT1). Advanced ZLD processes are used to enhance the disinfection of effluents. ABC undertook substantial modifications to its existing water pipelines across its entire infrastructure and engaged the services of an external contractor. Thus, some tacit and socially complex resources were used to implement GI-1. Post-implementation, ABC achieved a 9% reduction in fresh water usage in 2022–2023, reducing manufacturing expenses. ETP not only enhanced river water cleanliness but achieved zero effluent discharge, fortifying water safety amid changing climates.

GI-2 Installation of Slag Processing Plants for Recycling of Wastes ABC company places consistent emphasis on waste retrieval and recycling, notably targeting steel-making waste called slag. By using slag processing plants (SPP) established within ABC, the company effectively converts waste steel slag into valuable slag balls, a sought-after abrasive blast material. GI-2 aligns with "PP-Resource conserving" technology (PP-GT2), embodying resource conservation and efficiency, particularly in recovering costly metals from slag. SPP implementation led ABC to expand its workforce, introducing a dedicated managerial role for slag processing oversight and establishing a multidisciplinary team for flexibility. Retrieving and utilizing previously land filled slag resulted in a substantial reduction in production expenses, recovering approximately 2.8 tonnes of iron metal worth Rs. 3 Lakhs in 2022–23. SPP implementation showcased improvements in resource preservation, waste recovery, material utilization, and manufacturing quality and efficiency.

GI-3 Implementation of ISO 50001 Energy Management System ABC prioritizes energy savings, implementing ISO 50001-aligned strategies for reduction, conservation, and technical measures. Aligned with the "National Steel Policy," ABC focuses on eco-conscious and energy-efficient steel manufacturing. The decision to implement GI-4 corresponds to "MS-Environmental infrastructural support" (MS-GT4), aiming for national recognition and compliance with energy regulations. ABC establishes a "power portfolio" division, integrating continuous energy monitoring. External consultants aid in software setup, accreditation, and training. ISO 50001 leads to a 17% reduction in ABC's net energy usage in 2022–2023, minimizing wastage, yielding cost advantages (Rs. 25 Lakhs saved in 2022–23), and enhancing competitiveness. Findings for GIs are summarized in Table 3 under suggested GT categories.

4 Findings and Conclusions

This research significantly advances theoretical understanding by establishing a framework grounded in NRBV principles, elucidating the selection of GTs (GTs) under the influence of drivers and the development of enabling key resources. A case study of a steel

Table 3. Observations for GT based GIs from the case company

GIs	Drivers of GTs	Key resources	Realized competency
GI-1 (in line with PC-GT1)	Adherence to legal constraints under the zero liquid discharge policy	Using advanced ZLD equipments, technology and operations, *modifications to current pipelines, hiring other entity for ETP management*	Raising purity of river water, achieving minimal water contamination, *reduction in usage of fresh water, cost savings*
GI-2 (in line with PP-GT2)	Minimizing material expenses, waste handling expenses, improving cost competition	Hiring skilled operators, employee education and skill training, *restructuring the departmental setup, formation of multidisciplinary team*	Enhancements in manufacturing quality, recycling of waste material, environmental stewardship in preserving resources, reduced production expenses
GI-3 (in line with MS-GT4)	Senior leadership's progressive approach, *government regulations involving preservation of energy,* national recognition for green initiatives	Creating a specific division, using tracking and monitoring systems, forming unit level focused group, *hired outside consultant, employee awareness and education*	Lowering total energy use, energy footprint, improving sustainability performance, *creating a competitive edge, cost advantages*

Note The underlined terms indicate deviations from the theoretical framework

manufacturing company, considering four green initiatives, serves as a rare empirical contribution to the field. While the study broadly aligns with the theoretical framework, notable exceptions emerge. The End-of-pipe technology (PC-GT1), initially perceived as an expense, is observed in GI-1 to yield cost savings through tacit and socially complex resources, deviating from the theoretical framework. Similarly, PP-resource conserving technology (PP-GT2) demonstrates incongruity as scarce/firm-specific and socially complex resources, beyond tacit ones, contribute to manufacturing quality and environmental stewardship in GI-2. Paradoxically, MS-GT4, represented by GI-3, deviate from expected behaviors, and are employed to comply with laws on energy conservation by enabling tacit resources which can then result in competitive edge and cost savings.

This paper underscores the adoption of GTs as a solution to sustainability challenges in manufacturing firms, focusing on a single steel manufacturing company. The generalizability of outcomes for aligning GIs with specific GTs to diverse industrial contexts is constrained, warranting further studies for broader applicability. The study explores a limited number of GTs, suggesting future research to encompass additional GT categories, such as green product design, sourcing, logistics, marketing, and standards. For top managerial decision-making and strategic planners, this study offers crucial insights for selecting valuable GTs and field studies, enhancing organizational performance and learning.

References

1. Alayon C, Safsten K, Johansson G (2016) Conceptual sustainable production principles in practice: do they reflect what companies do? J Clean Prod 141:693–701
2. Pande B, Adil GK (2023) An enquiry into competitive value of sustainable manufacturing capabilities. J Manuf Technol Manage
3. Shukla GP, Swarnakar V, Singh SJ (2023) Assessment of drivers and barriers of green manufacturing practices in Indian manufacturing companies. J Inst Eng (India): Ser C 104(1):45–54
4. Du Plooy S, Neethling K, Nel A, Nel JD (2022) Drivers of and barriers to green manufacturing in South Africa. J Contemp Manage 19(1):260–298
5. Shankar KM, Kumar PU, Kannan D (2016) Analyzing the drivers of advanced sustainable manufacturing system using AHP approach. Sustainability 8(8):824
6. Karurkar S, Unnikrishnan S, Panda SS (2018) Study of environmental sustainability and green manufacturing practices in the Indian automobile industry. OIDA Int J Sustain Dev 11(06):49–62
7. Shang KC, Lu CS, Li S (2010) A taxonomy of green supply chain management capability among electronics-related manufacturing firms in Taiwan. J Environ Manage 91(5):1218–1226
8. Potrich L, Cortimiglia MN, de Medeiros JF (2019) A systematic literature review on firm-level proactive environmental management. J Environ Manage 243:273–286
9. Shukla GP, Adil GK (2021) A four-stage maturity model of green manufacturing orientation with an illustrative case study. Sustain Prod Consumption 26:971–987
10. Shukla GP, Adil GK (2021) A conceptual four-stage maturity model of a firm's green manufacturing technology alternatives and performance measures. J Manuf Technol Manag 32(7):1444–1465
11. Aboelmaged M (2018) The drivers of sustainable manufacturing practices in Egyptian SMEs and their impact on competitive capabilities: a PLS-SEM model. J Clean Prod 175:207–221
12. Hart SL (1995) A natural-resource-based view of the firm. Acad Manag Rev 20(4):986–1014

Corporate Climate Protection Measures to Improve the Carbon Footprint in Production: A Systematic Development of a Toolbox with Climate Protection Actions for the Reduction of Greenhouse Gas Emissions in Small and Medium-Sized Enterprises

Felix Budde[1][(✉)] and Holger Kohl[1,2]

[1] Fraunhofer Institute for Production Systems and Design Technology IPK, Pascalstraße 8-9,
10587 Berlin, Germany
`felix.budde@ipk.fraunhofer.de`
[2] Technische Universität Berlin, Pascalstraße 8-9, 10587 Berlin, Germany

Abstract. Individual climate protection measures are already being successfully implemented by many companies. But often, there is a lack of comprehensive information for selecting and implementing the right measures that fit the individual company's emissions profile. This paper presents the systematic development of a toolbox of climate protection actions for small and medium-sized enterprises in order to reduce greenhouse gas emissions. The development process followed a four-step methodological approach: conducting intensive research on climate protection measures in applied and practical literature, reducing and grouping the measures into their corresponding emission categories, structuring and categorizing the measures into specific types of areas, and providing detailed profiles and descriptions of each measure. The resulting list of climate protection actions follows underlying basic principles (i.e. improvement of the emission factor, reduction of emission-generating activities and behavioral changes) which allow an individual derivation of measures. In its final version, the toolbox provides small and medium-sized enterprises with a comprehensive and accessible resource for identifying and implementing effective climate protection actions according to their emissions profile.

Keywords: climate protection management · toolbox · climate protection measures · SMEs

1 Introduction

The latest Intergovernmental Panel on Climate Change (IPCC) synthesis report highlights that without drastic reductions in climate-damaging greenhouse gas emissions before the end of this decade, the 1.5-degree global warming target will already be

© The Author(s) 2025
H. Kohl et al. (Eds.): GCSM 2023, LNME, pp. 480–487, 2025.
https://doi.org/10.1007/978-3-031-77429-4_53

exceeded in the 2030s. The report indicates that climate change is progressing faster and the consequences are more severe than previously assumed [1]. Human-induced warming has been increasing at an unprecedented rate, caused by greenhouse gas emissions being at an all-time high over the last decade [2].

Together, world´s nations must cut greenhouse gas emissions 60% by 2035 to limit warming to 1.5 degrees Celsius over preindustrial levels [3]. For businesses, this means above all that they must save greenhouse-gases efficiently and sensibly within the same timeframe. This cannot be achieved without small and medium-sized enterprises (SMEs) as the cornerstone of the European economy. In 2021, about 22.8 million SMEs were active in the EU-27 and these SMEs accounted for 99.8% of all enterprises in the non-financial business sector. Subsequently, SMEs are critical to the success of the green transition in the EU since SMEs are currently responsible for around 60% of all greenhouse gas emissions by enterprises [4]. Most SMEs are being prevented to take action on climate change by their lack of skills and knowledge regarding their levers for a reduction in emissions [5].

2 Climate Management to Improve the Carbon Footprint in Production

The German funded project "KliMaWirtschaft - Nationwide Climate Protection Management for the Economy" focuses on a company's ecological impact on the climate as part of the corporate sustainability performance. Over 250 German SMEs are being supported in implementing a corporate climate management system to proactively engage in climate protection measures and thus reducing greenhouse gas emissions to minimize their environmental impact.

Climate management can be defined as the "identification, recording, active reduction and avoidance of relevant emission sources and emissions at the site as well as from upstream and downstream activities along the value chain [...]" [6]. This includes the preparation of a greenhouse gas footprint, the development of climate protection targets and measures to achieve the targets, and reporting and communication on these [7].

The basic structure of corporate climate management can therefore be divided into four steps: (1) assessment of the carbon footprint, (2) development of corporate climate protection goals, (3) implementation of actions to reduce the carbon footprint and (4) reporting and communication internally and externally [7]. Step 2 and 3 form the climate strategy. Participating SMEs are supported with a three-part interactive workshop including phases for application in the companies.

Narrowing down on step (3), a toolbox in form of a web application with resources for corporate climate protection management was developed aiming to strengthen the action competence of companies for their own climate protection management. The toolbox allows interactive use and a multi-perspective approach. Multi-criterial filters were developed to provide companies with tailored access to measures (e.g., sector-, process- or topic-oriented structuring of content). The knowledge base is continuously expanded through the experience of companies (good practices). In this way, companies that are not directly involved in the project are also empowered to improve their operational climate protection. This builds up planning, decision-making and action competence, promotes the implementation of measures and strengthens the effective-ness of the measures beyond the project.

This paper describes the systematic development as well as the resulting toolbox.

3   Methodology of the Toolbox Development

The development of the climate protection toolbox followed several systematic steps:

1. Extensive research of climate protection measures
2. Reduction through clustering
3. Structuring and typologization of measures
4. Detailing and description of measures
5. Correction of wording
6. Technical implementation as a web application

Firstly, extensive research of climate protection measures was conducted. The resulting catalogue included measures originating from scientific and practice-oriented literature as well as from companies and network partners.

Existing guidelines on corporate climate and sustainability management provided a first overview on climate protection measures [8] as well as best practice [9, 10]. The screening of individual enterprise sustainability reports, e.g. from German companies introduced further measures that were already applied in practice [11, 12]. Studies on societal level [13–15], allowed for an overarching view on necessary climate protection measures and actions. Most measures were found in best practice collections from the following sources: extensive catalogues on energy saving and efficiency measures with best practices from German enterprises [16–21], a checklist for energy efficiency measures for the retail industry [22], best practices from local enterprises from Berlin [23] as well as guides and tips for energy saving and climate protection for private households [24]. Additionally, a large number of measures were collected from over 250 participating companies in the project in form of good practices and plans for future application during several workshops.

Secondly, the resulting list of measures collected was then shortened by clustering the measures, eliminating doubles and summarizing similar measures into equally detailed groups. This helped to reduce the original list of approx. 180 measures to around 80 measures.

For step 3, the remaining measures were subsequently structured and attributed by the following criteria: emission category according to the GHG Protocol [25], type of measure (i.e. actions affecting electricity, heating, vehicle fleet, procurement, employee involvement, production processes and more), relevance to existing building stock (e.g. does the measure apply to rented or owned premises) as well as applicability in the sectors of manufacturing, services, trade. These criteria subsequently allow for setting multi-criterial filters to provide companies with tailored access to the measures in the toolbox.

The result is a structured list of categorized measures. In step 4, descriptions were added to each measure. The most important as well as most widely applicable measures were detailed in the form of fact sheets.

In step 5, the final fact sheets were revised linguistically in order to improve readability, user-friendliness as well as practical relevance for SMEs.

Eventually, the climate protection toolbox consisting of the catalogue of measures including fact sheets and best practice examples will be made available as an nation-wide online database in step 6. The resulting toolbox and measures catalogue will be initially used and evaluated by the participating pilot companies.

4   Results

4.1   Toolbox of Climate Protection Actions for the Reduction of Greenhouse Gas Emissions in Medium-Sized Enterprises

The resulting toolbox of climate protection measures for the support of SMEs in reducing their greenhouse gas emissions contains a catalogue of over 80 measures. All measures are listed with a short description. Just under 40 percent of the measures are further detailed with fact sheets including further information. Overall, most of the measures are primarily aimed at Scope 1 and 2 (e.g. electricity, heating, vehicles, production processes, product or recycling), even though most Scope 3 emission categories (e.g. procurement, employee integration, logistics, business travel) are covered as well. This could stem from the current focus on energy efficiency and the availability of practical short-term options for energy saving. The measures are mostly applicable in all three industry sections, manufacturing, services as well as trade. Manufacturing enterprises are able to find general measures as well as measures specific to production.

4.2   Fact Sheets for Measure Description

Widely applicable measures are accompanied with fact sheets. The fact sheets include short and focused description of all relevant information about a specific climate protection action to support the operational decision-making process. The fact sheets are typically between one and two pages long and include information as ID, measure name, short description, impact on GHG Protocol emission category, quantitative or at most qualitative information on potential greenhouse gas emission reductions, practical steps for action implementation, associated efforts and costs, hurdles and possible set-backs, further background information as well as links to federal subsidizing. Some measures of high importance include a best practice example too.

4.3   Basic Underlying Principles

During the development of the toolbox it came clear that all measures followed similar ideas and approaches. Especially during the structuring and typologization of the measures, basic underlying principles which allow categorizing most corporate climateactions were found. These principles can be used for descriptive categorization of existing measures. Alternatively, the found principles provide a systematic way of creatively deriving new climate protection measures.

The basic underlying principles are shown in Fig. 1. These basic principles define measures as with either direct or indirect impact on reducing greenhouse gas emissions. Actions with direct impact effect emissions in one of two ways following the basic calculation method for emissions (activity data x emission factor = emissions). One option is to improve the emission factor of an activity via substitution. This can be realized through the use of alternative technology mostly, i.e. electrification of processes that required fossil fuels before, building up own production of renewable energy, choosing alternative means of transport. The second option of directly impacting measures is the reduction of emissions-generating activities. This can be achieved again in two ways: by reducing the consumption of energy by an enterprise, i.e. through in-creased maintenance and cleaning of heating equipment and production machinery, dimensioning of the companies' infrastructure according to given needs and realistic requirements instead of over-dimensioning or by avoiding unnecessary activities like flights, business trips or non-regional procurement. Measures with an indirect effect are mostly directed to behavioral changes and fall into the four categories of communication, training of staff, implementation of guidelines, incentives and monitoring. Measures falling into these categories build up a holistic understanding in the organization and support changes with direct impact on emission reductions.

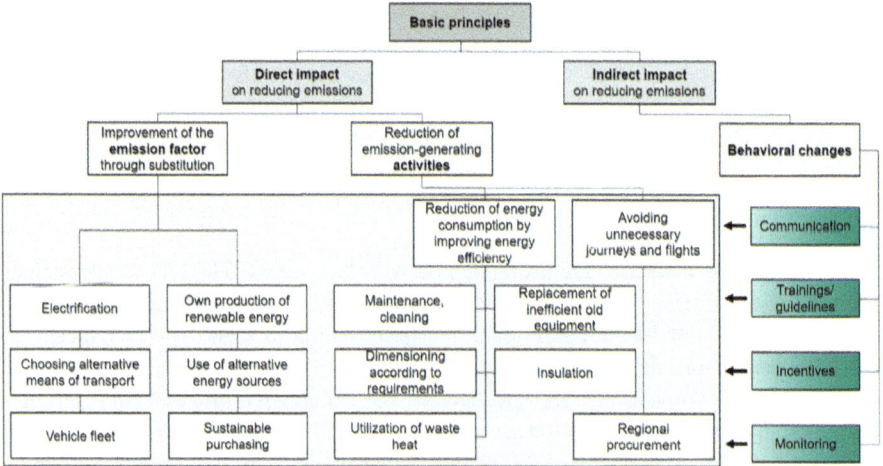

Fig. 1. Basic underlying principles of climate protection measure development.

5   Discussion of Application in Enterprise Practice

This section will explore feedback and experiences of pilot companies that have worked with the toolbox, discuss its wider application potential, and highlight the plans for publication and public use beyond the project's completion.

The participating companies expressed their satisfaction with the toolbox, particularly with the slimmed-down version of the action list provided in the workshops. Companies with several years of experience in corporate climate management found inspiration in the toolbox, discovering measures they hadn´t previously considered. As a result, actions from the toolbox were integrated into the company's action programs, reinforcing its practical value.

To assess the toolbox's efficacy, it is important to conduct a critical review of the actions included to ensure no crucial actions are not included in the toolbox. Through the workshops and documentation, it became evident that the toolbox already covers a broad spectrum of actions. Few new actions were suggested by companies during the workshops, indicating that the toolbox has been comprehensive in its coverage.

Additionally, the toolbox was developed with a holistic approach and integrates various industry-oriented measures make it applicable to all types of companies. The sources used for the catalogue development included a wide range of measure and best-practices from different industry sections as well as heterogenous company sizes. Furthermore, measures with specific use for special industry application were pooled into general yet specific groups of measures.

6 Conclusion

6.1 Summary

This paper presents the development of toolbox of climate protection actions for small and medium-sized enterprises in order to reduce greenhouse gas emissions. The development process followed the methodological approach of conducting intensive research on various climate protection measures, reducing and grouping the measures into their corresponding emission categories, structuring and categorizing the measures into specific types of areas, and providing detailed profiles and descriptions of each measure. The resulting list of climate protection actions follows several underlying basic principles which allow an individual derivation of measures. In its final version, the toolbox provides small and medium-sized enterprises with a comprehensive and accessible resource for identifying and implementing effective climate protection actions according to their emissions profile.

6.2 Outlook

The application of the toolbox in enterprise practice has yielded positive feedback and experiences from pilot companies. With plans for wider application, publication, and public use, the toolbox is posed to contribute to climate action beyond the project's completion.

At the point of submission, wider application is still upcoming, but the initial results were promising. The toolbox is set to be published on the project website, making it accessible for public use, even by non-participating companies. Moreover, the platform will remain online after the project's conclusion, allowing enterprises and multiplicators to continue utilizing it. The effectiveness of the toolbox is projected with an 8%

implementation rate, indicating the share of users of the platform that might implement climate measures as a result to their use of the climate protection toolbox. It has to be noted, that the effectiveness additionally depends on external factors, mainly SME motivation to engage in climate actions and their internal resources and capacities.

To further improve the toolbox, a survey will be conducted to gather feedback from users. Additionally, companies will be required to document the actions taken and subsequent emission reductions as part of the KliMaWirtschaft project.

Keeping the use after the project's completion in mind, the continuation of the toolbox is crucial. The catalogue of measures allows for updates in the future. New measures could come from the pool of companies participating in the project as well as from newly released guidelines and publications covering technological advances. Systematic updates of the toolbox need to guarantee up to date information in the fact sheets, e.g. on emission values, efficiency of actions and funding opportunities.

Additionally, the measures included in the toolbox relate significantly to Germany. In the further development, the toolbox should be expanded with practices from enterprises literature from other countries. Opening up to practices from other countries will lead to a gain of knowledge and best practices examples and the geographical expansion of its application.

Finally, an expansion of the toolbox's platform with an analytics module for direct identification of potential for emission reduction and derivation of measures from greenhouse gas data is possible.

Acknowledgments. The joint project between the Fraunhofer Institute for Production Systems and Design Technology IPK and the Bundesverband Der Mittelstand. BVMW is funded by the German Federal Ministry of Economics and Climate Protection (BMWK) as part of the funding call for innovative climate protection projects of the National Climate Protection Initiative (NKI) under the code 67KF0166B.

References

1. Intergovernmental Panel on Climate Change IPCC (2023) Ar6 synthesis report: headline statements. https://www.ipcc.ch/report/ar6/syr/resources/spm-headline-statements. Accessed 21 Mar 2023
2. Forster PM, Smith CJ, Walsh T et al (2023) Indicators of global climate change 2022: annual update of large-scale indicators of the state of the climate system and human influence. Earth Syst Sci Data 15:2295–2327. https://doi.org/10.5194/essd-15-2295-2023
3. Miller E (2023) Time is running out to Curb Climate Change, IPCC Report says. U.N. panel of scientists say limiting global warming requires a massive and rapid shift in the world´s energy supply
4. European Commission (2022) Annual report on European SMEs 2021/2022: SMEs and environmental sustainability, SME performance review 2021/2022, Luxembourg
5. SME Climate Hub (2023) Small business climate action: barriers & bridges
6. Kube M, Rhiemeier J-M, Stern F et al. (2016) Unternehmerisches Klima management entlang der Wertschöpfungskette - eine Sammlung guter Praxis
7. Budde F, Gellert B, Orth R et al (2023) Klimamanagement zur Verbesserung der CO_2-Bilanz in der Produktion: Systematische Analyse von Emissionsmin-derungspotenzialen und Ableitung

von Reduktionsmaßnahmen für Industrieun-ternehmen. Zeitschrift für Wirtschaftlichen Fabrikbetrieb ZWF 118:40–44

8. UN Global Compact Netzwerk Deutschland (2022) Einführung Klimamanage-ment: Schritt für Schritt zu einem effektiven Klimamanagement in Unternehmen

9. Galeitzke M (2021) Leitfaden Nachhaltigkeitsbenchmarking, Berlin

10. WWF, CDP (2016) Unternehmerisches Klimamanagement entlang der Wert-schöpfungskette: eine Sammlung guter Praxis

11. Gelsenwasser AG (2021) Nachhaltigkeitsbericht 2021

12. Evangelische Jugendbildungsstätte Neckarzimmern (2018) Umwelterklärung 2018

13. McKinsey & Company (2009) Pathways to a low-carbon economy: version 2 of the global greenhouse gas abatement cost curve

14. McKinsey & Company (2021) Net-Zero Deutschland: Chancen und Herausfor-derungen auf dem Weg zur Klimaneutralität bis 2045

15. Prognos AG (2021) Klimaneutrales Deutschland - In drei Schritten zu null Treibhausgasen bis 2050 über ein Zwischenziel von -65 % im Jahr 2030 als Teil des EU-Green-Deals

16. BMWi (2021) Energieeffizienz in Unternehmen: Das rechnet sich: Mehr aus Energie machen

17. VEA-Initiative Klimafreundlicher Mittelstand (2023) Schnell realisierbare Maß-nahmen zur Energieeffizienz und Erdgassubstitution, Version 9.0

18. Initiative Energieeffizienz- und Klimaschutz-Netzwerke (2022) Liste für Kurz-fristmaßnahmen für Energieeinsparung und Energiesubstitution in Unternehmen

19. Deutsche Energie-Agentur GmbH (2015) Energieeffizienz in kleinen und mitt-leren Unternehmen: Energiekosten senken. Wettbewerbsvorteile sichern

20. Klimaschutz Unternehmen (2022) Vordenken: Innovative Anregungen für Un-ternehmen. https://www.klimaschutz-unternehmen.de/erfolgsrezepte/. Accessed 26 Jul 2022

21. Stiftung KlimaWirtschaft, Better Earth (2022) Von Haltung zu Handlung: Was Unternehmen im Hier und Jetzt für die Dekarbonisierung tun können

22. Klimaschutzoffensive Handelsverband Deutschland (HDE) e. V. (2019) Einfach Energies-paren: Arbeitsbuch für Kaufleute

23. Forum CEOs FOR BERLIN (2021) Gemeinsam für eine klimaneutrale Metro-polregion

24. CO₂Online (2023) Energiesparen & Klimaschutz zuhause. https://www.co2on-line.de/. Accessed 17 Mar 2023

25. The Greenhouse Gas Protocol Initiative (2004) The greenhouse gas protocol: a corporate accounting and reporting standard, revised edn.

Data and Simulation

Optimizing GHG-Emissions in Milling by Integrating Electricity Mix Data into Manufacturing Parameter Decisions

Sebastian Felix Karnapp⬤, Magnus von Elling$^{(\boxtimes)}$ ⬤, Erkut Sarikaya⬤, Astrid Weyand⬤, and Matthias Weigold⬤

Department of Mechanical Engineering, Institute of Production Management, Technology and Machine Tools (PTW), Technical University of Darmstadt, Otto-Berndt-Str. 2, 64283 Darmstadt, Germany
m.vonelling@ptw.tu-darmstadt.de

Abstract. Industry is responsible for approximately 25% of global greenhouse gas (GHG) emissions, contributing substantially to climate change. Sustainable manufacturing has therefore become a significant topic of discussion, resulting in a better understanding of key GHG emitters in milling processes. This paper investigates the influence of different compositions of the German electricity mix on the CO_2-eq. Emissions of a milled steel disc defined by the Product Carbon Footprint (PCF). Using more wind and solar energy reduces CO_2-eq. Emissions significantly in comparison to an energy mix almost without green energy types. The cradle-to-gate PCF of the investigated steel workpiece ranges from 8.44 to 11.91 kg CO_2-eq., corresponding to a 41.18% increase caused solely by alternations in the energy mix composition. Feed rate was selected as the parameter for the optimization of the PCF, suggesting maximizing the feed rate to shorten production times. Emissions directly attributable to the milling process can be reduced by 11.59% and 16.93%, respectively in the two scenarios. Expected increase in cost and varying surface quality due to increased tool wear were not part of this investigation. This approach demonstrates the potential for environmentally sustainable manufacturing strategies by integrating electricity mix data into manufacturing parameter decisions.

Keywords: Sustainable machining · CO_2 emission · Optimization

1 Introduction and Motivation

With its substantial energy demand, the industrial sector holds significant potential to mitigate GHG emissions as the second largest emitter worldwide. Industry alone accounts for approximately 25% of total global GHG emissions [1]. Among industrial activities, the metal production and processing sector stands out as the second largest consumer of electrical energy in Germany. With 23.4% of the total electrical energy demand, the sector is one of the main contributors to industrial GHG emissions [2]. In order to meet the ambitious targets, set by the Federal Government of Germany, including achieving

H. Kohl et al. (Eds.): GCSM 2023, LNME, pp. 491–499, 2025.
https://doi.org/10.1007/978-3-031-77429-4_54

climate neutrality by 2045 and reducing emissions by 65% until 2030 compared to 1990, the discrete manufacturing sector must continue to intensify its efforts to minimize GHG emissions. The first step towards reducing and optimizing GHG emissions involves analyzing and understanding them. One tool commonly used for this purpose is the PCF. It serves to determine the specific GHG emissions associated with the entire lifecycle of a particular product, therefore also known as life cycle assessment (LCA). PCFs are a specialized form of LCA that focus solely on the global warming impact category, using global warming potentials (GWP) as a basis. The result of a PCF is typically expressed in equivalent CO_2 emissions (CO_2-eq.), allowing the standardized comparison of the climate change impact of different products [3]. Due to the major energy consumption of machine tools, research has focused on their energy efficiency in recent years [4]. Electrical energy savings of 40% and more are achieved by modernizing cooling systems and pumps of machine tools or by changing from two-point-controlled to speed-controlled components [5–7]. Intelligent standby management systems show potential savings in electrical energy of 20% and more [8]. Besides that, multi-parameter optimization models have shown that power consumption can be reduced by around 6.5% through optimization of process parameters like feed rate, cutting speed and depth [9]. Higher feed rates or cutting speeds result in faster machining times and therefore reduced energy consumption but reduce the lifetime of tooling at the same time. This indicates the existence of an emission-optimal set of process parameters [10, 11]. Increased efforts to integrate renewable energy sources into the electricity mix have reduced the carbon footprint of generated electrical power on average, however the share of renewable energy fluctuates significantly over time. This paper investigates the influence of fluctuating shares of renewable energy in the German electricity mix on the PCF of a milled workpiece. By analyzing the hourly direct electricity emission intensity data for the year 2022, this research aims to provide insights into the environmental implications of different energy mix compositions. Furthermore, an experimental investigation is conducted to analyze the influence of the feed rate on the PCF and identify optimized process parameters that effectively minimize GHG emissions in the process.

2   Methodology

2.1   Experimental Setup and Data Collection

The basis for the calculation of a PCF of the investigated milled steel disc (X14CrMoS17) is a thorough understanding of the process and a precise measurement of relevant operational data. The process consists of three individual NC programs, each representing a complex 5 axis-machine operation with clamping operations in between. Three tools are utilized, referred to as T_1, T_2 and T_3 as can be seen in Fig. 1.

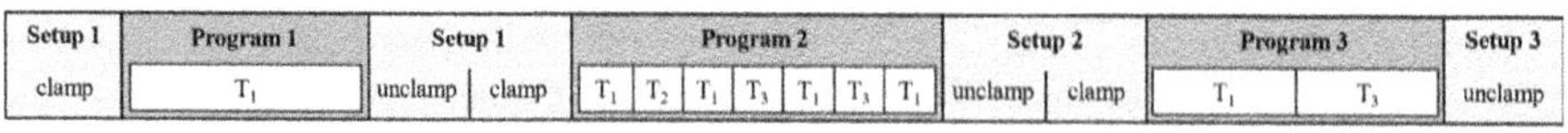

Setup 1	Program 1	Setup 1		Program 2							Setup 2	Program 3		Setup 3
clamp	T_1	unclamp	clamp	T_1	T_2	T_1	T_3	T_1	T_3	T_1	unclamp	clamp	T_1	T_3	unclamp

Fig. 1. Schematic representation of the manufacturing process.

Input and output factors necessary to operate the machine tool like electrical power, compressed air as well as necessary operating materials must be measured precisely. Measurement and storage of the active power data is realized through custom-built measuring devices. These include a voltage analyzer that allows voltage measurements at the main connector with three-phase alternating current. In combination with three current transformer clamps on the three phases, it is possible to determine the active power consumption of the machine tool at a frequency of 1 Hz. For the compressed air measurement, an external flow sensor is installed. Data acquisition and storage from the flow sensor is realized through a self-developed LabVIEW program. Furthermore, the machine tool is equipped with an edge computer that allows recording and processing of up to 100 Programmable Logic Controller (PLC) variables. In this case, the PLC data is used to acquire and store the power, voltage and speed data from the spindle motor as well as the feed drives. Process information such as tool changes and cutting times are derived from the acquired spindle speed. A spindle speed greater than 0min^{-1} indicates cutting time and cutting fluid usage. All collected datasets have been processed and harmonized to a common time zone and frequency of 1 Hz. Lubricating oil, grease and filter fleece consumption are derived from the machine tool manual.

2.2 Carbon Accounting Methodology

Several methods to determine a product specific carbon footprint exist [3, 12]. Due to its broad adoption in science and industry, ISO 14040/14044 is used as the accounting standard in this study. The scope of the standard includes a full LCA. For the purpose of a PCF solely the indicator climate change expressed in CO_2-eq. is relevant. A crucial step in carbon accounting is the definition of the system boundary. It separates inputs, which are necessary for the milling process, from the outputs, which are generated in the form of waste. It is commonly separated between cradle-to-grave, cradle-to-gate and a gate-to-gate approach. For this study, a cradle-to-gate and a gate-to-gate approach are investigated. Figure 2 gives an overview of both relevant system boundaries and the corresponding input and output factors. Emissions from material sourcing, transport and production of the input factors are neglected in the gate-to-gate approach.

2.3 Carbon Emission Factors

Calculating the PCF of the investigated workpiece requires a variety of carbon emission factors (CEF). For the purpose of this study, CEFs are obtained from the LCA database ecoinvent 3.8 according to the ReCiPe-2016 (H) allocation method [13]. CEFs for the electrical energy, tungsten carbide tools and the cutting fluid were not available in the database and therefore derived from other data sources or own calculations. The fluctuating share of renewable energies in the German electricity mix has a direct impact on its CEF. In the year 2022 the hourly emissions of one kWh ranged from $CEF_{kWh,min}$ of $165.70 \text{ g CO}_2 \text{ eq./kWh}$ to $CEF_{kWh,max}$ of $757.44 \text{ g CO}_2 \text{ eq./kWh}$ with a median CEF_{kWh} of $484.57 \text{ g CO}_2\text{eq./kWh}$ [14].

The milling process requires three different tungsten carbide cutting tools. A two-step production process is assumed that involves sintering and grinding of tungsten carbide rods [15]. The material composition of the rods is assumed to be 92% tungsten and 8%

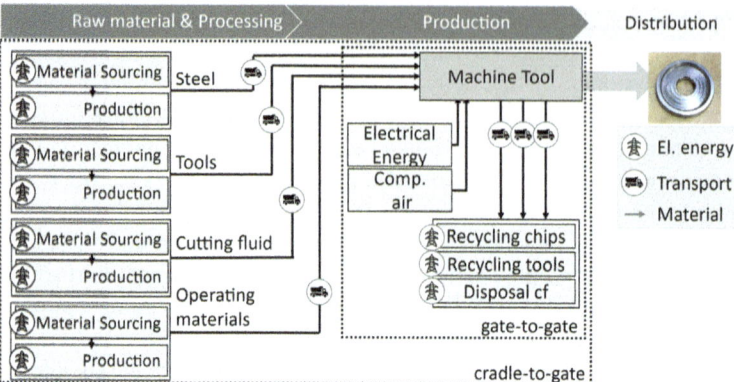

Fig. 2. System boundaries and flow of material.

cobalt [16]. The CEFs for sintering and grinding the tungsten carbide are approximated using the corresponding CEFs for the sintering and grinding of steel. Tool recycling is considered with $10\,kg\,CO_2eq./kg$ [17], resulting in total carbon emissions (CE) of 3.71, 3.65 and 0.98 kg CO_2eq. for the three tools. The proportional allocation to individual workpieces is based on Eq. 1 where the tool life T is approximated using the Taylor model [18]. The cutting fluid is a mineral oil-based cooling lubricant with a CEF of $1.34\,kg\,CO_2eq./kWh$ [19].

$$CE_{tool} = \sum_{i=1}^{3} \frac{t_{c,i}}{T_i} CEF_{tool,i} \tag{1}$$

3   Results

The cradle-to-gate PCF of the steel disc is composed of the emissions of the individual CE occurring before and during production. These include material sourcing, production, electrical energy, transport, wear and recycling or disposal of the corresponding input factor. Since the CE resulting from the electrical energy demand are considered in this paper as time dependent, the PCF is determined according to Eq. 2.

$$PCF_{c2g}(t) = CE_{kWh}(t) + CE_{air} + CE_{cf} + CE_{tools} + CE_{chips} + CE_{lub} + CE_{steel} \tag{2}$$

3.1   Product Carbon Footprint

As mentioned before, the carbon emissions resulting from electrical power vary significantly over time. Therefore, the time of production is considered to determine the PCF of each steel disc. The cradle-to-gate PCF for the minimum and maximum daily electricity CEF for 2022 are displayed in Fig. 3. Other carbon emission factors are kept constant.

The minimum cradle-to-gate PCF equals 8.44 kg CO_2eq., while the maximum PCF equals 11.91 kg CO_2eq. per steel disc. This corresponds to a 41.18% increase in the

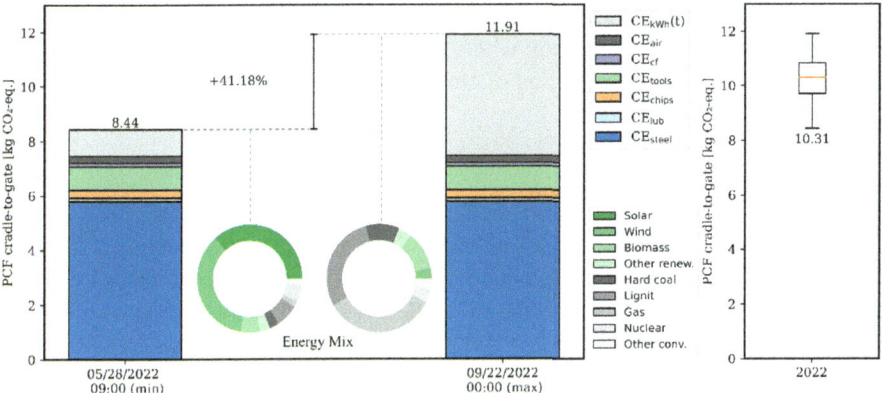

Fig. 3. Electricity emission intensity dependent cradle-to-gate PCF.

cradle-to-gate PCF, solely caused by a varying electricity emission intensity at the time of production. The median PCF equals 10.31 kg CO_2-eq. per steel disc. Steel represents the biggest emission factor with 68.51% in the minimum scenario and 48.53% in the maximum scenario. Electricity demand is responsible for 11.62 and 37.40% respectively, followed by the tools with 10.21 and 7.23%. These three factors combined account for more than 90% of the PCF. Besides the cradle-to-gate PCF a closer look at the gate-to-gate PCF is given in Fig. 4. This approach solely addresses emissions that are directly caused by the manufacturing process itself. Emissions from upstream and downstream activities of the steel disk are neglected. The results show a PCF of 2.05 kg CO_2eq. emissions per disc in the minimum case and 5.52 kg CO_2eq. in the maximum case. Emissions directly accountable to the machining process therefore vary by 169.85%. Median gate-to-gate emission equal 3.91 kg CO_2eq. per disc.

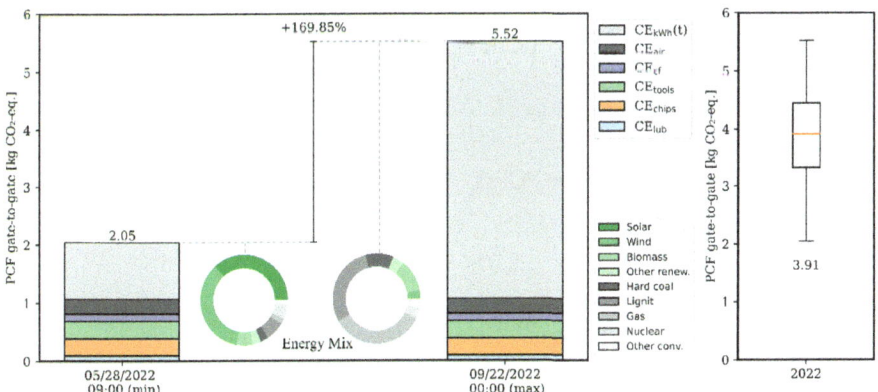

Fig. 4. Electricity emission intensity dependent gate-to-gate PCF.

In the gate-to-gate approach, the electricity demand is responsible for 47.93% in the minimum scenario and 80.70% in the maximum scenario. Tools account for 14.52

and 5.38%. These three factors combined account for more than 90% of the PCF. The chip recycling becomes noteworthy with a 14.37 and 5.33% share. The results confirm the importance of timely accounting of emissions resulting from electricity demand in machine tools.

3.2 Controlling Feed Conditions

The results of the gate-to-gate PCF reveal that the electrical energy demand is the main source of GHG emissions during production followed by emissions caused by tool wear. As mentioned in the introduction, several authors have shown a significant correlation between the feed rate and the energy efficiency of machine tools. Therefore, a simplified model of the machining process is derived, which only utilizes one single tool and neglects tool changes during the machining process. The end mill (T1) is chosen since the tool holds the largest share of cutting time in the original process. Initially, the process runs with a feed rate of $v_f = 1159 \frac{mm}{min}$ and a total cutting time of 1281 s. The gate-to-gate emissions for power consumption, compressed air, cutting fluid and lubricating oil are now distributed proportionally over the cutting time, resulting in emission per second factors. Emissions resulting from setup times are thus averaged over the cutting time. These emissions are combined and indexed with avg in Eqs. 3. Emissions resulting from the chips are independent of the cutting time and therefore added as a constant factor. Tool wear is calculated accordingly to the emission minimal feed rate. Equations 4, 5, 6 and 7 formulate the problem as a minimization problem with the objective of minimizing the gate-to-gate PCF. The decision variable feed rate is subject to the boundary condition that it cannot exceed $1500 \frac{mm}{min}$. A full production day (24 h) is investigated as displayed in Fig. 5. The results show an 11.59% decrease in carbon emissions for the minimum scenario and a 16.93% decrease in the maximum scenario. This is achieved by maximizing the feed rate to $1500 \frac{mm}{min}$, minimizing the cutting time and the electricity demand per workpiece.

$$PCF_{g2g,i}(t, v_f) = t_c(v_f) \cdot \left(CE_{kWh,avg}(t) + \frac{CE_{tool,g2g}}{T_{tool}(v_f)} + CE_{other, avg} \right)$$
$$+ CE_{chips} \tag{3}$$

$$\text{Minimize } PCF_{g2g} = \sum_i PCF_{g2g,i}(t, v_f) \tag{4}$$

$$0 \leq t \leq 24 \tag{5}$$

$$450 \frac{mm}{min} \leq v_f \leq 1500 \frac{mm}{min} \tag{6}$$

$$t_0 = 00 : 00 \tag{7}$$

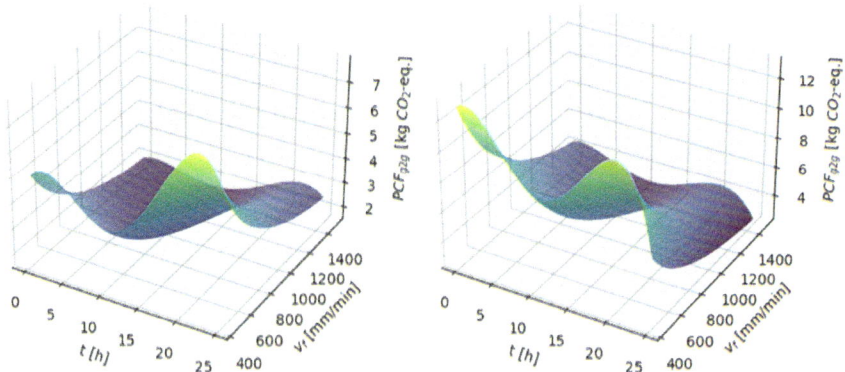

Fig. 5. Visualization of $PCF_{g2g}(t, v_f)$ for scenario 1 (left) and 2 (right)

4 Conclusion and Outlook

In conclusion, this study underscores the importance of considering the emission intensity of electrical energy on the PCF during milling at the time of production. Both the cradle-to-gate and the gate-to-gate PCF show significant variation due to a varying electricity mix. Emissions directly connected to the machining process (gate-to-gate) show an even greater dependency on the emission intensity of electricity at the time of production. To ensure the comparability of these results, it is important to use comparable CEFs. For the scope of this study, the CEF have been derived from the ecoinvent 3.8 LCA database. Additionally, it is worth noting that, in the context of this study, compressed air was measured using a flow sensor and evaluated using a CEF. The findings of this research highlight the significant impact of the electricity mix on the PCF. This suggests that directly measuring the electrical energy demand of the compressor during production, while considering the corresponding electricity emission intensity, would result in an even greater variation in PCFs. Currently there is no database or data exchange available that would allow the exchange of product specific CEF based on the time of production of the tools or the cutting fluid. CEFs for the tools have therefore been derived from a two-step production process. In reality, a variety of production processes exist, which can vary between manufacturers and even between tools. Increased accuracy of the PCF can be achieved by validating the process model with the tool manufacturers. Besides that, this study shows the possibility of reducing GHG emissions resulting from milling processes by adjusting the feed rate of the machining center. In accordance with previous studies investigating the energy efficiency of machine tools, increased feed rates reduce the carbon footprint by reducing the overall electrical energy demand for producing a workpiece.

Looking into the future, the trajectory of the electricity mix indicates a continued shift towards greener sources. It is important to note that while GHG emissions associated with electrical power demand cannot be entirely eliminated, due to the emissions generated during the production and recycling of renewable energy sources such as wind, solar, and hydro, they are expected to gradually diminish over the lifespan of these technologies. Consequently, the overall GHG footprint of electrical energy and the products derived from it will be significantly reduced. This reduction will have a profound impact on the CEFs used for calculating the carbon footprint of tools, cutting fluids, and lubricants. Therefore, it is crucial for future studies to examine the implications of this evolving trend on the PCF and consider the necessary adjustments to the optimization model.

References

1. Ritchie H, Roser M (2020) CO_2 and greenhouse gas emissions. Our world in data
2. Federal Statistical Office. https://www.destatis.de/DE/Presse/Pressemitteilungen/2022/12/PD22_530_435.html. Last accessed 23 Jun 2023
3. Deutsches Institut für Normung e.V. DIN EN ISO 14044:2021-02 (2021) Umweltmanagement-Ökobilanz-Anforderungen und Anleitungen. Deutsche Fassung EN ISO 14044:2006 +A1:2018 + A2:2020. Beuth Verlag GmbH, Berlin
4. Denkena B, Abele E, Brecher C, Dittrich M, Kara S, Mori M (2020) Energy efficient machine tools. CIRP Ann Manuf Technol 69:646–667
5. Helfert M, Kohne T, Petruschke L, Burkhardt M, Abele E (2018) Energieeffiziente Kühlung durch Einsatz innovativer Aggregate. Werkstatt + Betrieb 151(9):218–220
6. Brecher C, Triebs J, Heyers C, Jasper D (2012) Effizienzsteigerung von Werkzeugmaschinen durch Optimierung der Technologien zum Komponentenbetrieb—EWOTeK. Apprimus-Verlag, Aachen
7. Campitelli A, Cristóbal J, Fischer J, Becker B, Schebek L (2019) Resource efficiency analysis of lubricating strategies for machining processes using life cycle assessment methodology. J Clean Prod 222:464–475
8. Abele E, Sielaff T, Beck M (2013) Maximierung der Energieeffizienz spanender Werkzeugmaschinen: Schlussbericht zum Projekt Maxiem. Darmstadt
9. Kant G, Kuldip S (2014) Prediction and optimization of machining parameters for minimizing power consumption and surface roughness in machining. J Clean Prod 83:151–164
10. Zhang H, Deng Z, Fu Y, Lv L, Yan C (2017) A process parameters optimization method of multi-pass dry milling for high efficiency, low energy and low carbon emissions. J Clean Prod 148:174–184
11. Zhou G, Lu Q, Xiao Z, Zhou C (2019) Cutting parameter optimization for machining operations considering carbon emissions. J Clean Prod 208:937–950
12. World Resource Institute (WRI), World business council for sustainable development (WBSC). https://ghgprotocol.org/product-standard. Last accessed 10 Jul 2023
13. Ecoinvent, ecoinvent 3.8. https://ecoinvent.org/the-ecoinvent-database/. Last accessed 10 Jul 2023
14. Electricity Maps. https://www.electricitymaps.com/data-portal/germany. Last accessed 10 Jul 2023
15. Ceratizit. https://www.ceratizit.com/int/de/company/passion-for-cemented-carbide-/production.html. Last accessed 10 Jul 2023
16. Chemie.de. https://www.chemie.de/lexikon/Wolframcarbid.html. Last accessed 10 Jul 2023
17. Furberg A, Arvidsson R, Molander S (2019) Environmental life cycle assessment of cemented carbide (WC-Co) production. J Clean Prod 209:1126–1138
18. Taylor FW (1906) On the art of cutting metals: an address made at the opening of the annual meeting in New York. American Society of Mechanical Engineers
19. Technologieland Hessen. https://redaktion.hessen-agentur.de/publication/2021/3442_ArePron_Ressourceneffizienz-Produktion_2021_web.pdf. Last accessed 10 Jul 2023

Combined Evaluation of Digitalization and Sustainability Maturity of Small and Medium Sized Enterprises

Marc Münnich[1]([✉]) [ID], Annabell Schönfelder[1], Maximilian Stange[1] [ID],
Marian Süße[1] [ID], and Steffen Ihlenfeldt[1,2]

[1] Fraunhofer Institute for Machine Tools and Forming Technology IWU, Chemnitz, Germany
marc.muennich@iwu.fraunhofer.de
[2] Institute of Mechatronic Engineering, Technische Universität Dresden, Dresden, Germany

Abstract. Sustainability and digitalization are of high importance for companies to stay competitive and cope with market requirements from internal and external stakeholders. Especially the current changes in reporting standards and frameworks increase pressure to quickly adapt internal processes and acquire the relevant data. In this context, Industry 4.0 (I4.0) is discussed to enable companies to also foster sustainability. But many small and medium sized enterprises (SME) lack of experience and capacity and need support from experienced partners to develop suitable strategies and implement technologies. Appropriate support comprises general guidance through these domains and an evaluation of current and desired state of expertise to choose suitable actions. Maturity models (MM) are well-established to initiate and support the transformation of corporate structures. There are many models available that support digital and sustainable transformation, but almost no models are available that include both dimensions what results in decoupled developments in these domains. Hence, this contribution introduces a combined model for both domains and systematically describes the extension of an existing digitalization MM (DMM) that also considers rebound effects. The MM is tested within an industrial context use case. The results are evaluated and discussed for further development of the model and workflow.

Keywords: Digitalization · Sustainability · Maturity model

1 Challenges in Sustainability Consideration for SME

Man-made climate change is an undeniable fact and one of the main drivers of new environmental regulation for businesses. In addition, there are many more pressing environmental issues, such as deforestation and loss of biodiversity, that are forcing countries to impose tighter restrictions on companies. Therefore, the European Parliament and the Council of the European Union adopted the Corporate Sustainability Reporting Directive (CSRD) in November 2022. Consequently, larger businesses and increasingly more SMEs in the EU will have to implement or extend their sustainability reporting practices. Core requirements of the directive are formulated in specific principles. For

H. Kohl et al. (Eds.): GCSM 2023, LNME, pp. 500–508, 2025.
https://doi.org/10.1007/978-3-031-77429-4_55

example, 'double materiality' involves assessing the impact of business operations on sustainability issues, but also requires an assessment of sustainability risks to business success.

There are also indirect routes that demand for greater transparency and data availability from a sustainability perspective. The Sustainable Finance Disclosure Regulation (SFDR) introduces comprehensive reporting standards for financial market participants, which in turn raise the requirements for sustainable investments in industry and manufacturing. This is directly linked to the EU taxonomy that includes a set of criteria for aligning economic activities and investments with environmental sustainability requirements. All these regulatory standards constitute important building blocks of the European Green Deal. Not only at the European level, but also at the global level, the required sustainability competencies and capacities of companies will have to be increased due to the reporting standards of the International Sustainability Standards Board announced in 2021.

SMEs may struggle to comply with these new regulations because of their limited resources. Moreover, they operate in an increasingly challenging environment with a shortage of skilled labor, rising material prices and market competition. To address these issues, SMEs need to adopt digital tools and I4.0 solutions. But there is also growing interest in exploring how I4.0 can contribute to sustainability efforts by reducing emissions, minimizing material and energy consumption, and fostering a more environmentally conscious approach [1]. Despite the potential benefits, there a lack of knowledge on how to integrate sustainability in corporate strategic planning [2]. Additionally a significant number of companies do not perceive I4.0 and digitalization as opportunities to enhance sustainability [3].

2   Role of Cyber-Physical Approaches in Sustainable Manufacturing

When it comes to research in manufacturing, digitalization measures are often touted as automatically benefiting sustainability in companies. As cyber-physical production systems (CPPS) rely on the integration of information and communication technologies, sustainability assessment is directly linked to both and, in parallel, is part of an ongoing scientific discussion. As stated by Lange et al., the effect of digitalization on energy consumption on a more macroeconomic scale depends on four factors: direct effects (such as sector growth and its energy intensity), changes in energy efficiencies, economic growth, and changes in the sectoral structure [4]. Hence, the selection of appropriate system boundaries and cut-off criteria finally affects the evaluation of the advantageousness of CPPS from a sustainability point-of-view. This interrelation should therefore be considered together to avoid any negative impact of digitalization projects on sustainability. Rebound effects, that are characterized by partially offsetting efficiency improvement, are caused by increased spending of resources in other areas. Backfire effects describe a complete offset of the initial efficiency gains [5]. Hence, consideration is viable when planning and implementing digitalization measures.

Nevertheless, digitalization is a key component in environmental efforts. Methods such as corporate carbon footprinting or life cycle assessment (LCA) require a comprehensive database to account for all relevant environmental impacts. A specific example is the assessment of Scope 3 emissions, that occur in a company's value chain, including activities such as raw material extraction, transportation, product use and disposal. The complexity of value chains, involving multiple suppliers and stakeholders, complicates data collection on emissions. Varying reporting practices and limited transparency also make it difficult to obtain accurate information. In addition, many companies lack the necessary systems to track and report emissions data, and external entities may be reluctant to disclose relevant information.

Moreover, manufacturing companies find themselves in increasingly complex environments as they turn into prosumers in a more volatile energy grid. Especially when energy supply is composed by flexible tariffs and local renewable facilities optimization of production and facilities becomes a challenging task, which requires reliable data, fitted algorithms and management systems. As described by Salonitis and Ball [6] and also adopted by Schulz et al. [7], the classical goal triad in manufacturing is expanded by flexibility and therefore contributes to sustainability as an overarching goal. As an extension, they present an IoT-based concept for energy-flexible factories that incorporates simulation for the control of factory operations. Thus, improved energy and sustainability performance require appropriate input data and computational power.

These examples demonstrate the strong connection of digitalization and sustainability. To also address a stronger connection in corporate strategic planning, a combined maturity model (CMM) is proposed that integrates sustainability and digitalization equally. This model aims to raise awareness among companies, highlighting the potential benefits of sustainability and enabling a holistic approach that considers both technological advancements and sustainable practices. To reduce implementation barriers for sustainability aspects in SME, it is essential to emphasize benefits for involved companies [8].

3   Extended View on Sustainability in Existing Digitalization Maturity Models (DMM)

As elaborated, MM are perceived to support corporate strategic planning. According to Ref. [9], a MM is a supportive tool for transformation processes. Various domain-specific models have been developed, e.g., for digitalization [10], corporate sustainability strategies [11], or for sustainability in remanufacturing [12]. The authors extend an existing DMM, that is introduced in Refs. [10, 13], by specific categories for sustainability, boundary conditions of the DMM are considered and explained in model development.

3.1   Development Process of the Sustainability Maturity Model (SMM)

To create a valid and useful MM for corporate transformation processes, a systematic process is applied. Knacksted et al. [14] gave an overview on the principles of MM development. They analyzed a wide range of models to understand the underlying development processes and derived a structured approach. The approach was used and adopted to include the existing DMM. The resulting approach is shown in Fig. 1 with the boundary conditions, e.g., the target and scope, categorical framework and transfer strategy of the existing DMM as additional input data.

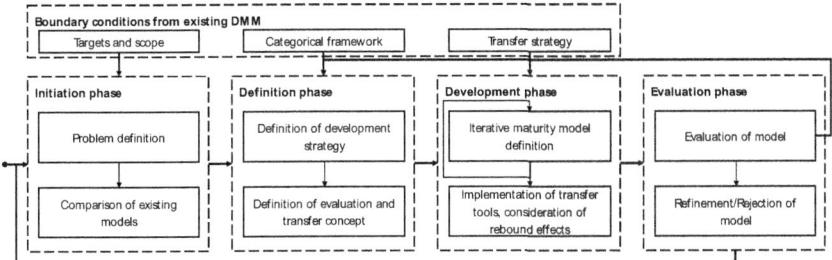

Fig. 1. MM development process (adapted from [14])

The process is structured in four major steps and aims to first define the model boundaries and to screen existing approaches before defining a strategy for development and transfer. The development itself is an iterative process to sharpen the identified maturity categories and integrate them into a transfer concept. Oftentimes, negative effects of digitalization on sustainability are discussed, e.g., by Refs. [15] or [16] in terms of additional resource use through increased efficiency of processes or waste streams by additional hardware usage, but not considered in MMs yet. Hence, the developed model also implements a guideline to foster awareness of rebound effects during evaluation phase as an additional module of the transfer tools. Before the model can be accepted and further refinement can be planned, an evaluation needs to be conducted.

3.2 Initiation Phase

This model development requires it to define goal and scope of the model to ensure the correct selection of a development strategy and transfer tools in later phases. The existing DMM has been tested and refined many times and proven to be an useful tool. Hence, to maintain these qualities, the target and scope have been adopted for the SMM. Consequently, SME that produce goods remain the target group and the CMM intends to inspire for sustainable transformation in connection with the adoption of digitalization strategies into corporate strategic planning and to identify mutual impacts.

The comparison of existing models reduces redundancy between existing MM. Similar approaches need to be investigated whether they can be reused or adapted for the own development process. This process was conducted by a literature review of MM for either sustainability or both domains following the PRISMA model by Moher et al. [17]. From that, the identified categories of SMM were analyzed and clustered to subject areas and main categories that are based on the existing DMM.

In sum, 28 identified studies were relevant, 21 gave a detailed view on a MM with sustainability aspects. The vast number of studies does not cluster the maturity categories further. 64% of the studies list one to nine categories, the majority between ten and 20 categories except for one study that lists 62 categories. Sustainability is the main goal of all the models with varying foci. Exemplarily, one focuses on sustainable project management whilst others emphasize corporate sustainability. Some refer specifically to recycling management, supply chain sustainability, energy management as cross-sectional area or software development processes. Eisner et al. include digitalization requirements into the sustainability assessment [16] what constitutes the closest related approach to the herein proposed one. The 28 identified sources were analyzed and resulted in nine main categories for the SMM that are clustered in three subject areas based on the DMM, as shown in Fig. 2. The DMM also includes the main category Energy as a separate level that was excluded in the SMM an integrated into the remaining categories. This ensures a better balancing of categories in the model due to the large amount of energy-related aspects that were identified in literature review.

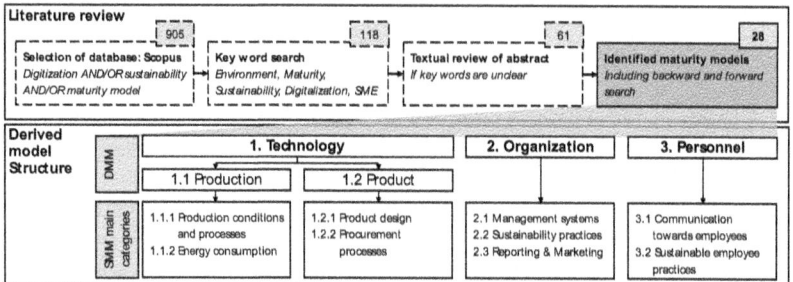

Fig. 2. Results of literature review and derived main categories for the SMM

3.3 Definition and Development Phase

During the **definition phase**, the development strategy was influenced by the existing DMM. However, the goal is to use the SMM in connection with the DMM but also independently from each other. This was addressed by the mapping of specific maturity categories to main categories and subject areas that can be excluded if they do not apply to the specific goals or requirements of the SME. Exemplarily, service-oriented companies that do not produce own goods would exclude area 1.2 Product of the MM completely or only the main category 1.2.1 Product design if procurement processes still apply to the scope of the SME. The definition of evaluation and the transfer concept is also defined by the transfer concepts of the existing DMM. Hence, the workshop concept and transfer tools for assessment of ideas and project concept were adopted as well. This process will be extended by the consideration of rebound effects between digitalization activities and sustainability goals.

During the **development phase** the MM was designed iteratively based on the findings of the literature review and included the feedback from experts and evaluation results. In total, 37 categories for the maturity model were identified and classified in the nine main categories. Each of the 37 categories consists of five maturity levels based on the DMM where each of them is described separately. The overall list of (main) categories is shown in Table 1.

The implemented transfer tools are based on the DMM, as elaborated in Ref. [13] in detail. Furthermore, a procedure to foster the awareness of rebound effects was implemented. Despite the rough planning state of possible digitalization activities, negative impacts on sustainability need to be considered. Specific assessment methods lack the

Table 1. Detailed subcategories of the maturity model

Main category	Categories within the main categories with five maturity levels each
1.1.1	Production site; lighting; machines; combustibles and raw materials; technologies and production processes
1.1.2	Energy data and measuring equipment; energy supply systems
1.2.1	Material input; circularity; repairability; packaging
1.2.2	Environmental properties of suppliers; cooperation with stakeholders
2.1	Heating; finance; consumption; waste; water; environment; data; innovation; energy; mobility
2.2	Sustainability practices; stakeholder requirements; life cycle assessment of products; assessment of the organization; laws and certifications
2.3	Sustainability reporting; customer communication; environmental marketing
3.1	Internal communication management; idea management of employees
3.2	Job design; selection and recruiting of employees; sustainability training and continuing education; sustainability incentives for employees

availability of data in the current planning phase. Hence, a rough estimation to create awareness of the stakeholders is the goal of the procedure that is shown in Fig. 3.

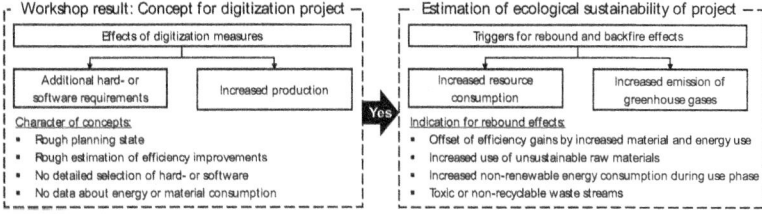

Fig. 3. Consideration of possible rebound effects during concept phase of digitalization projects

The procedure raises the question of potential additional hard- or software requirements or increased production of goods by the digitalization activities. If either of these exist, then potential triggers such as increased resource consumption or emissions can indicate on rebound or backfire effects. The created awareness can then be included in the planning and implementation of the designed projects to reduce ecological impacts.

3.4 Evaluation Phase and Application in Industrial Context

The model was tested with and industrial service supplier for cleaning, surface treatment and logistics. The company has buildings with facilities and workplaces but is not producing own products. The main customers are from the automotive sector that is putting more pressure on suppliers to improve and report about production processes and internal sustainability strategies. After a tour through the production facilities, the

DMM categories Production, Organization, Energy and Personnel were assessed. The result was low to mediocre maturity levels in digitalization and very low levels in sustainability. Among others, projects for digital document storage and increased energetic self-sufficiency and measuring equipment were identified and detailed to project concepts. The latter resulted in the selection of measuring points and equipment as well as supply strategies by renewables. Also, load management strategies as one category of the DMM were discussed in more detail. Due to load limits in the energy supply system, the current production site cannot be expanded with new machines. A well-established load management could enhance energetic flexibility what could result in the procurement of a new production facility. Here, the considerations of rebound effects showed a potential drawback of the load management strategy on absolute sustainability since new facilities could result in additional energy and material consumption and therefore, among others, increased greenhouse gas emission. Hence, a potentially drawback of the load management towards sustainability should be considered when implementing the strategy and necessary equipment.

The SMM itself was evaluated as well-designed and helped the company to develop a general understanding of the overall opportunities for sustainable business practices. Despite the currently low levels in sustainability, a general roadmap of concrete tasks could be defined by the target maturity levels. Some comments for improvement in terms of understanding of the main and subcategories as well as minor changes in the maturity levels were recommenced. Hence, the model can be seen as validated in general, specific refinements were considered and implemented though and will be evaluated again in the future.

4 Outlook

Climate change-related directives lead to more pressure on SME to implement ecological sustainability practices. Future reporting requirements force them to raise transparency in terms of their own contribution to emissions into the environment. This data acquisition is often connected to digitalization activities and requires a general strategy in both, digital and sustainable transformation. Hence, a CMM has been introduced that integrates bot dimensions and indicates on rebound effects. The development and validation of the CMM in an industrial use case has been shown and resulted in good overall results, but also feedback that was implemented accordingly. Nevertheless, the current opaque in reporting and assessment standards shows improvement potentials for the CMM. Here, positive impacts of sustainability on various aspects such as cost reduction, enhanced reputation and increased competitiveness needs to be highlighted more to motivate companies to embrace sustainable practices within digitalization strategies.

Acknowledgement. The authors gratefully acknowledge the financial support of the "Mittelstand Digital Zentrum Chemnitz" (01MF21007B) by the Federal Ministry for Economic Affairs and Climate Action (BMWK) and the project supervision by the project management organization Projektträger Deutsches Zentrum für Luft- und Raumfahrt e.V. (DLR).

References

1. Ghobakhloo M (2020) Industry 4.0, digitization, and opportunities for sustainability. J Clean Prod 119869
2. Mitchell S, O'Dowd P, Dimache A (2020) Manufacturing SMEs doing it for themselves: developing, testing and piloting an online sustainability and eco-innovation toolkit for SMEs. Int J Sustain Eng 3:159–170
3. Brozzi R, Forti D, Rauch E et al (2020) The advantages of industry 4.0 applications for sustainability: results from a sample of manufacturing companies. Sustainability 9:3647
4. Lange S, Pohl J, Santarius T (2020) Digitalization and energy consumption. Does ICT reduce energy demand? Ecol Econ 106760
5. Despeisse M, Ball P, Evans S et al, Zero carbon manufacturing through process flow modelling. Tagungsband. In: Proceedings of the 11th MITIP conference, Bergamo
6. Salonitis K, Ball P (2013) Energy efficient manufacturing from machine tools to manufacturing systems. Procedia CIRP 634–639
7. Schulz J, Popp RS-H, Scharmer VM et al (2018) An IoT based approach for energy flexible control of production systems. Procedia CIRP 650–655
8. Cantele S, Zardini A (2020) What drives small and medium enterprises towards sustainability? Role of interactions between pressures, barriers, and benefits. Corp Soc Responsib Environ Manag 1:126–136
9. VDI 4000-Blatt 1 (2022) Auswahl von Industrie-4.0-Reifegradmodellen zur digitalen Transformation produzierender Unternehmen - Grundlagen. Beuth-Verlag
10. Langer T, Singer A, Wenzel K et al (2017) Modulbaukasten Digitalisierung. Zeitschrift für wirtschaftlichen Fabrikbetrieb 12:902–906
11. Baumgartner RJ, Ebner D (2010) Corporate sustainability strategies: sustainability profiles and maturity levels. Sustain Dev 2:76–89
12. Golinska P, Kuebler F (2014) The method for assessment of the sustainability maturity in remanufacturing companies. Procedia CIRP 201–206
13. Münnich M, Stange M, Süße M et al (2023) Integration of digitization and sustainability objectives in a maturity model-based strategy development process. Tagungsband: manufacturing driving circular economy. In: Lecture notes in mechanical engineering. Springer International Publishing, Cham, pp 918–926
14. Knackstedt R, Pöppelbuß J, Becker J (2009) Vorgehensmodell zur Entwicklung von Reifegradmodellen. Tagungsband: Wirtschaftsinformatik Proceedings
15. Chen X, Despeisse M, Johansson B (2020) Environmental sustainability of digitalization in manufacturing: a review. Sustainability 24:10298
16. Eisner E, Hsien C, Mennenga M et al (2022) Self-assessment framework for corporate environmental sustainability in the era of digitalization. Sustainability 4:2293
17. Moher D, Liberati A, Tetzlaff J et al (2009) Preferred reporting items for systematic reviews and meta-analyses: the PRISMA statement. PLoS Med 7:336–341

Data-Driven Approach for a Continuous Information Flow in a Closed-Loop Supply Chain

Celine Letzgus[1(✉)], Jennifer Alica Kirsch[1], and Thomas Bauernhansl[1,2]

[1] Fraunhofer Institute for Manufacturing Engineering and Automation IPA, Nobelstraße 12, 70569 Stuttgart, Germany
`celine.letzgus@ipa.fraunhofer.de`

[2] Institute of Industrial Manufacturing and Management, University of Stuttgart, Allmandring 35, 70569 Stuttgart, Germany

Abstract. Due to the large variety of actors, channels, and materials involved, the processes of the circular economy are very complex. This complexity makes it difficult to maintain consistent data quality throughout the supply chain. Circular processes can only be applied extensively if enough data, for example supplier information, is available. Data and information from suppliers, manufacturers, end-users, and recyclers are not shared transparently through circular processes. One reason is the insufficient support from IT systems for data exchange between the different parties. To improve data exchange, a product-independent process view of the closed-loop supply chain based on the supply chain operations reference model is introduced in this work. Then, an approach is presented to improve data quality in closed-loop supply chain processes through data-based models. The human-in-the-loop method is used to capture data and enable integration of this approach in IT systems for closed-loop supply chains. Therefore, a more transparent data flow laying the foundation for improved data quality throughout the supply chain can be achieved.

Keywords: Sustainable supply chain · Human-in-the-loop · Supply chain management

1 Introduction

The central problem of the linear form of value creation is the monodirectional use of limited resources. Such a form of development requires a constantly growing influx of raw materials and not only leads to the deterioration of global livelihoods but also has irreversible economic and ecological effects [1].

An established alternative model for decoupling economic growth, resource extraction, and environmental degradation is the transition to a circular economy. This involves moving from a linear supply chain to a closed-loop supply chain (CLSC). The recovery of materials and their reintroduction into the value chain provides significant added

© The Author(s) 2025
H. Kohl et al. (Eds.): GCSM 2023, LNME, pp. 509–515, 2025.
https://doi.org/10.1007/978-3-031-77429-4_56

value for companies and consumers [2]. The path to a resource-conserving and resource-efficient economy leads primarily through the conscious limitation, saving, and closing of material and energy flows [3].

Despite numerous approaches to the transition to a circular economy, the path from a linear to a CLSC has barriers. These barriers make it difficult to recycle products and raw materials and prevent the integration of recycling into business planning [4]. The main challenge is the complexity of supply chains resulting from the opacity of material status, the diversity of actors, channels, and materials and take-back logistics. Lack of knowledge, the inability to share information, and the resulting low availability of information lead to poor data quality, which can currently be seen as one of the main obstacles to the transition to a circular economy and thus a CLSC. Existing IT systems do not sufficiently support this complexity, making it almost impossible to apply and plan reuse strategies [5].

Existing models for representing the circular economy focus on the product view and thus do not adequately cover the processes taking place within a circular economy [6]. The current state of research shows that there are still considerable gaps in a comprehensive mapping and evaluation of the circular economy. A holistic view of the entire system, including cross-company processes, is necessary to close this gap [7]. Intelligent IT components can help reduce the complexity of process design and minimize barriers in the transition to a circular economy [8].

2 Related Work

The barriers for companies to adopt CLSCs can be divided into the categories of finance, management, infrastructure, networking, and technology. On closer examination, the networking category emerges as the area with the most barriers to the transition to a CLSC. This is mainly due to a lack of competence in data collection and data exchange. This can be attributed to various problems, such as data sensitivity, the retention of knowledge by actors, and the lack of data consistency [4]. The management processes of the supply chain operations reference model (SCOR) can be used to describe both simple and complex supply chains consistently. From the source of supply to the point of consumption, the reference model covers the entire supply chain, from suppliers through production and distribution to the end customer. The model is dynamic and, therefore, responsive to paradigm shifts. The introduction of return adds an additional paradigm to the model essential to the CLSC [9]. SCOR is a hierarchical model consisting of several levels, each more detailed [10, 11]. Figure 1 shows the classic SCOR model with the process categories plan, source, make, deliver, and return.

To improve data provision in CLSC modelling using SCOR, various digitalisation technologies can be used. These digitalisation technologies can be divided into IoT, big data, blockchain, and artificial intelligence [12]. In particular, using artificial intelligence can help to create the necessary infrastructure to ensure the dynamic flow of information required for data integrity [13]. The integration of artificial intelligence creates the ability to track, analyse, and provide meaning to large amounts of data in situations of varying complexity and context. This will enable rapid and agile learning processes in data analysis and, thus, more flexible action, opening up new possibilities for the

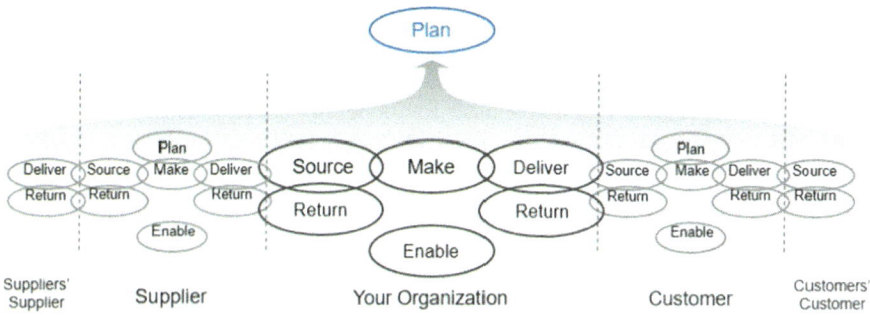

Fig. 1. SCOR model based on [9]

flow of information [14]. Machine learning (ML) is a branch of artificial intelligence. ML algorithms can monitor processes and detect and predict failures [15]. In this way, knowledge about the characteristics of CLSCs can be gathered and expanded [12].

3 Method for Improving Data Quality in Closed-Loop Supply Chains

A CLSC consists of various interconnected process levels. This results in a large amount of data with a variety of different process parameters. To ensure the reliability of the data, it is necessary to provide sufficient input information to update the data [16]. ML algorithms can help, but even these methods have limitations. Incorporating prior knowledge into the model is a common technique to reduce the required training data, as deriving the knowledge from the data itself is not necessary [17]. Here, humans have a wealth of prior knowledge that can be used to compensate for the lack of training data [18]. The approach that involves humans by integrating their expertise into the training, testing, and adaptation of ML models is also referred to as human-in-the-loop (HITL) in the literature. The human can provide feedback to the algorithm to correct errors and make improvements. In this way, they are integrated into the continuous feedback loop of the model [19].

In addition, due to their resilience, humans have the ability to cope with change and adapt quickly. This enables both the correct interpretation of novel combinations of existing concepts and a quick and good generalisation in complex situations in which humans have little previous experience. Furthermore, humans can derive new causal relationships on the basis of a few observations [20]. In this way, distortions of the algorithm that arise through training and inference, for example, can be significantly reduced through human feedback [21]. Prior knowledge about the probability of occurrence of various causal relationships and the type of mechanisms for linking cause and effect is particularly important in inference [20].

4   Integration of a Data-Driven Approach into the Closed-Loop Supply Chain

The SCOR model provides a suitable basis for modelling CLSC processes as it is capable of representing internal and external company processes. In addition, the HITL method can be used in conjunction with a ML algorithm to achieve better data quality for implementing the CLSC.

Figure 2 shows the integration of the HITL method into the SCOR model. In addition to the classical process categories, the HITL method is located at the planning level of the SCOR model. Information from different experts with implicit knowledge is stored in a database. Different methods, such as natural language processing, text classification, topic modelling, or sentiment analysis, can be used for this purpose [21]. The HITL method should be able to continuously adapt the process structure to ensure the efficiency and effectiveness of the system through feedback from the real world, consisting of actual business data, human feedback data, and specific metrics. A continuous flow of information is essential for a functioning CLSC. At the planning level, humans can continuously correct, validate, and provide feedback on the data. In this way, model training and inference can be improved. In the example, a database is integrated into the planning area of the SCOR model. Based on relevance, all areas can provide input and knowledge from here and thus feed the database. Conversely, information can be made available to all stakeholders in the system, and more transparency is created. The approach aims to automatically design the process structure using company data and human input.

There are different approaches for implementing HITL by ML. One approach is active learning (AL), where the human acts only as an oracle. In interactive machine learning (IML), the human plays a more important role and provides more information to the system. In Machine Teaching, on the other hand, the human takes the leading role in the learning process and deliberately limits the knowledge of the ML model [22]. In the model presented in this paper, a combination of the different approaches is suggested. Thus, if necessary, the human can limit which information is allowed to reach certain parts of the CLSC and which information is only relevant within the company.

To model the HITL approach with ML, an appropriate method has to be chosen. Case-based reasoning (CBR) is a method for implementing HITL. The method derives conclusions from experience in previous cases. It can respond to new situations and provide flexible, contextual solutions to problems [23]. This method provides a practical approach to decision support and knowledge management in various application domains by incorporating human expertise and experience. It enables the extraction of relevant information from past events and the application of this information to the current situation.

The synergy between human expertise and CBR enables optimised decision-making by integrating human expertise and experience into the model. This ensures effective decision-making support. The integration of HITL and CBR approaches can lead to improved decision-making and increased efficiency throughout the process. Context is vital when using HITL, as human experts provide the model with specific information about the use case. The inclusion of context can be made possible in decision-making by the CBR. This way, the model can recognise the situation and incorporate relevant

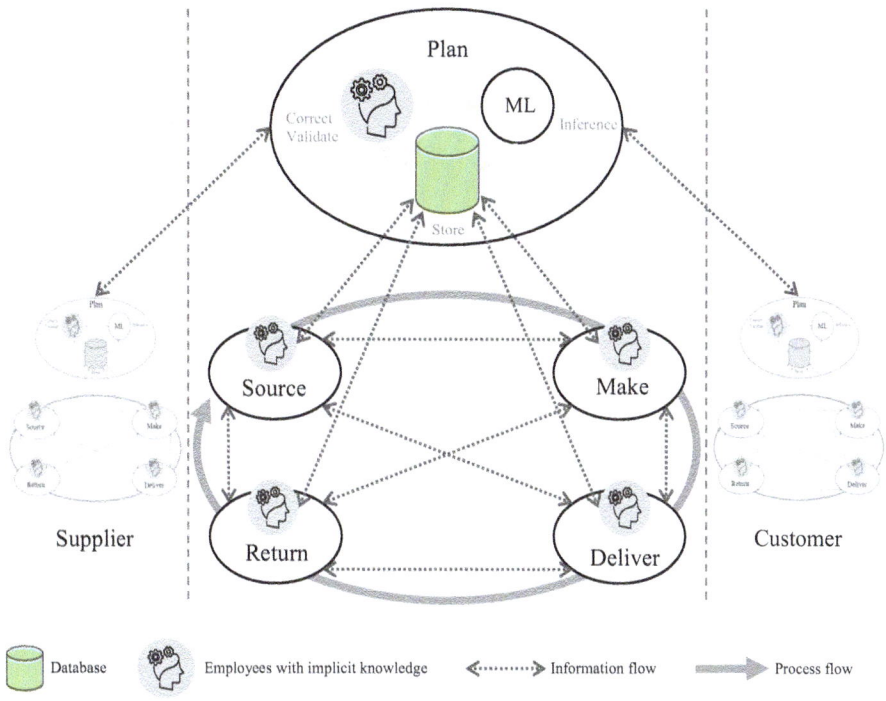

Fig. 2. Integration of the HITL method with ML into the SCOR model

cases and experiences to make appropriate decisions. By combining both approaches, the model can adapt to new requirements or conditions.

5 Conclusion

When moving from a linear supply chain to a CLSC, barriers prevent implementation. The return of products and materials back into the supply chain and the integration of returns into business planning are hampered. One problem is the limited ability to share information and the lack of available data.

The use of artificial intelligence, especially through the integration of the HITL approach, contributes to the improvement of data quality. The integration of humans allows feedback from reality, such as current company data, human feedback, and specific key figures, to continuously adapt the process structure and ensure a continuous information flow. Therefore, the developed process model is designed to collect reliable, complete, and consistent data. Close cooperation of experts and continuous evaluation of the process model are of major importance to meet the requirements and barriers of a CLSC. A structured and comprehensive basis for the process model development is provided using the SCOR model. This allows for defining the most important process elements and their relationships, providing a clear orientation and a solid basis for further modelling. The application of the CBR methodology integrates the HITL approach into the process model. Through the holistic approach, both current information and past best practices flow into the feedback loop, leading to effective knowledge utilisation and improved data quality.

References

1. Klenk F, Potarca M, Häfner B et al (2020) Kreislaufwirtschaft in Produktionsnetzwerken. In: Zeitschrift für wirtschaftlichen Fabrikbetrieb, pp 668–672
2. Franzò S, Urbinati A, Chiaroni D et al (2021) Unravelling the design process of business models from linear to circular: an empirical investigation. Bus Strat Env 2758–2772
3. Pieroni M, McAloone T, Pigosso D (2019) Business model innovation for circular economy and sustainability: a review of approaches. J Clean Prod 215:198–216
4. Lobo A, Trevisan A, Liu Q et al (2022) Barriers to transitioning towards smart circular economy: a systematic literature review. In: Scholz SG, Howlett RJ, Setchi R (eds) Sustainable design and manufacturing, vol 262. Springer Singapore, pp 245–256
5. Wilts H (2017) The digital circular economy: can the digital transformation pave the way for resource-efficient materials cycles?
6. Bianchini R et al (2019) Overcoming the main barriers of circular economy implementation through a new visualization tool for circular business models. Sustainability 6614
7. Elia V, Gnoni M, Tornese F (2017) Measuring circular economy strategies through index methods: a critical analysis. J Clean Prod 2741–2751
8. Rosa P, Sassanelli C, Urbinati A et al (2020) Assessing relations between circular economy and industry 4.0: a systematic literature review. Int J Prod Res 1662–1687
9. APICS (2017) APICS supply chain operations reference model: version 12.0
10. Werner H (2013) Supply chain management. Springer Fachmedien Wiesbaden, Wiesbaden
11. Wannenwetsch H (2014) Integrierte Materialwirtschaft, Logistik und Beschaffung. Springer Berlin Heidelberg, Berlin, Heidelberg
12. Chauhan C, Parida V, Dhir A (2022) Linking circular economy and digitalisation technologies: a systematic literature review of past achievements and future promises. Technol Forecast Soc Change
13. Roberts H, Zhang J, Bariach B et al (2022) Artificial intelligence in support of the circular economy: ethical considerations and a path forward. AI & Soc
14. Kristoffersen E, Blomsma F, Mikalef P et al (2020) The smart circular economy: a digital-enabled circular strategies framework for manufacturing companies. J Bus Res 241–261
15. Su D, Wu Y, Chai Z (2019) Advanced integrated manufacture by application of sustainable technology through product lifecycle. In: Proceedings of the 2019 international conference on artificial intelligence and advanced manufacturing. Association for computing machinery, New York, NY, United States, pp 1–4
16. Wu X, Xiao L, Sun Y et al (2022) A survey of human-in-the-loop for machine learning. Future Gen Comput Syst
17. Diligenti M, Roychowdhury S, Gori M (2017) Integrating prior knowledge into deep learning. In: 2017 16th IEEE international conference on machine learning and applications (ICMLA). IEEE, pp 920–923
18. Zhang R, Torabi F, Guan L et al (2019) Leveraging human guidance for deep reinforcement learning tasks
19. Xin D, Ma L, Liu J et al (2018) Accelerating human-in-the-loop machine learning: challenges and opportunities
20. Holzinger A (2021) The next frontier: AI we can really trust. In: Kamp M, Koprinska I, Bibal A et al (eds) Machine learning and principles and practice of knowledge discovery in databases, vol 1524. Springer International Publishing, Cham, pp 427–440

21. Dong X, Sarker S, Qian L (2022) Integrating human-in-the-loop into swarm learning for decentralized fake news detection. In: 2022 international conference on intelligent data science technologies and applications (IDSTA). IEEE, pp 46–53

22. Mosqueira-Rey E, Hernández-Pereira E, Alonso-Ríos D et al (2023) Human-in-the-loop machine learning: a state of the art. Artif Intell Rev 56:3005–3054

23. Richter M, Weber R (2013) Case-based reasoning. Springer, Berlin Heidelberg, Berlin, Heidelberg

Digital Technologies Enabling Resilience in Manufacturing Networks

Nikolaos-Stefanos Koutrakis[1]([✉]), Maria Chiara Magnanini[2], Eckart Uhlmann[1,3], Julian Polte[1,3], Eujin Pei[4], Foivos Psarommatis[5], and Alexandra Brintrup[6]

[1] Fraunhofer Institute for Production Systems and Design Technology IPK, Pascalstraße 8-9, 10587 Berlin, Germany
nikolaos-stefanos.koutrakis@ipk.fraunhofer.de

[2] Department of Mechanical Engineering, Politecnico di Milano, Via La Masa 1, 20156 Milano, Italy

[3] Institute for Machine Tools and Factory Management (IWF), Technische Universität Berlin, Pascalstraße 8-9, 10587 Berlin, Germany

[4] Brunel University London, Kingston Lane, Uxbridge UB8 3PH, UK

[5] University of Oslo, Gaustadalleen, 23B, 0373 Oslo, Norway

[6] Department of Engineering, Institute for Manufacturing, University of Cambridge, Cambridge CB3 0FS, UK

Abstract. Unforeseen events have the potential to cause disruptions throughout the entire manufacturing value chain, ranging from interruptions in production processes on the shop floor level to shutdowns in the supply chain and logistics. These disruptions increase the necessity for the establishment of manufacturing networks that prioritize cooperation and circularity to strengthen resilience, predict and counteract such impacts. This paper provides an overview of the main digital technologies required to create a resilient and sustainable manufacturing network and the implementation of the Manufacturing as a Service (MaaS) approach. For each digital technology, a synthetic characterization is provided, with respect to requirements, exemplary applications and involved standards. Challenges on the use of such technologies are presented and suggestions for future developments on the integration and deployment of digital technologies with the aim of achieving resilience in manufacturing networks is described.

Keywords: Digitally integrated production · Resilience · Supply chain · Circularity · Manufacturing networks · Sustainability · Manufacturing as a service

1 Introduction

In 2020, EU trade was hit by the COVID-19 pandemic, resulting in a significant fall for exports (-9.3%) and imports (-11.5%) [1]. In a positive light, EU trade has recovered strongly in 2021 and 2022, with imports of goods growing by 23.8 and 41.3% respectively, while exports of goods increased by 12.8% in 2021 and by 17.9% in 2022. In 2021, the value of sold production in the EU amounted to €5209 billion, an increase by almost

H. Kohl et al. (Eds.): GCSM 2023, LNME, pp. 516–523, 2025.
https://doi.org/10.1007/978-3-031-77429-4_57

14% compared with €4581 billion in 2020 [2]. The manufacturing sector is critical to the EU, employing around 29.4 million persons in 2020 and generating €1880 billion [3]. The region is now seeing a strong rebound in global demand for manufactured goods. However, some sectors are still hit by severe supply shortages, particularly in high value products, with supply struggling to accommodate this surge in demand. Disruptions in the logistics industry remain and have exacerbated supply bottlenecks. Companies may face severe shortages, particularly in high value process, with manufacturers struggling to accommodate this incremental surge in demand.

Manufacturing as a Service (MaaS) will play a critical role in terms of economic and societal impact. Bruegel, the European think tank specializing in economics, highlighted the importance of making the EU manufacturing sector more competitive and this is vital to restore economic growth in Europe in a recent report. In particular, the adoption of service business models in terms of servitization or "servinomics" will enable a wider portfolio of services for manufacturers and the fact that such services will be more resilient to business cycle fluctuations and to create a differentiation. The report also found that the externalization of services is more prevalent in the USA and in the Far East such as South Korea and increasingly in China. Therefore, it is imperative that EU manufacturers close this digital gap to increase its competitiveness and to reshape the way that manufacturing firms produce and compete [4].

As a result, the establishment of manufacturing networks that prioritize cooperation and circularity to strengthen resilience and counteract such impacts becomes all the more necessary [5]. The efficient utilization of resources by combining and deploying digital technologies for different processes across the manufacturing value chain has the potential to enhance resilience and raise the level of competitiveness of the manufacturing industry, as well as improving the sustainability of production processes [6]. Increased asset utilization along with circularity and minimized waste at their core will foster sustainability and increase the utilization of resources in manufacturing supply chains across various industries. Furthermore, in the literature has been pointed out that isolated solutions offer lower efficiency compared to holistic approaches [7]. Therefore, this paper provides a holistic framework for the synthetic characterization for each digital technology along with examples, standards, challenges and suggestions for future developments on their integration and deployment.

2 Enabling the Establishment of Manufacturing Networks

Disruptions described in the context of discrete manufacturing or other production industries can be mitigated if the system is able to quickly react to external or internal influences, and to also predict and anticipate these disruptions. The disruptions can trigger specific actions covering different processes and inform the supply chain network which in turn can offer alternative solutions. Relevant stakeholders need to be provided with information and assisted for optimal decision making on how to overcome the problems when they arise. These solutions require technologies to ensure that best solution can be applied for the adapted manufacturing strategy, such as dynamic rescheduling tools for accommodating the changes as efficient as possible. Additionally, by relying on synthetic System Digital Twins (DT) of the manufacturing plant to evaluate the effect of reconfigurations, process adaptation and rescheduling in a fast and efficient way [8, 9].

As described in Fig. 1, the architecture of a manufacturing network needs to be constructed in a way that the data can be shared horizontally and vertically in each layer. Starting from the bottom, the resources of the network participants, e.g. machine tools, materials or manufacturing capacities etc., must be able to provide actual information to all manufacturing entities. The data can be used to effectively communicate available resources across other network participants and at the same time inform decisions at an application level, e.g. for decision making, business approaches etc.

2.1 Dynamic Models and Digital Twins

Functional and geometrical models of products/components will need to be developed to contribute to the product-process matchmaking. The process planning should consider:

1. physical constraints such as process workspace or payload requirements,
2. operational capabilities and capacities,
3. level of automation available and
4. technological constraints.

The provision of models and simulators based on physics, technological features and real-time data to describe process capabilities and product requirements in the network is essential. It allows the implementation of actionable Digital Twins (DTs) that provide the exchange of information flow of manufacturing processes at network level and the identification of robust manufacturing strategies in the context of MaaS.

2.2 Networked Distributed Components

Industrie 4.0 with the implementation of Cyber Physical Systems (CPS) gives the ability to create highly automated digital factories, closing the gap between Operation Technology and Information Technology [10]. This integration ensures that information from different entities across the manufacturing value chain can reach potential or intended parties efficiently. Modular architectures incorporating concepts of DT and Asset Administrations Shells (AAS) assist in creating virtual representation of physical assets in a standardized way and will bring homogeneity in a heterogeneous interface field [11]. Such modular distributed Industrial Internet of Things (IIoT) architectures of production systems are often made of self-contained production modules. Every module must have the required capability for local control, to manage all operations that need to be performed within the entity, e.g. part reception, transformation operations start, termination etc. Furthermore, monitoring of process quality, equipment health status, faults, energy consumption, as well as part tracing, are all capabilities the module should include. Approaches that deploy Multi Agent Systems (MAS) can provide methods for distributed production control that enables production entities in the form of agents to securely exchange information within the network and to lead towards optimal decision-making regarding capacity utilization, complying with delivery dates and to account for real-time disruptive events and simulating different scenarios affecting production [12].

2.3   Data Driven Approaches and Analytics

Data driven Artificial Intelligence (AI) has created unprecedented opportunities for improvements in predictive analytics. This in turn helps improve efficiency in planning, forming the backbone of Industrie 4.0. Accurate predictions on quality outcomes help optimize process parameters and buffers, energy use, optimizing scheduling, suggesting predictions on machine operational lifespan, recommending times for service interventions, and making reliable predictions on supply stock and supply chain configuration. The implementation of effective analytical approaches is critical to the operation of a manufacturing network. Network participants must deploy various models across the entire organization to enable the generation of accurate information for stakeholders.

2.4   Human-Centered Value Chains

Businesses seek to boost efficiency in the value chain while also enhancing the quality and uniformity of their products. Human operators are being used for both quality control and at the same time are expected to be responsible for decision-making processes at the production level as well as at the value chain level. In some scenarios such as dealing with single or multiple unexpected disruptions, human operators may be unable to cope efficiently with the assigned tasks without any human centric assistive tools, purely because of human limitations in terms of cognition and mental processing. The new value chains should be characterized through cooperation between humans and machines, and they should be created to enhance and optimize, rather than to fully replace the knowledge and skills of human workers [13]. Beyond the traditional human factors that only help organizations to manage the workload safely and healthily, the design of these technologies must be centered around the human worker. Therefore, the design solution should advance to a more humanistic level, considering aspects of human experience, individual expertise and job satisfaction. This includes ensuring that the interface between Man and Machine, or the Graphic User Interface should be easy to understand and sufficiently intuitive to operate [14].

2.5   Cross Organization Data Exchange

The Internet and the services delivered by it are increasingly centralized on available platforms. Current frameworks are created to cater for increasing returns to scale and to deploy a business model with asymmetries between users and services. Existing multi-agent and agent architectures have seen no significant adoption outside niche applications. Consequently, the necessary agent frameworks need to be designed to allow for decentralized approaches, where each individual and organization can be represented by an autonomous entity with its own independent agency. Such an approach would bridge the gap between existing applications and new advancements in this field, such as by employing distributed ledger technologies as core parts of its construction. Other approaches such as decentralized Web 3.0 and Distributed Ledger Technologies, specifically Bitcoin or Ethereum, can be deployed to create decentralized systems for manufacturing networks.

2.6 Business Decisions

MaaS is part of a family of business models that are broadly summarized under the term servitization. In this context, "as-a-service" refers to the provision of a specific capability, and in the case of manufacturing industries, it means having an all-inclusive fee compared to selling equipment, solutions and services as separate transactions. Servitization therefore allows manufacturing companies to go beyond just selling products and instead offering a range of services that add value for customers. These services can include maintenance, spare parts management, process optimization, system integration, financing etc. [15]. In today's increasingly complex technological landscape, product differentiation alone may not suffice for a competitive advantage [16]. By leveraging their existing knowledge and expertise, companies can be faster to react to the market, develop and offer solution packages that set them apart from their competitors. This requires not only technical capabilities such as maintenance and repair, but also service-related skills such as data processing, risk assessment, and financing [17].

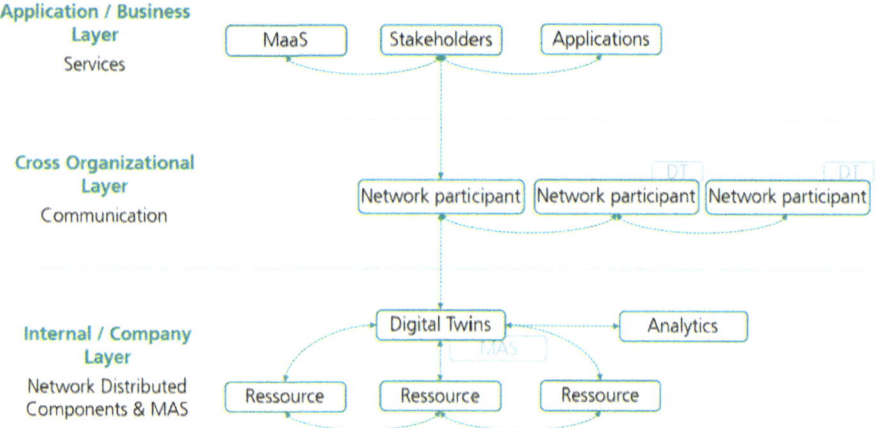

Fig. 1. Reference architecture for manufacturing networks

3 Challenges and Suggestions

Effective cross organizational communication requires a common language for all participating entities. Each organization, from the supply chain to the factory floor level uses different ontologies to describe common scenarios, applications and problems. Information included in various architectures of DTs differs significantly, thus making it increasingly difficult to relay the appropriate information when required. In manufacturing scenarios, common descriptions between machine tool capabilities and product requirements will need to be established, in order for appropriate matching to take place. This includes standardized translations of product designs to manufacturing specifications.

Key issues also arise in the data driven approaches. Deployed models typically suffer from data imbalance, due to the limited availability of historical data. Increased data sharing and collective decision making could consequentially improve predictive power. Such approaches have been extremely limited as organizations fear that sharing sensitive information such as process vulnerabilities, capacity and excess stock can be used opportunistically by others for cost reduction or for unfair competition [18].

4   Conclusion

This paper presents an overview of the main digital technologies required to create resilient and sustainable manufacturing networks. This starts from building upon the networked aspects of Industry 4.0, where all devices and components already possess cyber-physical capabilities and using agile resource exchange to solve real-world industry problems. A holistic and structured approach is necessary to build an organized hierarchical structure of assets that can be utilized within the manufacturing network. Data captured from sensors and machine tools during production processes should be integrated and translated into useful dynamic models, to be proactively used to control subsequent production processes and to flexibility adapt to existing production supply chains. The interconnected elements of the described architecture include dynamic models and digital twins, networked distributed components, data driven approaches and analytics, human-centered value chains, cross organization data exchange and business decisions. Cooperation among companies will improve asset utilization, thus reducing expected material and energy waste, reduce production costs by identifying optimal solutions at network level and foster proactive circular approaches. These technologies will allow the adaptation of logistics and production to varying external conditions, improving the resilience of the industrial systems and value chains. Such a shift from linear approaches to closed-loop, circular processes that also minimize waste will therefore ensure resilience, sustainability and relevance of manufacturing supply chains.

Acknowledgments. We thank our colleagues from TTTech Computertechink AG, Fetch.ai Research and Development GmbH, Servitize UG, M&M Group, EIT Manufacturing East GmbH and EIT Manufacturing South East who provided insight and expertise that greatly assisted this research.

References

1. EUROSTAT International trade in goods. https://ec.europa.eu/eurostat/statistics-explained/index.php?title=International_trade_in_goods#EU_trade_increased_strongly_in_2022. Last accessed 06 Apr 2023
2. EUROSTAT Industrial production statistics. https://ec.europa.eu/eurostat/statistics-explained/index.php?title=Industrial_production_statistics#Overview. Last accessed 06 Apr 2023
3. EUROSTAT Businesses in the manufacturing sector. https://ec.europa.eu/eurostat/statistics-explained/index.php?title=Businesses_in_the_manufacturing_sector#Structural_profile. Last accessed 29 Mar 2023
4. Bruegel Services in European manufacturing: servinomics explained. https://www.bruegel.org/blog-post/services-european-manufacturing-servinomics-explained. Last accessed 27 Mar 2023

5. Psarommatis F, May G (2022) Achieving global sustainability through sustainable product life cycle. In: IFIP international conference on advances in production management systems. Springer Nature Switzerland, Cham, pp 391–398

6. Psarommatis F, Sousa J, Mendonça JP, Kiritsis D (2022) Zero-defect manufacturing the approach for higher manufacturing sustainability in the era of industry 4.0: a position paper. Int J Prod Res 60(1):73–91

7. Psarommatis F, Bravos G (2022) A holistic approach for achieving sustainable manufacturing using zero defect manufacturing: a conceptual framework. Procedia CIRP 107:107–112

8. Magnanini MC, Tolio TA (2021) A model-based digital twin to support responsive manufacturing systems. CIRP Ann 70(1):353–356

9. Psarommatis F, May G (2022) A literature review and design methodology for digital twins in the era of zero defect manufacturing. Int J Prod Res 1–21

10. Gowtham V, Willner A, Pilar von Pilchau W, Hähner J, Riedl M, Koutrakis N-S, Polte J, Uhlmann E, Tayub J, Frey V (2021) A reference architecture enabling sensor networks based on homogeneous AASs. Automation 5–16

11. Koutrakis N-S, Gowtham V, Pilar von Pilchau W, Jung TJ, Polte J, Hähner J, Corici M-I, Magedanz T, Uhlmann E (2022) Harmonization of heterogeneous asset administration shells. Procedia CIRP 107:95–100

12. Psarommatis F, Kiritsis D (2022) A hybrid decision support system for automating decision making in the event of defects in the era of zero defect manufacturing. J Ind Inf Integr 26:100263

13. Romero D, Noran O, Stahre J, Bernus P, Fast-Berglund Å (2015) Towards a human-centred reference architecture for next generation balanced automation systems: human-automation symbiosis. In: Umeda S, Nakano M, Mizuyama H, Hibino H, Kiritsis D, von Cieminski G (eds) APMS 2015, vol 460. IAICT. Springer, Cham, pp 556–566

14. Peruzzini M, Pellicciari M, Gadaleta M (2019) A comparative study on computer-integrated set-ups to design human-centred manufacturing systems. Robot Comput Integr Manuf 55:265–278

15. Vandermerwe S, Rada J (1988) Servitization of business: adding value by adding services. Euro Manag J 6(4)

16. Minarsch D, Favorito M, Hosseini SA, Turchenkov Y, Ward J (2021) Autonomous economic agent framework. https://fetch.ai/content-hub/publication/autonomous-economic-agent-framework. Last accessed 10 Apr 2023

17. Benedettini O (2022) Structuring servitization-related capabilities: a data-driven analysis. Sustainability

18. Psarommatis F, Dreyfus PA, Kirtsi D (2022) Chapter 9—the role of big data analytics in the context of modeling design and operation of manufacturing systems. In: Design and operation of production networks for mass personalization in the era of cloud technology. Elsevier, pp 243–275

Data and Learning

Deep Learning-Based Optical Character Recognition for Identifying On-Label Printed Part Numbers of Used Automotive Parts: A Comparative Study of Open Source and Commercial Methods

Marian Schlüter[✉], Christian Tepper, Clemens Briese, Ole Kroeger, Raul Vicente-Garcia, and Jörg Krüger

Fraunhofer Institute for Production Systems and Design Technology IPK, Berlin, Germany
marian.schlueter@ipk.fraunhofer.de

Abstract. This paper explores the use of deep learning-based optical character recognition (OCR) to identify part numbers for used automotive parts. It compares open source and advanced AI methods to commercial tools from Google, Amazon, and Microsoft. The study finds that fine-tuned open source models outperform commercial services, especially for complex part numbers unrelated to any language structure. The preferred open source method, MaskedTextSpotter, is fine-tuned with image data from old vehicle and electrical parts, captured by a smartphone and 2D barcode scanner. Additionally, a new data augmentation method, CharChan, is introduced, replacing detected characters with random examples for better character recognition. The experiments demonstrate the efficacy of deep learning-based OCR for automotive part number identification.

Keywords: Optical character recognition · Circular economy · Artificial intelligence · Reverse logistics

1 Introduction

At the end of a product's service life, there are various ways to reuse, recycle or dispose of the product [1]. For reuse, repair or remanufacturing, the identification of a product during the reverse logistics handling is crucial. Recently, AI-enhanced identification systems [2, 3] addressed the challenges to achieve a high recognition rate of part numbers despite the condition of nameplates, printed or embossed serial numbers and barcodes, which can deteriorate to the point of illegibility due to dirt and wear [2]. Nevertheless, the majority of products in circular economy still have - at least partly - legible part numbers which are typically read out and entered into a logistics system manually with keyboard inputs. Workers often have to wear gloves because - especially in the automotive aftermarket - old parts are very dirty and oily. The accurate identification of part numbers on dirty car parts is hindered by obscured characters and smudging caused by accumulated dirt and oil, presenting significant challenges for OCR systems striving to interpret partially illegible text accurately. An elegant solution for supporting the worker in the identification process would be to automatically read out and enter part numbers, captured by imaging systems such as handhelds, barcode scanners or stationary cameras. This paper analyzes state-of-the-art methods of AI-based text recognition methods for reading out part numbers plates of used parts. The objectives include the analysis and evaluation of different text detection and text recognition models. A comparison between commercially available and pre-trained as well as fine-tuned open source text recognition models will be performed finally.

2   Literature Review

Optical Character Recognition (OCR) converts images into analyzable, editable, and searchable data [4]. Scene Text Recognition (STR) is a challenging aspect of OCR, given the diverse environmental factors in image capture. OCR involves two steps: text or character detection, where the region of Interest (RoI) containing the text is extracted, and text recognition in the extracted region using a specific method. Open source models for experiments were chosen based on their performance in competitions like ICDAR2019 Robust Reading Challenge on Arbitrary-Shaped Text (ArT) and Out of Vocabulary Scene Text Understanding.

Cloud-based text recognition systems The companies Google, Amazon and Microsoft each offer their own commercial cloud-based computer vision service. These three services can recognize objects, faces or text in images. The services of all three manufacturers offer pre-trained text recognition services that output the characters found for an image entered. All services incur costs that depend on the number of images to be recognized. In the context of the paper, the models are referred to as Google, Amazon, and Azure.

Text detection Deep Learning algorithms have made significant progress in text detection, particularly for scanned documents. However, detecting text in real environments with varying shapes, colors, and complexities remains a challenge. Text detection methods can be divided into word-based and character-based approaches. Among these methods, TextFuseNet (TFN) performed exceptionally well in the ICDAR2019 ArT competition. In this paper, TextFuseNet [5] is used as an end-to-end model. Another effective text detector considered is the Pixel Aggregation Network (PAN) [6], which showed promising results on the SCUT-CTW1500 and ICDAR2015 datasets.

Text recognition comes after text detection and individual symbols are recognized and arranged correctly. Baek's design pattern [7] represents modern STR models. Additionally, there are Transformer-encoder-decoder-based text recognition models like ViT-STR [8]. For text recognition in text boxes, the TRBA model (TPS-ResNet-BiLSTMAttn) [7], which achieved third place in the ICDAR2019 ArT competition, was selected. The Vision Transformer for Image Recognition (ViT-STR) model, developed by Rotenzia [8], is another text recognition approach based on the decoder-encoder model.

End-to-End models integrate text detection and recognition into one pipeline, training them simultaneously for improved results. TextFuseNet (TFN) [5] is an example of such, and combines character-based and word-based text detection approaches. Another end-to-end model is MaskTextSpotter (MTS) [9], which excels in recognizing text with arbitrary shapes and was best at ICDAR2019 ArT competition. DEER [10] achieved the best results in the OOV Competition [11], offering reduced dependency on detection and recognition modules by using a single reference point for each text instance instead of region-based recognition.

3   Materials and Methods

3.1   Dataset Description

The dataset consists of images of old parts from the automotive industry and old printer cartridges. In each case, a type plate of a component is present in the images. Part of the images were taken with a ZebraScanner DS 2200 barcode scanner and the other part with a standard smartphone. The images have different resolutions from 640 × 480 pixels up to 4032 × 3024 pixels. The main dataset consists of a total of 441 images, with 75 images representing old printer cartridges. Of the 441 images, 230 were captured in color with a smartphone and 211 in grayscale with the ZebraScanner.

Fig. 1. Typical core label information

The distribution of characters is even between test and training dataset. However, the letters "q" and "z" are rarely represented in the dataset, so only small improvements in the recognition of these letters can be expected. Figure 1 shows a typical label of an automotive core. It contains information about the manufacturer, country of manufacture, date, voltage, power and current values and other designations. In Fig. 1 the manufacturer is marked in yellow, text parts are marked in red and power/voltage values are marked in blue. The serial numbers highlighted in green are the most important information in this context and should be recognized by the procedures. These serial numbers belong to three different manufacturers: BOSCH, Mercedes and DENSO. It is a starter motor that is installed in different types of vehicles (Fig. 2).

Data augmentations: Augmentation techniques involve applying various transformations to the existing data, such as rotation, scaling, flipping, and adjusting brightness or contrast. These transformations create new variations of the original data, enabling the model to generalize better when faced with diverse and challenging real-world scenarios, including dirty and oily surfaces. To extend the main data set and generalize the training data, two augmentation procedures were applied to the main data set.

Fig. 2. Top: STR Augmentation, Bottom: CharChan Augmentation

STR Augmentation: Rowel Atienza developed the STR augmentation model, which contains 36 different image augmentation functions to emulate real-world perturbations [12]. The application of STR augmentation has significantly improved the accuracy of some models. For example, the accuracy of the Rosetta model was increased by 2.10%, the R2AM model by 1.48%, and the TRBA model by 1.06% [12]. Atienza's augmentation was developed specifically for text boxes that have already been cut out, so not all the framework's features are applied to the present dataset. Augmentations that change the shape of the text boxes are therefore not considered. Of the total 36 available functions of the model, 28 are applied, resulting in an augmented dataset of 1174 images. We introduce **CharChan augmentation** to exchange characters within the text boxes of the ground truth. The objective is to hinder the model interpreting semantic information within the characters and to achieve better generalization. The characters on the images are cut out using the TFN model [5], which allows both word-based detection and character-based detection. Within the ground truth text boxes, the letters are rearranged, with the number of letters to be distributed being dictated by the original text box. The height of the new letters is adjusted to the height of the text box to avoid overlapping with other text boxes. Then the color or grayscale values of the individual character images are adjusted. This augmentation created additional 1126 images. Character recognition is particularly challenging for the model on these images.

3.2 Model Selection for Focused Analysis in Text Recognition

To focus the remainder of the paper and not include all models, they were evaluated using common evaluation metrics, as shown in Fig. 3. Based on this evaluation, certain models were selected that showed promise and performance. This targeted selection of promising models allows the paper to dive deeper into the study of these specific models and gain detailed insights into their strengths and weaknesses. This facilitates a more precise and focused analysis of text recognition and allows relevant aspects to be examined in more detail. The models were evaluated using the main dataset according to different metrics as shown in Fig. 3. The end-to-end models achieved significantly better results in this evaluation. For this reason, stand-alone detection or recognition models are not pursued further in this paper. Due to the long evaluation time, the DEER model is also not considered further. Of the end-to-end models, the MTS model has the highest recall value and is therefore considered further. Of the cloud-based services, the Google and Azure services show the best values.

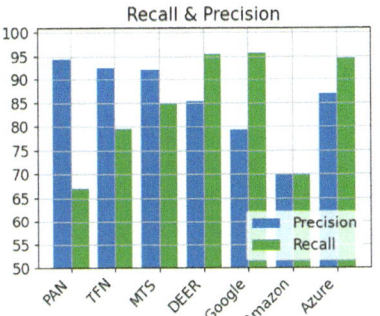 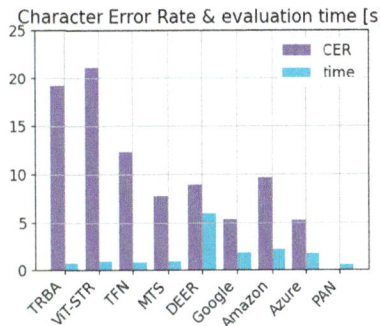

Fig. 3. Evaluation of models on main data set

3.3   Fine-Tuning the MTS Model

To enhance results on the main dataset, the MTS model undergoes fine-tuning using a transfer learning approach. Pre-trained on datasets like SynthText, ICDAR2013, ICDAR2015, Total-Text, and SCUT-CTW, the MTS model leverages its prior knowledge of various characters and text recognition. The training is conducted in two variants. In the first variant, the model is trained in three steps: STR augmentation dataset, CharChan dataset, and finally, the main dataset. Each step undergoes 25,000 iterations. The second variant also consists of three steps, with 30,000 iterations each. It starts with training solely on the main dataset, followed by gradual inclusion of STR augmentation and CharChan datasets.

3.4   Evaluation Metrics

The evaluation primarily emphasizes the recall metric when assessing predicted text boxes. This emphasis stems from the crucial need to recognize all pertinent text boxes accurately. In this context, the precision metric holds less significance because identifying more than the necessary relevant boxes is not as critical. Any inaccurately identified text boxes can be rectified later by filtering them out from the ground truth dataset. The accuracy of the recognized text is assessed using the accuracy value. For this, the Character Error Rate (CER) is calculated. And the accuracy (Acc) value is calculated as follows: $Acc = 1 - \frac{CER}{textlength}$

4 Results

In this chapter, we conduct an evaluation of the models using the datasets presented in the previous chapter. We begin by evaluating the fine-tuned MTS model and analyzing its performance. Next, we compare the MTS model's results with those of the Google Vision Cloud model.

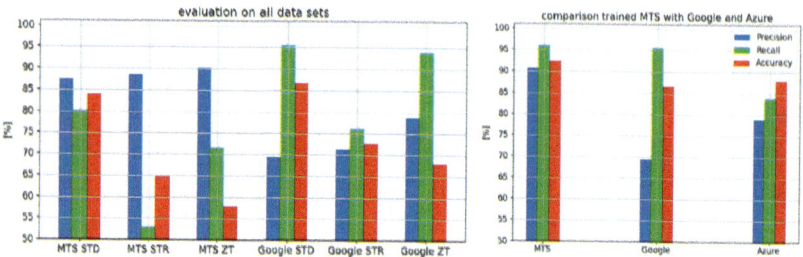

Fig. 4. Pretrained MTS (left) fine-tuned MTS (right) versus Google, Azure

Pre-trained model: The MTS and Google models exhibit varying performance in text recognition across different datasets, as illustrated in Fig. 4. The abbreviations used are STR for the STR augmentation dataset, ZT for the CharChan dataset, and STD for the main dataset. The Google model demonstrates notably superior results in both text detection and recognition compared to the pre-trained MTS model. This improvement can possibly be attributed to the Google model's enhanced generalization capabilities. It is likely that the Google model was trained on a larger and more diverse dataset, enabling it to recognize various text patterns and styles present in the input images more effectively. On the other hand, the MTS model may be less robust when encountering slightly different images, since it might not have encountered sufficient variation in the training data to handle such changes optimally. Fine-tuned model: Following the 3rd stage of training, the model achieves outstanding Accuracy, Recall, and Precision values on the main dataset, all surpassing 90%. A comparison between the Google, Azure and the MTS model after training (refer to Fig. 4) reveals that the MTS model outperforms both cloudbased models across all evaluation metrics. Specifically, the MTS model detects fewer text boxes that are not present in the ground truth and successfully identifies nearly all ground truth text boxes. Moreover, the detected text demonstrates a very low error rate, with an accuracy of 94%. Comparison of the recognition rates for different classes indicates minimal differences between both models. The training in the 2nd variant did not yield any improvement in the results, and therefore, it is not further explored in this paper.

5 Discussion

There is a concern regarding the MTS model's results after the second stage of the first variant's training. In this stage, few test dataset images from the main dataset were mistakenly included in the training dataset of augmented images. This situation is known as "data leakage", where the model benefits from test dataset information during training, potentially affecting its performance when encountering new, unseen test images. To make sure that the method did not overfit on the accidentally leaked data, these were sorted out in a following test. However, the values in Precision, Recall and Accuracy did not change significantly. Hence, data leakage was not found to be severe.

6 Conclusion

This paper analyzed and compared various text detection and recognition models using datasets of old parts from the automotive industry and old printer cartridges. End-to-end learned models outperformed stand-alone text detection models, showing at least a 15% improvement in recall values. Without fine-tuning, cloud-based models from Azure and Google achieved the best results in text detection with around 95% recall values, and in text recognition on the main dataset. In contrast, with fine-tuning for part number recognition, open source models outperformed commercial solutions, achieving more accurate and reliable text box contents with a 4 points lower character error rate (CER). While the DEER model showed exceptional performance potential, its high evaluation time made it unsuitable for time-sensitive reverse logistics operations. Such models can significantly support workers in part identification processes. By leveraging the text recognition's ability to accurately identify and categorize parts, workers can streamline their tasks, leading to increased efficiency and productivity. The CharChan Augmentation proved to be beneficial in dismissing the contextual information of characters for the models. Instead of attempting to identify complete words in the character sequence, fine-tuning with CharChan augmentation method enables pretrained models to improve text recognition. This approach proves advantageous in applications where contextual information is not crucial for accurate text detection. In the future, better AI models for text recognition will be released, mainly trained for natural language recognition. By applying techniques like CharChan to those, they will likely perform better for part number recognition on-label or any other surface.

Acknowledgements. This paper was written as part of the research and development project "Biological Transformation 4.0: Further development of Industry 4.0 by integrating biological principles (BioFusion 4.0)", which is funded by the German Federal Ministry of Education and Research (BMBF) and supervised by the Project Management Agency Karlsruhe (PTKA), as well as part of the EIBA project 033R226 in the ReziProK programm, which was funded over the FONA platform for sustainable research.

References

1. Mast J, von Unruh F, Irrek W (2022) R-Strategien als Leitlinien der Circular Economy, RETHINK. Impulse zur zirkulären Wertschöpfung - Enabling the Circular Economy, Bd. 2022 03, 25. Mai 2022. https://prosperkolleg.ruhr/wp-content/uploads/2022/05/rethink22-03r-str ategien.pdf
2. Schlüter M, Niebuhr C, Lehr J, Krüger J (2018) Vision-based identification service for remanufacturing sorting. Procedia Manuf 21:384–391. https://doi.org/10.1016/j.promfg. 2018.02.135

3. Schlüter M et al (2021) AI-enhanced identification, inspection and sorting for reverse logistics in remanufacturing. Procedia CIRP 98:300–305. https://doi.org/10.1016/j.procir.2021.01.107

4. Memon J, Sami M, Khan RA, Uddin M (2020) Handwritten optical character recognition (OCR): a comprehensive systematic literature review (SLR). IEEE Access 8:142642–142668. https://doi.org/10.1109/ACCESS.2020.3012542

5. Ye J, Chen Z, Liu J, Du B (2020) TextFuseNet: scene text detection with richer fused features. In: Proceedings of the twenty-ninth international joint conference on artificial intelligence, IJCAI-20, pp 516–522

6. Wang W, Xie E, Song X, Zang Y, Wang W, Lu T, Yu G, Shen C, Efficient and accurate arbitrary—shaped text detection with pixel aggregation network. https://doi.org/10.48550/ARXIV.1908.05900

7. Baek J, Kim G, Lee J, Park S, Han D, Yun S, Oh SJ, Lee H, What is wrong with scene text recognition model comparisons? Dataset and model analysis. https://doi.org/10.48550/ARXIV.1904.01906

8. Atienza R (2021) Vision transformer for fast and efficient scene text recognition. https://doi.org/10.48550/ARXIV.2105.08582

9. Liao M, Lyu P, He M, Yao C, Wu W, Bai X (2021) Mask TextSpotter: an end-to-end trainable neural network for spotting text with arbitrary shapes. In: IEEE transactions on pattern analysis and machine intelligence, vol 43, Nr. 2, pp 532–548. https://doi.org/10.1109/TPAMI.2019.2937086

10. Kim S, Shin S, Kim Y, Cho H, Kil T, Surh J, Park S, Lee B, Baek Y, DEER: detection-agnostic end-to-end recognizer for scene text spotting. https://doi.org/10.48550/ARXIV.2203.05122

11. Bitenand S, Bordilsand A, Delgadoand A (2022) ECCV 2022 Out of Vocabulary Scene Text Understanding. https://rrc.cvc.uab.es/?ch=19,2022

12. Atienza R (2021) Data augmentation for scene text recognition. In: CoRR abs/2108.06949. https://arxiv.org/abs/2108.06949

A Peak Shaving Approach in Manufacturing Combining Machine Learning and Job Shop Scheduling

Eddi Miller[1]([✉]), Anna-Maria Schmitt[1], Tobias Kaupp[1], Rafael Batres[2], Andreas Schiffler[1], and Jan Schmitt[1]

[1] Institute of Digital Engineering, Technical University of Applied Sciences Würzburg-Schweinfurt, Ignaz-Schön-Straße 11, Schweinfurt, Germany
eddi.miller@thws.de

[2] Tecnologico de Monterrey, Epigmenio González 500, San Pablo 76130, Queretaro, Mexico

Abstract. Computerized Numerical Control (CNC) plays an important role in highly autonomous manufacturing systems with multiple machine tools. The necessary Numerical Control (NC) programs to manufacture the parts are mostly written in standardized G-code. An a priori evaluation of the energy demand of CNC-based machine processes opens up the possibility of scheduling multiple jobs according to balanced energy consumption over a production period. Due to this, we present a combined Machine Learning (ML) and Job-Shop-Scheduling (JSS) approach to evaluate G-code for a CNC-milling process with respect to the energy demand of each G-command. The ML model training data are derived by the Latin hypercube sampling (LHS) method facing the main G-code operations G00, G01, and G02. The resulting energy demand for each job enhances a JSS algorithm to smooth the energy demand for multiple jobs, as peak power consumption needs to be avoided due to its expense.

Keywords: CNC · Machining · Energy consumption · Machine learning · Design of experiments

1 Introduction

Computer Numerical Control (CNC) is a form of automation in which the mechanical actions of a machine tool or other equipment are controlled by a program containing a list of standardized instructions. CNC has become an important technology in construction, automotive, aerospace, and medical industries, both for part manufacturing and product design [1].

The energy crisis in Europe and, as a consequence, the increasing energy cost force manufacturing companies to analyze their processes in terms of energy efficiency. In Germany, the manufacturing industry causes around 24 % of nationwide CO_2 emissions. According to Anderberg and Kara [2], the energy cost of

© The Author(s) 2025
H. Kohl et al. (Eds.): GCSM 2023, LNME, pp. 535–543, 2025.
https://doi.org/10.1007/978-3-031-77429-4_59

CNC machining accounts for 1–6% of the total machining expense, excluding the escalation rate of energy costs. Energy consumption in a CNC machine tool originates from the spindle operation, axis feed, servos system, tool change system, and cooling equipment. Traditionally, in a company that deploys a large fleet of CNC machines and in production planning, customer orders are scheduled by optimizing their makespan, lead times, or downtime of resources. Scheduling a series of orders over a defined production period (e.g. one day) can lead to electricity load peaks and can, therefore, cause higher costs. This is because, in addition to the costs for electricity consumption, there is also a considerable performance price, which the respective energy supplier calculates from the measured load peaks. Therefore, energy consumption in Job-Shop-Scheduling (JSS) has gotten a lot of attention in recent times and is a promising way of adapting to volatile electricity prices [3]. Hence, in this contribution, we present an ensemble Machine Learning (ML) model for the standard G-code operations G00 (rapid movement), G01 (linear movement), and G02 (circular movement), that predicts the energy demand for an entire Numerical Control (NC) program. The coordinates X, Y, and Z, which result in the distance traveled, f (feed rate), and s (spindle speed) are selected as input features. Additionally, for the G02 movement, the radius R is added. The aim is to estimate the energy demand of a series of jobs that should be evenly distributed over a defined production time to avoid load peaks.

In the scientific literature there are some approaches to utilizing ML to predict the energy consumption of machine tools by G-code evaluation: Borgia et al. [4] developed a Neural Network model to predict energy consumption given axis velocity, axis acceleration, spindle speed, spindle torque, and other machining parameters. They use Latin hypercube sampling (LHS) and Sobol sampling to generate data for training a two-layer Neural Network model. Bhinge et al. [5] proposed a Gaussian-Process regression model for energy prediction with three quantitative inputs (feed rate, spindle speed, and depth of cut) and two categorical inputs (cutting direction and cutting strategy). They report Normalized Mean Absolute Errors between 9.5% and 13.5% for estimating the energy for all the blocks in a given NC-code. A different approach was taken by [6]. The authors implemented a component-based energy-modeling methodology to ensure online optimization and real-time control. These energy models are able to forecast energy consumption up to the tool-path level. The effectiveness of the proposed methodology was proven on a milling machine example. Contrary to ML approaches, researchers have also provided analytical concepts to estimate the energy demand of numerical control machining. One such approach was explored by [7]. Not relying on G-code commands directly but rather on motions like auxiliary, air-cutting, or removal motions, the authors were able to estimate the power consumption of these movements. By carrying out a Design of Experiment, different lathes and CNC machine tools were tested. In the end, the authors were able to distinguish the power consumption between machines using the proposed motions. Edem et al. [8] investigated analytical models for predicting the energy demand of toolpaths. Their models aimed at estimating the energy demand for the commands G00, G01, G02, and G03. They conclude that shorter linear path lengths with G01 and circular path segments with G02 and G03 were highly energy intensive. Another analytical method was proposed by [9]. The authors first established a correlation between numerical control codes and energy consumption components of machine tools. Subsequently, the total energy consumption of an NC machine was estimated by accumulating the energy of all machine components, like spindle, axis feed, and others. To test the prediction of their analytical model, the authors compared its results to the actual energy measurements of two workpieces. The results were very conclusive and can help to make process planning more energy efficient.

In our ML approach, we minimize the number of experiments through LHS and then train models to be ensembled to estimate the energy consumption of NC programs. As a novelty, we integrate the findings through ML into a JSS algorithm to smooth the energy required for a series of jobs planned over a production period in a factory. Due to this, the contribution of this paper can be summarized as follows:

- Pioneer work to link ML and JSS to smooth energy consumption of a fleet of NC-machines during the production planning phase
- Provision of a tool to utilize order data and NC-data to optimize the production schedule
- Results of various production order scenarios indicate that our JSS algorithm is able to schedule jobs with two objectives—minimize makespan and variance of energy consumption of all working machines in a fixed time interval (e.g. 2 shift system) to reduce cost during the production planning phase by performing peak shaving as a form of load management.

The paper is structured as follows: Sect. 2 describes the research methodology and experimental setup. This includes the Design of Experiment and the generation of training data for the ML models. Based on that, Sect. 3 gives an overview of the used models and the validation results. After the ML models are established, Sect. 4 demonstrates the results these models have on the energy balancing aspect of production orders in JSS. Section 5 concludes this paper by discussing limitations and further research.

2 Research Methodology and Experimental Setup

Figure 1 shows the chosen research approach for energy assessment from G-code and JSS integration. First, a minimum of experimental sample points for ML model training are obtained by means of an optimal Latin Hypercube Sampling (LHS) design. LHS is a Design of Experiments method that produces a sampling plan in which points are spread across a bounded input space. A space-filling method such as LHS is useful when one of the objectives is to carry out as few experiments as possible. LHS divides each dimension of the input space into equally-sized bins and randomly generates a sample in each bin. Based on the work of Morris and Mitchel [10], we use Simulated Annealing for sample generation [11]. Then the experiments to obtain training data are conducted for the base G-code commands G00, G01, and G02 by varying the features X, Y, and Z for the motion path, which results in the distance as a feature. Additionally, f for the feed rate, S for spindle speed, and R for the radius of the circular interpolation of G2 were chosen as input parameters. The predictor is the energy consumption E for each G-code line. When the ensemble model is validated, the proposed method parses the G-code, calculates the energy consumption rpm for each G-code block, and then sums up the total energy consumption for the whole NC program. For each command, the model predicts the electric energy and provides the data for each job in the JSS algorithm shown in Sect. 4.

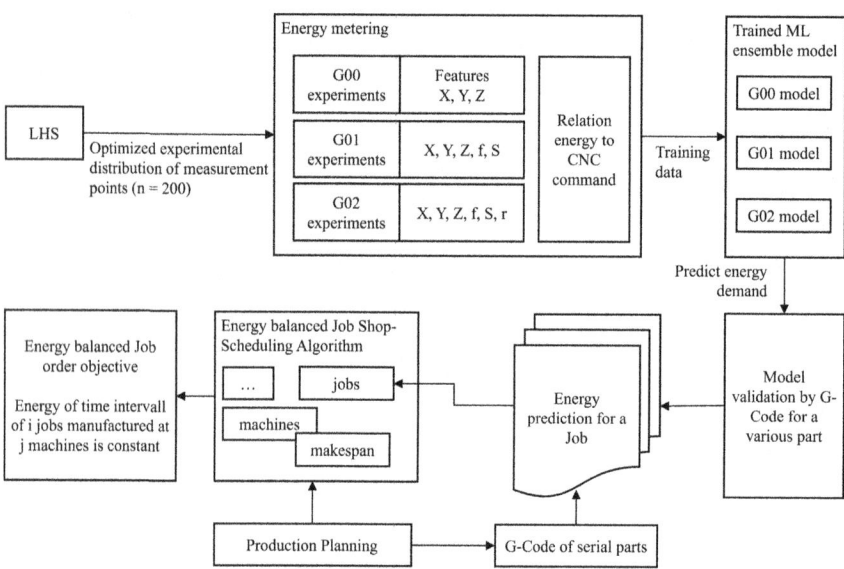

Fig. 1. Principle scheme of the combined approach of ML and JSS to conduct electricity peak load shaving in manufacturing utilizing production data (NC programs with G-code and order data)

The experiments are carried out with a Stepcraft M1000 CNC milling machine. It is a three-axis machine equipped with a spindle motor. The maximum travel distance for the X-axis is 678 mm, 1043 mm in the direction of the Y-axis, and 30 mm on the Z-axis. The distance in Z is kept shorter to account for the mounted tool. The maximum feed rate f of the machine, used with the command G00, is 5100 mm/min or 85 mm/s. For commands G01 and G02 the feed rate was set between 200 and 2000 mm/min. These feed rates are enabled by the NEMA 23 stepper motor. The spindle speed is capped out at 25000 rpm and has to run at a minimum of 10000 rpm. The machine is controlled via the UCCNC machine control software. The experimental search space is shown in Table 1. The experiments are conducted without machining a material, as we tend to

Table 1. Machining limits defining the experimental search space

	G00		G01		G02	
	min	max	min	max	min	max
X [mm]	0	678	0	300	0	200
Y [mm]	0	1043	0	500	0	300
Z [mm]	0	30	0	30	0	25
f [mm/min]	5100	5100	200	2000	200	2000
S [rpm]	0	0	10000	25000	10000	25000

validate the entire approach on the conceptual level in this contribution. Hence, the variations in milling different materials, e.g. aluminum or steel, are left out and the spindle speed is varied to in-/ decrease the energy consumption while executing G01 and G02. As previously mentioned, the Design of Experiment is based on LHS, which provided 200 sample points for each G-code command. These sample points range from 0 to 1 and are multiplied with the maximum of each feature to convert them into values the G-code can work with. A Python script created all necessary G-code lines for all three commands G00, G01, and G02 by using the values LHS provided. During all experiments, the energy of the CNC machine and spindle is measured using a time interval of 0.1 s. The values are then matched to the specific lines of G-code that correspond to the movement or activation of the component. The last step is to train the ML models, which will be explained in Sect. 3.

3   ML-Model Training and Validation

Three separate ML models for G00, G01, and G02 are trained with 80% of the available data. The different model types are trained and evaluated with the metric root mean square error (RMSE) on the remaining 20% of the data. The overall ML approach is shown in Fig. 2 (left). The RMSE of the three best models for each command are shown in Fig. 2 (right). For G00, the three best-performing models regarding the RMSE are Support Vector Regression (SVM-6.25), Bayesian Ridge Regression (BRR-6.38), and Linear Regression (LR-6.4). The best models for predicting the energy consumption of G01 are CatBoost Regression (CatBoost-539.28), Artificial Neural Network (ANN1-587.25), and Nearest Neighbor Regression (NNR1-692.83). For G02, the models with the lowest RMSE are again Artificial Neural Network (ANN2-499.44), Random Forest Regression (RF-721.9), and Nearest Neighbor Regression (NNR2-726.71).

The RMSE of G01 and G02 is significantly higher than the RMSE of G00 as the spindle speed has a high impact on the energy consumption, and for rapid movement commands (G00) the spindle is normally off. After combining the three best models to get the ensemble model, the total RMSE is 478.08.

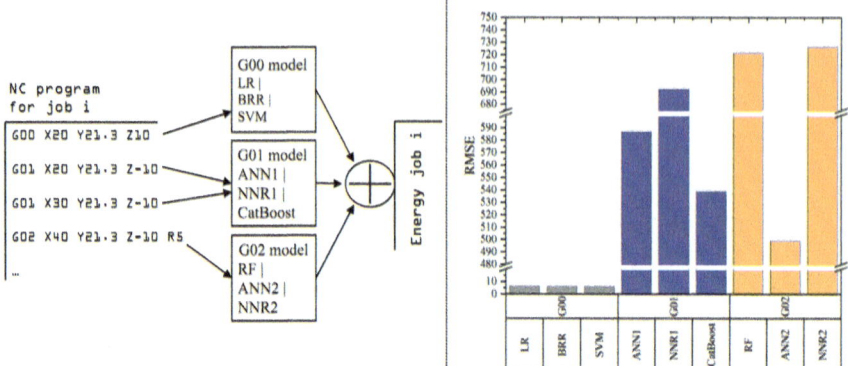

Fig. 2. G-code evaluation principle using an ensemble model approach (left); performance comparison of different ML models by RMSE measure

4   Scheduling Results and Discussion

The ML ensemble model results in an additional data point regarding the specific energy consumption of each machining job. Based on this, we developed a JSS algorithm with two objectives (see Eqs. 1 and 2) with J_i being the processing times of each production order, n the total number of orders, E the total energy consumption, and μ the mean value of the total energy consumption:

$$\min{}_{\text{Makespan}} = \sum_{i=1}^{n} J_i \tag{1}$$

$$\min{}_{\text{Var(E)}} = \frac{\sqrt{Var(E)}}{\mu} \tag{2}$$

The JSS algorithm is based on our previous research presenting a cascaded JSS approach for highly autonomous production systems [12] utilizing CP-SAT (constraint programming and Boolean satisfiability), which has a high computational effort. The current JSS algorithm further develops the approach from [12] by a new metaheuristic called Micro Evolutionary Particle Swarm Optimization (MEPSO) [13]. MEPSO is a modified Particle Swarm Optimization algorithm that includes Micropopulations, which lead to a decrease in computational complexity, an accelerated convergence rate, simplified implementation, and prevention of premature convergence. To verify the proposed JSS algorithm, we created seven scenarios SC1 to SC7, each with 16 production orders containing data on processing times, order quantity, due dates, and energy consumption. SC1, SC2, and SC3 do not consider energy consumption and are only optimized by makespan. To flatten out the energy demand, a time interval of 1000 min was chosen, which roughly translates into two work shifts in a manufacturing company. For SC4 to SC7, the JSS algorithm considers energy output for all machines and minimizes the variance between all intervals. The results show peak shaving for SC4 to SC7 under an exemplary limiting peak load of 4 kW, whereas SC1, SC2, and SC3 form a peak in different time intervals (see Fig. 3, left). The algorithm flattens the energy curves of scenarios 4 to 7 to reach a standard deviation of less than 15% from the mean energy consumption of all time intervals. The minimum distance to the limit value is at $SC4(6000$ minutes$) = 11.25$ % (see Fig. 3, right).

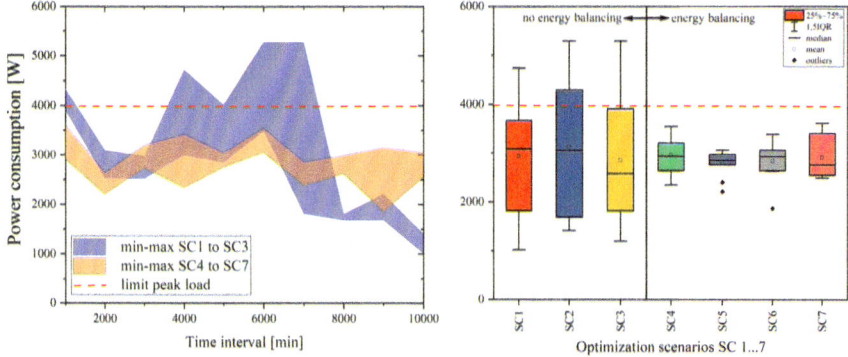

Fig. 3. Results of the energy consumption over all time intervals and scenarios SC1 to SC7 (left); box-plot of the assigned jobs and their energy consumption in each scenario

5   Conclusions, Limitations and Further Research

The presented contribution faces an innovative mix of methods regarding energy balancing or peak shaving in CNC machining for autonomous machine fleets by combining G-code analysis with ML and JSS. At this moment, an ensemble ML model is trained and validated with a minimum of experimental effort to predict the energy demand of an NC program. As a result, a customer order/machining job contains the energy consumption as additional information, which is then used to perform a JSS algorithm. The objective of this algorithm is to smooth the energy consumption of a multitude of jobs to avoid electricity load peaks over a specified production period. This will help to adapt to varying energy prices set by the power supplier to increase energy output when prices are lower and decrease power output when prices are higher. The approach presented is currently limited to the three base G-code operations G00, G01, and G02 as well as spindle on/off and motion parameters. As of now, other parameters or external boundary conditions such as material to be machined or different M-commands e.g. the cooling pump on/off are not included in the model. This is not a problem, since the chosen ML models are capable of training and validating energy consumption even without actual machining. This makes a good baseline to further enhance the ensemble ML model by steadily adding more features generated by machining actual parts. Also worth noting is the expansion of different machining equipment like robots e.g. which also contribute to power consumption. These resources can not only be optimized by power but also integrated into the JSS to further enhance other variables like makespan or cycle time.

Acknowledgment. The authors acknowledge the "Bayerisches Staatsministerium für Wirtschaft, Landesentwicklung und Energie" for funding the project "Autonomer Bereitstellungs- und Bearbeitungsassistent", Grant no. DIK-2104-0013//DIK0248/03.

References

1. Negi PK, Ram M, Yadav OP (2018) 1 CNC machine and its importance. In: Basics of CNC Programming. River Publishers, pp 1–16
2. Anderberg S, Kara S (2009) Energy and cost efficiency in CNC machining. In: The 7th CIRP Conference on Sustainable Manufacturing
3. Gong X, De Pessemier T, Martens L, Joseph W (2019) Energy-and labor-aware flexible job shop scheduling under dynamic electricity pricing: a many-objective optimization investigation. J Clean Prod 209:1078–1094
4. Borgia S, Pellegrinelli S, Bianchi G, Leonesio M (2014) A reduced model for energy consumption analysis in milling. Proced CIRP 17:529–534. Variety Management in Manufacturing
5. Bhinge R, Park J, Law KH, Dornfeld DA, Helu M, Rachuri S (2016) Toward a generalized energy prediction model for machine tools. J Manufact Sci Eng 139(4)
6. Shin SJ, Woo J, Rachuri S (2017) Energy efficiency of milling machining: component modeling and online optimization of cutting parameters. J Clean Prod 161:12–29
7. Lv J, Tang R, Jia S, Liu Y (2016) Experimental study on energy consumption of computer numerical control machine tools. J Clean Prod 112:3864–3874
8. Edem IF, Mativenga PT (2017) Modelling of energy demand from computer numerical control (CNC) toolpaths. J Clean Prod 157:310–321
9. He Y, Liu F, Wu T, Zhong F, Peng B (2012) Analysis and estimation of energy consumption for numerical control machining. Proceed Institut Mech Eng, Part B: J Eng Manuf 226(2):255–266
10. Morris MD, Mitchell TJ (1995) Exploratory designs for computational experiments. J Statis Plann Infer 43(3):381–402
11. Pholdee N, Bureerat S (2015) An efficient optimum Latin hypercube sampling technique based on sequencing optimisation using simulated annealing. Int J Syst Sci 46(10):1780–1789
12. Miller E, Kaupp T, Schmitt J (2022) Cascaded scheduling for highly autonomous production cells with agvs. In: Global Conference on Sustainable Manufacturing, pp 383–390. Springer
13. Solano-Rojas BJ, Villalón-Fonseca R, Batres R (2023) Micro evolutionary particle swarm optimization (MEPSO): a new modified metaheuristic. Syst Soft Comp, 200057

Towards Sustainable Machining: Synthetic Data Generation for Efficient Optical Tool Wear Monitoring via Generative Adversarial Networks

Markus Friedrich[1]([✉]), Engjëll Ahmeti[1], Jonas Dumler[2], and Frank Döpper[1,2]

[1] Chair Manufacturing and Remanufacturing Technology, University of Bayreuth, Bayreuth, Germany
{markus.friedrich,engjell.ahmeti,frank.doepper}@uni-bayreuth.de
[2] Fraunhofer IPA, Process Innovation Group, Bayreuth, Germany
jonas.dumler@ipa.fraunhofer.de

Abstract. To increase sustainability in the field of machining with regard to energy efficiency improvement and resource conservation, accurate tool wear classification is of a paramount importance. In particular, optical wear monitoring approaches that use artificial intelligence and computer vision techniques have shown enormous potential in the recent past. A critical point to these approaches is the requirement of a lage image dataset of worn tools. The generation of such dataset is associated with a waste of resources, since recording of the data has to be done elaborately during ongoing production and classified by an expert. In this paper, the possibilities of generating synthetic training images with the help of generative adversarial networks (GANs) were investigated. A limited set of images was acquired for training GANs which then were used to increase the number of images for the training of our tool wear classificator. The approach enables a resource-efficient approach for the training process such that the accuracy and robustness of tool wear classification systems through the use of GANs can be ensured.

Keywords: Generative Adversarial Networks · Machine Sustainability · Tool Wear · Condition Monitoring

1 Introduction

As public awareness of the effects of climate change and resource depletion has increased in recent years, so too have the requirements for sustainable manufacturing processes. This is reflected, for instance, in the increasing number of companies reporting on sustainability manners, including material assessments and decarbonization goals [1]. The field of machining with processes such as milling, turning and drilling represents a large part of the manufacturing industry and thus offers significant leverage with opportunities to respond to these requirements. One aspect to contribute to the sustainability of production is the

© The Author(s) 2025
H. Kohl et al. (Eds.): GCSM 2023, LNME, pp. 544–552, 2025.
https://doi.org/10.1007/978-3-031-77429-4_60

optimization of the tool replacement rate. The tool condition significantly affects the manufacturing quality and the energy consumption of the manufacturing process. Therefore tool replacements are often carried out rather conservatively and independently of the actual remaining useful lifetime, leading to a waste of resources in terms of tools, which are often made from critical raw materials.

In order to enable maintenance and replacement of tools as needed, thus reducing operational costs and environmental impact, efficient and reliable systems for Tool Condition Monitoring (TCM) and wear classification are required. Recent advancements in artificial intelligence (AI) and computer vision (CV) opened up new possibilities for optical tool wear monitoring and classification, where machine learning algorithms analyze images of worn tools automatically and provide valuable insights for optimizing machining processes and aid in using the tool's full lifespan. However, the most challenging and critical requirement of these approaches is the need of a large dataset of worn tool images for the training of the AI models. Generation of a comprehensive dataset of worn tool images requires the acquisition of images during production and manual classifications throughout the labeling-process by domain experts, which is very resource-intensive and time-consuming. This is why a more efficient solution for the generation of the dataset needs to be explored.

This research investigates the possibilities and limitations of synthetically generating a training dataset for tool wear classification using generative adversarial networks (GANs).

2   Related Work

2.1   Tool Condition Monitoring (TCM)

A variety of approaches have already been developed in the area of TCM. Distinctions can be made between direct and indirect approaches and additionally between online and offline methods. Indirect approaches use secondary metrics, like forces, vibrations or acoustic emissions, that are affected by tool wear. Such methods offer the advantage of simpler implementation, but accuracy often suffers due to the longer measuring distance [2]. Direct approaches typically use optical sensors for TCM and are able to provide very accurate information about the wear condition. Major challenges are the monitoring of individual cutting edges, the cost and effort of camera technology or image segmentation methods.

With online TCM approaches, the process is continuously monitored to obtain information in real time for process control, for example. These methods are often very cost-intensive and special attention must be paid to the effects of the running process. Offline approaches are based on data acquisition and evaluation after the machining process. Here, very detailed analyses can be carried out, although a direct influence on process control is rather not possible [3–8].

2.2   Optical Wear Monitoring

Decades of research have explored optical wear monitoring and classification. Initially focused on digital image processing [9–11], advancements in AI led to the integration of Computer Vision (CV) methods. These methods offer the advantages of direct methods, in particular providing very high accuracies. By using it as an online method, the wear condition of the tool can be recorded within the machine's installation space, without disrupting the process or requiring tool unclamping. Despite their benefits, CV methods are often underrepresented in publications summarizing and categorizing TCM techniques compared to other methods [5–8]. In particular, the summary publications by the authors in [3,4] specifically address wear monitoring and classification methods using CV. In terms of recent research, reference should be made in particular to the work of authors [9,12,13], individually generating good results for their respective use case.

2.3   Synthetic Data Generation for Tool Wear Monitoring

The primary drawback of using CV for TCM is the need for a large training data set. Traditional data collection involves the recording of a huge number of worn tools during production and manual classification by domain experts, which is resource-intensive and time-consuming. The latter classification introduces human biases, leading to dataset inconsistencies. Although this approach is viable for individual tools, an alternative approach to data collection is needed, particularly for generalizing the TCM method.

The generation of synthetic data is a promising solution to overcome these challenges. Based on only a few real images, huge volumes of realistic images can be generated representing different wear conditions. Therefore the dependency on manual data collection is reduced. Additionally, it provides the ability to augment data sets with underrepresented wear patterns, improving the accuracy and robustness of tool wear classification models.

Generative Adversarial Networks (GANs) have gained significant attention for their ability to generate synthetic data. First a generator network learns to generate synthetic data resembling the training data. Then a discriminator network tries to distinguish between real and synthetic data samples. During the adversarial training process, both, the generator and discriminator networks improve iteratively, leading to increasingly realistic data samples. [14]

For TCM, GANs have been applied in previous research mainly on time series data such as vibrations or acoustic emissions or their image representation in plots of spectral data [15,16]. However, applications from other fields show the potential of GANs in terms of generating synthetic image data. Therefore, in the following, GANs are used to generate synthetic images of worn tools for resource-efficient training of wear classification models.

The Deep Convolutional GAN (DCGAN) adds convolutional layers in both networks, which enable the networks to effectively capture spatial dependencies and learn hierarchical representations, making it suitable for generating complex images of worn tools.

3   Methodology

3.1   Image Acquisition for Training

The camera system and algorithms for image data acquisition and preprocessing have been developed during earlier research and are described in more detail in the corresponding publication [17]. The tool studied is a face milling tool used for machining of tungsten. One tool has 14 individual cutting edges. The images of the cutting edges were divided into four wear states (new, light, medium and heavy) by manual and subjective labeling by domain experts. To use the image data for training the DCGAN, the images were further cropped to a size of 64×64 pixels.

3.2   Architecture and Training of the DCGAN

The same model with the same architecture as in [18] was built and trained in accordance with the design of the generator and discriminator of DCGAN. The hyperparameters were adjusted. A DCGAN model was trained for each class using 50 images per wear state, i.e. New, Light, Medium, and Heavy tool wears, and after hyperparameter tuning, each of the models had a different set of hyperparameters. The learning rate (lr) and epochs are the two key hyperparameters that were tweaked for each class. Learning rates of 0.0002, 0.0005, 0.0009, and 0.00005 and epoch values of 500, 1000, 2200, 5000, 10000, and 15000 were investigated. By generating images at the end of each training with respective models and assessing the quality of images for tool wear with the naked eye, the evaluation of the hyperparameters was completed. Table 1 shows the ideal set of hyperparameters for each class-DCGAN.

Table 1. DCGAN hyperparameters per each class.

	Learning rate	Epochs
New	0.00005	10000
Light	0.00002	10000
Medium	0.0005	500
Heavy	0.0009	2200

3.3   Architecture and Training of the Tool Wear Classifier (TWC)

A CNN was used to classify tool wear. The basis for the CNN used is ToolWear-Net (see [9]), a model developed for a similar application and adapted from our previous research [17]. Based on the GANs output of image (64×64), the CNN model's input shape was changed to meet the image size requirement. There were three models trained with three different image datasets. Images used on this paper were described in Sect. 3.1. The first model (M_1) was trained on 1345 real images. The second model (M_2) was trained on 200 real images (50 real images per class). The third model (M_3) was trained on 400 images (200 real images and 200 generated images). Hence, 50 real images and 50 generated images per class were used. All models were tested on 200 real images.

4   Results and Analysis

This section first explains the metrics used to evaluate the GAN and TWC models and then analyzes the performance of the aforementioned models.

4.1   Evaluation Metrics

The performance of GANs is commonly evaluated using metrics that evaluate quality and diversity of the generated data. In the context of this paper, we will focus on two widely used metrics: the Inception Score (IS) and the Fréchet Inception Distance (FID).

With the help of the IS, the quality and diversity of the generated image data can be evaluated through the evaluation of generative model G [19]. This particular score is used due to its ability to correlate with human judgements of image quality. Here, the generator model G is evaluated based on its generated images x with their distributions p_g compared to the real ones y.

$$IS(G) = exp\left(\mathbb{E}_{x \sim p_g} D_{KL}\big(p(y|x)\|p(y)\big) \right) \tag{1}$$

FID is a metric that maps a set of samples to a shared feature space, where by treating the feature space as a continuous multivariate Gaussian distribution, it becomes possible to examine the difference between the mean ($\mu_y, \mu_{\hat{y}}$) and covariance ($Cov_{y,y}$ and $Cov_{\hat{y},\hat{y}}$) of two different datasets [20]. The distance between the two distributions can be calculated by the Fréchet distance (Wasserstein-2 distance), where Tr stands for trace and is the sum calculation of a matrix along the diagonals:

$$FID = \| \mu_y - \mu_{\hat{y}} \|^2 + Tr\left(Cov_{y,y} + Cov_{\hat{y},\hat{y}} - 2\sqrt{Cov_{y,y} \cdot Cov_{\hat{y},\hat{y}}} \right) \tag{2}$$

To assess the TWC's performance, F1-score is used as a balanced measure for precision and recall.

4.2   Performance of GAN and TWC

The performance of the DCGAN model is evaluated using the two metrics IS and FID. Table 2 shows the respective results for the four individual wear classes.

The relatively low values of IS suggest a low diversity of variants in the generated images, which is due to the previous sorting by wear classes. The differences within the respective class are also marginal in the training images.

Table 2. Descriptive statistics for the generated images using DCGAN.

	Inception score (IS)	Fréchet inception distance (FID)
New	1.31	192.79
Light	1.412	236.67
Medium	1.44	208.19
Heavy	1.3	256.12

FID values between 192.79 and 256.12 indicate a moderate distance between the distribution of generated images and the distribution of real images. This could indicate that the underlying patterns of wear properties in the images could not be fully captured.

In general, these metrics show that, after adjusting the parameters, the model can be used to generate images that capture and reflect the corresponding properties of the real images. However, it is also shown, that there is further room for improvement, which can only be exploited to a limited extent by further tuning of the hyperparameters. Nevertheless, after manual analysis of the generated image data, the quality was found to be sufficient and the generated images were used to train the TWC.

Figure 1 shows a comparison of one real image from each of the four wear classes with three generated images each. It can be seen that the generated images show very high similarities to the real images and represent new images, thus providing a good extension of the limited real dataset.

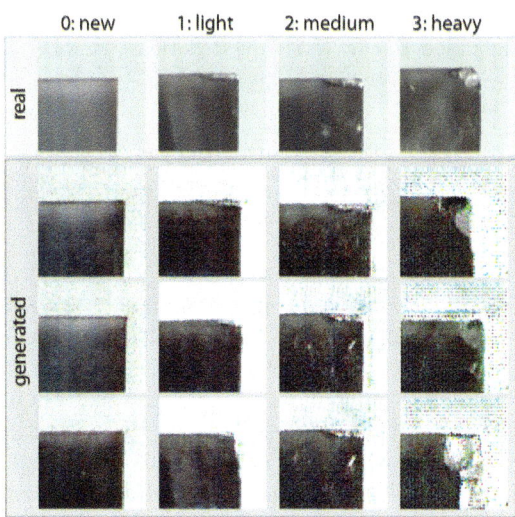

Fig. 1. Display of the real images (top row) versus the generated images (rows 2–4) of the respective wear class

To analyze the performance of the TWC trained using synthetic data, the f1-scores (Table 3) and confusion matrices (Table 4) for each run will be shown.

Table 3. F1-Scores of the three models for the individual classes

	M_1	M_2	M_3
New	1.00	0.82	0.99
Light	0.64	0.08	0.59
Medium	0.77	0.52	0.54
Heavy	0.89	0.67	0.76

5   Discussion

With the help of only 50 real images per wear class, synthetic images could be generated that significantly improve the classification results. This is evident when comparing the corresponding metrics (see Table 3) and the confusion matrices (see Table 4). In comparison with the original TWC (M_1) trained with 1345 real images, the weak points are evident after analysis of the confusion matrices.

Table 4. Confusion matrices of the three models

	M_1				M_2				M_3			
	New	Light	Med.	Heavy	New	Light	Med.	Heavy	New	Light	Med.	Heavy
New	40	0	0	0	28	6	4	2	39	0	0	1
Light	0	3	36	1	0	2	22	16	2	25	7	6
Med.	0	0	37	3	0	0	24	16	0	21	12	7
Heavy	0	0	1	39	0	0	3	37	0	1	1	38

One of the key advantages of the proposed approach is its scalability. By using GANs for synthetic data generation, it could be demonstrated that a TWC can achieve promising results with a significantly reduced number of real images. The presented approach shows promising results, however, certain limitations need to be acknowledged. The performance of the TWC with synthetic data relies heavily on the quality and diversity of the initial set of real images, which explains the confusion of wear classes "light" and "medium". The initial labeling process was done subjectively by an expert, introducing bias to the original dataset. Secondly, metrics such as Inception Score and Fréchet Inception Distance could provide some quantitative insights to the quality of generated images, the GANs were trained in isolation for each wear class, limiting the diversity and variations in the training dataset. Nonetheless the ability to generate realistic tool wear patterns through GANs offers substantial contributions to sustainable machining, since the dependence on extensive data collection can be greatly reduced. In this case, the necessary amount of training data was reduced from 1345 to 200 real images, which corresponds to a reduction of about 85 % and thus also promises a reduction of the required resources (time, energy etc.) in a similar order of magnitude during data acquisition. Moreover, the resource-efficient training process enabled by synthetic data aligns with the principles of sustainable manufacturing, promoting energy efficiency and reducing environmental impacts.

6   Conclusion and Outlook

In this paper, a novel approach utilizing synthetic data generation with DCGAN has shown promising results in tool wear monitoring. By using only 50 real images per wear class, good classification results could be shown. The scalability of this approach opens up possibilites for future developments. However, challenges such as ensuring dataset diversity and quality, as well as the bias introduced by human expert evaluation should be addressed to enhance the effectiveness of the synthetic data generation process. Future research should focus on refining this process, including the exploration of advanced GAN architectures and integrating objective metrics while labeling the initial dataset. In summary, the presented approach represents a significant step towards sustainable machining by reducing resource consumption and improving tool wear monitoring systems. With further research and development, synthetic data generation using GANs has promising potential for more efficient and environmentally responsible maching processes.

References

1. KPMG Int. (2022) Big shifts, small steps—survey of sustainability reporting
2. Karpuschewski B (2001) Sensoren zur Prozeßüberwachung beim Spanen, Fortschr.-Ber. VDI Reihe 2, Nr. 581. Düsseldorf: VDI Verlag
3. Dutta S, Pal S, Mukhopadhyay S, Sen R (2013) Application of digital image processing in tool condition monitoring: a review. CIRP J Manuf Sci Tech Jg 6(3). https://doi.org/10.1016/j.cirpj.2013.02.005
4. Kurada S, Bradley C (1997) A review of machine vision sensors for tool condition monitoring. Jg 34(1):55–72. https://doi.org/10.1016/S0166-3615(96)00075-9
5. Rehorn AG, Jiang J, Orban PE (2005) State-of-the-art methods and results in tool condition monitoring: a review. Int J Adv Manufact Tech 26(7–8):693–710. https://doi.org/10.1007/s00170-004-2038-2
6. Jantunen E (2002) A summary of methods applied to tool condition monitoring in drilling. Int J Mach Tools Manuf 42(9)
7. Nath C (2020) Integrated tool condition monitoring systems and their applications: a comprehensive review. Proced Manuf 48:852–863

8. Kuntoğlu M, Aslan A, Pimenov DY, Usca ÜA, Salur E, Gupta MK, Mikolajczyk T, Giasin K, Kapłonek W, Sharma S (2021) A review of indirect tool condition monitoring systems and decision-making methods in turning: critical analysis and trends. Sensors 21:108. https://doi.org/10.3390/s21010108

9. Bergs T, Holst C, Gupta P, Augspurger T (2020) Digital image processing with deep learning for automated cutting tool wear detection. Proced Manuf

10. Fernández-Robles L, Sánchez-González L, Díez-González J, Castejón-Limas M, Pérez, H (2021) Use of image processing to monitor tool wear in micro milling. Neurocomp

11. Bagga PJ, Makhesana MA, Patel K, Patel KM (2021) Tool wear monitoring in turning using image processing techniques

12. Hou Q, Jie S, Panling H (2019) A novel algorithm for tool wear online inspection based on machine vision [online]. Int J Adv Manuf Techn Jg 101(9-12):2415–2423. https://doi.org/10.1007/s00170-018-3080-9

13. Wu X, Yahui L, Xianliang Z, Aolei M (2019) Automatic identification of tool wear based on convolutional neural network in face milling process. Sensors Jg 19. https://doi.org/10.3390/s19183817

14. Goodfellow I, Pouget-Abadie J, Mirza M, Xu B, Warde-Farley D, Ozair S, Courville A, Bengio Y (2014) Generative adversarial nets. Adv Neural Inform Process Syst, 2672–2680

15. Shah M, Borade H, Sanghavi V, Purohit A, Wankhede V, Vakharia V (2023) Enhancing tool wear prediction accuracy using Walsh-Hadamard transform. DCGAN and Dragonfly algorithm-based feature selection. Sensors

16. Chen B-X, Chen Y-C, Loh C-H, Chou Y-C, Wang F-C, Su C-T (2022) Application of generative adversarial network and diverse feature extraction methods to enhance classification accuracy of tool-wear status. Electronics 11:2364

17. Friedrich M, Gerber T, Dumler J, Doepper F (2023) A system for automated tool wear monitoring and classification using computer vision. Proced CIRP

18. Radford A, Metz L, Chintala S (2016) Unsupervised representation learning with deep convolutional generative adversarial networks. http://arxiv.org/abs/2016

19. Barratt S, Sharma R (2018) A note on the inception score. ArXiv

20. Dowson D, Landau B (1982) The Fréchet distance between multivariate normal distributions. J Multivariate Anal 12(3):450–455

Language Models for Functional Digital Twin of Circular Manufacturing

Haluk Akay[1,2(✉)], Antonio J. Capezza[3], Maryna Henrysson[1,2],
Iolanda Leite[1,4], and Francesco Fuso-Nerini[1,2]

[1] KTH Climate Action Centre, KTH Royal Institute of Technology,
Stockholm, Sweden
[2] Department of Energy Technology, KTH Royal Institute of Technology,
Stockholm, Sweden
haluk@kth.se
[3] Department of Fiber and Polymer Technology, KTH Royal Institute of Technology,
Stockholm, Sweden
[4] Department of Intelligent Systems, KTH Royal Institute of Technology,
Stockholm, Sweden

Abstract. A key challenge for implementation of a circular economy model in manufacturing systems is the functional dependence of downstream processes on upstream byproducts. Design principles provide a framework for mapping goals to solutions by decomposing complex engineering problems into structured sets of requirements to be satisfied and embodied by design parameters and process variables. Large Language Models can computationally represent such textually-described design elements to quantify interconnections between problems, solutions, and processes. We present a Functional Digital Twin concept, powered by AI language modeling and guided by principles of manufacturing systems design, to identify functionally coupled process variables in an industrial symbiosis and automatically push alerts to stakeholders in a circular manufacturing system. Changes in byproduct composition are pushed downstream, and upstream decision-makers are guided to balance satisfying their design requirements with maintaining circularity of the system. The presented method is demonstrated in a case study of bio-based absorbent materials for intended use in disposable sanitary articles developed from byproducts of the agro-food industry.

Keywords: Language Models · Digital Twin · Industrial Symbiosis · Circular Economy

1 Introduction

To achieve sustainable manufacturing, industry must introduce solutions that maximize satisfaction of human needs while minimizing negative externalities affecting society and the environment, optimising resource flows and minimizing dependency on nonrenewable resources. Manufacturing systems must be

H. Kohl et al. (Eds.): GCSM 2023, LNME, pp. 553–561, 2025.
https://doi.org/10.1007/978-3-031-77429-4_61

designed to uphold boundaries of planetary health [1] while preserving economic competitiveness [2]. As demand increases with growing population, so does consumption of materials and energy in production. The circular economy proposes an alternative to prevailing linear "take, make, waste" resource consumption by closing loops in industrial ecosystems to minimize wastage and maximize value of resources, materials and products [3,4]. Circularity can be implemented at the systems level, connecting processes through material, energy, and information flow with an industrial symbiosis framework [5]. A mechanism for facilitating such industrial interconnections is input-output matching [6] which involves using the byproducts of one process as the inputs to another, illustrated in Fig. 1. Input-output matching is an efficient mechanism for decreasing and optimizing resource consumption and waste but in reality, industrial processes continuously experience adjustments according to new process technologies, regulations, and scale requirements. Flexibility is a key value of advanced manufacturing [7], and coupling upstream outputs to downstream inputs challenges the ability of a manufacturing system to be agile under disruption.

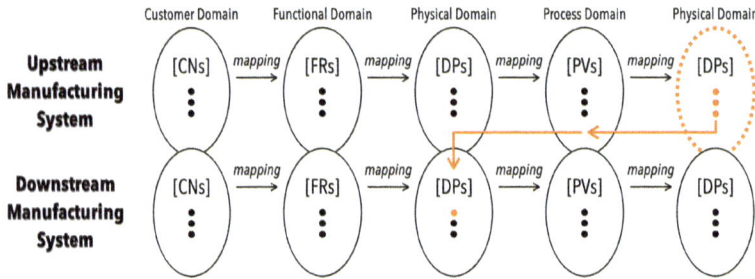

Fig. 1. System representation of connection between physical byproducts of one process coupled (orange line) to the physical inputs of another downstream.

In circular manufacturing, downstream processes are meticulously designed through resource-intensive and time-consuming innovation, such as for the bio-based absorbent materials [8] used in this work's case study. Adjustments to the upstream outputs can interrupt circularity for months resulting in capital loss and compromise in material quality as downstream design adapts. While not feasible to control all external factors, decision processes involved in adapting to new conditions can be cooperative by sharing design information bidirectionally upstream and downstream. Digital twins of manufacturing systems are a concept using virtual representations of an environment [9], representing physical processes to experiment and monitor in a lower-cost simulation setting. However, the physical domain of a process does not necessarily encompass information from society, the environment, or early-stage design. The concept of a *functional* digital twin presented in this paper instead represents the functional domain of a product's development, tracking how societal needs map to design requirements which must be addressed and produced.

2   Digital Representations of Manufacturing Systems

Design methodology such as Axiomatic Design (AD) [10] provide framework for representing mapping from customer needs in society to functional requirements that define the concrete goals of a design, to design parameters of the physical embodiment of a product, to the process variables governing production, as shown in Fig. 1. AD is useful in digital transformations of manufacturing because such mappings between domains are quantified in vector space representing function. Design methodology often lack consistent adoption in industry [11], particularly because requirements are expressed textually, a challenge to quantify for computation. Recent progress has been made in AI technology for representing language for machine understanding. The shift from rule-based algorithms relying on keywords to probabilistic models which learn language meaning by training on examples of context [12] and the application of neural networks [13] accelerated this field. Improvements on learning architecture, including the development of the Transformer [14] which decouples contextual relationships from the distance between words in a sequence, provide a framework for training language models such as Google's BERT or OpenAI's GPT that have transformed the value of textual data in many fields but not yet extensively in manufacturing.

Widespread digitalization in production has resulted in massive quantities of data, and also various concepts for representing such data in manufacturing systems. Digital Product Passport (DPP) provides information exchange in the supply chain to facilitate stakeholder decision-making in remanufacturing [15]. The digital twin concept has been proposed for tracking energy and material flows in small volume furniture production systems [16]. Digital modeling of material and information flow in the timber value chain provides data-driven decision-making in building prefabrication. [17]. In the specific case of input-output matching, ontological approaches have been explored to develop algorithms for valorizing resources in industrial symbioses [18]. Prior work representing functional information in manufacturing systems by applying statistical language models has demonstrated how to extract *what-how* structures from process documentation [19], and estimate metrics for functional coupling between requirements and design parameters [20,21]. In this work, such representations are used to identify coupled functions between processes in an industrial symbiosis, so that such information could be pushed to decision-makers [22] to strategically preserve circularity.

3   Method: Functional Digital Twin Concept

The functional digital twin concept models the evolution of problem-solving from the customer domain through design and manufacturing to the process domain, mapping the effect of decisions along this sequence, and alerting stakeholders when dependent processes in an industrial symbiosis are affected by upstream adjustments, as illustrated in Fig. 2. Axiomatic Design (AD) provides a framework for relating customer needs (CNs) from society, functional requirements

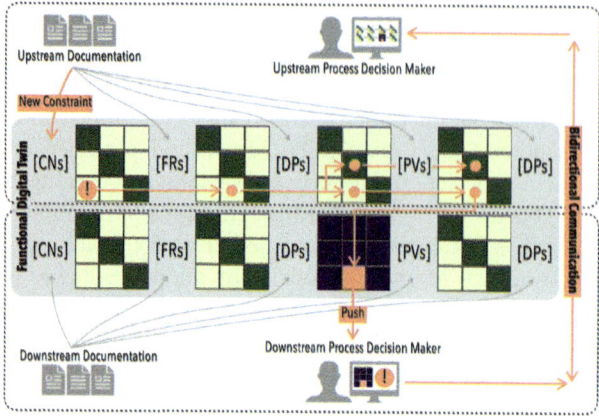

Fig. 2. Functional digital twin maps effects of new upstream adjustments to downstream processes and pushes actions to decision-makers.

(FRs) as the minimum set of goals that collectively exhaust the necessary functionalities of a product, design parameters (DPs) describing the physical embodiment of the solution, and process variables (PVs) defining production. To model input/output resource interdependence in a circular economy, we reintroduce the physical domain after the process domain, populated again by physical DPs but now mapped from process variables, post-process domain. The mapping between each domain can be expanded element-wise for a system of any size. This is described in Eq. 1 for a 2×2 system mapping the functional domain to the physical domain.

$$\begin{bmatrix} \mathbf{FR_1} \\ \mathbf{FR_2} \end{bmatrix} = \begin{bmatrix} b_{11} & b_{12} \\ b_{21} & b_{22} \end{bmatrix} \begin{bmatrix} \mathbf{DP_1} \\ \mathbf{DP_2} \end{bmatrix} \tag{1}$$

Elements b_{ij} where $i = j$ reflect the degree to which ith requirement is satisfied. Other elements, reflect functional relationships or *couplings* which can lead to inefficiencies and difficulties when adjusting the system [10]. The element-wise values of any matrix mapping between manufacturing domains b_{ij} are approximated using distributed vector representations from a statistical language model [23], and computed as the cosine similarity being the dot product of the vectors normalized by their magnitude, described in Eq. (2). Previously we demonstrated semantic similarity as a metric of functional coupling in manufacturing systems [21]. The textual descriptions can be extracted automatically from documentation [19], surfaced from decision-makers at the point of decision [22], or input manually.

$$b_{ij} = \frac{vec(\mathbf{FR}_i) \cdot vec(\mathbf{DP}_j)}{\|vec(\mathbf{FR}_i)\| \|vec(\mathbf{DP}_j)\|} \tag{2}$$

Table 1. Upstream agro-food process domains (adjustment in red text)

Customer	Functional	Physical in	Process	Physical out
CN_1:must not dispose concentrated liquid byproducts into surface water	FR_1:byproducts must be soluble in water for environmental health	DP_{i1}:dry liquid protein concentrate into powder form	PV_1:apply steam and high temperature to dry protein concentrate	DP_{o1}:dry protein concentrate
CN_2:minimize energy consumption of process	FR_2:minimize treatment of byproducts after processing	DP_{i2}:clean starch and gluten products	PV_2:clean with centrifuge washing machine	DP_{o2}:wet high water-content protein suspension

Table 2. Downstream sanitary process domains prior to upstream adjustment

Customer	Functional	Physical in	Process	Physical out
CN_1:the product should feel comfortable at all times while being worn	FR_1:the product must be able to uptake liquid at all times while being worn	DP_{i1}:bimodal distribution of pores in absorptive dry protein concentrate to uptake liquid during use	PV_1:create pores in dry protein concentrate with continuous extrusion process	DP_{o1}:waste water from extrusion process
CN_2:the product should remain the same aesthetic color at all times during use	FR_2:the product must remain inert to preserve color at all times during use	DP_{i2}:inert gases in product packaging to prevent color change through oxidation before use	PV_2:nitrogen gas injection into product packaging	DP_{o2}:excess plastic packaging material
CN_3:the product should remain usable for a long time	FR_3:the product must maintain functionality over time	DP_{i3}:cross link proteins to maintain functionality over time by delaying material degradation	PV_3:add cross linking agent Genipin during continuous extrusion process	DP_{o3}:leftover pulp and peel from Genipin extraction

Table 3. Downstream sanitary process domains after upstream adjustment (in red text)

Customer	Functional	Physical in	Process	Physical out
CN_1:the product should feel comfortable at all times while being worn	FR_1:the product must be able to uptake liquid at all times while being worn	DP_{i1}:bimodal distribution of pores in absorptive wet high water-content protein suspension to uptake liquid during use	PV_1:create pores in high water-content protein suspension with foaming agent batch-wise in oven	DP_{o1}:excess protein suspension material
CN_2:the product should remain the same aesthetic color at all times during use	FR_2:the product must remain inert to preserve color at all times during use	DP_{i2}:inert gases in product packaging to prevent color change through oxidation before use	PV_2:nitrogen gas injection into product packaging	DP_{o2}:excess plastic packaging material
CN_3:the product should remain usable for a long time	FR_3:the product must maintain functionality over time	DP_{i3}:cross link proteins to maintain functionality over time by delaying material degradation	PV_3:add cross linking agent Genipin batch-wise in oven	DP_{o3}:leftover pulp and peel from Genipin extraction

When an adjustment to customer needs, the design, or process occurs at any domain in the industrial symbiosis, the change is captured by the functional digital twin representing the mappings between domains. The propagated effects of the change downstream are used to notify decision-makers to facilitate bidirectional communication as an early-warning tool to alert threats to circularity the moment an adjustment is documented.

3.1   Case Study: Bio-Based Absorbent Material Development

In this case study we focus on an example in bio-based material development, specifically the case of utilizing outputs from the agro-food industry to construct absorbent materials which may be used as an alternative to single-use plastic products (Table 2). To apply circular principles within this value chain economy, starch byproducts obtained from potato processing undergo a continuous process to be transformed into an absorptive material by creating pores. We examine the effect of a hypothetical cost-saving measure by the upstream potato manufacturer to stop drying the byproducts into powder resulting instead in a liquid suspension (Table 1). This triggers the need for downstream process adjustment (Table 3), one option of which is from a continuous to batch-wise process. We create a functional digital twin of both the upstream and downstream processes and measure the effect of the upstream change on the downstream mappings.

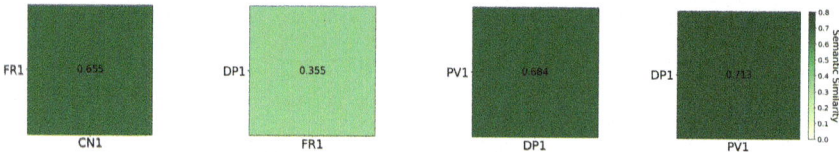

Fig. 3. Upstream domain mappings prior to adjustment

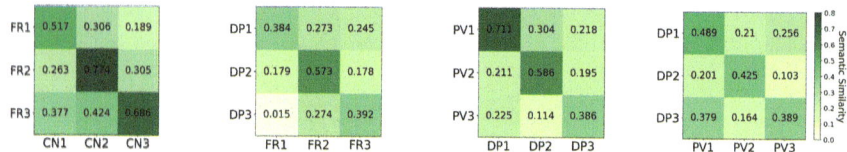

Fig. 4. Downstream domain mappings prior to adjustment

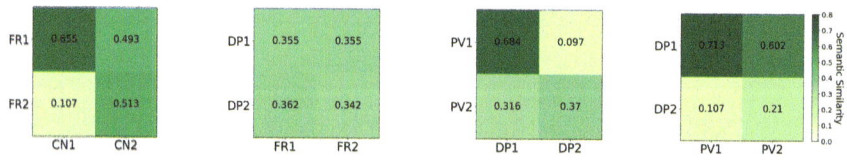

Fig. 5. Upstream domain mappings after adjustment

4   Discussion of Results

Figures 3 and 4 visualize functional coupling between domains measured by semantic similarity (Eq. 2) prior to the introduction of a new upstream customer need. Figure 5 shows the upstream domains after this adjustment, and

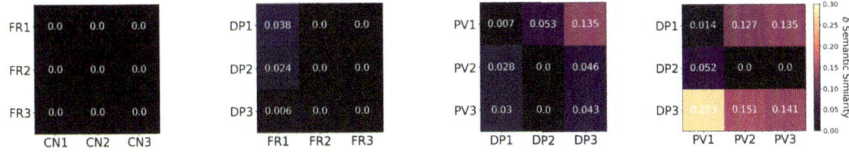

Fig. 6. Change to downstream domain mappings due to adjustment

Fig. 6 shows the *change in* (rather than absolute value of) the value of the downstream mappings due to the adjustment. In each downstream 3×3 domain the mapping is highest in value along the diagonal, which is expected. The introduced upstream customer need's effects propagate through the symbiosis to the downstream process, and the change in value of the mappings is captured in Fig. 6. The highest change in mappings is measured for the downstream process to physical outputs, due to the continuous process switching to a batch-wise process resulting in different byproducts. The effects of the adjustment propagate to influence two of the three downstream functions embodied by PV_1 and PV_3. Because this circular case is physically input-output coupled, no change is observed to mappings from customer needs in the downstream process.

A limitation of the direct application of statistical language modeling is the possibility to introduce bias from vector representations of textually described customer needs. Language is often gendered and using data-driven models to compute mappings from societal needs into design requirements will perform more accurately on descriptions of needs more commonly occurring in training data [24], which often skews to English-language sources in the Global North. Before integrating this concept at scale, thorough risk analysis of bias is required.

5 Conclusion

A Functional Digital Twin concept was presented, using descriptions of manufacturing systems quantified by language models to construct a model tracing the satisfaction and embodiment of function along product development domains. For a circular economy, it was shown how this concept can measure the propagating effects of an upstream change through an industrial symbiosis. This case study was performed on high-level functions through product development domains. The effects measured may be intuitive for an expert, but if such a tool is deployed at scale for a manufacturing system with hundreds of variables, then human decision making can be significantly augmented to preserve circularity. This concept can also be applied to any large system of stakeholders to enhance cooperation and guide decision making resulting in mutually beneficial outcomes in the face of complex functional challenges.

Acknowledgment. Digital Futures Postdoc Fellowship (Data-Driven Design for Climate Action), KTH Sustainability Office (Environment and sustainability without boundaries grant), Swedish Research Council (BioRESorb).

References

1. Johan Rockström et al (2009) A safe operating space for humanity. Nature 461.7263:472–475
2. Stark R, Seliger G, Bonvoisin J (2017) Sustainable manufacturing: challenges, solutions and implementation perspectives. Springer Nature
3. Kirchherr J et al (2023) Conceptualizing the circular economy (revisited): an analysis of 221 definitions. Resou, Conserv Recyc 194:107001
4. Stahel W (2016) The circular economy. Nature 531:435–438
5. Boons F, Spekkink W, Mouzakitis Y (2011) The dynamics of industrial symbiosis: a proposal for a conceptual framework based upon a comprehensive literature review. J Clean Prod 19:9–10
6. Halstenberg F, Lindow K, Stark R (2017) Utilization of product lifecycle data from PLM systems in platforms for industrial symbiosis. Proced Manuf 8:369–376
7. Wheelwright SC (1985) Restoring the competitive edge in US manufacturing. In: California Manage Rev 27(3):26–42
8. Capezza A et al (2023) Greenhouse gas emissions of biobased diapers containing chemically modified protein superabsorbents. J Clean Prod 387:135830
9. Grieves M, Vickers J (2017) Digital twin: mitigating unpredictable, undesirable emergent behavior in complex systems. In: Transdisciplinary perspectives on complex systems: new findings and approaches
10. Suh NP (1990) The principles of design. Oxford University Press, p 6
11. Nordlund M, Tate D, Suh NP (1996) Growth of axiomatic design through industrial practice'. In: 3rd CIRP Workshop on Design and the Implementation of Intelligent Manufacturing Systems, pp. 77–84
12. Bengio Y, Ducharme R, Vincent P (2000) A neural probabilistic language model. In: Advances in neural information processing systems, p 13
13. Mikolov T et al (2013) Distributed representations of words and phrases and their compositionality. In: Advances in neural information processing systems, p 26
14. Vaswani A et al (2017) Attention is all you need. In: Advances in neural information processing systems, p 30
15. Gallina V et al (2022) Reducing remanufacturing uncertainties with the digital product passport. In: Global Conference on Sustainable Manufacturing, pp 60–67. Springer
16. Krommes S, Tomaschko F (2022) Conceptual framework of a digital twin fostering sustainable manufacturing in a brownfield approach of small volume production for SMEs'. In: Global Conference on Sustainable Manufacturing, pp 519–527. Springer
17. Schramm N, Oertwig N, Kohl H (2022) Conceptual approach for a digital value creation chain within the timber construction industry—potentials and requirements. In: Manufacturing driving circular economy, p 595
18. Trokanas N, Cecelja F, Raafat T (2014) Semantic input/output matching for waste processing in industrial symbiosis. Comp Chem Eng 66:259–268
19. Akay H, Kim S-G (2021) Reading functional requirements using machine learning-based language processing'. In: CIRP Ann 70(1)
20. Akay H, Kim S-G (2020) Design transcription: deep learning based design feature representation. In: CIRP Ann 69(10):141–144
21. Akay H, Kim S-G (2020) Measuring functional independence in design with deep-learning language representation models. Proced CIRP 91:528–533
22. Akay H, Lee SH, Kim S-G (2023) Push-pull digital thread for digital transformation of manufacturing systems. CIRP Ann

23. Reimers N, Gurevych I (2019) Sentence-BERT: sentence embeddings using Siamese BERT-networks. In: Proceedings of EMNLP-IJCNLP
24. Rudinger R et al (2018) Gender bias in coreference resolution'. In: Proceedings of NAACL-HLT, pp 8–14

Process- and Material-Specific Modeling to Use in Simulation-Based and Resource-Oriented Decision Support Systems Using the Example of Calcium Silicate Brick Production Planning

Tobias Schrage[1]([✉]), Peter Schuderer[1], and Jörg Franke[2]

[1] Technische Hochschule Ingolstadt, Esplanade 10, 80584 Ingolstadt, Germany
tobias.schrage@thi.de

[2] Department of Mechanical Engineering, Institute for Factory Automation and Production Systems, Friedrich-Alexander-University, Egerlandstr.7, 91058 Erlangen, Germany

Abstract. As part of the fourth industrial revolution, data analysis and artificial intelligence are being integrated into production processes. In addition, energy consumption and CO_2 costs are becoming decisive factors in the resource-oriented management of companies. For energy intensive and hybrid production processes a simulation-based decision support system (DSS) for production planning is validated and further developed for the sand lime brick industry to support the production planning process. The integration of empirical knowledge in the energy-intensive control of steam processes, in which quality-critical product parameters are set via thermodynamically complex relationships, is still part of current research approaches. In this paper, an approach for the mapping of an energetic system behavior in the energy-intensive and hybrid production processes will be discussed using the example of calcium silicate brick (CSB) production. Possibilities for using Discrete event simulation (DES) to increase the energy efficiency of steam processes are summarized and linked to formalized empirical knowledge in artificial Intelligence (AI) approaches.

Keywords: Decision support system · Resource-orientation · Calcium silicate brick production

1 Introduction

The economic activities of companies are increasingly determined by the competitive factor energy in the resource-oriented control of companies [1, 2]. At the same time, digitalization offers new opportunities for value chains and the optimal design of production by means of analysis and use of operating data as well as the use of simulation for production planning and control [3]. The manufacturing processes of building materials, foodstuffs, fiber composites and other products are characterized by a transition from bulk to piece goods and feature at least one process step with thermal energy input. Autoclaves in a steam system are used for the curing process, which requires a large

© The Author(s) 2025
H. Kohl et al. (Eds.): GCSM 2023, LNME, pp. 562–568, 2025.
https://doi.org/10.1007/978-3-031-77429-4_62

amount of energy. At the same time, the challenge is to increase efficiency in this process step. It should be possible to integrate solutions close to the application, especially for small and medium-sized companies with limited financial resources and know-how.

2   State of the Art

Based on the process description for calcium silicate brick (CSB) production, the possibilities for process-oriented energy determination are researched. Furthermore, the difficulty to describe the autoclaving process becomes clear and possibilities for artificial intelligence (AI) modeling of steam processes are summarized.

2.1   Process Description for Calcium Silicate Brick Production

The construction material CSB is an important building block for a sustainable construction industry due to its load-bearing capacity and sound-insulating properties and is produced locally due to the geological conditions. For this work CSB serves as an exemplary application. The production process of CSB consists of five process steps (Fig. 1).

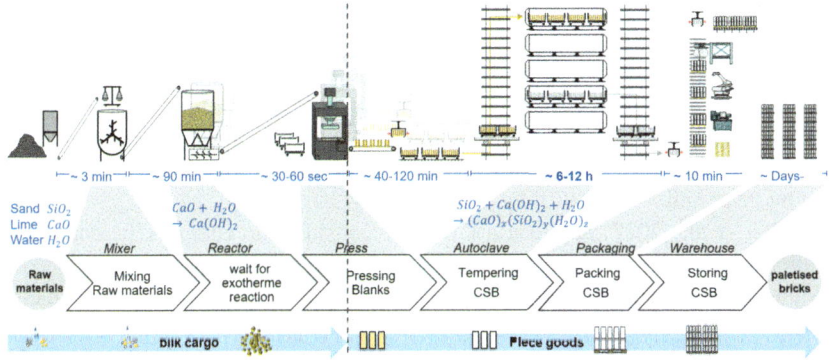

Fig. 1. Production process for calcium silicate bricks (CSB)

In the first step, the raw materials lime, sand, water and possibly aggregates are dosed according to a grade-specific recipe and then mixed. Water, sand and quicklime (CaO) react exothermically to form hydrated lime [Ca(OH)$_2$]. After completion of this process, the mixture is pressed into blanks. These are then autoclaved in a steam atmosphere to achieve the required strength. Finally, the finished bricks are bundled into transportable brick packs in the packaging area and finally stored [4].

Only the German sand-lime brick industry alone has an annual energy demand of about 800 gigawatt hours (GWh), making the calcium silicate brick industry responsible for 0.7% of all industry-related greenhouse gas emissions in Germany [5]. In 2021, the CSB-industry produced around 8.0 million tons of CSB in 77 plants in Germany. In total, about 800,000 tons of CO_2 equivalents were emitted [6]. Approx. 80% of the CO_2

emissions are attributable to the raw material quicklime due it's production ($=$ 640,000 tons CO_2-eq) and approx. 20% to the production process ($=$ 160,000 tons CO_2-eq). The CO_2 emissions from quicklime and production the total emissions in the life cycle of the calcium silicate brick [6]. In addition, to the raw material components used, the steam curing process is the most energy-intensive, which also applies to other hybrid production processes, such as in the manufacture of fiber composites [7].

The regulations on CO_2 pricing under the national emissions trading system, which came into force at the beginning of 2021, place an additional burden on manufacturing costs. Against this background, there is a need for ecological and economic action to make the manufacturing process more productive and resource oriented. This will help to sustainably reduce manufacturing costs and achieve the required climate neutrality by 2045 [8]. The control of the manufacturing process also currently requires specific expertise, which is subject to imminent loss due to the increasing shortage of skilled workers.

2.2   Energy Prediction in Production Processes with Simulation and AI

A widely discussed method for predicting the energy demand in production is determination by means of simulation. Discrete-event material flow simulation is used in a concept by Donhauser (2015) as the basis for a decision support system in CSB manufacturing for resource-oriented production control [4].

In addition to this overall concept, Schmidt and Pawletta (2014) restrict themselves to a linking of DES under consideration of differential equations and numerical modeling [9]. The benefit lies in the consideration of time and temperature dependencies of the workpieces and production plants, although this is accompanied by a high modeling effort. The possibility of thermal modeling of standardized, discrete-event processes is shown by Peter and Wenzel (2015), among others [10].

Schlüter et al. (2017) also use a DES and consider the energy efficiency via the utilization of the production facilities [11].

Waltersmann et al. (2021) see the potential in the use of AI-based analysis methods using pattern recognition to make industrial production processes more energy-efficient by supporting decision-makers and automating analyses [12]. In the production of fiber-reinforced composites, optimization by means of AI is available and increasingly relates to the prediction of electricity costs and CO_2 emissions [7]. A concrete implementation approach is not shown. The integration of a machine learning (ML) method for precise prediction of pressure vessel conditions into material flow simulation is possible and supports production planning [12].

In the following, the focus will be left on pressure vessels and autoclave systems in the production environment. Ogugua et al. (2023) show a Fourier-series based modeling for energy prediction in an autoclave model and validate their models [13]. With the help of a parameter adjustment, the models are transferable, but show the strong dependence on the temperature cycle and autoclave volume insulation thickness and conductivity as most influential parameters [13]. In addition to the design-related influencing factors, the autoclave loading has a significant impact on the process and energy efficiency [14].

Overall, production systems can be modeled and simulated using various simulation methods. Only a linkage of evaluation possibilities by means of AI algorithms makes the potential visible and usable.

3   Prediction of Energy Consumption in CSB-Production

Innovative solutions for the reduction of energy consumption and avoidance of CO_2 emissions are becoming a decisive competitive factor in the small and medium enterprises-oriented industry. For the present case of steam control of autoclave systems using the example of the CSB-industry, the autoclaving process will first be shown in detail and then a simulation-based modeling will be described.

For steam curing, autoclaves are used which are connected via a piping system and connected to a central steam generator [12]. Autoclaves are available in different sizes, one popular model in CSB production has a volume of 123 cubic meter and an diameter of 2.4 m. This one can produce 50 to 130 tons of CSB, depending on the density, in one charge. A central steam control system regulates the supply of live steam from the steam generator to the autoclave system. The sand-lime brick blanks are usually heated by means of saturated steam of about $p_e = 16$ bar gauge pressure. The associated saturated steam temperature until completion of curing is $T = 203$ °C. The pressure and temperature curves are stored in the steam control system over time by means of target hardness curves (Fig. 2). The hardness parameters depend on the format and raw material composition of the product, as well as on the temperature and pressure of the process [4]. The process steam and thus a large part of the process energy is then discharged from the autoclave. To increase energy efficiency, while starting and stopping, autoclaves are connected—where available—to enable steam transfer and thus further utilization of the heat energy.

Figure 2 shows a typical and real hardness curve. The sole temperature shown is measured in the autoclave sole. The overpressure is recorded via a manometer in relation to the ambient pressure. In the start-up phase, the sole temperature increases continuously, while the overpressure can only be increased with a delay. In the holding phase, the steam control system controls the above parameters in accordance with the target hardness curve via fully automatic addition of live steam from the steam generator. The slow cooling at constant pressure at the end of the holding phase is shown when the live steam supply is shut off and waiting for transfer. Subsequently, two steam transfers take place before the residual steam is discharged in this example. In the lower part of Fig. 2, the states of the donor autoclave (1) and the two associated receiver autoclaves 2,3 are shown over the time axis to illustrate an example transfer behavior. Autoclave 1 is started up by transfer (light orange) and live steam (dark orange) and the holding phase is controlled (red). At the end of the holding phase, the live steam supply is terminated, and the autoclave waits for a transfer partner (light violet). At 8:30, autoclave 2 is ready for the start-up phase and steam is transferred from autoclave 1 (dark violet) to autoclave 2 (light orange). Autoclave 2 continues to be started up with live steam (dark orange) and then enters the hold phase (red). Autoclave 1, on the other hand, waits again for a transfer partner and the process described is repeated between autoclave 1 and 3. The example describes only sequential transfer processes. Parallel transfer from one

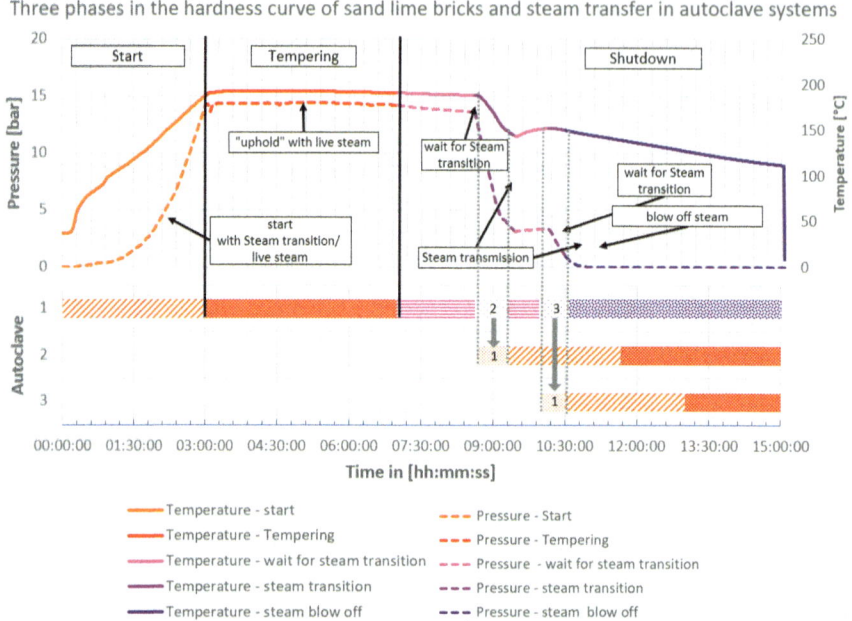

Fig. 2. Hardening curve of CSB (Pressure on the primary axis, Temperature on the secondary axis, Status of autoclaves during steam transition)

donor autoclave to several recipient autoclaves and the waiting of recipient autoclaves for transfer are also possible. The interaction of several autoclaves as transfer partners reveals a complex overall system behavior.

This system behavior could be modeled in a DES for CSB-plants and provides the basis for operational and simulation-supported decision-making in the CSB industry.

In a simulation toolbox, standardized hardness curves can be selected by means of an experiment manager or product-specific hardness curves can be imported. Currently, the pressure and temperature over various discredited points in time are used to describe the condition of the autoclave. The energy requirement of an autoclave is derived from the energy required to produce the steam, which is determined by means of enthalpy considerations. Subsequently, the critical pressure gradients are determined, above which a steam transfer reaches physical limits, and these steam quantities are transferred into energy quantities. In the context of energy balancing, heat conduction via the pressure vessel to the environment is considered. Within the autoclave, there is the assumption of complete loading, with the starting temperature before heating being set equal to the ambient temperature. In addition, a discretization is carried out over the course of time as well as the subdivision into heating, curing, and transfer/bleeding. Over the simulation time, the energy consumption is continuously determined and compared with the standard or real hardening curve.

Available plant parameterizations of production plants, hardening curves for the production program and energy consumption in hourly resolution have resulted in a deviation of max. 15% of the simulated energy consumption compared to the real energy consumption for an observation period of one week. Finally, the specific curing times in the real system cannot be predicted exactly, due to transition. Therefore, the operative system operation to maximize the efficiency is difficult.

4 Conclusion

The energy consideration in the production of resource-intensive products and materials has arrived in the focus of research. For the CSB application example, decision support is available by means of material flow simulation and DES.

Overall, there are two approaches for increasing the efficiency of autoclave systems.

1. The proposed solution considers the increase in efficiency through improved transfer times for steam reuse.
2. A second approach is to shorten curing times. Previous attempts have failed due to a complex modeling of the chemical-thermodynamic property change of CSB in the autoclave.

Future approaches may allow a modeling of the individual autoclave behavior, e.g. by means of physical neuronal networks. In a further step, the interaction of a modeled hardening behavior of CSB and the simulated energy consumption can be used to derive an optimization component that has not been available so far. With the storage of hardening curves and a product-specific neural network, a transfer to all steam hardening processes is conceivable.

References

1. Dyckhoff H (2018) Produktion und Umwelt, Bd. 2018. In: Handbuch Produktions-und Logistikmanagement in Wertschöpfungsnetzwerken, S 949–975
2. Rackow T, Schuderer P, Franke J (2013) Green ontrolling – Ressourcenorientierte Steuerung von Unternehmen. Zeitschrift für wissenschaftlichen Fabrikbetrieb 108(10):773–777
3. Wenzel S (2018) Simulation logistischer Systeme, Bd. 2018. In: Modellierung logistischer Systeme, S 1–32
4. Donhauser T (2020) Ressourcenorientierte Auftragsregelung in einer hybriden Produktion mittels betriebsbegleitender Simulation. Erlangen, FAU University Press. Dissertation
5. Deutsche Emissionshandelsstelle (DEHSt) (2021) Im Umweltbundesamt: Treibhausgasemissionen 2020 : Emissionshandelspflichtige stationäre Anlagen und Luftverkehr in Deutschland (VET-Bericht 2020). Mai
6. Institut Bauen und Umwelt e.V. (2021) Umwelt-Produktdeklaration: Kalksandstein. Bundesverband Kalksandsteinindustrie e.V. 11.10.2021
7. Adomat V, Ahanpanjeh M, Kober C, Fette M, Wulfsberg JP (2022) Helmut-Schmidt-Universität Hamburg (Mitarb.): Interdisziplinäre Forschungsperspektiven auf die Digitalisierung in der Leichtbauproduktion und Anwendungsmöglichkeiten in der LaiLa Modellfabrik, Bd. 1. In: Beiträge der Helmut-Schmidt-Universität / Universität der Bundeswehr Hamburg: Forschungsaktivitäten im Zentrum für Digitalisierungs-und Technologieforschung der Bundeswehr dtec.bw - Band 1 · 2022, S 235–241
8. Geres R, Lausen J, Weigert S, Roadmap für eine treibhausgasneutrale Kalksandsteinindustrie in Deutschland : Eine Studie für den Bundesverband Kalksandsteinindustrie e. V. von FutureCamp Climate GmbH

9. Schmidt A, Pawletta T (2014) Hybride Modellierung fertigungstechnischer Prozessketten mit Energieaspekten in einer ereignisorientierten Simulationsumgebung. Proceedings ASIM 2014 - Symp. Simulationstechnik 22:109–116

10. Peter T, Wenzel S (2015) Simulationsgestützte Planung und Bewertung der Energieeffizienz für Produktionssysteme in der Automobilindustrie. In: Rabe M, Clausen U (Hrsg.) Simulation in production and logistics. Tagungsband zur 16. ASIM Fachtagung. Stuttgart: Fraunhofer Verl. (ASIM-Mitteilung, 157), S 535–544

11. Schlüter W, Henninger M, Buswell A, Schmidt J (2017) Schwachstellenanalyse und Prozessverbesserung in Nichteisen-Schmelz- und Druckgussbetrieben durch bidirektionale Kopplung eines Materialflussmodells mit einem Energiemodell. In: Wenzel S, Peter T (Hrsg.) Simulation in Produktion und Logistik. Tagungsband zur 17. ASIM-Fachtagung. Kassel University Press GmbH, Kassel, S 19–28

12. Donhauser T, Kisskalt D, Mayr A, Scholz M, Schuderer P, Franke J (2019) Integration maschineller Lernverfahren in eine Materialflusssimulation zur Verhaltensabstraktion und -vorhersage komplexer Fertigungssysteme. In: Simulation in Produktion und Logistik

13. Ogugua CJ, Anton SV, Tripathi AP, Larrabeiti MD, van Hees SO, Sinke J, Dransfeld CA (2023) Energy analysis of autoclave CFRP manufacturing using thermodynamics based models. Compos Part A: Appl Sci Manufact 166:107365

14. Mirzaei S, Krishnan K, Al Kobtawy C, Roberts J, Palmer E (2021) Heat transfer simulation and improvement of autoclave loading in composites manufacturing. Int J Advanced Manufact Tech 112:11–12, S 2989–3000

Technical Processes

Residual Stresses Generation in Multi-stage Rotary Swaging of Steel Tubes

Lasse Langstädtler[1,2,3](✉), Kasra Moazzez[1,3], Dhia Charni[4], Christian Schenck[1,2,3], Jeremy Epp[2,4], and Bernd Kuhfuss[1,2,3]

[1] University of Bremen, Bibliothekstr. 1, 28359 Bremen, Germany
langstaedtler@uni-bremen.de
[2] Center for Materials and Processes—MAPEX, 28334 Bremen, Germany
[3] Bremen Institute for Mechanical Engineering, Badgasteiner Str. 1, 28359 Bremen, Germany
[4] Leibniz Institute for Materials Engineering, Badgasteiner Str. 3, 28359 Bremen, Germany

Abstract. Forming induced residual stresses give the potential to improve the application conditions which are dependent on the whole material flow history along the production line. A better knowledge of the generation of residual stresses leads to a more sustainable manufacturing process. Furthermore, hollow parts offer reduced weight thus saving energy and material consumption especially in the transportation sector. Thus, a sustainable swaging process needs to be designed with the understanding of multi-critical effects and interactions. In this study, the diameter reduction of annealed E355 steel tubes by infeed rotary swaging was exemplarily investigated. The reduction was performed by three different routes—one stage, two stages and three stages swaging using several swaging dies to reduce a diameter from 20 to 8 mm. The resulting residual stresses and geometric properties were discussed. It could be observed that an increased number of forming stages contributed to improved part properties like better surface quality and compressive residual stresses. Furthermore, the integration of multiple forming stages in a new swaging die design was presented. The improved part properties and new die design enhanced the processes and parts sustainability.

Keywords: Free forming · Rotary swaging · Residual stresses

1 Introduction

Rotary swaging is an incremental forming process used industrially to reduce and profile the cross section of axisymmetric parts that bear high mechanical loads such as axle shafts and steering spindles [1]. Additionally, it has gained more relevance as a manufacturing process of lightweight components across the automotive and aerospace industry due to its adaptability [2, 3]. Especially, the advantage of producing complex geometries in tight tolerances enables a cost-effective production of lightweight components with a sustainable use of material resources.

In industrial applications, several swaging machines—each with a special set of swaging dies—are used sequentially in multi swaging stages to form the final geometry

© The Author(s) 2025
H. Kohl et al. (Eds.): GCSM 2023, LNME, pp. 571–579, 2025.
https://doi.org/10.1007/978-3-031-77429-4_63

stage by stage to the workpiece. During the process, the forming dies—as a part of the pressure column together with cylinder roller, base jaw and spacer—rotate around the workpiece and strike simultaneously towards the center of the swaging axle, while the workpiece is axially fed into the swaging unit (Fig. 1). In infeed rotary swaging, after the reduction zone, the workpiece enters the calibration zone where the part properties are homogenized. This homogenization additionally takes place while retraction of the workpiece.

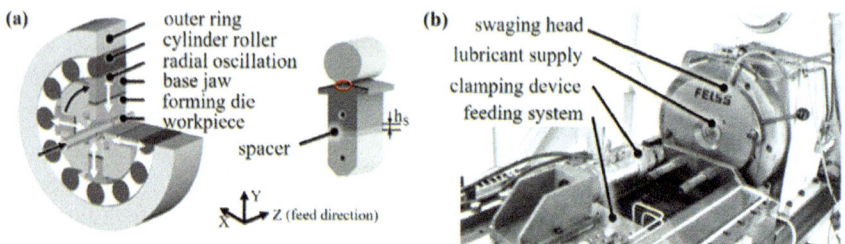

Fig. 1. Infeed rotary swaging: (a) swaging head and pressure column, (b) picture of a swaging machine

As a first approximation, the total forming degree φ_{sum} can theoretically be calculated independent from the number of interim forming stages with Eq. 1 [4].

$$\varphi_{sum} = \sum_{i=1}^{n} \varphi_i \tag{1}$$

As a result, the ideal forming work W_{id} for forming the material volume V when assuming a constant k_{fm} as the average yield stress can be calculated to be theoretically constant for all numbers of forming stages (Eq. 2 [4]).

$$W_{id} = V \int_{0}^{\varphi_{sum}} k_f d\varphi \approx V k_{fm} \varphi_{sum} \tag{2}$$

Based on this equations, in the swaging process, the forming work per stroke results from the actual forming stage respectively its forming degree as well as the fed material volume. Hence, using different forming degrees with constant feed rate, the forming work and by this the radial forming forces change. Furthermore, due to the material flow and the process kinematics, a changed axial back-pushing arises per stroke and the friction conditions can change. This causes different losses and wear of the dies which finally results in the total forming energy $W_{sum} \approx W_{id} + W_{loss}$ that can differ for different number of forming stages [4].

In detail, rotary swaging induces a locally distributed material flow bringing material modifications such as work hardening [5] and residual stresses in the workpiece [6] incrementally accumulating with each forming stroke. In previous studies, the final residual stress fields at the surface were investigated in steel tubes and bars and were found to be sensitive to process parameters [7] as well as their fluctuations leading to residual stresses evolving at different length scales [8].

The development of residual stresses along multiple swaging in correspondence with the development of geometrical features has so far not been investigated. According to forming simulations as described in [9], the residual stresses are reduced to zero in the reduction zone which results in the expectation, that residual stresses are in a comparable range after swaging to the final reduction stage. However, the material flow history can have an influence on forces, friction and heat generation in the process and on the development of geometry and surface properties as well as the stability of the residual stresses due to relaxation effects.

2 Materials and Methods

2.1 Experimental Setup

The experiments were conducted on a swaging machine Felss HU 32V. Annealed (700 °C; 2h) E355 steel tubes (Ø20 × 3 mm^2) were clamped in a hydraulic clamping device and axially fed with linear direct drive to the swaging head. Along the multi-stage swaging different swaging dies were used (Fig. 2a) with the same spacer height h_s of 5.08 mm (cp. Fig. 1a). The opening angle of the reduction zone was $\alpha = 10°$ and the length l_c of the calibration zone was 20 mm in the used dies, Fig. 2b.

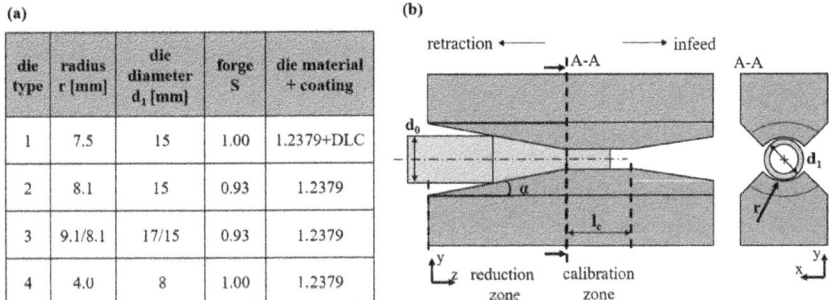

(a)

die type	radius r [mm]	die diameter d_1 [mm]	forge S	die material + coating
1	7.5	15	1.00	1.2379+DLC
2	8.1	15	0.93	1.2379
3	9.1/8.1	17/15	0.93	1.2379
4	4.0	8	1.00	1.2379

Fig. 2. Swaging dies: (a) used die types (b) schematic drawing of one-step die and related die parameters (only two dies out of four are shown)

The reduction zone and the calibration zone were connected by a transition radius of 20 mm. In the exit zone, the workpiece was gradually unloaded. The zones were equipped with a round shape crosswise to the infeed direction, whereas the die radius r varied for the used die types. According to the swaged workpiece diameter d_1 which was defined by the set of dies, the forge value S was calculated, see Eq. 3 [10].

$$S = \frac{d_1/2}{r} \tag{3}$$

The workpiece was swaged to the final diameter of 8 mm ($\varphi_{20\&\#xF0E0;8} \approx 0.92$) with different routes using different sets of dies as given by Fig. 3a. Three stage swaging (III) was performed with a newly designed two-step die along the swaging route that consists of an interim stage to 17 mm while reducing to 15 mm, Fig. 3b. The infeed rate was set to 1000 mm/min for all reduction stages from 20 mm to $d_1 = 15$ mm ($\varphi_{20\&\#xF0E0;15} \approx 0.29$) and to an infeed rate of 100 mm/min for all reduction stages to $d_1 = 8$ mm ($\varphi_{15\&\#xF0E0;8} \approx 0.63$). Here, due to high back-pushing forces—due to material flow within each stroke—and vibrations with higher infeed rate, a reduced infeed rate was required to stabilize the process. The low retraction speed of 500 mm/min was used in all stages due to its positive effect on residual stresses homogenization. The machine was equipped with different measurement systems to measure forces on the outer ring and displacement at the feeding axis.

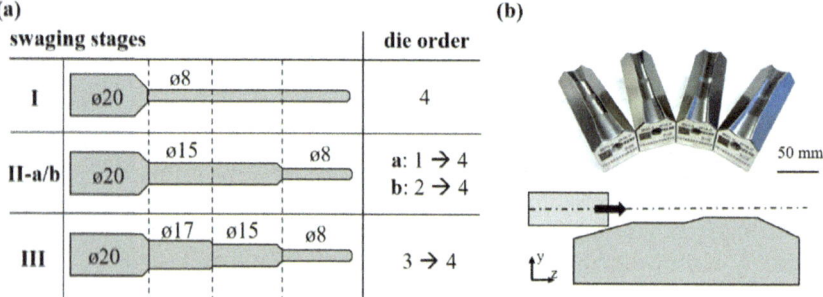

Fig. 3. Multi-stage swaging: (a) swaging stages and die order, (b) two-step die (die type 3)

2.2 Measurement Setup

The diameter was measured with a screw gauge at in 60° steps around the workpiece at two different points. Mean values and standard deviations were calculated separately at both points.

The roundness of the workpiece was measured with a Taylor Hobson TalyRond 252 at a distance of 30 mm from the reduction zone due to process stability issues and afterwards the roundness deviation RONt was calculated. The surface roughness was measured with a laser scanning microscope Keyence VK-X210. The image for roughness average Ra was measured with 50x magnification, filtered and evaluated over the image area of 500x500 μm^2.

X-ray diffraction (XRD) was used for residual stress analysis. Due to its low penetration depth in steel of a few micrometers, this method was applied for surface residual stress measurement as well as by local material removal to assess gradients within the first few mm (depth profile). For this, the residual stress measurement was performed with a commercial 8-axis diffractometer Type ETA 3003 from GE inspection technologies, equipped with a position sensitive detector. A beam of 1 mm diameter of Vanadium filtered Cr-Kα radiation was used. All residual stresses were determined using the $sin^2\psi$-method with the {211} diffraction peak of α-iron along 13 χ-angles between $-45°$ and $+45°$ and X-ray elastic constant $\frac{1}{2} S_2 = 5.81 \ 10^{-5}$ MPa^{-1} [11]. Since bending fatigue was considered, axial residual stresses were determined to reduce measurement time. In addition, previous studies show that the residual stress values in the tangential direction correlate with the axial ones.

3   Results

All swaging experiments were conducted with stable (vibrations and back-pushing within the allowed range) process behavior. Workpieces were reduced in diameter and increased in length due to volume constancy. The following workpiece properties could be identified.

3.1   Geometry and Surface Measurement

Examples for the multi-stage swaged workpieces are given in Fig. 4. The upper part was milled to have an access to the inside of the workpiece. According to an uninterrupted material, smooth transition between the different diameters appeared. Furthermore, with each forming stage the wall thickness slightly increased due to radial material flow. This in the end led to a closing of the tube and increased pressure in the forming zone to 8 mm.

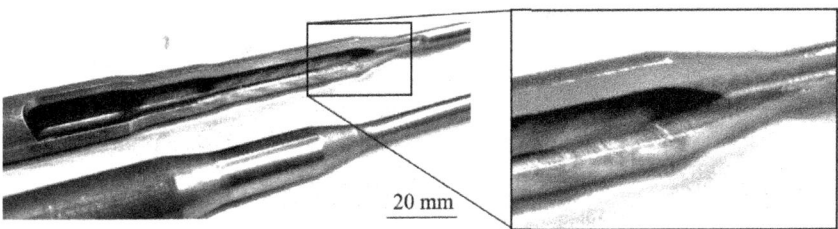

Fig. 4. Example of a multi-stage swaged workpiece (top/zoom III; bottom II-a)

The final diameter slightly varied from one-stage to three-stage swaging, Fig. 5a. The difference can be explained by differences in forming history and strain hardening. One-stage and two-stage swaging II-a resulted in a changed diameter and a conical shape. Two-stage swaging II-b and three-stage swaging III showed a constant diameter and a cylindrical shape.

The roundness deviation RONt improves with the number of forming stages, Fig. 5b. This on the one hand side was due to the increased number of calibration strokes by additional stages. On the other hand, regarding the different dies (die type 1 and 2) in the first stage of the two-stage swaging, the die shape contributed to a better roundness deviation.

The surface roughness average Ra was also improved with the number of forming stages which again can be explained by the increased number of calibration strokes, Fig. 5c. Furthermore, the high roughness values for one-stage swaging are due to heat generation which is related to strain hardening. A partially blue coloring of the workpiece indicated this heat generation during swaging. Beneath inner effects on the workpiece like residual stress relaxation, the heat caused an evaporation of oil thus a change in lubrication.

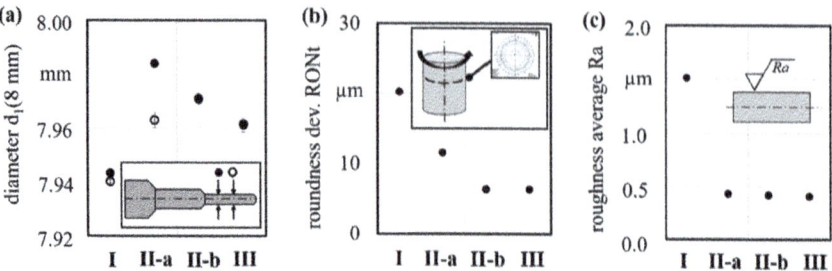

Fig. 5. Geometry and surface measurement after final stage to 8 mm diameter: (a) workpiece diameter d_1, (b) roundness deviation RONt, (c) roughness average Ra

3.2 Residual Stresses

Stress-free annealed E355 steel tubes developed different residual stress depth profiles after rotary swaging with different dies to a diameter of 15 mm (Fig. 6). This can be explained by the different die geometries and shapes which lead to different material flow. Swaging with die type 1 (II-a) resulted in high residual compressive stresses at the surface whereas swaging with die type 2 (II-b) and 3 (III) caused residual stresses close to zero at the surface. In the depth, the residual stresses where in the tensile range for die type 1 (II-a) and close to zero slightly shifting to the compressive range for die type 2 (II-b) and 3 (III).

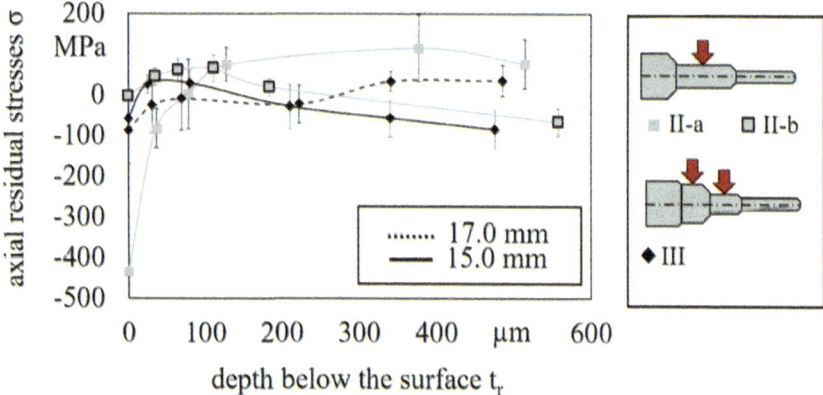

Fig. 6. Residual stresses depth profile in multi-stage swaging for different swaging sequence II-a, II-b and III (measured on different stages—red arrow)

The last reduction stage to 8 mm led—independent of the number of previous stages and previous residual stress state—to comparable residual stresses at the surface as well in the depth profile (Fig. 7). This can be explained by the effect of residual stress redistribution within the reduction zone and mainly setting the residual stresses in the last forming stage. However, with an increased number of stages from route I to route III, the residual stresses slightly increased in the compressive range below the surface.

Here, depending on the previous forming stages, a difference in the radial material flow while closing of the tube was assumed that caused small differences in residual stresses under the surface. Furthermore, the highest residual stresses at III can be explained by the effect of increased number of interim stages while combining two stages in one die hence reducing the number of calibration strokes that can lead to relaxation of stresses.

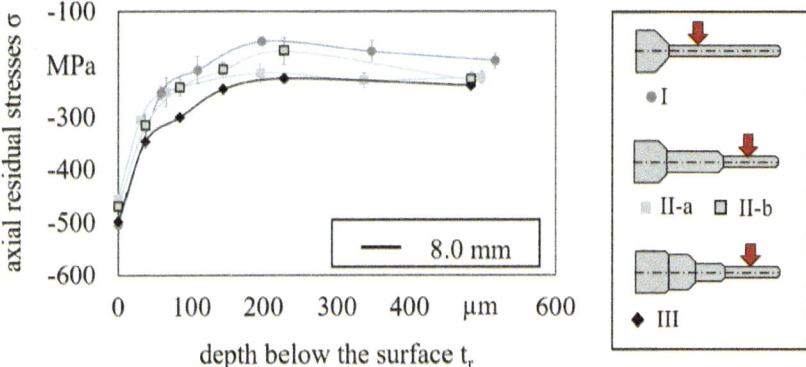

Fig. 7. Residual stresses depth profile in multi-stage swaging (after final stage to 8 mm)

4 Discussion and Conclusion

Rotary swaging as a forming process features a high material efficiency of industrial components. Due to the increased real process time, the consumption of electrical energy by the swaging machine was increased with the number of forming stages. Furthermore, with an increased number of stages, additional swaging dies were required with increased manufacturing expense. The combination of multiple stages in one die—as proven by die type 3—would be a possible way to reduce the manufacturing expense as well as process time. However, by combining multiple stages in one die, an increase of forming degree within one die results in higher forces as observed in one-stage (I) swaging. It needs to be considered in design, that the material flow is less localized and the pressure on the workpiece and the die are limited. Otherwise, an increased load to the dies increases the development of heat and wear. Furthermore, the process would behave less stable while increasing back-pushing and vibrations.

Regarding the workpiece properties, multi-stage swaging improved the geometrical features by correlating values of roundness deviation and surface roughness average value which was related to the number of calibration strokes. Additionally, residual stresses at the surface and in the depth were shifted to slightly higher compressive values with increased number of forming stages as aimed to improve the part properties. Thus, due to the influence on material flow leading to a conical shape of the workpiece, better results were found by multi-stage swaging with a reduced forge value S in the first reduction stage.

As a summary:

- Process and die design of inner and outer features had to be considered, whereas an increased number of forming stages contributes to better roundness, smoother surface and higher compressive residual stress values.
- The combination of multiple forming stages located in one die was possible.
- A reduction of forming stages was possible, which led to a more sustainable process, whereas effects on workpiece properties need to be considered.

Acknowledgement. This research is funded by the Deutsche Forschungsgemeinschaft (German Research Foundation)–374789876 within the sub-project (EP 128/5-2 and KU 1389/16-2) "Control of component properties in rotary swaging process" of the priority program SPP 2013 "The utilization of residual stresses induced by metal forming."

References

1. Rauschnabel E, Schmidt V (1992) J Mater Process Tech 35:371–383
2. Piwek V, Kuhfuss B, Moumi E (2010) Light weight design of rotary swaged components and optimization of the swaging process. Int J Material Forming 3:845–848
3. Lim S-J, Choi H-J, Lee C-H (2009) Forming characteristics of tubular product through the rotary swaging process. J Mater Process Technol 209(1):283–288
4. Doege E, Behrens B-A (2010) Handbuch Umformtechnik. Maschinen. Springer-Verlag, s.l, Grundlagen, Technologien
5. Ameli A, Movahhedy MR (2007) A parametric study on residual stresses and forging load in cold radial forging process. Int J Adv Manuf Technol 33:7–17
6. ASM Committee on Rotary Swaging (1988) Rotary swaging of bars and tubes. In: Metals handbook Vol. 14 forming and forging, 9th edn. Metals Park, Ohio, American Society for Metals, pp 268–309
7. Charni D, Ishkina S, Epp J, Herrmann M, Schenck C, Zoch HW, Kuhfuss B (2018) Residual stress generation in rotary swaging. In: Proceedings of the 5th International Conference on New Forming Technology, Bremen, Germany
8. Ishkina S, Charni D, Herrmann M, Liu Y, Epp J, Schenck C, Kuhfuss B, Zoch H-W (2019) Influence of process fluctuations on residual stress evolution in rotary swaging of steel tubes. Materials
9. Ortmann-Ishkina S, Charni D, Herrmann M, Liu Y, Epp J, Schenck C, Kuhfuss B (2021) Development of residual stresses by infeed rotary swaging of steel tubes. Arch Appl Mech
10. Kienhöfer C, Grupp P (2003) Rundknettechnik - Verfahren, Vorteile, Möglichkeiten. Die Bibliothek der Technik, Band 252
11. Noyan IC, Cohen JB (1987) Residual stress measurements by diffraction and interpretation. Springer, New York

Multi-stage and Multi-technology Forming of Paper for Sustainable Packaging

Nicola Jessen$^{(\boxtimes)}$ ⓘ and Peter Groche ⓘ

Technical University of Darmstadt, 64287 Darmstadt, Germany
nicola.jessen@ptu.tu-darmstadt.de

Abstract. Forming of paperboard enables great design freedom, high productivity and sustainable packaging solutions. The limiting factor in forming processes is the low ductility of the natural fiber material. Not only does this often lead to unwanted cracks, but also to an insufficient control of wrinkle formation. Compared to packaging manufactured from plastics, visual deficiencies have been unavoidable so far. In order to extend the process limits and improve the product appearance, two-stage processes as well as the combination of different forming methods are promising process routes and are therefore discussed in this paper. So far, research in paperboard forming has mainly focused on forming with rigid tools. Recently, forming with active-media is gaining interest. Investigations of process combinations of active-media based forming followed by deep drawing show that both maximum forming depth as well as the wrinkle compression are enhanced. In order to explain the findings scientifically and to develop a numerical model, the materials used have been characterized. Subsequent investigations were carried out numerically and validated experimentally using optical analysis methods. The findings open up new approaches in design and process planning for sustainable, formed packaging.

Keywords: Paper-forming · Multi-stage-process · Media-based forming

1 Introduction

Climate change is one of the greatest challenges facing society today. Not only is it politically relevant, with new regulations [1] being introduced on a regular basis, but it is also of great social interest and public concern. To comply with legislation, but also to retain customers, industry must move towards sustainable materials and processes.

A survey of consumers in Europe and the Americas in 2023 found that 82% of respondents are willing to pay more for sustainable packaging. 71% of consumers have made a product choice in the last 6 months based on its sustainability credentials [2]. To meet these customer and legislative demands, more and more packaging is being made from paper.

Paper packaging can be produced by various techniques, including folding [4], molding [5], and forming [6]. Each method has distinct advantages and limitations, impacting the geometric freedom, production time, energy efficiency, and visual properties of the packaging.

© The Author(s) 2025
H. Kohl et al. (Eds.): GCSM 2023, LNME, pp. 580–588, 2025.
https://doi.org/10.1007/978-3-031-77429-4_64

Additionally, paper as a material presents challenges due to its inhomogeneity, anisotropy and limited strain [7], which means that paper packaging is still often inferior to plastic packaging. To obtain higher quality sustainable products from paper, its material capacities must be exploited as much as possible.

2 State of the Art and Challenges

There are a number of processes that can be used to manufacture paper packaging, the most popular being folding, casting and forming. Production by folding has limited geometric freedom, allowing only rectangular geometries with linear bending lines. The often necessary use of gluing and cutting processes results in waste production and less sustainability. Nevertheless, folding is a well-established multi-stage process leading to wrinkle free products that facilitate printability for marketing and informational purposes [4].

Casting as an alternative allows a high degree of geometric freedom, but requires expensive tooling, long process times and high-energy processes. The resulting surfaces exhibit irregularities and moderate optical properties, which result in marketing and labeling limitations due to restricted printability [5].

Forming, on the other hand, offers greater geometric freedom than folding while requiring shorter production time and less energy compared to casting [3]. The highly productive deep drawing process is well established for forming of 3D geometries from sheet metal. Yet, paper, as an inhomogeneous natural fiber material, presents new challenges and, consequently, limitations to the process. The unique properties of paperboard, such as its limited elongation of 15% maximum and the significant variation in local properties depending on fiber orientation and arrangement, result in significant technological differences compared to metals. On the other hand, customer perception is also different, as exemplified by the general acceptance of wrinkles [6].

In addition to deep drawing with rigid tools, active media based forming (AMF), which allows "free forming" but with slower processing and higher energy requirements than conventional deep drawing, is gaining interest. The processes differ greatly in their frictional properties in the forming area. Rigid tools result in high friction between tools and paperboard, constricting movement, while media-based forming using pneumatic pressure results in air resistance until the material contacts the forming mold.

Several techniques, such as increased process humidity and temperature control, have been used to extend the formability of paperboard [10]. The primary objective of this work is to develop a paperboard-specific extension of the deep drawing process that ensures the exploitation of the full strain potential of the material. The aim is to enhance the drawing depth, which leads to the production of packaging with increased volume. In addition, efforts are being made to optimize wrinkle compression for superior optical properties and to expand the wrinkle free areas to enable printing and improve overall aesthetics by limiting spring-back. Homogenizing the strain distribution is critical to effectively exploit the material's capacity for maximum drawing depth and utilization.

3 Approach and Set-Up

The aim of the current work is to investigate new multi-stage technologies for paperboard forming. One potential approach to address the challenges of realizing the full potential of paper forming is by homogenizing the material strain in the forming process. The demonstrator is a cuboid with corner radii of 20 mm, as shown in Fig. 1. The numerical results are outlined showing a quarter of the model as shown in Fig. 2.

Fig. 1. Exemplary geometry, Dimensions: 110 mm x 80 mm x drawing depth

Figure 2b) shows the strain resulting from the conventional deep drawing process. Only minimal strain is introduced into the deep drawn product, as the material flow from the flange is primarily used to form the material. Friction at the bottom of the punch limits the material strain in the bottom area. In comparison, active media-based forming can be employed to achieve a more homogeneous distribution of strain in the paper, as shown in Fig. 2a). This technique involves the use of an active medium, such as a fluid or gas, to apply pressure and form the material into a mold. By applying a high blank holder force in the active media-based forming process, material flow is restricted and less friction affects the free forming process. As a result, a comparison at the same forming depth in Fig. 2 shows a much higher strain in the results of AMF than in deep drawing.

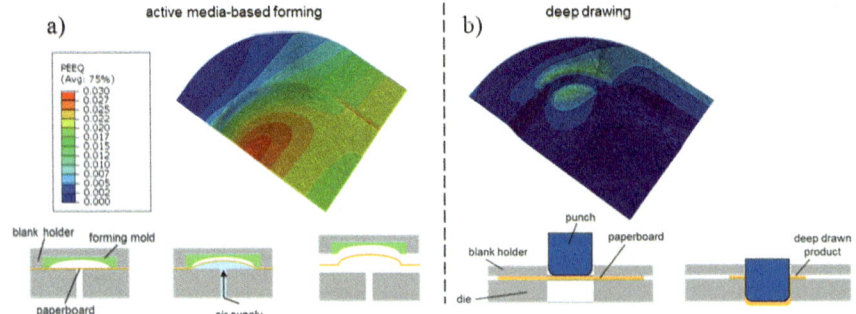

Fig. 2. Comparison of the plastic-elastic-strain-equivalent (PEEQ) at the same depth – a) active media based, b) deep drawing

In addition, Fig. 3 shows the blank and the formed product superimposed for both processes. Figure 3a) displays a smaller, red dashed outline of the deep drawn product, compared to the blank due to material flow. In Fig. 3b), the component formed with active media shows almost identical outer contours before and after forming, with no apparent flow as the forming process uses the elongation of the material instead.

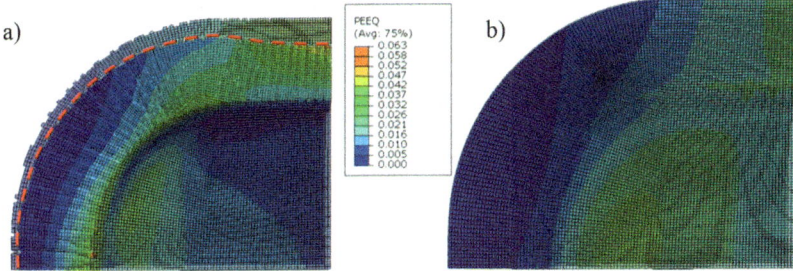

Fig. 3. Superimposition of deformed components and blanks showing the material flow – a) deep drawing and b) AMF

However, AMF provides a more homogenous strain, as a single process it cannot provide sufficient depth in the final packaging structure, due to lack of material flow from the flange and simultaneous strain limitations of the paperboard. The process window of deep drawing processes could be expanded by adding a second process step - deep drawing. Deep drawing allows material flow, but limits elongation in the wrinkle areas of the products and the bottom of the part. Therefore, the goal of an optimized process design is an appropriate mold geometry in the AMF, which leads to a homogenized strain distribution after the deep drawing while minimizing buildup of wrinkles in the flange area.

4 Results

In the following, a two-stage process is studied, consisting of an up-stream stretch forming process performed by active media-based forming and a subsequent deep drawing process. Results have been obtained through numerical simulations as well as experimental investigations. Details of the numerical model are described in previous publications of the authors [8, 9]. While the numerical model shows great accuracy in terms of the timing of wrinkle appearance and achievable drawing depth depending on moisture content, the wrinkle locations are not as accurate compared to experimental results. In all experiments, TrayForma [7] with a grammage of 350 g/mm^2 was used with an initial moisture content of 15%, and no temperature control or additional moisture was applied during the process. The blank holder force was determined by analyzing the maximum applicable blank holder force at the moment of wrinkle formation without exceeding the maximum tolerable strain in the work piece.

4.1 Wrinkle Formation

Figure 4 shows the deep drawing process and the multi-stage process with the same drawing depth. In Fig. 4b), wrinkles have already formed in the deep drawing process, while the multi-stage process does not show any wrinkles. This is a direct result of the material flow shown in Fig. 3.

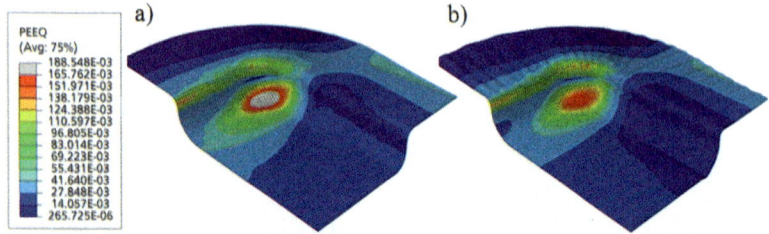

Fig. 4. Wrinkle formation at the same drawing depth: a) deep drawing after an upstream AMF and b) deep drawing

The formation of wrinkles can also be verified by analyzing the kinetic process energy of the simulation. The red as well as the blue line show the kinematic energy during the deep drawing process. Since both simulations are performed at the same speed, the simulation time can be considered equivalent to the drawing depth. The peaks of the kinematic energy can be explained as follows: a: Relaxation of the blank after the previous active media based forming process, b-c: Drawing punch bottom reaches the drawing depth of the pre-stretched blank, d: Wrinkling without pre-stretching, e: Wrinkling with pre-stretching. The difference between d and e shows the difference in the achievable drawing depth before wrinkle occurrence in the two processes (Fig. 5).

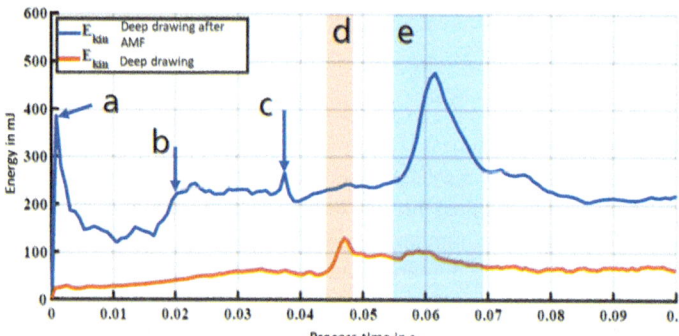

Fig. 5. Comparison of kinetic process energy: I) deep drawing (red) and II) deep drawing after active media based forming (blue), d) (red) and e) (blue) showing the moment of wrinkle formation for the respective process routes

4.2 Active-Media-Formed Geometries

After verifying the positive effect in numerical simulations, an experimental approach was chosen to compare three different geometries (Fig. 6) for the AMF process design.

The active media based geometries were chosen to evaluate different options. Geometry a) corresponds to the entire bottom geometry of the drawing punch, including the corner radii. b) corresponds to the bottom geometry of the drawing punch, excluding the corner radii and c) is a slightly larger than a) and has an octagonal geometry.

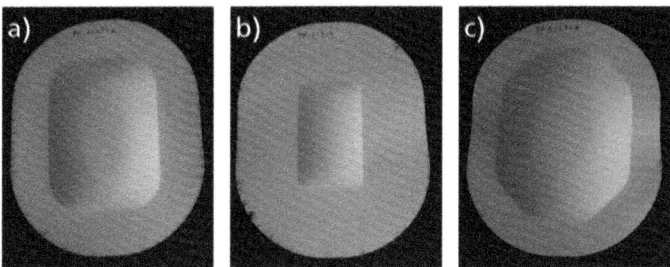

Fig. 6. Various paperboard geometries formed by AMF; a) including corner radii, b) excluding corner radii and c) shifted corner positions

While geometry a) pre-stretches a larger part of the material, it runs the risk of overstressing the material in the radii, as this part will also be stretched in the drawing process. Therefore, geometry b) only stretches the area that will not be stretched in the deep drawing process due to the friction between punch and material. Geometry c) is used to evaluate the possibilities to control the locations of wrinkle occurrence and the continuation of the AMF geometry in the subsequent drawing process.

In the following deep drawing process of geometry b), it becomes evident, that the area selected in the AMF process is too small. It does not overlap with the area stressed in the deep drawing process. As a result, the bottom is not flattened, the AMF geometry is still visible and the pre-process stretching has no positive impact on the deep drawn product (see Fig. 7).

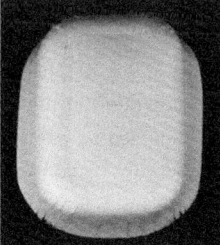

Fig. 7. Geometry b) of Fig. 6 after deep drawing

Geometry c) shows a high continuation of the geometric features even after deep drawing. The octagonal geometry is still visible, resulting in a poorer overall geometric accuracy (see Fig. 8). In terms of wrinkle formation, the radii show fewer wrinkles, but the wrinkles appear between the radii. The geometry has shown, that the AMF geometry has a high relevance to the final product geometry in a multi-stage forming process.

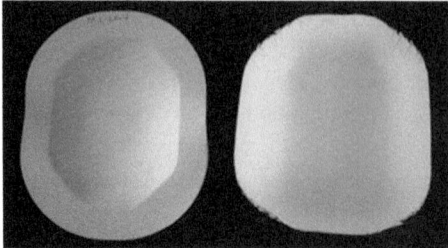

Fig. 8. Geometry c) features after the deep drawing process

Geometry a) includes the radii that results in overlapping strain. Therefore, the final geometry does not show any traces of the AMF process. Figure 9b) shows the deep drawing process without previous forming, while Fig. 9a) shows the results after previous forming with geometry a), the drawing depth is the same in both images. It can be seen, that more wrinkles are formed in Fig. 9a) in accordance with the results described in Sect. 4.1. Deep drawing in one step shows wrinkles on the long side of the product, while the multi-step product shows wrinkles only in the upper and lower part of the radii.

a)
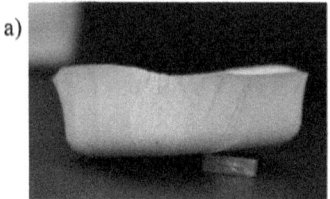

b)
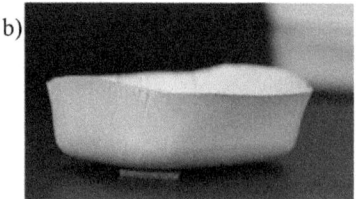

Fig. 9. Formed parts: a) deep drawing in one step, and b) deep drawing after AMF

4.3 Spring Back and Drawing Depth

Spring back is an issue in all forming processes, but shows especially when the excess material in wrinkles allows them to open up after the process. Consequently, deep drawing of paperboard is typical performed using conical punches that have an angle of 89° or less between punch bottom and side faces to counteract the spring-back [10]. If neither increased moisture nor temperature control is applied, the spring back is easily noticeable in the conducted experiments.

Figure 10 compares the spring back of three identical multi-stage formed products and three deep drawn products through a section of their laser scan with the same fiber orientation. Both processes result in high repeatability, represented by the aligned curves of the three tests. Yet, spring back is much lower using a multi-stage process, as the previous strain limits the elastic strain and thus the potential for spring back.

Ultimately, the drawing depth was analyzed by comparing the punch path to the depth at which the paperboard products showed failure. Depending on the fiber orientation along (MD) or perpendicular (CD) to the long side of the product geometry, the average drawing depth could be increased by 11.16% to 36 mm (MD) or 13.5% to 39 mm (CD) by the multi-stage process.

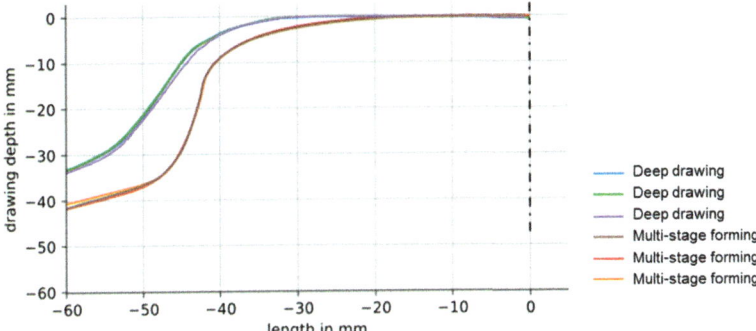

Fig. 10. Spring back comparison between the deep drawing and the investigated multi-stage forming process

5 Conclusion

It has been shown, that the multi-step process allows to control wrinkle areas by selecting the active media based forming geometry. The geometry should be chosen to overlap with the area that will experience strain in the deep drawing process. If the geometry selected for AMF has radii in areas that are not present in the target geometry, the wrinkle areas are moved there, which can be used to even out the wrinkle distribution.

By combining the stretching capabilities of active media based forming in the first stage with the material flow and stretching achieved by deep drawing in the second stage, the overall depth of the formed paper package can be increased. In addition, fewer wrinkles are created at the same drawing depth and less spring back occurs, resulting in a higher quality product. The two-stage process capitalizes on the benefits of both techniques and maximizes the utilization of the material's strain capacity while ensuring controlled folding areas and improved structural integrity. Due to that circumstance, a more homogenous overall strain in the material can be achieved and process limits can be shifted to higher levels.

References

1. CMS   Homepage.   https://cms.law/en/int/expert-guides/plastics-and-packaging-laws/eur opean-union. Last accessed 07 Dec 2023
2. Trivium Packaging, Buying Green Report 2023. https://www.triviumpackaging.com/media/ pe5hfxsp/2023buyinggreenreport.pdf. Last accessed 07 Dec 2023
3. Groche P, Huttel D (2016) Paperboard forming—specifics compared to sheet metal forming. In: BioResources, vol 11, no 1

4. Coles R, McDowell D, Kirwan MJ (2009) Food packaging technology. Wiley-Blackwell, London
5. World Paper Mill. https://worldpapermill.com/pros-and-cons-molded-fiber/. Last accessed 07 Dec 2023
6. Hauptmann M, Kaulfürst S, Majschak J-P (2016) Advances on geometrical limits in the deep drawing process of paperboard. In: BioResources, vol 11, no 4
7. StoraEnso. https://www.storaenso.com/-/media/documents/download-center/documents/pro duct-specifications/paperboard-materials/trayforma-en-15112021.ashx. Last accessed: 07 Dec 2023
8. Jessen N, Groche P (2022) Numerical analysis of the deep drawing process of paper boards at different humidities. In Production at the Leading Edge of Technology, Stuttgart
9. Jessen N (2023) Final report, paper forming with steam, IGF project 21562 N
10. Franke W (2021) Umformung naturbasierter Faserwerkstoffe unter Einflussnahme von Wasserdampf, 2021, Shaker Verlag, Berichte auf Produktion und Umformtechnik Band 126

Method to Simulate the Dynamic Behavior of Micro Ball End Mills

Steffen Globisch[1](✉), Markus Friedrich[1], and Frank Döpper[1,2]

[1] University of Bayreuth, Universitätsstraße 30, 95447 Bayreuth, Germany
steffen.globisch@uni-bayreuth.de
[2] Fraunhofer-Projectgroup Process Innovation, Universitätsstraße 9, 95447 Bayreuth, Germany

Abstract. Instable process behaviour is one of the biggest challenges in micromilling and has a significant impact on tool wear. This leads to reduced workpiece quality and higher scrap. Therefore, it increases the resource and energy consumption. This paper presents a new simulation method and uses the example of a micro ball end mill for machining X37CrMoV5-1 to show how tool deflection and process forces can be determined as a function of individual process parameters and how optimal process kinematics can be achieved. For this purpose, the individual modelling steps are presented and it is explicitly shown how the necessary characteristic values for the simulation can be determined experimentally.

Keywords: micro milling · process stability · energy efficiency

1 Introduction

Sustainable processes have become a real competitive factor in chipping processes, and not only because of the recent increase in costs for raw materials and energy. High process stability leads to less tool wear, higher component quality and thus lower scrap. This can significantly reduce the need for resources and energy. The energy demand of chipping processes depends significantly on the process performance and the machining time [1]. The process performance depends on the cutting speed, the feed rate and the process forces. Pratap et al. analysed the process forces along the tool life in micro milling with micro ball end mills and found that these are mainly caused by accelerated tool wear [2]. In micro milling, accelerated tool wear is mainly due to process instabilities caused by tool vibrations. The most common form of tool vibrations are self-excited chatter vibrations. These are triggered by unfavourable cutting conditions. The occurrence of such chatter vibrations often depends on the selection of process parameters. These also have a significant effect on the process time. A longer use of the milling tools and thus a longer tool life through optimized process parameters also has a positive effect on the CO_2 balance. While milling tools in the macro range can usually be reground several times, milling tools in the micro range are fed directly into the material cycle. In specialized recycling plants, the carbide is crushed and melted so that it can then be reused for the production of new carbide products. Although this reduces the need for primary raw materials, recycling is still very energy intensive. In order to increase

H. Kohl et al. (Eds.): GCSM 2023, LNME, pp. 589–597, 2025.
https://doi.org/10.1007/978-3-031-77429-4_65

energy efficiency and improve the CO_2 balance, optimal process parameters are therefore essential. For this purpose, a new simulation method is presented in this paper, which makes it possible to determine the dynamic process behavior during the machining of complex component geometries as a function of tool geometry, the material, the process kinematics and the process parameters. This enables companies to determine optimum process parameters, taking into account the process forces and tool deflection, without the need for time-consuming experimental trials. This can be used to increase process efficiency. Furthermore, it is possible to use the simulation method in tool development and thus reduce the duration of the development process.

2   Background and Related Work

2.1   Influence to an Efficient and Sustainable Milling Process

Milling can generally be seen as a system with input and output operands (Fig. 1) [3]. The input operands can be differentiated according to system and manipulated variables. While the system variables (machine, tool, material) are unchangeable over a long period of time, the manipulated variables (cutting speed, feed and depth of cut) must be individually adapted to the process in order to ensure stable process behavior [3]. The input operators influence the process and effective variables. Favorably selected input operators can thus reduce unpredictable tool wear and increase workpiece quality. This contributes to more efficient and sustainable milling processes by reducing scrap.

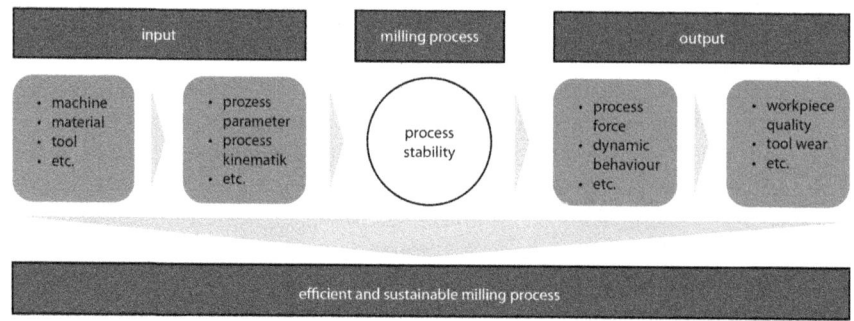

Fig. 1.   Influence of input and output operators to an efficient and sustainable milling process.

2.2   Methods to Analyze the Dynamic Behavior

To analyse the dynamic behaviour, dynamic models are required to consider the relationships between the system variables. Methods for predicting dynamic behaviour can be found in literature and can be divided into analytical and numerical approaches [4]. Analytical methods are often limited to the calculation of stability lobes. They are also often limited to specific applications. Numerical methods are called material removal simulations (MRS). Their main advantage over analytical models is their flexibility with respect to process kinematics and workpiece geometry. In addition, the accuracy of the stability prediction is higher, since non-linear effects can be taken into account by discretised simulations in the time domain. MRS are also advantageous because they respect the necessary flexibility with regard to the complex workpiece geometries and the calculation of the meshing parameters, taking into account the real process kinematics. Surmann et al. developed a numerical time domain simulation in which the tool dynamics can be simulated [5]. Kersting et al. extended this approach by taking workpiece vibrations into account [6].

3 Workflow and Experimental Parameter Determination

In the following, it's shown how the dynamic behaviour of a micro ball end mill can be analysed with the approach developed by Kersting et al. and how optimal process parameters can be determined in order to reduce the tool deflection and the process forces as well as the process load. The approach is dexel-based and takes into account real process kinematics as well as tool and workpiece geometries and material properties.

3.1 Workflow

The creation of the simulation model requires several steps (Fig. 2). First, some characteristic values must be determined experimentally. The procedure for this is described in the following section.

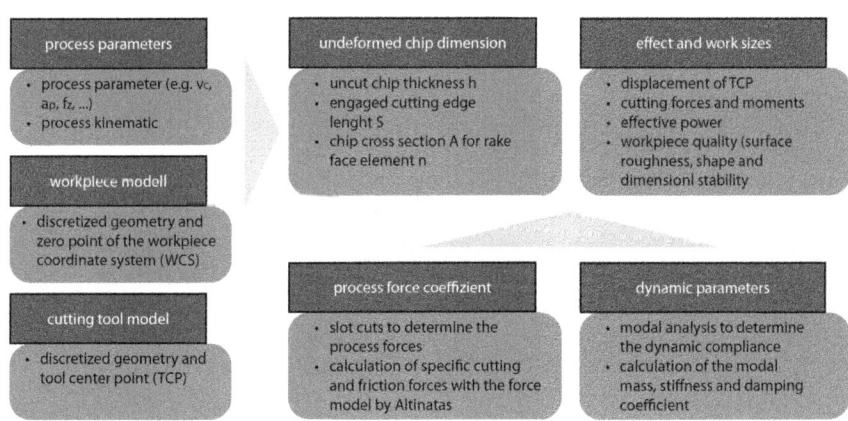

Fig. 2. Representation of the workflow of the material removal simulation (MRS)

3.2 Experimental Parameter Determination

In order to take into account the tool geometry, the material to be machined and the structural dynamics, experimental tests must be carried out to determine characteristic values. The necessary test procedure is described below. For the example in this paper, a micro ball end mill with diameter d = 3 mm and free length l = 15 mm is used.

Modal parameter. The calculation of the tool deflection requires the modelling of the structural dynamics of the tool at the tool center point. This requires knowledge of the compliance frequency responses in the different axial directions of the tool coordinate system. These can be determined by means of experimental modal analysis or numerically. The experimental modal analysis can be carried out, for example, with a laser vibrometer (Fig. 3).

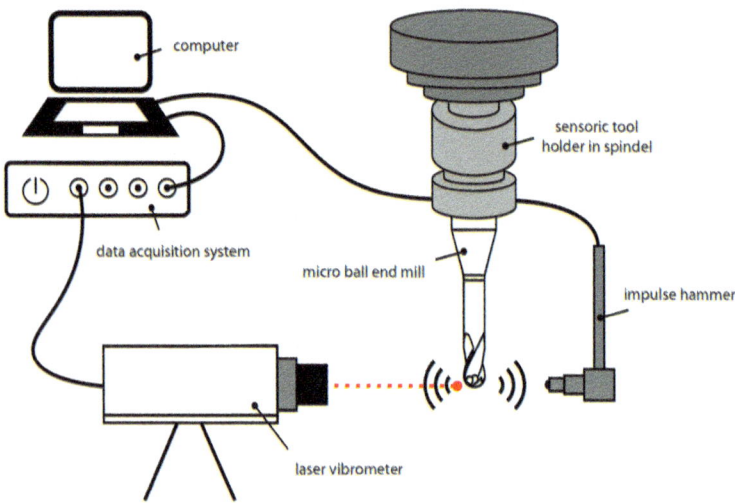

Fig. 3. Experimental setup of an experimental modal analysis to determine the compliance frequency response

The compliance frequency responses must then be replicated by superposition of single-mass oscillators, for which curve-fitting tools are used. The single-mass oscillator consists of its modal mass m [kg], a damping constant d [Ns/m] and a stiffness k [N/m]. Each single-mass transducer represents one mode of the underlying compliance frequency response (Fig. 4).

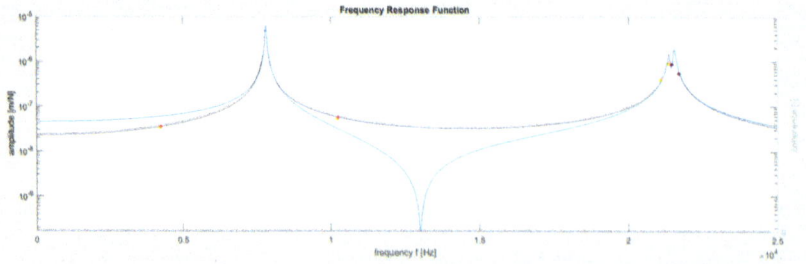

Fig. 4. Compliance frequency response with the curve fitting function

The dynamic characteristic values can be determined from the compliance frequency response by curve fitting. Particle Swarm Optimization (PSO) is used for this purpose. The resulting dynamic characteristic values as a function of the natural frequency can be found in Table 1.

Table 1. Dynamic parameters of the micro ball end mill

	Frequency [Hz]	Damping [N/m/s]	Stiffness [N/μm]	Lower boundary [Hz]	Upper boundary [Hz]
1	7816.52	3.02	29.48	4260	10250
2	21348.1	6.25	203.74	21120	21380
3	21503.69	4.08	138.11	21440	21690

Force coefficient. To predict the process forces, coefficients of the force model must be determined experimentally. This takes into account the influence of the workpiece to be machined as well as the macro and micro geometry of the tool. Figure 5 shows the force model used, where dF_i are the differential process forces in tangential, axial and radial direction of each rake face element. For each force component, a specific friction force $K_{e,i}$ and a specific cutting force $K_{c,i}$ are used for the calculation. This shows that influences such as the rake or clearance angle as well as the rounding of the cutting edge are represented.

To determine the force coefficients, the micro ball end mill is used to produce groove paths with a defined axial cutting depth a_p and varying feed rate f_z. The cutting forces and moments are measured directly at the cutting edge with a rotating dynamometer. Figure 5 shows the experimental setup.

The forces required for the force model according to Altintas in tangential (F_t), radial (F_r) and axial direction (F_p) can be determined from the recorded forces F_x, F_y and F_z as well as the spindle torque M_z [7]. Based on this, a specific friction force coefficient $K_{e,i}$ and specific internal force coefficient $K_{c,i}$ can be determined for each force component.

4   Simulation Model

4.1   Create Simulation Model

Several steps are necessary to create the simulation model. First, the geometry model of the tool must be imported. Afterwards, the TCP can be defined and the cutting edge can be defined by selecting the clearance and rake faces. This must then be meshed. The workpiece can be imported or created manually. This must be defined workpiece coordinate system (WCS). The process parameters and the process kinematics are read in via G-code. This can be generated on a CNC control, for example. Finally, the previously determined dynamic parameters and the specific friction and cutting force coefficients can be specified.

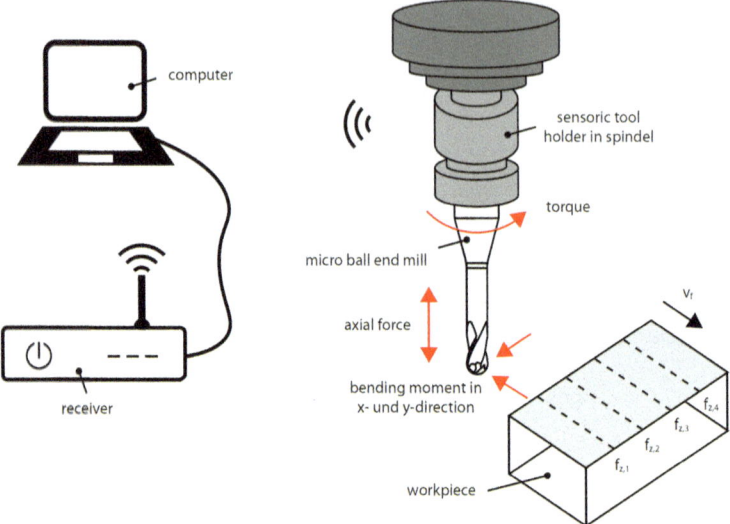

Fig. 5. Experimental setup for the force measurement to determine the process force coefficients

4.2 Modell Validation

Before the deployment behaviour can be reliably investigated, the simulation model must first be validated. For this purpose, several trajectories are generated and the results are compared with the forces already determined experimentally (Fig. 6).

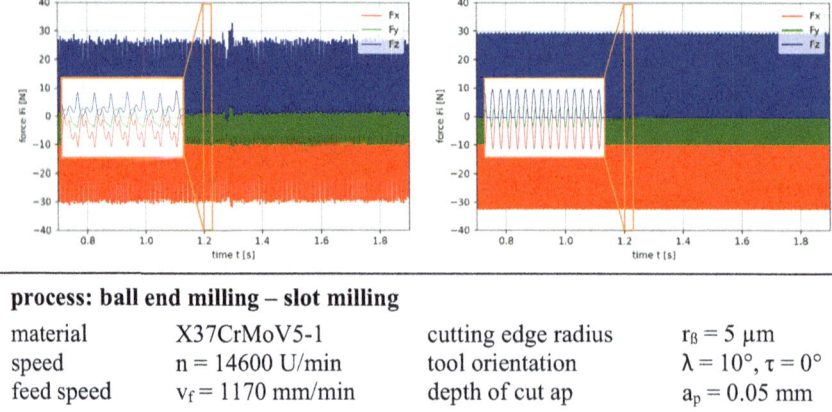

process: ball end milling – slot milling

material	X37CrMoV5-1	cutting edge radius	$r_\beta = 5\ \mu m$
speed	n = 14600 U/min	tool orientation	$\lambda = 10°, \tau = 0°$
feed speed	v_f = 1170 mm/min	depth of cut ap	$a_p = 0.05$ mm

Fig. 6. Experimental (left) and simulated (right) process forces to validation

It can be seen that the simulated process forces agree well with the measured data. An enlargement of a section shows that there are differences in the maximum forces and in the fluctuation of the signal, as the radial runout of the signal was not taken into account in the simulation. In addition, the process damping of the flank surface was not modelled. The tooth meshes show comparable frequency curves. Similar harmonics are visible. The different decay of the experimental tooth mesh can be attributed to the natural frequencies of the dynamometer.

4.3 Case Study to Evaluate the Dynamic Behavior and Cutting Power

To derive an optimal tilt angle, the effects on tool deflection and cutting forces of varying feed rates were simulated. The parameters of the case study can be found in Table 2.

Table 2. Parameter of the case study

Parameter	Value	
Material	X37CrMoV5–1	
Speed n	14300 U/min	
Feed speed v_f	1170 / 2340 mm/min	
Depth of cut a_p	0.035 / 0.07 / 0.2 mm	
Cutting edge radius r_β	5 μm	
Lead angle λ	10° / 20° / 30°	
Tilt angle τ	0°	

Figure 7 shows the results of the case study. It clearly shows that at an angle of inclination of 10°, the cutting force and, derived from this, the cutting performance is the lowest, especially with increasing cutting depth. A similar trend can be seen at double the feed rate, but the results at an angle of inclination of 10° and 20° hardly differ. Overall, it is clear that the cutting torque increases with increasing angle of inclination and therefore requires higher energy consumption.

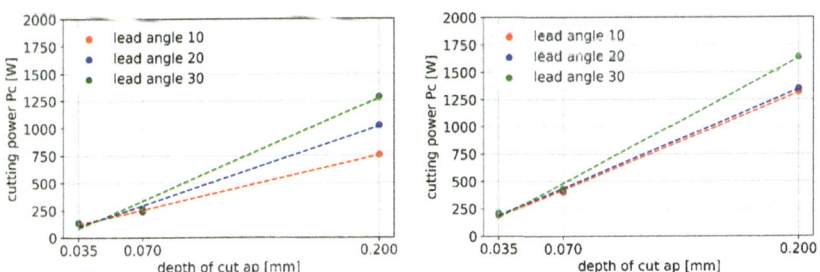

Fig. 7. Influence of feed rate, depth of cut and tilt angle on cutting power (left: $v_f = 1170$ mm/min; right: $v_f = 2340$ mm/min)

Figure 8 shows the tool deflection with reference to this. It can be seen that the lowest tool displacement occurs with an angle of inclination of 10°. It can also be seen that the feed rate has only a minor influence and that it is therefore possible to increase it and thus reduce the process time.

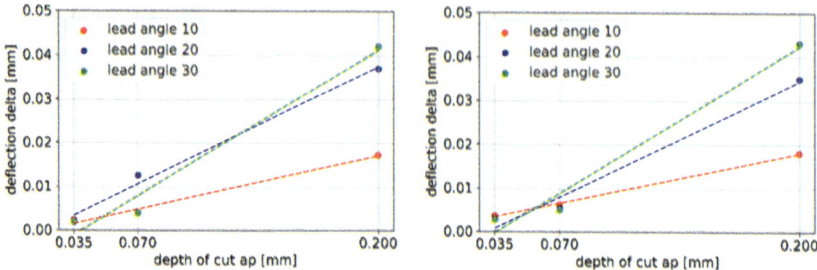

Fig. 8. Influence of feed rate, depth of cut and tilt angle on tool deflection (left: $v_f = 1170$ mm/min; right: $v_f = 2340$ mm/min)

5   Conclusion and Outlook

In this paper, it was shown how the tool deflection and cutting force and the resulting cutting performance of a micro ball end mill can be analysed with a MRS in the time domain depending on process parameters and process kinematics. The workflow was described in detail and in particular the procedure for determining the required characteristic values was shown. Overall, it can be concluded that the described method can be used to reliably determine optimal process parameters and process kinematics. The possibility to assess the component quality leads to the prediction of scrap and thus contributes to more sustainable machining processes.

In future work, it will be investigated to what extent tool wear can also be taken into account with the presented method. For this purpose, it is planned to import real tool geometry with defects into the simulation. It will be investigated how this affects the process and effective variables.

References

1. Mühlberger A (2022) AMB-Trend-Lounge: Bei der Zerspanung CO2 und Kosten sparen. Fertigung. https://www.fertigung.de/news/aus-der-branche/amb-trend-lounge-bei-der-zerspanung-co2-und-kosten-sparen-803.html
2. Pratap T, Patra K (2015) Modelling cutting force in micro-milling of Ti-6A1–4V titanium alloy. In: International conference on industrial engineering, pp 1345–139
3. Denkana B, Tönshoff H (2010) Spanen – Grundlagen. Springer, Heidelberg
4. Mamedov A (2021) Micro milling process modelling: a review. In: Editor F, Editor S (eds) Manufacturing review 2021. EDP Science
5. Surmann T, Biermann D (2008) The effect of tool vibrations on the flank surface created by peripheral milling. CIRP Ann 57(1):375–378
6. Wiederkehr P, Biermann D (2012) Modeling workpiece dynamics using sets of decoupled oscillator models. Mach Sci Technol 16(4):564–579
7. Altintas Y (2001) Analytical prediction of three dimensional chatter stability in milling. JSME Int J Ser C 44(3)

Feasibility Analysis of Carbide Broaching Tools in the Machining of Inconel 718

Leonardo Sastoque-Pinilla[1,3(✉)], Ander del Olmo[1], Endika Tapia[1], Unai López-Novoa[1,2], and Luis Norberto López de Lacalle[1,3]

[1] Advanced Manufacturing Centre for Aeronautics, University of the Basque Country, 48170 Biscay, Zamudio, Spain
edwarleonardo.sastoque@ehu.eus
[2] Department of Computer Languages and Systems, University of the Basque Country, 48013 Bilbao, Biscay, Spain
[3] Department of Mechanical Engineering, University of the Basque Country, 48013 Bilbao, Biscay, Spain

Abstract. Broaching is a highly efficient process for producing high-quality parts with precise dimensions. However, process efficiency alone does not guarantee sustainability. To address this, several key factors need to be considered, including process parameters, machine configuration, material and design of the workpiece, and tool characteristics. Several studies have focused on investigating the sustainability of the broaching process and its environmental impact, but the broaching tool has been less studied, even though it determines the geometric and sur-face quality of the broached grooves. To address this gap in the literature, this study compares the performance of three carbide tools for broaching Inconel 718 under identical machining conditions. The aim is to evaluate the selection of tools that align with the machine and workpiece material conditions to maintain a highly efficient process while improving process sustainability.

Keywords: Broaching Process · Sustainability performance · Tool wear analysis

1 Introduction

To date, there has been a lack of studies related to sustainability in the broaching process. Despite its effectiveness, speed, and reliability, broaching is a highly valued manufacturing process that finds wide application in industries such as energy, automotive, and aerospace, especially for producing complex parts with high dimensional accuracy and optimal surfaces [1]. Although many studies have been carried out in recent years related to sustainability in machining processes, broaching has been left aside. This is likely because the energy consumption of the broaching machine is relatively low, and the use of lubricants can be minimized by efficient cleaning and recycling systems. However, broaching has a unique characteristic: the tools manufactured for this process have a very limited range of use, as they are designed for specific geometries and materials. This lack of versatility can make broaching less sustainable than other machining processes.

H. Kohl et al. (Eds.): GCSM 2023, LNME, pp. 598–605, 2025.
https://doi.org/10.1007/978-3-031-77429-4_66

During broaching, wear, chipping, or fracture of the tool's teeth can negatively affect the surface quality of the workpiece, resulting in a finish lower than expected, and even lead to the rejection of expensive components such as aerospace turbine discs. Unlike other machining processes, there has been little research on the control and analysis of tool wear in broaching, despite the considerable economic investment in broaching tools and the high value of the machined parts [2, 3]. Some research [4, 5] is being developed to reduce costs and waste by reusing cutting tools or extending their lifespan. This is more sustainable and cost-effective than frequent tool replacement.

Moreover, controlling the tool's life cycle is an important factor in determining the sus-tainability of the broaching process. Therefore, the materials used in broaching tools directly influence their behavior during use, affecting process costs. A tool with lower wear results in reduced costs for re-sharpening, fewer tool replacements, and a de-creased risk of breakage or damaging the broached piece during the process. However, most studies on broaching tools are conducted using single-tooth tools to reduce testing costs. In contrast, the importance of this study lies in the fact that it uses three sets of broaching tools composed of different materials, and it takes place in a broaching ma-chine located at a research center with a Technology Readiness Level of 5 to 7. This ensures that the research is conducted under conditions closely resembling those of the end-user, for faster knowledge transfer. Consequently, real tools designed for roughing the fir trees on aerospace turbine discs were used in the study. Given that tests were performed on tools specifically designed for a particular material and with specific components, the conclusions obtained may not be fully representative when applied to other tools or piece materials with similar characteristics. To address this, the cutting conditions were deliberately kept the same for the three tools (cutting speed (Vc), cool-ant use, material to be broached, and angle of inclination of the lifting platform) to facilitate later data analysis and clarity in conclusions. The differences between the tools lie in variations in the rise per tooth and their physical components. The document is structured as follows: the next section provides details about the machine setup, workpiece, and tools used in the experiments. Subsequently, the results of the tests will be presented and analyzed for their implications, followed by the conclusions of the study.

2   Experimental Setup

The objective of the trial is to verify the technological suitability of carbide broaching tools for machining titanium and nickel base superalloys, and to determine the compo-sition that yields the best performance. Carbide broaching tools are available in a vari-ety of grades, which are related to the binder content within. Binder content can be measured in a range of 3 to 10 wt% and affects the properties of the tool. Grain size is another important factor to consider, as it confers resistance and high reliability against breakage. Grain size can be classified into three different groups according to a range of values: nano (<0.2 μm), ultrafine (2.0–0.5 μm), and submicron (0.5–0.9 μm). Three different tools are used in the trial, with specific weight and atomic fraction as describe in Table 1.

As explained earlier, parity was sought during the trial by keeping the machining con-ditions constant (Vc $=$ 18 m/min) and selecting a similar machining configuration (cool-ant: Yes; cutting oil: Yes). The only differences identified were in the variation of the rise per tooth, as described in Table 2.

Table 1. Weight and Atomic fraction of used tools.

	Tool 1		Tool 2		Tool 3	
	Grain size: submicron		Grain Size: ultrafine		Grain size: submicron	
	Weight fraction		Weight fraction		Weight fraction	
Element	Atomic (%)	Weight (%)	Atomic (%)	Weight (%)	Atomic (%)	Weight (%)
Tungsten	77,98	91,7	73,14	89,47	80,78	92,91
Cobalt	22,02	8,3	26,86	10,53	19,22	7,09

Table 2. Broaching Conditions and Tool Configuration

Tool	Z (Edges)	h (rise per tooth) [mm]	ap (Cutting depth) [mm]
Tool 1	24	0,064	1,49
Tool 2	24	0,069	1,49
Tool 3	24	0,065	1,49

The broached slots were performed on two Inconel 718 discs with a diameter of 489.8 mm and a height of 33.3 mm using an Ekin A218/Rashem in the machine as shown in Fig. 1(a). The disc to be broached was set at a 15° inclination on the electromechanical broaching machine for roughing in full slotting conditions (Fig. 1(b)). A total of at least 400 slots were planned to be machined with each tool, positioned in the machine as shown in Fig. 1(c). Approximately every 10 slots, tool wear was measured on three teeth of the broaching tool: Z08 (at the initial stage of the tool), Z16 (in the middle part of the tool), and Z24 (the final tooth of the tool). Notably, the Z24 tooth is the most critical part of the tool, as it determines the final quality of the broached surface.

Tool wear is one of the most important parameters to control during the cutting process, as it affects the cutting forces, the finish of the workpiece, and the overall efficiency of the process. Flank tool wear is the gradual deterioration or wear that occurs on the side of a cutting tool due to the abrasive action of the workpiece material during the cutting operation. Flank tool wear is the most common and important type of tool wear, as it has a direct impact on the quality of the machined surface. The maximum allowable tool wear values for different machining processes, such as turning [6] and

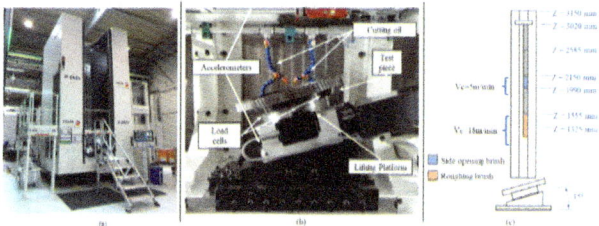

Fig. 1. (a) Broaching machine (b) Machine assembly (c) Tool assembly.

milling [7], are well-defined in the standards. However, the broaching process does not have its own standards for tool wear evaluation, so it is proposed to use the turning standards (Fig. 2).

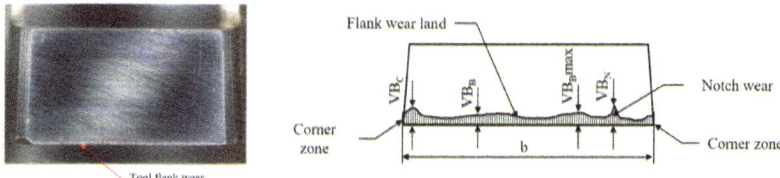

Fig. 2. Broaching tool wear description.

The initial working range for the cutting speed was Vc = 18 m/min. However, if excessive vibration was detected through the accelerometers, the cutting speed could be in-creased up to 20 m/min.

3 Results Description

Minimal wear was observed in tools 01 and 03, allowing for additional slots to be machined beyond the originally planned number. A total of 528, 440, and 497 slots were broached with tools 01, 02, and 03, respectively.

3.1 Tool 01

Figure 3 (a) shows the results of the tool wear (Vb) vs machined length measurement. The wear rate is nearly the same for all three cutting edges under observation. During the initial phase of the trial, fractures were observed in the teeth, each exhibiting different behaviors (Fig. 3 (b)). The teeth in the middle section of the tool (Z13, Z14, Z15, Z17, and Z23) experienced fractures right from the beginning. Similarly, the teeth at the beginning of the tool (Z04, Z05, and Z11) also suffered fractures starting from slot 120.

Additionally, tooth Z07 also presented fractures from slot 250 onward.

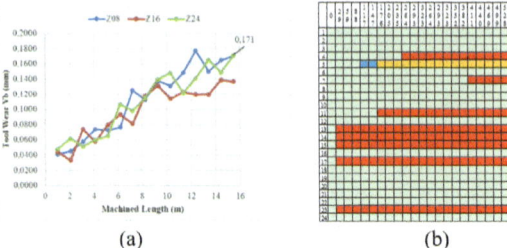

(a)   (b)

Fig. 3. (a) Tool Wear vs Machined Length – Tool 01. (b) Tool Breakage [Green: usual wear. Blue: break to the right of the edge. Red: break to the left of the edge. Orange: break on both sides of the edge]

3.2 Tool 02

Tool 2 also exhibited fractures, but in this case, they occurred right from the broaching of the first 20 slots. The tool's wear remained stable throughout the trial (Fig. 4 (a)). However, with a total of 440 broached slots, larger fractures occurred compared to those observed in Tool 01, and they occurred earlier in the tool's use (Fig. 4 (b)). The middle section of the tool experienced fractures in the teeth (Z13, Z17, Z18, and Z23) from the beginning of the process, and in the teeth (Z10, Z11, Z12, Z15, and Z16) from the first 20 slots. Starting from slot 30, the teeth in the initial part of Tool 02 (Z05 and Z06) also experienced fractures.

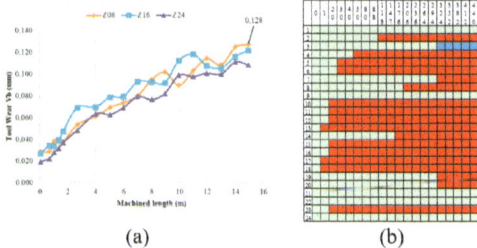

(a)   (b)

Fig. 4. (a) Tool Wear vs Machined Length – Tool 02. (b) Tool Breakage [green: usual wear. Blue: break to the right of the edge. Red: break to the left of the edge. Orange: break on both sides of the edge]

3.3 Tool 03

When the first broaching operations were carried out with Tool 03, fractures were observed in all sections of the tool (Fig. 5 (b)), including the first contacting teeth (Z3 and Z4), the middle section (Z15), and the final section (Z22) of the tool, starting from the first 10 slots broached. Subsequently, starting from slot 120, fractures occurred in the contacting and middle sections of the tool (Z5, Z9, Z11, Z12, and Z13). Additionally, teeth Z6, Z7, Z9, Z14, Z16, and Z18 experienced fractures from the broaching of slot

250. Tool wear (Fig. 5 (a)) of Tool 03 was more uniform along the tool, presenting a linear trendline.

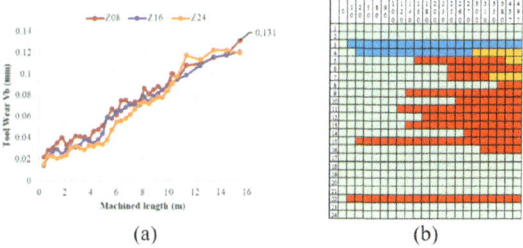

(a)   (b)

Fig. 5. (a) Tool Wear vs Machined Length – Tool 03. (b) Tool Breakage [green: usual wear. Blue: break to the right of the edge. Red: break to the left of the edge. Orange: break on both sides of the edge]

4 Discussion of the Results

Once the trials were completed, the results were summarized, and the findings were discussed. Tool 01 exhibited slightly higher wear (Vb = 0.171 mm) than Tool 03 (Vb = 0.131 mm) and Tool 02 (Vb = 0.128 mm) for an equivalent machined length (see Fig. 6 (a)). However, Tool 02 did not reach such high and uniform wear levels. Wear was more significant in the contacting area of the initial teeth of the tools, but more fractures occurred in the middle section of the tool. This may be because as the contacting area of the tool wears more, the middle section must work harder to remove the required amount of material, resulting in more fractures. Additionally, Tool 02, with a higher rise per tooth, did not generate greater wear in the teeth (Fig. 6 (b)), but it did experience a higher number of fractures right from the beginning of the broaching process. Regarding the material composition of the used tools, Tool 03 had a longer lifespan than the tools made with other material percentages.

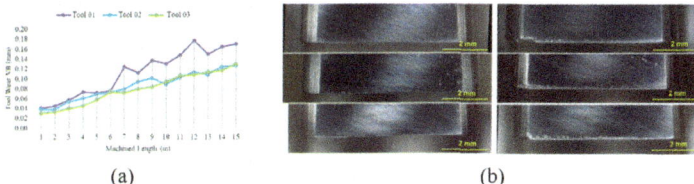

(a)   (b)

Fig. 6. (a) Tool Wear vs Machined Length – Comparison Tool 01, 02 and 03. (b) Tool 02 Wear and breakage example.

The analysis of these three tools demonstrates that the variation of the percentage of the components in the tools produces a significant effect on their performance. However, there seems to be no linear relationship between the percentage of elements from which the tools are made and their performance, but there could be a relationship between the chemical balance of the two main components. It is important to consider that other factors such as workpiece eccentricity or tool wear can cause a tooth to remove more or less material than is necessary, or what it is designed to support, which can seriously affect both the durability of the tool as well as the final quality of the broached piece.

Additionally, it can be determined that wear alone does not allow clear conclusions to be drawn about the behavior of the tool. Therefore, and as stated by Axinte and Gindy [3], measures of energy consumption were considered to complement the analysis. For this investigation, the readings of the torque consumed by the servomotors (derived from the energy consumption of the two servomotors of the machine) were used. In Figs. 3(b), 4(b), and 5(b), it is observed that the teeth that suffered the most during the test were those found in the middle section of the tool (approximately tooth 13). This may be due to the accumulation of mechanical stress that the tool suffers as it advances through the piece. This can be confirmed thanks to the information in Fig. 7, in which three broaching zones can be determined. In the first zone (Fig. 7(a)), broaching begins, and the cutting edges come into contact with the workpiece, so the measured torque gradually increases as more cutting edges come into contact with the disc. In a second zone (Fig. 7(b)), the maximum number of teeth comes into contact with the workpiece and the torque acquires a maximum value that becomes constant since the number of teeth in contact also remains stable. Finally, in a third zone (Fig. 7(c)), the torque de-creases as the cutting edges leave the workpiece until broaching stops.

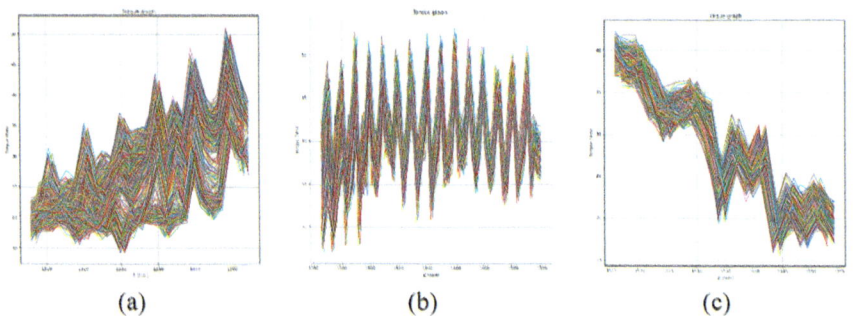

| (a) | (b) | (c) |

Fig. 7. Torque Signal (a) First Zone, (b) Second Zone, (c) Third Zone.

Another important aspect of the research lies in being able to determine how the variation in the rise per tooth of each tool affects its durability. In the case of the tools studied, Tool 02 (with a rise per tooth of 0.069 mm) was the one that suffered the most breakages during the process (79.17% of the tool teeth suffered breakages). However, and despite the fact that Tool 03 suffered breakages during broaching (62.50% of the tool's teeth showed breaks) like the other two tools, and under the test conditions, it turned out to be the most suitable tool for broaching of Inconel 718, since it was the one that exhibited the least wear and showed that the radius of machined slots versus the breaks presented was smaller for this tool.

5   Conclusions

The analysis of breakages and wear in tools made with different materials is important when evaluating the feasibility of broaching tools. It is not possible to draw a definitive conclusion about which of the three analyzed tools is the best, but several conclusions can be drawn about their performance. For example, Tool 01 showed the most wear per machined length, but it also experienced the fewest breakages (only 37.50% of the teeth broke). Additionally, the greater wear on this tool is likely because it had a larger sur-face area classified as wear rather than breakage. It is important to note that there is no technical standard for defining these concepts in broaching. On the other hand, and as mentioned in the analysis of the results, there does not seem to be a linear relationship between the materials that make up the broaching tool and its performance. This suggests that the success of a broaching tool may lie in the chemical balance of its compo-nents. An important area of future research is to analyze and determine the effect of tool breakage and wear on the surface quality of broached slots (roughness and dimensionality).

References

1. Davim JP (ed) (2010) Surface integrity in machining (Vol. 1848828742). Springer, London
2. Budak E (2001) Broaching process monitoring. In: Proceedings of third international conference on metal cutting and high-speed machining, pp 251–260
3. Axinte DA, Gindy N (2003) Tool condition monitoring in broaching. Wear 254(3–4):370–382
4. Bediaga I, Ramirez C, Huerta K, Krella C, Munoa J (2016) In-process tool wear monitoring in internal broaching. In: UMTIK 2016–17th international conference on machine design and production
5. Del Olmo A et al (2022) Tool wear monitoring of high-speed broaching process with carbide tools to reduce production errors. Mech Syst Signal Process 172:109003
6. International Organization for Standardization (1993) International standard ISO 3685: tool-life testing with single-point turning tools, 2nd edn
7. International Organization for Standardization (1989) ISO 8688–2:1989—tool life testing in milling—part 2: end milling, 1st edn

Sustainable Cutting Fluid Usage in Bandsawing of Additively Manufactured 316L Stainless Steel: Surface Integrity Evaluation

Kevin Wolters[1], Mohammad Sayem Bin Abdullah[3], Nithin Rangasamy[3],
James Caudill[2], C. S. Rakurty[3(✉)], and I. S. Jawahir[2]

[1] RWTH Aachen University, Aachen, Germany
[2] Institute for Sustainable Manufacturing, University of Kentucky, Lexington, USA
[3] The M. K. Morse Company, Canton, OH, USA
rakurtys@mkmorse.com

Abstract. Metal additive manufacturing (MAM) components often require post-processing for end applications. Bandsawing is an alternative process established in industry to remove parts from the build plate in wire arc additive manufacturing (WAAM). The removal of WAAM components from the build plate by bandsawing shows high potential, especially in the sustainable manufacturing of large-scale MAM parts. This work aims to study the influence of cooling conditions on the surface integrity of band sawed WAAM and conventionally manufactured 316L stainless steel. Sustainable cutting fluid strategies could impact surface integrity and their effectiveness in bandsawing MAM parts. The investigated cooling strategies are compressed air, minimum quantity cooling (MQC), and flood cooling. The surface integrity metrics of the band-sawed surfaces have been evaluated. The work aims to gain a deeper understanding of bandsawing AM components for industrial applications. Overall, the sustainable cutting fluid solution, MQC, has a similar effect on the surface roughness, microstructure, and cutting forces, compared to flood cooling.

Keywords: Additive manufacturing · Sawing · Sustainable cutting fluid · WAAM

1 Introduction and Background

Metal additive manufacturing (MAM) is gaining importance in major manufacturing industries due to its advantages, such as waste reduction, cost savings, complex and unique design solutions, and sustainability [1, 2]. MAM components require postprocessing, such as machining, coating, and heat treatments, before end applications [1, 2]. Because the parts come out of the machine fused to the build plate in the WAAM process, a postprocessing step is required to remove the parts and the support structures. Two main processes, wire electrical discharge machining (wire-EDM) and bandsawing, have been established in industry to remove parts from the build plate [2, 3]. The major advantages of bandsawing over wire-EDM are attributed to their costs [2].

© The Author(s) 2025
H. Kohl et al. (Eds.): GCSM 2023, LNME, pp. 606–614, 2025.
https://doi.org/10.1007/978-3-031-77429-4_67

Besides the cost-intensiveness of the process, many potential hazards exist in EDM [4]. Regarding removing additively manufactured (AM) components from its build plate, bandsawing shows high potential, especially in terms of sustainable manufacturing of large-scale MAM parts. Bandsawing is a pre-machining process to cut bar stock off into the desired length for postprocessing steps such as milling, turning, etc. Due to the additional surface processing, just a few efforts in research have been made to investigate the surface integrity (SI) of band-sawed components. Rangasamy et al. reported a literature review of the usage of cutting fluids in bandsawing regarding sustainable and economic aspects [5]. Most industrial applications in bandsawing metal parts use flood cooling for bandsawing.

Machining operations like bandsawing generate a new surface and, thereby, new surface characteristics depending on the process parameters and cutting strategies. Surface integrity can be summarized as the surface and sub-surface condition of a manufactured part, which influences the product properties such as fatigue strength, resistance to corrosion or wear, etc., and is, therefore, the key to the components' functional performance and part lifetime. As such, to achieve the ideal functional performance of machined components, the influence of machining parameters and cutting strategies surface integrity characteristics must be elucidated [6, 7]. In this context, bandsawing, a machining process, is becoming increasingly important for process optimization to reduce costs and waste of resources.

In metal sawing, the choice of blade, cutting parameters, cutting fluid application, and type of bandsaw depends on the workpiece material properties and geometries. A review of the literature reported on bandsawing has highlighted that the details of the blade, cutting parameters, and saw type are documented, with very few studies focusing on cutting fluid usage [5]. The lack of a comprehensive understanding of cutting fluids and their application in bandsawing solids motivated authors to study the cutting fluid application for bandsawing in previous studies [5, 8, 9]. Rakurty et al. investigated the four cutting fluid applications dry (no cutting fluid), minimum quantity lubrication (MQL), minimum quantity cooling (MQC), and industrial standard flood cooling [8, 9]. The authors emphasized that a comprehensive understanding must be developed of tribo-chemical phenomena which arise due to cutting fluid interactions at the tool tip-workpiece interface. Additionally, they argue that the bandsaw tooth acts as a microorthogonal cutting edge with a depth of cut on the order of microns, which would explain the major tool wear mechanism in bandsawing, namely adhesive wear along the flank surface and edge radius [8, 9]. Cutting fluid application has the potential to affect the machining dynamics, such as vibration and saw blade deflections, which directly affect the geometrical accuracy, the surface roughness, and the overall SI of the sawed component. By improving the SI parameters, it is possible that some finishing operations can be avoided [10]. The use of the right cutting fluid and the optimized strategy for specific cutting conditions show high potential for SI enhancements is the focus of this paper. Apart from that, no study discusses cooling strategies in bandsawing additively manufactured metals, while a few studies discussed the effects of machining (milling, turning, facing) on the mechanical and microstructural features of WAAM produced stainless steel components. Feldhausen et al. studied the microstructure and porosity of WAAM parts (as built and machined) and their effects on mechanical performance

[11]. Rodrigues et al. investigated the evolution of the microstructure of WAAM 316L stainless steel during the annealing process [12].

This work aims to study the influence of sustainable cooling conditions on the surface integrity of band-sawed WAAM components compared to a band-sawed conventionally manufactured (CM) material. This fundamental work is intended to provide the basis for further research in the field of bandsawing AM components. The investigated cutting fluid strategies include compressed air, MQC, and industry-standard flood cooling. Finally, the surface integrity of the band-sawed surfaces has been evaluated for surface roughness, microstructure, microhardness, and the force date during sawing.

2   Experimental Plan

The work piece material is a chromium-nickel-molybdenum austenitic stainless steel (AISI 316L), widely used in several AM technologies [13]. The used base material for WAAM 316L stainless steel is the ER316LSi wire feedstock.

Table 1. Experimental plan.

Ex No	Work material manufacturing process	Cooling strategy	Cutting speed (m/min)	Feed (mm/min)	Depth of cut (μm)
1/4	AM/CM	Flood	48.8	27.9	4.06
2/5	AM/CM	MQC	48.8	27.9	4.06
3/6	AM/CM	Compressed Air	48.8	27.9	4.06

A total of six experiments were performed - three for the AM material and three for the conventionally manufactured (CM) material, as shown in Table 1. A new uncoated carbide-tipped M-Factor GES (41 mm × 1.2 mm × 2/3 TPI) [14] saw blade was used for all six experiments, and all the blades were made from the same coil to reduce process variation. The diameter of CM and AM bars were, respectively, ~ 101 mm and ~ 113 mm. Since AM components are going to be annealed before cutting them off the base plate, both materials (AM and conventional) have been annealed before conducting the experiments to ensure comparability. The semi-synthetic water-soluble coolant (Nanotech 6800) was used for flood cooling and MQC at a concentration of 4%. Flood cooling and MQC flow rates were respectively 6 l/min and 0.01 l/min. The air pressure for MQC and compressed air was held constant at 0.689 MPa. The round bar stock was cut on the Amada HA400W saw, as shown in Fig. 1. The cutting forces were measured using a Kistler 9129A 3-axis dynamometer. Unist saw blade lube system was used as the delivery system for MQC and compressed air. The system consists of two outlets, one focused on the back edge of the blade to cool the back guides (left outlet) and the other focused towards the tooth to cool the tooth (right outlet) [9]. Both outlets are in front of the cut, and the volume flow of coolant is evenly distributed. Three outlets are used for flood cooling, as shown in Fig. 1. One is placed towards the blade in front of the cut, one is directed into the kerf, and one is placed towards the blade at the end of the cut. Slices with a thickness of 8 mm were cut off the bar stock. Figure 1 (right) shows the location for measuring surface roughness using Zygo NewView 7300 and microhardness using Shimadzu HMV-2TE tester. The microstructures have been inspected by cutting 10 x 10 mm samples using abrasive water jetting to minimize any heat influence. Each specimen was etched by V2A-etchant (47.6% H2O, 47.6% HCl, 4.8% HNO3) for ~ 90 s until the microstructure was revealed. The microstructures were investigated using the Keyence VHX 7000 digital microscope.

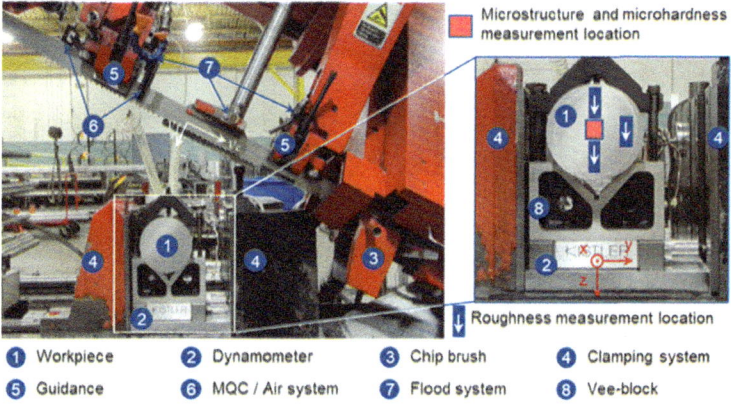

Fig. 1. Experimental Setup on Amada HA400W Scissor Style Bandsaw.

3   Results and Discussion

3.1   Force Analysis

The influence of the material and the cooling strategy on the average cutting forces from the middle portion of the workpiece can be seen in Fig. 2. Overall, the feed forces F_f are higher than the cutting forces F_c. This is probably because the undeformed chip thickness is of the same order of magnitude as the cutting-edge radius. The ploughing effect occurs if the depth of cut is of the same order as the cutting-edge radius. Ploughing increases friction processes around the cutting-edge radius and on the flank surface, particularly increasing the feed force. This effect depends on the machined material and the cutting conditions, as was described by Klocke et al. [15]. In comparing the different cooling conditions it is observed that the lowest cutting and feed forces for both the additive and the conventional material occur when MQC is employed. For the conventional material, the average cutting forces are higher when using compressed air cooling than using flood cooling. This is very different from the additive material, where the difference in cutting forces between flood cooling and compressed air cooling is negligible. Flood cooling leads to lower comparative cutting temperatures, which minimizes the thermal softening effect of the material and thereby results in higher cutting forces [15, 16]. In addition to the increased prevalence of workpiece thermal softening, the higher cutting temperatures generated during machining with compressed air cooling induces higher chemical affinity between the workpiece and the tool material. This leads to the built-up-edge (BUE) phenomenon, which can change the cutting tooth's effective rake angle and result in higher cutting forces [16]. Comparing the forces for the different cooling strategies, there is a high potential for using a more sustainable cooling strategy than the current industry-standard flood cooling. Specifically, when bandsawing the additive material, compressed air results in the same cutting forces as flood, and MQC can minimize the forces for both AM and CM material.

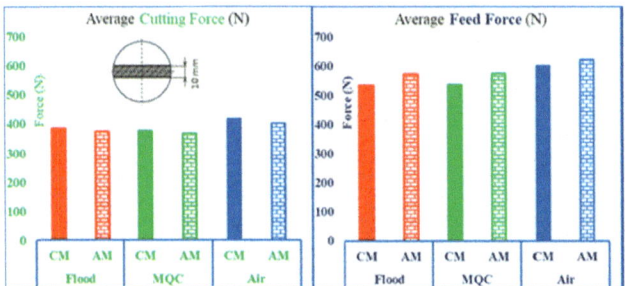

Fig. 2. Average cutting force (left) and feed force (right) at the center of the workpiece for AM and CM material for different cooling strategies.

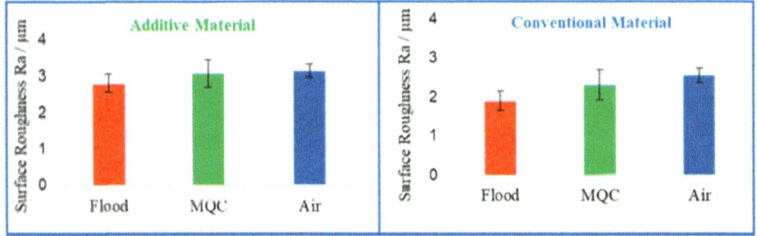

Fig. 3. Average surface roughness values for AM and CM material with different cooling strategies. Error bars indicate standard deviation

3.2 Surface Roughness

Figure 3 shows the measured average surface roughness R_a. The roughness values for bandsawing the additive parts are higher than for the conventional part. This could be explained from the higher feed force data while sawing AM parts. Increased feed force could lead to increased vibration, affecting surface roughness. The surface roughness in bandsawing depend on the stability of the cut, and is influenced by the cutting parameters, the properties and geometry of the work material, the saw blade pitch and geometry, and the blade tension. Additionally, the cooling strategy can positively affect surface quality [8]. A minimal statistical difference exists between the average surface roughness values with flood and MQC for both AM and CM materials. However, flood cooling (as the industry standard) delivers the lowest average surface roughness values for both materials. Also, the difference between the cooling strategies is usually of the order of the standard deviation. Comparing the flow rates of flood cooling and MQC, in this experimental design, flood cooling utilizes 600 times the amount of coolant as MQC. Thus, in this case, MQC, a sustainable cooling strategy has a high potential since the reduction of coolant is equivalent to the cost reduction [9].

3.3 Microstructure and Microhardness

Figure 4 depicts the microstructures of the annealed CM and annealed WAAM specimens. Both microstructures are mostly composed of FCC Austenite (Υ) matrix with BCC δ-Ferrite and possible Cr-rich carbides [12, 17]. Austenite is the bright phases shown in the optical micrographs, while ferrite and carbide phases are dark phases. While the microstructure of annealed CM 316 stainless steel is composed of fully equiaxed grains, the microstructure of annealed WAAM is composed of a mixture of columnar dendritic structures and equiaxed-like grains. The microstructure of as-built WAAM 316L stainless steel is entirely dendritic and columnar [12, 17]. In this experiment, the as-built WAAM 316L specimens were annealed at 1065 °C (1950 °F) for an hour before sawing. The columnar dendritic structures is less noticeable due to the dissolution of dendritic arms. However, the columnar structure is not fully dissolved probably due to shorter annealing time at 1065 °C. Rodrigues et al. reported complete dissolution of dentritic δ-Ferrite into Austenitic $\Upsilon + \delta$-Ferrite when annealed at 950 °C and 1050 °C for 2 h [12]. Thus, the microstructure of annealed WAAM 316L stainless steel in this experiment falls between wrought and as-built WAAM. The distribution of ferrite in AM air-cooled and AM flood specimens is consistent with annealed WAAM microstructure reported by Rodrigues et al., except the presence of some columnar features [12]. In this experiment, the dissolution mode is also ferrite solidification mode as Cr/Ni ratio > 1.55 [12, 18] and ferrite volume fraction would be 5–10% [12]. The microhardness measured within 100 μm from the sawed surface. The microhardness in MQC sawed workpieces is lower than flood cooling due to lower cooling rate in MQC than flood. Overall, the effects of flood coolant and MQC on the observed microstructure are similar for both the AM and CM processes. Thus, a sustainable solution (MQC) can be an effective strategy for sawing both AM and CM based parts.

Acicular α'-martensite is also observed on the microstructures near the sawed edge. Martensite is composed of distorted body centered cubic or body centered tetragonal that can more effectively dissolve carbon. The formation of martensitic phase occurs due to quenching effects from coolant or lubricant while sawing and possible strain hardening from sawing/machining [18, 19]. Feldhausen et al. also reported martensitic phase formation in hybrid manufacturing of 316L stainless steel. Martensite is less pronounced in AM air cooled specimens since it has a slower cooling rate than flood coolant and MQC. During the heat treatment process, Cr-rich carbides are precipitated in austenitic steel if it is cooled between 550 °C and 950°Cat a low enough rate and for a sufficient duration. Depending on cooling rate, δ-Ferrite can be precipitated as σ-phase (intermetallics) in 316L stainless steel during cooling. The triangular or polygonal

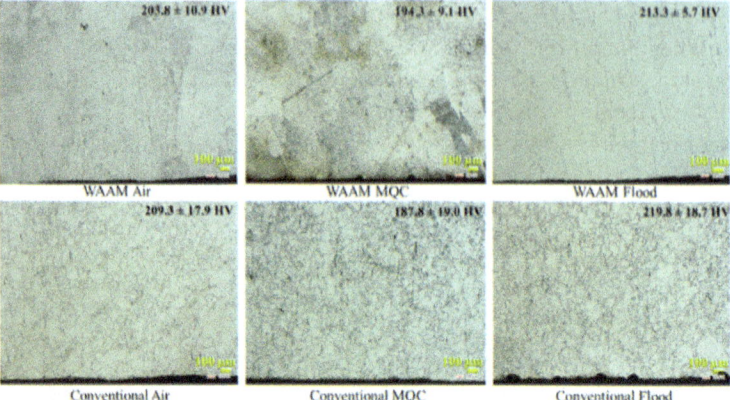

Fig. 4. Microstructure and average microhardness of sawed 316 stainless steel, both annealed AM and annealed conventionally manufactured (casting) in all cooling conditions.

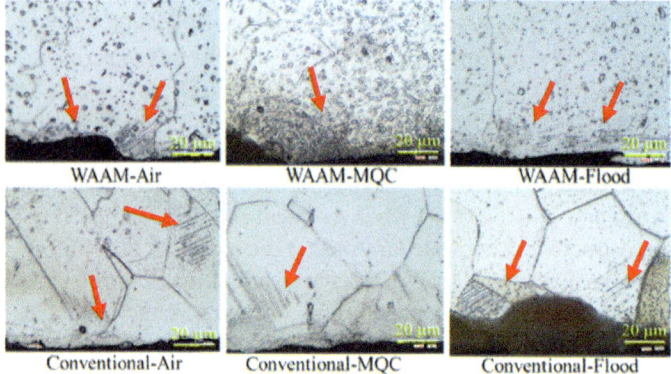

Fig. 5. Microstructure near sawed surface showing formation of martensite (indicated by arrow) and carbide in all cutting conditions.

feature in Fig. 5 resembles carbide [17]. The grains boundaries are often deep dark/black, which is probably due to carbide and intermetalics diffusion into the boundaries.

4   Conclusions

Comparing the forces for the different cooling strategies, there is a high potential for using a more sustainable cooling strategy than the current industry-standard flood cooling. Specifically, when bandsawing the additive material, compressed air cooling results in the same cutting forces as flood cooling, and MQC can minimize the forces for both AM and CM materials. A minimal statistical difference exists between the average surface roughness values with flood and MQC for AM and CM materials. The difference between the cooling strategies is usually of the order of the standard deviation. Comparing the flow rates of flood cooling and MQC, sustainable cooling strategies have a high potential since the reduction of coolant is equivalent to the cost reduction. Moreover, the influence of flood cooling and MQC on the observed microstructure are similar for both the AM and CM processes. Thus, a sustainable solution (MQC) can be an effective strategy for sawing both AM and CM based parts. Overall, the sustainable cutting fluid solution, MQC, has a comparable effect on the surface roughness, microstructure, and cutting forces, compared to flood coolant. Thereby, MQC cooling in WAAM parts could be a sustainable alternative to the industry standard flood cooling for bandsawing.

References

1. Kumar LJ, Pandey PM, Wimpenny DI (2019) 3D printing and additive manufacturing technologies. Springer
2. Gibson I, Rosen D, Stucker B, Khorasani M (2021) Additive manufacturing technologies, 3rd edn. Springer, Cham
3. Moylan S, Slotwinski J, Cooke A, Jurrens K, Donmez MA (2013) Lessons learned in establishing the NIST metal additive manufacturing laboratory. NIST
4. Rajurkar KP, Hadidi H, Pariti J, Reddy GC (2017) Review of sustainability issues in non-traditional machining processes. Procedia Manuf 7:714–720
5. Rangasamy N, Kirwin R, Rakurty CS (2023) A comparative sustainability assessment of cutting fluids usage in band sawing. In: Kohl H, Seliger G, Dietrich F (eds) GCSM 2022. Lecture Notes Mech Eng 21–29. https://doi.org/10.1007/978-3-031-28839-5_3
6. Jawahir IS et al (2011) Surface integrity in material removal processes: recent advances. CIRP Ann 60(2):603–626
7. Balaji AK, Rakurty CS, Ghatikar V (2014) A novel multiple cutting fluid dispensing system for sustainable manufacturing. Proceedings of the STLE AM&E, Lake Buena Vista, pp 18–22
8. Rakurty CS, Rangasamy N (2021) Cutting fluid application for bandsawing: a sustainable solution for cutting solids. In: Proceedings of ASME IMECE Vol 2B p V02BT02A065. https://doi.org/10.1115/IMECE2021-73127
9. Rakurty CS, Rangasamy N (2021) An experimental study on sustainable bandsawing solutions for structural applications. In: Proceedings of the ASME 2021 IMECE Vol 2B p V02BT02A066 (2021). https://doi.org/10.1115/IMECE2021-73133
10. Rakurty CS (2019) Targeted and variable minimum quantity cutting fluid application for finish machining of steels. Ph.D. dissertation, The University of Utah. ark:/87278/s6vq93x8
11. Feldhausen T, Raghavan N, Saleeby K, Love L, Kurfess T (2021) Mechanical properties and microstructure of 316L stainless steel produced by hybrid manufacturing. J Mater Process Technol 290
12. Rodrigues TA et al (2021) Effect of heat treatments on 316 stainless steel parts fabricated by wire and arc additive manufacturing: microstructure and synchrotron X-ray diffraction analysis. Addit Manuf 48
13. Sandmeyer Steel (2014) Specification sheet: alloy 316/316L
14. Rakurty CS, Vandervaart PC (2019) Ground set saw blade, no US 10, 279, 408 B2
15. Klocke F (2018) Fertigungsverfahren 1. Springer, Berlin
16. Pervaiz S, Deiab I, Rashid A, Nicolescu M (2017) Minimal quantity cooling lubrication in turning of Ti6Al4V: influence on surface roughness, cutting force and tool wear. Proc Inst Mech Eng B J Eng Manuf 231(9):1542–1558
17. Setyowati VA, Widodo EW, Hermanto SA (2019) Normalising of 316L stainless steel using temperature and holding time variations. In: IOP CSMSE
18. Jin W, Zhang C, Jin S, Tian Y, Wellmann D, Liu W (2020) Wire arc additive manufacturing of stainless steels: a review. Appl Sci 10(5)

19. Kim C (2016) Nondestructive evaluation of strain-induced phase transformation and damage accumulation in austenitic stainless steel subjected to cyclic loading. Metals 8(1), 2018 Author, F.: Article title. Journal 2(5):99–110

Frontloading Approach for Energy Efficiency in Forming Technology

L. Kluy[1(✉)], M. Becker[1], A. Wächter[2], and P. Groche[1]

[1] Institute for Production Engineering and Forming Machines (PtU), Technical University of Darmstadt, Otto-Berndt-Str. 2, 64287 Darmstadt, Germany
lukas.kluy@ptu.tu-darmstadt.de
[2] Institute for Production Management, Technology and Machine Tools (PTW), Technical University of Darmstadt, Otto-Berndt-Str. 2, 64287 Darmstadt, Germany

Abstract. The early stages of forming process development offer a high degree of design freedom and therefore cost advantages in implementing energy efficiency measures. However, this potential often remains unused in industrial applications due to the lack of a cost-efficient and simple approach.

This frontloading approach addresses the specifics of forming process development in existing production environments and bridges the gap between new development and retrofitting in SMEs. Four steps from analysis to modelling and measures to implementation provide an easy-to-follow guidance. Key challenges in data collection and uncertainty management are addressed.

Using two forming processes as examples (bulk and sheet metal forming), a reduction in total energy consumption is achieved while the productivity remains unchanged. Therefore, the proposed frontloading approach opens up new opportunities for energy efficiency in forming technology.

Keywords: energy efficiency method · forming technology · frontloading

1 Introduction to Energy Efficiency in Forming Technology

Societal interest in ecologically compatible products and rising costs for energy sources are motivating the development of energy-efficient production processes in industry. The industry sector account for 38% (169 EJ) of total global final energy use in 2021 with around 2% annual increase since 2010. This makes industry the consumer group with the highest energy consumption [1]. The provision of this energy is extremely resource-intensive. As a result, global resource consumption and emissions of climate-active gases are closely linked to production. Improving energy efficiency in industrial processes helps to reduce resource requirements and emissions of climate-active gases.

In addition to contributing to climate goals, improving energy efficiency offers economic benefits to companies. Over the past 20 years, the gross price of electricity for

Take Advantage of the High Degree of Freedom in the Early Stages of Process Design

© The Author(s) 2025
H. Kohl et al. (Eds.): GCSM 2023, LNME, pp. 615–622, 2025.
https://doi.org/10.1007/978-3-031-77429-4_68

industrial companies has risen in the U.S., China, and the EU [2]. In Germany in particular, there has been a significant price increase, not least due to taxes. The successful implementation of efficiency measures thus becomes a competitive advantage for companies [3].

Today's energy efficiency approach in manufacturing is predominantly focused on measures on existing processes (retrofitting). In particular, the efficient use of energy is often insufficiently addressed in the definition of requirements for new production processes [4]. The measures taken on existing manufacturing equipment usually culminate in process control, heat recovery, optimized compressed air generation, use of frequency converters for variable power requirements and use of new energy-efficient drives [5]. This makes a major contribution to energy efficiency in production, as many old systems used in small and medium-sized enterprises (SMEs) can be upgraded with little effort to achieve great benefit. The disadvantages are that functional components have to be replaced during downtime and implementation is only carried out if the payback is acceptable.

On the other hand, some publications deal with methodologies to consider energy efficiency initially in the planning and construction of new production plants and systems as well as new machines [6, 7]. These can be characterized as replanning.

For most SMEs, however, the application-oriented middle course is useful, as new processes are often implemented in existing production environments. Metal forming, in particular, is often characterized by individual machines (not series machines); high interconnect power, servo technology with high peak power, and a high degree of flexibility. Compared to other manufacturing processes like machining, the cold, warm and hot forming processes have comparatively high energy requirements. However, due to the high productivity and material utilization during forming, the environmental footprint of each finished part is comparatively small. Because of these aspects, in many cases neither the retrofit nor the new design of energy efficiency is beneficial for forming technology in SMEs.

The novelty of the approach presented here is to address the specifics of forming process development in existing production environments and to bridge the gap between replanning and retrofitting in SMEs. Following the approach, the energy consumption of the forming process is linked to the material flow and mapped in a model. The high degree of design freedom in the early phases of development allows efficiency measures to be implemented cost-efficiently.

2   Frontloading Approach for Energy Efficiency in Forming Technology

The major challenge in implementing energy efficiency measures in metal forming, especially in SMEs, is dealing with the complexity of the systems and uncertainties in the data without risking productivity and quality losses. The aim of this approach is therefore to provide an easy-to-follow guide for the energetic analysis of the process chain, to identify potentials and to derive recommendations for the development of energy-efficient process design. For this purpose, the forming process is divided into subsystems, abstracted, aggregated to a total system and concretized. The approach consists of four steps: process abstraction, model development, identification of the field of action and implementation of the process (Fig. 1).

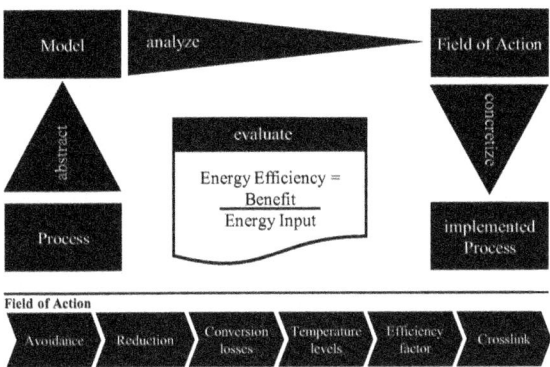

Fig. 1. Frontloading approach for energy efficiency in forming technology.

2.1 Process Abstraction and Model Analysis

In order to model a production process chain in terms of energy demand, it is useful to link the energy source consumption with the material flow. The complexity of the forming processes makes it difficult to estimate energy carrier consumption using physical models. For example, peripheral equipment (cooling, lubrication circuits), mass accelerations and operating scenarios must be included in the models. Therefore, empirical data are needed. On the other hand, energy data are hardly available in the early stages of process development.

To solve this dilemma, data collection begins with individual consumers to allocate energy source consumption. Based on the power of the forming machine and the aggregates specified by the manufacturers, the consumers can be relatively sorted into potential large and small consumers [6]. The measurement effort can be reduced by measuring only one among several similar consumers and considering the power consumption as representative of all similar consumers and operating points.

Current clamps and Rogowski coils provide a practical way to measure the conductor current if the devices are already available and can be operated with similar modes. By evaluating the induced current, a quantitative statement about the conductor current is possible. Another possibility for energy determination is the use of heat measurements (e.g. contact-free via thermographic cameras), which is particularly interesting for hot forming, since the provision of heat is energy-intensive.

In addition, manufacturer information as well as related processes can serve as data sources. Data that do not show significant variations in energy demand are assumed to be constant during data processing. For validation, the operating points are compared with measured real data. To create a functional model, the individual subsystems are linked to a total system.

2.2 Field of Action

According to Hesselbach [8] and Schnellbach [9], the fields of action avoidance, reduction, conversion losses, temperature levels, efficiency and crosslink are suitable for energy efficiency and are transferred here to forming technology. The aim of the measures is to reduce power consumption while the productivity remains unchanged.

Avoidance aims to determine whether the process step must necessarily be operated, whether it can be omitted or integrated into another process step. This can be done, for example, by optimizing the design or the functional integration of components.

Reduction aims at lowering the energy level by reducing e.g. feed rates, blankholder forces as well as friction due to lubrication and hydraulics. The analysis aims at downsizing the drives, actuators and periphery.

Conversion loss affect the demand-responsive energy supply both in the timing of useful energy provision and in the dimensioning of the energy supply, e.g. the reduction of the system pressure beyond the actual demand without throttling.

Adjustment of the temperature level addresses the targeted reduction of the temperature in the process, taking into account the yield stresses, as well as the heating of material before and after forming. If heat treatment is carried out after forming, the warm material can be fed into the furnace directly after forming to match the temperature levels more efficiently.

Efficiency aims at energetically optimal operating points and focuses on efficiency losses due to non-optimal operating parameters, e.g. the processing speed. This determines the time utilization of operating resources and can be adjusted with the aim of reducing idle times.

Crosslink means connecting at least two process steps to exchange energy and information. For example, the time required for transferring sheet blanks in process chains can be reduced, the duration of power consumption in standby mode can be shortened or waste heat from forging can be used to preheat components.

Once the scope for action has been opened up via the fields of action, measures are transferred to the process in the next step.

2.3 Implementation of Energy Efficiency Measures in the Process

The energy efficiency measures identified in the fields of action are transferred to the process in the next step. In hot forging processes, a major energy requirement is attributable to heating. From the point of view of an energy-efficient process design, it must be clarified whether lower forming forces at increased material temperatures justify the energy expenditure of heating. The measures implemented are incorporated in the design, process sequences and parameters, and the components in the trade-off between energy efficiency and technical framework conditions such as safety factors, formability and tool wear.

2.4 Evaluation of Energy Efficiency Measures

In production processes, energy efficiency can be defined as the ratio of goods produced per unit of energy used [10]. The reference value for the output must be chosen individually for the process. The measurement in kWh/kg is recommended for forming processes.

If machining is integrated into the process chain, it is advisable to use kWh/piece as the component mass changes significantly.

For simplicity, this approach does not take into account energy-related effects in the operational environment. These can lead to further savings elsewhere or even reduce the desired savings as a rebound effect.

3 Examples in Forming Technology

The potentials of the frontloading approach are discussed based on cold sheet and hot bulk metal forming (Fig. 2).

As an example of cold sheet forming the incremental bending process roll forming is considered. A large proportion of the annual steel production is processed by roll forming and therefore energy efficiency measures offer a great lever to improve the environmental footprint of profiles. In roll forming, the sheet strip is transported through several forming stages by means of rotating rolls and is formed at the same time. Since the diameters of the top and bottom rolls in the contact zone with the sheet vary according to the forming contour, the circumferential speed deviates locally from the sheet speed. The resulting slip ratio and the local contact stresses often result in inhomogeneous drive states with even decelerating torques at certain forming stages and therefore increased energy requirements [11].

In hot forging, the newly developed Equal Channel Angular Swaging (ECAS) process is analyzed for the production of medical titanium implants on a high-speed press. This process produces nanostructured titanium bars from a coarse-grained starting material, which is used for improved bone healing and reduced bacterial wetting. A titanium rod is fed into the high-speed press by drive rolls. In the die, energy-intensive heating of the bars takes place by induction continuously and the bar is formed incrementally [12].

3.1 Process Abstraction and Model Analysis

The first step of the approach is the abstraction of the process (Fig. 2A). In roll forming, the drive motor, the cardan shafts and the rolls are considered, as they represent the decisive subsystems for the process. In the hot forging process, the focus is on the feed unit, induction heating and the drive of the press. The second step is to analyse the power consumption and connect it to the material flow (Fig. 2B). In roll forming, drive torque and motor current are identified as essential process parameters. In the hot forging, the feed rate, the forming temperature and the strokes per minute are the relevant parameters. For the decoupled energy demand measurement, a current clamp, a thermal camera and sensory cardan shafts are used.

However, not all data can be mapped by preliminary tests. Thus, analytical models for heat transfer are used to calculate the heat losses in the dies as well as the deformation work. Polynomials describing the energy-related behavior of the components are fitted to the measured data. The models used for cold sheet metal forming are discussed in more detail in [11, 13] and for bulk metal forming in [12, 14, 15].

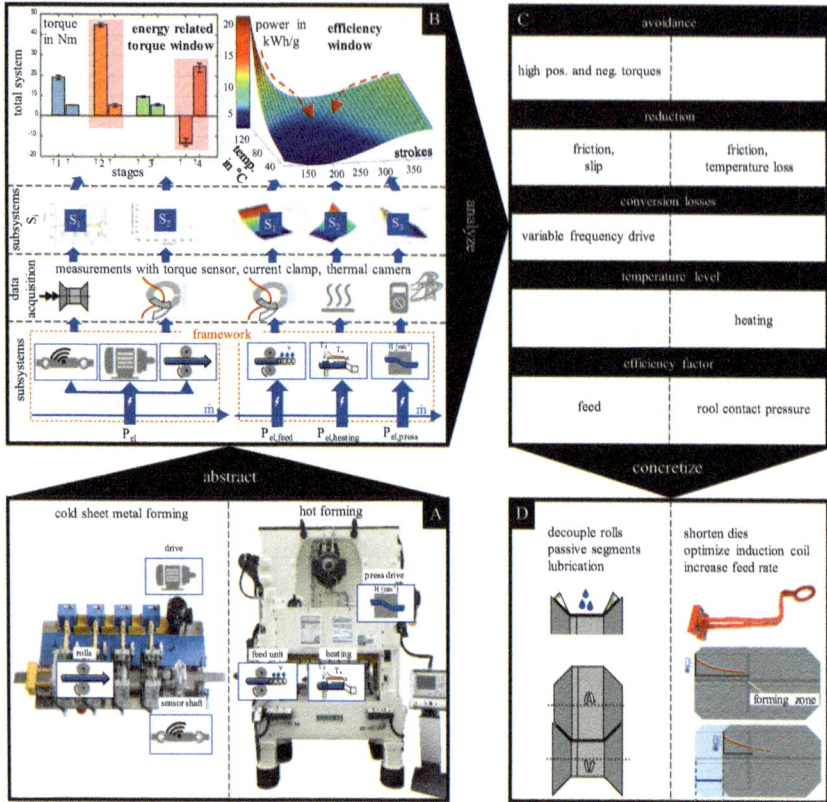

Fig. 2. Practical procedure according to the frontloading approach for energy-efficient process development in forming technology, cold sheet metal forming (left) and hot forming (right).

3.2 Field of Action

In roll forming, the positive and negative torques of the rolls are linked to the power demand of the motor (Fig. 2C). Here, both the direct power demand of the motor and the applied drive torque of the sensory driveshafts at the forming stages are considered. From numerical results [13] and experimental tests, it can be deduced that the torque correlates with the power demand and a homogeneous feed of all forming stages is targeted *(avoidance of high pos. And neg. Torques)*. During the FEM design of the process, the required torques of the forming stages can be optimized by tool design and tribological adjustments can be made by surface and lubricant modifications *(reduction of friction and slip)*. Energy efficiency potentials through variable frequency drives *(conversation losses)* and optimization of the feed *(efficiency factor)* are implemented.

In hot forging, the power consumption of the press, the induction heating and the feed is investigated (Fig. 2C). The forming temperature depends on the die length, the time between heating and forming, and the heating temperature *(temperature level of the heating)*. The time between heating and forming is directly related to the feed rate and the die geometry at the die inlet. At low stroke rates and high temperatures, the highest energy intensity can be expected. The higher the number of strokes, the less quickly the material cools down in the inlet channel of the tools and the more material is processed, which lowers the energy intensity *(reduction of temperature loss)*. Limitations arise from the fact that the process leads to material failure at low temperatures, i.e. there is a technical limit here.

3.3   Implementation and Evaluation of Energy Efficiency Measures

In the given roll forming process, negative torques occur at the rolls of the fourth stage due acceleration. Consequently, push and pull effects of the profile arise, leading to increased energy requirements. As a solution, a partial decoupling of roll segments and the use of passive side rolls are evaluated experimentally for a representative example (Fig. 2D). Here, the energy requirement of the motor averages 576.4 W before and 547.1 W after the measure, which corresponds to an optimization of $5\% \pm 0.7\%$. Calculations of the optimized process energy based on the sensor-measured torques even show a reduction of the energy requirement by 16%. After quantifying these measures, a systematic integration into the FE design of the process at an early stage is targeted by means of frontloading [13].

In hot forging, the frontloading approach shows heat dissipation into the die as the dominant energy efficiency potential. The focus is therefore on optimized inductor coil geometry and shortened dies (Fig. 2D). Again, an example of certain operating parameters is used. The power consumption of the inductor to reach the target forming temperature is 3.6 kW before optimization and 3.1 kW afterwards, so that the energy efficiency of the process is improved by $13\% \pm 2\%$ for the same productivity.

4   Summary and Outlook

The early stages of forming process development offer a high degree of design freedom and therefore cost advantages in implementing energy efficiency measures. However, this potential often remains unused in industrial applications due to the lack of a cost-efficient and simple approach. The presented frontloading approach addresses the specifics of forming process development in existing production environments and bridges the gap between new development and retrofitting in SMEs. Four steps from analysis to modelling and measures to implementation provide an easy-to-follow guidance. Key challenges in data collection and uncertainty management were addressed. Using two forming processes as examples (cold sheet metal forming and hot bulk forging), a reduction in total energy consumption of up to 10% is achieved while the productivity remains unchanged. Therefore, the proposed frontloading approach opens up new opportunities for energy efficiency in forming technology.

Acknowledgements. The frontloading approach was awarded second place in the Hessian State Prize for Energy (Germany). The work was funded by the German Federal Ministry of Education and Research (BMBF) and the German Federal Ministry for Economic Affairs and Climate Action (BMWK) as part of IdentiTI and Mittelstand Digital Center Darmstadt as well as the Joachim Herz Stiftung.

References

1. International Energy Agency: Tracking Industry 2021, International Energy Agency IEA, www.iea.org, Paris (2022)
2. Commission E (2020) Study on energy prices, costs and their impact on industry. Report from the European Commission, Brussels
3. Fresner J, Morea F, Krenn C et al (2017) Energy efficiency in small and medium enterprises. J Clean Prod 142:1650–1660
4. VDI 2221: Design of technical products and systems, model of product design. Association of German Engineers (VDI), Beuth Verlag, Berlin (2019)
5. Gao M, He K, Li L, Wang Q, Liu C (2019) A review on energy consumption, energy efficiency and energy saving of metal forming processes from different hierarchies. Processes
6. Kornes M (2017) A method to identify energy efficiency measures for factory systems based on qualitative modeling. Springer, Wiesbaden
7. Denkena B, Abele E, Brecher C et al (2020) Energy efficient machine tools. CIRP Ann 69:646–667
8. Hesselbach J (2012) Energy and climate-efficient Production. Springer, Wiesbaden
9. Schnellbach P (2015) Methodology for the reduction of energy waste under consideration of target values of holistic production systems. Technical University Munich
10. DIN EN ISO 50001:2018–12 (2018) Energy management systems—requirements with guidance. Beuth Verlag, Berlin
11. Traub T, Güngör B, Groche P (2019) Measures towards roll forming at the physical limit of energy consumption. Int J Adv Manuf Technol
12. Klinge L, Kluy L, Siemers C et al (2022) Nanostructured Ti–13Nb–13Zr for dental implant applications produced by severe plastic deformation. J Mater Res
13. Becker M, Groche P (2022) Optimization of slip conditions in roll forming by numerical simulation. In: NUMISHEET 2022. The minerals, metals & materials series. Springer
14. Kluy L, Becker M et al (2021) Frontloading for energy efficiency, ZWF Gruyter
15. Klinge L, Kluy L, Siemers C et al (2023) Nanostructured Ti-13Nb-13Zr alloy for implant application. Front Bioeng Biotechnol

Polycrystalline Diamond Tool for Milling of Granite

A. Ahsan$^{(\boxtimes)}$, M. Kahlmeyer, E. Fedotov, and S. Böhm

Department for Cutting and Joining Manufacturing Processes, University of Kassel,
Kurt-Wolters-Str. 3, 34125 Kassel, Germany
a.ahsan@uni-kassel.de

Abstract. State of the art in tooling for cutting granites are sintered grinding tools having synthetic diamond grains bound together in metal matrix usually a cobalt copper alloy. Routers for CNC machining of stone surfaces come in various forms such as face milling tools, end milling cutters, edge profiling tools etc. and have either segmented or continuous grit. Characteristic of the grinding processes the material removal rates are relatively low and inefficient compared to geometrically defined tools such as milling tools for metal or wood. In this study, a milling tool with four vacuum brazed polycrystalline diamond inserts have been used to machine Nero Assoluto Granite. The cutting edge of the brazed PCD tips were laser ablated using ultrashort pulsed lasers to form round and chamfer micro-geometries. Wet milling of granite plates was carried out at constant cutting and feed speed to study the effect of various edge geometries on the tool wear as well as cutting forces. The results showed that the cutting edges with laser ablated chamfer resulted in the smallest tool wear as well as the lowest average cutting force and therefor make a decisive contribution to energy and resource efficient machining processes.

Keywords: resource efficient milling · PCD tool development · milling of granite

1 Introduction—Application and Properties of PCD Cutting Tools

Granite is a hard, heat resistant, igneous rock composed of 10–65% feldspar, 20–60% quartz and 5–15% mica depending on the origin [1]. The hardness of granite ranges from 6–7 on Mohs scale and makes it a superior quality product for flooring, facades and tombstones. Due to its superior heat resistance and vibration damping capability, granite is also used for making beds and frames for high precision machines [2]. State of the art of cutting tools for machining granites are diamond tools which contain segments having synthetic diamond grains sintered in a metallic binder. A common feature of machining with sintered diamond tools is high cutting speed (30–40 m/s) with low cutting feed and depth as well as the need for water cooling [3]. The size of diamond grains, grit distribution, binder composition, sintering process of the inserts as well as number and geometry of them vary significantly for different granite compositions and hardness values. This compels the tool manufacturers and stone processors to manufacture and maintain a large inventory of tools.

© The Author(s) 2025
H. Kohl et al. (Eds.): GCSM 2023, LNME, pp. 623–631, 2025.
https://doi.org/10.1007/978-3-031-77429-4_69

This study is aimed to test the feasibility of geometrically defined tools having brazed polycrystalline diamond inserts as an alternative to the sintered diamond tools for milling of granites. After monocrystalline diamond, polycrystalline diamond (PCD) is the hardest cutting tool material with a hardness between 50–70 GPa [4]. Due to their high cost, PCD tools are used in high end applications of cutting wood, non-ferrous metals and fiber reinforced plastics. PCD is produced by high pressure high temperature (HPHT) sintering of synthetic diamond grains on tungsten carbide (WC-Co) substrates. Under pressures of around 5 GPa and temperatures of about 1400 °C, Cobalt infiltrates from the substrate to fill the voids between diamond grains, binding them together [5]. Typical PCD compositions contain 4–6% of cobalt with the rest being diamond. Cobalt also promotes strong diamond—diamond bonding resulting in the superior hardness and low fracture toughness [6]. PCD tools are characterized by low thermal stability as cobalt can accelerate graphitization rates of diamond at temperatures above 700 °C. In addition, cobalt binder has about 13 times higher thermal expansion coefficient than diamond which can lead to thermal cracks [7]. Due to the strong affinity of carbon to iron, cobalt and nickel, which results in accelerated phase transformation of diamond to graphite followed by dissolution of carbon (from PCD) into the work piece, the application of PCD as a cutting tool material is at present limited to machining of aluminum, wood and fiber composites. First tests on the application of PCD tools for cutting natural stones were carried out by the University of Kassel on band saw blades equipped with brazed and resistance welded PCD saw teeth [8, 9]. The results showed an increase in feed rate to up to 100 mm/min at cutting speed of 50 m/min compared to typical feed of 2–4 mm/min of diamond wire saws.

2   Investigation Method and Experimental Set-Up

2.1   The PCD Milling Tools

The cutting tools used in this study were 25 mm diameter end milling cutter made of tungsten carbide (WC) having four vacuum brazed PCD cutting inserts. Table 1 shows the tool with corresponding geometrical data. The PCD consists of mixed grains with sizes of 30 μm and 2 μm. PCD inserts of length 30 mm and width 5 mm with a corner radius of 2 mm were wire eroded out of 2 mm thick discs having 0.5 mm thick PCD layer with 1.5 mm thick WC backing layer. The inserts were vacuum brazed to the WC tool body in a furnace after sandblasting to remove the white layer from carbide using a silver-based sandwich brazing filler BrazeTec 49/Cu. For this study, three milling tools were manufactured. The first PCD inserts were tested in the 'as-eroded' condition, i.e. without further edge preparation. For the second and the third inserts, picosecond pulsed laser ablation was used to create 50 μm chamfer and 50 μm radius on all four PCD cutting edges of the tool (see Fig. 1).

Due to its superior hardness, mechanical grinding of PCD tools becomes uneconomical. With ultra-short pulsed laser ablation, the high pulse energy evaporates the material with a small heat affected area and almost no formation of the liquid phase. This results in a smooth ablated surface. In addition, by using five-axis laser heads, complex contours and fine geometries can be created.

Table 1. Properties of the end milling cutter used in this study.

Property	Value	
Diameter	25 mm	
Number of inserts	4	
Rake/relief angle*	−20°/2°	
Tool clamping	Weldon	
Max. Cutting depth	30 mm	
Angular pitch of the cutting inserts (variable pitch tool design is used for reduced chatter)	85°/95°/85°/95°	

* The cutting inserts did not have a safety chamfer. The inserts were wire eroded such that the wedge angle was 108°. Seats for the inserts on the tool body were made such that the rake face of the inserts was at −20° from the radial line. This resulted in a rake angle of −20° and a relief angle of 2° while providing the inserts with added stiffness due to a large wedge angle.

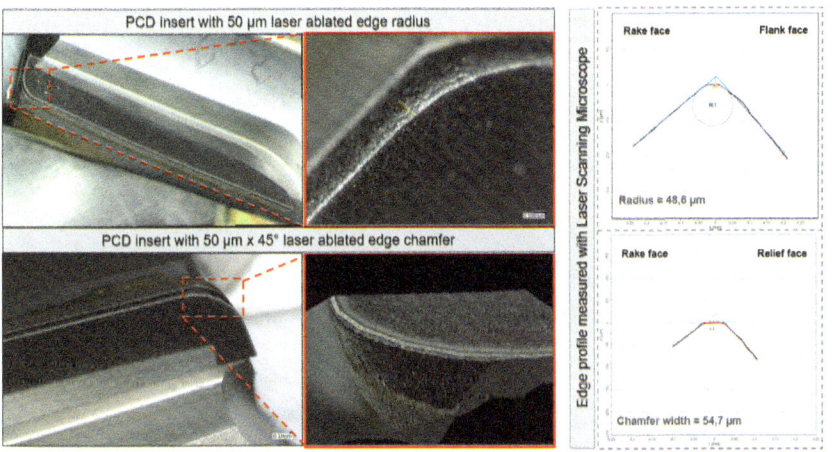

Fig. 1. Edge preparation of the PCD inserts using Laser ablation technique with 50 μm radius (top) and 50 μm x 45° chamfer (bottom).

2.2 Granite Properties

African black (Africa Schwarz) granite plates with dimensions 250 x 100 x 20 mm^3 were used as work piece. Mechanical and physical properties are given in Table 2.

2.3 Cutting Process and Test Set-Up

The cutting tests were performed on a Deckel FP2 milling machine driven by a 2.2 kW spindle. Test setup is shown in Fig. 2. The rotational speed for the milling tests was set

Table 2. Physical properties of African black granite [8].

Property	Value	Property	Value
Density	3.1 g/cm^3	Compressive strength	$200-300$ MPa
Bending strength	$20-30$ MPa	Poisson's ratio	$0.2-0.3$
Elasticity modulus	77 GPa	Thermal conductivity	-3.5 W/(m·K)

to 1.250 min^{-1} for all three tools and tooth feed (f_z) to 0.05 mm. Axial depth of cut (a_p) was the full thickness of the granite plates (20 mm) and radial depth of cut (a_e) was the full tool diameter that is 25 mm. The cutting parameters resulted in a material removal rate (Q) of $125 \text{ cm}^3/\text{min}$. Cutting tests were performed under flood cooling by water. The test setup on the milling machine had to be enclosed and isolated from the rest of the machine mainly because water was being used as the coolant which would otherwise cause rusting of the machine components.

Fig. 2. Test set-up for cutting granite on Deckel FP2 milling machine.

The granite plate was mechanically clamped to a piezoelectric dynamometer of type Kistler 9129AA, which measured force components in three linear directions (Fig. 3). The measurement of forces during the cuts was performed using a program written in LabVIEW, whereby unfiltered force data were captured at 1 kHz sampling rate from the sensor. On each granite plate a total of six cuts were made. Each cut had a total length (including the tool radius) of 80 mm resulting in the amount of material volume removed per cut of 40 cm^3. At any given point there are two cutting edges in contact with the work piece. The selection of the cutting parameters was based on the results of the single tooth cutting test study carried out at the University of Kassel on a similar granite type

[8]. Flank wear on each of the four cutting edges was measured after every 280 cm³ of removed volume. The tool life criteria were set as doubling of the machining forces (F_x or F_y) compared to the first cutting test or fracture of one or more of the brazed PCD inserts.

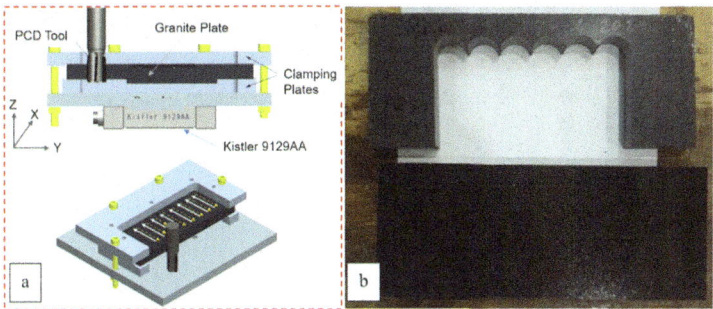

Fig. 3. Experimental set-up including the Kistler 9129AA force dynamometer; (a) configuration and force measurement set-up; (b) granite plate before and after machining.

3   Results and Discussion

Cutting forces in three directions namely feed force (F_x), radial force (F_y) and axial force (F_z) were measured during the cuts. In this section only the feed force measurement results are discussed as the other two components were in the range of 5 to 10 times smaller than the feed force. In addition, the tool wear on the flank face of the four inserts on each tool was measured using optical microscopy after each cutting cycle amounting to a removed material volume of 300 cm³. The two results are discussed below.

3.1   Influence of Cutting-Edge Geometry on Cutting Forces

For the analysis of the force data, the raw force data was filtered using a Butterworth low pass filter with 10 Hz limit frequency. Figure 4A shows the force component along the tool feed direction for one single cut for the three tools. As has been previously observed in the case of machining brittle materials like stones the feed forces are by a factor of 2–3 higher than the cutting forces hence the most important in terms of load on the tool and energy consumption. The feed force is higher at the beginning of the first cut of each tool owing to the unevenness and notches on the cutting edge from the wire erosion process. In almost all inserts some of the notches left over from the erosion process were larger and deeper than the material removed by laser ablation to form chamfers and edge radius. As the uneven cutting edge comes in contact with the stone as the tool enters the workpiece, the notches get removed resulting in edge chipping. Consequently, the succeeding insert experiences higher loads and thus higher feed forces result. Gradually, the grinding action of the rock particles hone the cutting inserts resulting in a smooth cutting edge and the feed forces drop down to a stable level for the rest of the cut.

The cutting tests were carried out until the failure of the tool. Mean feed force during each cut (represents 300 cm³ removed volume) with the three tools is shown in Fig. 4B. The results demonstrate that the difference in the feed force of the three tools is minimal at the beginning. Whereas, the inserts with sharp and rounded edges fail after 3 and 4 cuts respectively, the chamfered inserts last until 6 cuts which means about 1.800 cm³ of removed material after which the tests were stopped.

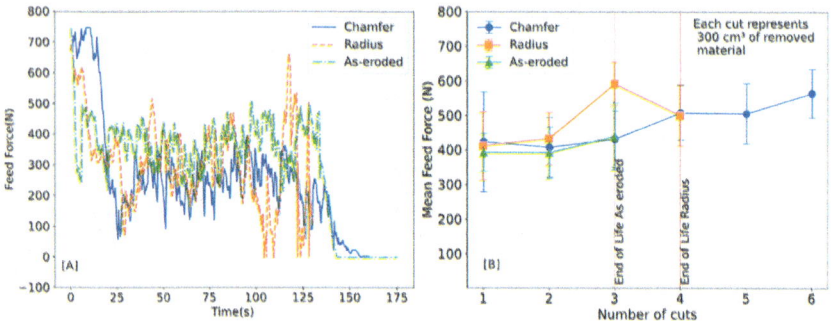

Fig. 4. [A] Comparison of feed force during the first cuts with each of the three tools; [B] Average feed force for the three cutting edge configurations.

3.2   Tool Wear and Tool Life

Figure 5 shows the average of maximum widths of flank wear land of the four inserts on each tool. The observation of tool wear pattern shows that the initial mode of wear is edge chipping on the chamfered and rounded tools, specifically on the laser ablated region. This may be attributed to the line pattern of material removal by the picosecond pulsed laser. Optical microscope images of each tool at the end of life are also shown in Fig. 5.

On the as-eroded tool, the wear pattern observed is primarily abrasion. This is probably due to the wire erosion process of cutting PCD inserts out of the disc shaped preforms. The wire cuts only through the binder regions as the diamond itself is a poor conductor leading to a very uneven edge. In the as-eroded tools, the relatively large, protruding regions on the cutting edges are (ground) removed during the first few seconds of the cut due to rubbing effect of the stone mud (particles mixed with water). This is leading to a smooth edge which wears under abrasion afterwards. The chamfered inserts showed the longest tool life and were able to cut further. For this study a chamfer width and radius of only 0.05 mm was used. A larger chamfer width would result in complete elimination of the uneven edge caused by the wire EDM process. However, it would increase the secondary shear zone size thus resulting in enhanced abrasion on the flank face (Fig. 5).

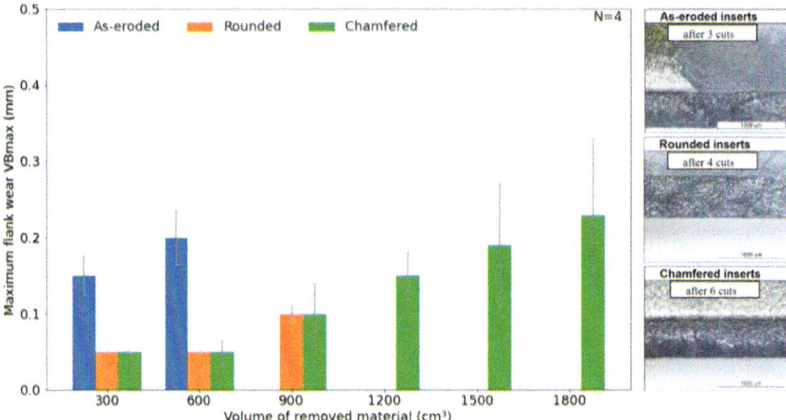

Fig. 5. Maximum flank wear (average value of four cutting edges) for the three edge geometries and the images of one cutting edge of each tool at the end of life.

3.3 The Chip-Separation Process

The abrasion observed on PCD in these tests is typical to what was observed during sawing of granite with PCD in previous studies at the University of Kassel [8, 10]. Besides primarily shearing, crushed granite particles remain at the surface once the tool has passed and rub against the flank face of the tool the leading to a secondary chip formation (Fig. 6). This rubbing effect, coupled with the high pressing force, not only wears the binder (cobalt) in the PCD but also rips out the individual diamond grains. However, if the chamfer width is large enough such that it is larger than f_z and if the corners of the chamfer are even and have small radius then this could result in significant reduction of the flank wear. Additionally, approaches like internal high pressure cooling with nozzles directed at the flank face of the tool could also reduce the abrasion caused by the secondary shearing of the stone work piece.

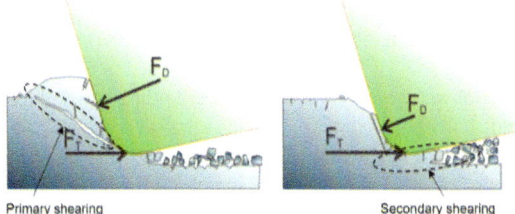

Fig. 6. Primary and secondary shearing of natural stones when machined with geometrically defined tools (as proposed by [8, 10]).

4 Conclusion and Outlook

In this study milling tests were carried out using a PCD router on granite. The results show that the milling tools with geometrically defined PCD are able to machine granite, at removal rates which are up to 30 times higher than the sintered diamond tools of the same size. However, the tool life observed in these tests is about ten times less than that of the sintered diamond tools and thus needs further research and improvement on edge quality and tool finishing. The primary wear mechanism during the start of the cuts is edge chipping after the initial phase is abrasion. The wear is primarily caused by the rubbing of crushed stone particles against the flank face of the tool. Therefore, in author's opinion, larger chamfer size along with greater relief angles perhaps also with high pressure internal cooling could solve the wear problems in future studies. These adjustments could contribute to energy and resource-saving machining processes of mineral materials through the use of PCD tools [11, 12].

References

1. Castro A, Fernández C, Vigneresse JL (1999) (eds) Understanding Granites: integrating new and classical techniques. Geol Soc 168 London
2. Reitz Natursteintechnik https://blog.reitz-natursteintechnik.de/thermischer-ausdehnungskoef fizient-granit-stahl. Last Accessed 2020/7/20
3. Carrino L, Polini W, Turchetta S (2003) Wear progression of diamond mills. Diam Relat Mater 12(3–7):728–732
4. Wilks J, Wilks E (1994) Properties and applications of diamond, 1st edn. Elsevier, Boston
5. Astakhov VP, Stanley A (2015) Polycrystalline diamond (PCD) tool material: emerging applications, problems, and possible solutions. In: Davim J (eds.) Traditional machining processes. Materials forming, machining and tribology. Springer, Heidelberg, pp 1–32
6. Laurindo QMG et al (2023) Molybdenum as a new binder for polycrystalline diamond (PCD) prepared by HPHT sintering. Ceram Int 49(11):17313–17322
7. Chen Z et al (2020) Effects of graphene addition on mechanical properties of polycrystalline diamond compact. Ceram Int 46(8):11255–11260
8. Reichenbächer H (2010) Trennen mineralischer Werkstoffe mit geometrisch bestimmten Schneiden. Kassel Univ. Press, Kassel
9. Ahsan A, Böhm S (2020) Noch effizienter Trennen mittels PKD-Sägeband. Naturstein 8:48–51
10. Böhm S, Scherm W, Schwarte S, Heise C (2012) Bearbeitung von Stahlbeton mit geometrisch bestimmter Schneide aus PKD. Diamant Hochleistungswerkzeuge 2:28–35
11. Vagnorius Z, Sørby K (2011) Effect of high-pressure cooling on life of SiAlON tools in machining of Inconel 718. Int J Adv Manuf Technol 54:83–92
12. Masek P, Maly J, Zeman P, Heinrich P, Alagan NT (2022) Turning of titanium alloy with PCD tool and high-pressure cooling. J Manuf Process 84:871–885

Initial Study on Milling in an Ultrasonically Excited Metalworking Fluid Bath

Daniel Gross, Miriam Eichinger, Jan Harald Selzam(✉), and Nico Hanenkamp

Institute of Ressource and Energy Efficient Production Systems,
Friedrich-Alexander-Universität Erlangen-Nürnberg, Dr.-Mack-Str. 81, 90762 Fürth, Germany
jan.selzam@fau.de

Abstract. Energy and resource efficiency are seen as key drivers of the industrial transformation towards a low carbon economy. In particular, machine tools, which are essential to value creation, have a significant share of industrial greenhouse gas emissions due to their high-energy consumption. The energy demand for the provision and supply of metalworking fluids (MWF) alone can account for up to 25% of the total energy demand. The aim of this study is to investigate and fundamentally design an alternative cooling lubricant concept. This concept does not require continuous pumping of the coolant supply. Instead, the milling operation takes place in an immersion bath filled with metalworking fluid. An additional ultrasonic (US) actuator causes the coolant to vibrate, allowing the coolant to penetrate deeper into the cutting zone. The machining tests include a comparison of classic flood cooling, immersion bath milling and immersion bath milling with US stimulation. To compare the processes, the flank wear, the workpiece surface roughness and the bending moments due to the cutting forces were analyzed. The energy assessment allows a final evaluation of the processes.

Keywords: Milling · energy efficiency · immersion bath · ultrasonic

1 Motivation and State of the Art

Machining centres are an essential part of the manufacturing sector. For this reason, their contribution to industrial sustainability is crucial. Machine tools consume between 10 and 25% of the total energy for the supply of metalworking fluids (MWF) alone [1]. One approach minimizing this energy-intensive consumption is to convert wet machining to minimum quantity lubrication or dry machining. However, the wide range of materials and workpieces in industry often makes it impossible to completely eliminate wet machining. This study therefore investigates whether wet machining can be carried out without a permanently running coolant pump in order to reduce the energy consumption of the machine tool. The machine concept of electric discharge machining is used as an approach. The workpiece and tool are submerged in an immersion bath filled with a dielectric, usually deionised water or oil. By filling the tank once, the workpiece is completely submerged. In machining, this approach is used firstly to ensure a deeper wetting of the cutting zone, secondly to bind the dust and thirdly to improve energy efficiency by ensuring that the coolant pump does not run continuously.

© The Author(s) 2025
H. Kohl et al. (Eds.): GCSM 2023, LNME, pp. 632–640, 2025.
https://doi.org/10.1007/978-3-031-77429-4_70

Temperatures of up to 1000 °C in the cutting zone cause evaporation of the coolant during the cutting process. The evaporation and the challenge to access the contact zone between the tool surface and the chip prevents complete wetting with liquid coolant. To reduce the formation of insulating vapour bubbles, known as the Leidenfrost effect, ultrasonic (US) sensors are used when quenching workpieces. Ultrasound reduces film boiling, allowing the coolant to get closer to the workpiece surface. This can increase the uniformity of the quenching [2]. Within the machining process, this effect should allow the coolant to penetrate deeper into the process zone and thus have a positive effect on tool wear.

The use of submerged baths in machining has been investigated to a limited extent. A liquid nitrogen bath has been used in the preparation of biological samples. The medium is used to freeze the sample material and thus enable machining [3].

An US bath installed in a machine tool was used by [4] to clean the tool during micro-milling. The purpose of the bath is to remove contamination from the tool in order to make the laser measurement of the milling cutter more accurate.

One study describes the use of a US actuator to improve wetting of the cutting zone. The contact between the tool and the workpiece is repeatedly interrupted to allow the coolant to flow in. As a result, tool life can be increased up to three times compared to conventional flood cooling [5].

A comparative study concluded that for the same cutting parameters, milling under flood cooling with ultrasonically excited tools resulted in 27% less wear than without ultrasonic assistance [6]. They assume that ultrasound allows the cutting fluid to reach deeper into the cutting zone and dissipates the heat during the disengagement period.

2 Experimental Set-Up

The tests were carried out on a DOOSAN DNM 500 3-axis machining centre. A special tank was constructed for the immersion bath (Fig. 1). The tank is used to hold water or cooling lubricant during the experimental investigation. A pull-down vise is used to clamp the workpiece. A circulating pump with a nozzle removes the chips from the cutting zone.

The tank is mounted to the machine table with sliding blocks. Additional seals prevent the fluid from leaking at the joints. A ball valve allows the used fluid to be drained. There is a base plate in the tank. The individual components, such as the US actuator and the vise, are attached to this. Both the conventional machining and the filling of the tank are carried out with the Blaser MWF Blasocut BC 935 Kombi. The concentration is 10% for all processes examined.

Tullker's T-1006Z actuator and FJD-1006 generator were used for the US excitation. The actuator has a power of 300 W and an operating frequency of 40 kHz. Frequency modulation was not used.

Machined material is a high alloy austenitic stainless steel. It forms titanium carbonitrides due to the addition of titanium, resulting in increased tool wear. The external dimensions of the pre-machined material are 68 mm × 68 mm.

The tool used is a solid carbide end mill with a diameter of d = 6 mm and a TiAlN coating. The length of the cutting section of the tool is 12 mm with a total tool length

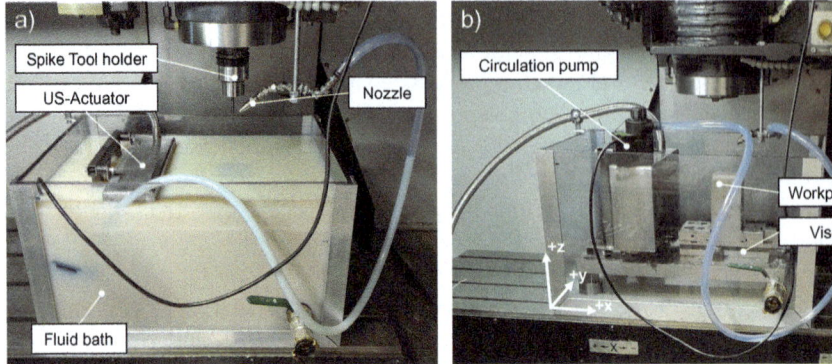

Fig. 1. a) Bath filled with metalworking fluid b) empty bath with vise mounted on base plate

of 100 mm. The short cutting edge in relation to the total length is to ensure that the rotating tool does not carry out too much coolant of the tank during the submerge process. The tests are carried out using a Pro-Micron Sensory tool holder spike. This allows the bending moments and therefore the cutting forces to be measured during the process. The test parameters are given in Table 1.

Table 1. Cutting parameter of the experiments

Parameter	Cutting speed v_c	Feed f	Radial width of cut a_e	Axial depth of cut a_p
Value	51 m/min	0.15 mm	0.3 mm	6 mm

3   Results

3.1   Flank Wear

Flank wear was recorded using a Keyence VHX 5000 measuring microscope. The flank wear values shown in Fig. 2 are the average values of the four cutting edges for all three test repetitions.

There is an approximately linear increase for all MWF concepts. In the reference test, after a cutting length of 30.2 m, the mean value of the flank wear is 87.3 µm. It is noticeable that when machining with conventional flood cooling, the wear is lower at all cutting lengths considered than in the submerged bath tests with and without US. When milling in the submerged bath without US, the mean value of flank wear after machining is 116.25 µm, an increase of 33.1% compared to conventional machining.

The US submerged milling tools have a mean flank wear value of 104.3 µm at this point. This is 19.5% higher than for conventional machining. The standard deviations of the mean values vary between the tests and are 15.4 µm after a cutting length of 7.3 m for conventional machining, which is 36.5% with a mean value of 42.1 µm. It is therefore difficult to determine the influence of the submerged bath or the US. However, there is a noticeable tendency for increased flank wear when milling in a submerged bath with or without US. The use of US seems to reduce this effect slightly.

The wear images show more wear on the clearance surface with conventional machining. In contrast, the submerged milling with and without US shows little wear. It is also noticeable that during the submerged tests there was material build-up on the leading chamfer of the tool. This could be caused by insufficient chip removal from the cutting zone. The formation of built-up edges was observed in all tests.

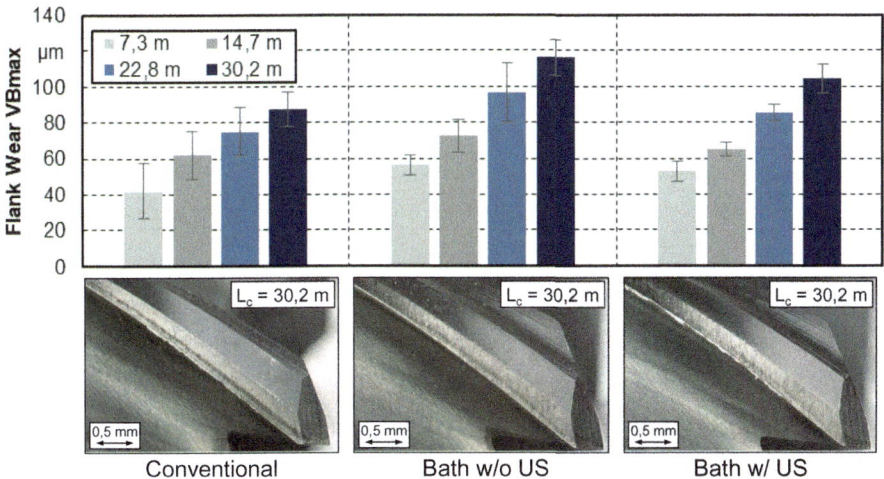

Fig. 2. Flank wear of conventional, bath w/o US and bath w/US for different cutting lengths

3.2 Bending Moment

With flood cooling as a reference, the average bending moment is $M_B = 9.3$ Nm after a cutting length of 7.3 m (Fig. 3). It then increases to 10.3 Nm ($L_c = 14.7$ m) and 11.2 Nm ($L_c = 30.2$ m). This curve is similar for all the tests. The increasing moment is due to increasing tool wear during the machining time. When milling under submerged conditions with the US actuator deactivated, the moment after a cutting length of $L_c = 14.7$ m is 9.6 Nm, which is an increase of 2.9% compared to conventional machining. This trend continues with an average increase of 1.9% in the bending moment from 10.5 Nm to 11.4 Nm after a cutting length of 14.7 m to 30.2 m.

With the US submerged bath, a bending moment of 11.0 Nm was obtained after a cutting length of 30.2 m. The moments at the end of the test series are therefore approximately the same.

The values are also compared with those of flood cooling at the same machining time. The submerged bath increases the process forces insignificantly, while those of the US submerged bath are comparable to those of conventional machining. The standard deviations are between 9.0% and 11.4% of the measured mean value for all experimental setups. There is no pattern to the variation. No valid conclusions can be drawn about the influence of the US or the submerged bath, as the standard deviation exceeds the differences by a multiple.

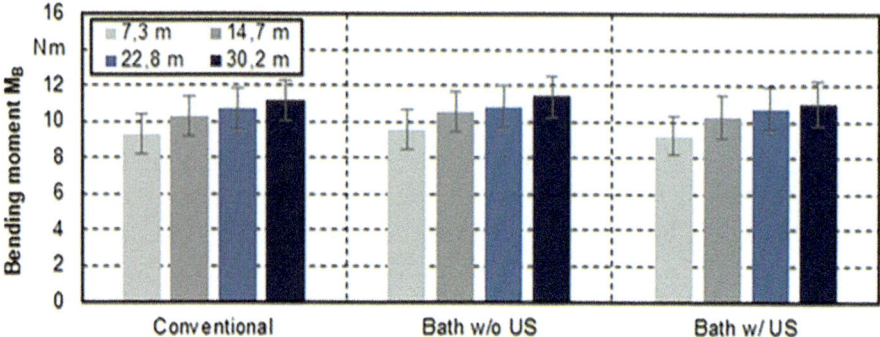

Fig. 3. Bending moment of conventional, bath w/o US and bath w/US for different cutting lengths

3.3 Surface Roughness

With conventional milling, the roughness R_z is 3.68 μm after a cutting length of 7.8 m (Fig. 4). This decreases to 2.85 μm as machining continues.

It can be seen that after a cutting length of 7.3 m, the R_z for milling in the submerged bath without US is 4.47 μm higher than that of the conventional process. Surface roughness is similar for submerged US milling. After a cutting length of 14.2 m, R_z decreases to 4.10 μm. This is an increase of 32.7% compared to flood cooling. The standard deviation is between 6.2% and 11.2%. After 30.2 m, the values are increasingly similar. In the reference process, R_z is 2.85 μm. During milling in the submerged bath, R_z increases to 3.14 μm and in the US submerged bath to 3.22 μm. This corresponds to increases of 10.2% and 13,0% respectively.

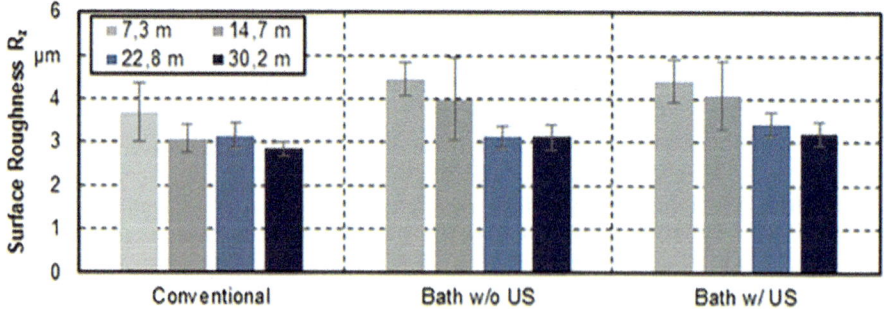

Fig. 4. Surface Roughness of conventional, bath w/o US and bath w/US for different cutting lengths

In all tests the surface roughness decreases with increasing cutting length. This is due to increased wear in the form of edge rounding. The rounding of the cutting edges leads to an increase in the cutting edge radius and the onset of so-called "ploughing". As a result, the workpiece surface is smoothed and the surface roughness is reduced.

Based on the surface roughness test results, no direct influence of the submerged bath and the US can be determined. It is noticeable that when looking at the surface of the workpiece in the submerged bath, material particles were more often welded to the surface. One reason for this may be inadequate chip removal.

4 Energy Assessment

In order to evaluate the energy demand, the energy consumers are considered, which are different for each process. For conventional machining, the pumps for internal and external cooling were considered. For US machining, these pumps are needed to fill the tank. The filling time is 5.94 min for the tank used in the test with a volume of 13.2 dm^3. Constructions that reduce the filling volume were not taken into account.

The consideration refers to a functional unit (FU) with a processing time of 60 min. To determine the energy requirements, the US actuator and the circulation pump for the submerged bath were first measured using the C.A. 8336 power analyser from Chauvin Arnoux. The actuator consumes the most power with 249 W for the US bath treatment. The circulation pump consumes 64 W. Next, the coolant pumps for 91 W for the external feed and 740 W for the internal feed. In relation to the filling of the tank for which the coolant pumps are required, this results in an energy requirement of 82 Wh for the submerged bath with a filling time of 5.94 min. Including the other consumers, the total energy requirement for the US bath machining is $E_{US} = 395$ Wh/FU. Submerged bath processing without US activation therefore has a total energy requirement of $E_{Bath} = 146$ Wh/FU (Fig. 5).The energy requirement for conventional flood cooling is mainly influenced by the choice of coolant supply. For external cooling, the energy requirement is 91 Wh. For internal cooling, the energy requirement is 740 Wh. A combination of internal and external cooling results in a total energy requirement of $E_{Con} = 831$ Wh/FU. Figure 6 shows the energy demand as a function of the process duration. As expected, the energy requirement increases linearly with increasing process duration. For the submerged bath with and without US there is an initial energy demand of 82 Wh/FU even before the process starts due to the filling of the tank. The slope of the straight line gives an indication of the efficiency of the process. The flatter the line, the more efficient the process. The process comparison shows that the external coolant supply is most efficient up to a process time of 3.036 h. After 3.036 h, the external coolant line and the submerged bath line intersect.

Submerged milling is the most energy efficient process. Internal and external coolant supply surpassed the other strategies in terms of energy demand after 0.159 h and is therefore the most energy inefficient process.

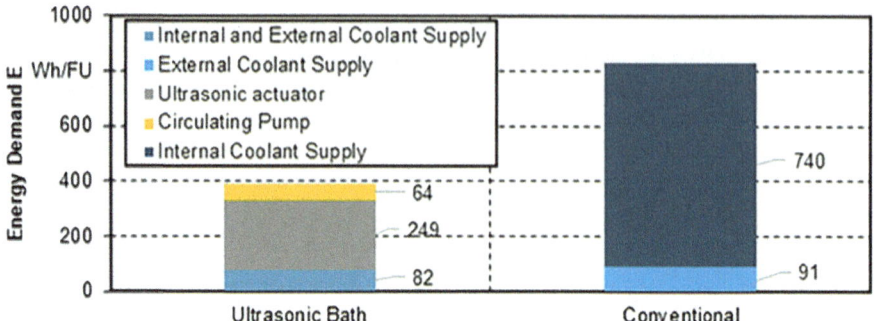

Fig. 5. Comparison of the energy demand of different cooling strategies

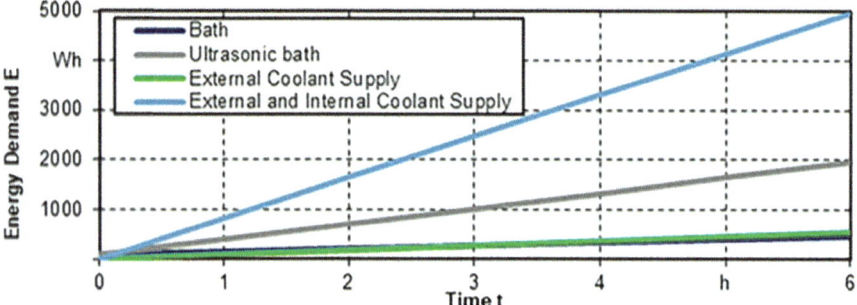

Fig. 6. Energy demand of different cooling strategies over increasing processing time

5 Discussion

Submerged milling in the US bath is a coolant supply method that has not been sufficiently investigated. A disadvantage is the machine design where the bath is mounted on a moving machine table. The movements of the table cause acceleration and build up of the fluid. Firstly, this places unfavourable loads on the machine. Secondly, it reduces the level in the tank. As a result, the tests could only be carried out at reduced feed rates. It is therefore recommended that tests are carried out on portal machines.

It was also observed that the liquid formed a standing wave due to the US. The reason is the reflection of the wave on the walls. As a result, some chips were not removed from the machining zone and were entrained by the tool. The chips were pushed across the surface of the workpiece, resulting in surface welds.

The standard deviations of the recorded bending moments are a multiple of the differences. Therefore, no conclusions can be drawn about the influence of the submerged bath or the US on the process forces. However, there is a tendency for these to be higher in the submerged bath without US than in the other setups.

The lowest surface roughness was obtained when milling with flood cooling. When milling in the submerged bath without US, they are approximately the same as when milling in the submerged bath with US. Again, the values are partly influenced by high standard deviations.

To determine tool wear, the maximum flank wear was measured. On average, this is the lowest over the entire machining process with conventional cooling; the submerged bath results in higher tool wear, with or without US. The overall wear can be considered irregular, which is reflected in the sometimes high standard deviations of flank wear. All tools remained below the maximum flank wear of 300 μm.

The energetic consideration of the different consumers in the process shows that submerged milling can contribute to a reduction of the energy requirement after a certain process time. If the process allows, an internal coolant supply should be avoided as the highest energy demand is caused by the high pressure pumps. The energy required to supply the US actuator exceeds the energy required with an external coolant supply.

6 Conclusion

This study examines submerged milling with and without US support and compares it to conventional flood cooling. The study included the bending moments, surface roughness, tool wear and energy efficiency of the processes. The results can be summarised as follows:

- The applied forces are similar in all test series. Wear tests show only minor differences, with conventional flood cooling showing the lowest wear.
- Surface roughness does not show any clear differences either. It is noticeable that there is more welding of the chips to the surface of the material during submerged machining with US. The reason for this is that the chips remain in the cutting zone due to a standing wave.
- Excitation of the fluid by ultrasound did not result in any significant improvements in the characteristics investigated. The effect of suppressing bubble formation by exciting the fluid does not appear to be strong enough due to the lower amount of heat to be dissipated compared to quenching.
- In terms of energy efficiency, the submerged bath without US has the lowest energy consumption for a machining time of more than 3 h.
- For shorter machining times, internal cooling should be eliminated where possible to reduce energy consumption.
- In general, with submerged cooling, care should be taken to ensure that the submerged bath is used on gantry type machines in order to reduce fluid build-up. The lack of transport, particularly of coarse chips, limits the application to the machining of small components made from high temperature materials, such as those used in medical technology.

Acknowledgements. This article is part of the work of the Nuremberg Campus of Technology, a researchcooperation between Friedrich-Alexander-Universität Erlangen-Nürnberg and Technische Hochschule Nürnberg Georg Simon Ohm supported by the state of Bavaria.

References

1. Denkena B, Abele E, Brecher C, Dittrich M-A, Kara S, Mori M (2020) Energy efficient machine tools. CIRP Annals—Manufacturing Technol 69:–667

2. Redmann R, Kessler O (2012) Ultrasonic assisted water quenching of aluminium and steel cylinders. Int Heat Treat Surf Eng 6(3):115–121
3. Disselhorst JA, Krueger MA, Minhaz Ud-Dean SM et al. (2018) Linking imaging to omics utilizing image-guided tissue extraction. PNAS 115(13):E2980-E2987
4. Popov KB, Dimov S, Pham DT, Ivanov A, Gandarias E (2010) New tool-workpiece setting up technology for micro-milling. Int J Adv Manuf Technol 47:21–27
5. Zhang X, Peng Z, Wang D, Liu L (2023) Theoretical analysis of cooling mechanism in high-speed ultrasonic vibration cutting interfaces. Int J Therm Sci 184:108033
6. Airao J, Nirala CK, Bertolini R, Krolczyk GM, Khanna N (2022) Sustainable cooling strategies to reduce tool wear, power consumption and surface roughness during ultrasonic assisted turning of Ti-6Al-4V. Tribol Int 169:107494

Eco-sustainable Improvement of Super Duplex Stainless Steel Turning Trough CryoMQL Technique

O. Pereira[1], N. Villarrazo[1], A. Rodríguez[1], L. N. López de Lacalle[1(✉)], and D. Martínez-Krahmer[2]

[1] CFAA—Aeronautics Advanced Manufacturing Center, University of The Basque Country, Biscay Science and Technology Park, Ed. 202, 48170 Zamudio, Spain
norberto.lzlacalle@ehu.eus

[2] INTI—National Institute of Industrial Technology, Avenida General Paz 5445, 1650 Miguelete, Provincia de Buenos Aires, Argentina

Abstract. In machining processes, the environmental footprint reduction is a fact that is been addressed in recent years. In this line, the use of recycled CO_2 as cryogenic cutting fluid in those processes is taking advantage in comparison with other lubricooling alternatives to replace oil emulsions in workshops. This technique is characterized by its high cooling capability and its cleaning efficiency. Besides, it is environmentally innocuous due to the CO_2 is captured form a primary process to be used as cutting fluid. However, the CO_2 has low lubricant properties what implies a challenge to be used in heat resistant alloys because combine both high hardness and low thermal conductivity. Therefore, in this work is explored the use of CO_2 combined with biodegradable oil spray under CryoMQL technique for dealing with super duplex stainless steel during turning operations. In particular, the CryoMQL technology was optimized for achieving a successful process through tool life improvement. The results shown that the optimization carried out implies that the use of CryoMQL technique extend tool life a 45% in comparison with the use of conventional oil emulsions, obtaining in this way not only an environmental improvement but also a technical one in this kind of alloys.

Keywords: CryoMQL · Eco-sustainable machining · Difficult to cut alloy

1 Introduction

In the current industry dealing with environmental issues with the aim of reducing their impact processes and being more ecofriendly is mandatory due to not only more restrictive laws but also society's greater environmental awareness [1]. In this line, from machining industry the reduction or suppression of cutting fluids is the path to answer those requirements because they are the major source of contamination. In particular, in machining processes, such as turning and milling operations, traditionally oil emulsions are used as cutting fluids. However, despite from a technical point of view those oil emulsions works and assist machining processes lubricating and cooling the cutting zone,

H. Kohl et al. (Eds.): GCSM 2023, LNME, pp. 641–649, 2025.
https://doi.org/10.1007/978-3-031-77429-4_71

from a environmental and health point of view have several disadvantages that cause damages in the environment and workers' health. In one hand, regarding environmental issues, oils emulsions are petroleum derivatives in which 30% are lost through leaks and workpieces cleaning processes, etc. [2]. This implies that these emulsions reach the food chain through the drain that led to the rivers and these in turn to the sea in an uncontrolled way [3]. On the other hand, in relation to workers' health it is know that continuous exposure to these emulsions by workers means that in the short term they may present skin and respiratory irritation and in the long term even lung cancer [4, 5]. Therefore, as was mentioned above, its elimination implies achieving processes that are not only safer for workers but also more environmentally friendly.

In this line, in the literature several alternatives were proposed with the aim of suppressing oil emulsions from machining processes, being the alternatives most efficient the Minimum Quantity Lubrication (MQL) and Cryogenic cooling. MQL consists of using an biodegradable oil spray as cutting fluid, which has excellent lubricating properties [6]. In the case of Cryogenic cooling, nitrogen or carbon dioxide liquified is used as cutting fluid, presenting both gases a high cooling capacity [7]. These techniques were widely analyzed in several materials such as aluminum and Ti6Al4V alloys obtaining satisfactory results [8, 9]. However, in difficult-to-cut alloys such as nickel alloys and super duplex stainless steels where cooling and lubricating is needed at the same time to deal with their high hardness and ductility MQL and Cryogenic cooling need to be combined under CryoMQL technology where lubricating and cooling properties are achieved at the same time due to the cryogenized biodegradable oil particles that are sprayed in the cutting zone [10]. For example, CryoMQL technique was successfully applied to turning AISI 52100 bearing steel. In particular, in this case the biodegradable oil spray was injected through the toolholder and the CO_2 was used to cool the workpiece. The results showed that in comparison with the use of oil emulsions the tool wear was similar, being therefore an alternative to be applied in this kind of alloys [11]. Other research deals with Inconel 625 in turning operations in which also CryoMQL in comparison with MQL and cryogenics "stand-alone" mode obtained and improvement in tool life of 80% [12]. Other recent research combined CryoMQL lubricooling technique with nanoparticles to turning Haynes 25 superalloy obtaining among the ecofriendly alternatives the longest tool life [13].

In summary, the utilization of CryoMQL technology proves to be a viable option for difficult to cut alloys. Nevertheless, to implement it in an industrial setting, it is imperative to optimize the consumption of liquefied gas in order to effectively manage the associated costs and it is here where there is lake of information in the researches. Therefore, this work steams from the idea of applying an optimized cryogenized biodegradable oil spray under CryoMQL conditions to turning AISI 440C super duplex stainless steel to be applied in industrial processes. Besides, once CryoMQL was optimized the results obtained were compared with the use of oil emulsion and CO_2 in "stand alone mode". The results shown that optimization is mandatory, as the injection of additional CO_2 does not yield improved outcomes. Thus, achieving the appropriate flow rate emerges as a critical parameter for attaining a reliable and effective CryoMQL technique suitable for practical implementation.

2 Experimental Setup

Experimental tests were carried out in a CMZ TC25BTY lathe with 35 kW of power. The alloy employed was AISI 440C super duplex stainless steel with 100 mm of initial diameter and length of 400 mm. This steel is characterized by presenting a high chromium content ($\approx 25\%$) followed by nickel ($\approx 5\%$). Additionally, its microstructure consists of a roughly equal proportion of austenite and ferrite, making it challenging to machine.

The lubricooling technologies tested in these experiments were CryoMQL, CO_2 and oil emulsions, respectively. In particular, for the application of CryoMQL and CO_2 as cutting fluid, the BeCold® system was used. The system injects CO_2 at a pressure of 12 bar, and the flowrate of biodegradable oil was 100 ml/h. Finally, Houghton® Horocut 4260 oil emulsion with a concentration of 9% was used as a reference.

The utilized inserts were Sandvik® CNMG120408 MM 2025 (TiN-coated carbide) with cutting conditions of a cutting speed of 100 m/min, a feed rate of 0.15 mm/rev, a depth of cut of 1.5 mm, and an initial cutting length of 300 mm. The employed tool holder was an Iscar® PCLNR 2525M-12X-JHP which during the tests to optimize the process its nozzle outlet diameter was modified. Cutting forces were measured during the experiments using a Kistler® dynamometer equipped with piezoelectric sensors. Additionally, after each pass, the cutting edge was examined using a PCE-200 microscope. The test stop criterion was established according to ISO 3685, which specifies that the test should be stopped upon reaching a flank wear of 0.3 mm, severe chipping, or general chipping. Figure 1 illustrates the experimental setup and summarizes the employed cutting conditions.

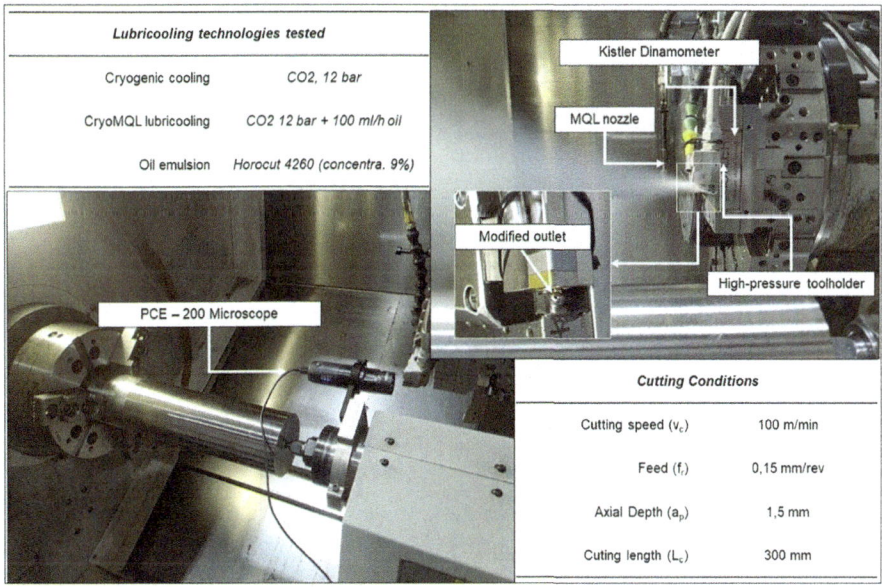

Fig. 1. Experimental setup and cutting conditions used.

3   CryoMQL Optimization and Validation.

3.1   CryoMQL Optimization Process

The optimization began by applying cryogenic cooling using the commercially available high-pressure tool holder (without modifying the outlet), with CO_2 directed towards the chip and rake face, at a flow rate of 5 kg per 8 min of machining. In this case, the insert successfully machined a length of 458 mm but experienced catastrophic failure, as depicted in Fig. 2.

Fig. 2. Insert once CO_2 test was finished.

As can be observed in the figure, the failure was due to mechanical causes due to lack of lubrication. In this case the insert did not show any wear due to thermal causes, which indicates that CO_2 is able to solve this problem but not the mechanical stress to which the insert is subjected. Therefore, subsequently it was proceeded to combine cryogenic cooling with CO_2 combined with biodegradable oil aerosol under CryoMQL technology as cutting fluid. In this second test, the nozzle was placed on the rake face parallel to the CO_2 nozzle as is shown in Fig. 3.

Fig. 3. Initial CryoMQL setup.

In this case, the biodegradable oil particles were absorbed by the Venturi effect due to the CO_2 flow. In this test, the CO_2 flank face outlet of the tool holder was also closed to reduce the CO_2 consumption. However, this test was aborted for several reasons. Firstly, the chips were getting entangled in the MQL system nozzle, causing movement. Additionally, the CO_2 consumption remained substantial. Finally, the insert exhibited premature wear due to thermal and mechanical effects after a single pass, as depicted in Fig. 4.

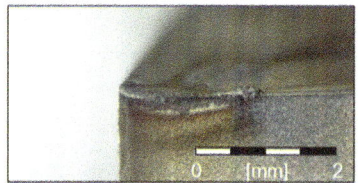

Fig. 4. Insert tool wear once CryoMQL test without optimization was finished.

In this case, the wear due to mechanical effects on the rake face is significant, along with the high CO_2 consumption. This prompted a reconsideration of how CryoMQL technology should be applied. Consequently, the MQL nozzle was repositioned on the flank face of the insert, and an external CO_2 nozzle was used on the rake face. However, in this case, the issue arose with chip entanglement in the nozzles, rendering their use ineffective based on the proposed setup (Fig. 5).

Then, based on this result, the commercial high-pressure tool holder was modified so that the CO_2 outlet diameter was suitable for this application. Specifically, the rake face outlet diameter was reduced to 1.5 mm. The use of this diameter was based on previous experiences carried out by the authors [14]. The optimized CryoMQL setup was as shown in Fig. 6.

Fig. 5. Optimized CryoMQL setup

With this optimized assembly it was possible to reduce the CO_2 consumption up to 1 kg of CO_2 every minute and forty seconds, that is, the CO_2 flow rate was reduced 40%. In this case, the aerosol injected by the MQL nozzle at 6 bar travels by Coanda effect through the wall of the tool holder and once it reaches the CO_2 flow, it is absorbed by it and injected much more efficiently into the cutting zone.

In this case, it was able to machine 1.570 mm of cutting length, stopping the test due to the formation of a ridge around the workpiece due to the loss of cutting edge of the tool.

3.2 Validation of CryoMQL Optimization

Once CryoMQL was optimized, other tests were conducted using oil emulsions as cutting fluid with the aim of obtain reference values and analyze the feasibility of employing the optimized CryoMQL technology in this kind of alloys. The cutting length achieved in this test before insert failure was 1.540 mm.

Afterwards, once these tests were finished, an analysis and comparison were conducted between the different lubricooling techniques were carried out taking into account the machined lengths (number of passes), machining time, volume of chips removed and cutting forces obtained. Figure 6 presents the obtained results related with the process productivity, as well as the condition of the inserts used with coolant and CryoMQL in their final stages of useful life.

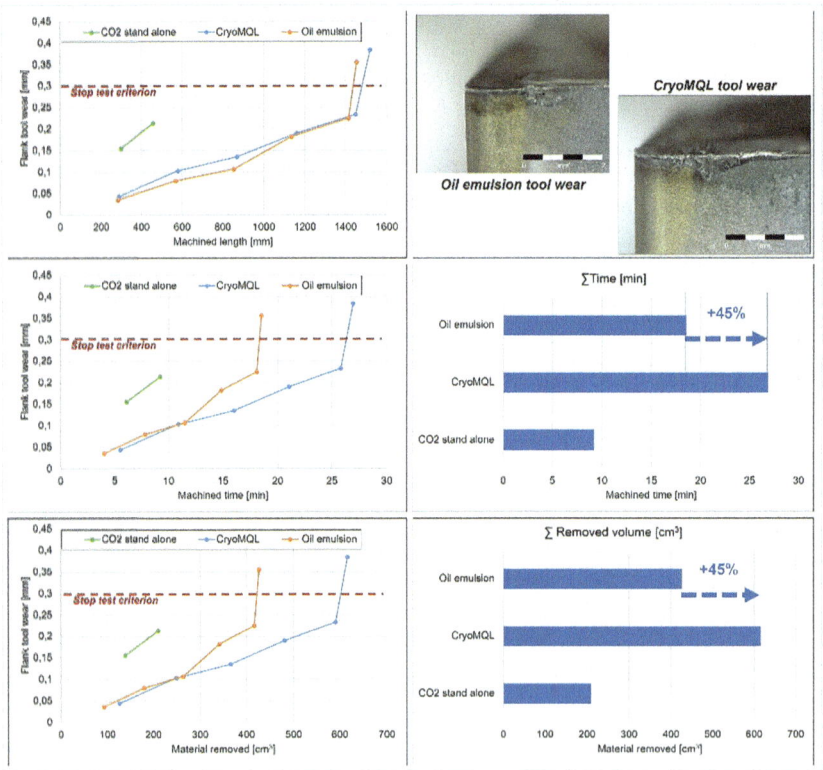

Fig. 6. Productivity results obtained with the different lubricooling technologies.

From the analyzed results, it can be deduced that the use of cryogenic cooling with CO_2 is not viable compared to the use of cutting fluid or CryoMQL, ruling out the use of CO_2 in "stand alone" mode. Taking the oil emulsion results as a reference, in the case of the linear length machined by CryoMQL, the tool life is comparable for both technologies. However, it should be noted that as the diameter of the AISI 440C steel workpiece decreases, the machining time per pass and the volume of chips removed also decrease, as the cutting speed remains constant during machining. Therefore, when analyzing the feasibility of CryoMQL technology, these two parameters should be taken into account.

Observing the machining time, which is the actual tool durability parameter, using oil emulsion as cutting fluid, the technology achieves 18.52 min of machining. However, utilizing CryoMQL as cutting fluid results in a 45% increase in this value, allowing for 26.93 min of machining before the insert reaches the end of its useful life. Regarding the volume of chips removed, the result is analogous, with CryoMQL evacuating 45% more chip volume (616.62 cm^3) compared to cutting fluid (427.05 cm^3).

Furthermore, upon analyzing the images of the insert's edge condition at the end of its useful life, it can be observed that the wear morphology in both cases is similar, exhibiting the typical notch characteristic of materials with austenite in their microstructure. This notch is caused by the hardening of the surface layer generated during machining due to the compression exerted by the insert tip on the material surface. Consequently, this is the area where the insert experiences the most stress, resulting in the formation of the aforementioned notch on the edge (Fig. 7).

Finally, regarding cutting forces, Fig. 7 shows the average cutting force modulus. In this figure it can be observed that in all technologies the values and the trends are analogous and that the value of the average force increases linearly without presenting values that indicate destabilization of the machining and could cause the breakage of the insert in an uncontrolled manner.

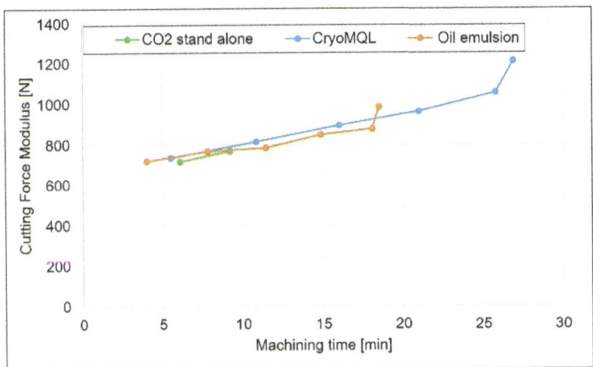

Fig. 7. Average cutting force modulus obtained.

4   Conclusions

In this work a CryoMQL optimization was carried out with the aim of obtaining a feasible technology to be applied to super duplex stainless steels industrially. Besides, several tests were carried out to validate the optimization in comparison with the use of conventional oil emulsions and CO_2 in "stand alone" mode. The conclusions obtained are listed below:

- The optimization of CO_2 flowrate to apply CryoMQL technology is mandatory. In this case, the reduction of the initial flowrate used was a 40%, being this the optimum point to deal with super duplex stainless steel.
- The use of CO_2 in "stand alone" mode does not sufficiently reduce the mechanical wear effects on the insert, resulting in premature wear of the insert.
- The use of CryoMQL as lubricooling technology implies an increase of tool life compared to oil emulsions of 45% and therefore a similar increase in the volume of chips dislodged.
- The cutting forces do not reveal unstable machining for any of the three technologies analyzed.

References

1. Pereira O, Rodríguez A, Fernández-Abia AI, Barreiro J, López de Lacalle LN (2016) Cryogenic and minimum quantity lubrication for an eco-efficiency turning of AISI 304. J Clean Prod 139:440–449
2. Lawal S, Choudhury I, Nukman Y (2012) Application of vegetable oil-based metalworking fluids in machining ferrous metals: a review. Int J Machine Tools Manufact 52:1–12
3. Byrne G, Dornfield D, Denkena B (2003) Advancing cutting technology. CIRP Ann – Manuf Technol 52(2):483–507
4. Cetin MH, Ozcelik B, Kuram E, Demirbas E (2011) Evaluation of vegetable based cutting fluids with extreme pressure and cutting parameters in turning of AISI 304L by Taguchi Method. J Clean Prod 19:2049–2056
5. Park KH, Olortegui-Yume J, Yoon MC, Kwon P (2010) A study on droplets and their distribution for minimum quantity lubrication (MQL). Int J Mach Tools Manuf 50:824–833
6. Sharma VS, Dogra M, Suri N (2009) Cooling techniques for improved productivity in turning. Int J Mach Tools Manuf 49(6):435–453
7. Pereira O, Rodríguez A, Calleja-Ochoa A, Celaya A, López de Lacalle LN, Fernández-Valdivielso A, González H (2022) Simulation of Cryo-cooling to improve superalloys cutting tools. Int J Precis Eng Manuf-Green Techn 9:73–82
8. López de Lacalle LN, Angulo C, Lamikiz A, Sánchez JA (2006) Experimental and numerical investigation of the effect of spray cutting fluids in high speed milling. J Mater Process Technol 172(1):11–15
9. Khanna N, Shah P, López de Lacalle LN, Rodríguez A, Pereira O (2021) In pursuit of sustainable cutting fluid strategy for machining Ti-6Al-4V using life cycle analysis. Sustain Mater Technol 29:e00301
10. Pereira O, Celaya A, Urbikaín G, Rodríguez A, Fernández-Valdivielso, A, López de Lacalle LN (2020) CO_2 cryogenic milling of Inconel 718: cutting forces and tool wear. J Mater Res Technol 9:8459–8468
11. Çetindağ HA, Çiçek A, Uçak N (2020) The effects of CryoMQL conditions on tool wear and surface integrity in hard turning of AISI 52100 bearing steel. J Manuf Process 56:463–476
12. Yildirim ÇV, Kivak T, Sarikaya M, Sirin S (2020) Evaluation of tool wear, surface roughness/topography and chip morphology when machining of Ni-based alloy 625 under MQL, cryogenic cooling and CryoMQL. J Mater Res Technol 9(2):2079–2092
13. Sirin S (2022) Investigation of the performance of cermet tools in the turning of Haynes 25 superalloy under gaseous N_2 and hybrid nanofluid cutting environments. J Manuf Processes 76:428–443

Improving Chip Curling with Reduced Amount of Cutting Fluid in Circular Sawing with Internal Coolant Supply

Christian Menze[1]($\boxtimes$), Jan Stegmann[2], Stephan Kabelac[2], and Hans-Christian Möhring[1]

[1] Institute for Machine Tools (IfW), University of Stuttgart, Holzgartenstraße 17, 70174 Stuttgart, Germany
christian.menze@ifw.uni-stuttgart.de
[2] Institute of Thermodynamics, Leibniz University Hannover, An Der Universität 1, 30823 Garbsen, Germany

Abstract. Circular sawing is an important intermediate step in industrial production. Due to the increasing demand for high-performance materials such as titanium alloys, efficient manufacturing for such difficult-to-machine materials is becoming more and more important. Traditionally, large quantities of cutting fluid are supplied to the sawing process by means of flood lubrication. However, the application of cutting fluids is expensive and energy-intensive. Additionally, cutting fluids pollute the environment and are harmful to health. The aim should therefore be to develop resource-saving machining processes and to reduce the need for cutting fluids in a sustainable manner. During sawing, a saw kerf is created which makes it difficult to supply the cutting process with cutting fluid. Chip formation plays an important role in the circular sawing process. This paper shows that by using a transparent kerf, the chip deformation can be studied and evaluated using a high-speed camera. By investigating the chip diameter from the camera recordings, the filling of the saw tooth gullet can be evaluated. As main result of the investigation an internal coolant supply in circular sawing can reduce the volume flow by 50% and improve the chip deformation process at the same time. This allows a sustainable and resource-efficient design of sawing processes.

Keywords: Chip curling · Cooling · Efficiency

1 Introduction

Metal cutting is an important machining process and circular sawing plays a significant role in the manufacture of semi-finished products. During metal cutting, high thermal and mechanical loads occur at the tool [1]. Therefore, a cutting fluid is often added to the cutting process. Cutting fluids cool and lubricate the cutting process and serve as a transport medium for the produced chips [2].

Higher technical requirements but also the demand for sustainable mobility by land and air have pushed the use of high-performance materials such as titanium alloys. Due

© The Author(s) 2025
H. Kohl et al. (Eds.): GCSM 2023, LNME, pp. 650–658, 2025.
https://doi.org/10.1007/978-3-031-77429-4_72

to the excellent density to strength ratio, the material has a great potential for future designs of aircrafts and cars [3]. However, the supposedly favorable properties of the material lead to problems in machining. Especially due to the small product of thermal conductivity, specific heat and density as well as the relatively low elongation and the low Young's modulus, titanium alloys are considered to be difficult to machine [4].

A reliable supply of cutting fluids is indispensable when machining materials such as titanium alloys and comparable materials [5–7]. In general, cutting fluids cause considerable operating costs [8–10]. These vary from 7 to 17% and can increase up to 20% for difficult-to-machine materials such as titanium alloys [11]. Such costs include the cutting fluids as well as the operating resources, personnel, energy, and waste disposal required for the operation. Thus, the use of cutting fluids requires a considerable amount of energy and is also harmful to the environment and health [12].

In circular sawing, the cutting fluid is traditionally supplied by means of external flood lubrication. In this case, large quantities of cutting fluid are supplied to the machining process to flood a large area around the component and the cutting zone. However, during the sawing process, a kerf is created which covers up the actual cutting operation. The cutting edge enters the saw kerf, and the chip is formed in the saw tooth gullet. This makes it difficult for the cutting fluid to reach the cutting operation in the necessary quantity by means of conventional flood lubrication. One possibility to improve this situation is an internal coolant supply (ICS). With this technology, the cutting fluid is delivered through the tool directly to the cutting zone. In various machining processes such as turning [13], milling [14] and drilling [15], an improvement of the machining operation could be achieved, and the cutting fluid could be supplied in a purposeful and resource-efficient way. For circular sawing, an improvement of the machining quality and a reduction in cutting temperature could also be demonstrated [16].

The state of the art shows the increasing importance of the machining of titanium alloys. In [17, 18] it was shown that a saving potential of more than 60% is possible by a specific design of the cutting fluid system. The use of cutting fluids is associated with high operating costs and pollutes the environment. The circular sawing operation is an important intermediate step in the production of titanium components, as it is used to provide semi-finished products for the further machining process. A resource-saving machining chain of difficult-to-machine materials such as titanium should be carried out in all intermediate steps. In this research work, the chip formation during circular sawing was investigated by means of an artificial kerf. It was distinguished between dry machining and machining with external and internal coolant supply. For the first time, a high-speed camera was used to visualize the chip movement under these different conditions and to evaluate the chip filling of the saw tooth gullet. This methodology allows tool and process developers to improve the design of the saw tooth gullet and the process itself regarding an efficient and resource-saving sawing process. The investigations showed how an internal supply of cutting fluid during sawing can reduce the amount of cutting fluid required and improve chip formation at the same time. This allowed the realization of a resource-saving sawing process with better cutting conditions.

2 Experimental Set-Up

During the circular sawing process, a single cutting edge rotates and translates in a combined motion, creating a kerf. This makes it difficult to observe the chip deformation process during sawing. Therefore, the actual circular sawing process was abstracted into a system that allows an examination with a high-speed camera (Fig. 1). For the investigation, a circular saw blade (type: Tube Cut by AKE Knebel) was used with a diameter of 350 mm, a width of cut of 2.7 mm, a blade thickness of 2.5 mm and a number of teeth of $z = 80$. The tool used and the cutting parameters correspond to the recommendations of the tool supplier. Two-toothed segments were cut out of the circular saw blade. These were clamped into a tool holder in a test rig with linear cutting motion. This test rig was designed to investigate orthogonal and oblique cutting and reached a maximum cutting speed of $vc = 120$ m/min. During the experiment, the tool was clamped tightly and the workpiece executed the cutting motion. The workpiece had a width of 2.8 mm (including width of cut of 2.7 mm and a safe distance between tool and sapphire glass of 0.05 mm on both sides) and a length of cut of 100 mm. Ti6Al4V was used as test material. The workpiece was clamped between two sapphire glass panel. This created an artificial kerf, allowing the chip deformation process to be observed and maintaining the conditions of the saw kerf. An external pump was used to supply the cutting fluid in the experiment. In the case of external flood lubrication, the cutting fluid was delivered through a channel directed to the two-toothed segment above the kerf. During the machining test, the two-toothed segment was 20 mm deep in the artificial kerf. The internal cooling supply was realized through a Ø1.5 mm diameter hole in the two-toothed segment. To ensure a transparent visibility, water was used as cutting fluid. The chip forming process was recorded with a high-speed camera (type IDT OS8, 8,000 fps). A three-component dynamometer (by Kistler, type 9129AA) was used to measure the cutting forces acting on the workpiece.

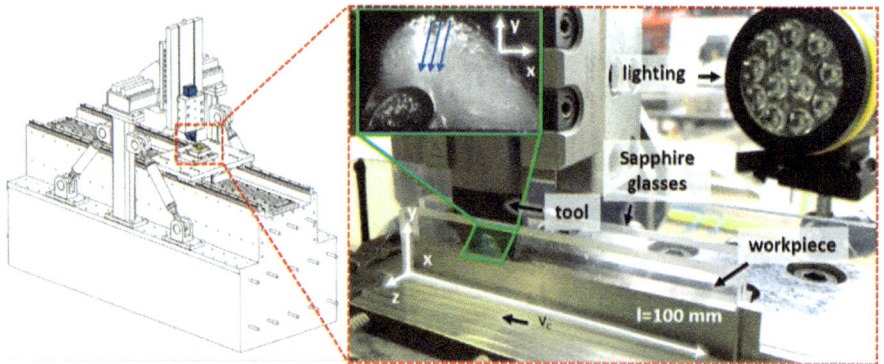

Fig. 1. Experimental set-up to investigate the chip forming process in the saw kerf.

3 Results

The form of the chips produced during cutting is an important criterion for the evaluation of a machining process [19]. If a chip forms unfavorably, this can lead to a damage of tool or workpiece and requires unnecessary downtime for the removal of the chip material. A favorable chip form is important particularly in sawing processes. Since the chip is in the saw kerf during the entire time of cutting action, unfavorably shaped chips can damage the machining surface and thus reduce the machining result. Furthermore, chip jamming can occur and result in damage to the tool and workpiece.

A cutting fluid can have a significant effect on the resulting chip form. On the one hand, the cutting fluid lubricates the contact partners, and on the other hand, the cooling effect can directly influence the chip form. Moreover, a fluid jet causes a mechanical force that can influence the chip curling. As a result of the formed saw kerf, it is difficult for the cutting fluid to reach the machining process by means of external flood lubrication. Therefore, large quantities of cutting fluid have to be used. The ICS was designed to circumvent this problem and transport the cutting fluid directly into the saw tooth gullet in the kerf.

To investigate the influence of an ICS on the cutting process, a high-speed camera was used to record the chip formation in the artificial kerf as shown in Fig. 1. The cutting speed was $v_c = 50$ m/min in all tests. The uncut chip thickness was h = 0.025 and 0.05 mm. The volume flow was $\dot{V} = 60$ l/h for the external flood lubrication and $\dot{V} = 20$ and 40 l/h with ICS. Figure 2 shows an exemplary sequence of pictures for the chip formation in dry cutting, with external and internal coolant supply. In dry cutting, the chip curled up and filled most of the chip space. Finally, the friction between the chip and the surrounding geometry caused the chip to buckle and led to uncontrolled chip movements. In the case of external flood lubrication, the chip formation was initially similar to dry cutting. It was visible that the chip was slightly watered with the cutting fluid. However, the cutting fluid flowed into the saw kerf only slowly. Yet even with external flood lubrication, the total space in the saw tooth gullet was quickly filled and the chip folded up. In the case of ICS, the chip space was continuously flooded with cutting fluid. The incoming fluid flow thus exerted a force on the chip, which curled up tightly as a consequence. This change in the macroscopic chip shape is also shown by investigations in drilling and turning [6, 7, 18]. Even though the chip finally filled the space between the tooth and the saw kerf, the ICS achieved significantly better friction conditions and thus prevented uncontrolled buckling. This leads to an improvement in process reliability as described in [19].

Figure 3 shows the chip diameter depending on the length of cut (l = 100 mm) from the high-speed recordings for dry cutting, external flood lubrication and ICS. The evaluation was carried out by taking the chip diameters at the discrete distances in steps of 10 mm between 0 mm (tool entry) and 100 mm (tool exit). The chip space of the tool had a maximum diameter of 6 mm. If the chip diameter reached this value, the chip space was filled. Shortly afterwards, as the chip continued to pour in, it buckled in an uncontrolled manner in most cases of dry cutting and external flood lubrication. Figure 3 shows that dry machining and the use of external flood lubrication led to a significantly faster filling of the chip space than when using an ICS. Due to the increase in volume flow with an ICS, the chip was curled up more tightly. However, the effect

Dry cutting	External coolant supply	Internal coolant supply

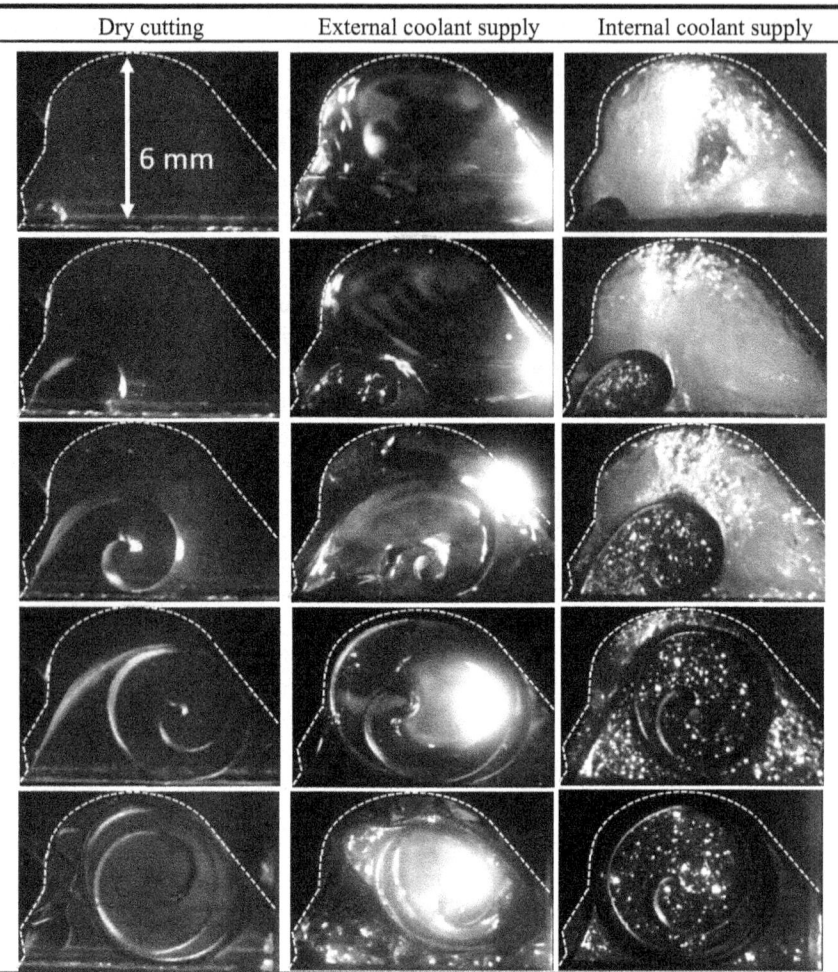

Fig. 2. Exemplary comparison of picture sequences from a high-speed film for dry cutting, external flood lubrication and ICS at $v_c = 50$ m/min and h = 0.025 mm [16].

was only minor even though the volume flow used was doubled. Furthermore, it could be seen that with an uncut chip thickness of h = 0.05 mm, the chip space filled up significantly faster regardless of the coolant supply strategy. This could be explained by the chip being thicker and taking up more volume. Additionally, the chip was stiffer and therefore curled up more extensively. However, the use of an ICS showed a noticeably better chip movement in the chip space of the two-toothed segment. This demonstrates for an ICS an improvement of chip curling with a simultaneous reduction of the volume flow by 50%. Thus, the known effects of cutting fluids [2] and a special coolant supply can be used to significantly reduce the cost-intensive and harmful amount of coolant [12].

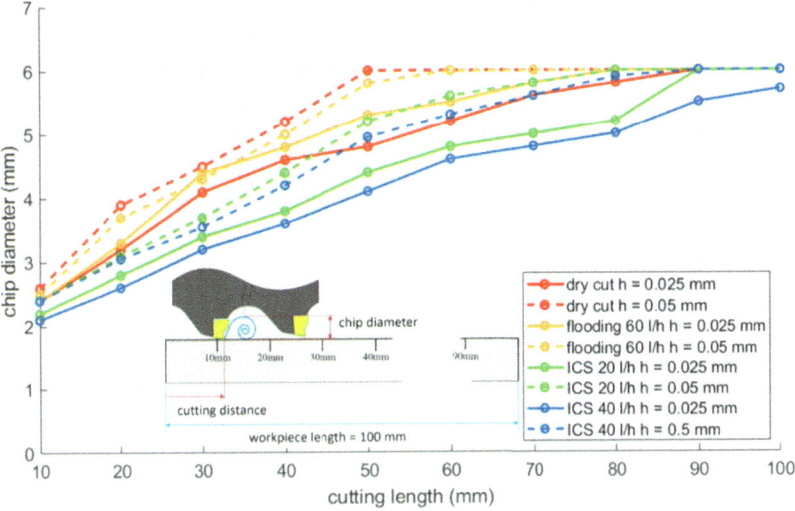

Fig. 3. Filling of the saw tooth gullet at $v_c = 40$ m/min by the chip for different cooling strategies.

Figure 4 illustrates the comparison of the cutting forces in the experiment. The results showed that the cutting force decreased with the application of a cutting fluid. With an ICS, this effect was slightly more pronounced than with external flood lubrication. This could be explained by a better cooling of the cutting process. This effect was more noticeable with an increase in chip thickness to h = 0.05 mm. It was caused by the greater deformation work and the higher temperatures resulting from that. As a consequence, titanium alloys tend to have a stronger chip segmentation. The cutting fluid counteracts this effect and the friction conditions are generally improved.

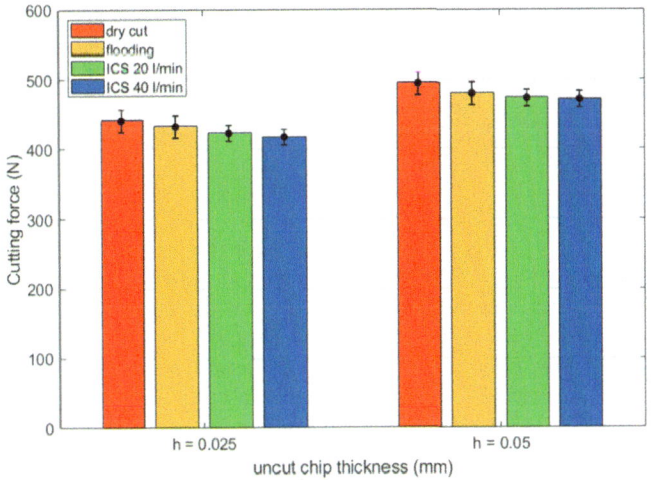

Fig. 4. Cutting force at $v_c = 40$ m/min for different cooling strategies.

4 Conclusion

This paper shows how a sophisticated coolant supply strategy can be used to improve the chip forming process in circular sawing and reduce the required amount of cutting fluid at the same time. Particularly materials that are difficult to machine, such as titanium alloys, require a reliable supply of cutting fluid, otherwise the thermal load on the tool and the wear increase considerably. At the same time, it is a global challenge to design sustainable manufacturing processes. This requires the optimization of the existing manufacturing processes and the reduction of the resources used, such as the application of cooling lubricants. The experiments showed that, even under the simplified conditions in the test (lateral safe distance between tool and sapphire glass) and with a comparatively low depth of engagement (20 mm in the saw kerf), the cutting fluid flowed only scarcely into the saw kerf and the saw tooth gullet. The ICS, on the other hand, reliably supplied the cutting process with cutting fluid, even though the volume flow was reduced by 50%. Thereby, a sustainable circular sawing process can be designed and considerable amounts of cooling lubricant can be saved.

- Chip formation could be influenced and improved by an ICS.
- The cutting force decreased slightly with ICS compared with dry cutting and external flood lubrication.
- Increasing the volume flow from 20 to 40 l/min when using ICS improved the chip formation marginally.
- With ICS, the volume flow could be reduced by 50% with better chip formation.

The presented investigations allow to study the chip curling and the fluid-structure interaction between cutting fluid and chip in the chip space. In future, this will enable to optimize the tool as well as the process with regard to a sustainable and efficient production. Therefore, design studies will be carried to improve the design of the saw tooth gullet with regard to a good flow behavior inside the saw kerf. As a result, the amount of cutting fluid can be further reduced and the sawing process can be improved.

Acknowledgement. The authors would like to thank the German Research Foundation (DFG) for their funding – project number 439925537.

References

1. Kato S, Yamaguchi K, Yamada M (1972) Stress distribution at the interface between tool and chip in machining. J Eng Ind 94(2):683–689. https://doi.org/10.1115/1.3428229
2. Singh J, Gill SS, Dogra M, Singh R (2021) A review on cutting fluids used in machining processes. Eng Res Express 3(1):12002. https://doi.org/10.1088/2631-8695/abeca0
3. Oyesola M, Mpofu K, Daniyan I, Mathe N (2022) Design and simulation of a bearing housing aerospace component from titanium alloy (Ti6Al4V) for additive manufacturing. Acta Polytech 62(6):639–653. https://doi.org/10.14311/AP.2022.62.0639
4. Davim JP (2014) Machining of titanium alloys, 1st edn. Springer, Heidelberg, Germany
5. Venugopal KA, Paul S, Chattopadhyay AB (2007) Growth of tool wear in turning of Ti-6Al-4V alloy under cryogenic cooling. Wear 262:1071–1078

6. Courbon C, Kramar D, Krajnik P, Pusavec F, Rech J, Kopac J (2009) Investigation of machining performance in high-pressure jet assisted turning of Inconel 718: an experimental study. Int J Mach Tools Manuf 49(14):1114–1125

7. Shah P, Khanna N, Maruda RW, Gupta MK, Krolczyk GM (2021) Life cycle assessment to establish sustainable cutting fluid strategy for drilling Ti-6Al-4V. Sustain Mater Technol 30. https://doi.org/10.1016/j.susmat.2021.e00337

8. Sanchez JA, Pombo I, Alberdi R, Izquierdo B, Ortega N, Plaza S, Martinez-Toledano J (2010) Machining evaluation of a hybrid MQL-CO_2 grinding technology. J Clean Prod 18(18):1840–1849. https://doi.org/10.1016/j.jclepro.2010.07.002

9. Klocke F, Lung D, Krämer A, Cayli T, Sangermann H (2013) Potential of modern lubricoolant strategies on cutting performance. KEM 554–557:2062–2071. https://doi.org/10.4028/www.scientific.net/KEM.554-557.2062

10. Denkena B, Abele E, Brecher C, Dittrich M-A, Kara S, Mori M (2020) Energy efficient machine tools. CIRP Ann 69(2):646–667. ISSN 0007-8506. https://doi.org/10.1016/j.cirp.2020.05.008

11. Pereira O, Rodríguez A, Fernández-Abia AI, Barreiro J, López de Lacalle LN (2016) Cryogenic and minimum quantity lubrication for an eco-efficiency turning of AISI 304. J Clean Prod 139:440–449. https://doi.org/10.1016/j.jclepro.2016.08.030

12. Wu X, Li C, Zhou Z, Nie X, Chen Y, Zhang Y et al (2021) Circulating purification of cutting fluid: an overview. Int J Adv Manuf Technol 117(9–10):2565–2600. https://doi.org/10.1007/s00170-021-07854-1

13. Fang Z, Obikawa T (2020) Influence of cutting fluid flow on tool wear in high-pressure coolant turning using a novel internally cooled insert. J Manuf Process 56:1114–1125

14. Lakner T, Bergs T, Döbbeler B (2019) Additively manufactured milling tool with focused cutting fluid supply. Procedia CIRP 81:464–469

15. Evans R (2012) Selection and testing of metalworking fluids. In: Astakhov VP, Joksch S (eds) Metalworking fluids (MWFs) for cutting and grinding. Fundamentals and recent advances. Woodhead Publ, Oxford, pp 23–78

16. Möhring H-C, Menze C, Werkle KT (2023) Internal coolant supply in circular sawing. CIRP Ann

17. Denkena B, Helmecke P, Hulsemeyer L (2014) Energy efficient machining with optimized coolant lubrication flow rates. Procedia CIRP 24:25–31

18. Wang X, Li C, Zhang Y, Said Z, Debnath S, Sharma S et al (2022) Influence of texture shape and arrangement on nanofluid minimum quantity lubrication turning. Int J Adv Manuf Technol 119(1–2):631–646. https://doi.org/10.1007/s00170-021-08235-4

19. Klocke F, Lung D, Essig C (2008) Kontrollierter Spanbruch. ATZ Prod 1:28–33. https://doi.org/10.1007/BF03224085

Machining Cycle Detection Based Expert System for Improving Energy Efficiency in Manufacturing

Borys Ioshchikhes$^{(\boxtimes)}$ ⓘ, Paul Heller, and Matthias Weigold ⓘ

Institute for Production Management, Technology and Machine Tools (PTW), Technical University of Darmstadt, Otto-Berndt-Str. 2, 64287 Darmstadt, Germany
`b.ioshchikhes@ptw.tu-darmstadt.de`

Abstract. The transformation of manufacturing companies towards a carbon-neutral economy requires energy transparency, energy analyses and the implementation of energy efficiency measures. Given the continuing skills shortage, the need for automated analysis methods to gain insights from measurement data is increasing. Expert systems that combine the knowledge of multiple experts, analyze load profiles, and derive energy efficiency measures are one approach to tackle this challenge. This paper presents an expert system that quantifies energy efficiency potentials based on the detection of machining cycles and derives promising measures. For this purpose, a new algorithm for the detection of machining cycles is introduced, which shows an accuracy between 76.7% and 94.3% on a representative production day for electrical load profiles of different types of production machines. Since the detected machining cycles are in a form impractical for further processing, information is extracted as energy performance indicators. The expert system utilizes this aggregated information to identify energetic hotspots and derive appropriate energy efficiency measures. The machining cycle detection based expert system is demonstrated on a typical production chain for the metal-working industry within the ETA research factory at the Technical University of Darmstadt.

Keywords: Pattern recognition · Energy analysis · Sustainable manufacturing

1 Introduction

As manufacturing companies increasingly face the challenges of transitioning to a carbon-neutral economy, rising energy costs and a shortage of skilled workers, the need for innovative solutions is becoming pressing [1]. One promising approach to solve these challenges by improving energy efficiency in manufacturing are expert systems, which are artificial intelligence applications that are designed to emulate the decision-making ability of a human expert in a specific domain. Expert systems can help manufacturers reduce energy costs and increase productivity through the analysis of large amounts of data and recommendations for energy optimization of production machines. Moreover, expert systems can contribute to solving the issue of skills shortages in manufacturing

H. Kohl et al. (Eds.): GCSM 2023, LNME, pp. 659–667, 2025.
https://doi.org/10.1007/978-3-031-77429-4_73

by providing a way to leverage the knowledge and expertise of skilled workers, even as they retire [2].

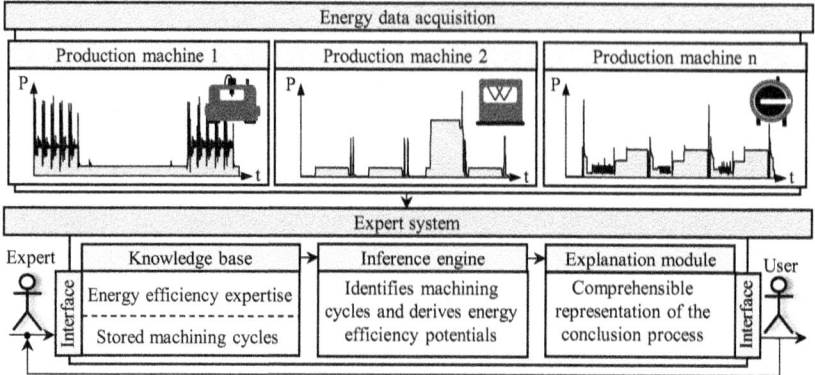

Fig. 1. Methodological approach.

Previous expert system approaches have focused on energy optimization of individual machines, possibly overlooking energy hotspots of other machines in the same production system. In addition, existing approaches have been limited to specific machine types such as machine tools or cleaning machines [3, 4]. By extending the system boundary to multiple production machines, a machine-type independent and cross-machine prioritization of energy efficiency potentials can be realized, with machining cycles serving as the comparison baseline.

Following the introduction, Sect. 2 outlines the overall methodology, which is composed of the energy data acquisition and the expert system itself. In this context, the main elements of the expert system - the knowledge base, the inference engine and the explanation module - are covered. Section 3 describes the implementation of the underlying algorithm for detecting machining cycles. Subsequently, in Sect. 4, the proposed approach is evaluated using a real production chain. Finally, Sect. 5 provides a summary and a conclusion for future research.

2   Methodology

The general approach shown in Fig. 1 aims to increase the energy efficiency of production machines within a production system. This approach involves the automatic identification and quantification of energy hotspots and inefficiencies based on cycle detection in electrical load profiles, followed by proposals for prioritized measures to unlock potential improvements. The methodology can be divided into two major parts: The energy data acquisition and the expert system. The energy data acquisition in this paper focuses on electrical energy, due to its high share in the industrial final energy consumption and the relatively low metering effort compared to non-electric energy flows [5, 6]. The prerequisite of the presented approach is the measurement at the main power supply of the considered production machines, which can be accomplished by stationary, temporary or virtual metering [6]. Important characteristic features of the expert system are the knowledge base, inference engine and the explanation module [7]. The knowledge base contains expertise related to the problem domain [7]. In this approach, the knowledge base includes stored machining cycles as well as facts and rules to increase the energy efficiency of production machines. Experts and users of the expert system can expand the knowledge base, e.g., when machining cycles for new production machines need to be defined. The inference engine analyzes the electrical load profile, identifies machining cycles and uses the stored knowledge to derive conclusions. The explanation module explains how the system comes to certain conclusions or recommendations, which helps the user understand the system's decision-making process [7]. Both the knowledge base and the explanation component provide a user interface, which serves as the communication channel between the expert and the expert system, and the expert system and the user.

2.1 Machining Cycle Detection

Dehning et al. describe the two value-based consumer states *unproductive* and *productive* [8]. The unproductive state covers the states off, standby, operational and powering up/down according to VDMA 34179 [9]. No value is created during this state. The productive state refers to a value-adding process and corresponds to the state working according to VDMA 34179 [9]. The productive state is typically characterized by cyclically recurring manufacturing operations and has the highest average power and energy demand [10]. For the presented expert system, we focus on the automated identification of machining cycles during the productive state.

In general, there are many approaches for pattern recognition in time series based on machine learning [11] and motif methods [12, 13]. However, these approaches are not robust to the effects of a dynamic production environment, varying setup times and other process anomalies [10]. To take these effects into account, Seevers et al. presented their own method, which, however, is based on the assumption that typical machining cycle times for machine tools have a time range from minimum 20 s to a maximum of 120 s [10]. Since this is not a valid assumption for all machine tools and other machine types, the algorithm developed in this paper does not follow this premise.

2.2 Energy Efficient Manufacturing

Given the necessity to minimize the energy demand of machine tools in the use phase, measures have been collected [14] and classified [15]. We propose that these measures are mostly applicable to other types of electric actuated production machines. Figure 2 shows the classification into the three main levels of energy recovery, energy input reduction and energy reuse. Energy recovery is generally approached by thermal management, while energy reuse aims to operate motors as generators during deceleration processes [15]. Energy input reduction can further be divided into machine-related and process-related measures [16]. Machine-related measures are all measures that require a technical change, while process-related measures require organizational intervention. For the expert system, two measures are considered as examples. Demand-oriented-control aims at (partially) switching off or adjusting components when they are not required, while energy-oriented planning targets minimizing the energy demand of machining cycles, e.g., by reducing process times.

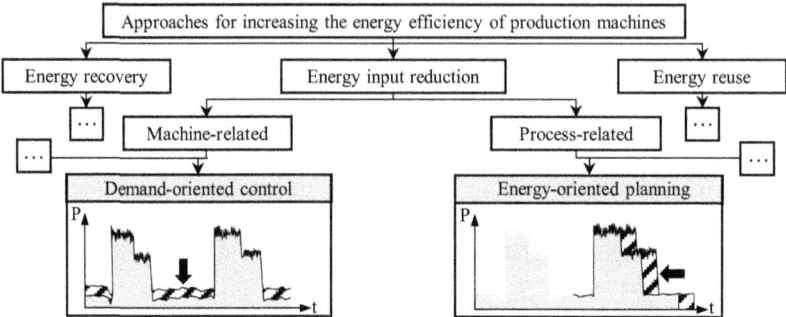

Fig. 2. Approaches to increase the energy efficiency of production machines bases on [15].

To identify the first measure, the unproductive energy factor (*UPEF*) is considered according to Eq. (1). It is defined as the ratio of the energy demand in unproductive times E_{np} and the total energy demand E_{total} [8]. For the second measure, the potential energy savings during the productive state (*PESP*) is calculated following Eq. (2). The *PESP* reflects the potential energy savings during the productive state if the cycle with minimum energy demand $E_{cycle,min}$ would correspond to all identified cycles k instead of the measured energy demand E_p.

$$UPEF = \frac{E_{np}}{E_{total}} \cdot 100\% \tag{1}$$

$$PESP = \frac{E_p - k \cdot E_{cycle,min}}{E_p} \cdot 100\% \tag{2}$$

3 Machining Cycle Detection Algorithm

Expert knowledge is needed to identify a representative machining cycle, hereafter referred to as a sample cycle, within each machine's load profile. These sample cycles are stored within the knowledge base. Subsequently, the expert system leverages this selected sample cycle to detect additional machining cycles using our algorithm.

The first step of the algorithm is to calculate the z-normalized Euclidean distance between the sample cycle and the load profile to generate a matrix profile using the Python library STUMPY [17]. The matrix profile contains the start indices of all identified putative patterns and the computed values of the z-normalized Euclidean distance. However, relying solely on STUMPY to detect machining cycles has two major drawbacks: Firstly, not every identified start index found by STUMPY corresponds to the actual beginning of a machining cycle. Secondly, STUMPY provides only the start indices, but not the end indices of the detected cycles. Our algorithm not only identifies incorrect start indices but also accounts for varying cycle lengths. To identify start indices that do not correspond to the actual beginning of machining cycles, we compare the identified cycles with the sample cycle based on additional statistical properties such as mean and standard deviation. If the deviation of these metric values surpasses a predetermined relative threshold, our algorithm labels the start index identified by STUMPY as one that does not correspond to a true machining cycle. Consequently, this index is removed from the matrix profile. To determine the lengths of the cycles, the length of each identified cycle is first calculated as the difference between two consecutive start indices. Subsequently, a comparison of the statistical properties and a condition check is conducted. If the discrepancies between the mean and standard deviation values are below the predetermined threshold, our algorithm designates the start index associated with the newly calculated length as a true machining cycle. In such cases, the algorithm continues with the next iteration and repeats this process for the subsequent start index. However, if the statistical properties differ significantly, the length of the identified cycles is adjusted by adding or subtracting a small percentage of the length of the sample cycle. The differences between these metrics are then recalculated. This process is repeated until the algorithm classifies the cycle as a true machining cycle, or until a predefined number of iterations is reached. If the specified number of iterations is reached without a true machining cycle classification, the cycle is identified as a false machining cycle, and the corresponding start index is removed from the matrix profile. The algorithm then continues with the next iteration, repeating the entire process for the next start index in the matrix profile. Finally, our algorithm returns the start indices of the classified true machining cycles, along with a list of their corresponding lengths. It is important to note that the specific percentage of length adjustment and the predetermined number of iterations are hyperparameters in this algorithm. These hyperparameters provide the flexibility to tailor and fine-tune the algorithm to suit specific data and analysis requirements, ensuring more accurate identification of machining cycles.

4   Use Case

To ensure the practical applicability of the developed expert system, the approach is evaluated on a representative production chain for the metalworking industry in the ETA research factory at the Technical University of Darmstadt. The production chain manufactures parts for axial piston pumps and begins with lathing raw pieces on the machine tool EMAG VLC-100Y (EMAG Y), followed by the first cleaning process on the cleaning machine MAFAC JAVA. Subsequently, a tempering process is conducted on the gas nitriding furnace IVA RH65. Next, a grinding process is performed on the EMAG VLC-100 GT (EMAG GT), which is followed by the second cleaning process on the machine MAFAC KEA [18].

4.1   Application

Figure 3 shows a section of the identified machining cycles after the expert system was applied to measurement data of a representative production day. The highest accuracy of the machining cycle detection algorithm is achieved at EMAG Y with 94.3% and the

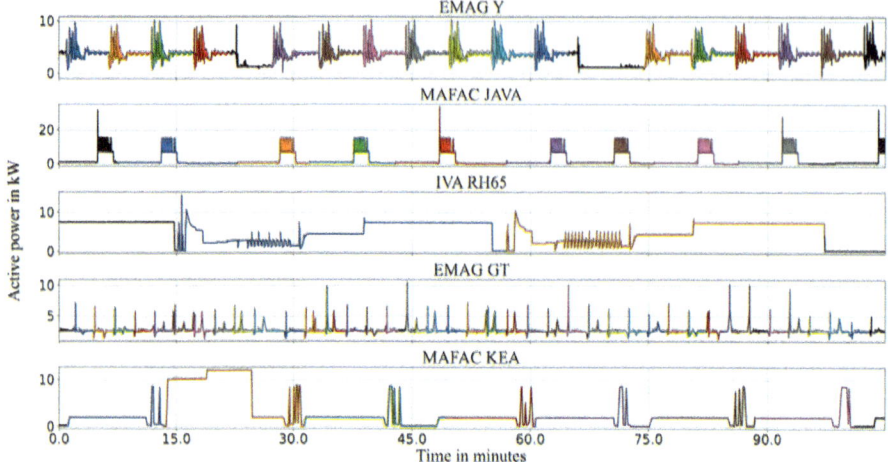

Fig. 3. Identified machining cycles.

lowest accuracy at MAFAC KEA with 76.7%. The accuracy corresponds to the ratio of correctly identified to total cycles. The lower accuracy on MAFAC KEA can be explained by the acyclic pulsing of a decentralized tank heating which occurs irregularly.

The expert system extracts further information using the detected machining cycles, some of which can be found in Table 1. According to the results in Table 1, the lathing process at the machine EMAG Y is identified as an energetic hotspot for the produced part. Thus, this machine could have a significant impact on reducing energy consumption. In addition, with a *UPEF* of 16.7%, the same machine has the greatest potential for energy savings in the unproductive state. Consequently, the measure of demand-oriented control is prioritized here highly by the expert system. Furthermore, with a *PESP* of 9.5%, the measure of energy-oriented planning is given considerable priority at the EMAG Y. The cleaning machines MAFAC JAVA and MAFAC KEA show significantly higher *PESP* values, which are caused by larger deviations in the cycle times, but also offer a lower energy savings potential due to the lower average energy demand per part.

Table 1. Results of the expert system.

Machine	Average parts per cycle	Average energy demand per cycle in kWh	Average energy demand per part in kWh	UPEF in %	PESP in %
EMAG Y	1	13849.6	**13849.6**	16.7	9.5
MAFAC JAVA	42	8883.1	211.5	2.8	29.7
IVA RH65	42	17653.8	420.3	1.7	2.2
EMAG GT	1	9655.7	9655.7	2.9	9.5
MAFAC KEA	42	9875.6	235.1	1.3	29.1

4.2 Evaluation

The use case reveals that the accuracy of the developed cycle detection algorithm can vary considerably with different machines. Especially for machines like MAFAC KEA, with a load profile strongly characterized by acyclic events, the algorithm performs inferior. As with all expert systems, it should be noted that the output is significantly determined based on the knowledge stored by the experts. Accordingly, the pattern cycles must be correctly defined for a useful utilization of the system. Furthermore, only one cycle pattern per machine, i.e., the production of identical parts, was considered in the use case. In industrial practice, multiple different parts could also be produced on one machine. Consequently, a sample cycle would have to be defined not only for one machine, but also for each part that is produced on a machine.

5 Summary and Conclusion

The developed expert system enables automated data analysis with the aim of improving energy efficiency in manufacturing. In our use case, machining cycles within a production chain were identified, energy hotspots revealed, and exemplary energy efficiency measures derived. The algorithm for the detection of machining cycles takes into account two factors: Firstly, the possibility that there may be a pause between cycles, i.e., that cycles do not have to follow each other directly. Secondly, that cycles detected by STUMPY do not have to be true machining cycles. The algorithm achieves up to 94.3% accuracy for similar machining cycles, while performing inferiorly for machines with acyclic events. Future research activities are expected to include further energy efficiency measures and deeper refinement of the developed algorithm. This optimization might involve comparing potential machine cycles with the sample cycle based on additional statistical properties, to increase the accuracy of the algorithm. Moreover, the algorithm can be utilized for semi-automated data labeling for machine learning applications. Some of the product's individual carbon footprint could also be calculated automatically by linking the current electricity mix with the identified cycles. By integrating forecasts of renewable energies, the expert system could also be extended to include energy flexibility measures.

Acknowledgements. The authors gratefully acknowledge financial support by the German Federal Ministry of Economic Affairs and Climate Action (BMWK) for the project *KI4ETA* (grant agreement No. 03EN2053A) and project supervision by project management Jülich (PtJ).

References

1. European Investment Bank (EIB) (2021) EIB investment survey 2021: European Union overview. European Investment Bank (EIB), Luxembourg
2. Basden A (1994) Three levels of benefits in expert systems. Expert Syst. https://doi.org/10.1111/j.1468-0394.1994.tb00003.x
3. Petruschke L, Elserafi G, Ioshchikhes B, Weigold M (2021) Machine learning based identification of energy efficiency measures for machine tools using load profiles and machine specific meta data. MM SJ. https://doi.org/10.17973/MMSJ.2021_11_2021153

4. Ioshchikhes B, Elserafi G, Weigold M (2023) An expert system-based approach for improving energy efficiency of chamber cleaning machines. In: Herberger D, Hübner M, Stich V (eds) Proceedings of the conference on production systems and logistics: CPSL 2023 - 1. Publishing, Hannover, pp 1–11. https://doi.org/10.15488/13419

5. Posselt G (2016) Towards energy transparent factories. Springer International Publishing, Cham

6. German Environment Agency (UBA) (2022) Energieverbrauch nach Energieträgern und Sektoren. Accessed 6 Feb 2023. https://www.umweltbundesamt.de/daten/energie/energieverbrauch-nach-energietraegern-sektoren#allgemeine-entwicklung-und-einflussfaktoren

7. Schäfer KF (2020) Netzberechnung. Verfahren zur Berechnung elektrischer Energieversorgungsnetze. Springer Vieweg, Wiesbaden

8. Dehning P, Blume S, Dér A, Flick D, Herrmann C, Thiede S (2019) Load profile analysis for reducing energy demands of production systems in non-production times. Appl Energy. https://doi.org/10.1016/j.apenergy.2019.01.047

9. VDMA (2019) Messvorschrift zur Bestimmung des Energie- und Medienbedarfs von Werkzeugmaschinen in der Serienfertigung. Verband Deutscher Maschinen- und Anlagenbau e.V., 2019th edn. 25.080.01 (34179)

10. Seevers J-P, Jurczyk K, Meschede H, Hesselbach J, Sutherland JW (2020) Automatic detection of manufacturing equipment cycles using time series. J Comput Inf Sci Eng. https://doi.org/10.1115/1.4046208

11. Keogh E, Lin J (2005) Clustering of time-series subsequences is meaningless: implications for previous and future research. Knowl Inf Syst. https://doi.org/10.1007/s10115-004-0172-7

12. Gao Y, Lin J (2018) Exploring variable-length time series motifs in one hundred million length scale. Data Min Knowl Disc. https://doi.org/10.1007/s10618-018-0570-1

13. Linardi M, Zhu Y, Palpanas T, Keogh E (2018) Matrix profile X. In: Das G, Jermaine C, Bernstein P (eds) Proceedings of the 2018 international conference on management of data. SIGMOD/PODS '18: international conference on management of data, Houston, TX, 10–15 June 2018. ACM, New York, NY, pp 1053–1066. https://doi.org/10.1145/3183713.3183744

14. CECIMO (2005) Roadmap for CECIMO's self-regulative initiative (SRI) for the sector specific implementation of the directive 2005/32/EC (EuP directive) for 2009 to 2011. Brussels

15. Zein A, Li W, Herrmann C, Kara S (2011) Energy efficiency measures for the design and operation of machine tools: an axiomatic approach. In: Hesselbach J, Herrmann C (eds) Glocalized solutions for sustainability in manufacturing. Springer Berlin Heidelberg, Berlin, Heidelberg, pp 274–279

16. Flum D, Sossenheimer J, Stück C, Abele E (2019) Towards energy-efficient machine tools through the development of the twin-control energy efficiency module. In: Armendia M, Ghassempouri M, Ozturk E, Peysson F (eds) Twin-control: a digital twin approach to improve machine tools lifecycle. Springer International Publishing, Cham, pp 95–110

17. Law SM (2019) STUMPY: a powerful and scalable Python library for time series data mining. J Open Source Softw 4:1504

18. Abele E, Schneider J, Beck M, Maier A (2018) ETA – the model factory. Technical University of Darmstadt, Darmstadt

Effects of Drill Point Geometry on Cutting Forces and Torque When Drilling AA1050

A. Simoncelli[1,2(✉)], L. Buglioni[1,3], G. Abate[1,2], P. Gayol[2], A. J. Sánchez Egea[4], and D. Martínez Krahmer[1,5]

[1] Centro de Mecánica, Instituto Nacional de Tecnología Industrial (INTI), Buenos Aires, Argentina
asimoncelli@inti.gob.ar
[2] Facultad de Ingeniería, UNLZ, Buenos Aires, Argentina
[3] Facultad de Ingeniería, Universidad de Buenos Aires, Buenos Aires, Argentina
[4] Departament d'Enginyeria Mecànica, Universitat Politècnica de Catalunya, Cataluña, Spain
[5] Instituto de Tecnología e Ingeniería, Universidad Nacional de Hurlingham, Buenos Aires, Argentina

Abstract. Reducing energy consumption in drilling operations is crucial for achieving sustainability goals. A study examined 36 drill bits with different geometries and conditions on AA1050. It assessed thrust forces and torque in two machining conditions (Cmin and Cmax) while considering mesh density, tool geometry, and boundary conditions. The results show that finer mesh models exhibit lower thrust forces, while mass scaling primarily influences torque. The pilot hole configuration decreases force, consistent with experiments. Torque decreases by increasing mesh density, matching with the experimental results. Finally, temperature and chip shape are mesh-dependent, affecting torque and force. As a result, our FEM model effectively predicted thrust force and torque, emphasizing the role of the pilot hole configuration in temperature and plastic strain results.

Keywords: Twist drill · Point geometry · Cutting forces · Computer simulation · Pilot hole

1 Introduction

Sustainability in drilling processes requires a multifaceted approach that considers four aspects [1, 2]. Firstly, the material being drilled plays a significant role. Some materials are easy to cut, while others pose challenges. Even with the possibility of using free machining elements added to the material in very small proportions, they act as tribofilms and chip breakers, facilitating chip evacuation and reducing cutting forces [3, 4]. Secondly, the cutting tool used is crucial. Coated drill bits can improve the coefficient of friction, resulting in easier chip sliding, reduced cutting forces, prevention of adhesion on contact surfaces, and decreased heat generation due to chip sliding [2]. Thirdly, lubrication is essential. Lubrication and cooling systems are employed to consume less energy and prevent tool breakage. While dry machining is the ideal technique, implementing it poses difficulties, especially in drilling aluminum and its alloys. However,

© The Author(s) 2025
H. Kohl et al. (Eds.): GCSM 2023, LNME, pp. 668–675, 2025.
https://doi.org/10.1007/978-3-031-77429-4_74

sustainable alternatives such as compressed air or Minimal Quantity Lubricant systems exist [5]. Finally, finite element simulation methods, once validated with experimental results, prove to be an effective approach for verifying tool designs aimed at achieving minimal energy consumption [6]. Most studies focusing on drilling processes mainly address determining cutting forces based on cutting conditions and drill diameter [7], while a few consider the prediction of shear stresses in predrilling processes [8]. It is known that the drill point geometry is primarily defined by the point angle (ε), the clearance angle (α), and the web thickness (s). Each geometric variable has specific standards that provide a minimum-maximum range for their values. Since optimizing the drill geometry is a fundamental step toward developing a sustainable drilling process, dynamometric tests are done to correlate drilling forces (feed force and torque) with the drill point geometry.

This work introduces several innovations in the field of machining and drilling modeling. Firstly, it explores no-conventional geometric conditions of the tooltip (α, ε, and s) on a material of wide industrial application, including empirical tests and results, which challenge the established norms in the industry, thus opening up new possibilities and avenues for precision machining techniques. It also underscores the limited body of work dedicated to drilling with a pilot using FEM models by using a split coupled Eulerian-Lagrangian, an area that remains relatively unexplored. We are aware that a pilot hole is used to determine the constants for a mechanistic model, but it is also true that further studies are required to relate the tool geometry, material and cutting conditions.

2   Methodology

2.1   Material, Drill Bits and Drilling Conditions

The experimental drill tests were conducted on 19 mm square cross-section of AA1050 aluminum bars and material hardness of HV10 106 $\pm$ 2. Aluminum bars of approximately 400 mm in length were used during the experimental tests. A total of 72 drilling tests were done, based on combining 9 groups of drills, 4 repetitions per group and 2 cutting conditions. Consequently, 36 helical drill bits (DIN 338 helical twist drills) with a diameter of 7.5 mm were used during the experimental tests. These drills were of type N and constructed from AISI M2 high-speed steel, with the following nominal chemical composition (% by weight): 0.9% C; 6% W; 5% Mo; 4% Cr; 2% V and balance Fe. To analyze the influence of drill point geometry on drilling forces, these drills were manufactured outside the standard production program to provide four values for the clearance angle (α) (12°, 14°, 16°, and 18°), three values for the web thickness (s) (0.88 mm, 1.18 mm, and 1.38 mm), and four values for the point angle (ε) (110°, 118°, 130°, and 140°). All of these values fall within the range specified by the relevant standard for the three mentioned variables. Table 1 presents the nine tested drill groups and their associated geometries. Drilling tests were done in AA1050 with two configurations: minimum configuration (Cmin) and maximum configuration (Cmax). Cmin is defined as a 40 m/min cutting speed, 1698 rpm of spindle speed and 212 mm/min of feed speed. The Cmax is 60 m/min cutting speed, 2546 rpm of spindle speed and 318 mm/min of feed speed. All of the drilling tests were carried out in dry conditions.

Table 1. .

Drill group	Web thickness [mm]	Measured w.t. [mm]	Clearance [*]	Point [*]
G1	1.38	1.43 ± 0.02	14	130
G2	1.38	1.41 ± 0.02	14	118
G3	1.18	1.22 ± 0.02	16	118
G4	1.18	1.23 ± 0.04	14	140
G5	0.88	0.88 ± 0.02	14	118
G6	1.18	1.19 ± 0.01	14	118
G7	1.18	1.23 ± 0.04	12	118
G8	1.18	1.20 ± 0.01	18	118
G9	1.38	1.45 ± 0.10	14	110

2.2 Experimental Tests

A PROMECOR CNC milling machine with 8000 rpm and 10 kW of power was used for drilling. The machining forces were measured using a two-channel Kistler model 9271A piezoelectric dynamometer with their respective charge amplifiers, and the test data was recorded with a Labjack T7-Pro acquirer. The data processing, filtering and analysis of the signals were performed by scripts in Python programming language. All holes were drilled at a single tool depth of 10 mm (without chip withdrawal). Using the data acquirer, the torque (T) and feed force (F) signals were recorded for each drill in each test condition (Cmin/Cmax) at a sampling rate of 4 kHz. F_{RMS} and T_{RMS} average values were calculated per test with a 0.1 s window. Each signal was segmented into five equal parts corresponding to the records obtained at depths from 2.5 and 8.5 mm, to avoid transitions and filter smoothing. Table 2 shows the experimental test summary.

Table 2. .

Drill group	Thrust F max [N]	Thrust F min [N]	Torque max [Nm]	Torque max [Nm]
G1	438.7 ± 28.1	513.7 ± 33.6	0.86 ± 0.022	0.90 ± 0.023
G2	398.6 ± 20.0	430.6 ± 24.0	0.86 ± 0.026	0.89 ± 0.014
G3	373.2 ± 14.7	400.1 ± 22.2	0.84 ± 0.028	0.86 ± 0.017
G4	657.5 ± 58.3	705.7 ± 64.7	0.87 ± 0.037	0.90 ± 0.029
G5	316.4 ± 9.1	322.0 ± 14.9	0.82 ± 0.021	0.83 ± 0.010
G6	366.0 ± 18.9	394.9 ± 26.6	0.84 ± 0.024	0.85 ± 0.018
G7	378.9 ± 32.0	416.9 ± 36.4	0.84 ± 0.022	0.87 ± 0.027
G8	362.1 ± 20.09	393.8 ± 31.2	0.84 ± 0.028	0.86 ± 0.012
G9	362.1 ± 26.0	393.0 ± 24.0	0.89 ± 0.035	0.91 ± 0.031

2.3   FEM Formulation and Description

Several explicit dynamics with coupled temperature displacement FEMs have been developed in ABAQUS/Explicit to characterize the drilling process. The numerical approaches have been done in an Intel I7 10700 6-core and 12-thread CPU. These models consist of an elastic tool represented with a Lagrangian mesh (fixed with the material) and an Eulerian mixed void and material domain representing a workpiece (WP) in which material can flow through the mesh elements. This approximation of the workpiece mesh is known as Coupled Eulerian-Lagrangian and has been used before to simulate drilling processes [9, 10]. Tool geometries correspond to G4 and G5 from Table 1. For all the Eulerian cubic shape domains, the material was initially removed in a conical shape following the tip angle of the tool, in order to reduce the transient phase, such as done in [9]. Moreover, hexahedral 8 nodes and reduced integration elements were used for meshing the WP, according to [9, 10], while the tool was meshed with tetrahedrons with a higher density in the tooltip. Table 3 details model parameters, such as E_s, which stands for element length, Mass Scaling (MS) factor, H represents maximum domain height, being included the "void" space which is filled with chip formation, Pil represents the pilot hole configuration, and MecBC and ThBC mechanical and thermal boundary conditions tests to study influence over results. Mesh densities are detailed below group model IDs. ID 1 and 3 are the coarse mesh ones (with and without pilot hole, respectively), and specifically, ID 1 is also used for BC and MS testing, and 2 and 4 are the dense mesh models.

Table 3. Model parameters. MS and BC tests were done only with G4 tool.

Mesh ID	Tool	E_s [μm]	MS	t [ms]	H [mm]	Pil	Cond	MecBC	ThBC
1	4.5	100	100	30	3/5	No	min, max	Sides	No
1	4	100	25	30	3/5	No	max	Sides	No
1	4	100	100	30	3/5	No	max	Sides/All	No/20 °C
2	4.5	50	100	20	3	No	max	Sides	No
2	4	50	400	20	3	No	max	Sides	No
3	4.5	100	100	30	5	Yes	max	Sides	No
4	4.5	50	100	20	3	Yes	max	Sides	No

Two mesh densities have been developed and studied. The first one was adopted based on the work of [10], considering the solving time. It consists of 100 μm element length, with an initial domain total height of 5 mm, which provides sufficient clearance to maintain the chip intact. The second mesh density was adopted later when it was observed that the thrust force obtained by the model exceeded the experimental values, considering that mass scaling did not affect it. The mesh consists of 1.5 million elements for the WP, with an element length of 50 μm. In this model, consistent with the aforementioned clearance, the overall domain height has been set to 3 mm. Regarding the boundary conditions, it have been applied to the domain sides, where all displacements are fixed.

In particular, in the first model, a test was run with the bottom also fixed. Another test has been done by applying a constant temperature of 20 °C to the sides and bottom surfaces, according to [9], which constrains a convective heat transfer of the coolant. Another domain with a 2 mm pilot hole has been included to analyze the influence of this hole both in thrust force and torque. Cutting conditions Cmin and Cmax are 1698 rpm–212 mm/min and 2546 rpm–318 mm/min, respectively, and were applied as a linear function from initial time to 10 ms, to avoid oscillations. The overall runtime goes from 15 to 17 h for coarse mesh models (1 and 3) with 100x mass scaling (30 ms model time), about 22–24 h for model 1 with 25x mass scaling, and about 40 h for the dense mesh models (2 and 4) with 100x mass scaling (20 ms model time) and 20 h for the 400x mass scaling dense mesh models. Tool material has been adopted as elastic carbon steel (7850 kg/m^3, 200 GPa, 0.3 of Poisson ratio, 45 W/m°C and 420 J/kg°C) and WP is assumed as AA1050 (2710 kg/m^3, 69 GPa, 0.3, 160 W/m°C and 899 J/kg°C). A plastic flow Johnson-Cook material model has been adopted for the WP:

$$\sigma = \left[A + B(\varepsilon_P)^n\right]\left[1 + Cln(\varepsilon_p)/\varepsilon_0\right]\left[1 - (T - T_0)/(T_m - T_0)^m\right]$$

The WP parameters are defined as A-110 MPa, B-150 MPa, C-0.014, ε_0-1, n-0.36, m-1, T_m-645 °C and T_0-20 °C. The proportion of heat flux produced at the contact interface that was conducted to the tool and WP is determined by a heat partitioning coefficient, which is adjusted to 0.5 (equally distributed heat). Contact between tool and workpiece is based on the Coulomb friction model, which assumes a constant friction coefficient of 0.2, following [10]. Contact conductance was kept constant at 40 kW/(m^2 K), according to [9]. Finally, 90% of the heat generated by plastic strain is converted to heat (inelastic heat fraction). Shear stress has not been limited for this contact model and the material damage has not been considered in these models.

3   Results and Discussion

The thrust forces and torque for the two machining conditions, Cmin and Cmax, are then analyzed. Figure 1 shows at left mean values of thrust force obtained from different models. All values are obtained by averaging data from 15 ms when values stabilize.

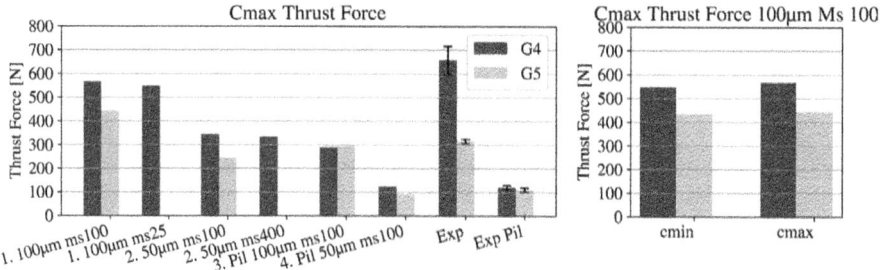

Fig. 1.   Thrust force. Cmax condition (left). Cmin vs. Cmax (right).

The influence of the mesh density shows that the thrust forces are found below experimental ones in the finer mesh density models, particularly in G4 tool. Several factors could influence this low thrust force. According to [11], contact friction formulation and tool wear play a fundamental role. Among other material parameters, like Johnson Cook parameters, thermal expansion coefficient and material damage, have been neglected (although damage usually even decreases thrust force according to [9]). The second and fourth bars indicate mass scaling influence, showing that 100x is a good mass scaling selection since its results are similar to 25x (which requires 2x calculation times), even if it could be chosen a value of 400x, in particular for a finer mesh, considering solving times, and thrust force difference which is negligible. Bars with models 3 and 4 show thrust force in pilot hole configuration for the two mesh densities. For the finer mesh density, a numerical axial force is almost exactly the same value as the experimental one. It is seen that thrust force diminishes considerably in this configuration against WP without a pilot hole, due to the tool core. This is also observed in experimental measures, giving the major capacity of the model to predict this behavior. The right side of Fig. 1 shows differences in thrust force for both cutting conditions and tools for coarse mesh. It is important to note that thrust forces are smaller for Cmin conditions, whereas the experimental responses behave the opposite. However, these differences are 3.7% and 1.8% for G4 and G5, respectively, which is small considering the experimental dispersions. Figure 2 shows torque for all conditions and compares cutting conditions. The torque decreases significantly with mesh density, but in this case, those obtained from finer mesh are very similar to the experimental ones. The prediction capacity is a crucial advantage of the model, specially considering that power consumption is directly determined by torque. For all cases, G5 tool results in a greater torque than G4, but with small differences. Thrust force and torque values have also been evaluated for Cmax conditions, coarse mesh, G4 tool changing both mechanical (by adding displacement restrictions at the bottom) and thermal (by the constraining temperature at the bottom and sides to 20 °C) boundary conditions. The results seem not to be affected by these constraints. The two bars on the left show torque in the pilot configuration. Regarding the mass scaling, from 25x to 100x is an increase in torque of 7%, whereas from 100x to 400x is an overestimation of 16%. Here, the torque is almost the same with or without a hole, which is coherent with experimental observations, and it makes sense since what contributes to torque is mainly the cutting part of the tooltip. It is also denoted by a slight increase in the value from Cmin to Cmax. These variations are small, 4.5% and 1.4% for G4 and G5, respectively. One of the causes is that friction could have been underestimated and, if increased as in [9], could produce more heat in the case of Cmax due to greater rotational velocity. Another friction-sliding mechanism could be analyzed, allowing limitation of shear stress like [9, 10].

Temperatures after 19 ms are shown in Fig. 3. It is evaluated at this particular time and not when the simulation is over due to the chip shape, which rises up until it flows above and outside the upper face of the domain, losing its continuity and giving the resulting chip remains, which is difficult for a proper view. The maximum values for coarser mesh are nearly the same for each tool, at about 200 °C. Also, the mesh density has a crucial impact on both the thickness and shape of the chip, having a smaller thickness and narrow curvature in the case of finer mesh, and is directly related to the impact on both torque and thrust force. The cut chip shape in the Cmax is due to a small overall domain height and does not affect the results.

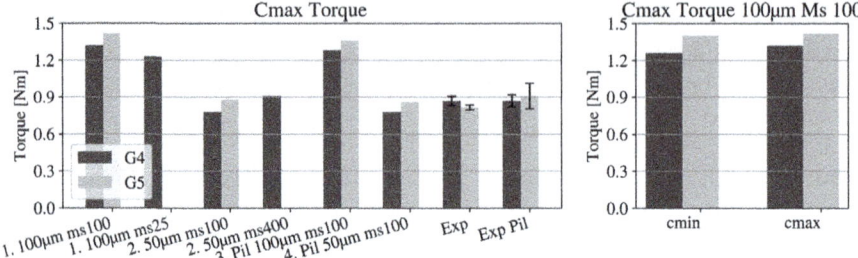

Fig. 2. Thrust force. Cmax condition (left). Cmin vs. Cmax (right).

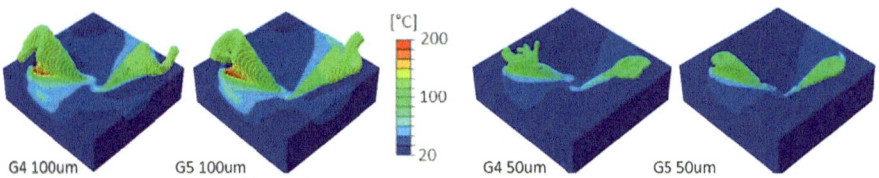

Fig. 3. Temperatures after 19 ms. Models 1 and 3. 100x MS, Cmax conditions.

Finally, the ratio of kinetic energy against internal energies should be maintained above 5%, in order to be sure that the mass scaling magnitude is not too excessive (mass scaling also affects forces, as seen before). Also, artificial energy (which is involved in solving element reduced integration) should be below 1–2% of internal energy. It is observed that the kinetic vs. internal energy relation is significantly low, about 1.3%, which has not been diminished even with finer mesh.

4 Conclusions

Several finite element models have been developed to investigate and predict outcomes in drilling processes. Various model parameters have been studied, such as mass scaling, mesh density and both thermal and mechanical boundary conditions. For most refined mesh models and considering the simplifications adopted, torque values are notoriously accurate with respect to the experimental tests. This is a major achievement of the model, bearing in mind that power consumption is directly related to torque values. The model also shows thrust force and torque extremely similar for the pilot hole configuration, in which the influence of the tool core is neglected. Another major capacity of the model is the prediction of magnitudes that are difficult or even impossible to measure, such as temperature or plastic strain.

References

1. Pervaiz S, Kannan S, Kishawy HA (2018) An extensive review of the water consumption and cutting fluid based sustainability concerns in the metal cutting sector. J Clean Prod 197:134–153

2. Astakhov VP (2014) Drills: science and technology of advanced operations
3. Krahmer DM, Urbicain G, Egea A (2020) Dry machinability analyses between free cutting, resulfurized, and carbon steels. Mater Manuf Process 35(4):460–468
4. Krahmer DM, Hameed S, Sánchez Egea AJ (2019) Wear and MnS layer adhesion in uncoated cutting tools when dry and wet turning free-cutting steels. Metals 9(5)
5. Berzosa F, Rubio EM, Agustina B, Davim JP (2020) Geometric optimization of drills used to repair holes in magnesium aeronautical components. Metals 10(11)
6. Pervaiz S, Samad WA (2021) Drilling force characterization during Inconel 718 drilling: a comparative study between numerical and analytical approaches. Materials 14:4820
7. Tamura S, Matsumura T (2022) Cutting force in drilling with flat bottom drill. Key Eng Mater 926:1636–1642
8. Kim KS (2015) Analysis of drilling of a metal plate with pilot hole. Arch Metall Mater 60(2):1–4
9. Priest J, Ghadbeigi H, Avar S (2021) 3D finite element modelling of drilling: the effect of modelling method. J Manuf Sci Technol 35:158–168
10. Abdelhafeez AM, Soo SL, Aspinwall D (2016) A coupled Eulerian Lagrangian finite element model of drilling titanium and aluminium alloys. CIRP J Manuf Sci Technol 9:198–207
11. Laakso SVA, Agmell M, Ståhl J-E (2018) The mystery of missing feed force – the effect of friction models, flank wear and ploughing on feed force in metal cutting simulations. J Manuf Process 33:268–277

Trends and Resource-Efficient Application of Solid End Mills in Plastics Machining: A Comparative Market Study

J. Schmidt[1(✉)], B. Thorenz[1,2], and F. Döpper[1,2]

[1] Chair Manufacturing and Remanufacturing Technology, University of Bayreuth, Universitätsstraße 30, 95447 Bayreuth, Germany
Julian.Schmidt@uni-bayreuth.de
[2] Fraunhofer IPA Project Group Process Innovation, Universitätsstraße 9, 95447 Bayreuth, Germany

Abstract. Due to their high share of value creation in manufacturing, the influence of machining processes and the therein applied cutting tools on profitability and sustainability is often high. Plastics machining differs significantly from metal machining and requires special tools or machining strategies. However, innovation and research usually address the area of metal machining, while plastics machining is often based on in-house experience and the use of tools that were originally developed for metal machining. Thus, potentials regarding energy and resource efficiency of those cutting processes often remain unused. This paper analyses the market situation and needs for required innovations of companies in the plastics machining industry. Therefore, a market study was conducted in which both tool manufacturers and users were asked about relevant aspects regarding the production and application of tools as well as assessments of trends and desired innovations. The results help to identify future research needs so that potentials in the area of plastics machining can be raised with regard to energy and resource efficiency. An important finding of the study is that both tool manufacturers and users see a need for vibration-damped tools and for a wider range of tools with more special tools for plastics machining.

Keywords: Milling · Plastics machining · Market study

1 Introduction

In order to achieve internal and external environmental goals, the energy- and resource-efficient design of manufacturing processes is becoming increasingly important. Due to the very high specific energy consumption compared to other manufacturing processes, machining processes like milling in particular play a decisive role here [1]. Machining of plastics is for example necessary when the number of parts to be manufactured does not justify the production of an injection mold or extrusion die, or when the parts to be manufactured must have a very high dimensional accuracy [2]. While research work on machining processes is mainly concerned with the machining of metals, only a few

© The Author(s) 2025
H. Kohl et al. (Eds.): GCSM 2023, LNME, pp. 676–683, 2025.
https://doi.org/10.1007/978-3-031-77429-4_75

papers deal with the machining of plastics – and these in turn mainly with the machining of fiber-reinforced plastics [3]. For this reason, plastics machining processes are often only experience-based or are carried out with conventional tools that were originally developed for metal cutting.

This market study focuses both on tool manufacturers and industrial users of solid end mills in engineering plastics machining and thus might contribute to answer urgent questions regarding operating conditions, current trends and developments in the field of machining processes. The results are discussed in order to identify future research and development needs for improving the resource efficiency in manufacturing enterprises which are applying solid end mills in plastics machining.

2 State of the Art

2.1 Plastics Machining

The numerous approaches regarding milling processes from science and practice relate to a large extent to the application case of metal milling. Plastic milling represents a deviating special case with some peculiarities in comparison to conventional metal milling.

Plastics can basically be produced by non-cutting processes, for example by injection molding or extrusion. Nevertheless, machining is often necessary, for example when reworking molded parts, for the production of small batches [4] or for the production of complicated individual parts. In general, plastics have lower strength properties compared to metals. As a result, the cutting force F_c to be applied in the process is lower [5]. The thermal properties of plastics in particular set them apart from metals in their processing. Due to their low thermal conductivity, the heat generated in the cutting process must be dissipated mainly via the tool, which negatively affects the stability of the cutting edge. Heat dissipation by cooling is often controlled by compressed air supply, since some plastics exhibit an increased swelling effect when water and emulsions are used [6]. The process heat generated can also have a negative effect on the material being machined. Plastics with low temperature resistance can start to smear and stick at too high temperatures and, due to thermal or chemical degradation, impair the dimensional accuracy of the manufactured geometries [4]. In order to counteract these circumstances, high speeds and corresponding feed rates are used when milling plastics, so that the heat can be dissipated as far as possible via the chips and thus an annealing of the tool cutting edge can be prevented [7]. Furthermore, the geometric shapes produced in plastics machining differ greatly from those in metal machining, which is why the tools are subjected to different stresses due to the machining strategy [8].

The milling of fiber-reinforced plastics poses further challenges for the machining process [9], but will not be included in the scope of this paper.

For these reasons, existing research and practical investigations for metal milling cannot be transferred to the application of plastics milling. This paper therefore addresses this still insufficiently investigated topic area.

2.2 Market Studies

In order to discover new insights and collect purposeful data properly scientifically based research methods and a planned research process are required. These two aspects can be taken into account by focusing on an ideal market research process [10]. For this purpose, different types of process models which are essentially similar are described in literature. Depending on how many individual steps are summarized in a process phase, these process models are divided into five [11], seven [12] or ten phases [10]. The sequence of a study in a five-stage process model according to Magerhans is shown in Fig. 1 [11]. This sequence is based on a seven-stage process model such as from Mooi [12] while partly aspects of Homburg are introduced [10].

Fig. 1. Five-phase model of the market study

First, the object of investigation and there research objectives are defined. Then, the examination design is determined and the survey is designed. In the data collection phase, suitable addresses are selected, and the scope of the survey is determined. In the data analysis phase, the collected data is evaluated. The documentation phase includes the summary of the results, suggestions for further market research activities and recommendations for future research and development.

3 Approach and Study Design

A standardized survey was developed and sent to a variety of enterprises by email. The collected data has been analyzed and summarized. Subsequently, a comparative analysis of different groups of participants has been carried out.

The survey consists of two different aspects. One aspect is the description of the status quo, the other aspect are current trends as well as the identification of potentials for further development. The implementation of these two research approaches is carried out within the framework of a standardized survey by using different question formats. A distinction is made between closed and open questions. Furthermore, combined forms of these two questions formats are used. Closed questions serve to collect quantitative results, whereas open questions are apt to gain new insights in the course of qualitative research.

The survey is aimed at both tool manufacturers for plastics processing and their users. To ensure that the questions assigned to each are asked, the interviewee must indicate which of these two groups he or she belongs to [13].

Three types of questions are used: Non-specific questions that are relevant to all participants, manufacturer-specific questions that only concern the manufacturers of tools for plastics milling and user-specific questions that only concern users of tools for plastics milling. Figure 2 shows the distribution of the three types of questions among the various topic blocks into which the survey is divided.

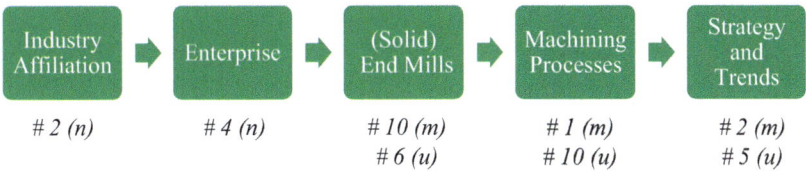

Industry Affiliation		Enterprise		(Solid) End Mills		Machining Processes		Strategy and Trends
# 2 (n)		# 4 (n)		# 10 (m)		# 1 (m)		# 2 (m)
				# 6 (u)		# 10 (u)		# 5 (u)

 # number of questions; (n) nonspecific; (m) manufacturers; (u) users

Fig. 2. Study design

With the help of associations, industry directories and chambers of commerce, a total of 54 companies in Germany were identified that could be considered as manufacturers and were contacted accordingly. Due to the small number of potential companies, they were contacted both in writing and by telephone. 18 of these companies completed the survey, which corresponds to a response rate of 33%. On the user side, 207 companies were identified and contacted, of which 25 responded to the survey, representing a response rate of 12%. Online surveys are always expected to have lower response rates than other formats such as a telephone survey [14]. A single-digit response rate can be expected, with the maximum achievable value being up to 20% [15]. Thus, the response rate of the users is above what can be expected, and that of the manufacturers is even above the maximum achievable value according to the literature. The latter can be explained by intensive follow-up and repeated reminding of the companies contacted. This was necessary in order to obtain a reliable number of responses to the significantly smaller number of companies contacted.

4   Results and Discussion

4.1   Surveyed Enterprises

In order to obtain an overview of the structure of the companies surveyed, the first questions asked for general information such as company size and industry affiliation - regardless of whether they were manufacturers or users.

According to EU Recommendation 2003/361 [16], the majority of the companies surveyed (86%) are SMEs, i.e. companies with fewer than 250 employees. 32% are even classified as small companies with fewer than 50 employees. Figure 3a illustrates this relationship graphically. The manufacturers surveyed all assigned themselves to the tool and die making industry. Therefore, Fig. 3b continues to show only the industry affiliation of the user companies surveyed. At 48%, just under half of the companies surveyed have assigned themselves to the machinery and plant engineering industry. Medical technology and automotive and vehicle manufacturing each account for 12%.

4.2   Machining Processes and Tools

The first question asked in this section was how high the share of value added and thus of sales can be attributed to processes from plastics machining. For the manufacturers,

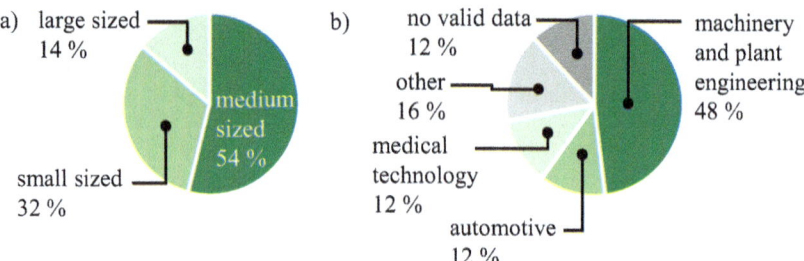

Fig. 3. Size (a) and industry sector (b) of surveyed companies

this is the tools produced, and for the users, the share of components produced. As shown in Fig. 4a, the share of most manufacturers is less than 25%, from which it can be concluded that none of the manufacturers specializes explicitly in the production of tools for plastics machining. Among the users, 20% of the companies surveyed generate more than 75% of their sales with plastics machining, which indicates a higher degree of specialization of users for this industry. Furthermore, the users were asked about the most frequently machined plastics in their company. Three answer options were permitted in each case. Polyoxymethylene (POM) was named most frequently with 68%, followed by polyethelene (PE) and polyurethane (PU) with 44% each, see Fig. 4b.

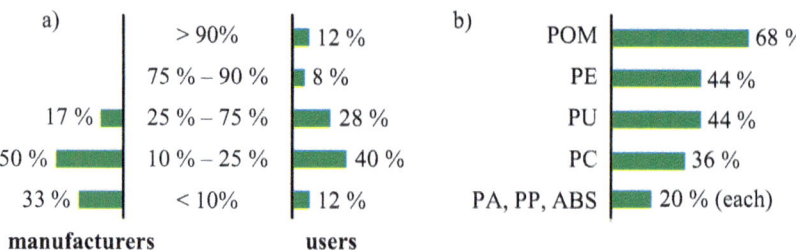

Fig. 4. Share of sales related to plastics machining (a) and share of machined plastics among users (b)

Furthermore, users indicated that on average 74% of their tools used are made of carbide. In addition, 87% of the tool designs used are solid material tools. This can be explained by the fact that carbides allow higher cutting speeds than high-speed steel (HSS), which is often required for plastics machining. In addition, the cutting edges of carbide tools can be subjected to higher thermal loads, which is highly significant due to the lack of thermal conductivity of plastics [6]. While 69% of the tools used are uncoated, the largest group of tool diameters used is in the medium range between 10 mm and 20 mm (46%). Due to the different material properties and behavior of plastics, companies usually use several cooling strategies - the most frequently mentioned are the use of cooling emulsion (76%) and compressed air (72%). All these statements are summarized graphically in Fig. 5.

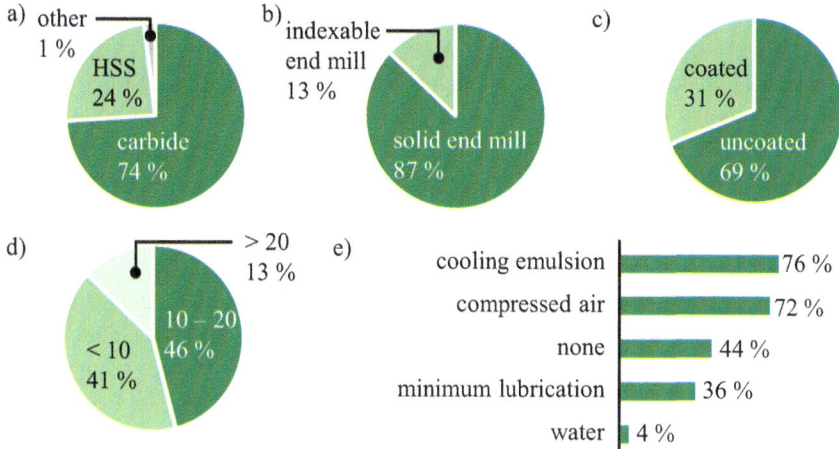

Fig. 5. Cutting material (a), end mill type (b), coated tools (c), tool diameters in [mm] (d) and cooling strategies (e) deployed by the users

4.3 Trends and Room for Improvement

While users were asked to rate five selected target dimensions in relation to their business, manufacturers were asked to provide an estimate of how their customers rate them. For each target dimension, a score between 0 (unimportant) and 5 (very important) could be given. It is noticeable that the manufacturers estimate that the users rate all target dimensions as very important. With a score of 4.7, energy efficiency is rated as the most important target dimension, see Fig. 6. The users take a much more differentiated view of the target dimensions and rate tool life as the most important criterion with a score of 4.1. Energy efficiency, which is considered by the manufacturers to be the most important target dimension, is even rated lowest by the users with a score of 2.1. This discrepancy can be explained by the fact that users take the view of their specific application case of plastics machining, while manufacturers unconsciously take into account the view of all their customers, most of whom come from the application case of metal machining. In the latter, energy efficiency plays a far greater role due to the higher energy input when machining high-strength materials [1].

Fig. 6. Target dimensions from the perspective of manufacturers and users

With regard to innovations in demand, manufacturers estimate the needs of users very accurately in three of the categories, see Fig. 7. Only with regard to tool variety is there a large discrepancy, as manufacturers rate this as the least important (2.1) and users rate it as the most important (4.4). The high demand for flexible molds can be explained by the wide range of plastics processed and their widely varying material properties. Furthermore, while large quantities are produced with injection molding or extrusion processes, machining of plastics is often done in small batches [4], which is why a tool is more often exposed to varying machining requirements. It can also be noted that innovations on vibration-damped tools are rated as very important and advances in cooling strategies and energy efficiency are rated as important with scores of 3.2 and 3.0, respectively. The discrepancy in users' perceptions of energy efficiency between the current importance of the target dimension and the need for innovation suggests that users are aware that aspects of energy efficiency and sustainability will become much more important in the near future.

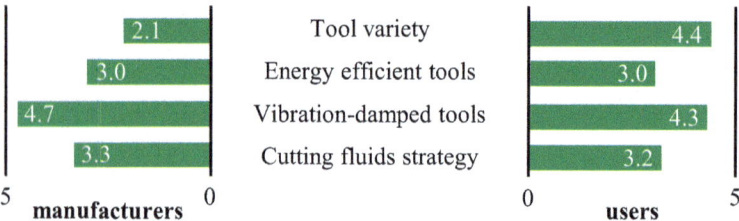

Fig. 7. Need for innovations from the perspective of manufacturers and users

5   Conclusion and Outlook

In summary, this paper describes the successful design and execution of a market study on end mill operating conditions as well as trends and need for innovation in plastics machining. 261 companies were contacted, of which 43 returned a usable questionnaire. In the first specific block of questions, the companies surveyed provided information on the tools used and the machining processes applied in plastics machining. For example, POM was indicated as the most frequently machined plastic as well as carbide solid end mills with medium diameters of 10–20 mm as the most frequently used tools. The assessment of manufacturers and users regarding the importance of target dimensions diverge in some points, for example in the assessment of the significance of energy-efficient processes. However, the need for further research was identified in several categories, especially for vibration-damped tools.

The results can be used as an orientation for research work dealing with the field of plastics machining, e.g. in the context of innovative tool or sustainable process development. The information on the current conditions of use of solid end mills in plastics machining can help to select the right references and to address the corresponding cross-company target groups and market segments. This helps to maximize the benefits of newly developed solutions. Furthermore, the survey serves as an overview for users who are only sporadically confronted with the machining of plastics in order to get an impression of the commonly used tools.

References

1. Westermann HH (2016) Entwicklung einer energieverbrauchsoptimierten Schneidengeometrie für Vollhartmetall-Schaftfräser. Shaker, Bayreuth
2. Xiao KQ, Zhang LC (2002) The role of viscous deformation in the machining of polymers. Int J Mech Sci 44(11):2317–2336
3. Aruna M (2020) Optimization of cutting parameters in machining polyoxymethylene using RSM. IOP Conf Ser Mater Sci Eng 893:012005
4. Baur E, Osswald TA, Rudolph N (2019) Plastics handbook, 5th edn. Hanser, Munich
5. Schwarz O (2016) Kunststoffverarbeitung, 11th edn. Vogel, Würzburg
6. Degner W, Lutze H, Smejkal E, Heisel U, Rothmund J (2019) Spanende Formung, 18th edn. Hanser, Munich
7. Hopman C, Michaeli W (2017) Einführung in die Kunststoffverarbeitung, 8th edn. Hanser, Munich
8. Schmidt J, Thorenz B, Döpper F (2023) Design of a reference part for the comparison of tools in plastic milling operations. In: Liu A, Kara S (eds) 33rd CIRP design conference 2023, Sydney, vol 11. Springer, Heidelberg, pp 981–986
9. El-Hofy MH, Soo SL, Aspinwall DK, Sim WM, Pearson D, Harden P (2011) Factors affecting workpiece surface integrity in slotting of CFRP. Procedia Eng 19:94–99
10. Homburg C (2017) Marketingmanagement, Strategie – Instrumente - Umsetzung – Unternehmensführung. Springer Gabler, Wiesbaden, pp 252–253
11. Magerhans A (2016) Marktforschung, Eine praxisorientierte Einführung. Springer, Wiesbaden, p 47
12. Mooi E, Sarstedt M, Mooi-Reci I (2018) Market research. Springer, Singapore
13. Sreejesh S, Mohapatra S, Anusree MR (2013) Business research methods. Springer, Cham
14. Daikeler J, Bosnjak M, Manfreda KL (2020) Web versus other survey modes: an updated and extended meta-analysis comparing response rates. J Surv Stat Methodol 8(3):513–539
15. Scholl A (2018) Die Befragung, 4th edn. UVK/Lucius, Munich
16. European Union Commission (2003) Commission recommendation of 6 May 2003 concerning the definition of micro, small and medium-sized enterprises. Off J Eur Union 46(L124):36–41

Sustainable Dry Turning of Aluminum Alloys Using Pulsed High-Pressure Cryogenic Jet Cooling

Jan Harald Selzam[✉], Niklas Enslein, Daniel Gross, and Nico Hanenkamp

Institute of Resource and Energy Efficient Production Systems, Friedrich-Alexander-Universität Erlangen-Nürnberg, Dr.-Mack-Str. 81, 90762 Fürth, Germany
Jan.Selzam@fau.de

Abstract. As well as producing contamination-free parts, dry machining is a sustainable alternative to conventional machining processes using metalworking fluids. Despite these benefits, chip breaking and chip removal are challenges in dry machining processes, particularly when turning ductile materials, which are known for continuous chip formation and therefore long chips. Due to their tendency to become entangled around machine components, these chips have to be removed manually, posing a risk to productivity and operator health. Recent studies have demonstrated the effectiveness of pulsed high-pressure emulsion cooling for chip breaking. This paper aims to transfer these findings to sustainable dry machining and investigates the influence of pulsed high-pressure cryogenic jet cooling on chip breaking and chip removal. Therefore, liquid CO_2 at pressures up to 200 bar is used as a cooling medium in turning processes of two aluminum alloys and is applied continuously or pulsed to the machining area. It has been shown that high-pressure cryogenic jet cooling improves chip breaking and chip removal, allowing a wider range of cutting parameters, widening the process window and providing a higher level of sustainability, productivity and safety in dry machining.

Keywords: Sustainable machining · Pulsed high-pressure cryogenic jet cooling · Dry turning · Aluminum · Carbon dioxide

1 Motivation and State of the Art

Most metal machining processes are performed using oil- or water-based metalworking fluids (MWF). The three main tasks of these MWF are to cool and lubricate the zone between the tool and the workpiece and to remove chips from this area. While mostly fulfilling these tasks adequately, conventional MWF come with noticeable drawbacks in the field of sustainability. They pose a risk to operator health and the environment, require a high maintenance effort and are difficult and expensive to dispose.

By producing contamination free parts, dry machining offers a more sustainable alternative to conventional machining methods. Most commonly, no medium is applied to the zone between tool and workpiece in dry machining, but this lack of cooling often causes excessive tool wear and impedes chip breaking and removal.

© The Author(s) 2025
H. Kohl et al. (Eds.): GCSM 2023, LNME, pp. 684–692, 2025.
https://doi.org/10.1007/978-3-031-77429-4_76

Cryogenic cooling, which covers a temperature range from 0 K to 0 °C, offers a way to compensate for this lack of cooling. In contrast to conventional MWF, different cryogens such as liquid nitrogen (LN_2), carbon dioxide (CO_2) or helium (He) are applied to cool the interaction zone. As these media evaporate without residue immediately after use, this cooling method provides technically clean and dry parts. It can therefore be considered dry machining in the same way as media-less processes, while still providing the cooling required for most machining operations.

The shortcomings in the field of chip removal on the other hand cannot be addressed as easily. This proves to be particularly critical in turning of ductile materials, which are known for continuous chip formation and therefore long chips.

Recent studies have demonstrated the effectiveness of pulsed high-pressure (HP) emulsion cooling for chip breaking and removal of Inconel 718 resulting in higher levels of productivity and quality [1, 2]. To combine these positive effects also with a higher level of sustainability, the concept of a pulsating coolant jet is transferred to cryogenic cooling using HP liquid CO_2, which cools down to − 78.5 °C when relaxed to atmospheric pressure due to the Joule-Thomson-effect.

In addition to the positive impact on sustainability, there are further reasons for dry or cryogenic cooling. These specific use cases include difficult-to-cut materials like titanium alloys, where the cryogen provides a better cooling than conventional emulsion and therefore facilitates longer tool life [3, 4].

Several publications show improved chip breaking in cryogenic machining. This is often claimed to be due to the embrittlement of the chips produced by the low temperatures in cryogenic machining [4–7].

2 Experimental Setup

2.1 Materials

The two aluminum alloys EN AW 2007 and EN AW 6082 were selected for the experiments, both in the shape of round extruded bars with a diameter of d_{bar} = 140 mm. EN AW 2007 has a tensile strength of 340 MPa, an elongation at break of 8% and a Brinell hardness of 95 HBW. EN AW 6082 has a tensile strength of 205 MPa, an elongation at break of 14% and a Brinell hardness of 70 HBW. Therefore, EN AW 2007 can be considered stronger and harder while EN AW 6082 is tougher [8, 9].

2.2 Machine, Process and Tools

All experiments were carried out on a DMG Mori CLX 350 lathe performing a longitudinal external cylindrical turning operation. The machine has a maximum spindle power of P = 11 kW and a maximum speed of n = 5500 1/min.

The tool holder SVJCL 2020 K 16-M-A was equipped with the insert VCGT 160402F-AL:T0315 (PRAMET). Specialized on non-iron metals, this polished insert allows for a cutting depth a_p = 0.3–4.0 mm, a cutting speed v_c = 195–630 m/min and a feed rate f = 0.06–0.12 mm.

2.3 Cryogenic Medium Delivery

The general setup of the delivery system for the cryogen is shown in Fig. 1. Depending on the examined application strategy the delivery system was adjusted.

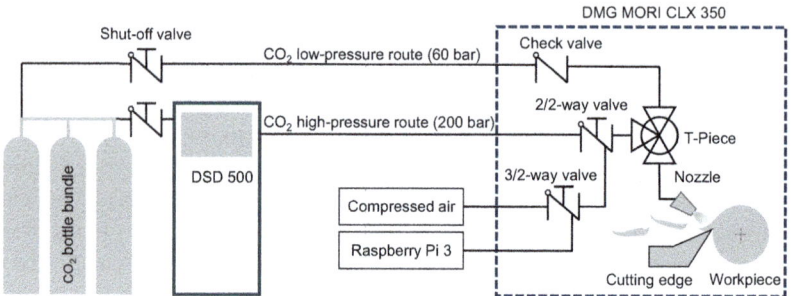

Fig. 1. Cryogenic medium delivery system

As cryogenic medium CO_2 was chosen. It is delivered via a riser bottle bundle at a pressure of $p_{bundle} = 60$ bar. From there the liquid CO_2 can take the high- and/or low-pressure route. While the low-pressure route ($p_{low} = p_{bundle} = 60$ bar) feeds the CO_2 to the T-piece just before the nozzle, the HP route supplies the CO_2 to a Maximator DSD 500 pressure booster. The booster system further compresses the CO_2 to an average pressure of $p_{high} = 200$ bar. The DSD 500 is connected to the 2/2-way valve Swagelok SS-HBS4-C that is normally closed. To open the pneumatically actuated SS-HBS4-C an electrically actuated 3/2-way valve is controlled by a Raspberry Pi 3 running a Python script. As can be seen in Fig. 2a, the switchable valves are integrated into the lathe to ensure the application of the CO_2 close to the interaction zone. The nozzle with a diameter of $d_{nozzle} = 0.5$ mm is positioned at a distance of 10–15 mm behind the cutting insert (Fig. 2b).

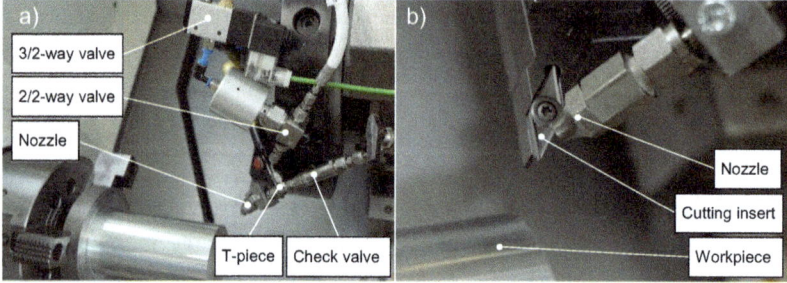

Fig. 2. a) Pulsing valve setup inside the lathe. b) Cutting insert and nozzle position

3 Experimental Design and Chip Classification

For the experiments, a full factorial design was chosen. While the cutting speed was kept constant at $v_c = 200$ m/min, the cutting depth a_p, the feed rate f and the cooling setup were varied. The cutting depth a_p was set to 0.5/1.0/1.5 mm resembling a finishing operation. The feed rate f was set to 0.06/0.08/0.10/0.12 mm to cover the full range of tool capability. For the cooling setup five different dry machining cases were varied. For reference, machining was carried out without any media. Furthermore, the CO_2 was applied continuously, firstly with a low pressure $p_{low} = 60$ bar and secondly with a high pressure $p_{high} = 200$ bar. Finally, the CO_2 was applied in a pulsed manner by continuously opening and closing the pulsing valves throughout the machining process. The opening time was set to $t_{open} = t_{pulse} = 0.1$ s, while the closing time was set to $t_{close} = t_{base} = 0.2$ s. In the first case, the pulse pressure was $p_{pulse} = p_{low} = 60$ bar and there was no base flow. In the second case, the pulse pressure was $p_{pulse} = p_{high} = 200$ bar with a base flow of $p_{base} = p_{low} = 60$ bar. Each factor combination was repeated three times to ensure a statistically relevant result.

In order to allow a consistent assessment of the influence of pulsed cryogenic jet cooling on the chip breakage and removal, the chips produced were examined and classified. The length of the chips is determined by taking them out of the machine, arranging them on a grid and taking a photo, as shown in Fig. 3. Additionally the cutting process is recorded by a GoPro Hero7 Black and the slow motion videos provide information about the moment of chip breakage.

As shown in Table 1 and Fig. 3 five different chip morphology categories were selected. Category 1 are short chips with a length of l < 10 mm. They are considered usable and would rather be expected for brittle materials. Category 2 are medium sized chips with a lengths of 10 mm $\leq$ l $\leq$ 100 mm. As they are unlikely to get entangled around machine components they are of optimum length. Chips with a length of l > 100 mm are described as long and belong to Category 3. Category 4 is for snarled chips. In this case, some of the chips become entangled around the tip of the tool and thereby influence the morphology of all subsequent chips randomly. Both categories 3 and 4 are equally unfavorable. Category 5 is specifically designed for chips that are likely to be generated during pulsing. It is a combination of short chips produced during the pulse phase and long chips produced in between pulses. These long chips produced in between pulses are each of approximately equal length with their length being influenced and adjusted by the time in between pulses [1]. If any factor combination produces a mixture of different chip categories, this will be marked accordingly [10].

4 Results

4.1 EN AW 2007

The chip morphology from machining without CO_2 is taken as reference and compared to the results of machining with CO_2 (Fig. 4). Improvements (+), no changes (o) and deteriorations (−) are marked according to Table 1, with usability "good" and "adjustable" being on the same level.

Table 1. Chip morphology and property categorization [10]

Category	Description	Length	Chip space	Usability
1 (Fig. 3a)	Short	< 10 mm	≥ 3	Usable
2 (Fig. 3b)	Medium	10–100 mm	≥ 25	Good
3 (Fig. 3c)	Long	> 100 mm	≥ 50	Unfavorable
4 (Fig. 3d)	Snarled	–	≥ 90	Unfavorable
5 (Fig. 3e)	Pulsed	Adjustable	Adjustable	Adjustable

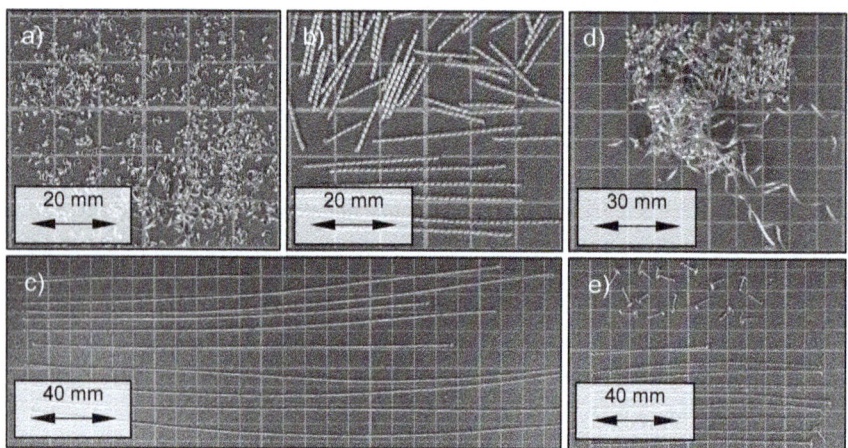

Fig. 3. a) Short chips. b) Medium sized chips. c) Long chips. d) Snarled chips. e) Chips with the typical pulsed pattern

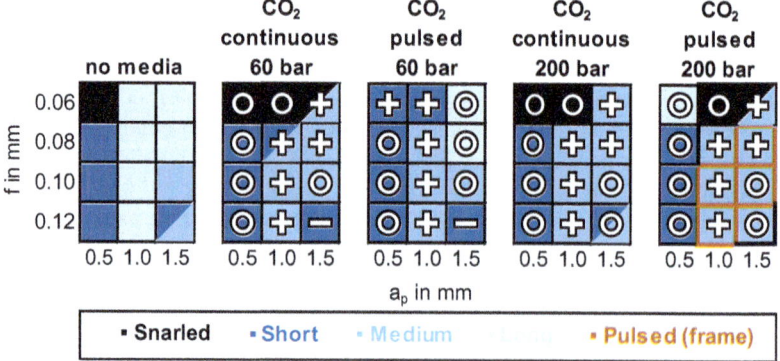

Fig. 4. Influence of the cooling strategy on chip breaking for EN AW 2007

While the cutting depth a_p has a major influence on the chips produced, the influence of the feed rate f appears to be less significant. For the smallest cutting depth of $a_p = 0.5$ mm the chips produced are predominantly short. There are only a few exceptions at low feed rates (f = 0.06 mm). These results are independent of the cooling strategy. At a cutting depth of $a_p = 1.0$ mm the influence of the cooling strategy becomes clear. When machining without coolant, long chips are produced at all feed rates. When CO_2 was used, chips were almost always broken either for continuous or pulsed application and for both pressures. Short chips can be seen to a large extend with continuous and pulsed application, while for HP pulsing the video recordings partially indicate the chip breaking being initiated by the pulse. Again, the only exceptions are at the lowest feed rate. For the highest cutting depth, the results without media application are inconsistent. Lower feed rates result in long chips; higher feed rates result in medium or short chips. With the use of cryogenic cooling, all strategies lead to a shortening of the chips produced. The only exception is noticeable at low feed rates using 60 bar pulsed CO_2.

4.2 EN AW 6082

Similar to EN AW 2007 the influence of the cutting depth is generally higher than that of the feed rate for EN AW 6082 (Fig. 5). Without CO_2 application, the chip length for the lowest a_p shrinks from long to short as the feed rate increases. As soon as CO_2 is applied, only short chips are produced. At $a_p = 1.0$ mm mostly long chips are produced when machining without media. This remains basically the same for CO_2 use with low pressure, regardless of continuous or pulsed flow. When used at high pressure, the cryogen is able to shorten the chips in most cases, producing medium sized chips in the continuous case and the typical chip pattern for pulsed application. At maximum cutting depth of 1.5 mm, machining with no media produces medium sized chips with the exception of $f = 0.06$ mm. For CO_2 application with 200 bar mainly medium sized chips can be observed for continuous flow and chips of Category 5 for pulsed flow. With low pressure CO_2 the chips tend to be longer than without media use.

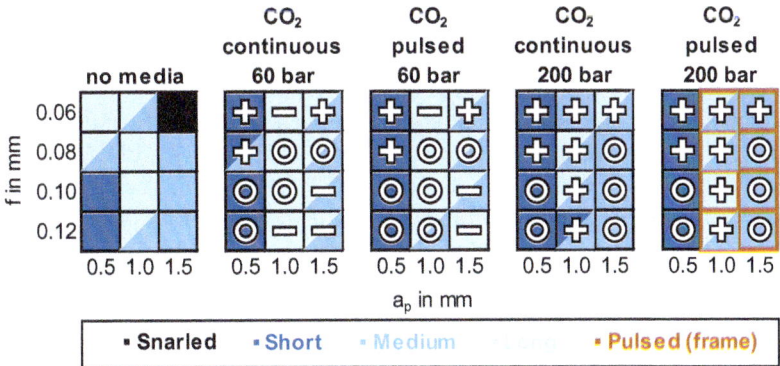

Fig. 5. Influence of the cooling strategy on chip breaking for EN AW 6082

The chip breaking mechanism for pulsed CO_2 application is shown in Fig. 6. This specific case shows the machining of EN AW 6082 with a cutting depth $a_p = 1.0$ mm and a feed rate of $f = 0.10$ mm and a cryogenic jet with $p_{pulse} = 200$ bar and $p_{base} = 60$ bar. Figure 6(a) shows the long outgoing chip as the CO_2 cools the interaction zone with the base flow. The bright flare around the nozzle shows the pulse setting in (Fig. 6b). The impact of the pulse causes the chip to break. After the initial impulse, the pulse weakens, allowing a new chip to emerge. Meanwhile the previous chip falls down (Fig. 6c).

Fig. 6. a) Outgoing chip during base flow. b) Chip breaking by pulse. c) New chip emerging with previous chip falling down

5 Discussion

When comparing the chip breakability of the two selected materials, a noticeable difference can be observed. Whereas for EN AW 2007 the chip breaking capabilities of the cryogen manifest for high and low pressure, for EN AW 6082 it only does so at a pressure of 200 bar. As the pressure is largely responsible for the impulse generated, it can be seen, that for tougher materials a higher impulse and therefore higher pressure is required to initiate the chip breaking. This might also be an indication that chips are not being torn off by the pulse, but are rather being bent beyond their maximum deformability.

As mentioned above, the cutting depth has a larger influence on the chip morphology than the feed rate. Only the snarled chips produced at $f = 0.06$ mm are particularly noteworthy. At the lowest cutting depth, the chips are mainly short. This applies for almost all setups. The influence of the cryogenic cooling on the chip morphology is rather small. Only little improvement can be achieved for EN AW 6082. For the highest cutting depth, mixed results and mainly medium sized chips are observed. While for EN AW 2007 improvements are achieved at lower feed rates, for EN AW 6082 the chip morphology deteriorates with low-pressure CO_2 application. The largest effect of cryogenic cooling is seen for the cutting depth of $a_p = 1.0$ mm. Without media, long chips are always produced for both alloys. Depending on the pressure applied, these chips can almost always be broken, widening the possible process window. In most cases, it does not matter whether the CO_2 is applied continuously or in a pulsed manner.

In the experiments, HP application always resulted in equal or improved chip breakage, whereas low pressure had variable effects depending on the material and cutting parameters. This indicates that with correctly chosen pressure cryogenic jet cooling is capable of improving chip breakage. At the correct pressure setting, both continuous flow and pulsed CO_2 had the same positive effect on chip morphology. This allows for shorter and/or length-adjusted chips that improve productivity and safety while additionally offering a higher level of sustainability compared to conventional MWF.

The pulsed CO_2 application increases the sustainability even further. Comparing the flow rate of 200 bar continuous flow with the pulsed flow ($p_{pulse} = 200$ bar, $t_{pulse} = 0.1$ s, $p_{base} = 60$ bar, $t_{base} = 0.2$ s), 30.1% of CO_2 can be saved. By adapting the pulsing times, even larger savings are possible.

6 Conclusion

Different HP cryogenic jet cooling strategies were applied during a turning operation on two different aluminum alloys. The aim was to positively influence the chip morphology. The results of this study can be summarized as follows:

- In most cases, HP cryogenic jet cooling improves chip breakage, depending on several factors such as material, pressure and cutting parameters.
- This allows for a wider process window, opening up dry machining, with cryogenic cooling, for further machining operations.
- The shorter and/or length-adjusted chips are much less likely to become entangled in machine components or the workpiece. This results in higher productivity, quality and safety, which all together improve sustainability.
- With the right pressure setting, pulsed cryogenic jet cooling delivers the same result as continuous flow cooling, while improving sustainability even further.

Acknowledgements. This article is part of the work of the Nuremberg Campus of Technology, a research cooperation between Friedrich-Alexander-Universität Erlangen-Nürnberg and Technische Hochschule Nürnberg Georg Simon Ohm supported by the state of Bavaria.

References

1. Splettstoesser A, Schraknepper D, Bergs T (2021) Influence of the frequency of pulsating high-pressure cutting fluid jets on the resulting chip length and surface finish. Int J Adv Manuf Technol 117:2185–2196
2. Bergs T, Splettstoesser A, Schraknepper D (2019) Pulsating high-pressure cutting fluid supply for chip control in finish turning of Inconel 718. MM Sci J
3. Biermann D, Abrahams H, Metzger M (2015) Experimental investigation of tool wear and chip formation in cryogenic machining of titanium alloys. Adv Manuf 3:292–299
4. Jerold BD, Kumar MP (2013) The influence of cryogenic coolants in machining of Ti–6Al–4V. ASME J Manuf Sci Eng 135(3):031005
5. Jerold BD, Kumar MP (2012) Experimental comparison of carbon-dioxide and liquid nitrogen cryogenic coolants in turning of AISI 1045 steel. Cryogenics 52(10):569–574
6. Kanyak J, Gharibi A (2018) Progressive tool wear in cryogenic machining: the effect of liquid nitrogen and carbon dioxide. J Manuf Mater Process 2(2):31
7. Hong SY, Ding Y (2001) Micro-temperature manipulation in cryogenic machining of low carbon steel. J Mater Process Technol 116(1):22–30
8. Kloeckner metals. Accessed 18 July 2023. https://facts.kloeckner.de/werkstoffe/aluminium/3-1645/

9. Kloeckner metals. Accessed 18 July 2023. https://facts.kloeckner.de/werkstoffe/aluminium/3-2315/
10. Stahl-Eisen-Prüfblatt 1178-90

Impact of Modified Clamping System on Rotary Swaging Workpiece Surface Quality and Sustainability

Kasra Moazzez[1,3](✉), Lasse Langstädtler[1,2,3], Christian Schenck[1,2,3], and Bernd Kuhfuss[1,2,3]

[1] University of Bremen, Bibliothekstr. 1, 28359 Bremen, Germany
moazzez@bime.de
[2] Center for Materials and Processes - MAPEX, 28334 Bremen, Germany
[3] Bremen Institute for Mechanical Engineering, Badgasteiner Str. 1, 28359 Bremen, Germany

Abstract. Rotary swaging is a sustainable cold forming process that reduces the radius of axis-symmetrical workpieces. The clamping system is a critical part of this process that affects the behavior of the workpiece during the process. The dynamic characteristics of the clamping-workpiece system can affect the process dynamics, such as vibrations in the whole structure, dynamic forces and contact time between the tools and the workpiece, and angular displacement of the workpiece. This research aims to modify the dynamics of the process by changing the contact characteristics between the clamping system and the workpiece. Instead of using a quasi-rigid clamping head, an elastic rubber spacer was tested as a spring-damper element. The study evaluated the impact of this clamping-workpiece system modification on the sustainability of the rotary swaging process in different aspects and surface quality of the final product via roundness and cylindricity measurements. The results could lead to parts that require less post-processing and are closer to a ready-to-use state, depending on the desired application.

Keywords: Sustainability · Modified clamping · surface quality

1 Introduction

Fixtures and clamping systems are vital for ensuring machining stability in manufacturing processes by positioning and stabilizing the workpiece. Consequently, extensive research has been conducted on clamping and fixturing in manufacturing processes from various perspectives. Modifying and adjusting the clamping status can effectively eliminate undesirable phenomena, such as vibrations during the process, depending on the specific process type [1–3]. The known findings from machining processes like turning and milling highlighted the potential of a well-designed clamping system to enhance work quality.

H. Kohl et al. (Eds.): GCSM 2023, LNME, pp. 693–700, 2025.
https://doi.org/10.1007/978-3-031-77429-4_77

It is worth noting that while the majority of research investigations have primarily focused on clamping design and modifications in machining processes to enhance process and workpiece quality [4–7], a few studies have explored similar aspects in forming processes [8, 9]. Consequently, the design of the clamping system is recognized as having significant importance in a wide range of both machining and forming processes.

The aim of this study is to assess how modifications to the stiffness and damping properties of the clamping system in the rotary swaging process affect the quality of the workpiece. The primary modification implemented involves using clamping jaws with a larger diameter than the workpiece and filling the resulting gap with a rubber spacer to enhance damping effects and eliminate direct metal-to-metal contact between the workpiece and clamping jaws (see Fig. 1). Results indicate that this modification improves the surface quality of the workpiece, reduces energy consumption, and mitigates undesired forces during the process. Importantly, the final geometric profile of the workpiece remains almost unchanged regardless of the clamping system used. Consequently, this approach enables the attainment of higher quality final products with reduced energy requirements and lower machine loads, thereby contributing to a more sustainable forming process.

2 Experiment Setups

2.1 Rotary Swaging

Rotary swaging is an incremental cold-forming process that involves reducing the radius of the samples through the repeated strokes of dies on the surface of the workpiece. In the used rotary swaging machine (Felss HU 32V), the dies rotate around the axially fed workpiece and during the closing of the dies, the workpiece is forced to follow the rotation of the swaging head. The process was done using Diamond-Like-Carbon Coated (DLC) dies. Cold drawn E355 steel tubes with an initial length of $l_0 = 300$ mm, initial diameter of $d_0 = 20$ mm, and initial wall thickness of $s_0 = 3$ mm were reduced to a final diameter of $d_1 = 15$ mm. The process parameters were feeding velocity of $v_f = 1000$ mm (feed per stroke of $l_{st} = 0.46$ mm/stroke), rotation speed of the motor of $r_s = 999$ rpm which resulted in a stroke frequency of 35 Hz.

The rubber spacer used to fill the gap between the sample and the clamping jaws was Ethylene Propylene Diene Monomer (EPDM) rubber with a shore hardness A (DIN 53505) of 5–10°, elongation after fracture of > 150% (DIN 53571), and tensile strength of 0.6 MPa (as specified by the manufacturer). The samples number ST15-01, ST15-02, and ST15-03 were formed using the normal clamping system and the modified system were used to swage the samples number ST15-05, ST15-06, ST15-07.

Process data can be obtained by analyzing signals from displacement sensors on the swaging machine. These signals provide valuable information about the feed per stroke (the progress of the workpiece between consecutive strokes), back-pushing reverse motion (the displacement of the feeding system opposite to the feeding direction caused by each stroke), feeding increment (fed length per stroke which is determined by summation of the feed per stroke and the back pushing displacement), and the touching time (the duration of tool-workpiece contact when the dies are closed).

2.2 Roundness and Cylindricity Measurements

Roundness and cylindricity of the swaged samples were measured using the Talyrond 252 from Rank Taylor Hobson, a computer-controlled software-based instrument for precise geometric form measurements. The roundness measurements were taken at 50 mm and 90 mm from the tip, while cylindricity measurements covered a longer path on the surface, starting at 60 mm and 130 mm from the tip (see Fig. 2).

Fig. 1. Normal and Modified clamping status.

3 Results and Discussion

The adjustment of the clamping system's stiffness and damping characteristics directly influences the machining process dynamics. For instance, it affects the angular displacement of the workpiece, where the workpiece rotates with the dies during die closure, and, due to a rubber spacer acting as a rotational spring, it returns to its original position. This alteration in dynamics is crucial for workpiece quality. Notably, there are differences in surface patterns between workpieces produced using conventional and modified clamping methods.

In Fig. 3, we observe a spiral pattern in workpieces produced with rigid clamping, while this pattern is absent in samples produced with the modified clamping system. Furthermore, the typical wavy pattern between the deformed and undeformed regions of the workpiece, seen in normal clamping, is eliminated with the modified clamping.

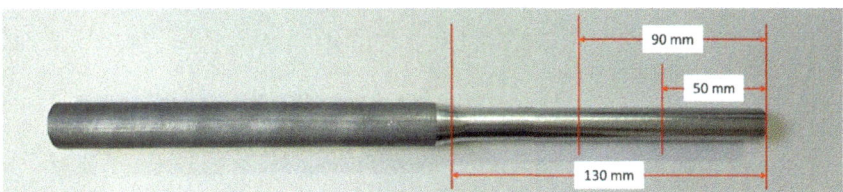

Fig. 2. Cylindricity and roundness measurement points

Figure 4 illustrates the reverse motion of the feeding system for different samples. It should be noted that this displacement occurs in the opposite direction of the feeding, indicated by negative values. The modified clamping results in a lower back-pushing displacement, reducing the load on the machine. In the figures, samples labeled by "Normal" refer to the application of normal clamping system and those labeled by "Modified" refer to the application of modified clamping system during the process.

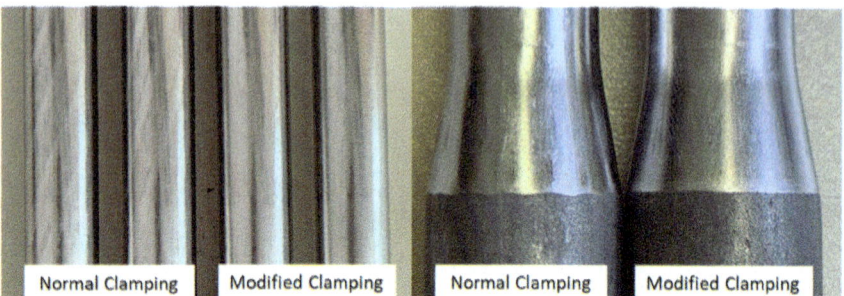

Fig. 3. Samples from the swaging process with different clamping status

Another factor impacting process parameters and workpiece quality is the touching time, as shown in Fig. 5. Despite fluctuations, the average touching time is generally lower in the process with modified clamping compared to the normal clamping system. The reduced touching time results in a shorter overall process duration, which is particularly significant for achieving higher productivity in mass production.

Next, an analysis was undertaken to determine the average feed per stroke for four samples under varying clamping conditions, utilizing data acquired from sensors. Surprisingly, the results remained consistent, even with reduced backward motion in the feeding system. This means that by employing the modified clamping system, we can maintain a constant feed per stroke at around 0.468 mm, leading to reduced energy consumption due to the decreased backward force.

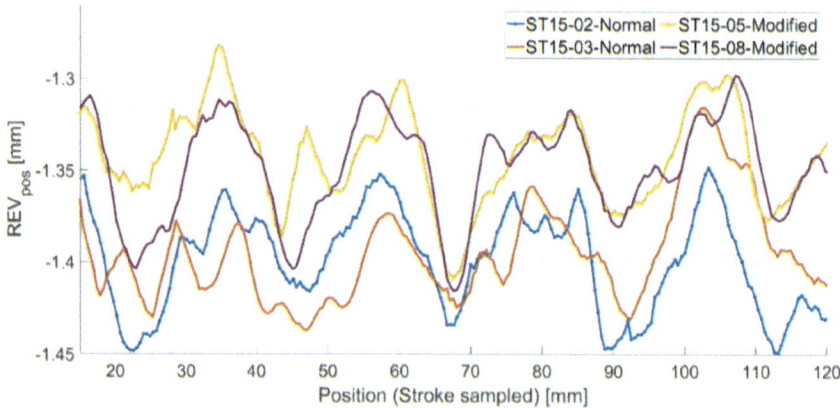

Fig. 4. Back-pushing reverse motion of the feeding system in oppose to the feeding direction

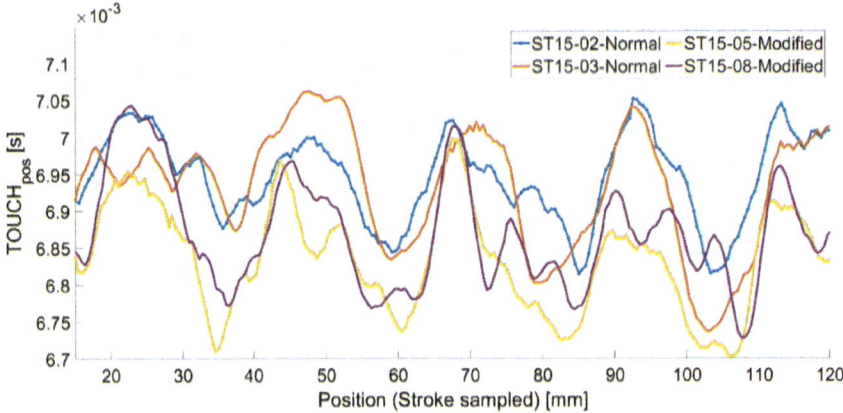

Fig. 5. Touching time during the process

Figure 6 provides a visual representation of the feeding increment across the four samples subjected to diverse clamping conditions. Notably, rigid clamping cases exhibit a more significant feeding increment, primarily attributed to the increased backward displacement. Consequently, despite maintaining consistent feeding velocity and feed per stroke, rigid clamping leads to a higher volume of material introduced into the machine during each stroke. This, in turn, results in elevated energy consumption and a decline in the quality of the final products.

The benefits of clamping system modification become evident when assessing workpiece properties, particularly surface quality. As seen in Fig. 7, a distinct contrast in surface patterns is apparent between samples produced using the modified clamping system and their conventionally clamped counterparts. The modified system yields a smoother and more uniform surface, free from the prominent marks typically caused by forming tools.

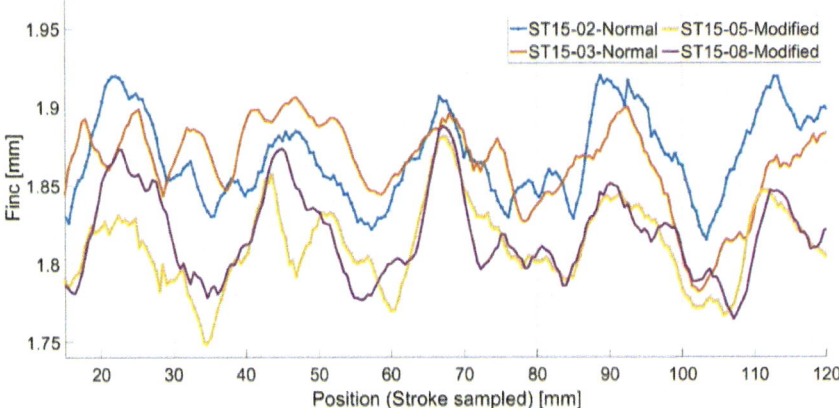

Fig. 6. Feeding increment during the process

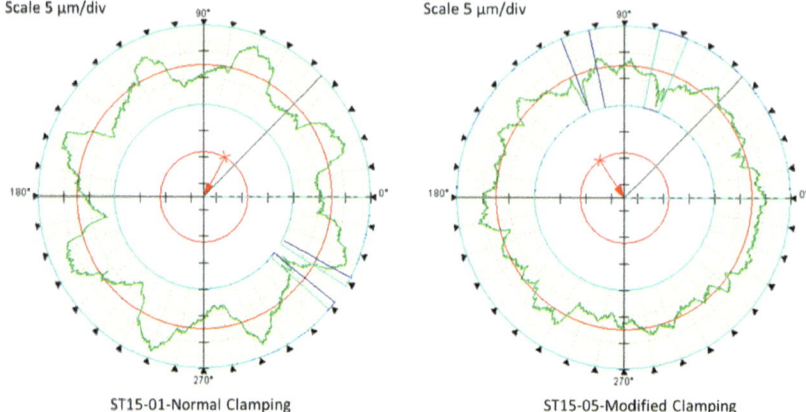

Fig. 7. Roundness data measured at 90 mm from the tip on ST15-01 and ST15-05

Figure 8 presents the total roundness value (RONt), indicating the overall deviation from the mean circle at 90 mm and 50 mm from the tip. Additionally, Fig. 9 displays the total cylindricity value (CYLt) measured at 60 mm and 130 mm from the tip, covering a 10 mm length on the workpiece surface. The clamping system modification leads to reduced error values in both CYLt and RONt, indicating an improved surface quality.

This improvement can be attributed to several factors. Firstly, the rubber spacer's function as a rotational spring ensures that the workpiece consistently returns to its initial orientation after each stroke, maintaining a relatively constant stroke following angle of approximately 51°. Stroke following angle is the rotational displacement between the wokpiece and the dies between each stroke. This angle minimizes the appearance of the prominent structure observed in Fig. 7 (left).

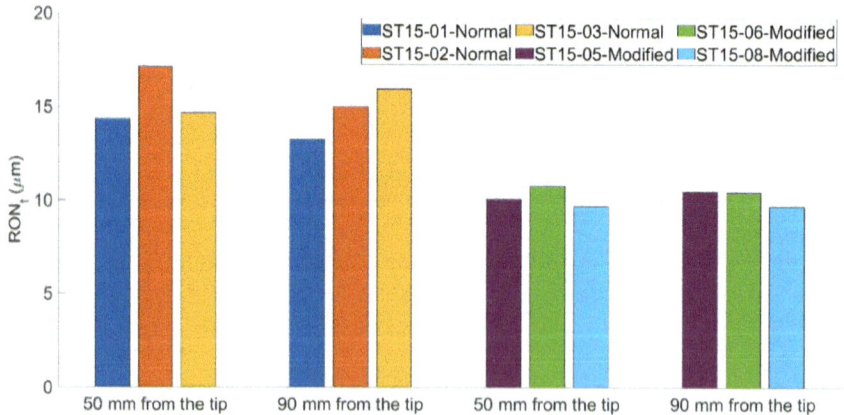

Fig. 8. RONt for different clamping status at 90 mm and 50 mm from the tip

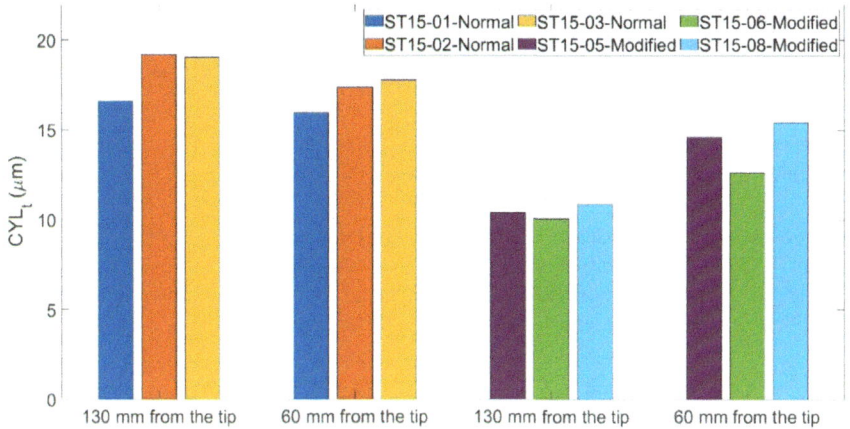

Fig. 9. CYLt for different clamping status at 60 mm and 130 mm from the tip

Secondly, the spacer serves as a planar spring oriented perpendicular to the feeding direction, mitigating forces generated between the tools and the workpiece due to any misalignment between the clamping device's centerline and the swaging head caused by machine vibrations.

4   Conclusion and Outlook

Based on the obtained results, the application of the modified clamping system in rotary swaging offers significant sustainability advantages as below:

- **Efficient Energy Utilization**: The feed per stroke remains constant, yet the reduction in reverse back-pushing motion leads to lower fed material volume per stroke and reduced forming forces. This translates to less energy consumption to form the samples while maintaining the same geometrical properties, ultimately enhancing sustainability and efficiency.
- **Enhanced Productivity**: Shortened touching times in the modified clamping process reduce overall process duration. This particularly benefits mass production scenarios by improving productivity and overall resource efficiency.
- **Improved Workpiece Quality**: The modified clamping system enhances workpiece surface quality by reducing prominent patterns on its surface, and ensuring consistent roundness and cylindricity properties by maintaining a constant stroke following angle. This reduces or eliminates the need for post-processing and product finishing, depending on the workpiece's intended application.
- **Machine Longevity**: The planar spring function of the rubber spacer reduces forces arising from misalignment between the clamping device and the swaging head due to machine vibrations. This not only ensures stable machining processes but also extends the lifespan of machine tools, reducing the environmental impact associated with machinery replacement.

In conclusion, the process with modified clamping system not only conserves energy, improves productivity, enhances product quality, and extends machine longevity but also contributes to the overall sustainability of the rotary swaging process. These encouraging outcomes underscore our overarching objective to develop an industrial-grade clamping system with similar attributes suitable for large-scale and series production, further advancing sustainable manufacturing practices.

Acknowledgement. This research is partially funded by the Deutsche Forschungsgemeinschaft (German Research Foundation)–374789876 within the sub-project (EP 128/5-2 and KU 1389/16-2) "Control of component properties in rotary swaging process" of the priority program SPP 2013 "The utilization of residual stresses induced by metal forming."

References

1. X Dong W DeVries M Wozny 1991 Feature-based reasoning in fixture design CIRP Ann 40 1 111 114
2. C Costa FJG Silva RM Gouveia RP Martinho 2018 Development of hydraulic clamping tools for the machining of complex shape mechanical components Proc Manuf 17 563 570
3. A Al-Habaibeh N Gindy RM Parkin 2003 Experimental design and investigation of a pin-type reconfigurable clamping system for manufacturing aerospace components Proc Inst Mech Eng Part B J Eng Manuf 217 12 1771 1777
4. J Li B Li N Shen H Qian Z Guo 2017 Effect of the cutter path and the workpiece clamping position on the stability of the robotic milling system Int J Adv Manuf Technol 89 2919 2933
5. J Munoa M Sanz-Calle Z Dombovari A Iglesias J Pena-Barrio G Stepan 2020 Tuneable clamping table for chatter avoidance in thin-walled part milling CIRP Ann 69 1 313 316
6. HT Sanchez M Estrems F Faura 2006 Fixturing analysis methods for calculating the contact load distribution and the valid clamping regions in machining processes Int J Adv Manuf Technol 29 426 435
7. JH Yeh FW Liou 1999 Contact condition modelling for machining fixture setup processes Int J Mach Tools Manuf 39 5 787 803
8. Birnbaum AJ, Cheng P, Yao YL (2007) Effects of clamping on the laser forming process, pp 1035–1044
9. Nirala HK, Agrawal A (2017) Design and fabrication of electromagnetic fixture for incremental sheet metal forming

Hybrid Polishing Tool for Internal Finishing of Complex Part Interiors in Magnetic Abrasive Finishing

Adriel Magalhães Souza[1]([✉]), Justin Rietberg[2], Eraldo Jannone da Silva[1], and Hitomi Yamaguchi[2]

[1] São Carlos School of Engineering, University of São Paulo, São Carlos, SP, Brazil
adrielmagalhaes2@gmail.com
[2] Herbert Wertheim College of Engineering, University of Florida, Gainesville, FL, USA

Abstract. This paper describes the development of a new hybrid polishing tool, which consists of magnetic particles and a water-soluble binder and transforms from a bonded-tool phase to a particle-brush phase during polishing. During the bonded-tool phase, the binder remains intact, and material is removed from the peaks of the target surface where abrasive is pressed against the surface. Once the binder dissolves in the lubricant, the magnetic particles separate, form a brush tool (particle-brush phase), and polish the surface while conforming to the surface geometry. The phase transition timing is determined by the lubricant type. The magnetic particles can be collected using magnetic force and reused; the hybrid tool thus contributes to resource utilization and waste reduction. The effects of tool geometry, tool binder content, and lubricant type on the tool-phase transformation and polishing characteristics are clarified by applying the new hybrid tool to internal finishing of stainless steel tubes, made using directed energy deposition, using magnetic abrasive finishing (MAF). MAF mechanically removes material by the action of abrasives pressed against a workpiece with magnetic tools suspended in a magnetic field. Compared with conventional MAF, the hybrid tool achieves a target surface finish with significantly less material removal, minimizing waste production.

Keywords: Magnetic Abrasive Finishing · Directed Energy Deposition · Reusable Polishing Tool

1 Introduction

Directed energy deposition (DED) is an additive manufacturing (AM) method for construction of metallic parts, which combines material by supplying energy layer-by-layer and consequently forms the part [1]. In general, components made using this method have low dimensional accuracy and poor surface quality, creating a need for post-processing [2].

© The Author(s) 2025
H. Kohl et al. (Eds.): GCSM 2023, LNME, pp. 701–709, 2025.
https://doi.org/10.1007/978-3-031-77429-4_78

Post-processing of AM parts has been done using chemical and hybrid methods such as electropolishing [3], chemopolishing [4], and chemical-abrasive flow polishing [5], however these processes generally create environmentally hazardous waste products. Magnetic abrasive finishing (MAF) is a mechanical process that been used for finishing AM components. MAF mechanically removes material by the action of abrasives pressed against the workpiece by magnetic tools (e.g., particles) suspended in a magnetic field [6]. Iron particles are commonly used as magnetic tools. Although the flexible motion of the magnetic particle brush is an advantage of MAF, it is sometimes advantageous to bond the magnetic tools to form a desired shape and construct a hybrid tool that transforms from an initial bonded-tool phase to a particle-brush phase during polishing. Figure 1 illustrates a schematic representation of the internal polishing process for cylindrical samples, showing the effect of the hybrid tool.

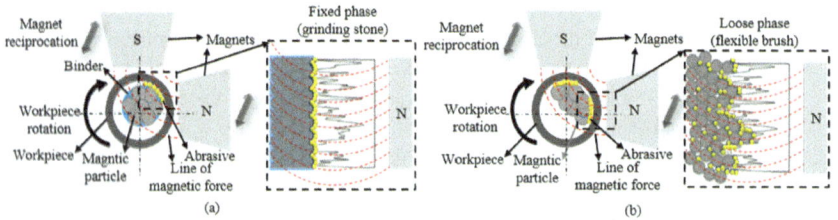

Fig. 1. Schematic of processing principle: (a) fixed phase and (b) loose phase.

During the bonded-tool phase, the binder in the hybrid tool remains intact, and material is removed from the peaks of the workpiece surface where the abrasive is mechanically pressed against it by the hybrid tool. Once the binder is dissolved, the magnetic particles separate and form a magnetic-particle brush (called the particle-brush phase). The particle brush causes the abrasive to remove material along the peaks and valleys of the surface, smoothing the surface [7, 8]. The lubricant type determines the phase-transition timing. The magnetic particles can be collected using magnetic force and reused; the hybrid tool thus contributes to resource utilization and waste reduction.

Earlier in this context, hybrid tools, which conform to the workpiece surface, were made using iron particles or magnetic abrasive with water-soluble binder. The feasibility of the tools was demonstrated for surface finishing of ceramic plates [7] and stainless steel tube interiors [8]. However, making hybrid tools to match the workpieces geometry can be tedious, can only address specific applications, but cannot accommodate different surface geometries. Moreover, it is not practical to make hybrid tools that conform to complex geometry made using DED.

This paper proposes a new form of hybrid tool—a ball-shaped hybrid tool—which is applicable to various geometries, including complex geometries made using DED. The development of the new hybrid tool is described, and the effects of tool geometry, tool binder content, and lubricant type on the tool-phase transformation and polishing characteristics are examined. The feasibility of the new hybrid tool was clarified for internal finishing of stainless steel tubes made using DED.

2 Methodology

2.1 Hybrid-Tool Fabrication

Two kinds of hybrid tools, semicircular and ball-shaped hybrid tools, were created. Figure 2 (a) shows the semicircular hybrid-tool fabrication process. An acrylic tube (Ø21.4 × Ø 18.2 × 92.3 mm) was cut in half (semicircular). This semicircular cutout was used as a mold, and three layers of tape (thickness 0.25 mm) were adhered to the tube internal surface, resulting in a mold with 16.7 mm diameter.

Magnetic particles (iron particles or steel grit) were mixed with binder (water-soluble polyvinyl acetate), and the mixture was cured in the semicircular mold. Table 1 shows conditions to fabricate hybrid tools with semicircular hybrid tools. Since the semicircular hybrid tools did not exactly match the workpiece geometry, a layer of magnetic particles was adhered to the tool surface using lubricant (see Fig. 2 (a)). When the hybrid tool is inserted into the tube, the loose particles relocate following the lines of magnetic force, generating a short particle brush. This helps the hybrid tool conform to the workpiece geometry and generate smooth relative motion against the workpiece surface.

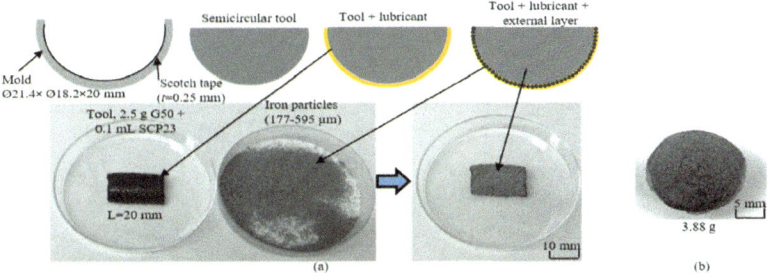

Fig. 2. (a) Semicircular and (b) ball-shaped hybrid tools.

Table 1. Specifications of the hybrid tools

Tool Code	HT1	HT2	HT3	HT4	HT5	HT6	B1/B2
Format	Semicircular						Spherical
Magnetic particles	177–595 µm iron particles: 1.3 g			177–595 µm iron particles: 2.5 g			177–595 µm iron particles: 2 g
Binder	0.26 g	0.65 g	1.3 g	1.25 g	1 g	1 g	1 g
Curing time	24 h				48 h		

For ball-shaped hybrid tools, a semispherical epoxy mold was made to create spherical halves. Two of them were bonded together with cyanoacrylate adhesive to form a ball-shaped hybrid tool (ellipsoid, Ø 9 × Ø 14 mm, Fig. 2 (b)). Tool B1 included a layer of magnetic particles (a mixture of 0.75 g of G25 steel grit and 0.5 g of 177–595 µm iron particles) adhered to the tool surface using lubricant. Tool B2 did not include this external layer.

2.2 Polishing Experiments

The effect of the size ratio between the binder and magnetic particles on the tool structures and their performance were tested using a machine developed for MAF. Experiments were conducted to polish commercially available 316 stainless steel tubes (Ø 19.1 × Ø 17.3 × 112.5 mm) and DED-produced 316L stainless steel tubes (Ø 20 × Ø 18 × 97 mm). The clearance between the magnets and 316 stainless steel tubes and DED-316L stainless steel tubes were 1.5 mm and 0.7 mm, respectively. The magnetic flux densities in those clearances were 363 mT and 451 mT, respectively.

Table 2 shows the experimental conditions for polishing 316 stainless steel tubes. Finishing time was 10 min, using a semicircular-hybrid tool (Ø13.2 mm × 20 mm long) with an external layer of 0.3 g of iron particles (177–595 μm). HT1–5 used linear magnet motion (1 mm/s rate, 10 mm feed), while HT6 used oscillation (2.5 mm amplitude, 1 Hz frequency) instead.

Table 2. Experimental conditions for polishing 316 SS using hybrid tools.

Tool Code	HT1	HT2	HT3	HT4	HT5		HT6
Abrasive	0.6 g, white alumina (#2500)					0.6 g diamond abrasive (D126)	
Lubricant	1 mL WBC* + 1 mL DI water	1.5 mL WBC + 0.5 mL DI water	2 mL WBC	2 mL Cutting oil	2 mL WBC		
Workpiece rotation	1500 min^{-1}						

*WBC: Water-soluble barreling compound

Prior to polishing using the hybrid tools, the DED-316L tubes were polished using a mixture of 2.5 g of G25 and 0.6 g of uncoated diamond (D 126) as abrasive to test baseline characteristics of the DED-316L stainless steel tubes. Table 3 shows the experimental conditions. 2 mL of water-soluble barreling compound was used as the lubricant (2 mL added every 10 min). Conditions 1 and 4 used linear magnet motion, while Conditions 2 and 3 used oscillation.

Table 4 shows polishing conditions for DED-316L stainless steel tubes using ball shaped hybrid tools. The finishing time was 30 min using linear magnet motion (1 mm/s rate, 10 mm feed).

Before and after each experiment, the workpieces were ultrasonically cleaned in ethanol for 10 min to ensure accurate measurement of material removal and prepare the surface for surface profilometry. Surface roughness was measured using a stylus-roughness tester (Mitutoyo SJ-400). Four parallel profile lines along the tube length (with a 90° gap between each) were evaluated and averaged. Material removal was obtained by measuring the difference in mass of the tubes before and after finishing using a microbalance (resolution 10 μg).

Table 3. Experimental conditions for polishing DED-316L tubes with loose particles.

Condition	#1	#2	#3	#4
Workpiece magnetic state	Non-magnetic	Magnetic	Magnetic	Non-magnetic
Magnet motion	1 mm/s feed rate, 10 mm feed	2.5 mm amplitude, 1 Hz frequency		1 mm/s feed rate, 10 mm feed
Workpiece rotation	1500 min^{-1}			

Table 4. Experimental conditions for polishing DED-316L using ball-shaped hybrid tools.

Tool Code	B1	B2
External layer	Yes	No
Abrasive	0.6 g diamond abrasive (D126)	
Lubricant	2 mL cutting oil	
Slurry addition	0.2 g abrasive + 2 mL lubricant every 10 min	
Workpiece rotation	1500 min^{-1}	

3 Results

3.1 Polishing of 316 Stainless Steel Tubes

Fundamental characteristics of the developed hybrid tools were firstly examined by polishing 316 stainless steel tubes. Figure 3 shows the results of polishing tests (under conditions reported in Table 2) using the semicircular hybrid tool with an external layer of loose particles. The roughness limit was Rz $\approx$ 0.6 μm. By changing the amount of lubricant and the ratio between binder and magnetic particles, the material removal increased from HT1 to HT2, and the hybrid tool was completely dissolved for those conditions. Since the phase-transition behavior of a hybrid tool is a function of the binder content, the lubricant was changed from water-soluble barreling compound to heavy duty metalworking fluid. For HT3, the surface roughness was improved, reaching the limit (see above) after polishing for 10 min. Also, the use of oil as a lubricant enabled the binder to remain intact after polishing. Thus, it was enough for the tool to remain throughout the process. As material was only removed from the peaks, low material removal occurred.

3.2 Polishing of DED-316L Stainless Steel Tubes

Although the developed hybrid tool with semicircular shape and an external layer of loose particles exhibited smooth rotation during polishing of 316 stainless steel tubes, irregular jumbling occurred with DED-316L stainless steel tubes. The irregular behavior of the tool during polishing of DED samples could be caused by geometric and dimensional

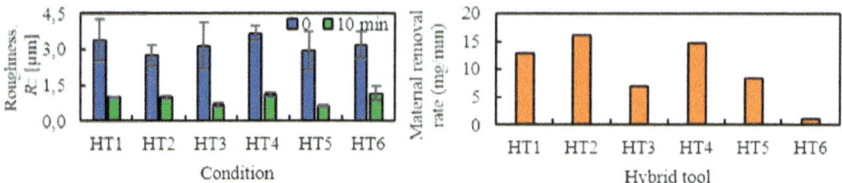

Fig. 3. Roughness and material removal rate after several polishing conditions using semicircular-shaped hybrid tools.

errors. Thus, semicircular hybrid tools could not be used to polish parts fabricated with DED. To overcome this limitation, ball-shaped hybrid tools were used. Showing the smooth rotation, polishing experiments were carried out to validate the efficiency of ball-shaped hybrid tools.

Prior to running the polishing experiments using the ball-shaped hybrid tools for the DED-316L tubes, polishing characteristics of the DED-316L tubes were tested using conventional MAF (i.e., using a mixture of loose magnetic particles and abrasive slurry) under the conditions shown in Table 3.

Figure 4 shows the results of polishing tests using DED-316L tubes. The initial roughness was reduced from over 160 µm to around 3 µm Rz (the limitation of the polishing conditions) in 60 min of polishing time. The material removal rate between Conditions #1 and #2 with the non-magnetic tube and Condition #3 with the magnetic tube were similar, with the magnetic tube having a slightly higher roughness. For Condition #1 followed by #2, the material removal was 1805 mg after polishing for 60 min. Using Condition #3, the material removal was 1501 mg after the same processing time.

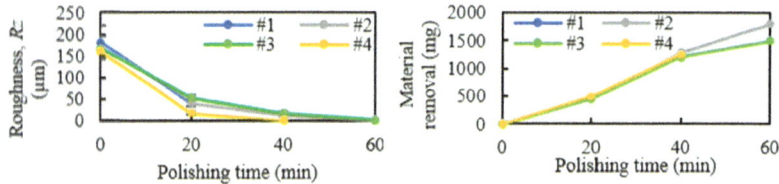

Fig. 4. Roughness and material removal after several polishing conditions using loose particles to polish DED-316L stainless steel tube.

Further investigation needs to be done to verify the saturation of magnetization and phase transformation during DED of 316L stainless steel tubes. However, these results demonstrated that MAF enables polishing of DED-316 tubes regardless of their magnetic properties. Based on these results, the ball-shaped hybrid tools were applied for polishing both 316 stainless steel tubes and DED-316 tubes.

Figure 5 shows the experimental setup and the results of polishing 316 stainless steel tubes using ball-shaped hybrid tools. B1 did not improve the surface quality, and the material removal was low. This resulted from the lack of smooth relative motion of the ball-shaped hybrid tool with loose particles against the tube inner surface. B2 (ball-shaped hybrid tool with no external layer) improved the surface roughness with more material removal. Thus, B2 was chosen for polishing DED-316l tubes.

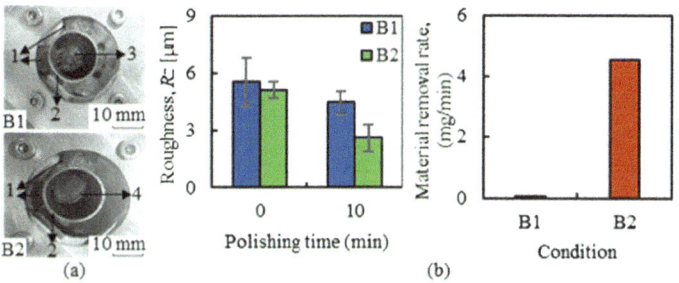

(a) (b)

Fig. 5. (a) Experimental setup for polishing with balls (1 = magnets, 2 = workpiece, 3 = ball-shaped hybrid tool with external layer, and 4 = hybrid tool without external layer); (b) polishing characteristics (roughness and material removal rate).

Figure 6 shows the roughness Rz and material removal over processing time while polishing DED-316L tubes with a ball-shaped hybrid tool. Two trials were carried out, referred as #1 and #2 in Fig. 6. The initial roughness was reduced from over 160 μm to around 3 μm R_z. Thus, the ball-shaped hybrid tool achieved the same roughness level compared to polishing with loose particles. However, 300–360 min of polishing time was required for the roughness reduction, 5–6 times longer than polishing with loose particles (see Fig. 4). Although the roughness limit was reached in only 40 min of polishing time for loose particles (Condition #4 in Fig. 4), the material removal was 2.2 times greater than using a ball-shaped hybrid tool. Considering that the goal of minimizing material removal is essential to reducing material waste for near-net-shape components produced with additive manufacturing, the use of the new hybrid polishing tools is a viable alternative for post-processing DED components. The hybrid tool can be reused in multiple applications, reducing the amount of iron-particle waste. Thus, the use of hybrid tools can be an alternative for resource utilization and waste reduction. Developing strategies to maximize the hybrid tool characteristics is a necessary next step. The effect of hybrid tool transition timing (from the fixed phase to the loose phase) on the material removal and surface roughness and on the total processing time required to meet a desired material removal and surface roughness will be studied in a future research work. In addition, a life-cycle assessment to quantify the overall sustainability of the proposed hybrid tool will be also studied in the future with a comprehensive cost-benefit analysis to assess the economic feasibility of the hybrid tool.

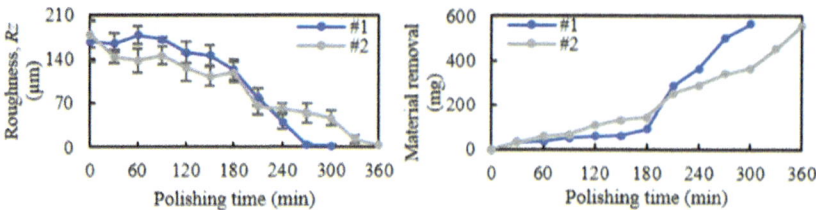

Fig. 6. Roughness and material removal after polishing of DED-316L SS with ball-shaped hybrid tool.

4   Conclusions

A new hybrid polishing tool, made by binding magnetic particles with glue, was developed with the ability to transition from a bonded-tool phase to a particle-brush phase during polishing. By adjusting the phase transition timing, the developed hybrid polishing tool can adapt to various polishing requirements, such as achieving desired surface finishes with specific material removal. When polishing DED-316L stainless steel tube, the hybrid polishing tool was able to achieve roughness values comparable to the loose-particle MAF method but with less material removal. This aligns with the objective of reducing material waste. In addition, the magnetic particles for the hybrid polishing tools can be collected after use and reused. Accordingly, the developed ball-shaped hybrid polishing tools are versatile and contribute to sustainable manufacturing by optimizing material utilization and reducing waste generation. Investigating ways to characterize the hybrid tool performance (e.g., effect of the hybrid tool transition timing from the fixed phase to the loose phase on the material removal and surface roughness), a life-cycle assessment of the hybrid tool, and a comprehensive cost-benefit analysis will be a future task in this study.

Acknowledgments. The authors acknowledge the São Paulo Research Foundation (FAPESP) (Grant numbers 2016/11309-0 and 2019/10758-4) for their financial support.

References

1. Thompson SM, Bian L, Shamsaei N, Yadollahi A (2015) An overview of direct laser deposition for additive manufacturing; Part I: transport phenomena, modeling and diagnostics. Additive Manuf 8:36–62
2. Maleki E, Bagherifard S, Bandini M, Guagliano M (2021) Surface post-treatments for metal additive manufacturing: progress, challenges, and opportunities. Additive Manuf 37
3. Chaghazardi Z, Wüthrich R (2022) Review—electropolishing of additive manufactured metal parts. J Electrochem Soc 169
4. B Wang J Castellana SN Melkote 2021 A hybrid post-processing method for improving the surface quality of additively manufactured metal parts CIRP Ann 70 1 175 178
5. N Mohammadian S Turenne V Brailovsky 2018 Surface finish control of additively-manufactured Inconel 625 components using combined chemical-abrasive flow machining J Mater Process Technol 252 728 738

6. T Shinmura K Takazawa E Hatano M Matsunaga T Matsuo 1990 Study on magnetic abrasive finishing CIRP Ann Manuf Technol 39 325 328
7. Stein M, Yamaguchi H (2017) Effect of binder content on hybrid magnetic tool behavior. In: Proceedings of the ASME 2017 international manufacturing science and engineering conference, Los Angeles
8. H Yamaguchi V Nteziyaremye M Stein W Li 2015 Hybrid tool with both fixed-abrasive and loose-abrasive phases CIRP Ann Manuf Technol 64 337 340

Comparison of an Additive with a Subtractive Method from the Perspective of Sustainability

Leopoldo De Bernardez[✉], Cristian Sandre, and Juan Sanguinetti

Instituto Tecnológico de Buenos Aires, Iguazú 341, C1437 Buenos Aires, Argentina
ldb@itba.edu.ar

Abstract. Additive Manufacturing (AM) is increasingly used for the manufacture of parts in different industrial sectors and, therefore, it becomes relevant to evaluate the mechanical performance that can be achieved with this process and the possible impacts on the environment compared to traditional processes. In this paper, two alternative production methods for fabricating a standard stainless steel tensile specimen are compared: Selective Laser Melting (SLM) and machining. Functional tests were carried out until the fracture of the pieces. The amount of material used was measured. In addition, the energy consumed to produce the pieces was estimated. Both production processes were compared concerning the measured tensile properties, material consumption, and additional data from the literature. The results obtained from generative design and topological optimization of a part are discussed, as well as the implications regarding the sustainability of the processes.

Keywords: Selective Laser Melting · Machining · Generative Design · Sustainability

1 Introduction

Additive manufacturing (AM), also known as 3D printing, has gained significant popularity and widespread adoption in various industry sectors due to its advantages in terms of design freedom, rapid prototyping, and customization capabilities. The advent of AM has revolutionized the manufacturing landscape by enabling the creation of high-precision complex geometries directly from digital models and reducing material waste. SLM uses a high-powered laser to selectively melt and fuse metallic powders layer by layer, resulting in the creation of complex three-dimensional (3D) structures. The powder material surrounding the rapidly moving laser spot undergoes selective melting due to radiation. Once the laser moves away, the melt pool solidifies rapidly and forms strong bonds with the layers beneath. Throughout the process, a controlled laminar gas flow is directed over the build platform to efficiently remove process waste. After completing one laser scan, the build platform is lowered and a new layer of powder is applied with a coater. This entire sequence of steps is repeated until the final part is completely manufactured.

H. Kohl et al. (Eds.): GCSM 2023, LNME, pp. 710–718, 2025.
https://doi.org/10.1007/978-3-031-77429-4_79

SLM enables the production of parts with excellent mechanical properties and intricate geometries, making it a versatile technique for various industries, including aerospace, automotive, and medical. However, as with any manufacturing process, it is crucial to evaluate the mechanical properties of AM-manufactured parts to ensure their suitability for specific applications. The printing conditions can affect the mechanical properties. Marattukalam et al. [1] studied the effect of laser scanning strategies on the mechanical properties of 316L SS, while Carassus et al. [2] discussed the specimen thickness and printing orientation on the dynamic response, Lee et al. [3] the effect of laser speed on mechanical properties and Wang [4] the microstructure and anisotropy obtained for the same material. Likewise, the effect of heat treatment after printing on the microstructure, and mechanical properties has also been studied [5]. On the other hand, it is essential to understand the potential environmental impacts of AM to make informed decisions for its adoption in the industry. Significant aspects related to sustainability in the additive manufacturing of 316L were previously considered, including consumption of materials and energy, consumables, and post-processes [6].

To reduce the amount of material used to build a piece or part, computer algorithms specifically developed to find optimised solutions can be used. Generative design (GD) is an innovative approach that enables the creation of optimized and highly efficient designs. Additionally, Topology optimization (TO) makes it possible to reduce the volume of previous designs, for instance, those that were developed to be used in the machining process, based on the stresses the piece has to resist when in use [7]. In both cases, GD and TO, it is advisable to continue refining the results so that the surfaces and shapes obtained are consistent with the application or use of the part or product.

This work contributes to the discussion of which cases the use of a particular 3D printing technology is appropriate compared to traditional machining processes, highlighting the much higher energy consumption but lower material requirements in the SLM printing process in comparison with machining. The mechanical performance and sustainability aspects of manufacturing stainless steel parts with different production processes are compared. The potential advantages of SLM over machining, especially when using GD and TO techniques are analysed.

2   Methods

The samples used for the tensile tests were manufactured according to the dimensions defined in the ASTM E8 standard. The dimensions of the specimens are presented in Fig. 1.

An EOS GmbH M290 machine was used to produce the samples by SLM. The laser power of the machine is 400W. For printing, a layer thickness of 20 microns was used. Three specimens were printed using three different scanning strategies, as shown in Fig. 2. In total, nine specimens were produced for testing. One of the scanning strategies followed the Z axis, which corresponds to the tensile the directions of the tensile test, the second strategy followed the Y axis, perpendicular to Z, and in the third, each layer was built by rotating the scan angle 67° related to that previously deposited [1].

Additionally, six specimens were obtained from a 5 mm thick stainless steel plate produced by rolling. First, a wire cut, following the approximate contour of the specimen,

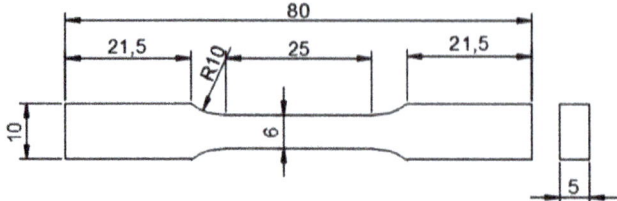

Fig. 1. Tensile test specimen according to ASTM E8 Standard. Dimensions are in mm.

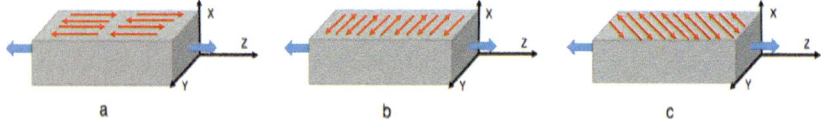

Fig. 2. Scanning strategies: a) along Z direction, b) along Y direction, c) rot-scan strategy

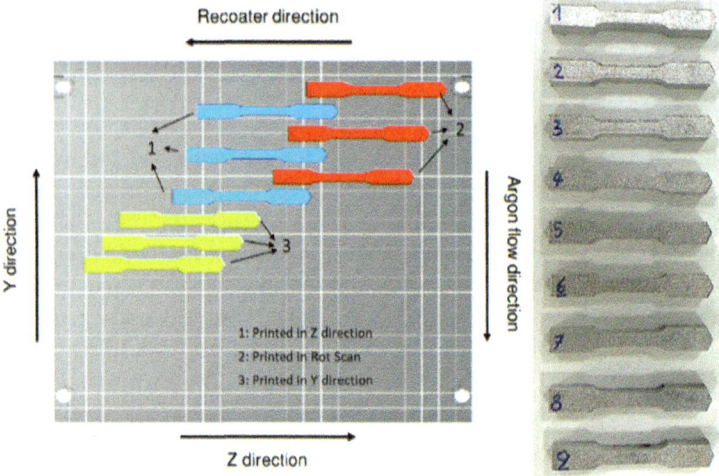

Fig. 3. Left: Specimens location in the chamber platform, recoating direction, and argon flow during printing. Yellow specimens are printed in the Y direction, blue specimens in the Z direction, and red specimens are printed following the rot-scan strategy. Right: Specimens printed with different scanning strategies: 1–3 correspond to Z scan, 4–6 Y scan, 7–9 to Rot-scan. All of them show the typical surface finish of the SLM process.

was made, and then milling was used to bring each specimen to the dimensions indicated in Fig. 1. In three of the specimens, the tensile direction coincided with that of the rolling of the plate, and in the other three the tensile direction was transversal to that of the plate rolling.

3   Results and Discussion

Physical properties. The samples were tested in accordance with ASTM E8 using a universal testing machine Instron 3382. Figure 4 shows the tensile curves of the test pieces manufactured by SLM for the three scanning strategies. The curves vary depending on the strategy used but are very similar within each strategy. Table 1 shows the Yield Strength, Ultimate Tensile Strength, and elongation values for each of the tested samples.

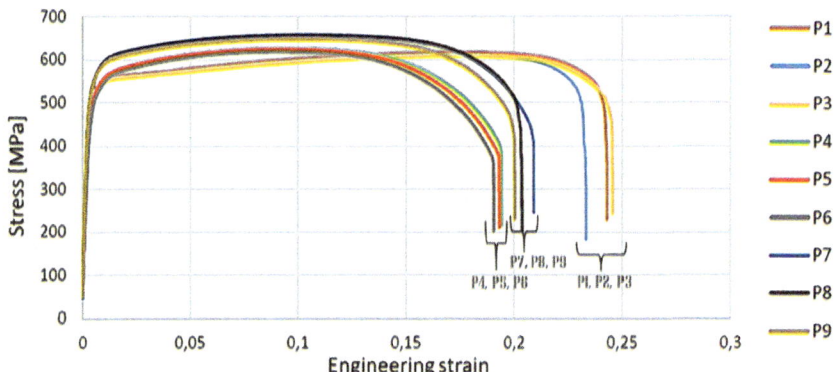

Fig. 4. Tensile test results for the samples obtained through the SLM process. Each specimen is identified in the figure with a P before the corresponding number defined in Fig. 3.

From Table 1 it can be seen that the Yield Strength is higher when the test piece is produced by the SLM process compared to rolling. This difference can be attributed to the fine-grained microstructure formed during the SLM process, resulting in enhanced grain boundary strengthening and improved mechanical properties [8]. The results show that the SLM process does not produce significant anisotropy in the material for different printing directions, corresponding to the different scanning strategies.

Consumed energy. The six specimens produced by SLM were obtained in the same run. For this reason, it was not possible to measure the print time for each scanning strategy. Instead, the print time for each of them was calculated using the EOS machine software. The values were quite similar, being 6050 s and 1.7% the time average and the relative standard deviation respectively for printing each specimen.

An estimation of the energy consumption was made from the analysis of previously reported data [6] for the same machine, although for a different material, in that case, Maraging Steel MS1. Table 2 summarizes the reported values.

The peripheral steps included in Table 2 refer to required processes other than printing, such as cleaning and sandblasting, as reported in reference 6. The consumed energy largely depends on the printing time, which in turn would depend on the scanning strategy and the material. The energy required when using SS316L can be estimated from the printing times for this material. Calculations were performed using the EOS printer software and it was found that the print times for the scanning strategies used in this work only differed between them by ±1.5%. To consider a case closer to real situations, the time to print 28 specimens was estimated, instead of the 9 printed for testing. These

Table 1. Physical properties of specimens obtained using SLM and machining.

SLM specimens				Machined specimens			
Specimen	Ys [MPa]	UTS [MPa]	A [%]	Specimen	Ys [MPa]	UTS [MPa]	A [%]
P1	486	616	46	L1	314.7	646	74
P2	486	611	46	L2	318.9	644	68
P3	491	609	46	L3	312.7	645	69
Average	487.7	612.0	46.0	Average	315.4	645.0	70.3
Std. Dev	2.9	3.6	0.0	Std. Dev	3.2	1.0	3.2
P4	494	626	37	T1	293	646	67
P5	505	624	37	T2	300	640	67
P6	488	618	34	T3	300	641	68
Average	495.7	622,7	36,0	Average	297.7	642.3	67.3
Std. Dev	8.6	4.2	1.7	Std. Dev	4.0	3.2	0.6
P7	539	658	38				
P8	542	654	40				
P9	534	647	40				
Average	538.3	653.0	39.3				
Std. Dev	4.0	5.6	1.2				

Table 2. Process time and consumed energy reported in ref. [6]

Process step	Process time [s]	Average power [kW]	Consumed energy [kWh]
Metal print	66384	2.02	37.3
Peripherals	6974	7.7	14.9

28 specimens would cover most of the chamber plate surface. The energy required was then calculated from the printing time. The results are presented in Table 3.

The specific energy per unit volume of printed material can then be calculated from the total consumed energy and the built volume. For the 28 specimens, the total built volume is 142 cm^3. Thus, the relative energy consumption is 1.6e + 06J/mm^{-3}. This value is in agreement with specific energy consumption (SEC) values previously reported [9]. The SEC should also include the energy necessary to produce the metal powder, typically by atomization, although this was not the case in the present work. This last process requires metal melting and is energy-intensive [10]. On the other hand, when the powder is recycled, only a small percentage of waste material is produced by screening it after printing to separate some particles welded together by the laser.

Table 3. Process time and consumed energy estimations for 28 SS316L specimens

Process step	Process time [s]	Average power [kW]	Estimated energy [kWh]
Metal print	84050	2.02	47.2
Peripherals	6974	7.7	14.9

Manufacturing processes comparison. To compare the advantages and disadvantages of additive manufacturing concerning machining, particularly regarding the material and energy consumption of both processes, it was decided to use a rocker arm as a reference part. By using a complex piece instead of a very simple test sample that has specified dimensions, the geometry of the part can be optimized, ensuring that mechanical requirements are met, to use a smaller amount of material. For design purposes, Fusion 360 software was used, as it incorporates generative design and topology optimization features.

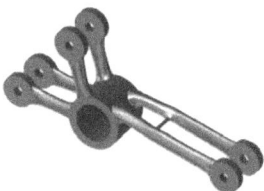

Fig. 5. Generative design of an SS316 rocker for an additive manufacturing process obtained using Fusion 360.

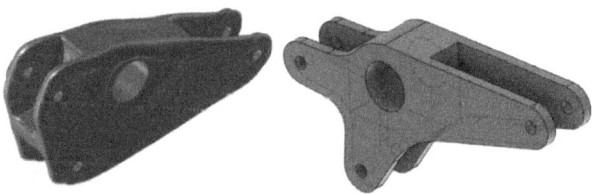

Fig. 6. Design of an SS316 rocker for a 3-axis milling process. Left: generative design obtained using Fusion 360. Right: further topology optimization of the original generative.

The design generated by the software seeks to maximize the rigidity of the piece. Fusion 360 uses default values of physical properties that were left unchanged because they do not significantly differ from the measured values. Figures 5 and 6 show the designs obtained for additive manufacturing and for 3-axis milling, respectively. Figure 6 right shows the design obtained after topology optimization of the initial generative design shown in Fig. 6 left. It is worth noting that Figs. 5 and 6 are perspective representations. In all three cases, the position of the connection points of the rocker's arm with the structure on which it is mounted was the same. Although the designs seem to be very different, all three of them satisfy the requirements as the software optimizes the geometry for each particular manufacturing technology.

In Table 4 Fusion 360 software results are shown. The table includes the results of GD for AM, applied in the present analysis to SLM. On the other hand, TO results correspond to a design for 3-axis milling. Additionally, it is worth noting that for all three designs, the maximum stresses calculated and reported by the software were well below the Yield Strength of the material.

Table 4. Fusion 360 software results and calculated energy

	GD AM	TO Milling
Part volume [mm^3]	30170	77272
Total volume [mm^3]	32000	332000
Part mass [g]	241	607
Total mass [g]	256	2650
Recycled mass [g]	15	2043
Removed volume [mm^3]	1800	255000
Consumed energy [J]	5,1E + 10	2,6E + 07

The consumed energy was calculated for both manufacturing processes and also included in Table 4. For AM, the energy consumption results from multiplying the volume of the part by the SEC previously estimated as 1.6e + 06 J/mm^3. On the other hand, the energy consumption in 3-axis milling was calculated by multiplying the removed volume by an SEC = 100 J/mm^3 SEC, which is representative of this process [11].

Table 4 also shows the material consumption for both manufacturing processes. The mass of the part resulting from a milling process is more than double that of AM, while the total mass involved in the milling process is about ten times the used in AM. It is worth noting that the volume removed by milling in the case of AM corresponds to the material that must be deposited before starting the printing of the part to allow subsequent sawing to separate it from the platform. In any case, the mass removed by milling could be recycled later. Conversely, the energy consumption in AM is more than a thousand times that of the milling because, even when the amount of material involved in milling is greater, the SEC for AM is several orders of magnitude higher than that for machining.

4   Conclusions

The differences in materials and energy consumption for additive and subtractive processes are significant. In the case of AM, although the energy consumption is much higher at the time of manufacture, the weight of the part is lower, and it would be advantageous if the energy associated with its use were lower, as would be the case in aerospace applications, compensating for the higher initial energy consumption.

In the case of machining, the weight of the part is greater, as is the total amount of material to be processed, but the energy used is significantly less. Presumably, this process would be more suitable for static applications. However, the energy required to produce the additional mass of the part compared to AM should be included, as well as the energy consumed in recycling the removed mass.

In brief, although a complete analysis of the factors involved is needed, from estimations of materials and energy consumption as previously described, it would be possible to have a first approach on the convenience of using an additive manufacturing process instead of a subtractive one before even starting to manufacture the first pieces.

References

1. Marattukalam JJ, Karlsson D, Pacheco V, Beran P, Wiklund U, Jansson U, Sahlberg M et al (2020) The effect of laser scanning strategies on texture, mechanical properties, and site-specific grain orientation in selective laser melted 316L SS. Mater Des 193:108852
2. Carassus H et al (2022) An experimental investigation into influences of build orientation and specimen thickness on quasi-static and dynamic mechanical responses of selective laser melting 316L stainless steel. Mater Sci Eng A 835:142683
3. Li Y, Ge Y, Lei J, Bai W (2022) Mechanical properties and constitutive model of selective laser melting 316L stainless steel at different scanning speeds. In: Advances in materials science and engineering
4. Wang L (2022) Microstructure and anisotropic tensile performance of 316L stainless steel manufactured by selective laser melting. Frattura ed Integrità Strutturale 16(60):380–391
5. Lei J, Ge Y, Liu T, Wei Z (2021) Effects of heat treatment on the microstructure and mechanical properties of selective laser melting 316L stainless steel. Shock Vib 1–12
6. Ochs D, Wehnert KK, Hartmann J, Schiffler A, Schmitt J (2021) Sustainable aspects of a metal printing process chain with laser powder bed fusion (LPBF). Proc CIRP 98:613–618
7. Barbieri L, Muzzupappa M (2022) Performance-driven engineering design approaches based on generative design and topology optimization tools: a comparative study. Appl Sci 12(4):2106
8. Zong X, Liu W, Yang Y (2020) Effect of different molding process on mechanical properties of 316L stainless steel. J Phys Conf Scr 1676(1):012097 (IOP Publishing)
9. Liu Z et al (2018) Investigation of energy requirements and environmental performance for additive manufacturing processes. Sustainability 10(10):3606
10. Jackson MA, Van Asten A, Morrow JD, Min S, Pfefferkorn FE (2018) Energy consumption model for additive-subtractive manufacturing processes with case study. Int J Precis Eng Manuf-Green Technol 5:459–466
11. Farooq MU, Anwar S, Ullah R, Guerra RH (2023) Sustainable machining of additive manufactured SS-316L underpinning low carbon manufacturing goal. J Market Res 24:2299–2318

Additive Manufacturing

A Proposal for an Extension of the Consumption Performance Sustainability Index to Machining Processes

Leopoldo De Bernardez[1], Giampaolo Campana[2(✉)], Rita Porcaro[3], and Sebastián Mur[1]

[1] Instituto Tecnológico de Buenos Aires (ITBA), Iguazú 341, C1437 Buenos Aires, Argentina
ldb@itba.edu.ar

[2] Department of Industrial Engineering, University of Bologna, Viale del Risorgimento 2, 40136 Bologna, Italy
giampaolo.campana@unibo.it

[3] Department of Civil, Chemical, Environmental and Mathematical Engineering, University of Bologna, Via U. Terracini 28, 40131 Bologna, Italy

Abstract. We previously proposed a Consumption Performance Sustainability Index to assess the sustainability of additive manufacturing products made by Fused Filament Fabrication. For simplicity and to get broader applications, the index was defined as dimensionless and composed of several factors related to the manufacturing process. The index included material and energy consumption and the performance of the transformed material, which, in the last case, was measured as the yield or the maximum resistance of the produced part in comparison with the characteristics of the feedstock material. The index can highlight the most sustainable among other solutions based on different print orientations and part densities.

We propose extending the Consumption Performance Sustainability Index's validity to additive manufacturing and Computerized Numerical Control (CNC) machining processes. For this purpose, each factor has been revised and, if necessary, expanded to include process parameters of machining processes. A new organization of all the factors has been introduced based on groups of parameters related to material properties and usage, part design, consumed energy, and machine time. Case studies have been investigated to evaluate this new index for assessing product sustainability preliminarily.

Keywords: Additive Manufacturing · Subtractive Manufacturing · Polymers · Sustainable Manufacturing · Consumption Performance Sustainability Index

1 Introduction

Life Cycle Impact Assessment (LCIA) is commonly used to evaluate environmental impacts (EIs) of products. This analysis is based on a method chosen among the numerous available ones. Then, the final results depend on the chosen method and need an

---

G. Campana is Corresponding author
L. De Bernardez is Co-corresponding author

© The Author(s) 2025
H. Kohl et al. (Eds.): GCSM 2023, LNME, pp. 721–729, 2025.
https://doi.org/10.1007/978-3-031-77429-4_80

interpretation [1]. The crucial aspects related to the Life Cycle Inventory (LCI) and the available database are not mentioned here.

This approach is beneficial in identifying critical aspects that can be improved by reducing the overall impacts of the following edition of the product under investigation. At the same time, it does not easily allow for direct comparisons among similar products because it neither necessarily nor explicitly considers all the parameters of manufacturing processes involved during their fabrication [2]. In addition, manufacturers can introduce specific local differences based on their production processes that cannot be easily included in the Life Cycle Assessment (LCA). Furthermore, the results of the LCA typically consist of multiple indicators, which cannot easily be interpreted in comparative issues nor easily aggregated in a lower number of indices, allowing for a more straightforward interpretation. Indeed, this approach poses serious complications when LCA is used for comparison purposes when the effects of manufacturing strategies on part properties are under consideration.

Over the past decade, the challenge has been identifying alternative sustainability performance indicators that can be complemental to the LCIA and directly related to production metrics, e.g. energy utilization as it relates to productivity. Mani et al., for example, identified the required elements and presented a methodology to evaluate the sustainability performance for manufacturing process characterization in order to ensure reliable and consistent comparisons [2]. Kibira et al. presented a procedure to select Key Performance Indicators (KPIs) for measuring, monitoring, and improving environmental aspects of manufacturing processes [3, 4]. Watson and Taminger presented a computational model for determining whether additive or subtractive manufacturing is more efficient for producing specific part geometries. In particular, they defined the total energy associated with the production of a part for both processes in terms of the volume fraction of the part [5]. Many other researchers have focused their research on the development of energy consumption models by associating them with different process parameters, and are not mentioned here for shortness. The challenge is still open, and several alternative indices complementing the LCIA are under development. LCA has been extensively adopted to quantify the Environmental Impacts of Three Dimensional Printing (3DP) and Computer Numerical Control (CNC) machining processes despite all the mentioned issues about using this approach, for example [6].

CNC machining is a mature and versatile subtractive production process to machine rigid materials such as metals, wood, polymers and composites. Substantial differences regarding cutting tools, cutting speeds, feed rates and coolant fluids depend on the architecture of machines and machined materials to achieve the desired surface finish and dimension accuracies of produced workpieces.

Additive manufacturing is often considered an alternative to subtractive processes, even if machining can be, on occasion, necessary to achieve the final part quality in terms of surface roughness and precision and accuracy of specific critical dimensions.

Manufacturers must evaluate and compare several factors when choosing between these two production methods for fabricating a specific product. In particular, they must consider the part size, production batches, expected mechanical properties, dimensional accuracy, surface quality, other issues and product life cycle sustainability.

In previous work, an innovative index named the Consumption Performance Sustainability Index (CPSI), which is complementary to LCA, was presented to compare production solutions. The index, initially developed on the additive process FFF (Fused Filament Fabrication), is calculated through a simple formulation that relates aspects relevant to the sustainability of the process, starting from energy and material consumption for specific sets of the manufacturing process parameters and by including the mechanical properties of the material and the final manufactured part. Furthermore, this index is dimensionless; thus, different designs, materials, and processes can be easily compared [7]. This article aims to extend the validity of the CPSI to CNC machining by appropriately modifying the index itself.

2 Methods

The Consumption Performance Sustainability Index (CPSI) has been recently introduced for additive manufacturing processes [7]. Its formulation has been modified here. All the formula elements have been revised, and new terms have been added to extend the use of the index to subtractive manufacturing parameters. It is defined as a multiplication of factors, as shown in Eq. 1:

$$CPSI = I_{prop} \times I_{design} \times I_{mat} \times I_{time} \times I_{s.e} \qquad (1)$$

In Eq. 1, each multiplication factor has its own physical and engineering meaning. I_{prop} accounts for material properties, I_{design} for those parameters related to the piece design, I_{mat} accounts for the amount of material in use, I_{time} for machine utilization in terms of time, and $I_{s.e}$ accounts for specific energy consumption.

By analyzing each multiplication factor individually, we can define the material properties index as in Eq. 2 and Eq. 3:

$$I^Y_{prop} = \frac{Y_m}{\rho_m} \qquad (2)$$

$$I^{UTS}_{prop} = \frac{UTS_m}{\rho_m} \qquad (3)$$

where Y_m, UTS_m and ρ_m are the Young moduli, the Ultimate Tensile Strength and density of the feedstock material, respectively. If using the International System of Units (ISU), the unit of measure of this index is m^2/s^2, representing the specific modulus of the material under consideration. Although this ratio is not dimensionless, its unit will be offset by those of the energy consumption index.

The formulation of the design index is shown in Eq. 4:

$$I_{design} = \frac{Y_s}{Y_m} \times \frac{\rho_m}{\rho_s} \times \frac{\sigma_{eq}}{\sigma_{eq,lim}} \qquad (4)$$

where the factor $\left[\frac{Y_s}{Y_m}\right]$ is related to the piece design and the material transformation processes occurring during the manufacturing technology; it accounts for the increase or reduction of the mechanical properties observed in the considered part, Y_s, in comparison with the feedstock material performance, Y_m; ρ_s is the part density; $\frac{\sigma_{eq}}{\sigma_{eq,lim}}$ represents the maximum value of the equivalent stress over the limit equivalent stress and describes material exploitation (e.g. evaluated by von Mises). It is worth mentioning that I_{design} is also dimensionless.

The material consumption index accounts for any material quantity required for the production process that is not a portion of the final piece. It is defined in Eq. 5:

$$I_{mat} = \frac{m_s}{m_t} \tag{5}$$

where m_s and m_t are the mass of the part and the total mass involved in the production process, respectively. I_{mat} represents the mass of the material wasted during the manufacturing process, m_m, that will not be part of the final piece (the support material in the case of 3DP or the removed material in the CNC machining processes). The total mass, m_t, can be calculated as the following sum: $m_t = m_s + m_m$. The material consumption index is also dimensionless, as is the design index. This factor always diminishes the index.

The machine utilization index is calculated as in Eq. 6:

$$I_{time} = \frac{t_p}{t_a} \tag{6}$$

where t_p is the average cycle time for the complete process (conditioning, loading, machining or printing, unloading), and t_a is the average time available for production. In particular, the latter is identified by t_c, the sum of the cycle time and t_{np}, the average non-productive time between runs: $t_a = t_c + t_{np}$. As for the previous factors, the machine utilization index is dimensionless and can be smaller or equal to 1.

The last multiplication factor represents the energy consumption index, and it is defined in Eq. 7:

$$I_{s.e} = \frac{m_s}{E_t} \tag{7}$$

where E_t is the total consumed energy, i.e. the sum of the energy used for material production, E_m, the energy consumed for the component production, E_f, and the energy consumed during non-productive hours due to specific requirements of production machines, E_{np}. Summarising: $E_t = E_p + E_{np} = (E_m + E_f) + E_{np}$, where E_p is the total energy consumed during productive hours. In addition, the energy consumed for the part production can be expressed as $E_f = m_p \times e_f$, where m_p is the processed mass; e_f is the used energy by a unit of the processed mass. If using the ISU, the unit of this index is s^2/m^2.

Then if the same system of units is used for the different variables, CPSI results in a dimensionless number. Eventually, by rewriting Eq. 1 by introducing all the mentioned factors, the following Eq. 8 and Eq. 9 can be obtained:

$$CPSI = \frac{Y_m}{\rho_m} \times \left(\frac{Y_s}{Y_m} \times \frac{\rho_m}{\rho_s} \times \frac{\sigma_{eq}}{\sigma_{eq,lim}} \right) \times \frac{m_s}{m_t} \times \frac{t_p}{t_a} \times \frac{m_s}{E_t} \tag{8}$$

$$CPSI = \frac{UTS_m}{\rho_m} \times \left(\frac{Y_s}{Y_m} \times \frac{\rho_m}{\rho_s} \times \frac{\sigma_{eq}}{\sigma_{eq,lim}} \right) \times \frac{m_s}{m_t} \times \frac{t_p}{t_a} \times \frac{m_s}{E_t} \tag{9}$$

It is evident that the CPSI will be higher if the design optimizes the mechanical performance, the specific modulus, the material use efficiency and the machine time utilization, while it will be lower if the energy consumption increases.

For the sake of simplicity and as a first estimate of the index, the following conditions were considered:

$$\frac{\overline{\sigma}_{eq}}{\sigma_{eq,lim}} = 1 \tag{10}$$

$$\frac{t_p}{t_a} = 1 \tag{11}$$

The final formulation of the CPSI extension can be obtained by making the necessary simplifications and by including Eq. 10 and Eq. 11 in Eq. 8 and Eq. 9. For both 3DP and CNC machining processes, if V_s is the volume of the feedstock material, the CPSI formulation can be reduced as shown in Eq. 12 and Eq. 13:

$$CPSI = \frac{Y_s}{\rho_s} \times \frac{m_s}{m_t} \times \frac{m_s}{E_t} = \frac{Y_s}{m_t} \times \frac{m_s}{E_t} \times V_s \tag{12}$$

$$CPSI = \frac{UTS_s}{\rho_s} \times \frac{m_s}{m_t} \times \frac{m_s}{E_t} = \frac{UTS_s}{m_t} \times \frac{m_s}{E_t} \times V_s \tag{13}$$

3   Discussion

A comparison between the original and the extended version of the proposed index is shown in the following Sect. 3.1. In Sect. 3.2, a comparison between the index calculated for FFF and CNC machining is introduced. The part design and material are the same as presented and discussed in [7].

3.1   Comparison Between Extended and Previous Formulations of CPSI in the Case of Additive Manufacturing Processes

In the case of additive manufacturing processes, the extended formulation of the proposed index introduces a simple modification that does not change the already examined final result but allows for extending the index to subtractive processes. It is worth mentioning that:

- m_t is a newly introduced term that includes the mass of the manufactured piece and the mass of the material that is required for the production process but is not part of the final piece, i.e. support material, extruded purges and disposable trays in the case of FFF machines investigated in the previous research work [7].
- Y_s could be different from Y_m and ρ_s could be different from ρ_m due to the part design (e.g. infill pattern and infill density in FFF) and the material transformation process (melting and solidification in FFF);
- E_t includes the energy used to print the specimen, the energy used to produce the raw material, the energy used to produce the filament from raw polymer, and the energy used to maintain the chamber at a temperature between runs or during standby time.

Figure 1 compares the calculation of the original and the extended version of CPSI. It is evident that the index depends on the infill orientation and part density. Moreover, the energy used for heating the chamber in the case of FFF industrial machines strongly affects the index. As a first comparative test, only the energy used during the printing process was considered (the energy needed to produce the raw material and the filament from it were not included in the calculations). As shown in Fig. 1, the CPSI depends on the infill orientation and density.

In addition, the energy used to heat the chamber in the case of FFF industrial machines strongly affects the index, being higher if the energy for heating is not included. Indeed, the chamber is heated to maintain process stability and optimize printing conditions, although it implies higher energy consumption. In the case of non-professional machines without a heated chamber, energy consumption is lower, but it can determine part defects due to the manufacturing process or, in the worst case, failure and, in general, produce non-consistent results over time. To evaluate the effect on the CPSI of keeping the chamber temperature constant, the energy for heating can be estimated, as discussed in the previous work [7], and discounted from the total energy consumed. Figure 1 compares the extended and previously defined CPSI for both the possible situations, with and without chamber heating. It is clear from Fig. 1 a) and b) that CPSI increases when heating energy is not included.

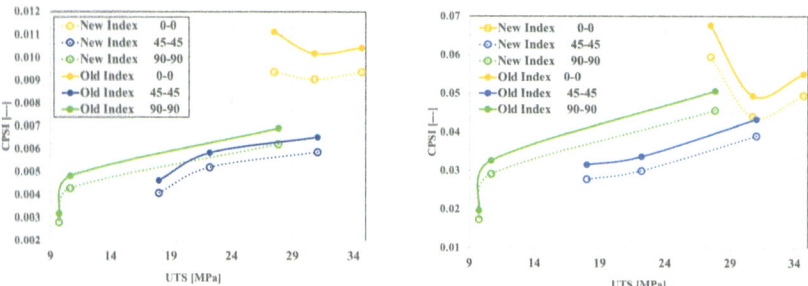

Fig. 1. *CPSI vs. UTS for FFF test samples with different infill densities and orientations. (left side) The consumed energy includes chamber heating. (right side) The consumed energy does not include chamber heating. (Experimental data used for CPSI calculation are published in [7])*

3.2 Estimation of CPSI for Machining Processes and Comparison Between Subtractive and Additive Processes

In the case of machining, it is worth mentioning that:

– m_t includes the mass of the manufactured piece, m_s, and the mass removed during the production process, m_m, which is not part of the final piece.
– E_t includes the energy used to produce the raw material, the one to produce the billet from the raw material, the energy used during the machining process and the energy consumed during the non-productive time, if applicable. The energy consumed during

the machining process strongly depends on the removal rate and can be defined as E_t = SEC x V_r, where SEC is the Specific Energy Consumption in J/mm^{-3}; V_r is the removed volume in mm^3.

On the other hand, the SEC is highly dependent on the material removal rate (MRR) [8]. For higher removal rates, the time required decreases, and correspondingly, the SEC decreases because the power involved in MRR is a fraction of the total milling power [8]. Figure 2 represents the index as a function of the total mass used in the final process for additive (FFF) and subtractive (CNC machining) processes.

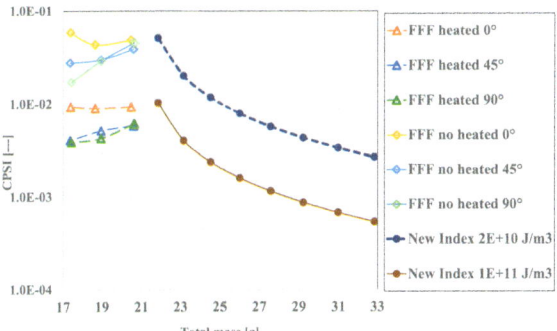

Fig. 2. CPSI vs total mass for FFF and CNC machining. (For CNC machining, CPSI is estimated based on the benchmark design in [7].)

In the case of the FFF results presented in Fig. 2, the calculations are based on experimental results previously reported in [7]. The total mass required increases if more support material is needed and decreases when the infill density decreases.

The amount of support material is relatively small in this case because the pieces arc printed horizontally. For this reason, the ratio between the mass of the part and the total mass is close to 1. If the parts are printed at a higher angle to the horizontal plane or complex parts requiring a lot of support material are printed, the relationship can be less than 1.

On the other hand, the relationship between Young's modulus and part density, which is part of the CPSI calculation (see Eq. 12), can increase depending on the infill density and print orientation. For this reason, Fig. 1a and Fig. 2 show that the index for FFF increases in some cases when the total mass decreases [7]. It can also be seen in Fig. 2 that there is a significant difference in CPSI for FFF when the heating energy of the chamber is not included.

In the case of CNC, the calculations of CPSI are based on some theoretical data. Two values of SEC were considered: SEC$_1$ corresponds to 20 J/mm^{-3}, a value in the high range of MRR, while SEC$_2$ corresponds to 100 J/mm^{-3}, in the low range of MRR, following data published by Yoon et al. [8]. Young's modulus was obtained from [9].

The higher total mass observed for the machining process than the FFF is because the total mass depends on the amount of removed material. The total mass increases as the billet mass increases and more must be removed by machining. Thus, the ratio between the part's mass and the billet's mass, or the total mass, decreases. Likewise, the energy required for machining also increases with the total mass because more material must be removed. For both reasons, the CPSI decreases as the total mass increases.

4   Conclusions

A dimensionless index, named CPSI, which was initially developed and validated for additive manufacturing processes (FFF), has been remodelled to extend its use to CNC machining processes. It is intended to be used for optimization purposes as a complemental index to the conventional LCA for comparisons between different manufacturing processes by considering those process parameters that are not typically introduced in an LCA. The present formulation is still under investigation for further extension to other manufacturing processes and, theoretically, could be extended to all the known processes by maintaining the same concept. In this work, some preliminary elaborations in graphical forms show that the results are coherent with the previous formulation in the case of FFF. Besides, some critical aspects related to the need for reliable data about the SEC for CNC machining are remarked to achieve a sound index evaluation.

References

1. Huijbregts MAJ et al (2017) ReCiPe2016: a harmonized life cycle impact assessment method at midpoint and endpoint level. Int J Life Cycle Assess 22:138–147
2. Mani M, Madan J, Lee JH, Lyons KW, Gupta SK (2014) Sustainability characterization for manufacturing processes. Int J Prod Res 52(20):5895–5912
3. Kibira D, Brundage M, Feng S, Morris KC (2017) Procedure for selecting key performance indicators for sustainable manufacturing. Proc ASME MSEC 2017:1–8
4. Kibira D, Brundage M, Feng S, Morris KC (2018) Procedure for selecting key performance indicators for sustainable manufacturing. J Manuf Sci Eng 140:011005, 1–7
5. Watson JK, Taminger KMB (2018) A decision-support model for selecting additive manufacturing versus subtractive manufacturing based on energy consumption. J Cleaner Prod 176:1316e1322
6. Landi D, Zefinetti FC, Spreafico C, Regazzoni D (2022) Comparative life cycle assessment of two different manufacturing technologies: laser additive manufacturing and traditional technique. Proc CIRP 105:700–705
7. De Bernardez L, Campana G, Mele M, Mur S (2023) Towards a comparative index assessing mechanical performance, material consumption and energy requirements for additive manufactured parts. In: Lecture notes in mechanical engineering, pp 302–310
8. Yoon HS, Lee JY, Kim HS, Kim MS, Kim ES, Shin YJ, Chu WS, Ahn SH (2014) A comparison of energy consumption in bulk forming, subtractive, and additive processes: review and case study. Int J Precis Eng Manuf Green Tech 1:261–279
9. Rahimi M, Esfahanian M, Moradi M (2014) Effect of reprocessing on shrinkage and mechanical properties of ABS and investigating the proper blend of virgin and recycled ABS in injection molding. J Mater Process Technol 214(11):2359–2365

Wire-Arc Additive Manufacturing Toolpath Optimization Using a Dexel-Based Temperature Prediction Model

G. Mauthner[1]([✉]), M. Stautner[2], S. Sell[2], M. Frings[3], A. Lorenz[3], D. Plakhotnik[3], and F. Bleicher[1]

[1] TU Wien, Institute for Production Engineering and Photonic Technologies, Karlsplatz 13, 1040 Vienna, Austria
mauthner@ift.at
[2] Hochschule Ruhr West, Duisburger Straße 100, 45407 Mühlheim an Der Ruhr, Germany
[3] ModuleWorks GmbH, Henricistraße 50, 52072 Aachen, Germany

Abstract. Wire-Arc Additive Manufacturing (WAAM) has been established as a new technology for industrial use-cases such as low-lot size manufacturing or part repair services. A key aspect when developing such WAAM processes is thermal management during the layer-by-layer metal deposition. To maintain a stable welding process in-depth knowledge about the heat distribution is required. Thus, predicting the heat flux for a given part geometry already in the process development stage using Computer-Aided-Manufacturing Systems (CAM) would be beneficial. However, current state-of-the-art approaches are computationally expensive and time intensive. Therefore, they are hardly applicable for WAAM applications. In this paper, a dexel-based metal cutting and deposition simulation is combined with a temperature prediction model, which is integrated in the toolpath planning algorithm when defining a build-up strategy for a given part geometry. The approach is based on a temperature prediction algorithm, that calculates temperature fields for deposited material volume considering basic material properties. Calculated temperature fields can be utilized for optimizing welding toolpath to achieve stable process conditions across the part geometry.

Keywords: Wire-Arc Additive Manufacturing · Toolpath Planning · Temperature Simulation · Computer-Aided-Manufacturing

1   Introduction

Wire arc additive manufacturing (WAAM) is an innovative metal additive deposition technique that employs the power of an electric arc to melt and deposit metal wire, thereby fabricating workpiece layer by layer [1]. WAAM supports trends towards more sustainability in production and consumption, due to its advantages over conventional manufacturing techniques regarding material and energy efficiency [2]. Campatelli et al. highlight the potentials integrating WAAM processes and conventional milling processes, demonstrating an energy saving of 34% compared to traditional manufacturing

H. Kohl et al. (Eds.): GCSM 2023, LNME, pp. 730–737, 2025.
https://doi.org/10.1007/978-3-031-77429-4_81

process chains [3]. One of the biggest challenges when utilizing WAAM technology is the transient temperature field created due to the inhomogeneous welding process. The temperature in the part is increasing during the layer-by-layer build-up phase and, if not managed properly, potentially leading to process quality issues such as weak connections between weld layers or geometrical errors. By carefully regulating temperature, one can optimize the entire process, enhancing its efficiency and ensuring the production of high-quality parts [4]. The induced temperature in the workpiece is mainly influenced by the welding parameters such as voltage or current and the speed of the welding torch manipulator. However, recent research indicates that toolpath strategies also play a major role for practical WAAM systems [5]. To further increase the industrial utilization of this important technology towards sustainable manufacturing systems, temperature related sources of defects need to be identified and eliminated already in early stages of process development.

2 State-of-the-Art Review

Temperature predictions are an essential part of the design of a stable WAAM process. This includes a series of tests and numerical simulations [6]. Various researchers have focused on developing finite element simulation (FEM) models for additive manufacturing processes to analyze thermo-mechanical performance of the WAAM process. Researchers have implemented models to simulate WAAM processes and analyze characteristics such as local and global heat development and the subsequent formation of residual stresses [7–9]. The impact of preheating the substrate and baseplates before building the actual geometry, as well as the impact of cooling periods is investigated [10, 11]. Various heat source models as basis for temperature simulation are implemented [12–14]. Mehnen et al. provide a design study, utilizing FEM to investigate different build-up strategies by implementing the welding toolpath as a moving heat source in the simulation software system [15]. The result of this work highlights the high dependency of the process quality on the selected build-up/toolpath strategy. Required toolpath information must be imported in numerical model by either (a) utilizing specialized tools from simulation software providers [16] or (b) implementing toolpath information by individual import scripts utilizing the application programmable interface (API) possibilities in the respective FEM software such as Ansys or Abaqus [17–19]. In contrast to FEM based models, Böß et al. developed a dexel-based simulation system to predict the weld seam geometry using a quadratic regression model [20]. Researchers also studied the effect of various toolpath strategies for optimized WAAM processes. According to Uyen et al., the choice of trajectory has a significant influence on the cooling behavior and therefore the grain size and mechanical properties of the workpiece [21]. To prevent final part distortion, studies about the impact of deposition toolpaths such as balanced building strategies are discussed [15, 22–24]. Algorithms enabling continuous welding toolpaths eliminating defects occurring from stopping the welding torch are investigated [5, 25]. An approach to optimize toolpaths based on a rapid thermo-mechanical field prediction model utilizing machine learning techniques is used by Zhou et al. [26].

To summarize, past research has mostly focused on optimizing final part defects such as distortions by developing simulation models for thermo-mechanical analysis as well as experimental studies analyzing varying toolpath strategies for the part build-up. While finite element analysis provides positive correlations when compared to experimental data, these kinds of simulation are computational expensive and operate on a fixed domain where initial temperatures and boundary conditions are known. However, in WAAM the domain is subject to change due to the constant material deposition. Additionally, current finite element approaches are limited in their use since toolpath and process information need to be integrated manually, utilizing provided software interfaces. Regarding toolpath strategies, research highlighted opportunities to optimize part quality. Current approaches are focusing on optimized build-up strategies for selective geometries and are based on large experimental data sets. In some approaches, experimental temperature data is collected and deployed as training set for machine learning approaches. However, no simulation data such as temperature prediction is used in the toolpath planning process directly to optimize the overall results. An integrated system, coupling a thermal deposition calculation, a material build-up simulation as well as a toolpath algorithm, could overcome the limitations of current approaches developed in academia and provide novel opportunities for virtual WAAM process optimization.

3 Concept for Wire-Arc Additive Manufacturing Toolpath Optimization

This paper presents an approach for a dexel-based temperature calculation model, enabling automated toolpath adaptions for optimized WAAM processes. The temperature simulation is coupled with the toolpath planning algorithm within one system, thus, providing novel opportunities for process optimization. The envisioned concept expands the functionality of traditionally used CAD/CAM systems along the horizontal process chain when generating numerical codes for WAAM machines. The concept proposed in this work builds on integrating the novel temperature calculation and toolpath planning algorithm by using only one dexel-based data model in the background. Figure 1 provides an overview of the concept and its main components. The proposed concept consists of two main modules: *(a)* the *temperature calculation module*, providing the time-dependent temperature in the deposited material during the welding process and *(b)* the *toolpath optimization module*, which consists of necessary temperature management rule sets to calculate an optimized toolpath. Both models interact and exchange information for an iterative optimization of the toolpath, based on simulated temperature values.

First, a draft toolpath is generated based on the standard CAM inputs such as welding geometry, layer height, layer width etc. In this step, conventional CAM algorithms are used to slice the geometries and provide necessary points and line segments as input for the machine movement commands.

Second, the current toolpath is then the basis for a material deposition simulation, utilizing modern dexel material modelling technique. This method discretizes the space by a three-dimensional dexel field. The boundary of the material is represented by intersection points on the dexels. During process simulation these intersection points are updated to represent the current material geometry. Third, the same dexel material model is now used for solving the respective heat equation in the temperature calculation model. The

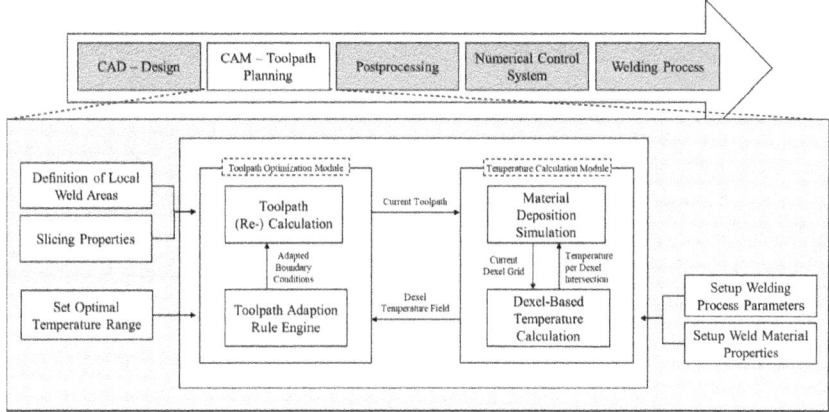

Fig. 1. Concept for WAAM toolpath optimization using dexel-based temperature prediction.

model requires standard material parameters such as heat transfer coefficient, room temperature and welding current/voltage as input and solves the transient heat equation using Lattice-Boltzmann-Method (LBM) [27]. Last, the calculated temperatures for a given toolpath position are provided to the toolpath optimization module. A rule engine consists of WAAM specific process rules, that determine respective actions if the temperature values are too high or low, thus, leading to a potential defect during or at the end of the welding process. The rule engine describes the boundary conditions and potential actions to be taken in such a case and initiate a re-calculation of the toolpath considering the changed situation. From here, the iterative optimization process starts again, given the new toolpath as input for the simulation.

4 Validation

The described concept for temperature dependent toolpath optimization has been implemented as a prototype system and was validated in the laboratory of *TU Wien Institute of Production Engineering and Photonic Technologies* in Vienna. The temperature calculation software has been implemented utilizing existing software frameworks of *ModuleWorks GmbH*. An existing toolpath calculation and deposition simulation environment has been expanded by the described modules in order to demonstrate a proof-of-concept implementation.

4.1 Experimental Setup

The novel functionality of the prototype is demonstrated for a simple two-wall weld geometry, utilizing a 6-axis welding robot and a Cold-Metal-Transfer (CMT) welding power source. For temperature measurement, an *Optris* infrared thermo camera was used. In the demonstration scenario, two walls with same geometry are deposited on a baseplate (H: 10 mm). An original welding toolpath for the robotic setup has been generated with *Siemens* NX CAD/CAM multi-axis-deposition software and was utilized as baseline for the optimization. This baseline toolpath aims to build the two wall geometries consecutively (S235 steel wire, L × W × H: 100 × 5 × 40 mm, number of layers: 16).

This toolpath does not consider information about the expected thermo-mechanical development during the deposition process. It is purely based on the wall geometries and the experience of the CAD/CAM programmer. During the welding process, the welding torch follows the toolpath while inducing heat energy in the baseplate and the already deposited wall structure material. Due to the layer-by-layer additive build-up process, temperature is continuously growing in the local toolpath welding area, potentially leading to defects such as uneven weld geometries, destruction of the material chemistry or postprocess deformations due to the induced stresses.

In the given demonstration scenario, the original toolpath builds both walls consecutively, reaching maximum temperatures of about 1300–1500 °C. With the developed optimization software, an additive manufacturing rule is added in the toolpath optimization module, limiting the maximum temperature in the local weld area to prevent expected welding defects. Whenever the simulation indicates that the temperature in the local weld area exceeds the defined maximum, a toolpath adaption is initiated. The manufacturing rule forces the system to reject the toolpath segment of the overheated layer. Instead, it adds a transfer move to the second wall geometry, which at this moment has a temperature below the defined maximum.

4.2 Results

The newly developed approach has been implemented and utilized in the described demonstration scenario. The original toolpath has been simulated, and respective temperature values for all dexels intersection are generated (c.f. Fig. 2).

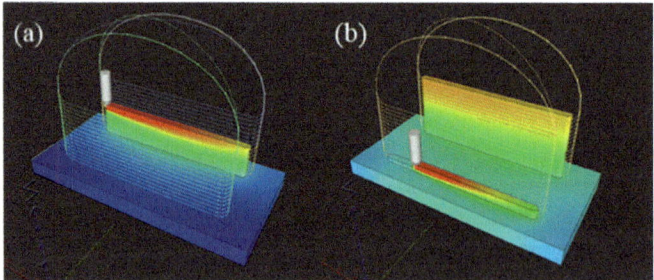

Fig. 2. Toolpath simulation (CLSF input from Siemens NX) with integrated temperature calculation module and colored visualization scheme. (a) Wall 1, Layer 10; (b) Wall 2, Layer 2.

The simulation data was compared to measurements to obtain information about the accuracy of the proposed LBM-based simulation approach. Depending on the camera calibration used, measured (maximum) temperatures in the welding zone vary significantly between 1100–1500 °C. In comparison, simulated temperature values in the currently deposited dexel intersection points illustrate temperatures between 1200–1400 °C. These first results indicate a positive correlation of simulated and measured temperatures of the developed model.

The temperature feedback provided is used as input for the toolpath optimization module. Based on developed optimization rules, the toolpath can be segmented and adapted to achieve optimal heat distribution across the entire toolpath and to prevent defects in the welding zone. In the presented experiment, a maximum has been defined as the initial point for a toolpath adaption. During the build-up simulation of the first wall, the temperature reaches this maximum for the first time right in the middle of deposition layer 10 (of 16). The deployed manufacturing rule forces the toolpath generator to reject the last toolpath layer segment on the current feature, and instead create a transfer move to another geometry whose current top layer is below the maximum temperature defined. From there, a new toolpath segment on the new geometry is created until the build-up is completed, or maximum temperature is reached, starting another optimization loop respectively. Figure 3 highlights the result of the adaptive toolpath planning, the created toolpath segments, and an updated simulation result during the welding process.

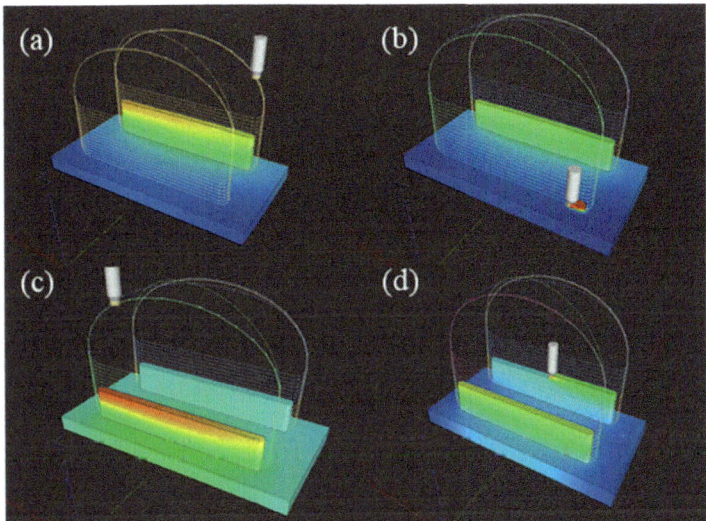

Fig. 3. Adapted toolpath strategy utilizing optimization module. (a) Wall 1, Layer 10 exceeds defined temperature maximum; (b) integration of transfer move to Wall 2, Layer 1 and start deposition simulation; (c) Wall 2, Layer 9 exceeds defined temperature maximum starting another optimization loop; (d) Continued welding on Wall 1, Layer 10 until geometry is finished.

5 Conclusion and Future Outlook

In this paper, a concept for an integrated dexel-based temperature prediction for WAAM toolpath optimization has been presented. The concept is based on a fast and reliable temperature simulation, which is fully integrated within an additive material deposition simulation and a toolpath planning algorithm. First prototype tests indicate that the proposed concept helps to give a quick understanding of the temperature behaviour during the process and demonstrates the usability for the optimization of WAAM path planning already in the virtual process planning stage.

Although this prototype gives a good estimation, further research is necessary in the future. To further increase the accuracy of the underlaying model, different models than LBM could be exploited. Additionally, increased accuracy could be achieved by optimizing the form and shape of the added material in the dexel simulation. More complex toolpath optimization strategies will be necessary leading to additional requirements for advanced collision control systems and dynamically adaptions of welding paths.

References

1. Singh S, Sharma S, Rathod D (2021) A review on process planning strategies and challenges of WAAM. Mater Today Proc 47:6564–6575. ISSN 2214-7853
2. Agnusdei L, Del Prete A (2022) Additive manufacturing for sustainability: a systematic literature review. Sustain Futures 4:100098. ISSN 2666-1888
3. Campatelli G, Montevecchi F, Venturini G et al (2020) Integrated WAAM-subtractive versus pure subtractive manufacturing approaches: an energy efficiency comparison. Int J Precis Eng Manuf Green Tech 7:1–11
4. Jafari D, Vaneker THJ, Gibson I (2021) Wire and arc additive manufacturing: opportunities and challenges to control the quality and accuracy of manufactured parts. Mater Des 202:109471
5. Ding D, Pan Z, Cuiuri D, Li H (2014) A tool-path generation strategy for wire and arc additive manufacturing. Int J Adv Manuf Technol 73(1–4):173–183
6. Santi A, Bayat M, Hattel J (2023) Multiphysics modeling of metal based additive manufacturing processes with focus on thermomechanical conditions. J Therm Stresses 46(6):445–463
7. Chen B, Hashemzadeh M, Guedes SC (2014) Numerical and experimental studies on temperature and distortion patterns in butt-welded plates. Int J Adv Manuf Technol 72(5–8):1121–1131
8. Ge J et al (2019) Thermal-induced microstructural evolution and defect distribution of wire-arc additive manufacturing 2Cr13 part: numerical simulation and experimental characterization. Appl Therm Eng 163(100):114335
9. Cadiou S, Courtois M, Carin M, Berckmans W, Le Masson P (2020) 3D heat transfer, fluid flow and electromagnetic model for cold metal transfer wire arc additive manufacturing (Cmt-Waam). Addit Manuf 36:101541
10. Xiong J, Lei Y, Li R (2017) Finite element analysis and experimental validation of thermal behavior for thin-walled parts in GMAW-based additive manufacturing with various substrate preheating temperatures. Appl Therm Eng 126:43–52
11. Mukherjee T, Zhang W, DebRoy T (2017) An improved prediction of residual stresses and distortion in additive manufacturing. Comput Mater Sci 126:360–372
12. Giarollo DF, Mazzaferro CCP, Mazzaferro JAE (2022) Comparison between two heat source models for wire-arc additive manufacturing using GMAW process. J Brazilian Soc Mech Sci Eng 44(1):1–13
13. Ding D, Zhang S, Lu Q, Pan Z, Li H, Wang K (2021) The well-distributed volumetric heat source model for numerical simulation of wire arc additive manufacturing process. Mater Today Commun 27:102430
14. Sampaio RFV, Pragana JPM, Bragança IMF, Silva CMA, Nielsen CV, Martins PAF (2023) Modelling of wire-arc additive manufacturing—a review. Adv Ind Manuf Eng 6:100121

15. Mehnen J, Ding J, Lockett H, Kazanas P (2014) Design study for wire and arc additive manufacture. Int J Prod Dev 19(1–3):2–20
16. Digital Engineering 247. https://www.digitalengineering247.com/article/simufact-additive-2020-ded-simulation-tool-announced-at-formnext. Last accessed 16 July 2023
17. Ansys. https://www.ansys.com/blog/what-is-apdl. Last accessed 16 July 2023
18. Zhao XF, Wimmer A, Zaeh MF (2023) Experimental and simulative investigation of welding sequences on thermally induced distortions in wire arc additive manufacturing. Rapid Prototyping J 29(11):53–63
19. Chergui A, Villeneuve F, Béraud N, Vignat F (2022) Thermal simulation of wire arc additive manufacturing: a new material deposition and heat input modelling. Int J Interact Des Manuf 16(1):227–237
20. Böß V, Denkena B, Dittrich MA, Malek T, Friebe S (2021) Dexel-based simulation of directed energy deposition additive manufacturing. J Manuf Mater Process 5(1):9
21. Uyen TMT, Minh PS, Nguyen V-T, Do TT, Nguyen VT, Le M-T, Nguyen VTT (2023) Trajectory strategy effects on the material characteristics in the WAAM technique. Micromachines 14:827
22. Li X, Lin J, Xia Z, Zhang Y, Fu H (2021) Influence of deposition patterns on distortion of H13 steel by wire-arc additive manufacturing. Metals 11(3):1–14
23. Lee Y, Bandari Y, Nandwana P, Gibson BT, Richardson B, Simunovic S (2019) Effect of interlayer cooling time, constraint and tool path strategy on deformation of large components made by laser metal deposition with wire. Appl Sci 9(23)
24. Chen X, Shang X, Zhou Z, Chen SG (2022) A review of the development status of wire arc additive manufacturing technology. In: Advances in materials science and engineering
25. Zhou Z, Shen H, Lin J, Liu B, Sheng X (2022) Continuous tool-path planning for optimizing thermo-mechanical properties in wire-arc additive manufacturing: an evolutionary method. J Manuf Process 83(August):354–373
26. Zhou Z, Shen H, Liu B, Du W, Jin J (2021) Thermal field prediction for welding paths in multi-layer gas metal arc welding-based additive manufacturing: a machine learning approach. J Manuf Process 64:960–971
27. Chen S, Doolen GD (1998) Lattice Boltzmann method for fluid flows. Annu Rev Fluid Mech 1998(30):329–364

Directed Energy Deposition of Functionally Graded Materials: Towards Resource- and Cost-Effective Manufacturing

Jacques Strauss[1,2], Natasha Sacks[1,2]([✉]) [iD], and Devon Hagedorn-Hansen[1,3]

[1] Department of Industrial Engineering, Stellenbosch University, Stellenbosch, South Africa
natashasacks@sun.ac.za
[2] DSI-NRF Centre of Excellence in Strong Materials, Cape Town, South Africa
[3] HH Industries, Cape Town, South Africa

Abstract. This paper presents an overview of the production of functionally graded materials (FGMs) using directed energy deposition (DED) where the advancements, challenges, and future directions in this field are explored. Functionally graded materials entail the joining of two or more different materials to produce a part of which the chemical and thus mechanical properties vary along one or more dimensions of a part. This study explored the benefits of FGMs, including enhanced multifunctionality, weight optimization and reduced cost. Additive manufacturing (AM), particularly DED, has emerged as a promising technique for fabricating complex FGM structures. By utilizing FGMs, designers can overcome the limitations imposed by expensive materials when developing critical components. Through strategic engineering, the material composition of the part can be tailored as a gradient between expensive materials and more affordable alternatives in non-critical regions. This approach optimizes costs while ensuring performance requirements are met. It is also widely acknowledged that utilizing AM techniques in place of or in addition to conventional manufacturing techniques can decrease material waste. The reduced material waste and enhanced resource utilization offered by DED make it a good solution for sustainable manufacturing. Furthermore, by harnessing the potential of FGMs and DED, industries could unlock new design possibilities, improve product performance, and achieve greater manufacturing efficiency.

Keywords: Directed energy deposition · functionally graded materials · waste reduction · resource efficiency · sustainable manufacturing

1 Introduction

Functionally graded materials (FGMs) are a class of advanced engineered materials that exhibit continuous spatial variations in composition, structure, and properties, enabling transitions between different functionalities within a single component [1]. These materials offer unique advantages over conventional homogeneous materials as their properties can be tailored to specific requirements, thereby enhancing overall performance and

H. Kohl et al. (Eds.): GCSM 2023, LNME, pp. 738–745, 2025.
https://doi.org/10.1007/978-3-031-77429-4_82

functionality. They have gained significant attention across various industries, including aerospace, automotive, biomedical, and energy sectors [2]. FGMs have shown significant promise in various domains, including thermal management, wear resistance, corrosion protection, and structural integrity. They achieve this by employing a gradient design, transitioning from costly materials to more economical alternatives in non-critical regions. This strategy enables cost optimization while guaranteeing the fulfillment of performance requirements [3, 4].

FGMs can be manufactured using techniques like fusion welding, powder metallurgy, ultrasonic welding, and vapor deposition, but they are limited to small gradient lengths (< 1 mm). Additive manufacturing (AM), specifically directed energy deposition (DED), is a promising method for creating intricate FGM structures with longer gradient zones on a larger scale. DED is an AM technology that employs the melting of a powder or wire material via a high-powered energy source (arc, plasma, laser, or electron beam) to produce a solid part [5]. The technology does present some advantages over other AM technologies, positioning it as a favorable option for specific applications. One of the notable benefits of DED is its ability to accommodate a wide range of feedstock materials, including those commonly utilized in welding and cladding procedures, making the required resources more readily available and cost-effective [6, 7]. Certain DED processes even permit the use of multiple materials, enabling the production of FGMs with continuous gradients. The combination of high build rates, material versatility, and accessibility contributes to the creation of FGMs by incorporating specific material properties in different regions of a component. Collectively, these advantages make DED an ideal manufacturing technology to produce FGMs in a sustainable manner.

To understand the current knowledge available on the production of FGMs using DED, an initial review was done of existing published literature with the findings presented in this paper. The review explores advancements, challenges, and future directions in this field. By harnessing the potential of FGMs and AM, industries could unlock new design possibilities, improve product performance, and achieve greater manufacturing efficiency.

2 Current State of Research

This section provides a comprehensive overview of the current state of research in DED for FGMs, focusing on three key subsections: Challenges and Limitations, Design Methodologies, and Material Selection. By exploring the complexities and opportunities within these domains, this paper aims to shed light on the potential of FGMs manufactured through DED as transformative solutions for sustainable, cost-effective, and high-performance engineering applications.

2.1 Challenges and Limitations

Careful consideration is required due to the presence of specific challenges and limitations associated with DED. The printing process involves significant thermal gradients, leading to a heightened risk of warping due to residual stress. Typically, additional steps such as heat treatments or machining are required for post-processing, which vary depending on the material and the intended application of the part [7].

The fabrication of FGMs can amplify these problems, as the process of combining two dissimilar materials introduces its own set of challenges and constraints. This is primarily due to variations in thermal expansion, melting point, and density between the materials [8]. Among the most commonly encountered issues are cracking, elevated residual stress, porosity, warping, undesired phase formations, intermetallic formation, and chemical segregation [2, 4, 9–11]. Fortunately, most of these challenges and issues can be minimized and addressed by employing suitable design methodologies.

2.2   Design Methodologies

The specific DED method and equipment utilized have a significant impact on the type of FGMs that can be created. Certain methods enable only immediate transitions, while others can produce either discontinuous gradients or continuous gradient structures. Figure 1 shows some of the different transition types.

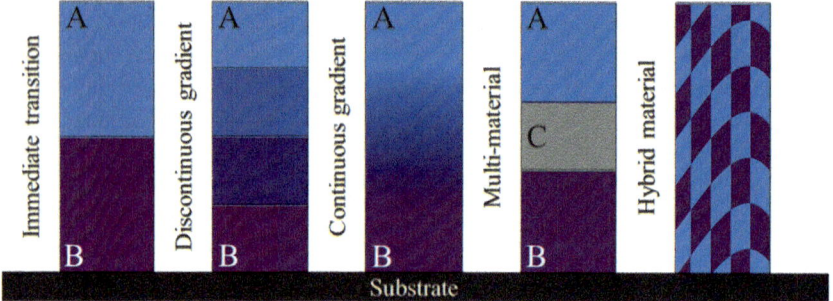

Fig. 1. Types of functionally graded materials

The production of successful immediate transitions can pose challenges as a significant difference in thermal expansion of the materials can result in cracks [12]. This risk can be mitigated by using continuous gradient structures. Powder-based DED systems with multiple powder feeders allow for seamless transitions between different powder compositions, ensuring a gradual adjustment and reducing the likelihood of cracks [5, 13]. Achieving this is not always straightforward or possible, as one must also consider the influence of machine capability, microstructure, formation of undesirable phases, unexpected precipitation, and intermetallic formation [8, 14]. Various approaches have been employed, incorporating different thermodynamic simulations and phase diagram data, to carefully plan the gradient structures.

Schaeffler diagrams, ternary phase diagrams, and thermodynamic phase simulations have been widely used to assist in the prediction of phases that may develop in FGMs. By connecting the nickel and chrome equivalents of various materials using a tie line, the Schaeffler diagram can provide a general estimation of the microstructure along the length of the gradient for ferritic systems [8, 14–16], as depicted in Fig. 2a. For the prediction of phases, including potential precipitation and intermetallic formation, ternary phase diagrams offer greater insight [17–19]. The procedure resembles that of Schaeffler diagrams, involving the use of a tie line to determine the phases present within the gradient section, as depicted by the linear path in Fig. 2b. Thermodynamic phase simulations have also been used to give more insight [14, 19]. Some authors have utilized ternary phase diagrams to strategically design a gradient path to avoid undesirable phases, as illustrated with the planned path in Fig. 2b [8, 20, 21].

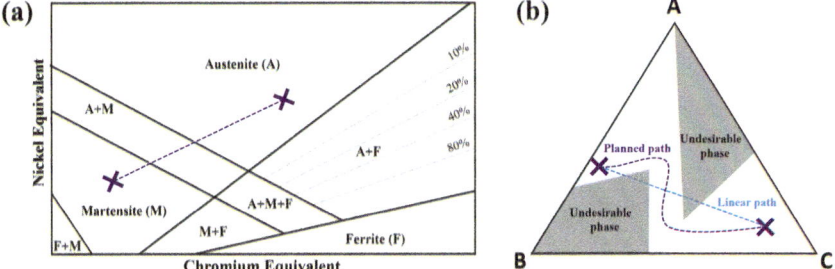

Fig. 2. Predicting microstructure of FGMs (a) Schaeffler diagram (b) Ternary phase diagrams

The primary constraint of the Schaeffler diagram, ternary phase diagram, and some thermodynamic phase simulations lies in their assumption of equilibrium solidification, which does not account for the rapid cooling rates typically encountered in DED [22]. As a result, these diagrams may not consistently provide a complete representation of the potential phases that can be formed. Hence, it can be necessary to employ more comprehensive thermodynamic approaches to account for the unique conditions associated with DED [19]. A. R. Moustafa *et al.* [22] and Z. Yang *et al.* [23], have proposed methods that incorporate Scheil calculations and data from ternary phase diagrams to generate phase diagrams resembling ternary phase diagrams for non-equilibrium cooling to address this issue.

2.3 Material Selection

FGMs rely on the careful selection of materials that possess: the required properties, thermodynamic feasibility, potential for cost savings, and availability. The cost of a material often reflects its level of complexity in beneficiation and processing. Opting for more economical materials not only results in a reduction of costs but also serves to mitigate the supplementary energy, environmental, and resource demands inherent in the manufacturing process of the product. Table 1 presents a comparison of the relative prices of various common materials, aiming to illustrate the potential cost savings achievable by utilizing FGMs. The majority of material systems utilized for FGMs are titanium, nickel, and iron-based systems.

Titanium and its alloys possess remarkable characteristics such as high strength-to-weight ratio, excellent corrosion resistance, and biocompatibility, making them ideal candidates for various engineering fields, including aerospace, automotive, and biomedical industries [19, 24]. Ti_6Al_4V (Ti64) is the most used alloy and is mostly joined

Table 1. Relative feedstock prices

Material	Ti$_6$Al$_4$V	Inconel 625	Inconel 718	316L stainless	17-4 PH	H11 Tool steel	Al-Si$_{10}$Mg$_5$	308L stainless
Powder	3.09	1.43	1.30	0.84	0.93	–	1.00	–
Wire	3.02	9.19	7.47	1.24	2.43	2.48	–	1.00

with Inconel 718/625 or stainless steel 316L/304L. The titanium systems tend to be the most difficult because of their tendency to form brittle intermetallic which results in cracks. L. D. Bobbio *et al.* observed critical cracking during the production of a non-continuous FGM while attempting to join Ti64 with 304L stainless steel. This cracking was attributed to the formation of σ-FeV [19]. However, in a subsequent study, L. D. Bobbio *et al.* successfully addressed this problem by incorporating an intermediate phase of vanadium into the FGM, which prevented cracking [24]. Similarly, B. Onuike *et al.* faced similar risks of cracks resulting from intermetallic formation. However, they were able to successfully avoid cracking by utilizing intermediate phases, specifically VC, to join Ti64 with Inconel 718 [25].

Nickel and its alloys possess high-temperature stability, excellent corrosion resistance, and exceptional mechanical properties, making them well-suited for various industries [4]. Inconel 718 and Inconel 625 are the most popularly used nickel-based alloys. While they are sometimes combined with Ti64, they are predominantly joined with stainless steel, with 316L being the most prevalent choice [3, 4, 26, 27]. Inconel exhibits a lower tendency to crack than titanium, but it remains a significant challenge. Various authors have noted that cracks occur due to the precipitation of Nb and Mo phases [3, 4, 27]. While titanium and nickel-based systems are commonly paired with iron-based alloys such as stainless steel, there are also other iron-based systems that are utilized. These systems are predominantly found in combinations of tool steels, copper, hard facing steels, and stainless steels, providing specific advantages and characteristics for various applications in the mold, tool and die industry [15].

In most of these research papers, stainless steel or low carbon steel is combined with a harder and more wear resistant steel, such as AISI H11, H13, M2, D2, or even vanadium carbide [9, 13, 15, 28]. These combinations have a lower risk of failure because the materials share similar chemical properties. Traditional welding techniques for these types of dissimilar metals have been extensively studied and often employ the Schaeffler diagram to predict the resulting microstructure. Considerable research has also been conducted on joining tool steel, primarily H13 tool steel, with copper. This is done to create tools and molds that benefit from both conformal cooling designs and copper's high thermal conductivity [29, 30]. X. Zhang *et al.* discovered that immediate copper and H13 FGMs are prone to cracking due to the significant difference in thermal expansion, resulting in solidification cracks. However, they found that incorporating a nickel-based D22 intermediate layer they could successfully produce a defect-free FGM [31].

3   Conclusion and Future Direction

The review found that FGM DED research is gaining increased attention due to the benefits of enhanced multifunctionality, reduced cost, and enhanced resource utilization, and DED can fabricate complex FGM structures. By utilizing FGMs, designers can overcome the limitations imposed by expensive materials when developing critical components. Moreover, the implementation of DED in FGM production offers significant cost and sustainability advantages. The reduced material waste and enhanced resource utilization offered by DED make it a good solution for sustainable manufacturing. The primary challenges encountered in the production of FGMs is the ability to create products without cracks, which arises from the intricate thermodynamic processes involved. To address this limitation, future research should prioritize the development and utilization of advanced thermodynamic models and simulations to aid in the design of defect-free FGMs.

Acknowledgements. The authors wish to acknowledge the financial support from the Department of Science and Innovation (DSI) and the National Research Foundation (NRF) in South Africa (Grant Nos: 41292 and 129313). Opinions expressed and conclusions arrived at are those of the authors and are not necessarily to be attributed to the CoE-SM, DSI or NRF.

References

1. Miyamoto Y, Kaysser W, Rabin B, Kawasaki A, Ford R (1999) Functionally graded materials. Materials technology series. Springer, Boston
2. Ghanavati R, Naffakh-Moosavy H (2021) Additive manufacturing of functionally graded metallic materials: a review of experimental and numerical studies. J Mater Res Technol 13:1628–1664
3. Zhang X, Chen Y, Liou F (2019) Fabrication of SS316L-IN625 functionally graded materials by powder-fed directed energy deposition. Sci Technol Weld Joining 24:504–516
4. Meng W, Zhang W, Zhang W, Yin X, Cui B (2020) Fabrication of steel-Inconel functionally graded materials by laser melting deposition integrating with laser synchronous preheating. Opt Laser Technol 131(106451) (2020)
5. Ahn D (2021) Directed energy deposition (DED) process: state of the art. Int J Precis Eng Manuf-Green Technol 8:703–742
6. Li Z, Sui S, Ma X, Tan H, Zhong C, Bi G, Clare AT, Gasser A, Chen J (2022) High deposition rate powder- and wire-based laser directed energy deposition of metallic materials: a review. Int J Mach Tools Manuf 181
7. Svetlizky D et al (2021) Directed energy deposition (DED) additive manufacturing: physical characteristics, defects, challenges and applications. Mater Today 49:1369–7021
8. Ben-Artz A, Reichardt A, Borgonia J-P, Dillon R, McEnerney B, Shapiro A, Hosemann P (2021) Compositionally graded SS316 to C300 Maraging steel using additive manufacturing. Mater Des 201(109500)
9. Baek GY et al (2017) Mechanical characteristics of a tool steel layer deposited by using direct energy deposition. Met Mater Int 23(4):770–777
10. Woo W, Kim D-K, Kingston E, Luzin V, Salvemini F, Hill MR (2019) Effect of interlayers and scanning strategies on through-thickness residual stress distributions in additive manufactured ferritic-austenitic steel structure. Mater Sci Eng 744:618–629

11. Sagong MJ, Kim ES, Park JM, Karthik GM, Lee B-J, Cho J-W, Lee CS, Nakano T, Kim HS (2022) Interface characteristics and mechanical behavior of additively manufactured multi-material of stainless steel and Inconel. Mater Sci Eng 847(143318)

12. Gandy D et al (2017) Design, fabrication, and characterization of graded transition joints. Weld J 96:295–306

13. Farren JD, Dupont JN, Noecker FF (2007) Fabrication of a carbon steel-to-stainless steel transition joint using direct laser deposition—a feasibility study. Weld J, 55–61

14. Xin D, Yao X, Zhang J, Chen X (2023) Fabrication of functionally graded material of 304L stainless steel and Inconel625 by twin-wire plasma arc additive manufacturing. J Mater Res Technol 23:4135–4147

15. Ostolaza M, Arrizubieta JI, Lamikiz A, Cortina M (2021) Functionally graded AISI 316L and AISI H13 manufactured by L-DED for die and mould applications. Appl Sci 11(771)

16. Zuback JS, Palmer TA, DebRoy T (2019) Additive manufacturing of functionally graded transition joints between ferritic and austenitic alloys. J Alloy Compd 770:995–1003

17. Li W, Chen X, Yan L, Zhang J, Zhang X, Liou F (2018) Additive manufacturing of a new Fe-Cr-Ni alloy with gradually changing compositions with elemental powder mixes and thermodynamic calculation. Int J Adv Manuf Technol 95:1013–1023

18. Bobbio LD, Otis RA, Borgonia JP, Dillon RP, Shapiro AA, Liu Z-K, Beese AM (2017) Additive manufacturing of a functionally graded material from Ti-6Al-4V to Invar: experimental characterization and thermodynamic calculations. Acta Mater, 133–142

19. Bobbio LD et al (2018) Characterization of a functionally graded material of Ti-6Al-4V to 304L stainless steel with an intermediate V section. J Alloy Compd 742:1031–1036

20. Reichardt A et al (2016) Development and characterization of Ti-6Al-4V to 304L stainless steel gradient components fabricated with laser deposition additive manufacturing. Mater Des 104:404–413

21. Eliseeva OV, Kirk T, Samimi P, Malak R, Arróyave R, Elwany A, Karaman I (2019) Functionally graded materials through robotics-inspired path planning. Mater Des 182(107975)

22. Moustafa AR, Durga A, Lindwall G, Cordero ZC (2020) Scheil ternary projection (STeP) diagrams for designing additively manufactured functionally graded metals. Addit Manuf 32(101008)

23. Yang Z, Sun H, Liu Z-K, Beese AM (2023) Design methodology for functionally graded materials: framework for considering cracking. arXiv Material Science

24. Bobbio LD, Bocklund B, Simsek E, Ott RT, Kramer MJ, Lui Z-K, Beese AM (2022) Design of an additively manufactured functionally graded material of 316 stainless steel and Ti-6Al-4V with Ni-20Cr, Cr, and V intermediate compositions. Addit Manuf 51(102649)

25. Onuike B, Bandyopadhyay A (2018) Additive manufacturing of Inconel 718 – Ti6Al4V bimetallic structures. Addit Manuf 22:844–851

26. Feenstra DR, Molotnikov A, Birbilis N (2020) Effect of energy density on the interface evolution of stainless steel 316L deposited upon INC 625 via directed energy deposition. J Mater Sci 55:13314–13328

27. Nannan C, Haris A, Zixuan W, John L, Hui S, Shun-Li S, Zi-Kui L, Jingjing L (2020) Microstructural characteristics and crack formation in additively manufactured bimetal material of 316L stainless steel and Inconel 625. Addit Manuf 32

28. Gualtieri T, Bandyopadhyay A (2018) Additive manufacturing of compositionally gradient metal-ceramic structures: stainless steel to vanadium carbide. Mater Des 139:419–428

29. Shin KH (2019) A method for representation and analysis of conformal cooling channels in molds made of functionally graded tool steel/Cu materials. J Mech Sci Technol 33:1743–1750

30. Articek U, Milfelner M, Anzel I (2013) Synthesis of functionally graded material H13/Cu by LENS technology. Adv Prod Eng Manage 8:169–176

31. Zhang X, Sun C, Pan T, Flood A, Zhang Y, Li L, Liou F (2020) Additive manufacturing of copper – H13 tool steel bi-metallic structures via Ni-based multi-interlayer. Addit Manuf 36(101474)

Investigations on the Use of Photodiodes for In-Situ Defect Detection in Laser-Based Powder Bed Fusion of Metals

Eckart Uhlmann[1,2], Gustavo Reis de Ascencao[1(✉)], Bernhard Hesse[3],
Jussi-Petteri Suuronen[3], David Carlos Domingos[1], and Jianlin Zhuang[1]

[1] Fraunhofer Institute for Production Systems and Design Technology (IPK), Pascalstraße 8-9, 10587 Berlin, Germany
gustavo.reis.de.ascencao@ipk.fraunhofer.de
[2] Technische Universität Berlin – Institute for Machine Tools and Factory Management (IWF), Pascalstraße 8-9, 10587 Berlin, Germany
[3] XPLORAYTION GmbH, Invalidenstr. 34, 10115 Berlin, Germany

Abstract. The sustainability of AM is strongly related to the ability of producing functional parts using minimal resources. In-situ monitoring saves energy and material by providing information to support process stops in case of anomalous events detected in critical regions. In this work, AlSi10Mg samples with different defect conditions were PBF-LB/M manufactured under the observation of photodiodes in three wavelength ranges. The samples were scanned through synchrotron-based micro-CT, and a defect score was attributed to every point in a non-destructive manner. The results were registered to the emissions acquired during the PBF-LB/M process. An exploratory data analysis has shown that the mean and standard deviation of the emissions of defect points do not differ from the ones found for points labelled as normal by micro-CT. A neural network was trained to infer the points' quality based on their emissions. The maximum F1-score was 4.79%, suggesting that the photodiodes in the studied set-up are not suitable for the in-situ identification of local porosities in PBF-LB/M parts.

Keywords: Additive Manufacturing (AM) · Laser-based Powder Bed Fusion of Metals (PBF-LB/M) · Process Monitoring · Computed Tomography (CT)

1 Introduction

AM is increasingly gaining industrial attention due to the flexibility and adaptability it might offer to products and production processes. The possibility of manufacturing parts with optimized lightweight design places AM as an important tool for the achievement of carbon neutrality. To mention one example of the aerospace sector, Lufthansa estimated fuel savings of 47 tons per aircraft and year for a general weight reduction of about 900 kg in an Airbus A340-600 [1]. However, it is widely accepted that the specific energy consumption of metallic AM processes is considerably higher than in traditional

© The Author(s) 2025
H. Kohl et al. (Eds.): GCSM 2023, LNME, pp. 746–754, 2025.
https://doi.org/10.1007/978-3-031-77429-4_83

manufacturing routes [2]. Moreover, trial-and-error approaches are still a common practice during the development of new AM applications, since closed-loop control is still in its infancy. Science and industry are increasingly putting effort into the strategies to enhance the sustainability of AM processes. In-situ process monitoring for the identification of defects during the build plays a significant role in economic and environmental terms, due to the saving potentials in time, energy and material usage. Grasso [3] presented a comprehensive review on PBF-LB/M in-situ monitoring methods. The goal of the current work is to investigate in how far photodiodes can be used to identify the formation of local defects already during the PBF-LB/M process. Samples with three defect conditions were analyzed: low porosity (samples "i" to "iii", with nearly no voids), key-holes (samples "iv" and "v", with spherical voids induced by high energy processing) and lack-of-fusion (samples "vi" to "viii", with irregular-shaped voids induced by low energy processing). In this study a dimensionless quantity called defect score φ is introduced. It is defined as the ratio between the number of voxels segmented as voids and the total number of voxels in a certain region of interest. It is used in this paper interchangeably with its synonymous expression local porosity.

2   Methodology

A RenAM 500Q HT system from Renishaw, UK, was used to conduct the PBF-LB/M experiments. The system is equipped with four laser units, each one containing three photodiodes to monitor the process. Figure 1a shows the schematic representation of one laser unit. The photodiodes PD_L, PD_P and PD_M were designed to capture laser ε_L (1080 nm), plasma ε_P (between 700 and 1050 nm) and melt pool ε_M (between 1100 and 1700 nm) emissions. PD_L is positioned in front of the laser source, whereas PD_P and PD_M detect the emissions generated in the processing zone that travel back in the same coaxial path of the laser beam. The fields of view of PD_P and PD_M in the build plane are approximately a square with a 6.0 mm side and a circle with 2.4 mm in diameter. The photodiodes have 12-bit resolution and their acquisition frequency is 100 kHz. Examples of the evolution of average emissions along layers are presented in Fig. 1d–e.

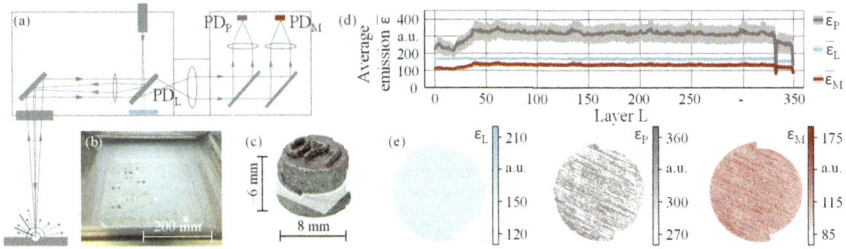

Fig. 1. The investigation process: monitoring system (a), off-axis image of build chamber (b), produced sample (c) and evolution of ε_L, ε_P and ε_M along layers (d) and within a single layer (e)

An off-axis image of the PBF-LB/M process is shown in Fig. 1b. A gas atomized AlSi10Mg powder from ECKART TLS, Germany was used in a non-virgin state in the experiments. The original particle size distribution was 20–63 μm. Based on the work of Gobert [4], a cylinder with an external ramp (Fig. 1c) was designed in order to enable the geometrical registration between data from different sources. All experiments were conducted in modulated laser modus meaning the laser operates sequentially exposing discrete points in order to form a scan line. For every discrete point, an average value of the PBF-LB/M emissions captured by the photodiodes is saved. A commercially available parameter set (supplied by Renishaw, UK) was used as basis for the experimental plan. Variations of laser power P_l, exposure time t_e and hatch distance h_d were designed in order to stochastically generate different defect conditions in AlSi10Mg, i.e. samples with low porosity, keyholes and lack-of-fusion. After the PBF-LB/M process, the samples were removed from the base plate via wire electrical discharge machining. The final height of the samples was approximately 6 mm. From all cylindrical samples, eight units were scanned via micro-CT in the biomedical beamline of the ESRF, each of them manufactured with a different combination of PBF-LB/M process parameters. The original voxel size achieved was approximately 3.5 μm for all three directions. The photodiodes in-situ measurements and the micro-CT results were used to develop a method for geometrical data preparation. This method enabled the comparison between PBF-LB/M emissions and local porosity values for every point of the sample in a non-destructive manner. The registered datasets served as basis for an exploratory data analysis approach and finally for the training and testing of a neural network designed to predict the local porosity based on laser ε_L and plasma emissions ε_P.

3   Results and Discussions

3.1   Developed Method for Geometrical Data Preparation

A method comprising six steps was developed in this work in order to enable the representation of data obtained from different sources in the same coordinate system. These steps are schematically illustrated in Fig. 2.

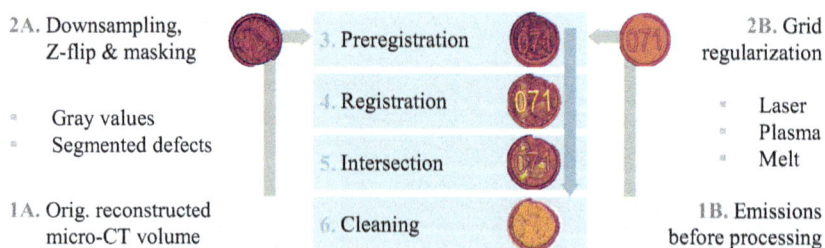

Fig. 2. The six steps for geometrical data preparation of micro CT and PBF-LB/M emissions

The starting point of the method are the original reconstructed micro-CT volume (1A) and emissions before processing (1B). Since the micro-CT has a much higher resolution than the grid of points whose emissions have been captured, a downsampling has to be conducted (2A). An Otsu thresholding is used to remove voxels corresponding to the sample surroundings during acquisition (e.g. plastic tube). In this step, the segmented defects volume generated by a U-NET, similarly to Hilbert [5], is also incorporated into the method. A real number between 0 and 1 is assigned to each voxel of the downsampled micro-CT volume, indicating its defect volume fraction or defect score φ (0: fully dense, 1: complete void). After the grid regularization (2B), which removes the texture of the emissions files arising from the acquisition along the scan vectors in each layer, all volumes exhibit an anisotropic voxel size of $90 \times 90 \times 30\ \mu m^3$. The preregistration (3) step consists of crops and rotations of the volumes in order to minimally position them in the same location in space. This reduces the divergence probability of the optimization algorithm run in the step (4) to register the volumes. The registration is conducted within the module ITKElastix for Python [6]. The laser emission volume was set as fixed image, while the micro-CT volume was used as moving image in the search for an affine transformation that minimizes a mutual information metric. Finally, an intersection (5) and a cleaning step (6) are conducted to isolate the sample's bulk for further analyses, thus removing the complex boarder region. This is plausible since PBF-LB/M parts usually undergo surface finishing prior to final application.

3.2 Correlation Between Photodiodes Measurements and Local Porosity

Once the cleaned datasets were generated for all samples according to the method presented in Fig. 2, an exploratory data analysis was conducted to investigate the correlation between the photodiodes measurements and local porosity. Firstly, scatter plots and histograms of the emissions in every single point of the cleaned datasets were used. Figure 3 presents these plots for sample "iv", which will be used for the discussions since other samples in this work presented similar behavior.

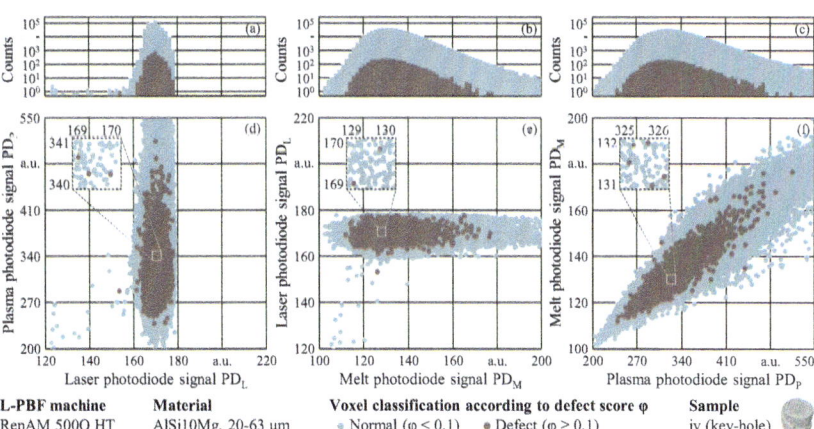

Fig. 3. Scatter plots and histograms of the emissions for all points of the cleaned dataset "iv"

Two different plotting colors were used in order to highlight possible clustering regions or histogram deviations of defect points with respect to normal points. The criteria used for the classification of a point as defect was a defect score φ higher or equal to 0.1, i.e. the ratio between the number of voxels classified as voids by the U-NET and the total number of voxels present in the kernel being analyzed higher or equal to 10%. The histograms in Fig. 3a–c show that the number of defect points is usually much lower than the number of normal points (between 2 and 3 orders of magnitude for each bin regardless of the emission type), revealing a high sparsity of defect events in the sample's bulk. Only sample "vii" is an exception to this rule due to the very low volumetric energy density E_v used for manufacturing, which caused a level of defect much higher than in the other seven analyzed samples (Fig. 4). In addition to the sparsity of defect events, one can note that the mean and standard deviation of the emissions distribution for defect points are virtually identical to the ones found in the emissions distribution for normal points. This provides an evidence for the absence of a simple correlation between emissions and local porosity.

The similarity between the distributions is also seen in Fig. 3d–f: the defect points are sparse and concentrated in the center of the scatter plots of the emissions, showing that a clear clustering region does not exist. The zoomed regions of the scatter plots highlight the complexity needed for a decision boundary to successfully classify the porosity level of a single point based on its emissions. A further takeaway from this analysis is the strong correlation between plasma ε_P and melt pool emissions ε_M (Fig. 3f). The average Pearson correlation coefficient r for all samples was 92.7%, showing that there would not have been a significant amount of information lost if only one of the photodiodes had been observed. This does not apply to the laser photodiode PD_L, since it does not capture emissions of the metallurgical process.

In order to further investigate the correlation between photodiodes measurements and local porosity, an analysis considering the average emission and defect score in each layer of all samples was conducted. Every circle plotted in Fig. 4 corresponds to a layer and the size of its diameter represents the average porosity φ_L of that layer, which is the average the defect scores of all points in that layer. The layers were qualitatively classified according to their types of defects. For the parameters studied in this work, the layers with lack-of-fusion defects (i.e. low volumetric energy density E_v) exhibit the larger porosities, followed by the layers with keyholes (i.e. high volumetric energy density E_v) and the low porosity layers, which have nearly no voids.

From Fig. 4a–b it can be seen that, although all lack-of-fusion layers exhibit a low average laser emission $\varepsilon_L < 125$, the keyhole and low porosity layers cannot be classified solely by this quantity. This is also the case for the average melt pool emission ε_M, since one can find layers with all types of defects in the range $100 < \varepsilon_M < 140$. The only quantity that could directly classify the type of defect and therefore also the average porosity of each layer was the average plasma emission ε_P (Fig. 4b). The layers with $\varepsilon_P < 200$ presented lack-of-fusion defects while keyhole was the predominant defect type for $\varepsilon_P > 300$. In-between were solely low porosity layers.

Although these classification thresholds were found, two facts limit their relevance for in-situ monitoring purposes: (a) there is a strong correlation between the emissions and the PBF-LB/M process parameters themselves ($R^2 = 98.8\%$ for the prediction of the plasma emission ε_P, for example), as exemplified in Fig. 4c for the volumetric energy density E_v. Therefore, one could directly use the PBF-LB/M process parameters to predict type and quantity of defects in each layer, removing the necessity of an extra

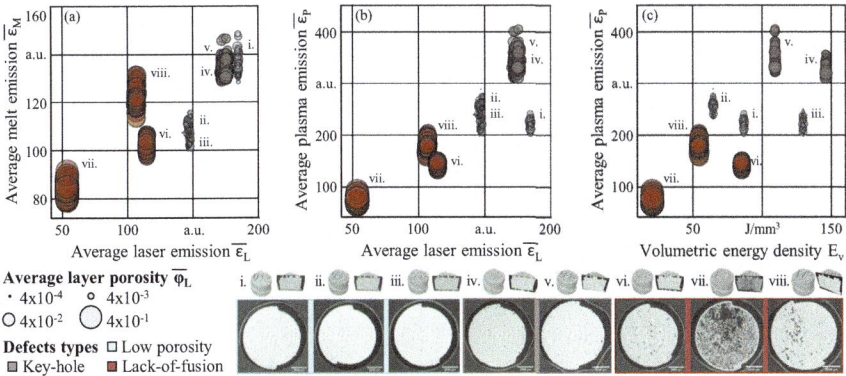

Fig. 4. Layerwise analysis of emissions (a–b) and correlation to volumetric energy density (c)

hardware for the in-situ monitoring. Additionally, (b) this analysis only considers average values along layers, thus removing its applicability for a local defect prediction.

3.3   Machine Learning (ML) Model Performance Assessment

A further attempt to predict the presence of defects in each point of the sample was conducted through the development of a ML model. Neural networks were chosen based on the complexity needed for the decision boundaries in this classification task. Instead of solely using the emissions of a single point, the emissions of neighboring points were also considered as input features. This approach is plausible since defects might also affect the emissions of their surroundings. Preliminary tests were conducted to determine the model's hyperparameters, focusing on the number of inputs and the structure of hidden layers. The best result was achieved with a fully connected neural network structure of (250-125-42-6-1). The number of inputs was 250, comprising the laser ε_L and plasma emissions ε_P of all neighbors in a cube $5 \times 5 \times 5$ centered in the analyzed point ($2 \times 5 \times 5 \times 5 = 250$). The melt pool emissions ε_M were not considered due to their strong correlation to the plasma emissions ε_P (Fig. 3f). The number of neurons in hidden layers gradually reduces down to six neurons in the last hidden layer, representing the six faces of a cube centered around the analyzed point. All points of the cleaned samples "ii", "iii", "iv" and "v" composed the dataset used to train the model. To ensure a more balanced number of normal and defect observations, a fraction of the good points with defect score $\varphi < 0.1$ was randomly removed from the dataset in order to keep the number of normal (i.e. $\varphi < 0.1$) and defect points (i.e. $\varphi \geq 0.1$) equal. Without the initial dataset balancing, due to the scarcity of defect points in the samples, the neural network would have probably been underexposed to this kind of situation, compromising its learning behavior. The training-testing ratio was set to 4:1 and MSE was used as loss function by the gradient descent optimization. The coefficient of determination R^2 was used as metric for the performance assessment and reached a maximal value of 0.003% for the test set at the final stage of the learning curve. A plateau was reached at a relatively early stage of the learning curve, showing that additional input data would not enhance the performance of the model. A similarly low value of R^2 for the training set was also exhibit, suggesting that the developed model suffers from high bias and still cannot capture the complex relationship between emissions and defect score in each point of the samples. The performance after a binarization of the actual and predicted defect scores using 0.1 as threshold is presented in Fig. 5.

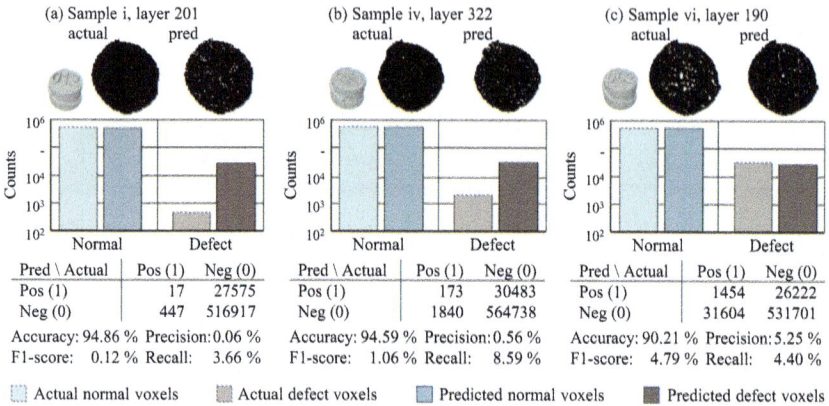

Fig. 5. Performance assessment of the ML model to predict defects based on emission values

The neural network was applied to all points of the samples "i", "iv" and "vi". Only the sample "iv" was known in forehand by the model, since it was included in the training set. One can notice the high accuracy values achieved by the model, greater than 90% for all samples. This is caused by the high number of true negatives: the vast majority of the samples' voxels are actually normal and classified by the model as so. The precision and recall values, however, were very low, yielding poor F1-scores for all samples. The lowest F1-score was 0.12%, achieved for sample "i". Sample "vi" has achieved a higher F1-scocre than sample "iv", which has previously been seen by the neural network. This is due to the fact that sample "vi" exhibits a high number of actual defects, which increases the probability of matching predicted defects along the analyzed volume. However, the F1-score of 4.79% achieved for sample "vi" still characterizes the model as inapplicable. In general, the model predicts more defects than the actual number, if applied to samples similar to the ones in the training dataset. This might be explained by the dataset balancing conducted prior to training.

4   Conclusions and Outlook

The results of this work suggest that the photodiodes are not a suitable tool for the in-situ identification of local porosities in PBF-LB/M parts, at least in the set-up used in the experiments. Assuming that (a) the quality of the micro-CT data is sufficiently high for this to be used as ground truth and also that (b) the developed method for geometrical data preparation is robust enough to avoid as mismatching between features and labels, one can state that the relationship between PBF-LB/M emissions and local defects is of random nature. The first assumption seems to be reasonable, since current state-of-the-art micro-CT techniques were used in this work. There is however on-going work to further development uncertainty metrics for 3D imaging methods. One limitation seen in the current work is the fact that defects that might have been created and destroyed by the layer remelting effect during the build are therefore not seen, introducing a labelling error. Regarding the second assumption, even if there was a slightly misregistration between micro-CT and emission volumes, the approach of additionally considering two neighbors in each direction as inputs for the neural network should absorb such as mismatch and minimize its eventual effects. An affine transformation was chosen due to its physical plausibility: it is the simplest transformation that can compensate for thermal shrinkage and other sort of volumetric deviations between designed and manufactured parts. In this sense, a possible explanation for the absence of correlation between the emissions and local defects relies on the large field of view of the photodiodes in the working plane. This might average out any eventual local emissions anomalies that could be related to the formation of micro defects (about 100 μm in equivalent diameter). Moreover, the larger the field of view, the greater the effect of sparks generated in outlying regions in the photodiodes signals for one point of interest. Due to the strong correlation of emissions to the PBF-LB/M process parameters, one possible use of the photodiodes could be system health monitoring. A strong drift in the laser power, for example, could be identified by the photodiodes since the actual volumetric energy density would change, thus affecting the emissions. The authors are working on the use of thermography for the identification of defects based on local heat accumulation.

Acknowledgements. This project was supported by the Federal Ministry for Economic Affairs and Climate Action (BMWK) on the basis of a decision by the German Bundestag. The authors acknowledge access to beamline ID17 of the European Synchrotron Radiation Facility (ESRF) and support of Michael Krisch and Luca Fardin.

References

1. Lufthansa Group (2012) Balance - Key data on sustainability within the Lufthansa Group
2. Priarone PC et al (2018) Laser powder bed fusion (PBF-LB/M) additive manufacturing: on the correlation between design choices and process sustainability. Procedia CIRP 78:85–90
3. Grasso M, Remani A, Dickins A et al (2021) In-situ measurement and monitoring methods for metal powder bed fusion: an updated review. Meas Sci Technol 32:112001
4. Gobert C, Reutzel EW, Petrich J et al (2018) Application of supervised machine learning for defect detection during metallic powder bed fusion additive manufacturing using high resolution imaging. Addit Manuf 21:517–528
5. Hilbert A, Madai VI, Akay EM, et al (2020) BRAVE-NET: fully automated arterial brain vessel segmentation in patients with cerebrovascular disease. Front Artif Intell 3
6. Klein S, Staring M, Murphy K et al (2010) Elastix: a toolbox for intensity-based medical image registration. IEEE Trans Med Imaging 29:196–205

In-Process Detection of Defects on Parts Produced by Laser Metal Deposition Using Off-Axis Optical Monitoring

Marco Mazzarisi[1]([✉]) [iD], Maria Grazia Guerra[1] [iD], Marco Latte[1] [iD], Fabrizia Devito[1,2] [iD], Luigi Maria Galantucci[1] [iD], Sabina Luisa Campanelli[1] [iD], Fulvio Lavecchia[1] [iD], and Michele Dassisti[1] [iD]

[1] Department of Mechanical, Mathematics, and Management (DMMM), Polytechnic University of Bari, Bari, Italy

{marco.mazzarisi,mariagrazia.guerra,marco.latte,fabrzia.devito, luigimaria.galanucci,sabinaluisa.companelli,fulvio.lavecchia, michele.dassisti}@poliba.it, fabrizia.devito@iusspavia.it

[2] Department of Sciences, Technologies and Society, Scuola Universitaria Superiore IUSS Pavia, 27100 Pavia, Italy

Abstract. Laser Metal Deposition (LMD) is emerging among metal Additive Manufacturing technologies due to its wide range of applications. This technique represents an evolution of laser cladding, currently used for fabricating and repairing complex metal components, promoting manufacturing sustainability. One of the main drawbacks hindering the widespread use of these technology is the complexity of implementing monitoring equipment on industrial LMD systems with limited modification setups. Therefore, it is essential to develop appropriate off-axis systems that allow effective monitoring of the deposition process. The present work proposes a prototype off-axis monitoring system consisting of a pair of specially set cameras capable of analyzing the evolution of the melt pool and discerning fundamental information on geometry, size and brightness intensity. By correlating this information with the process outcome, it could be possible to forecast the most frequent defects related to the deposition process. Experimental tests have been carried out, in which powder flow and laser alterations were specifically induced. The prototype system enabled the characterization of each type of process variation and the determination of specific indicators, serving as the basis for achieving a zero-waste sustainable manufacturing process.

Keywords: Additive manufacturing · Laser Metal Deposition · Sustainability · Process monitoring · Defect detection

1 Introduction

Additive manufacturing (AM) is nowadays recognized as the most suitable technology for developing a more sustainable industrial production [1]. The fabrication of components by selective addition of material results in a strong reduction of raw materials and

H. Kohl et al. (Eds.): GCSM 2023, LNME, pp. 755–762, 2025.
https://doi.org/10.1007/978-3-031-77429-4_84

energy consumption, which lowers the carbon footprint of the production process [2]. This approach aimed at producing components at the near net-shape drives the industrial manufacturing towards a zero-waste concept [3, 4].

Among the various AM processes, Laser Metal Deposition (LMD) plays a significant role thanks to the capability to work a wide range of metal materials [5] and ceramics [6]. The LMD is a particularly versatile process that is used both for the production of new component [7] and for remanufacturing purposes, such as the repair and modification of damaged tooling molds and dies components [8]. That makes the LMD one of the most promising technologies from a sustainability perspective, being capable of extending the lifespan of worn components and avoiding complete replacement [9].

However, such a sustainable approach implies a very high reliability of the manufacturing process that can minimize production defects. In fact, even the slightest problem during the deposition process can result in production failure or expensive rework [1]. Certain performances are still difficult to achieve because of some issues that plague LMD technology. The main barrier is given by the high complexity that characterizes the deposition process, which requires a precise control over process parameters and environmental conditions [10]. Furthermore, the LMD process can induce distortion, warping, or cracking of the component, which can result in dimensional inaccuracies [11]. This becomes even more pronounced when dealing with complex geometries or large components [12]. All these issues can have a significant impact on the final quality of the produced parts and, consequently, on the sustainability of the process [13, 14].

In order to overcome these challenges, researchers have developed various monitoring methods to improve the control and the quality of the deposition process. Monitoring methods for LMD process are based on the analysis of physical phenomena that occur in laser-material interactions and can be classified into two main categories: in-situ monitoring [15, 16] and post-process monitoring [17, 18]. In-situ monitoring techniques, whether coaxial or off-axis, are certainly the most useful for the identification of manufacturing defects during the production process and, in some cases, for their mitigation [19]. Coaxial monitoring utilizes the same optical path as the laser beam, precisely evaluating process alterations. Nevertheless, these systems have severe limitations regarding the field of view, which is restricted to a small region surrounding the process area, which may not be appropriate when assessing the quality of the entire layer.

On the other hand, off-axis techniques can be preferred when analyzing layer-wise information. In fact, these give the opportunity to evaluate several data ranging from the geometric characteristics of the melt pool to the thermal field obtained in the production of the entire component [15, 20]. Obviously, the choice between these approaches depends on specific requirements, priorities, and resources available. Hybrid approaches to correlate changes in melt pool emissions with alterations in process parameters was also developed [21].

The present work proposes a prototype off-axis monitoring system consisting of a pair of specially set cameras, providing data of the melt pool in video format, and of algorithms capable of detecting and analyzing the most frequent defects of the LMD process. Specifically, the system can discern fundamental information on geometry, size and brightness intensity of the melt pool throughout the deposition process. The potentials of the systems were investigated by means of experimental tests in which process flaws (powder and laser flow variation) were specially induced. Furthermore, specific indicators have been defined that can allow the identification of production defects in order to avoid unnecessary rework and achieve a zero-waste manufacturing process.

2 Experimental Setup

The off-axis optical monitoring system was applied on industrial LMD system. This is a prototype machine consisting of a 3 kW CO_2 laser source and a powder feeding system. The substrate and powders used are AISI 316L stainless steel. The latter were conveyed through a 3-way multi-jet nozzle into the melt pool by using a helium gas flow.

Fig. 1. The prototype monitoring system consisting of two cameras (front and side) implemented on industrial LMD instrumentation (a) and the sample produced for the experimental test (b).

The monitoring system consists of two high-resolution cameras (DSLR CANON EOS 760D) equipped with a 50 mm lens. These were placed at a distance of about one meter to avoid possible damage caused by the high temperatures generated during the laser process and dispersion into the environment of the metal powder. As shown in Fig. 1(a), the cameras were positioned with an inclination relative to the deposition plane of about 30° and with an angle between them of about 90°, so that a full field of view of the working area was achieved.

Videos were acquired with a resolution of 1920 × 1080 px and a frame rate of 25 frames per second. The ISO parameter was set to 100 and the camera aperture and the exposure time were specifically set according to the adopted laser process parameters, seeking to acquire only the melt pool area on each frame and to maximize the spatial resolution and the depth of field.

2.1 Experimental Test

The capabilities of the monitoring system were tested during the fabrication of a sample consisting of 7 layers with nominal dimensions of 50 × 25 mm, shown in Fig. 1(b). This was made adopting a bidirectional raster deposition strategy with 90° rotation at each layer. Table 1 shows the process parameters used:

Table 1. LMD process parameters.

Process parameter	Value
Effective laser power [W]	1600
Laser spot diameter [mm]	2.00
Scanning speed [mm/min]	200
Powder feed rate [g/min]	5.0
Carrier gas flow [l/min]	5.0
Shielding gas flow [l/min]	11.0

The effective laser power was evaluated using an analog laser power probe. Specific alterations in process conditions were induced during the deposition process, simulating some typical deposition problems. Specifically, in layer 3 the laser power supply was stopped, in layer 5 the powder flow was interrupted, and in layer 7 the laser head movement speed was varied. These events alter the regular deposition process, changing the size of the melt pool and its brightness.

3   Discussion and Results

The videos were analyzed using a specific algorithm developed in the MATLAB environment. This first extracts every single frame of the video related to the layer under consideration. Next, the algorithm allows for in-process reconstruction of the component geometry by comparing the melt pool size and brightness data. Figure 2 shows the grayscale reconstructions of the seven layers according to the two viewpoints. The false-color representation aids in distinguishing areas of higher brightness (red) from those characterized by lower light intensity (blue). In this way, the reconstruction of each layer of the constructed component can be obtained providing fundamental information about local defects and the homogeneity of the internal area can be also ascertained. The simultaneous implementation of the two cameras intended to limit problems due to the different points of view and possible reflections. It can be seen that the two reconstructions highlight different details of the deposition process, allowing a more in-depth analysis of the component. The layer reconstruction of the side camera is more affected by the perspective distortion along the prevalent dimension (the sample length). This highlights the directionality of this method when applied to a geometry with a prevalent dimension.

In order to highlight the effect of induced defects on the deposition process, it was selected the melt pool area as a reference indicator. Figure 3 shows the results of the detailed analysis performed on layers 3, 5 and 7. The layer reconstruction in Fig. 3(a) reveals how the shutdown of the laser beam results in the vanishing of the melt pool and

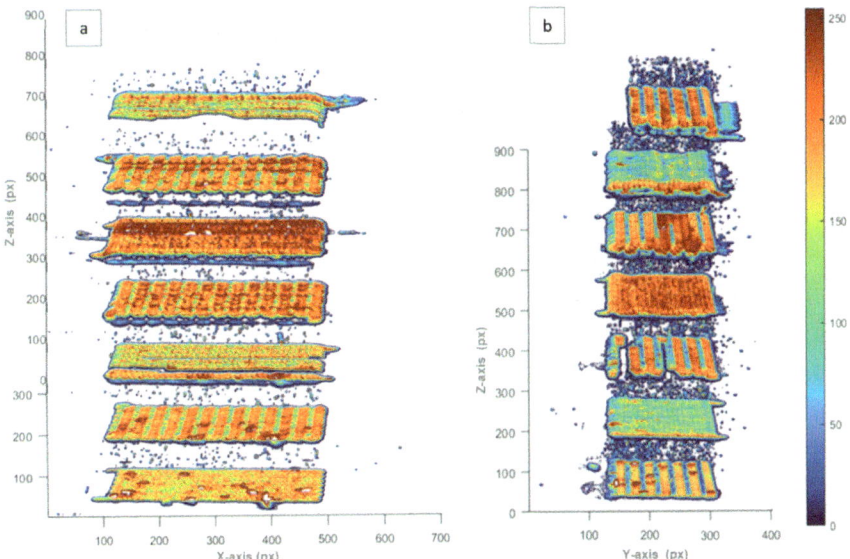

Fig. 2. Reconstruction of the component produced by front camera (a) and side camera (b), in which the brightness is represented in false color.

the lack of some traces. Consequently, the graph in Fig. 3(b) shows a sudden decrease in the melt pool size due to the impossibility of the algorithm to detect areas with sufficient light intensity. In Fig. 3(c) is shown the reconstruction of layer 5 in which the supply of the powder flow was interrupted in specific areas. An interruption of the powder flow generates an increase in the laser energy incident the component. These can be identified by the higher brightness of the affected area due to the lower shielding effect by the powders. This event can also be read in the graph in Fig. 3(d), in which the size of the melt pool increases significantly in the central area of the deposition process. Finally, Fig. 3(e) shows the reconstruction of layer 7 in which the speed movement of the nozzle was changed (first decelerated and then accelerated). This can be read in the reconstruction which, although it is carried out with a constant step, shows areas with more pronounced overlap (closer melt pools) and other areas showing the melt pools more widely spaced. This effect is readable, although in a less prominent form, in Fig. 3(f) in which the size of the melt pool varies substantially, showing the instability of the process in those areas.

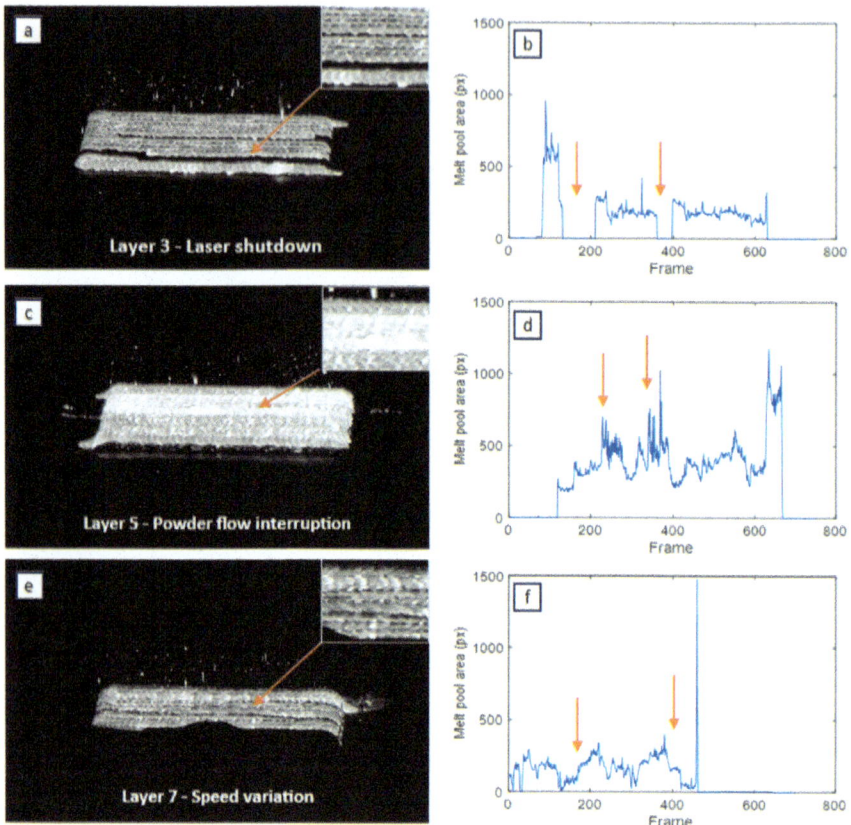

Fig. 3. Grayscale reconstruction of layers 3 (a), layer 5 (c) and layer 7 (e) and melt pool area trends of layers 3 (b), layer 5 (d) and layer 7 (f).

4 Conclusion

The proposed off-axis optical monitoring system exhibited several interesting features regarding defect detection in the Laser Metal Deposition (LMD) process. Indeed, it accurately defined the effects on the melt pool of the main process issues and successfully reconstructed the geometry of the component, precisely identifying and locating manufacturing defects. Although the system showed some points that require further development, particularly concerning issues of directionality of the system that may create blind spots in the analysis, it is proposed as a relevant contribution to optimizing AM processes for achieving a more sustainable production.

Acknowledgements. This work has been partially supported by Ministry of University and Research MUR, Italy, PON 2014/2020, Innovative Solutions for the quality and sustainability of ADDitive manufacturing processes – SIADD project.

This study was carried out within the MICS (Made in Italy – Circular and Sustainable) Extended Partnership and received funding from the European Union Next-GenerationEU (PIANO NAZIONALE DI RIPRESA E RESILIENZA (PNRR) – MISSIONE 4 COMPONENTE 2,

INVESTIMENTO 1.3 – D.D. 1551.11-10-2022, PE00000004). This manuscript reflects only the authors' views and opinions, neither the European Union nor the European Commission can be considered responsible for them.

References

1. Ghobadian A, Talavera I, Bhattacharya A, Kumar V, Garza-Reyes JA, O'Regan N (2020) Examining legitimatisation of additive manufacturing in the interplay between innovation, lean manufacturing and sustainability. Int J Prod Econ 219:457–468. https://doi.org/10.1016/j.ijpe.2018.06.001

2. Javaid M, Haleem A, Singh RP, Suman R, Rab S (2021) Role of additive manufacturing applications towards environmental sustainability. Adv Ind Eng Polym Res 4:312–322. https://doi.org/10.1016/j.aiepr.2021.07.005

3. Pratheesh Kumar S, Elangovan S, Mohanraj R, Ramakrishna JR (2021) Review on the evolution and technology of State-of-the-Art metal additive manufacturing processes. Mater Today Proc 46:7907–7920. https://doi.org/10.1016/j.matpr.2021.02.567

4. Singh S, Ramakrishna S, Gupta MK (2017) Towards zero waste manufacturing: a multidisciplinary review. J Clean Prod 168:1230–1243. https://doi.org/10.1016/j.jclepro.2017.09.108

5. Karmuhilan M, Kumanan S (2022) A review on additive manufacturing processes of Inconel 625. J Mater Eng Perform 31:2583–2592. https://doi.org/10.1007/s11665-021-06427-3

6. Pfeiffer S et al (2021) Direct laser additive manufacturing of high performance oxide ceramics: a state-of-the-art review. J Eur Ceram Soc 41:6087–6114. https://doi.org/10.1016/j.jeurceramsoc.2021.05.035

7. Xue L (2010) Laser consolidation: a rapid manufacturing process for making net-shape functional components. In: Advances in laser materials processing, pp. 492–534. Elsevier. https://doi.org/10.1533/9781845699819.6.492

8. Foster J, Cullen C, Fitzpatrick S, Payne G, Hall L, Marashi J (2019) Remanufacture of hot forging tools and dies using laser metal deposition with powder and a hard-facing alloy Stellite 21®. J Remanufactur 9:189–203. https://doi.org/10.1007/s13243-018-0063-9

9. Piscopo G, Iuliano L (2022) Current research and industrial application of laser powder directed energy deposition. Int J Adv Manuf Technol 119:6893–6917. https://doi.org/10.1007/s00170-021-08596-w

10. Parsazadeh M, Sharma S, Dahotre N (2023) Towards the next generation of machine learning models in additive manufacturing: a review of process dependent material evolution. Prog Mater Sci 135:101102. https://doi.org/10.1016/j.pmatsci.2023.101102

11. Yang S, Xu Z, Peng S, Cao S, Liu W, Wang F (2023) Investigation on the relationships among process parameters, molten pool characteristics, and fabrication quality of Ti-10V-2Fe-3Al by laser direct energy deposition. Optik 275:170521. https://doi.org/10.1016/j.ijleo.2023.170521

12. Saboori A, Aversa A, Marchese G, Biamino S, Lombardi M, Fino P (2019) Application of directed energy deposition-based additive manufacturing in repair. Appl Sci 9:3316. https://doi.org/10.3390/app9163316

13. Shamsaei N, Yadollahi A, Bian L, Thompson SM (2015) An overview of direct laser deposition for additive manufacturing; Part II: mechanical behavior, process parameter optimization and control. Addit Manuf 8:12–35. https://doi.org/10.1016/j.addma.2015.07.002

14. Selicati V, Mazzarisi M, Lovecchio FS, Guerra MG, Campanelli SL, Dassisti M (2022) A monitoring framework based on exergetic analysis for sustainability assessment of direct laser metal deposition process. Int J Adv Manuf Technol 118:3641–3656. https://doi.org/10.1007/s00170-021-08177-x

15. Mazzarisi M, Angelastro A, Latte M, Colucci T, Palano F, Campanelli SL (2023) Thermal monitoring of laser metal deposition strategies using infrared thermography. J Manuf Process 85:594–611. https://doi.org/10.1016/j.jmapro.2022.11.067
16. Latte M (2023) In process monitoring of geometrical characteristics in laser metal deposition: a comparative study, pp 101–110. https://doi.org/10.21741/9781644902479-12
17. Castellano A, Mazzarisi M, Campanelli SL, Angelastro A, Fraddosio A, Piccioni MD (2020) Ultrasonic characterization of components manufactured by direct laser metal deposition. Materials 13:2658. https://doi.org/10.3390/ma13112658
18. Pellegrini A, Palmieri ME, Guerra MG (2022) Evaluation of anisotropic mechanical behaviour of 316L parts realized by metal fused filament fabrication using digital image correlation. Int J Adv Manuf Technol 120:7951–7965. https://doi.org/10.1007/s00170-022-09303-z
19. Yi L et al (2022) Optical sensor-based process monitoring in additive manufacturing. Procedia CIRP 115:107–112. https://doi.org/10.1016/j.procir.2022.10.058
20. Garmendia I, Leunda J, Pujana J, Lamikiz A (2018) In-process height control during laser metal deposition based on structured light 3D scanning. Procedia CIRP 68:375–380. https://doi.org/10.1016/j.procir.2017.12.098
21. Zhang P et al (2021) Anomaly detection in laser metal deposition with photodiode-based melt pool monitoring system. Opt Laser Technol 144:107454. https://doi.org/10.1016/j.optlastec.2021.107454

Energy Generation and Efficiency

Potentials for Energy Savings and Carbon Dioxide Emissions Reduction in Cement Industry

Shoaib Sarfraz[1](✉) ⓘ, Ziyad Sherif[1] ⓘ, Michal Drewniok[2] ⓘ, Natanael Bolson[3] ⓘ, Jonathan Cullen[3] ⓘ, Phil Purnell[2] ⓘ, Mark Jolly[1] ⓘ, and Konstantinos Salonitis[1] ⓘ

[1] Sustainable Manufacturing Systems Centre, School of Aerospace, Transport and Manufacturing, Cranfield University, Cranfield MK43 0AL, Bedfordshire, UK
s.sarfraz@bham.ac.uk
[2] School of Civil Engineering, Faculty of Engineering and Physical Sciences, University of Leeds, Woodhouse Ln., Woodhouse, Leeds, UK
[3] Department of Engineering, University of Cambridge, Trumpington Street, Cambridge CB2 1PZ, UK

Abstract. Cement production accounts for 7% of global carbon dioxide emissions, 3 to 4% of greenhouse gas emissions, and 7% of global industrial energy use. Cement demand is continuously increasing due to the rising worldwide population and urbanisation trends, as well as infrastructure development needs. By 2050, global cement production is expected to increase by 12 to 23% from its current level. Following the net-zero carbon 2050 agenda, both energy and emissions must be significantly reduced. Different production routes exist to produce cement that differs in energy intensity as well as carbon intensity. Similarly, a range of values exists related to energy and emissions for the major cement production stages i.e., raw meal preparation, clinkerisation and cement grinding. The same is the case with cement types produced. This study presents a literature review-based investigation and comparison of cement production practices in terms of energy consumption and CO_2 emissions. This will provide perspectives to the cement industry by identifying approaches that are the least energy and emissions intensive.

Keywords: Cement Production · Clinker · Energy · CO_2 Emissions · Net-Zero Carbon

1 Introduction

The cement sector accounts for around 7% of worldwide CO_2 emissions (the second-largest industrial CO_2 emitter after the agriculture industry [1]) and is the third-largest industrial energy consumer [2]. The growth and development of societies have amplified the demand for construction materials, cement especially has been an essential material due to its superior binding capabilities required in making concrete. By 2050, global cement production is expected to rise by 12–23% from its present level, 4200 Mt [2].

Cement can be produced in large volumes from easily attainable materials from the earth's upper crust [3]. The most widely used type of cement is Ordinary Portland Cement (OPC), with Portland clinker content of 95%. Portland clinker cement-based concrete has come to be one of the most manufactured products in the world in terms of quantity and the most globally expended substance after water [4]. Conversely, the dependence of cement on the accessibility of natural resources is one of its limitations [5]. To put the matter in context, in order to manufacture a ton of Portland cement, 1.5 t of raw material is required [6], mainly limestone and clay (or marl) with lower amounts of gypsum. Clinker is the primary component of cement, and its use directly correlates to the CO_2 emissions produced during cement manufacturing as a by-product of fuel burning and limestone decomposition during clinker synthesis [7].

The cement industry is a massive consumer of raw materials, electrical power, and heat. Since the present study aims to review the literature and identify opportunities for the cement industry to get on track for the Net-zero agenda, it is essential to comprehend the cement manufacturing process and its principal mechanism in relation to energy utilisation and CO_2 emissions. The first process of OPC manufacturing is the production of Portland clinker. By adding up to 5% of gypsum, OPC is manufactured. By adding mineral additives (Supplementary Cementitious Materials – SCMs), different types of cement can be obtained. Therefore, it is also vital to highlight the different cement types and their substitutes that have a large impact on carbon reduction. In the next section, the research methodology followed is presented. Section 3 discusses the main cement production routes. Section 4 extensively examines the production process. Section 5 presents cement types and the latest advancements in binding materials with lower CO_2 footprints. Finally, Sect. 6 concludes the paper.

2 Methodology

The literature reviewed in this study was selected based on relevance to energy consumption and CO_2 emissions in cement manufacturing. Priority was given to recently published articles in scientific journals and technical reports from reputable organisations like the IEA. The aim was to compile comprehensive global perspectives on best practices, process enhancements, and efficiency opportunities across the main stages of cement production and the clinker substitutions. The main limitation is that some proprietary data on efficiency gains by individual manufacturers could not be accessed. The results should allow the industry to get an outlook on the current and emerging techniques for reducing energy consumption as well as carbon emissions.

3 Cement Production Routes

OPC is typically composed of calcareous and argillaceous minerals, with small amounts of iron-bearing elements and sand. The initial phase in making cement is to combine various raw materials that are readily available locally with fuel additives so that the final cement has the necessary chemical makeup.

Usually, the basic cement manufacturing process is divided into three production stages: 1) raw material preparation which involves mining, crushing, and grinding, 2) clinker production which entails calcining the materials in a rotary kiln then cooling the resulting clinker and mixing it with gypsum, and 3) cement finishing in which the cement is milled and then stored. Electrical motors, compressors, pumps, fans, conveyors, coolers, kilns, transportation systems, and lighting systems are the principal components of machinery used in the production of cement [8]. Cement can be manufactured using four methods, namely dry, semi-dry, semi-wet, and wet. The dry method is the most commonly used approach globally, with 90% of Europe employing this technique [9]. Generally, the state of the resources and raw materials have a major impact on the method selected.

The election of the manufacturing method is greatly impacted by the composition of raw materials and energy sources availability. Other factors also require consideration such as input materials quality and homogeneity, energy consumption, dust and CO_2 emissions, and operational cost and time. On average, the total energy cost of production using the wet process is approximately 50% higher in cement production than in the dry process [10, 11]. The energy efficiency can be improved from 26% to 58% by changing from a wet to a dry process [12]. Hence, the next section will focus on upgrades for the dry production route to further enhance its attributes.

4 Cement Production Process (Dry Route)

The production of one ton of cement generally requires 3.4–3.5 GJ of thermal energy and 110 kWh of electrical energy are needed (dry process) [13, 14]. Furthermore, manufacturing a ton of OPC releases 0.67–0.94 tons of CO_2 [11] which primarily depends on the fuel type and production route. Despite the superiority of the dry method in terms of production expenses and CO_2 emissions, it remains a highly energy-intensive process. Therefore, in order to analyse production characteristics and highlight points of possible improvements, a more profound outlook on the cement production process is crucial. The typical dry process of producing OPC consists of three major steps i.e., raw material processing, clinker production and finish grinding [14]. The majority of the carbon emissions are stemming from the clinker production stage which accounts for 85% of the total emissions. This stage is also the primary thermal energy consumer (99%) due to the kiln requiring a high temperature of up to 1450 °C [15]. Cement finishing is the most electrical energy-intensive stage (34%) due to the inherent inefficiency of the grinding process [16] as well as the small Blaine finesses required compared to the raw meal preparation stage [17].

Overall, energy conservation measures can range from straightforward ones like good housekeeping to complex ones with substantial capital expenses like replacing outdated equipment with energy-efficient ones. Several studies have generated a directory of possible measures and their associated possible energy savings per tonne of cement produced. Mokhtar and Nasooti [18] proposed several measures and the corresponding savings in terms of electrical and thermal energy of 0.1–18 kWh/t and 0.1–2 GJ/t. Hasanbeigi et al. [19] and Worrell et al. [20] presented possible technological advancements and processing measures concerning the three production stages in addition to general practices and plant wide procedures. Furthermore, they identified consequent fuel and electrical energy savings of up to 2.4 GJ/t and 24 kWh/t respectively along with CO_2

emissions reductions of up to 200 $KgCO_2$/t resulting from undertaking a specified action. The major production stages have been discussed in the following sections highlighting the key upgrades and prospective savings.

4.1 Grinding

The grinding process occurs at the beginning (raw material grinding) and the end of the cement making process (cement grinding) [21]. It is the main consumer of electrical power in cement production [22] which involves mills and air separators operating together in grinding circuits [23]. By analysing multiple dry cement production plants, Putra et al. [24] found that using vertical roller mills in the raw material grinding reduces the amount of energy required by 25% as well as the CO_2 produced by 57% when compared to ball mills. The efficiency of the roller mill can be further enhanced by utilising waste heat from the kiln for drying raw materials prior to grinding [25]. Upgradation of the mill for finish grinding has the potential to save up to 40% kWh/t electrical energy [11].

4.2 Clinker Production

In terms of clinker production, the most considerable improvements are related to energy efficiency measures in the kiln system which entails reducing the amount of fuel needed while maintaining clinker production volume [26]. Energy savings can be obtained through the realisation of a profitable trade-off between kiln fuel minimization and meal flow rate maximization, this also implies a significant emission reduction [27]. Dry-process kilns with a precalciner, a multistage cyclone preheater, and multichannel burners are considered to be at the forefront of current technology for clinker production. This results in the consumption of 3.0–3.4 GJ/t clinker which is deemed the best available energy level [2]. Operating the kiln with oxygen-enriched air can lead to up to 5% thermal energy savings. Grate clinker coolers are considered principal equipment that can enable greater heat recovery from hot clinker and can lead to up to 0.3 GJ/t clinker energy savings [28]. Beyond operating state-of-the-art kilns, increasing the burnability of raw materials can enhance the thermal efficiency of the process. Emission reduction of 112.61 $kgCO_2$/t can be achieved by implementing a preheater or precalciner kiln system [11].

4.3 General Measures

Afkhami et al. [14] found that general maintenance and plant optimisation can greatly enhance energy efficiency by minimising energy waste. They also indicated the importance of having the right auxiliary equipment which complement the systems in place. For instance, replacing outdated fans and pumps with more suitable models can reduce electrical energy consumption by 4%.

Assuming the best available energy efficiency technologies are fully implemented by 2030, Zhang et al. [29] quantified that 44% energy savings and 12% CO_2 emission reduction could be realised in China's major cement producing regions. This is mainly accomplished by upgrading outdated equipment to new high efficiency ones. An extensive variety of technologies, at different development stages, exist in the literature from research to commercialisation with the objective of reducing energy consumption and CO_2 emission. Advancing energy efficiency necessitates the strategic deployment of the Best Available Technologies (BAT) and the exploration of innovative methods to evaluate energy efficiency whenever possible.

By adopting relevant KPIs, the cement industry can successfully monitor their environmental impact at various levels of operations which can further inform decision-making and drive sustainable practices [30]. The energy benchmarking approach developed by Sarfraz et al. [31] revealed the elements responsible for actual energy use in the cement production (grinding) process. The proposed concept was applied to assess the efficiencies of grinding machines in relation to their theoretically required minimum energy, based on Gentani approach [32]. This analysis identified the most efficient equipment and highlighted the energy that is not directly utilized for production and could be utilised to assess retrofitting plans.

5 Cement Types

The global clinker-to-cement ratio was projected to have climbed from 2015 to 2020 at an average annual rate of 1.6%, reaching an anticipated 0.72 in 2020; this trend was the primary driver of the rise in direct CO_2 intensity of cement production during that time [7]. In contrast, the Net Zero Scenario sees a 1.0% annual decline in the clinker-to-cement ratio, with a global average of 0.65 by 2030 as a result of increased usage of blended cement and clinker alternatives.

Long-term replacements for clinker will be more crucial as the accessibility of by-products from other industries that are currently used as alternatives, such as fly ash from coal power plants and blast furnace slag (GGBS) from the steel industry, decrease due to the decarbonization of other sectors. According to the European standard EN 197 which defines and provides specifications to cement types, there are 5 families of cement products which are further divided into 32 common types of cement. The definitive difference between the types is the varying values of clinker content and SCMs. Due to the significant clinker content present in OPC, clinker substitution can significantly reduce CO_2 intensity from cement production even by 75% [33].

In 2017 the clinker-to-cement ratio in Europe was 77%. This indicates that, on average, 23% of clinker was exchanged for alternative materials [34]. India, after China, is the second largest producer of cement [35], with 9% of the global cement production in 2021, amounting to 300 Mt. It had a 30% share of blended cement production in total quantities of cement manufactured in 1995 which has increased to 73% in 2017, with overall clinker to cement ratio of 71%. This could be attributed to the boosted acceptability by the consumers, growing awareness of sustainability concepts, the obtainability of fly ash from power plants and the employment of advanced technologies [36, 37]. As a result, despite the increase in the overall production of cement, the total direct CO_2 emissions and energy utilisation have reduced over time due to the decline in clinker manufacturing. This is represented in Fig. 1. It can be seen that a considerable CO_2 reduction is possible in a scenario where OPC is completely replaced by Portland Pozzolana Cement (PPC). Further CO_2 reduction can also be achieved by replacing OPC with Portland Slag Cement (PSC) only.

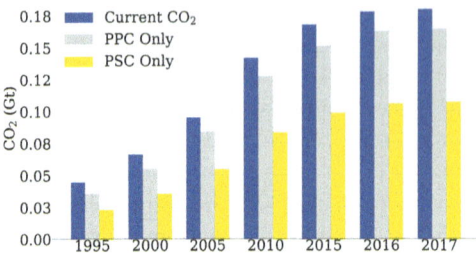

Fig. 1. CO_2 emissions for different scenarios of production in India based on cement type [36–38]

Saraswathy et al. [39] found that mechanical properties are not adapted by using blended cement with GGBS and pozzolana. More specifically, equivalent compressive strength values were obtained for PPC and PSC concrete types when compared with OPC concrete while the corrosion resistant properties for the blended cement were found greatly improved. However, the main limitation for mass production of blended cement is the availability of raw materials.

In recent years, new materials have been introduced to replace clinker with lower environmental impacts. One noteworthy material is calcined clay limestone cement (LC3), a promising new type of cement with 70% lower CO_2 emissions. It consists of clinker, limestone, and calcined clay [40]. Calcined clay being a manufactured product provides a much better opportunity for quality control [41]. Yu et al. [42] found that LC3 achieved the compressive strength that fulfils the requirements as per BS EN 197-1 standard. In comparison to regular Portland cement, blended cement with 50–60% LC3 has adequate compressive strength, lower hydration heat, less negative environmental effects, and lower material costs per unit strength, but less workability.

6   Conclusions

This study provides critical insights that can guide the cement industry practitioners towards enhanced sustainability and reduced environmental impacts. The research indicates multiple priority areas for the cement sector to target in pursuit of energy and emissions reductions.

Transitioning from wet to dry cement production processes is imperative, offering around 50% savings in energy consumption. Within dry processes, vertical roller mills should be adopted for raw material and cement grinding, reducing grinding energy intensity by 25% and the amount of CO_2 produced by 57%. For clinker production, best available technologies including precalciners, multistage cyclones, and modern kiln systems can minimise thermal energy use to 3.0–3.4 GJ/t clinker. Efficiency measures for clinker coolers, oxygen enrichment, and burnability improvements also hold promise.

However, process enhancements can only go so far. Fundamentally transitioning to blended cement products with higher proportions of fly ash, blast furnace slag and other supplementary cementitious materials is essential for deep decarbonization. Global clinker-to-cement ratios must decline, not rise further. Up to 75% reduction in CO_2 intensity can be obtained by substituting clinker with SCMs.

Cement companies must commit to phasing out outdated equipment, investing in efficiency upgrades, increasing clinker substitution, and developing innovative low-carbon solutions. With concerted efforts, the cement industry can overcome its inherent carbon-intensity and contribute positively to economy-wide decarbonization. Policymakers can assist through incentives for emissions reductions and sustainability measures.

The research is limited to commercialised technologies in which tools in development have not been considered, for example, carbon capture and alternative fuels. However, such tools have shown great potential in energy and CO_2 reduction and should have a worldwide implementation in the near future. Further work will be conducted on benchmarking consumption and production values across the cement industry in order to recognise best practices.

Acknowledgements. The authors would like to acknowledge the UK EPSRC-funded project "Transforming Foundation Industries Research and Innovation Hub (TransFIRe)" (EP/V054627/1) for the support of this work. All data supporting this study are provided in full in this paper.

References

1. Mittelman E (2018) The cement industry, one of the world's largest CO2 emitters, pledges to cut greenhouse gases. [Online]. Available: https://e360.yale.edu/digest/the-cement-industry-one-of-the-worlds-largest-co2-emitters-pledges-to-cut-greenhouse-gases
2. IEA (2018) Technology roadmap: low-carbon transition in the cement industry. [Online]. Available: https://www.iea.org/reports/technology-roadmap-low-carbon-transition-in-the-cement-industry
3. Gökcekuş H, Ghaboun N, Ozsahin DU, Uzun B (2021) Evaluation of cement manufacturing methods using multi criteria decision analysis (MCDA). In: 14th international conference on developments in esystems engineering (DeSE), pp 39–43
4. Sakai K (2009) Towards environmental revolution in concrete technologies. In: 11th annual international fib symposium, concrete: 21st century superhero
5. Gao T, Shen L, Shen M, Liu L, Chen F (2016) Analysis of material flow and consumption in cement production process. J Clean Prod 112:553–565
6. Elchalakani M, Aly T, Abu Aisheh E (2014) Sustainable concrete with high volume GGBFS to build Masdar City in the UAE. Case Stud Constr Mater 1:10–24
7. IEA (2022) Cement. [Online]. Available: https://www.iea.org/reports/cement
8. Madlool NA, Saidur R, Hossain MS, Rahim NA (2011) A critical review on energy use and savings in the cement industries. Renew Sustain Energy Rev 15(4):2042–2060
9. Kourti I, Sancho LD, Schorcht F, Roudier S, Scalet BM (2013) Best available techniques (BAT) reference document for the production of cement, lime and magnesium oxide: Industrial Emissions Directive 2010/75/EU (integrated pollution prevention and control). Publications Office. [Online]. Available: https://data.europa.eu/doi/10.2788/12850
10. Ohunakin OS, Leramo OR, Abidakun OA, Odunfa MK, Bafuwa OB (2013) Energy and cost analysis of cement production using the wet and dry processes in Nigeria. Energy Power Eng 05(09):537–550
11. Sahoo N, Kumar A, Samsher (2022) Review on energy conservation and emission reduction approaches for cement industry. Environ Dev 44:100767
12. Sousa V, Bogas JA (2021) Comparison of energy consumption and carbon emissions from clinker and recycled cement production. J Clean Prod 306:127277

13. Madlool NA, Saidur R, Rahim NA, Kamalisarvestani M (2013) An overview of energy savings measures for cement industries. Renew Sustain Energy Rev 19:18–29
14. Afkhami B, Akbarian B, Narges Beheshti A, Kakaee AH, Shabani B (2015) Energy consumption assessment in a cement production plant. Sustain Energy Technol Assess 10:84–89
15. Naranje V, Chidambaram TVS, Garg RB, Bachchhav BD (2021) Use of sustainable practices in cement production industry: a case study. In: Kumar S, Rajurkar KP (eds) International conference on recent advances in manufacturing (RAM 2020). Springer, pp 181–192
16. IEA (2006) Energy technology perspectives 2006: scenarios and strategies to 2050. Organisation for Economic Co-operation and Development
17. Hosten C, Fidan B (2012) An industrial comparative study of cement clinker grinding systems regarding the specific energy consumption and cement properties. Powder Technol 221:183–188
18. Mokhtar A, Nasooti M (2020) A decision support tool for cement industry to select energy efficiency measures. Energ Strat Rev 28:100458
19. Hasanbeigi A, Menke C, Therdyothin A (2010) The use of conservation supply curves in energy policy and economic analysis: the case study of Thai cement industry. Energy Policy 38(1):392–405
20. Worrell E, Galitsky C, Price L (2008) Energy efficiency improvement opportunities for the cement industry. Lawrence Berkeley National Lab. (LBNL), Berkeley, CA (United States)
21. Jankovic A, Valery W, Davis E (2004) Cement grinding optimisation. Miner Eng 17(11–12):1075–1081
22. Valderrama C, Granados R, Cortina JL, Gasol CM, Guillem M, Josa A (2012) Implementation of best available techniques in cement manufacturing: a life-cycle assessment study. J Clean Prod 25:60–67
23. Boulvin M, Wouwer AV, Lepore R, Renotte C, Remy M (2003) Modeling and control of cement grinding processes. IEEE Trans Control Syst Technol 11(5):715–725
24. Putra MA, Teh KC, Tan J, Choong TSY (2020) Sustainability assessment of Indonesian cement manufacturing via integrated life cycle assessment and analytical hierarchy process method. Environ Sci Pollut Res 27(23):29352–29360
25. Venkateswaran SR, Lowitt HE (1988) The US cement industry: an energy perspective. Energetics, Inc., Columbia, MD (USA). [Online]. Available: https://www.osti.gov/biblio/7224969
26. Salas DA, Ramirez AD, Rodríguez CR, Petroche DM, Boero AJ, Duque-Rivera J (2016) Environmental impacts, life cycle assessment and potential improvement measures for cement production: a literature review. J Clean Prod 113:114–122
27. Zanoli SM, Orlietti L, Cocchioni F, Astolfi G, Pepe C (2021) Optimization of the clinker production phase in a cement plant. In: Gonçalves JA, Braz-César M, Coelho JP (eds) Proceedings of the 14th APCA international conference on automatic control and soft computing. Springer International Publishing, pp 263–273
28. CSI/ECRA (2017) Development of state of the art techniques in cement manufacturing: trying to look ahead. [Online]. Available: https://docs.wbcsd.org/2017/06/CSI_ECRA_Technology_Papers_2017.pdf
29. Zhang S, Xie Y, Sander R, Yue H, Shu Y (2021) Potentials of energy efficiency improvement and energy–emission–health nexus in Jing-Jin-Ji's cement industry. J Clean Prod 278:123335
30. Sherif Z, Sarfraz S, Jolly M, Salonitis K (2022) Identification of the right environmental KPIs for manufacturing operations: towards a continuous sustainability framework. Materials 15(21):7690

31. Sarfraz S, Sherif Z, Jolly M, Salonitis K (2023) Towards framework development for benchmarking energy efficiency in foundation industries: a case study of granulation process. In: New directions in mineral processing, extractive metallurgy, recycling and waste minimization. Springer, Cham, pp 245–256

32. Sarfraz S, Jolly M, Salonitis K (2023) The use of Gentani approach for benchmarking resource efficiency in manufacturing industries. In: Manufacturing driving circular economy. Springer, Cham, pp 457–463

33. ICE (2022) Low carbon concrete routemap. [Online]. Available: https://www.ice.org.uk/eng ineering-resources/briefing-sheets/low-carbon-concrete-routemap/

34. CEMBUREAU (2019) Cementing the European green deal. The European Cement Association

35. U. S. G. Survey, Major countries in worldwide cement production in 2021. Statista. [Online]. Available: https://www.statista.com/statistics/267364/world-cement-production-by-country/

36. GCCA (2022) Blended cement - green, durable & sustainable

37. WBCSD (2018) Low carbon technology roadmap for the Indian cement sector: status review. World Business Council for Sustainable Development

38. USGS (2022) Cement statistics and information. [Online]. Available: https://www.usgs.gov/ centers/national-minerals-information-center/cement-statistics-and-information

39. Saraswathy V, Karthick S, Lee HS, Kwon S-J, Yang H-M (2017) Comparative study of strength and corrosion resistant properties of plain and blended cement concrete types. Adv Mater Sci Eng 2017

40. Wang DL, Chen ML, Tsang DDCW (2020) Green remediation by using low-carbon cement-based stabilization/solidification approaches. In: Sustainable remediation of contaminated soil and groundwater. Butterworth-Heinemann, pp 93–118

41. Díaz YC et al (2017) Limestone calcined clay cement as a low-carbon solution to meet expanding cement demand in emerging economies. Dev Eng 2:82–91

42. Yu J, Wu H-L, Mishra DK, Li G, Leung CKY (2021) Compressive strength and environmental impact of sustainable blended cement with high-dosage Limestone and Calcined Clay (LC2). J Clean Prod 278:123616

Development of a Minimalistic Smart Sensor System for Motion Measurement of Wind Turbine Towers

Johannes Rupfle[1,2](✉), Patricio Neffa[2], and Christian Grosse[1]

[1] Chair of Non-Destructive Testing, Technical University of Munich, Franz-Langinger-Str. 10, 81245 Munich, Germany
johannes.rupfle@tum.de
[2] Instituto Tecnológico de Buenos Aires, San Martín 202, 1004 Buenos Aires, Argentina

Abstract. The accurate measurement of wind turbine tower motion is crucial for assessing structural integrity, identifying damages, and estimating remaining useful life. In this study, the development of a smart sensor system is presented, which combines acceleration measurements, real-time kinematic measurements, and SCADA data to accurately determine the motion of slender structures. The proposed setup offers a cost-effective and easily deployable solution. Installed on the top of the wind turbine nacelle, the sensor system collects data to calculate the tower motion. Real-time kinematics measurements serve as correction input for a state estimator, to enhance accuracy and reliability. Data gathering and processing occur directly on the sensor node, allowing for efficient transfer. Test measurements on a wind turbine tower have demonstrated the effectiveness of the system for accurate motion measurements. Beyond wind turbine manufacturing, this sensor system holds potential for application in several domains, including tall buildings and bridges, where the detection of vibrational movements is valuable for structural health monitoring.

Keywords: sustainable manufacturing · life cycle assessment · smart sensor system · wind turbine tower motion · acceleration · real-time kinematic · damage assessment · remaining useful life · structural health monitoring · industry 4.0 · internet of things · sustainable development

1 Introduction

The rapid growth of wind energy as a sustainable power source has led to an increased demand for reliable and efficient wind turbine systems. The dynamic behavior of these tall structures, subjected to turbulent wind leads to tower vibrations and motion. Monitoring and analyzing tower motion is essential for assessing structural health, detecting potential damages, and estimating the RUL[1] of wind turbines, but may also allow extension of the overall operating range of the turbine.

---

[1] Remaining Useful life.

© The Author(s) 2025
H. Kohl et al. (Eds.): GCSM 2023, LNME, pp. 774–781, 2025.
https://doi.org/10.1007/978-3-031-77429-4_86

Traditional methods for tower motion measurement involve the use of complex and expensive instrumentation, such as laser-based displacement sensors or inclinometers as in [1]. While these techniques provide accurate measurements, their high cost and limited scalability hinder their widespread implementation for large-scale wind farms. Therefore, there is a need for a cost-effective and easily deployable monitoring solution that can capture accurate and reliable tower motion data.

We propose the development of a minimalistic smart sensor system for motion measurement of wind turbine towers. The system leverages the advancements in RTK[2] positioning technology and high-resolution acceleration sensors to enable precise and continuous motion monitoring. RTK is a high-precision positioning technique that uses reference stations to provide real-time corrections to GPS5[3] signals. It achieves sub-centimeter-level accuracy by comparing satellite measurements with known reference positions. RTK is widely used in various industries for applications requiring precise and continuous positioning information, such as agriculture, construction, and autonomous vehicles. The integration of an RTK module ensures accurate static positioning, while the high-resolution acceleration sensors capture the dynamic response of the tower.

The key advantages of the proposed smart sensor system include its ease of deployment, cost-effectiveness, and compatibility with existing wind turbine infrastructure. By utilizing commercially available components and wireless communication, the system minimizes the installation and maintenance costs associated with traditional monitoring methods. The compact size and lightweight design of the sensors allow for straightforward integration into the wind turbine towers without significant modifications.

For efficient data acquisition and analysis, a custom software platform has been developed. The software synchronizes data logging but also provides more complex extraction of modal parameters, tower movement analysis, and feature extraction from SCADA6[4] events.

The contributions of this study lie in the development and validation of a minimalistic smart sensor system for motion measurement of slender structures. A prototype installed in a large-scale onshore wind turbine utilized as test facility [2] for several years provides the depicted measurement data in the course of this publication. The system provides a practical solution for sustainable manufacturing in the wind energy sector. Through continuous monitoring and analysis of tower motion, operators can make informed decisions regarding maintenance, repair, and optimization of wind turbine performance.

2   Methodology

The following section describes the concept of the smart sensor system and gives insight into the development of both hard- and software parts of the system.

---

[2] Real-Time Kinematics.

[3] Global Positioning System.

[4] Supervisory Control and Data Acquisition.

2.1 Sensor System Development

Platform selection: A single-board computer is used as the platform for the sensor system. This choice allows for flexibility and integration with various sensors and communication protocols. The system also provides both wired and wireless internet connection and extension modules for cellular networks. The system includes separate storage for the operating system and the file storage.

Utilization of a high-resolution sensor: A high-resolution 3-axis MEMS7[5] accelerometer is utilized. The sensor element has a built-in 20 bit ADC8[6] and a sensitivity range from ± 2 g to ± 8 g. The sensitivity of the digital output is $256,000 \mathrm{LSB9}[7]$/g. Both high-pass and low-pass filter options can be applied internally. The spectral density of noise is specified as 22.5 µg/$\sqrt{\text{Hz}}$. The sensor node also provides an integrated temperature sensor.

Integration of the RTK module: The RTK module provides precise positioning information and an accurate GPS timestamp that is used for synchronization purposes with potential additional systems. The sample rate is adjustable from 1 Hz to 14 Hz. Correction input can be either provided by a self-hosted base station or publicly available reference station networks.

2.2 Data Acquisition Development

Platform for data acquisition: The measurement systems data acquisition consists of a Python-based framework for the basic functional modules and the optional modules for extended analysis of the measured data, data logging, and transmission. It is intended, that the essential modules work independently from non-essential analysis modules to improve reliability.

Connectivity: The MEMS element and the RTK module of the measurement system are connected to the single-board computer via serial interfaces. The external data sources SCADA and environmental data are implemented via programming interfaces.

Implementation of external data: SCADA data such as yaw angle, rotor speed, blade angle, power output, wind speed, wind direction, and status text messages of the turbines control system are used to complement the data. A further approach is to implement weather forecast data such as wind, wind gusts, new snow, temperature, dew point, humidity, and freezing altitude into the database.

2.3 Features of the Software

The proposed software consists of a central data buffer that is fed with data from different sources. The flowchart after initialization is depicted in Fig. 1. Independent modules can

---

[5] Micro-Electro-Mechanical System.

[6] Analog Digital Converter.

[7] Least Significant Byte.

then access the data and are also able to return data to the buffer. Besides the state estimator module which is seen as an essential module of the software package, optional modules can be added to fulfill specific tasks such as extraction of modal parameters according to [3], or ice detection as it is developed in [4].

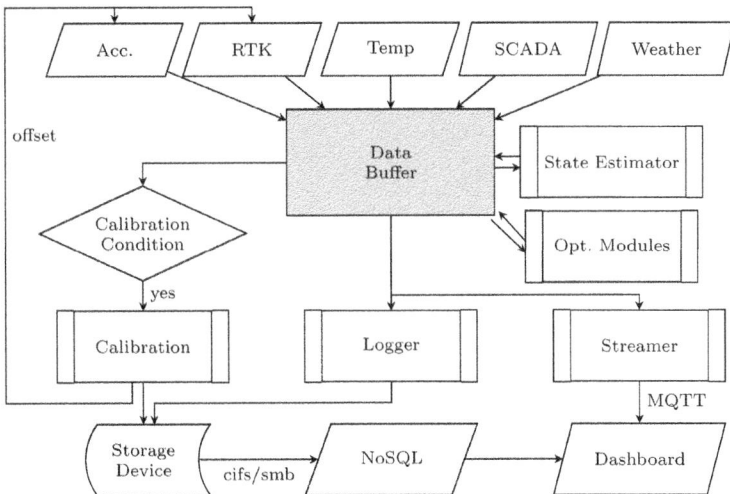

Fig. 1. Flow chart of proposed data acquisition and processing scheme.

Data acquisition with precise and unified timestamp: The clock of the system is given by the GPS module and is delivered with a precise time stamp with an accuracy of up to 30 ns, thus it is not necessary to have a real-time clock installed in the single-board computer.

State estimator: A state space model is a mathematical representation of the physical system, described by a set of differential equations known as the equations of motion. This model provides a foundation for fusing data from different measurements with different limitations to obtain a more precise estimation of the tower's state using a state estimator. The initial implementation of the state space model incorporates a Kalman filter as initially published by Kalman [5]. The Kalman filter is a recursive estimation algorithm widely used in control systems and signal processing. It optimally estimates the state of a system by combining predictions from a mathematical model (the state space model) and measurements.

Coordinate transformation: To be able to examine tower motion and modal features both in a tower fixed coordinate system but also in a nacelle fixed coordinate system, a continuous transformation between the coordinate systems is implemented. This allows for the examination of a subsequent module that analyzes direction-dependent variation of Eigenfrequencies. After simplification of the rotation matrix of the intrinsic sequence X–Y′–Z″ with the Tait-Bryan angles $\alpha = 180°$, $\beta = 0°$, and $\gamma = 0°$, respectively, the following rotation matrix $\mathbf{R}_{S \to N}$ as depicted in Eq. 1a results. For the intrinsic sequence

X–Y′–Z″ with the Tait-Bryan angles α = 0°, β = 0°, and γ = 270° − ϕ, where ϕ represents the yaw angle with positive angles for clockwise rotation from north direction, the rotation matrix $\mathbf{R}_{N \to T}$ as depicted in Eq. 1b results. Equally, the RTK data can be transformed into the nacelle coordinate system using the transposed rotation matrix $\mathbf{R}_{T \to N} = \mathbf{R}_{N \to T}$.

$$\mathbf{R}_{S \to N} = \mathbf{R}_Z(\gamma)\mathbf{R}_Y(\beta)\mathbf{R}_X(\alpha) \begin{bmatrix} 1 & 0 & 0 \\ 0 & \cos(\alpha) & -\sin(\alpha) \\ 0 & \sin(\alpha) & \cos(\alpha) \end{bmatrix} = \begin{bmatrix} 1 & 0 & 0 \\ 0 & -1 & 0 \\ 0 & 0 & -1 \end{bmatrix} \tag{1a}$$

$$\mathbf{R}_{N \to T} = \mathbf{R}_Z(\gamma)\mathbf{R}_Y(\beta)\mathbf{R}_X(\alpha) = \begin{bmatrix} \cos(\gamma) & -\sin(\gamma) & 0 \\ \sin(\gamma) & \cos(\gamma) & 0 \\ 0 & 0 & 1 \end{bmatrix} \tag{1b}$$

Automated offset determination and sensitivity tracking: Accurate measurement of wind turbine tower motion requires calibration of the sensor offset and orientation. The offset of MEMS accelerometers can arise from mechanical and electrical factors. The mechanical offset is caused by gap mismatch and residual stress, while the electrical offset is due to parasitic capacitance and electrical components [6]. Minimizing these influencing factors is crucial for precise data analysis. Additionally, determining the orientation of the sensor elements with high accuracy is essential. To achieve this, a calibration method utilizing the untwisting cable event is employed and will be discussed in future work. In [7] a method to determine both sensor offset values and relative sensitivity is proposed.

Data logger: The buffered data is logged on the local machine and securely transferred via MQTT10[8] to a central NoSQL11[9] database.

Real-time data stream: Utilization of MQTT enables real-time streaming of sensor data. This feature allows for integration with other systems and real-time motion monitoring. The MQTT performance is tested up to a sample rate of 20Hz which can be seen as the maximum RTK rate available on the market and sufficiently high resolved data for the expected Eigenfrequencies of up to 5.69Hz for the fourth bending Eigenfrequency of the tower as it was determined in [8].

Extraction of modal parameters: Modal parameters, such as natural frequencies and mode shapes, are extracted from the sensor data for structural analysis of the wind turbine tower. These parameters provide insights into the dynamic behavior and health condition of the tower. The extracted data can be clustered by operating ranges and directions to not only detect damage but also localize them.

3 Results

A measurement system prototype was installed at the tower top of the test facility with the GPS antenna on top of the nacelle.

---

[8] Message Queuing Telemetry Transport.
[9] Not only Structured Query Language.

Evaluation of displacement estimation: Fig. 2 presents measurement data (a–d) obtained from the RTK system (a, b) and the acceleration sensors (c, d) in the east (a, c) and north (b, d) directions. Subplots $Disp_x$ (e) and $Disp_y$ (f) illustrate the estimated displacement in the fore-aft and side-side directions of the nacelle, respectively. Subplots *Wind Speed* (g) and *Power Output* (h) display the wind speed and power output of the wind turbine. Notably, the fore-aft displacement of the turbine exhibits a direct correlation with the overall load on the tower, as evidenced by the similarities between the wind speed and power output curves and the fore-aft displacement curve in subplot $Disp_x$ (e). The power output corresponds to the thrust force leading to a tower displacement of the same order of magnitude. Conversely, the side-side displacement shows only minimal drift. Both displacement curves exhibit superimposed harmonics associated with the Eigenfrequencies of the tower and the rotor.

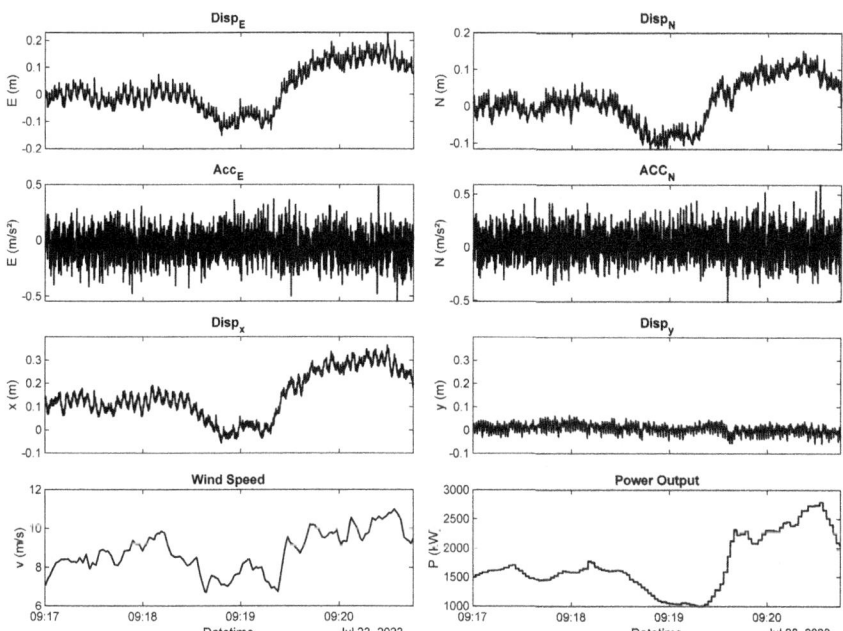

Fig. 2. Measurement results of both RTK (a, b) and acceleration (c, d) measurements and the estimated and rotated displacement in nacelle coordinates (e, f), compared with the wind speed (g) and power output (h).

Extension of operating range: An additional advantage of monitoring wind turbine tower motion lies in its potential to extend the operating range of the wind turbine. With precise estimation of the tension experienced by the tower under varying wind loads, the operational limits of the turbine can be expanded, provided that it remains within its structural safety limits. This increased flexibility allows the wind turbine to operate more efficiently across a wider range of wind speeds, leading to enhanced energy production and overall performance. As depicted in Fig. 2 (e), a notable rise in wind speed of approximately 3 m/s, from 7 m/s to 10 m/s, corresponds to an overall displacement of the tower of approximately 0.30 m. Both the static and dynamic displacements can then be employed as inputs for a finite element tower model, following the approach presented by Emiroglu et al. [9]. The resulting tensile stress can further be utilized as input for a rainflow counting algorithm, as previously developed in studies such as [10] and [11]. This comprehensive analysis enables a more accurate evaluation of the turbine's structural health and RUL.

4 Conclusion

The presented measurement system represents a significant advancement in accurately assessing tower motion. By employing a state estimator, the measurement system combines the dynamic displacement precision of acceleration measurement with the static accuracy of RTK data. Its superiority over purely vibration-based displacement calculations lies in the avoidance of drift resulting from the integration of acceleration values and thus enables precise determination of displacement of the tower.

Beyond its main use in monitoring tower motion, the system is capable of detecting damage and assessing RUL. Additionally, it can evaluate the initial state of turbines and monitor the construction phase, as resonance cases may occur at various stages of construction. Moreover, it facilitates estimation of the overall load on wind turbine supporting structures, allowing for an extended operating range under specific conditions where the cutout wind speed is already exceeded. The versatility of the system extends to various slender structures, such as skyscrapers, antenna towers, but also bridges, and dams. Its contributions can enhance structural integrity evaluation and sustainability in manufacturing. The usage of the system can vary from a single deployment in wind parks to an entire park, allowing for a better understanding of the turbine's target state and the detection of outliers.

Measured data provides reliable displacement values as long as a fixed solution of the RTK data is provided which relies on a stable internet connection for correction data and on tropospheric conditions. A loss of the fixed solution status results in a drift of the position data. This may be overcome in future iterations by considering the full covariance matrix of the RTK measurement.

The limited RTK sampling rate and event-triggered SCADA data, as well as the unavailability or low quality of RTK data, require imbalance measures. Firstly, the acceleration data can be adapted or downsampled to match the RTK data sampling frequency. In cases of unavailability, a mapped tilt sensor measuring the tower's top tilt can serve as a fallback system for static displacement. Furthermore, the Kalman filter can be enhanced to accommodate only the currently available data.

Acknowledgements. The authors express their gratitude to the AdV for providing the SAPOS data, which significantly contributed to the success of this research. Additionally, the authors extend their heartfelt appreciation to Prof. Dr.-Ing. Cecilia Smoglie, Director of Energy and Environment from the Instituto Tecnológico de Buenos Aires (ITBA) for her invaluable collaboration. The authors also acknowledge the TUM International Graduate School of Science and Engineering (IGSSE) for their support. Finally, special thanks to Ramzes Mosiej for his valuable contributions.

References

1. Kim K et al (2018) Structural displacement estimation through multi-rate fusion of accelerometer and RTK-GPS displacement and velocity measurements. Measurement 130:223–235
2. Botz M, Oberlander S, Raith M, Grosse C (2016) Monitoring of wind turbine structures with concrete-steel hybrid-tower design
3. Botz M, Zhang Y, Raith M, Pinkert M (2017) Operational modal analysis of a wind turbine during installation of rotor and generator
4. Wondra B, Rupfle J, Emiroglu A, Grosse CU (2022) Analysis of icing on wind turbines by combined wireless and wired acceleration sensor monitoring. Lecture notes in civil engineering. Springer International Publishing, pp 143–155
5. Kalman RE (1960) A new approach to linear filtering and prediction problems. Trans ASME–J Basic Eng 82:35–45
6. Dong X et al (2021) Research on decomposition of offset in MEMS capacitive accelerometer. Micromachines 12:1000
7. Rupfle J, Emiroglu A, Grosse C (2022) Investigation of the measurability of selected damage to supporting structures of wind turbines. J Phys Conf Ser 2151(1):012008
8. Botz M, Harhaus G, Grosse C (2019) Monitoring modal parameters and external loads of wind turbines for remaining useful life analysis
9. Emiroglu A et al (2017) Fe-modelling and analysis of a hybrid wind-turbine tower for fatigue analysis and remaining life-time prediction. In: WESC2017
10. Osterminski K, Gehlen C (2018) Echtzeitmodellierung der Ermüdung von Windenergieanlagentürmen
11. Loew S, Obradovic D, Bottasso CL (2019) Direct online rainflow-counting and indirect fatigue penalization methods for model predictive control. In: 2019 18th European control conference (ECC), pp 3371–3376

Optimisation of the Melting Furnace Unit in an Italian Aluminium Foundry to Reduce Gas Methane Consumption

Jessica Rossi[(✉)] and Augusto Bianchini

Department of Industrial Engineering, University of Bologna, Via Fontanelle 40, 47121 Forlì, Italy
{jessica.rossi12,augusto.bianchini}@unibo.it

Abstract. The foundry industry is one of the most energy-intensive industrial sectors. Consequently, the energy cost can reach 7–15% of the cost of the operations. Among all the types of energy used, the most significant part of energy consumption is associated in Italy with gas methane in different typologies of melting furnaces. According to the treated material (e.g., aluminium, steel, cast iron), the foundry process can vary; however, some operations characterize the entire sector, such as the metal melting phase, which is the most energy-intensive stage of the process (it can account up to 70% of the total energy consumption of the foundry). The energy crisis, which has affected companies in these years, determines instability and volatility in energy availability and costs and requires implementing some improvements to optimize energy efficiency and reduce consumption. With the aim of investigating the potential energy reduction in the foundry sector, an Italian aluminium foundry has been considered. The analysis consisted of three main activities: (i) Analysis of the process and mapping of energy and resource consumption at the factory level and in each unit; (ii) Quantification of energy and resource consumption at the factory level according to the ViVACE® method; and (iii) Addressing the critical points (energy consumption) to improve the environmental impact of the foundry. According to this methodology, the optimization of the melting furnace unit has been addressed, allowing the potential saving of gas methane up to 13%.

Keywords: Foundry · Methane · Aluminium · Optimization · Energy efficiency

1 Introduction

The foundry is a key sector able to product numerous typologies of products, which can interconnect different supply chains, such as automotive, mechanical and aerospace industry, agricultural machines, energy production plants and construction sector [1]. At the basis of foundry process, there is the material melting, which requires very high temperature, e.g., about 700 °C for aluminium and 1500 °C for steel, determining very high energy consumption. It makes the foundry sector one of the most energy-intensive industrial sector. In particular, in 2021, in Europe, the foundry sector consumed about

© The Author(s) 2025
H. Kohl et al. (Eds.): GCSM 2023, LNME, pp. 782–790, 2025.
https://doi.org/10.1007/978-3-031-77429-4_87

100,000 GWh (NACE 24.5, included 24.41 and 24.43), corresponding about to the 13.9% of the total energy consumption [2]. According to a previous analysis [3], it has been evaluated that energy consumption in a foundry can reach up to 15% of the total costs. In this context, energy consumption optimization and savings become crucial aspects, above all in the current historical period, characterized by a high and instable growth of energy price (both electricity and gas methane) that determines a huge economic impact on the industries.

In the entire pattern of energy consumption in foundries, the melting phase, where different types of furnaces can be used, generates the greatest part of energy consumption, reaching rates up to 70% of the total energy consumption [4, 5]. Consequently, paying the attention on the use of melting furnaces could generate relevant energy savings for foundries. In the scientific literature, the studies about the energy consumption reduction in foundry sector are limited: in [6], it has been explained that there could be numerous areas of investigation to reduce energy consumption, such as ventilation, and compressed air, and it is highlighted that a possible problem related to the lack of strong and repetitive procedure and techniques to use the furnaces; in [7], an experimental innovative burner unit is described with the potential to reduce energy consumption of about 36%, energy costs of 48% and CO_2 emissions of 41%; in [5], insights about correct feedstock condition, avoiding over-heating and heat losses; and, finally in [3], it has been investigated how small adjustments to the tolerance limits of multiple process variables in furnaces could make a saving of 60 kWh/ton of liquid metal. According to melting furnace constructor [8], there are four main activities that can be applied to reduce furnace energy consumption. (i) Checking and optimizing the fill: material typology and mix (ingots and foundry returns) and distribution could generate up to 4% of energy savings. (ii) Optimizing load efficiency during low utilization avoiding short melting periods and longer holding ones could reduce energy consumption by up to 20%. (iii) Adjusting the air-fuel ratio in the burner could save up to 2% of the energy used in a furnace with an excessive air utilization. (iv) Closed liquid metal holding and dosing system could avoid up to 66% of unnecessary energy consumption in foundry. Other studies on aluminium melting furnaces, such as [9–11], are focused on the identification of combustion performance according to some conditions, such as the type of fuel and the air-fuel ratio, however, this approach is more useful for furnace construction optimisation, since the users of the furnaces cannot optimise these parameters during the furnace usage. Finally, in [12] different solutions have been explained to optimise the consumption of gas methane in aluminium melting furnaces through flues gases heat recovery, which could be a downstream solution.

In this paper, with the aim of investigating the potential energy reduction in the foundry sector, an Italian aluminium foundry has been considered. The industrial process of the foundry has been analyses in detail and the consumption of the main resources (energy, waste generation, water and transport) were quantified, according to the ViVACE® method, an innovative quantitative tool developed by the authors [13, 14], able to assess the environmental sustainability of the companies. This methodology allowed to identify that, also in this case, melting furnace unit represents the most energy consumer unit of the foundry, proposing an improving way in using furnaces to reduce their energy consumption.

2   Methodology

The methodology applied in this paper consists of three main phases (Fig. 1): (i) Analysis of the process and mapping of energy and resource consumption at the factory level and in each unit; (ii) Quantification of energy and resource consumption at the factory level according to the ViVACE® method; and (iii) Addressing the critical points (energy consumption in melting furnaces) to improve the environmental impact of the foundry.

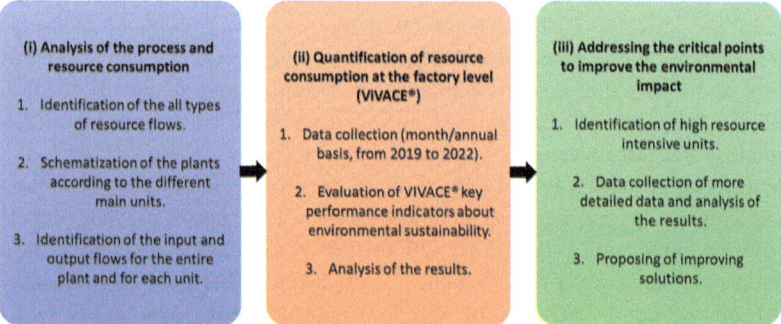

Fig. 1. Methodological approach applied to investigate the potential energy consumption reduction in the analyzed Italian aluminium foundry.

2.1   Analysis of the Process and Resource Consumption

The analyzed company is a die casting foundry, which produces aluminium components starting from their design and simulation, then their industrialization and production up to the delivery to their customers. The foundry production process is divided in five units: the melting unit, two units with the die casting machines (one unit contains old generation machines and the other contains new generation machines), quality lab, the unit which contains a specific mechanical operation for the final products and the unit which contains several types of general plants (compressed air, boilers, wastewater treatment plant, conditioning systems, solid/liquid particle emission filters). According to this structure, the entire plant has been schematized with twelve blocks. For each block, all the types of involved resources have been identified.

The main resource flows that involve the environmental sustainability and interest the typical industrial processes, independently on the specific sector, mainly fall in five categories, which are: (1) materials, (2) waste, (3) energy (electricity and thermal energy), (4) water and (5) fuels. In this specific case of the foundry, 39 resource flows have been identified and associated with their relative blocks and the with the entire factory. Each flow has been classified with a name, a numeric code, a colour and its relative category. Finally, the respective input and output flows have been associated to each single block, as in Fig. 2, which represents the block of the entire plant and consists of the reference for the following methodological phases.

2.2   Quantification of Resource Consumption at Factory Level by ViVACE®

According to the reference [9], in 2019, tools to proper quantify circular economy business models and company environmental sustainability lacked in literature (both scientific and grey literature), but were necessary to support the process management and company decision-making process. For this reason, ViVACE® has been defined. In these years, the ViVACE® tool have been further developed and it derives a practical and consolidated tool and software to assess environmental sustainability of companies in different sector. Now, the ViVACE® tool is applied and continuously updated and improved by a spin-off of the University of Bologna (Turtle Srl). Some IT management tools that manage data and indicators for sustainability exist on the market for 10–15 years [15], but their use is not widespread due to their complexity that binds them only to large companies. ViVACE® tool is simple and can be adapted to all sectors and company dimensions thanks also to the support in systematizing and interpreting the necessary data starting from specific company structures. The case study presented in this article represents the first application of ViVACE® tool to a foundry, consequently the academic approach has been necessary to address the adaptation of the tool to the sector.

According to ViVACE® tool steps and methodology, for each flow of resources, the quantitative value has been assessed through the data collection phase and registered according to: type of data, source(s), involved corporate function(s), recording methods, any potential critical issues.

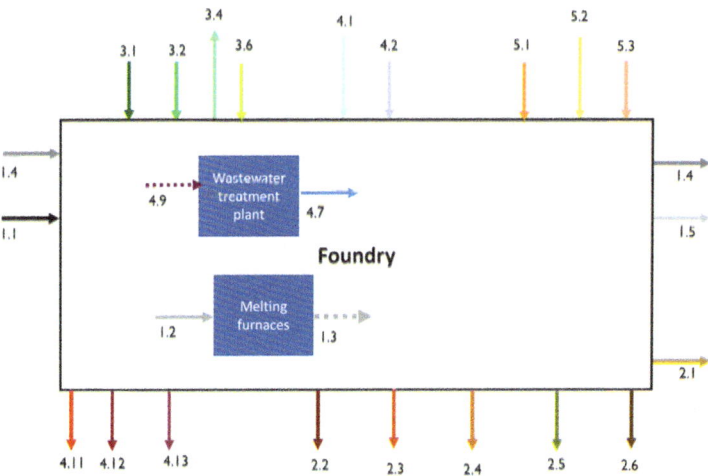

Fig. 2. Qualitative schematization of the foundry (plant level) and identification of all the codified input and output resource flows.

To understand the possible evolution of the company on the path towards sustainability, the data has been collected for the period 2019–2022. In this way it was possible to verify the situation before the Covid-19 pandemic, noting potential effects of the pandemic itself on production and evaluating the post-covid recovery in the two-year period

2021–2022. The data was collected on a monthly and/or yearly basis, depending on their availability. The data collection involved numerous company functions (administration, purchasing office, logistics office, industrial accounting, etc.) as the data necessary for quantifying sustainability are already present in the companies, but are currently managed for different purposes. Finally, Excel sheets have been prepared to continue collecting data also in the future: in particular, the Excel sheets have been structured to process the data in the form in which they are currently managed, but also to process the ViVACE® indicators and dashboard.

In fact, all the collected data have made it possible to calculate some operational KPIs relating to the consumption of the main resources (energy, water, fuel for transport on the vertical axis) and the flux of material and generation of waste (on orizontal axis) according to the ViVACE® method. Table 1 summarize the main indicators monitored through ViVACE® divided in four classes: (i) energy, which considers the energy (electrical and thermal) consumed by all the company's processes and services, and that self-produced internally, and it also includes the energy used for internal logistics or similar to it (electricity or produced from fuel); (ii) waste, which considers both industrial and urban waste, produced by the company, and it also includes materials valued as by-products; (iii) water, which considers the water consumed by all processes and services within the company and assesses the type of disposal, consequently special liquid waste falls into this category (and not in waste one); and finally (iv) transport, which considers the fuel consumption due to the transport of company vehicles, but it does not involve the logistics associated with the material purchase, the delivery of finished products and the disposal of waste.

Table 1. Table captions should be placed above the tables.

Category	Indicators	Unit
Energy	– Annual energy consumption	kWh/year
	– Rate of electricity/thermal energy	%
	– Rate of renewable energy	%
Waste	– Annual waste generation	kg/year
	– Rate of waste sorted by typologies	%
Water	– Annual water consumption	m^3/year
	– Rate of water sorted by wastewater treatment	%
Transport	– Annual travels sorted by fuels	km/year

2.3   Addressing Critical Points to Improve Environmental Sustainability

This last methodological phase is based on the concept to focus the attention on the resource consumption that emerges as critical from the analysis in previous methodological phase, in terms of CO_2 equivalent emissions. Consequently, the sub-methodological steps are always the same (identification of high resource intensive units/machines/processes, more detailed data collection and identifying improving solutions), but there are practically defined and applied in different ways, according to the resultant criticalities (e.g., in the energy field as in this specific case).

3   Results and Discussions

The results of the quantification of environmental sustainability of the analyzed foundry are expressed in terms of the main indicators listed in Table 1, sorted by category. Then the annual consumption of each resource is transformed in CO_2 equivalent emissions to identify the most impactful resource (Fig. 3 – 2022 data are the reference).

Energy. In 2022, the foundry consumed more than 35 GWh/year of energy. The thermal energy generated by the combustion of gas methane is the first energy source (73.8%), followed by electricity (26.1%). Fuel consumption for energy generation makes up a small part (0.1%). Starting from 2021, a portion of the electricity consumed comes from the photovoltaic system installed on the roof of the foundry plant. This portion constitutes 3.3% of the total electricity consumption, and is almost entirely self-consumed by internal processes (> 99%).

Waste. In 2022, the foundry generated about 1000 tons of waste. On average, the main type of waste is determined by aluminium waste (35%), followed by waste from production activities (24%) and finally by aluminium by-products (22%).

Water. In 2022, the foundry consumed about 18000 m^3/year of water. The main wastewater treatment methodology is its recovery through the internal wastewater treatment plant, which allows the reuse of the 30% of the incoming water. This is followed by civil wastewater (from bathrooms and changing rooms), equal to 15%. However, most of the water consumed (48%) is not found in the outgoing liquid flows, as it evaporates during the process.

Transport. In 2022, the foundry car park made more than 2000 km/year. The Covid-19 pandemic has certainly changed the methods of communication between people, determining procedures which then consolidated over time (e.g. telecommunications, video telephony), causing a drastic drop in travel and therefore in fuel consumption (− 65% from 2019 to 2022).

According to these data, it derives that the CO_2 equivalent emissions of the foundry production process are mainly generated by energy and waste, while water and transport have a negligible impact (Fig. 3). Surely, the energy determines almost all the emissions, consequently, in this specific case, it consists of the most critical resource to be deeply analyzed to improve the environmental impact of the foundry.

According to this evidence, a more detailed analysis of energy consumption has been conducted. Since the methane consumption consists of about the 74% of the entire energy consumption, it has been chosen to focus the attention on this resource. In particular, confirming the literature data [4, 5], it derived that, also in the analyzed foundry, the furnace melting unit is the greatest user of gas methane: in particular, the entire unit, constituted by 3 operating furnaces, generates the 48% of the gas methane consumption of the entire foundry.

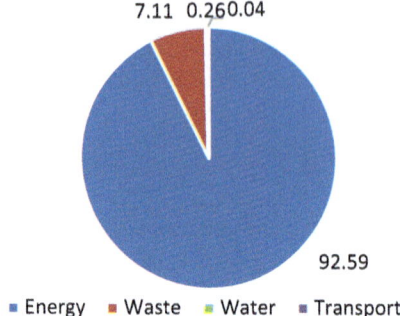

7.11 0.260.04

92.59

■ Energy ■ Waste ■ Water ■ Transport

Fig. 3. Composition of the CO_2 equivalent emissions generated by the production process of the analyzed foundry sorted by category of resources.

With the aim to deeply understand the composition of this consumption in the furnace melting unit, further data collection and analysis have been conducted. In particular, thanks to the availability of gas methane sensors on each furnace, it was possible to monitor the consumption over the time (data are available with a sample time of 20 min), and associating them with the number of aluminium spills (spill means the collection of a certain and fixed quantity of molten aluminium – 500 kg in this case – from the furnace) from each furnace, which is an information that is still manually registered. From this analysis, a correlation between the number of spills in a work shift and the corresponding average specific gas methane consumption of the furnace, assessed in mc/kg (Fig. 4), that is the quantity of gas methane (mc) associated to each kg of molten aluminium. This trend is opposite of the gas methane consumption in a work shift (measured in mc), which increases with the number of spills. Consequently, the specific gas consumption measure refers to the quality of the melting phase and its optimisation.

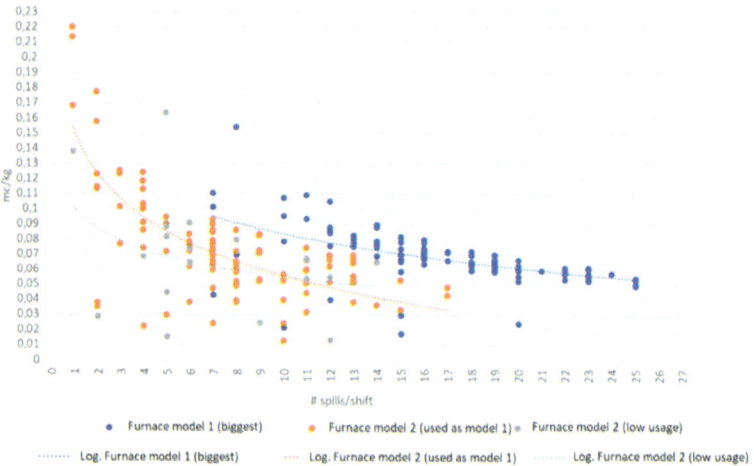

Fig. 4. Trends of the specific gas methane consumption (mc/kg) in each work shift for each of the three furnaces, according to the number of spills in the same work shift.

These results confirm the indications of furnace constructor about the optimization of the load efficiency during low utilization avoiding short melting periods and longer holding ones [8]. In fact, it derives that when a limited number of spills is made by each furnace, the specific gas methane consumption increases. It allows the identification of a minimum number of spills for each furnace in a work shift to guarantee to optimize its use and reduce the energy consumption. This operation, which does not involve any kind of investment, but only a different procedure in making spills from furnaces, could be reduce the average specific gas methane consumption of the entire furnace melting unit from 0.071 to 0.062 mc/kg ($-$ 13%). Considering that, in 2022, the foundry melted more than 11 Mkg of aluminium, it could determine an economic saving of about 100 k€/year (considering a gas methane price of 1 €/smc – dec-2022 data) and a reduction of CO_2 equivalent emissions of about 200 ton/year.

4   Conclusion

The foundry is one of the most intensive sectors, above all for its high gas methane consumption for metal melting, which is a fossil fuel determining high CO_2 equivalent emissions. However, potential solution to reduce this consumption are still lacking in literature. Following some rules given by the furnace constructor, it is possible to optimize the use of this equipment and hence reduce its energy consumption. To make it possible, it is necessary to have data and information, since quantify the resource consumption and some key performance indicators allows the design of the sustainability transition of the companies. Quantifying the sustainability and defining the improving steps are the main objective of the ViVACE® tool, which starts from the numerous data that the companies already manage. With these steps, in the analyzed foundry it was possible to find an improved procedure to use the furnaces without any other investment.

References

1. Riedel E, Ahmed M, Hellmann B, Horn I (2023) Foundry 4.0: digitally recordable casting ladle for the application of Industry 4.0-ready manual casting processes. Proc CIRP 116:95–100
2. Authors' elaboration from Eurostat dataset: NRG_D_INDQ_N
3. Arjunwadkar SH, Ransing MR, Ransing RS (2015) Seven steps to energy efficiency for foundries. Technical 143(3):24–29
4. Arasu M, Rogers Jeffrey L (2009) Energy consumption studies in cast iron foundries. In: Transactions of 57th Indian foundry congress. Recent Advances, Kolkata, India, pp 331–336
5. Lazzarin RM, Noro M (2015) Energy efficiency opportunities in the production process of cast iron foundries: an experience in Italy. Appl Therm Eng 90:509–520
6. Fayomi OSI, Agbola O, Oyedepo SO, Ngene B, Udoye NE (2021) A review of energy consumption in foundry industry. In: Proceeding of international conference on energy and sustainable environment. IOP Publishing, pp 1–6
7. Scharf S, Sander B, Kujath M, Richter H, Stein N, Felde J (2022) Sustainability potentials of an innovative technology and plant system in non-ferrous foundries. Proc CIRP 105:758–763
8. StrikoWestofen: 5 ways to save energy when melting and transferring metal. Available online at: https://www.strikowestofen.com/en-gb/foundry-efficiency/cut-furnace-costs, last access on 27 July 2023

9. Nieckele AO, Naccache MF, Gomes MSP (2011) Combustion performance of ana aluminium melting furnace operating with natural gas and liquid fuel. Appl Therm Eng 31(5):841–851
10. Brewster B, Webb BW, McQuay MQ, D'Agostini M, Baukal CE (2001) Combustion measurements and modelling in an oxygen enriched aluminium-recycling furnace. J I Energy 74:11–17
11. Adeniji TA, Waheed MA (2021) Evaluation of the energy efficiency of an aluminum melting furnace for a Nigerian cast-coiled plant. Fuel Commun 9:100027
12. Bratu V, Stoian EV, Florina VA (2016) Natural gas consumption reducing in aluminium melting furnaces by heat recovery of flue. Sci Bull Valahia Univ Mater Mech 14(11):17–22
13. Bianchini A, Rossi J, Pellegrini M (2019) Overcoming the main barriers of circular economy implementation through a new visualization tool for circular business models. Sustainability 11(23):1–33
14. Bianchini A, Rossi J (2021) An innovative visualization tool to boost and monitor circular economy: an overview of its applications at different industrial sectors. In: Product life cycle—opportunities for digital and sustainable transformation. IntechOpen
15. Startup Stash, Top 27 corporate sustainability tools, 1 Nov 2022. Available online: https://startupstash.com/corporate-sustainability-tools/, Accessed on 25 Sept 2023

Materials and Resource Efficiency

Per- and Polyfluoroalkyl Substances: An Environmental Challenge

Francesco Fontana[✉] and Giampaolo Campana

Department of Industrial Engineering, University of Bologna, 40136 Bologna, Italy
{francesco.fontana16,giampaolo.campana}@unibo.it

Abstract. Per- and poly-fluoroalkyl substances are a group of synthetic aliphatic compounds containing one or more perfluoroalkyl moieties (CnF2n + 1-). Due to their unique surface-active properties and high chemical and thermal stability, they are used in various consumer products such as cosmetics, food packaging, furnishings, and textiles. Their use dates back to the 1940s. Despite their peculiar performances, recent scientific works investigate their sustainability and safety deriving from environmental ubiquity, persistence and bioaccumulation. Indeed, these chemicals are not metabolized in animals and are eliminated mainly through the kidneys, with a low excretion capacity in the case of humans. Many studies conducted worldwide have reported significant serum levels of per- and poly-fluoroalkyl substances in most of the general population, associating them with several health conditions, including hepatoxicity, dyslipidaemia, and endocrine and immunotoxicity outcomes. Besides, new sources of emissions due to degradation phenomena are continuously identified. Consequently, long-chain per- and poly-fluoroalkyl substances have triggered supervising actions and restrictions that fostered research about possible alternatives. This article aims to screen current European scientific investigations conducted to evaluate the impact of these chemicals on the environment and human health, focusing on food packaging and textile production. Due to the paper's length limitation, not all the scientific literature is described because of the need to leave space to discuss possible substitutes. Relevant studies are reported about implementing sustainability and safety/hazard assessment for per- and poly-fluoroalkyl substances. Eventually, the need to develop new and specific analysis methods is discussed.

Keywords: PFASs · Food packaging · Textiles · Life cycle sustainability assessment · Safety assessment · Hazard assessment

1 Introduction

Perfluorocarbons' first appearance in the scientific field dates back to 1886, with Carbon tetrafluoride (PTFE) synthesis. The term *Per-Fluorinated Chemicals* was initially used as a reference for these substances. It was phased out after Per- and Poly-fluoroalkyl Substances (PFASs) with functional groups were introduced in the 1940s. The worldwide spread of PFASs was due to unique chemical and physical properties deriving from their molecular structure: indeed, the highly fluorinated portion of their molecules results in

© The Author(s) 2025
H. Kohl et al. (Eds.): GCSM 2023, LNME, pp. 793–800, 2025.
https://doi.org/10.1007/978-3-031-77429-4_88

amphiphilicity, whereas the functional group, which most have, allows for interaction with polar compounds. Moreover, their resistance to degradation makes them useful at high temperatures or pressure conditions or in corrosive environments. These features are appealing for a variety of industrial applications but lead to the persistence of these chemicals in the environment and even in living beings [1].

Recent studies on the environmental impacts and health effects of PFASs have significantly increased due to their predominantly diffuse use in the past decades without any restriction and a growing understanding of their unsustainable health risks. This concern has grown so much that, after proving and realising that bioaccumulation and likely health risks decrease with decreasing perfluoroalkyl chain, a novel classification was introduced to divide PFASs into long- and short-perfluoroalkyl chains. A few details of the most discussed substances are reported below [2] because the overall classification of PFASs is outside the scope of the present work. The long-perfluoroalkyl chain substances mainly refer to:

- PerFluoro-alkylCarboxylic Acids (PFCAs) with eight or more fluorinated carbons (whose PerFluoroOctanoic Acid, PFOA, is the most studied);
- Perfluoro-Alkane Sulfonates, PFSAs, with seven or more fluorinated carbons (whose PerFluoroOctane Sulfonic acid, PFOS, is the most studied);
- Substances (like Fluorinated Telomer alcohols, FTOHs) that have the potential to split off (e.g. due to atmospheric oxidation or biodegradation) to long-chain PerFluoroAlkyl Acids, PFAAs, giving rise to lower molecular weights products with higher mobility with respect to their precursors.

The short-perfluoroalkyl chain substances mainly refer to:

- PFCAs with seven or fewer fluorinated carbons; and
- PFSAs with six or fewer fluorinated carbons.

This short review attempts to excerpt some salient European case studies concerning food packaging or textile industrial production from the recent and profuse scientific literature. A preliminary list of possible substitutes under investigation for PFASs is also discussed (see Sect. 2). European investigations through Life Cycle Assessment (LCA) and Safety/Hazard Assessment have been recently addressed to evaluate the safety and sustainability of PFASs (see Sect. 3). Eventually, an evaluation of the need for new and specific analysis methods to understand the sustainability and safety of new chemicals before their introduction into the market is discussed.

2   European Case Studies and Possible Alternatives to PFASs

Since 2004, several per-fluorinated chemicals used in food packaging underwent an intense scrutiny by consumer groups and environmental authorities because of their relatively high toxicity and link to a variety of health disorders and cancer. Since PFOS and PFOA were regulated in 2004 and 2017, respectively, there has been a growing awareness that the usage of PFAS in textiles and outdoor wear increases human exposure at every stage of the product (i.e., during production, use and final disposal) [3].

An overview of the examined peer-reviewed articles addressing PFASs migration from food packaging is reported in Table 1.

Table 1. Studies addressing PFASs migration from food packaging and possible substitutes.

Chemicals	EU country/products	Methods/disadvantages	References
PFCAs	Spain/from pet food paper bags to whole and low-fat milk and Tenax®	Liquid chromatography (LC) coupled to tandem Mass Spectrometry (MS)/increased contaminant migration at high temperature (80–120 °C)	[4]
PFCAs	Sweden/fast-foods, pre-prepared meals, bake goods	LC-MS/significant detection in food samples	[5]
FTOHs	Germany/from packaging to dough and butter, Tenax®	Sliding spark spectroscopy/increased contamination at different humidity levels	[6]
FTOHs	Denmark/from dishes and cupcake packages to food	LC-MS/increased contamination at increased temperatures	[7, 8]
PE, PVA-co-PE, PET	Laminated/extruded coating paper/paperboard	Not compostable, not recyclable, not renewable resources	[9, 10]
PLA, PBS, PBAT, PHAs	Laminated paper	Low performances and high cost	[11]

Elizalde et al. observed PFCAs migration levels from packaging to the simulant Tenax®, after 10 days at 40 °C. Even if they did not detect any diffusion phenomenon at 40 °C, an increase of temperature in the range 80–120 °C gave rise to the migration of PFCAs [4]. Furthermore, when testing migration to real foods, such as lyophilized whole and low-fat milk samples, PFCAs migration into food samples were much higher than those obtained into Tenax®. Gebbink et al. tried to quantify PFCAs migration levels into several Swedish fast-foods, pre-prepared meals, and baked goods (before and after preparation), concluding that consumption of food packed in PFCAs precursors could be an indirect source of human exposure to PFCAs [5]. In a German study, Fengler et al. investigated the migration of FTOHs, from baking paper into real local food samples (dough and butter), as well as into Tenax® simulant, at defined parameters (temperature, time). The probability of FTOHs degradation increases with increasing temperatures, and, thus, baking paper with oleophobic surface treatment may be critical food contact materials. The authors came to justify the differences observed in the diffusion of these chemicals, when varying temperature and time, with the differences in samples humidity [6]. A study from Denmark, carried out by Granby et al., aimed to quantify FTOHs migration to food, from dishes and cupcake packaging provided by distributors and retailers in Norway. They also assisted to an increased migration with increasing temperatures [7].

As a consequence of that and other studies performed in Denmark, in 2020 the Danish Ministry of the Environment and Food has established a total fluorine indicator value of 0.1 μg/cm^2 in food packaging [8]. This represents the first measure taken by a European country in reducing PFASs in food packages and currently other European countries support this kind of restrictions after confirming the carcinogenic, endocrine disruptive and immunotoxic effects of these substances.

Concerning the present trends of substitution of PFAS in the food industry, lamination of paper and paperboard by extrusion coating or with preformed melted thin plastic films is the most common commercial method to provide long-term water and oil resistance. Even if PolyEthylene (PE), Poly(Vinyl Alcohol-co-Ethylene) (PVA-co-PE) and PolyEthylene Terephthalate (PET) represent good alternatives to PFASs, the major disadvantages of synthetic polymer laminated papers products are that they are neither compostable nor recyclable and they are derived from a nonrenewable, finite resource [9, 10]. As a result, there has been growing interest in laminating paper products with bio-based, biodegradable polymers, such as PolyLactic Acid (PLA), PolyButylene Succinate (PBS), PolyButylene Adipate Terephthalate (PBAT) and PolyHydroxyAlkanoates (PHAs). Drawbacks for these polymers are represented by their biodegradability encountered in commercial composting facilities, as well as their high cost and sometimes lower performances (in flexibility, heat sealability, adhesion to the paper, resistance to hot liquids), for which research efforts are being made and include also blending with other polymers, surface coating with sizing agents (like polysaccharides, proteins, polyesters, inorganics) or internal sizing (rosin, alkenyl succinic anhydride) [11].

An overview of the examined peer-reviewed articles addressing PFASs used in textiles is reported in Table 2.

Three different studies conducted by Greenpeace in several European nations (e.g., Italy, Sweden, Finland, Switzerland, etc.), found PFASs (in particular PFOA) in outdoor clothing and hiking gear coming in contact with human skin, as wells as FTOHs [12–14]. Furthermore, a recent study on the effect of weathering on Swedish water repellent clothing revealed an increase in concentrations of most PFAAs in weathered and laundered articles compared to the original samples [15]. An increase in volatile FTOHs was also observed in some samples after weathering and laundering, being it justified by the authors with the degradation of FTOHs precursors in time. Besides air, identified as direct exposure source for humans, these results also highlight water as an indirect one, since conventional wastewater treatment plants do not typically have technologies for PFASs capture and destruction, and so PFASs coming with precipitation and laundry water are emitted into the waterways with more risks posed for the environment and human health. A recent study from Sweden, conducted by Schellenberger et al., also revealed not only the loss of PFASs contained in textile fragments (e.g., microfibers), but also the formation and loss of low molecular PFASs, both the phenomena occurring throughout weathering [16].

To the current state of the art, very few alternatives to PFASs have been identified for the clothing market, since it is fiddly to replicate PFASs peculiar performances. Besides repellency to water, oil and dirt, the PFASs-based impregnation agents provide repellency to alcohol and a high level of washing and dry-cleaning durability. In recent years, many manufacturers of impregnating agents have developed non-fluorinated alternatives to

Table 2. Studies addressing PFASs migration from textiles and possible substitutes

Chemicals	EU country/products	Methods/disadvantages	References
PFOA, FTOHs	Several European countries/outdoor clothing and hiking gear	LC-MS/exposure through human skin contact	[12–14]
FTOHs	Sweden/water repellent clothing	Electrospray negative ionization; gas chromatography and MS/exposure through air and drinkable water	[15]
PFASs microfibers	Sweden/textiles for outdoor activities or personal protective equipment	Combustion ion chromatography and LC-TMS/increased exposure throughout weathering	[16]
Paraffin and silicon	Textile and clothing	No efficient repellency against oil and alcohol	[17]
Dendrimers	Textile and clothing	No efficient repellency against oil and alcohol; No data on health properties	[18, 19]
PVC, PU	Textile and clothing	Fabrics not breathable	[20]

PFASs-based finishing agents, in response to a demand for more "environmentally-friendly" solutions. Whilst many of these agents providing water repellency are marketed (e.g., those based on paraffin and silicone chemistries), none of the same agents provides efficient repellency against oil, alcohol and oil-based dirt [17].

Water repellent dendrimer-based impregnation agents have been recently introduced into the market, but they are still not able to guarantee repellency to oil, alcohol and oil-based dirt, even if research efforts on this line are being made. Furthermore, there are no data on health properties of the active substances and other components (since the product compositions were not sufficiently specified for an assessment), even if, according to the producer's information, these products should not be labelled or classified as harmful [18, 19]. Another alternative is represented by polymer coatings (e.g., PolyVinyl Chloride, PVC, or PolyUrethane, PU). They may provide repellency against water, oil and dirt, but the fabrics are not breathable, and have not been further assessed [20].

3   Life Cycle and Safety Assessments for Evaluation of PFASs' Impacts

LCA has been largely implemented to evaluate environmental impacts and sustainability of clothing, household textiles, industrial textile and geotextiles [17]. These works can be classified—based on the functional unit used as reference—in three categories: (1) Product-oriented: when the functional unit is a single item or the economic buying quantity of a target product; (2) Consumer-oriented: when references are the consumers/users;

(3) Use-oriented: when considering the number of wears, cleaning cycles or years of use of a target product. Impact category indicators can also be divided into three groups: (1) Footprint-based indicators: are focused on a single impact category (e.g. carbon footprint, water footprint); (2) Problem-oriented indicators: multi-dimensional assessment conducted at the midpoint level; (3) Damage-oriented indicators: reduced number of indicators conducted at the endpoint level.

It is worth mentioning that extensive studies on the impact of PFASs started in 2016 with a work about durable water-repellent chemistry for textile finishing, by Hanna Holmquist et al. [17]. Then, an original method has been proposed in 2020 to evaluate characterisation factors implemented in a standardised LCA with the aim of calculating the (eco)toxicity of several members of the PFASs family [21]. In particular, the author underlined the need to include an aspect that is not typically considered in a conventional LCA: the fate and degradation of these chemical products. By introducing a new step in the LCA approach—named "transformation"—located between the Life Cycle Inventory and the Life Cycle Impact Assessment (LCIA), they aimed at identifying relevant terminal degradation products from primary pollutant chemicals. Characterisation factors, defined as CF_i in expression (1), are modified by introducing a fraction factor "f" of the primary pollutant, so that they are finally converted into the active degraded chemicals:

$$IS_c = \sum_i (CF_i \times E_i) = \sum_i (CF_i \times f_i \times E_i) \tag{1}$$

where IS_c is the impact score of the category "c" and E_i is the mapped emission.

In 2021, a LCIA with this novel evaluation of the impact score was implemented [22]. As with other cases, authors overcame the lack of data by proposing an effective approach based on the combination of industry consultation (datasheets when available) and scientific or grey (patents repositories) literature research.

It is possible to mention some aspects that need further development: (a) concerning system boundaries, it is worth remarking a lack of studies focusing on a cradle-to-grave perspective and, in general, the need for more attention to textiles use stage by including use processes (washing, cleaning, drying, ironing, repairing, etc.). (b) Databases are often mentioned as the weak point, and missed data determine uncertainties that can only be avoided by reducing system boundaries. This issue is coherent with the previously mentioned concerns about the shortage of investigations based on a cradle-to-grave system. (c) An integrated approach—Safe-and-Sustainable-by Design (SSbD) material, process and products—is still under development and refinement.

4   Conclusions

PFASs are ubiquitous in the environment and our lives. Their chemical and thermal stability, as well as their hydrophobic and oleophobic properties, provide unique material benefits, but also high persistency, bioaccumulation, and toxicity. As studies about human health and environmental effects of PFASs continue to grow, and the controversy about the sustainability and safety of these chemicals is hopefully going to be sorted out, scientists, governments, product manufacturers, purchasing organizations, and consumers should work together to limit more and more the production and use of these substances globally and to develop safer and functional nonfluorinated alternatives.

Furthermore, given PFASs persistence in the environment, in absence of indisputably safer substitutes, an open discussion should be started, informed by the scientific evidence, to understand whether consumers would give up certain product functionalities to protect themselves against potential health risks.

It has been demonstrated the importance of the LCA approach and methods to assess the sustainability and safety of this class of chemicals. Despite the extensive use of the LCIA to evaluate and compare materials and products, scientists stressed the lack of studies regarding cradle-to-grave system boundaries, reliable and regional data inventories and, occasionally, the use of inconsistent functional units.

Eventually, in the case of PFASs, fate and degradation of these chemicals should be taken into account to correctly evaluate the environmental impact of their use. In particular, we examined a relevant case in which, for this purpose, a modification of characterisation factors calculation has been proposed and successfully implemented.

Acknowledgements. This work was promoted under the Project Horizon Europe ZEROF (Development of verified safe and sustainable PFAS-free coatings for food packaging and upholstery textile applications).

References

1. Wang Z, Dewitt JC, Higgins CP, Cousins IT (2017) A never-ending story of per- and polyfluoroalkyl substances (PFASs)? Environ Sci Technol 51(5):2508–2518
2. Wang Y et al (2019) A review of sources, multimedia distribution and health risks of novel fluorinated alternatives. Ecotoxicol Environ Saf 182:109402
3. Zheng G, Salamova A (2020) Are melamine and its derivatives the alternatives for per- and polyfluoroalkyl substance (PFAS) fabric treatments in infant clothes? Environ Sci Technol 54(16):10207–10216
4. Elizalde MP, Gómez-Lavín S, Urtiaga AM (2018) Migration of perfluorinated compounds from paperbag to Tenax® and lyophilised milk at different temperatures. Int J Environ Anal Chem 98(15):1423–1433
5. Gebbink WA, Ullah S, Sandblom O, Berger U (2013) Polyfluoroalkyl phosphate esters and perfluoroalkyl carboxylic acids in target food samples and packaging-method development and screening. Environ Sci Pollut Res 20(11):7949–7958
6. Fengler N, Schlummer R, Gruber M, Fiedler L, Weise D (2011) Migration of fluorinated telomer alcohols (FTOH) from food contact materials into food at elevated temperatures. Organohalogen Compd 73:939–942
7. Granby K, Håland J (2018) Per- and polyfluorinated alkyl substances (PFAS) in paper and board food contact materials—selected samples from the Norwegian market 2017. Technical University of Denmark
8. Schultes L et al (2019) Total fluorine measurements in food packaging: how do current methods perform? Environ Sci Technol Lett 6(2):73–78
9. Andersson C (2008) New ways to enhance the functionality of paperboard by surface treatment—a review. Packag Technol Sci 21(6):339–373

10. Glenn G, Shogren R, Jin X, Orts W, Hart-Cooper W, Olson L (2021) Per- and polyfluoroalkyl substances and their alternatives in paper food packaging. Compr Rev Food Sci Food Saf 20(3):2596–2625

11. Hubbe MA, Pruszynski P (2020) Greaseproof paper products: a review emphasizing ecofriendly approaches. BioResources 15(1):1978–2004

12. Greenpeace (2012) Chemistry for any weather—greenpeace tests outdoor clothes for perfluorinated toxins

13. Greenpeace (2013) Chemistry for any weather—part II excecutive summary—outdoor report

14. Greenpeace (2016) Leaving traces—the hidden hazardous chemicals in outdoor gear

15. van der Veen I, Hanning AC, Stare A, Leonards PEG, de Boer J, Weiss JM (2020) The effect of weathering on per- and polyfluoroalkyl substances (PFASs) from durable water repellent (DWR) clothing. Chemosphere 249:126100

16. Schellenberger S et al (2022) An outdoor aging study to investigate the release of per- and polyfluoroalkyl substances (PFAS) from functional textiles. Environ Sci Technol 56(6):3471–3479

17. Holmquist H, Schellenberger S, van der Veen I, Peters GM, Leonards PEG, Cousins IT (2016) Properties, performance and associated hazards of state-of-the-art durable water repellent (DWR) chemistry for textile finishing. Environ Int 91:251–264

18. Atav R, Bariş B (2016) Dendrimer technology for water and oil repellent cotton textiles. AATCC J Res 3(2):16–24

19. Lassen C, Jensen AA, Warmning M (2015) Alternatives to perfluoroalkyl and polyfluoroalkyl substances (PFAS) in textiles. In: Danish environmental protection agency. LOUS Survey of chemical substances in consumer products, vol 137

20. Özek HZ (2018) Development of waterproof breathable coatings and laminates. In: Waterproof and water repellent textiles and clothing, the textile institute book series, pp 25–72

21. Holmquist H, Fantke P, Cousins IT, Liagkouridis I, Peters GM (2020) An (eco)toxicity life cycle impact assessment framework for per- and poly fluoroalkyl substances. Environ Sci Technol 54(10):6224–6234

22. Holmquist H, Roos S, Schellenberger S, Jonsson C, Peters G (2021) What difference can drop-in substitution actually make? A life cycle assessment of alternative water repellent chemicals. J Clean Prod 329(129661)

Algae-Based Phlorotannins as a Sustainable Feedstock for Epoxy Resin Formulation

S. Böhm[1], A. Winkel[1], M. Kahlmeyer[1], B. Fazliu[2], M. Horn[2], and T. Fuhrmann-Lieker[2(✉)]

[1] Department for Cutting and Joining Manufacturing Processes, University of Kassel, Kurt-Wolters-Str. 3, 34125 Kassel, Germany
s.boehm@uni-kassel.de
[2] Physical Chemistry of Nanomaterials, University of Kassel, Heinrich-Plett-Str. 40, 34132 Kassel, Germany
th.fuhrmann@uni-kassel.de

Abstract. Bisphenol A is the most important chemical for producing epoxy resins, but as of today is not bio-based accessible. Furthermore, it is rated as a substance of very high concern and possesses reproductive toxic and endocrine-disrupting properties. Phlorotannins, a class of polyphenols, are structurally highly suited for serving as sustainable bisphenol A alternatives. They are largely found in brown algae, which are already being harvested for alginate production. Phlorotannins thus represent a promising marine raw material for the chemical industry which otherwise has received little attention in research to date, at least in the field of epoxy resin formulation. For this study, an epoxy-resin model compound based on phloroglucin, the simplest phlorotannin, was chosen to gain insight into reactivity and thermo-mechanical characteristics. As curing agents, well-established systems for ambient-temperature cure, e.g. isophorone diamine, as well as anhydrides for heat cure were applied. In all cases, thermosets with glass transition temperatures higher than 100 °C could be obtained under cross-linking conditions comparable to today's procedures. In the case of a phthalic acid anhydride derivative, even a T_g of 198 °C has been determined, proving the high potential of the cured systems for industrial usage, e.g. as impregnating resins for fiber-reinforced plastics.

Keywords: Phlorotannins · Epoxy resin · Algae

1 Introduction

1.1 General Discussion

The substitution of petroleum-based chemicals with sustainable alternatives has become ever so important as the worldwide community strives to reach CO_2 reduction goals and diminish the rate of global warming. Thus, in the past years, research focused on polymers that possess similar properties as well-established systems but with a smaller

H. Kohl et al. (Eds.): GCSM 2023, LNME, pp. 801–808, 2025.
https://doi.org/10.1007/978-3-031-77429-4_89

product carbon footprint, often in combination with high recyclability and biodegradability [1]. This is especially true for the thermoplastic sector where polymers made of starch [2], cellulose [3], proteins [4], etc. could be successfully developed, or, in the case of polylactic acid, are already in use [5]. Besides thermoplastics, thermosets, due to their unique properties such as high mechanical strength and chemical resistance, have proven to be of paramount importance for highly demanding industrial sectors such as automotive and aviation. Thus, when improving the sustainability of thermosets, the biobased synthesis of their precursors and components has become increasingly important. For epoxy resins, however, bisphenol A, the fundamental chemical for their production, is still unavailable from renewable sources [6]. Theoretical alternatives have emerged in recent years such as epoxidized vegetable oils (EVOs). Due to their size and limited epoxy content, thermosets made entirely of EVOs demand long curing times and high curing temperatures, even under catalytic conditions, or have, generally speaking, relatively low glass transition temperatures that limit the usability in high-temperature applications [7]. These drawbacks confine EVOs' use as thermoset resins for demanding industrial use cases. Therefore, the search for additional biological feedstock that can be readily converted into reactive resin components by well-established methods is crucial for sustainable manufacturing.

1.2 Phlorotannin-Based Epoxy Resins from Brown Algae as a Sustainable Feedstock

Phlorotannins are a special class of polyphenols that can be considered oligomers of phloroglucin (1,3,5-trihydroxy benzene), the simplest compound of the group. They are largely found in brown algae [8]—in their cells in organelles called physodes, and in the cell walls where they are cross-linked to alginates. Thus, in the long term, phlorotannin extraction could primarily be based on the same harvesting routes that are already established for alginate production. Sources assume an annually extracted alginate quantity of about 30,000–40,000 t worldwide [9, 10], associated with a maximum content of over 40% concerning the dry algae mass [11]. In terms of phlorotannins, taking into account the above values and a share of up to 20% by weight of the dry matter [12], this would ideally result in an algae-based phlorotannin quantity of 15,000–20,000 t p.a., which could be produced without algae farming specifically set up for phlorotannin extraction. However, although different procedures, mainly based on a solid-liquid extraction, are already described in the literature [13], efficient methods of extracting phlorotannins customized to the specific algal species on an industrial scale are yet to be developed, and the question of whether the exploitation of alginate extraction side streams is even possible remains unknown up to this point. Yet, because of phlorotannins' high hydroxyl content, they are structurally highly suited for the introduction of epoxy groups. The standard reactant for this is epichlorohydrin [14], which, e.g., can be sustainably obtained from bio-glycerol, a side product of biodiesel production. Since it is a carcinogenic substance, no matter the origin, alternative reaction routes are discussed, the most prominent being the reaction with allyl halides and the subsequent Prilezhaev reaction, e.g. in an enzymatic fashion [15].

1.3 Requisite Factors for Industrially Applicable Epoxy Resins

Due to numerous combinations of resins, curing agents, catalysts, and curing procedures, thermosets can be developed to suit a broad variety of applications. Thus, general valid statements about characteristic values of epoxy resins are difficult to make. However, certain remarks regarding the curing of bisphenol A-based epoxy resins with well-established curing agents like amines and anhydrides may be made at this point according to assertions cited in the literature [16]:

1. Curing with aliphatic amines readily occurs at room temperature but can be accelerated at elevated temperatures
2. Anhydride-cure is slow and demands heat for the cross-linking to proceed, in certain cases up to 200 °C
3. Anhydride-cure can be accelerated by adding catalysts such as tertiary amines or imidazoles
4. Anhydride-cure shows low curing enthalpies.

Among the many characteristics that give insight into the performance of epoxy resins, factors that describe the curing kinetics of thermosets are especially important for proving whether reactive resins are suitable for industrial purposes. For example, the gelation time for ambient temperature curing as well as the gelation temperature for heat cure depict the point in the curing process where the system transforms from a gel into a solid and thus possesses a certain degree of internal strength. In addition, the heat amount released during curing (curing enthalpy ΔH) should be low to prevent the mixture from self-combustion due to the exothermic nature of the cross-linking. Furthermore, time-efficient cross-linking should take place in lower-temperature regions, even for heat-curing systems since this greatly diminishes energy consumption of the manufacturing procedures. From a thermo-mechanical standpoint, the glass transition temperature is equally crucial since it limits the thermoset's applicability in higher-temperature regions due to the polymer becoming rubber-like.

In this study, these characteristics were determined for a glycidyl ether of the simplest phlorotannin, phloroglucin, in combination with well-established hardeners.

2 Experimentals

2.1 Materials and Methods

The phloroglucin glycidyl ether (PGE, (**1**)) with an epoxy content of 8.3 meq/g, which equals an average value of 3.57 epoxy groups per oligomer molecule, was purchased from Specific Polymers. As curing agents, amines for ambient-temperature cure have been tested: Diethylene triamine (**2**) (Sigma-Aldrich), isophorone diamine (**3**) (TCI Chemicals), and Jeffamine D230 (**4**) (Sigma-Aldrich). Admerginic acid (**6**), kindly provided by HOBUM Oleochemicals GmbH, Hamburg, Germany, and 4-methyl cyclohexane dicarboxylic acid anhydride (**5**) (TCI Chemicals) were used as hardeners for heat-cure, also catalyzed by 2-ethyl imidazole (**7**) and DMP-30 (**8**), both were supplied by TCI Chemicals. All chemicals were used as received (Fig. 1).

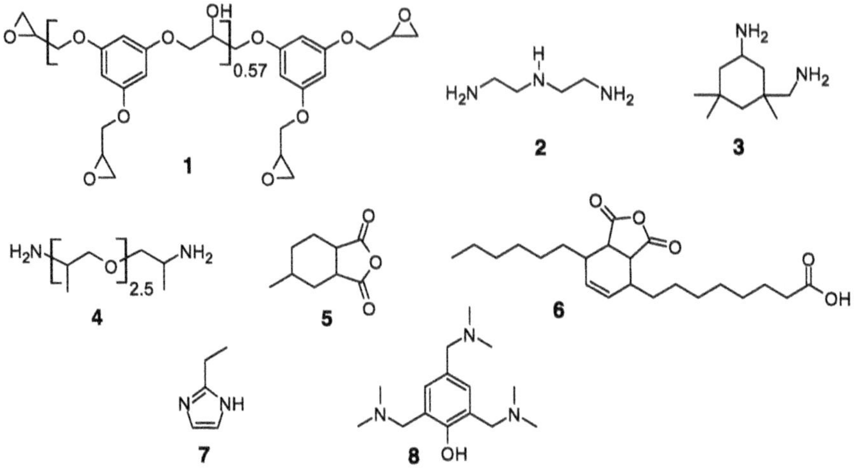

Fig. 1. Chemicals used in this study.

2.2 Preparing the Epoxy Resin Formulations

Around 2 g of reactive resin was prepared by mixing the epoxy resin **1** and the appropriate amount of curing agent, assuming a stoichiometric reaction of the functional groups involved. Catalysts **7** and **8** were added in a total of 5 weight% of the combined mass of resin and hardener. All mixtures were homogenized in a centrifugal mixer (SpeedMixer DAC 150 SP, Hauschild, Germany) before further use (Table 1).

Table 1. Epoxy resin compositions based on PGE (**1**) presented in this study.

Resin	Curing agent	Catalyst	Resin	Curing agent	Catalyst
R1	2	–	R6	5	8
R2	3	–	R7	6	–
R3	4	–	R8	6	7
R4	5	–	R9	6	8
R5	5	7			

2.3 Gel Point Determination

An Anton-Paar MCR 502 rheometer was used for gel point determination. Disposable aluminum plates with a diameter of the top plate of 15 mm were utilized for all experiments with a gap of 0.15 mm. The deformation value was 5% with a frequency of 1 Hz. Gel points were determined as the intersection of the temperature-dependent storage modulus and loss modulus. The resins containing amines **2**, **3**, and **4** were analyzed at 20 °C, whereas systems using anhydrides **5** and **6** were heated from 20 to 200 °C at a 5 K/min rate. Thus, the gelation point for the first group of resins is given as a time value, and as a temperature value for the latter.

2.4   Determination of Curing Enthalpy and Curing Kinetics

In a typical curing experiment, 5–10 mg of resin were heated in a 40 μL aluminum crucible from $-$ 25 °C to 300 °C in a Mettler Toledo DSC 1. For each system, four different heating rates were applied (2.5, 5, 10, and 20 K/min). The activation energy E_A of each curing reaction was calculated according to Kissinger [17] from the best-fit line's negative slope multiplied by the gas constant R when $\ln[\beta/T_P^2]$ is plotted against $1000/T_P$ (where β and T_P are the heating rate and the temperature associated with the corresponding curing peak, respectively). The exothermic curing peaks of the 5 K/min. Experiments are the basis for the associated peak temperature T_P and the individual curing enthalpies ΔH which themselves represent the basis for determining the conversion-temperature dependency (Fig. 2).

2.5   Determination of the Glass Transition Temperature

The cured sample of each 5 K/min-DSC experiment was heated from 25 to 250 °C, cooled, and heated a second time within the same temperature range. The heating and cooling rate was $\pm$ 20 K/min. The T_g was determined at half the individual step height.

3   Results and Discussion

The results of the different experiments are summed up in Table 2.

Table 2. Determined data for the curing of epoxy resins **R1** to **R9**.

Resin	Gel point	ΔH (J/g)	T_P (°C)	E_A (kJ/mol)	T_g (°C)
R1	65.4 min*	483.10	68.37	57.6	152.7
R2	128.4 min*	428.91	78.29	56.7	115.6
R3	404.6 min*	446.26	91.91	53.1	102.6
R4	135.6 °C	400.82	166.24	75.6	198.0
R5	98.9 °C	317.11	108.79	74.4	152.0
R6	90.6 °C	198.10	103.82	74.6	168.5
R7	169.8 °C	327.06	170.70	67.1	103.5
R8	105.6 °C	301.60	112.88	80.6	107.2
R9	110.0 °C	278.04	114.18	72.6	141.9

3.1 Amine-Curing Epoxy Resins

Of the amines used, especially diethylene triamine (**2**) proved to be highly reactive, up to the point, where the exothermic curing reaction was so violent that the mixture underwent self-combustion when working with bigger amounts, even without additional heating. In addition, **R1** possesses the highest curing enthalpy of all mixtures examined, and the gel point of **R1** could be determined to be 65.4 min, the lowest value of the amine-cured systems. In comparison, **R2** and **R3** have gel points of 128.4 and 404.6 min at 20 °C, respectively. These differences in reactivity are also mirrored by the various glass transition temperatures, that of **R3**, containing the biggest and thus the most slow-reacting amine, being the lowest (102.6 °C), whereas the highest T_g was obtained for **R1** (152.7 °C).

Although the amine-curing resins already convert at room temperature into hardened thermosets, curing can be enhanced by additional heating. The increase of conversion as a function of temperature is shown in Fig. 2 (left) for a 5 K/min heating rate. Under these conditions, the reaction partners' noticeable conversion already occurs at moderate temperatures of around 60–80 °C. In addition, a high conversion is first reached for amine **2**, followed by isophorone diamine (**3**) and Jeffamine D230 (**4**) which again illustrates the order of amine reactivity. However, in terms of the activation energy EA, resins **R1** to **R3** only differ marginally but exhibit lower values than the systems using anhydrides as curing agents, even when applying catalysts **7** and **8**.

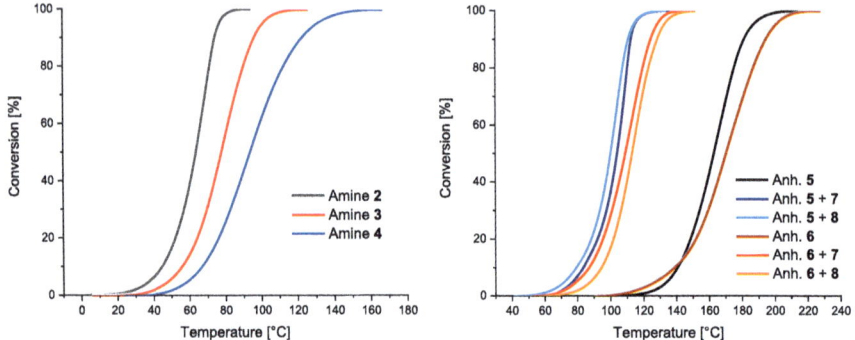

Fig. 2. Relation between conversion and temperature for the epoxy resins using amines **2** to **4** (left) and anhydrides **5** and **6** (right) as curing agents for 5 K/min heating experiments.

3.2 Anhydride-Curing Epoxy Resins

The anhydride curing agents **5** and **6** only make a marginal difference in terms of conversion-time dependency, as can be seen for resins **R4** and **R7** in Fig. 2 (right). These systems require higher temperatures for the curing reaction to proceed. Although the bespoke resins behave similarly in this regard, the smaller and thus more reactive anhydride **5** delivers a cured thermoset that reaches the gel point 35 °C earlier than the mixture using admerginic acid (**6**) (135.6 °C compared to 169.8 °C). In addition, the T_g of **R4** (198 °C) is nearly twice that of resin **R5** (103.5 °C).

Adding catalysts **7** and **8**, respectively, can drastically alter the curing behavior. E.g., the gel points of **R5** and **R6** are reasonably lower than compared to **R4**. Furthermore, the time for reaching 50 and 99% conversion at 160 °C, as deduced from the kinetic experiments, is only but a fraction of the value for **R4**, illustrating the ability of **7** and **8** to improve the curing rates drastically. However, the activation energies E_A for the hardening reactions of **R4** to **R6** do not differ to that extent as one could conclude the catalysts' activity. In addition, the T_g values for the resins containing a catalyst are lower in comparison to resin **R4** where no catalyst is applied.

This behavior is not found for the admerginic acid-containing resins **R7** to **R9**. With the addition of a catalyst, the T_g increases distinctly, with **8** being the most potent one in this regard (T_g of **R9** is 141.9 °C, compared to **R7** with a T_g of 103.5 °C). However, in terms of the gel point, the imidazole catalyst (**7**) proved to be more beneficial as compared to DMP30 (**8**), although the difference is not markedly pronounced.

The activation energies of **R7** to **R9** differ significantly without any obvious trend, thus it is not possible to derive any tendency about the reaction rate from this.

4   Conclusion

The phloroglucin-based epoxy resin was shown to provide duromers with glass transition temperatures higher than 100 °C by reacting with established curing agents. For some systems, these values were even above 140 °C and in individual cases approached 200 °C. In all cases, crosslinking took place in temperature ranges and within periods that are also found for established bisphenol A resins in the literature. The catalyzed blends with admerginic acid hardener turned out to be particularly promising since the hardener itself is based on sustainable sources (fatty acid). In summary, the authors therefore believe that phlorotannins extracted from algae and epoxy resins based thereof are not only promising alternatives for petrochemical reactive resins but also for those based on epoxidized vegetable oils.

5   Outlook

The authors presented results for a phloroglucin-based epoxy resin acting as a model substance for algae-based phlorotannin systems. The displayed curing behavior and high Tg demonstrate the general suitability for an industrial application, especially in the form of the anhydride-curing systems supported by a catalyst. However, numerous topics still need to be addressed before the production of epoxy resins from brown algae can be implemented on an industrial scale in the long term:

- Optimization of phlorotannin content depending on macroalgae species, age, size, tissue type, and abiotic factors
- Development of efficient and sustainable extraction methods
- Chemical characterization of the phlorotannin mixtures

- Separation of structurally promising phlorotannins and converting these into epoxy-bearing components for matrix resin production through a preferably sustainable chemistry
- Investigation of the crosslinking behavior of algae-based epoxy resins (kinetics, rheology) in dependence on industrially established or newly developed sustainable curing agents
- Mechanical characterization of the thermosets produced, also after aging
- Evaluation of the adhesion to a variety of substrates for use as adhesives, potting material, and the like, also regarding joint aging
- Introduction of advanced material functionalities such as self-healing or the incorporation of the vitrimer concept into the reactive resins.

References

1. Kharb J, Saharan R (2022) International conference on materials science and engineering (ICMSE 2022). IOP Conf Ser Mater Sci Eng 1248:012008
2. Surendren A, Mohanty AK, Liu Q, Misra M (2022) Green Chem 24:8606–8636
3. Immonen K et al (2021) Molecules 26:1701
4. Hernandez-Izquierdo VM, Krochta JM (2008) J Food Sci 73(2):R30–R39
5. Binti Taib NA, Rahman MR, Huda D, Kuok KK, Hamdan S, Bin Bakri MK, Bin Julaihi MRM, Khan A (2023) Polym Bull 80:1179–1213
6. Quian Z et al (2022) Chem Eng J 435(2):135022
7. Di Mauro C, Malburet S, Genua A, Graillot A (2020) Biomacromol 21:3923–3935
8. Imbs TI, Zvyagintseva TN (2018) Russ J Mar Biol 44(4):263–273
9. Saji S, Hebden A, Goswami P, Du C (2022) Sustainability 14(9):5181
10. Marburger A (2003) Alginate und Carrageenane—Eigenschaften, Gewinnung und Anwendungen in Schule und Hochschule. Dissertation. Philipps-Universität Marburg
11. Kane SN, Mishra A, Dutta AK (2016) International conference on recent trends in physics (ICRTP 2016). J Phys Conf Ser 755:011001
12. Steevensz AJ et al (2012) Phytochem Anal 23(5):547–553
13. Santos SAO, Félix R, Pais ACS, Rocha SM, Silvestre AJD (2019) Biomolecules 9(12):847
14. Genua A et al (2020) Polymers 12(11):2645
15. Aouf C, Lecomte J, Villeneuve P, Dubreucq E, Fulcrand H (2012) Green Chem 14(8):2328–2336
16. Sukanto H, Raharjo WW, Ariawan D, Triyono J, Kaavesina M (2021) Open Eng 11:797–814
17. Kissinger HE (1957) Anal Chem 29(11):1702–1706

Development of a Pre-treatment Process for EPS-ETICS to Enable a Solvent-Based Recycling

Sebastian Lumetzberger[1]($\boxtimes$), Andreas Lehner[1], Selina Möllnitz[2], Gernot Peer[3], Stephan Keckeis[1], Roy Huisman[1], Laura Kasinger[1], and Sebastian Schlund[1,4]

[1] Fraunhofer Austria Research GmbH, Theresianumgasse 7, 1040 Wien, Austria
sebastian.lumetzberger@fraunhofer.at
[2] Lindner-Recyclingtech GmbH, Manuel-Lindner-Straße 1, 9800 Spittal an der Drau, Austria
[3] SUNPOR Kunststoff GmbH, Tiroler Straße 14, 3105 St. Pölten, Austria
[4] TU Wien, Institute of Management Science, Theresianumgasse 27, 1040 Wien, Austria

Abstract. Large quantities of HBCD-containing Expanded Polystyrene (EPS) had been used in External Thermal Insulation Composite Systems (ETICS) from the 1970s until the flame retardant HBCD was banned in 2016 for being a persistent organic pollutant (POP). According to the Basel Convention, a solvent-based process is the best available technology for POP-containing waste. Such processes are subject to an impurity limit, thus requiring a pre-treatment of the input material. Therefore, a sequence of comminution and sorting technologies was designed and tested with real-life ETICS demolition waste. Three specimens were used as input material. Following an initial analysis, the demolition waste was comminuted, examined, and a sorting technology was selected. The resulting material fractions were weighted and analysed. The lightweight fraction (target: sole EPS) was additionally tested for its purity. The designed process achieved an efficient material disintegration and separation resulting in a purity of the lightweight fraction of up to 93 wt-%. This result represents a major step towards the circularity of EPS via solvent-based recycling.

Keywords: Pre-treatment · Solvent-based recycling · Expanded polystyrene

1 Introduction

The sustainable use of resources is a current issue that encompasses energy conservation as well as energy and resource efficiency. Insulating buildings with an External Thermal Insulation Composite System (ETICS) makes a significant contribution to energy conservation [1]. An ETICS is a multi-component composite, which may include expanded polystyrene (EPS) and has the following schematic structure: adhesive, insulation material (EPS), mechanical fixing devices, base coat, reinforcement and finishing layer [2]. Since 1970, EPS has been increasingly used in Austria, with up to 53 kt installed per year. The amount of EPS on Austrian facades was about 690 kt in 2016 [3]. Until then

© The Author(s) 2025
H. Kohl et al. (Eds.): GCSM 2023, LNME, pp. 809–817, 2025.
https://doi.org/10.1007/978-3-031-77429-4_90

EPS contained hexabromocyclododecane (HBCD) as a flame retardant, which was classified as a persistent organic pollutant (POP) by the European Commission in 2016 [4]. Increasing amounts of waste are expected in the future as many of these ETICS have an estimated service life of 40–80 years [5]. Studies have shown that almost all of the insulation materials collected in the construction sector are incinerated and thus removed from the material cycle [6]. Across Europe, EPS demolition waste amounted to 88 kt in 2019, of which 1.4% was recycled [7]. For EPS containing HBCD, either incineration or recycling with prior separation of HBCD is currently allowed. In this case, an HBCD content of less than 100 mg HBCD/kg must be achieved [8]. The solvent-based process has been included in the technical guidelines of the Basel Convention as the best available recycling technology for the treatment of HBCD-containing waste. The quality of the input material has a direct influence on the quality of the output material as well as the economic and ecologic efficiency. Pretreatment is required as the process allows up to 7 wt-% impurities [9].

2 State of the Art

There are three main methods for the deconstruction of ETICS:

- Non-selective deconstruction: Building materials are crushed without any pre-sorting. The result is a heterogeneous mixture, often contaminated with hazardous substances. Due to legislation and high disposal costs, this method is generally no longer used [10].
- Partial-selective deconstruction: This involves the removal of different components to minimize the mixing of construction waste [1].
- Selective deconstruction: This method means breaking down the building by function or material composition and is therefore very time-consuming [1].

2.1 Comminution

As there is currently no comminution technology specifically designed or optimised for ETICS waste, there have been several experimental implementations [10]. Fehn and Teipel [11], performed initial tests with the cutting mill and ball mill, but without satisfactory results. The main problem was the grinding mechanism of the cutting mill becoming heavily clogged, resulting in only 20% of the material being crushed. Additionally, there were still traces of plaster residue on the EPS material. Otherwise, the ball mill failed to achieve the desired disintegration and comminution. In addition, the required throughputs for an industrial application cannot be achieved with these aggregates. Heller [10] showed that by using an impact crusher, ETICS can be largely disintegrated and selectively crushed. However, Simons and Feil [12] showed that significant contamination remains in the same grain size range as most of the EPS. The hammer mill has been used in many experiments [1, 5, 10–13]. During these experiments, it was observed that without a sieve insert, the particle size varied significantly, and plaster residue was present on the EPS material [11]. The use of sieves has achieved a high degree of material separation [11–13], but EPS often accumulates on the walls of the grinding chamber of the hammer mill [13]. Single-shaft and twin-shaft shredders according to the cutting principle are more resistant to impurities, gentle on the material and produce a continuous output. With suitable screen inserts, a targeted particle size distribution can also be produced.

2.2 Sorting

Due to the large specific density difference between EPS and the expected impurities (e.g., plaster, adhesives, mechanical fixing devices), separation techniques which use the specific density as a separation characteristic, are used in the literature. Heller and Flamme [14] achieved a very precise result with float-sink sorting in water, but due to the high energy consumption during drying, this technology was not pursued further. In laboratory-scale trials from Heller [10], pure fractions were also obtained using a zigzag air classifier. This method was also found to be suitable for large-scale trials. However, the sorting of the light fraction was insufficient and additional screening was required. Other experiments have failed to separate the light fraction with the zigzag air classifier [11].

Table 1 shows the results of various comminution and sorting experiments. It should be noted that the impact crusher and hammer mill experiments were carried out on a laboratory scale and may not be readily applicable to industrial scale scenarios [10, 14].

Table 1. Overview of comminution and sorting experiments in literature

Comminution	Sieve opening (mm)	Sorting	Purity (wt-%)	EPS material output (wt-%)	References
Impact crusher	20.0	Sieve analyses	99.0	2.0	[12]
Impact crusher	31.5	Sieve analyses	41.0	85.0	[12]
Hammer mill	10.0	Sieve analyses	96.0	16.0	[12]
Hammer mill	31.5	Sieve analyses	64.0	45.0	[12]
Hammer mill	10	Zigzag air classifier	–	84.0	[11]
Flail drum	–	Zigzag air classifier	90.0	–	[10]

Simons and Lößner [15] used a hammer mill and a zigzag air classifier. In this experiment high purities and material yields were achieved, but only low throughputs. However, this trial is not included in the table as a two-stage sorting process was carried out with one sieve fraction.

3 Methodology

The methodology for designing a pre-treatment process for EPS-ETICS to enable solvent-based recycling is described in the following section. The initial input material as well as the target purity of ≥ 93 wt-% of the EPS were set as constraints and the yield rate was targeted to be as high as possible. In addition, the mechanical process steps were intended to be continuous rather than batch-based to boost the economic aspect. The development of the process steps required an iterative approach, as the outputs and results could not be anticipated prior to conducting the previous step. Therefore, each step was based and selected on the predecessors' results, as outlined in the following:

1. Initial analysis: The specimens which served as input material are analysed, characterised, and categorised. These include deconstruction method, fixing type, thickness of the EPS, characteristics, and a qualitative grading of impurity.
2. Comminution: Several experiments aiming at breaking up the interfaces between the EPS and the adhesive and the EPS and the base coat were conducted using hammering strain. Although the achieved purities were satisfactory, the corresponding yield rates were insufficient. This approach, therefore, involves the use of a single shaft shredder for comminution and disintegration. For possible scale-up and system design, the mass of the comminuted material as well as the machine running time are recorded to calculate the throughput. Additionally, the energy consumption is recorded during the test.
3. Post-comminution analysis: Representative samples are taken following ÖNORM EN 15442 for subsequent determination of bulk density per ONR CEN/TS 15401. A sieve analysis according to DIN 66165 is conducted to quantitatively determine the comminution and select an adequate sorting process.
4. Sorting: The output resulting from the comminution process is separated by using an air sorting table, which splits the material into three fractions: heavyweight (target: materials with density above the density of EPS), lightweight (target: EPS), and dust (target: dust). All fractions are weighed and the machine running time is recorded to calculate the throughput.
5. Post-sorting analysis: All resulting fractions are qualitatively analysed regarding their composition. For the lightweight fraction, samples were taken and tested for purity by weighing, dissolving in ethylbenzene, decanting, drying, and finally weighing the insoluble remains. This allows determining the weight percentage of insoluble impurities.

4 Case Study

This chapter presents the tests and analyses carried out according to the defined methodology with three different specimens to achieve an EPS fraction from ETICS that meets the requirements of the solvent-based recycling process.

4.1 Initial Analysis

ETICS with EPS from semi-selective mechanical and manual deconstruction were selected as specimens for the tests, as these are the two deconstruction methods most frequently used in practice. Figure 1 shows parts of the experimental material. Specimens 1 and 2 are relatively clean except for the adhesive residues and specimen 3 is mainly contaminated with plaster and reinforcement lattice.

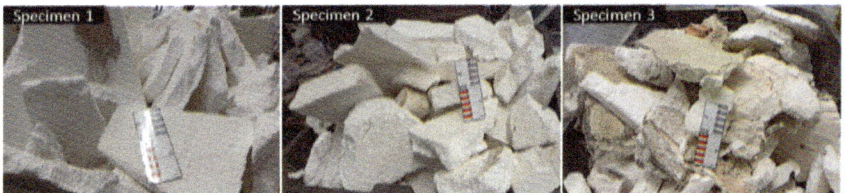

Fig. 1. Input material for comminution tests

The most important specifications recorded in an initial analysis of the samples are summarised in Table 2.

Table 2. Overview of specimen acquisition and characteristics

	Specimen 1	Specimen 2	Specimen 3
Deconstruction method	Manual deconstruction, complete stripping, chiselling with a chisel	Mechanical deconstruction, mainly stripped, scraping with sorting gripper	Mechanical deconstruction, mainly not stripped, scraping with sorting gripper
Fixing type	Glued	Glued and dowelled	Glued and dowelled
Thickness EPS (cm)	5	5–8	5–8
Characteristics	Very clean and dry, coarse (grain size > 200 mm), occasional adhesive and brick adhesions	Different adhesions and impurities (plaster, reinforcement grids, adhesive), dry, coarse (grain size > 200 mm)	Different impurities (plaster, reinforcement grids, adhesive), dry, coarse (grain size > 200 mm)

4.2 Comminution

A Lindner Antares 1600 single-shaft shredder with a 43P cutting system with minimal cutting gap and screens with a diameter of 50 mm was used to comminute the specimens. The speed of the shaft was 100 rpm and the condition of the knives, shaft and screens was as good as new. The shredder set-up was determined in preliminary tests. The specimens were successively filled into pallet boxes and fed into the hopper of the shredder. After comminution, the material was transported by a belt conveyor and filled into big bags, labelled, and weighed. The throughput was in the range between 680 and 1200 kg/h and the energy consumption was between 40 and 68 kWh. The tests showed that all specimens, even the heavily contaminated ones, could be comminuted without any problems. A first qualitative analysis of the crushed specimen indicated a good disintegration of the material, which is a basic prerequisite for further sorting. During operation, representative samples of each specimen were taken for bulk density determination and sieve analysis.

4.3 Post-comminution Analysis

The sieve analyses were carried out on an analytical vibratory sieve shaker Retsch AS 450 CONTROL. In addition to bulk density (20–55 kg/m³), material type and composition, a particle spectrum analysis supports the selection of sorting units and their setting parameters. Interpolations of the values provided by the sieve analysis for each specimen are given in Fig. 2. It allows determining the largest grain size (D95) for the specimens, lying between 17.16 and 38.54 mm. Consequently, the target grain size of < 40 mm was achieved without oversize grains and thus a good material disintegration should be achieved. The smallest grain (D10) is between 0.35 and 2.11 mm, which indicates an increased proportion of fines smaller than the average EPS pearl size (4–5 mm). The assumption that most of the impurities are brittle inorganic material (e.g., plaster), which is pulverised by the comminution process, is thus confirmed. Consequently, screening with a round hole of 5 mm (corresponds to approx. 90% < 4 mm) could theoretically separate between 23.2 and 62 wt-% of fine material, which is predominantly not EPS. Due to the low material density of EPS and the insufficient fraction separation, both screening and an air classifier were excluded.

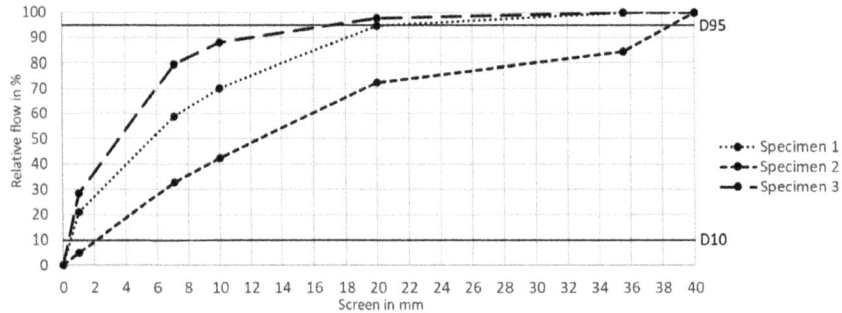

Fig. 2. Results of the sieving analyses (sieve curves displayed represent cumulative curves from several sieve analyses)

4.4 Sorting Process

The air sorting table IFE-Sort was used for sorting the comminuted specimens. The correct parameter setup was determined through preliminary tests. The impact angle (25°), rotational speed unbalance (1450 rpm), fan speed (600 rpm), suction power (400 m³/h) and barrage height (20 mm) were identical in all trials. Screening inclination and screening line were 15° and 0.9 mm for specimen 1 and 13° and 0.3 mm for specimens 2 and

3. A special dosing hopper was used for continuous, uniform feeding. The throughput was in the range between 300 and 900 kg/h. All fractions (heavyweight, lightweight and dust) per specimen were collected separately and weighed. In addition, the proportion that ends up in the filter of the suction unit was quantified. Table 3 shows the proportions of the respective fractions for each sample material examined.

Table 3. Output in different fractions after sorting

	Specimen 1 (wt-%)	Specimen 2 (wt-%)	Specimen 3 (wt-%)
Heavyweight fraction	79.9	67.1	42.1
Dust fraction	0.2	–	–
Lightweight fraction	17.3	29.3	52.6
Suctioned material	2.6	3.7	5.3

A qualitative analysis of the heavyweight fraction confirmed the good material disintegration by the shredder, as the proportion of EPS is very low. Only very sporadic, EPS with remaining adhesions could be found in this fraction. Representative samples of the lightweight fraction of each specimen were taken during operation for subsequent post-sorting analysis. The heavyweight and lightweight fractions produced from specimen 1 are shown in Fig. 3.

Fig. 3. Heavyweight and lightweight fractions after sorting of specimen 1

4.5   Post-sorting Analysis

In order to investigate the purity to which extent the EPS could be enriched in the light fraction, random samples of 6 L were taken from each specimen to analyse the remaining impurities. The weight percentage of insoluble components was determined as a measure of the purity of the lightweight fraction. To determine this, the samples were dissolved in 400 ml ethylbenzene, a specific solvent for polystyrene at these temperatures, which does not dissolve dowels, other plastics, etc. Subsequently, the suspension was decanted and the insoluble dust was dried at 70 °C for one week. The dried dust was weighed and represented the weight percentage of insoluble impurities. The share of insoluble impurities in the lightweight fraction was 10 wt-% in specimen 1, 13 wt-% in specimen 2 and 7 wt-% in specimen 3.

5 Conclusion and Outlook

In this paper, a novel pre-treatment process for EPS-ETICS using a single shaft shredder and an air sorting table was presented. By using this process design, it was demonstrated, that the comminution and sorting process works well for all specimens resulting in 7–13 wt-% impurity share. Contrary to previous attempts, these low impurity levels were achieved while having a continuous material flow and high throughput in combination with high yield rates of the target fraction EPS. Several machine parameters can be adapted to further improve the purity of the output material. However, the low impurity levels already indicate its suitability for economic solvent-based recycling. The differences in impurity levels are thought to have multiple causes. Firstly, fine dust lands on the lightweight fraction. This can be prevented by adding a sieving process as an additional step. Secondly, the sampling for post-sorting analysis was limited. Further tests with a wider range of samples are necessary to validate the process design for achieving the required purity. And finally, EPS-ETICS vary heavily through their wide range of compositions and fastening methods. The three specimens, therefore, represent only a small subset of the possible input material. Nevertheless, the results of this work have shown that pre-treatment according to the requirements of solvent-based recycling is possible, which represents a major step towards closing the loop of EPS from the construction sector. Additional work should focus on the economic and ecologic viability of the process as well as the development of suitable networks that enable the area-wide pre-treatment of EPS-ETICS.

Acknowledgements. This work received funding from the Austrian Federal Ministry for Climate Action (BMK) under grant number 889857 (EPSolutely). The Austrian Research Promotion Agency (FFG) has been authorized for program management.

References

1. Albrecht W, Schwitalla C (2015) Rückbau, recycling und Verwertung von WDVS: Möglichkeiten der Wiederverwertung von Bestandteilen des WDVS nach dessen Rückbau durch Zuführung in den Produktionskreislauf der Dämmstoffe bzw. Downcycling in die Produktion minderwertiger Güter bis hin zur energetischen Verwertung
2. European Association for ETICS, About ETICS [Online]. Available: https://www.ea-etics.com/etics/about-etics/. Accessed: 19 June 2023
3. Eibensteiner F, Paulik C, Stadlbauer W (2016) Endbericht STREC—EPS/XPS recycling im Baubereich
4. Commission Regulation (EU) 2016/460 of 30 March 2016 amending Annexes IV and V to regulation (EC) No 850/2004 of the European Parliament and of the Council on persistent organic pollutants. Off J Eur Union L 80:17, 31 Mar 2016
5. Netsch N et al (2022) Recycling of polystyrene-based external thermal insulation composite systems—application of combined mechanical and chemical recycling. Waste Manage 150:141–150
6. Kambeck N, Grunow M (2018) Recycling von HBCD-haltigen Dämmstoffen als Entsorgungsoption im Sinne der „circular economy". Z Recht Abfallwirtsch 289–292
7. Conversio Market and Strategy (2021) The EPS-industry's journey towards circularity: progress report

8. Federal Ministry for Climate Action (2020) Environment, energy, mobility, innovation and technology. EPS- und XPS-Dämmstoffabfälle ab der Baustelle
9. Demacsek C, Tange L, Reichenecker A, Altnau G (2019) PolyStyreneLoop—the circular economy in action. IOP Conf Ser Earth Environ Sci
10. Heller N (2022) Entwicklung und Bewertung von Entsorgungsstrategien für Wärmedämmverbundsysteme mit expandiertem. Shaker Verlag, Polystyrol
11. Fehn T, Teipel U (2019) Recycling von Wärmedämmverbundsysteme (WDVS): Abschlussbericht
12. Simons M, Feil A (2019) Gegenüberstellung Hammermühle und Prallmühle am Beispiel WDVS, vol 16. Münsteraner Abfallwirtschaftstage
13. Fehn T, Teipel U (2020) Werkstoffliche Aufbereitung von Wärmedämmverbundsystemen. Chem Ing Tech 92
14. Heller N, Flamme S (2020) Waste management of deconstructed external thermal insulation composite systems with expanded polystyrene in the future. Waste Manage Res 38(4):400–407
15. Simons M, Lößner C (2020) Anreicherung von Dämmstoffen aus Wärmedämmverbundsystemen mittels Windsichtung

Reducing the Down-Cycling of Carbon Fibre: An Observation on Preserving Different Woven Fibre Architectures

Di He$^{(\boxtimes)}$ [iD], Isaiah Pang, and Matthew Doolan [iD]

School of Engineering and Technology, UNSW Canberra, Campbell ACT 2612, Australia
di.he@adfa.edu.au

Abstract. The growing use of carbon fibre reinforced polymers (CFRP) in the aerospace and wind industries is leading to increases in waste carbon fibres. The current recycling practice for this waste results in significant reductions in material properties. Preserving the reinforcement architecture of carbon fibre has the potential to enhance the mechanical reinforcement capability of the recycled fibre. Nevertheless, the impact of this approach on different woven fibre architectures with varying material characteristics remains unclear. This study presents an observation on applying fibre architecture preservation to the recycling of carbon fibres in two woven fibre architectures, namely twill and satin weaves. Carbon fibres are recycled by a pyrolysis technique and remanufactured into recycled CFRP. Flexural tests indicate increases in the flexural properties of CFRP with twill and satin weaves after recycling, differing from the results for plain weave in a previous study. The results suggest potential influences of weave architectures on the material compositions of the recycled composites, resulting in different flexural properties. Further research is encouraged to investigate the potential correlations and provide deeper insights into the use of this approach to reduce the down-cycling of various types of carbon fibre.

Keywords: Carbon fibre recycling · Woven fabrics · Flexural properties

1 Introduction

The growing use of carbon fibre reinforced polymers (CFRP) is leading to increases in waste carbon fibres from both production and end-of-life stages. Carbon fibre has high specific strength, good thermal stability and corrosion resistance that are desirable properties in many applications. The global demand for carbon fibre is forecast to surge by 59% from 2021 to 2026, reaching over 180,000 tonnes [1]. This raises concerns regarding carbon fibre waste. The scrap rate of carbon fibre in various CFRP manufacturing processes, from conventional resin infusion to automated tape lay-up, is close to 30% [2]. Adding to this, an increasing amount of end-of-life CFRP waste is expected, with the aerospace and wind power sectors forecast to generate a total of nearly one million tonnes of CFRP waste by 2050 [3]. There is an increasing need for effective recovery and recycling of this waste to improve the sustainability of the carbon fibre industry.

© The Author(s) 2025
H. Kohl et al. (Eds.): GCSM 2023, LNME, pp. 818–826, 2025.
https://doi.org/10.1007/978-3-031-77429-4_91

The current recycling practices for carbon fibre waste result in significant reductions in material properties. The carbon fibre recycling sector primarily focuses on recovering carbon fibres from CFRP waste. Typically, the approach involves initially reducing the size of CFRP waste through chopping or shredding, followed by a thermal (e.g., pyrolysis) or chemical (e.g., solvolysis) process to remove the polymer matrix in CFRP and subsequently recover the carbon fibres [4]. Despite the ability to recover fibres from CFRP, the recycling process leads to fibre degradation. One aspect of this degradation involves the irreversible loss of compact fibre architectures caused by the shredding or chopping process. Recycled carbon fibres (rCF) are typically in a short and fluffy form, in contrast to the dense tow-based architectures of virgin carbon fibres (vCF) (Fig. 1). Incorporating these fibres into new CFRP products often results in low volume fraction of fibres [5], uneven fibre distribution, and increased porosity [6], compared to using vCF. The decline in material properties associated with rCF significantly limits their applications. The size reduction processes in the current recycling approach are implemented to facilitate material transportation and processing, following established practices for metal and plastic recycling. However, this approach does not take into consideration the distinctive functional characteristics of composites compared to materials such as metals and plastics, leading to significant down-cycling of carbon fibres.

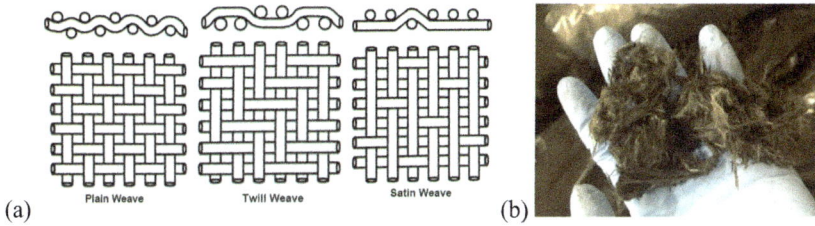

(a) Plain Weave   Twill Weave   Satin Weave   (b)

Fig. 1. Comparison of (a) tow-based woven fibre architectures and (b) current rCF [7].

Preserving the architecture of carbon fibre in the recycling process has the potential to enhance the mechanical reinforcement capability of rCF. The objective is to maintain the dense tow-based architectures of fibres without resorting to aggressive size reduction processes. Previous research efforts [8–10] have explored this approach to assess its potential in preserving the material properties of CFRP after recycling. In our earlier investigation [11], we demonstrated that preserving unidirectional and plain woven carbon fibres can reduce the degradation in flexural properties of CFRP by 51–65% compared to the commercial recycling approach. However, despite these previous studies, uncertainties still remain regarding the feasibility of this approach, particularly with respect to different woven fibre architectures. Various woven patterns, including plain weave, twill weave and satin weave (as depicted in Fig. 1a), are used in the market, each exhibiting distinct fibre characteristics. For instance, plain weave exhibits the most significant fibre crimp (Fig. 1a), resulting in the highest stress points at the contact areas between fibre tows. Consequently, plain weave offers the highest stability among the three weave patterns. Twill and satin weaves, as shown in Fig. 1a, possess reduced fibre crimp and thus exhibit lower fabric stability relative to plain weave (Twill > Satin). The implications of these differing characteristics on fibre degradation during recycling, and consequently on the material properties of the recycled CFRP, remain unclear. Understanding this aspect is important for assessing the applicability of this recycling approach for the various carbon fibre types currently used in the market.

Building upon our prior research, this study presents an observation on the impact of preserving fibre architecture on the flexural performance of recycled CFRP with various carbon fibre weave patterns. Specifically, the flexural properties of virgin and recycled CFRPs with twill and satin weaves are measured and compared to the findings from our previous investigation on plain weave. This comparative analysis will enhance our understanding of the potential associated with fibre architecture preservation in reducing the down-cycling of carbon fibres.

2   Methodology

This study presents a comparative experimental investigation on rCFRPs with different carbon fibre weave patterns. Two types of vCF fabrics, twill and satin weaves respectively, undergo virgin CFRP (vCFRP) manufacturing, recycling, and recycled CFRP (rCFRP) manufacturing with the same processing parameters (Fig. 2). This comparative experimental design provides the foundation for attributing variations in the material properties of CFRPs to the varying characteristics of carbon fibre fabrics.

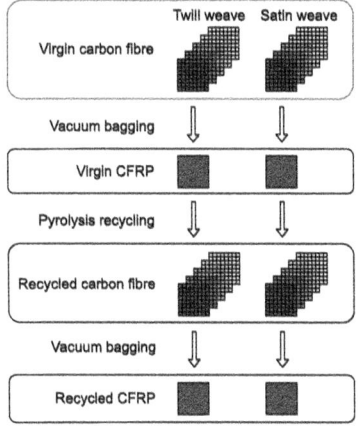

Fig. 2. Comparative experiment on the recycling of different carbon fibre woven fabric.

The experiment started with the manufacturing of CFRP using the two types of fibres via the wet layup technique and the vacuum bagging technique. This reflects the wide adoption of these techniques in the production of CFRP components for aeronautical and wind turbine applications [12]. The manufacturing procedure in this study closely follows the approach detailed in our previous investigations [11, 13]. It involves the laying of five sheets of 300 mm * 200 mm carbon fibre fabric and impregnating them with RIM935 epoxy resin (Hexion), serving as the polymer matrix. Each fibre layer was uniformly saturated with a resin-hardener mixture (100: 38 ± 2 by mass) and stacked upon one another. The resulting composite layups were cured overnight at ambient temperature under atmospheric pressure, followed by a post-curing step at 80 °C for 5 h. The identical manufacturing procedure was applied to both vCFRPs and rCFRPs. To assess any defects or porosity within the materials, the cross-sections of the CFRP samples were inspected utilising an optical microscope.

Subsequently, the vCFRPs underwent a thermal recycling approach to recover the carbon fibres. This approach, prevalent in the commercial recycling sector [4], starts with a pyrolysis process in nitrogen, aiming to thermally decompose the epoxy resin into gaseous byproducts and residues. The remaining residues were then removed through a subsequent treatment in heated air. It is important to note that in this study, the entire CFRP sample (300 mm * 200 mm) was placed in a LABEC SF-13-SD furnace each time for recycling, which differs from the current industry practice involving prior chopping or shredding of the materials. The parameters for the recycling process including the processing temperature were determined based on thermogravimetric analysis of the materials, as detailed in earlier studies [11, 14]. The parameters are listed in Table 1.

Table 1. Processing parameters for recycling carbon fibre.

Step	Temperature (°C)	Time (min)	Atmosphere
1. Pyrolysis	500	30	N_2
2. Residue removal	500	60	Air

The flexural performance of CFRPs holds great significance for their applications in various fields, such as aircraft components, wind turbine blades, and fishing rods. Hence, three-point bend tests were conducted on both vCFRPs and rCFRPs to measure and compare their flexural modulus and strength. The tests were conducted in accordance with the ASTM D790 standard [15]. Considering the potential variation in the amount of polymer residue across the five layers of rCF [7], the flexural tests of rCFRPs were conducted in two different configurations (Fig. 3) to evaluate the consistency of the flexural properties.

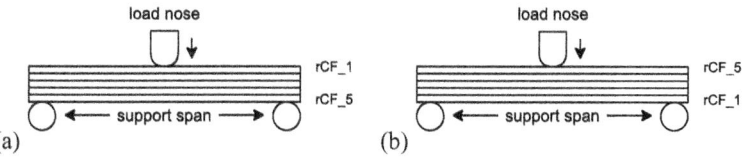

Fig. 3. Two flexural test configurations for rCFRP: (a) normal and (b) flipped (rCF_1 and rCF_5 were the top and bottom layers of fibre during recycling, respectively).

3   Results and Discussion

We start by assessing the quality of the recovered rCF through the recycling approach. As shown in Fig. 4b, c, both twill and satin weave architectures were successfully recovered with clean separations between fibre tows and fabric layers. The recycled woven fabrics have a similar appearance to their vCF counterparts, and are more compact and uniform compared to commercially available rCF mat (Fig. 4a). This suggests the different characteristics of the woven fabrics did not have a discernible impact on the recovery of carbon fibre weaves.

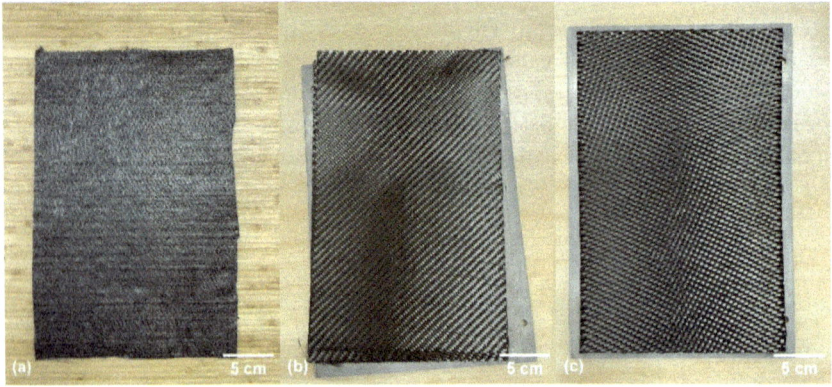

Fig. 4. Comparison of (a) commercial recycled carbon fibre mat [16], (b) recycled twill and (c) recycled satin carbon fibre woven fabrics.

At the composite level, some changes in the physical characteristics between rCFRPs and vCFRPs were observed. rCFRPs with twill weave and satin weave had slightly reduced but more uniform thickness compared to their virgin counterparts (Table 2). Whereas this change in thickness was not observed in the case of plain weave in our previous investigation [11]. Optical microscopic images of cross-sections of the CFRPs (Fig. 5) revealed that the rCF fabric appeared thinner than the vCF fabric, potentially contributing to the overall decrease in composite thickness. Further investigation is required to understand the reasons for the change in thickness and any potential correlation with the characteristics of different woven fabrics.

Table 2. Thickness of virgin and recycled CFRPs (figures for plain weave from previous study [11]; uncertainties from 15 measurements).

Fibre type	Virgin CFRP	Recycled CFRP
Plain weave	1.04 ± 0.02	1.05 ± 0.02
Twill weave	1.03 ± 0.02	0.98 ± 0.01
Satin weave	1.82 ± 0.06	1.78 ± 0.02

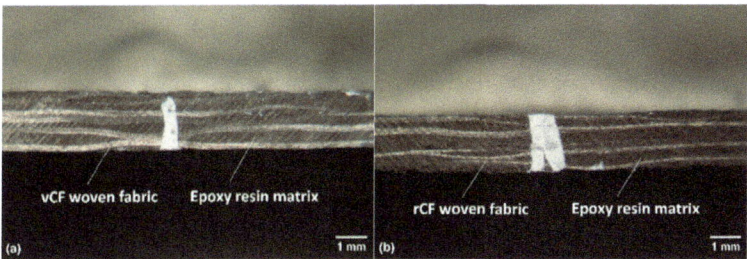

Fig. 5. Cross-sectional images of of (a) vCFRP and (b) rCFRP with satin weave (white mark in the middle indicates the thickness of the composite).

CFRPs containing twill and satin weave demonstrated varying changes in flexural properties in comparison to plain weave. In the cases of twill and satin weaves, flexural modulus of rCFRP increased compared to vCFRPs (Fig. 6). This contrasts with the results observed for plain weave in the previous study [11] that demonstrated a reduction in flexural modulus after recycling. These observed changes, coinciding with the observations on CFRP thicknesses mentioned earlier, suggests variations in the material compositions of CFRPs after recycling potentially influenced by the differing fibre architectures, which affects the mechanical properties of the rCFRP. It is plausible that twill and satin weaves experienced more significant flattening during the manufacturing-recycling-remanufacturing cycle, resulting in a smaller thickness and a higher fibre volume fraction of the rCFRPs. These factors contribute to the enhanced flexural properties of the composite. This hypothesis requires to be verified through further investigations, which will be addressed in the next section. In terms of the consistency of flexural properties, there were no statistically significant differences between the results from tests conducted in normal and flipped configurations.

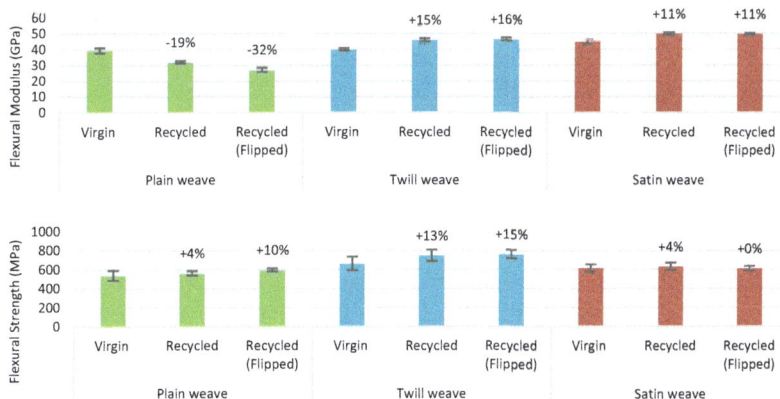

Fig. 6. Flexural modulus and flexural strength of CFRPs with twill weave and satin weave before and after recycling, compared with plain weave from previous study [13] (uncertainties from 5 measurements).

4   Future Work

This study presents intriguing observations regarding the potential relationships between different woven carbon fibre architectures and the material properties of rCFRP containing these preserved weaves. Future research should aim to substantiate these connections by delving into the underlying mechanisms that govern the mechanical properties of CFRP.

Based on the principles of composite mechanics, areas of further research can be divided into three distinct levels: the fibre level, fabric level, and composite level. First, at the fibre level, the potential impacts of woven fibre architectures on the mechanical properties of rCF should be investigated. As discussed in the Introduction, fibres within various woven fabrics experience varying constraints due to crimping. It is unclear whether this factor influences the degradation in fibre properties and subsequently affects the mechanical properties of the composite. At the fabric level, the varying fabric stability and constraints imposed on fibre tows may result in different variations in the physical characteristics of the fabric after recycling. These variations can include changes in fabric thickness, tow width and flatness. Investigating these aspects offer valuable insights into the overall thickness variations in rCFRPs observed in this study. Finally, at the composite level, the fractions of fibre, polymer, and porosity within rCFRPs with different woven fabrics are crucial parameters that directly impact their mechanical properties. Hence, it is imperative to quantify and compare these fractions to gain a comprehensive understanding of their influence. Given the multitude of factors that influence the material properties of CFRP, computational mechanical analysis can serve as a valuable tool for narrowing down the focus to key parameters in a controlled manner.

5   Conclusion

The current recycling approach for carbon fibre leads to significant down-cycling. Preserving the architecture of carbon fibres holds promise for improving the material properties of the recycled fibres and composites. This study presents an observation on the effects of recycling with fibre architecture preservation on carbon fibre reinforced polymers (CFRP) containing different woven fibre architectures. The experimental investigation focuses on comparing the changes in flexural properties of CFRP with twill and satin weaves after recycling. The results demonstrate an increase in the flexural modulus and strength of these CFRPs after recycling, which differs from our previous findings regarding plain weave where reductions in flexural modulus were observed. This suggests potential influences of the weave architectures on the material compositions of the recycled composites, resulting in different flexural properties. Further research effort should investigate the impacts of various characteristics of the woven fabrics on the material properties at the fibre, fabric and composite levels, as these factors directly affect the mechanical performance of the rCFRP. This will provide deeper insights into the potential of this recycling approach in enhancing the circularity and sustainability of carbon fibre.

References

1. CompositesWorld, The outlook for carbon fiber supply and demand 2021. https://www.com positesworld.com/articles/the-outlook-for-carbon-fiber-supply-and-demand, last accessed 13 June 2023
2. Rybicka J, Tiwari A, Alvarez Del Campo P, Howarth J (2015) Capturing composites manufacturing waste flows through process mapping. J Clean Prod 91:251–261
3. Lefeuvre A, Garnier S, Jacquemin L, Pillain B, Sonnemann G (2019) Anticipating in-use stocks of carbon fibre reinforced polymers and related waste generated by the wind power sector until 2050. Resour Conserv Recycl 141:30–39
4. Zhang J, Chevali VS, Wang H, Wang C-H (2020) Current status of carbon fibre and carbon fibre composites recycling. Compos B Eng 193:108053
5. Pickering SJ, Liu Z, Turner TA, Wong KH (2016) Applications for carbon fibre recovered from composites. IOP Conf Ser Mater Sci Eng 139:012005
6. Pimenta S, Pinho ST (2014) The influence of micromechanical properties and reinforcement architecture on the mechanical response of recycled composites. Compos A Appl Sci Manuf 56:213–225
7. He D, Compston P, Morozov E, Doolan M (2022) Reducing down-cycling of carbon fibre by fibre architecture preservation: multi-layer fibre surface quality investigation. Proc CIRP 105:637–641
8. Lopez-Urionabarrenechea A, Gastelu N, Jiménez-Suárez A, Prolongo SG, Serras-Malillos A, Acha E et al (2021) Secondary raw materials from residual carbon fiber-reinforced composites by an upgraded pyrolysis process. Polymers (Basel) 13:3408
9. Qazi H, Lin R, Jayaraman K (2021) Fibre structure preservation in composite recycling using thermolysis process. Resour Conserv Recycl 169:105482
10. Meredith J, Cozien-Cazuc S, Collings E, Carter S, Alsop S, Lever J et al (2012) Recycled carbon fibre for high performance energy absorption. Compos Sci Technol 72:688–695
11. He D, Kim HC, Sommacal S, Stojcevski F, Soo VK, Lipiński W et al (2023) Improving mechanical and life cycle environmental performances of recycled CFRP automotive component by fibre architecture preservation. Compos A Appl Sci Manuf 175:107749
12. Campbell FC (2003) Manufacturing processes for advanced composites. Elsevier
13. He D (2022) Life cycle strategies for improving mechanical and environmental performance of recycled carbon fibre composite in the automotive industry. Australian National University
14. He D, Soo VK, Stojcevski F, Lipiński W, Henderson LC, Compston P et al (2020) The effect of sizing and surface oxidation on the surface properties and tensile behaviour of recycled carbon fibre: an end-of-life perspective. Compos A Appl Sci Manuf 138:106072
15. ASTM International (2017) ASTM D790-03: standard test methods for flexural properties of unreinforced and reinforced plastics and electrical insulating materials
16. Gen 2 Carbon, G-TEX M. https://www.gen2carbon.com/product/g-tex-m/, last accessed 18 July 2023

A Circular Use of Oil by Double Separation Technology and Its Sustainable Impact Comparing to Conventional Industrial Oil Usage

Feng Qiu[1]([envelope]), Johanna Reimers[2], Annica Iseback[3], and Juan Ricardo Alvarez[4]

[1] SKF B.V., Meidoornkade 14, Houten 3992 AE, The Netherlands
feng.qiu@skf.com
[2] SKF RecondOil AB, Von Utfallsgatan 4, 415 05 Gothenburg, Sweden
[3] SKF AB, Sven Wingquists Gata 2, 405 15 Gothenburg, Sweden
[4] SKF RecondOil AB, Sven Hammarby Kaj 14, 120 30 Stockholm, Sweden

Abstract. Currently industrial oil is considered as a consumable material and used in a linear way: it gets discarded and replaced with new oil when the oil degrades and often loses its desired properties. This is unsustainable and can have hidden, indirect costs relating to productivity and product quality caused by wear on machinery due to degraded industrial oil properties. SKF RecondOil Double Separation Technology (DST) enables regeneration and reuse of industrial oil by removing contaminants and thereby increasing the lifespan of the oil. Such circular use of oil makes industrial oil an asset and can generate monetary and environmental benefits. A lifecycle-based study has been carried out to map and compare the climate impact of the SKF RecondOil DST process to a conventional industrial oil cycle. The results show that a reduction of more than 96% percent of CO_2-eq./m^3 oil can be achieved. Furthermore, the results also show that the SKF RecondOil DST processes use less fossil resources during the life cycle compared to a conventional industrial oil cycle. This paper will introduce DST, its applications, and discuss its future development based on the results obtained.

Keywords: Circular use of oil · Double separation technology (DST) · Life cycle assessment (LCA)

1 Introduction

1.1 Circular Use of Oil Enabled by DST

Industrial oil is currently considered as a consumable material and used in a linear way: it gets discarded and replaced with new oil when the oil degrades and often loses its desired properties, caused by contamination, oxidation and additive depletion [1]. This is unsustainable and can have hidden, indirect costs relating to productivity and product quality caused by wear on machinery due to degraded industrial oil properties.

© The Author(s) 2025
H. Kohl et al. (Eds.): GCSM 2023, LNME, pp. 827–834, 2025.
https://doi.org/10.1007/978-3-031-77429-4_92

As global concern for environmental sustainability grows, innovative technologies are sought to use industrial oil in a more sustainable way. Double Separation Technology (DST) from SKF RecondOil, is a two-stage combination of chemical separation and mechanical filtration. As shown in Fig. 1, a (chemical) booster is mixed into the (used) oil, and it attracts contaminants to lumps by surface polarity. The lumps then coalesce into larger lumps then can be easily mechanically separated from the oil via sedimentation and filtration. Depending on the oil analysis result, additive top-ups can be required to achieve the same additives level as in the virgin oil. A circular use of oil is achieved so that it can be used again.

 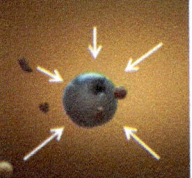 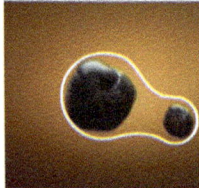 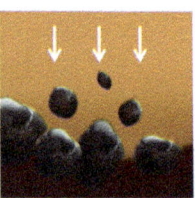

The booster is mixed with the oil

The booster attracts particles of all kinds and all sizes to lumps

The lumps coalesce into larger lumps

The lumps are now easy to separate from the oil via sedimentation and filtering.

Fig. 1. DST process.

1.2 Life Cycle Assessment (LCA)

LCA is a tool to assess potential environmental impacts throughout a product's (the term product includes goods, technologies and services) life cycle, i.e., from natural resource acquisition, via production and use stage to waste management (including disposal and recycling). Duda and Shaw [2] LCA framework is illustrated in Fig. 2, according to ISO 14040. To quantify the environmental benefits circular use of oil by DST processes, an LCA has been carried out, comparing to the conventional linear use of oil.

2 LCA on DST Systems

2.1 Goal and Scope

Environmental impact on climate change and fossil resource use were compared for the DST process and conventional industrial oil cycle. Two set-ups of SKF RecondOil DST process were assessed, an integrated system and a stand-alone system. In an integrated DST system, the oil undergoes a so called "continuous regeneration". This entails a DST unit being integrated into the customer's application, where it continuously regenerates the (same type of) oil (system boundary system Fig. 3). In the stand-alone DST system, the approach is instead a "batch regeneration", which happens either at defined intervals or based on the condition of the customer's oil. A used batch of oil is brought from the customer's system to a separate facility where the DST stand-alone unit is installed. The

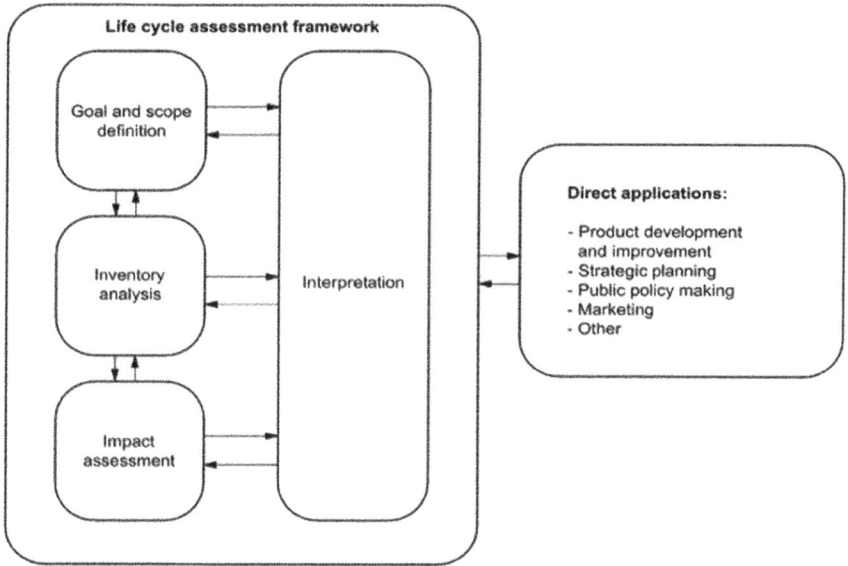

Fig. 2. Stages of LCA (ISO 14040, 2006).

treated oil is then brought back to the customer's application for re-use. The stand-alone system can treat several different types of oil. However, in between different oil types, the tanks in the stand-alone system need to be flushed with flushing oil (system boundary diagram Fig. 4). For both systems, new oil (additives) need to top up into the treated oil, depending on the needs followed by oil analysis. For the conventional industrial oil cycle, the used oil is replaced with new oil and is later discarded and incinerated (system boundary diagram Fig. 5).

The main purpose of this LCA is to analyze regeneration with the DST process and compare with conventional industrial oil cycle, not to compare the efficiency or impact of oil in use in its final application, thus the use phase of the lubricant (such as positive effects of regenerated oil by DST: reduced wear of machines, extended lifetime of equipment, reduced oil, improved production efficiency and product quality [3]) is excluded. Infrastructure, buildings (such as heat demand), process equipment (production and disposal), as well as packaging materials are excluded in the LCA.

The impact assessment method used is CML 2001, generating results in kg CO_2 equivalents for climate impact and MJ for fossil resource use. All systems were modelled using the software LCA FE (Life Cycle Assessment for Experts, formerly known as GaBi) version 10.6 and database content version 2021.1[4].

2.2 Life Cycle Inventory/Data Collection

Data for the DST Integrated system is based on a low viscosity honing oil used in SKF Airasca factory in Italy. The oil contains approximately 5% rapeseed methyl ester (RME). No other additives have been modelled. Input materials and energy, waste and transportation are included in the inventory shown in Fig. 3.

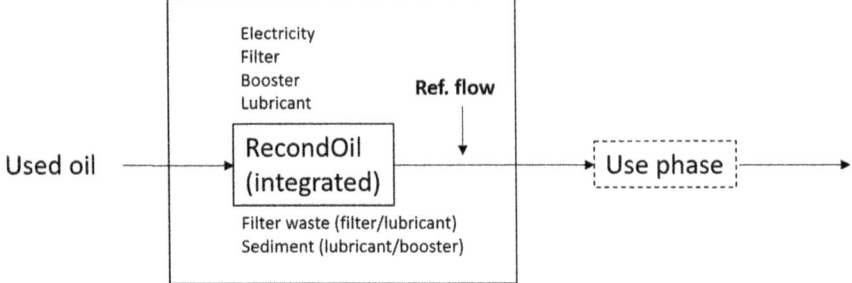

Fig. 3. System boundaries of the DST integrated system. The used oil enters the system boundaries without any burden from previous life cycles. The use phase is excluded.

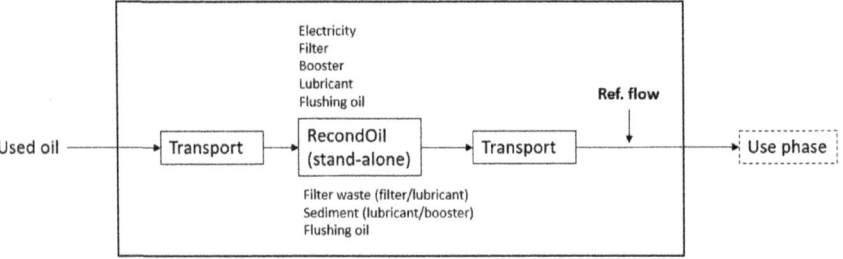

Fig. 4. System boundaries of the DST stand-alone system. The used oil enters the system boundaries without any burden from previous life cycles. The use phase is excluded.

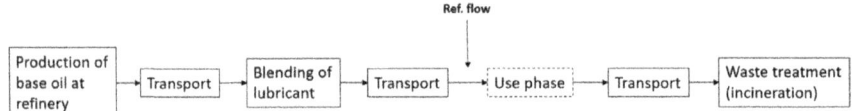

Fig. 5. System boundaries of the conventional lubricant cycle. The lubricant is produced from fossil resources. The use phase is excluded.

Data for DST stand-alone lubricant is processed in SKF Schweinfurt factory in Germany and is a zinc free hydraulic oil. The lubricant is assumed to contain 100% base oil with very low amounts of additives. The additives are therefore excluded from the analysis. The inventory includes input materials and energy, waste and transportation as illustrated in Fig. 4.

Conventional scenario consists of an average composition of lubricant assumed to be 100% base oil with no additives. Figure 5 describes the material, energy and transportation included in the inventory.

Due to space limitations, the detailed data for each scenario, such as lubricant properties, transportation distances, and amount of each material used are not presented in this paper. Complete data collection can be found in Reference [5, 6].

2.3 Results

The results of the climate impact potential per 1 m3 of lubricant for all studied three scenarios are presented in Table 1 and Fig. 6. Both DST systems have lower climate footprints than the conventional scenario, around 20 kg CO_2-eq for DST integrated, 154 kg CO_2-eq for DST stand-alone and 3830 kg CO_2-eq for the conventional scenario. DST stand-alone achieves 96% CO_2-eq reduction and DST integrated 99%.

The waste treatment, i.e. incineration of the oil for conventional scenario or lubricant, booster and filter for DST process, contributes to the largest share of climate impact in all cases. The production of lubricant also has a negative climate change for the conventional scenario. Both life cycle stages have minimal impact for the DST systems since the lubrication oil is reused and the emissions originating from production and waste treatment can therefore be avoided.

Table 1. Climate change impact potential of the studied systems per 1 m^3 lubricant.

	Conventional	DST integrated	DST stand-alone	Unit
Raw materials	1000	6.55	36.0	kg CO_2-eq
Transport of raw materials	4.98	0.59	13.8	kg CO_2-eq
Transport of waste	4.98	0.016	12.8	kg CO_2-eq
Waste treatment	2820	12.4	91.2	kg CO_2-eq
Total	3830	20	154	kg CO_2-eq

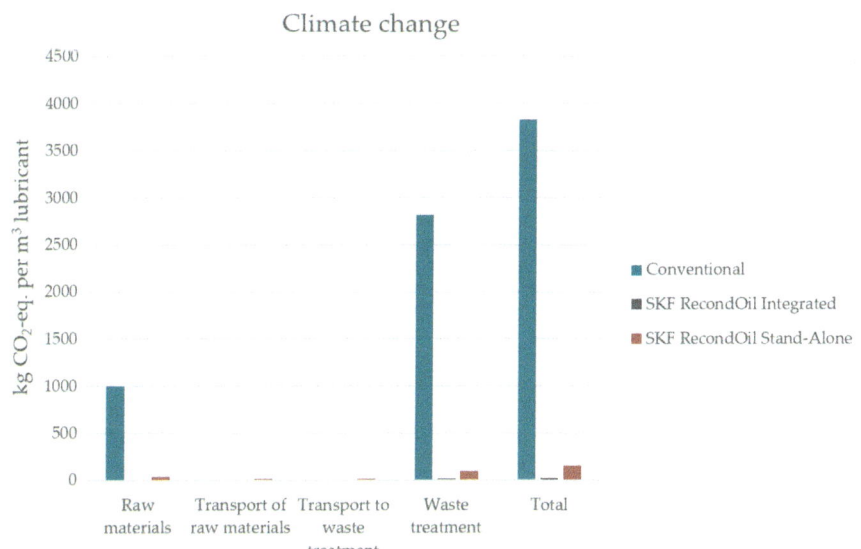

Fig. 6. Climate change impact of three studied systems.

The result of Fossil resource depletion potential per 1 m^3 of lubricant is presented in Table 2 and Fig. 7. Both DST systems have lower climate footprints than the conventional scenario, around 254 MJ for DST integrated, 1911 MJ for DST stand-alone and 46774 MJ for the conventional scenario.

Table 2. Fossil resource depletion potential of 1m3of lubricant.

	Conventional	DST Integrated	DST Stand-alone	Unit
Raw materials	45100	239	1503	MJ
Transport of raw materials	67.2	8.0	186	MJ
Transport of waste	67.2	0.21	173	MJ
Waste treatment	1540	6.5	49.5	MJ
Total	46774	254	1911	MJ

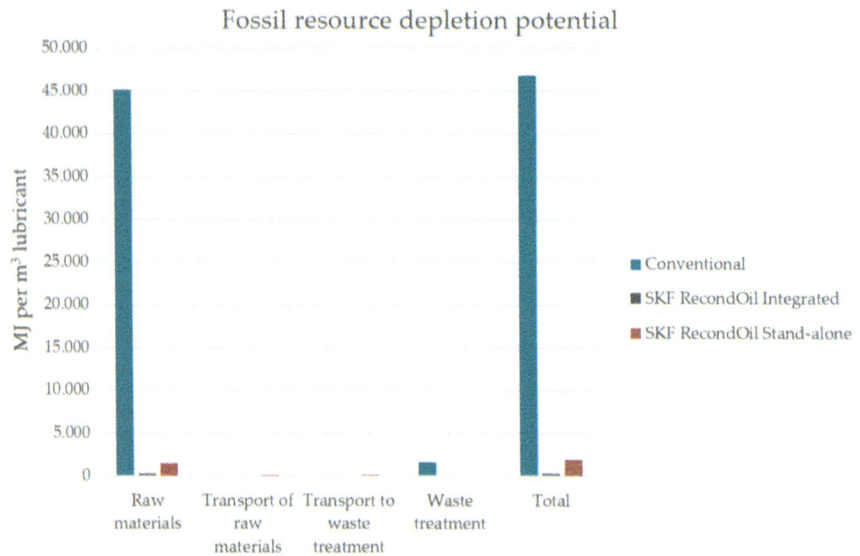

Fig. 7. Fossil resource depletion potential of the three studied systems.

2.4 Interpretation

In this LCA, it is assumed that the oil input to the DST processes enters the system boundaries without any burden from previous life cycles since the oil is regenerated from used oil, i.e. the impact from earlier life cycles is cut off. An alternative method to cut-off is to split, or to allocate, the initial environmental impact of the oil production evenly onto all oil regeneration cycles. For example, if the oil was to be reconditioned

10 times at an SKF site before the oil needed to be replaced all together and if we did not use cut-off between each regeneration cycle, the total climate impact of the DST processes would increase with 380 kg CO_2-eq./m^3 of oil (Since the climate impact of a conventional oil cycle corresponds to 3800 kg CO_2-eq./m^3 of oil, a tenth of this impact would be allocated to each oil regeneration cycle if we assume that the oil can be regenerated 10 times before it needs to be replaced completely). This would result in a total climate impact of 400 (20 + 380) kg CO^2-eq. For the DST Integrated and 534 (154 + 380) kg CO_2-eq. For the DST Stand-alone. However, the choice of allocation method does not affect the comparison between the DST processes and a conventional oil cycle – the conventional oil cycle still has a higher climate footprint independent of which allocation method is used.

The incineration of waste (oil) has potentially the highest climate change impact for both DST systems and conventional scenario. The production and incineration of the flushing oil contributes the most to climate change in the DST stand-alone system. The consumption of flushing oil is based on a worst-case scenario since it is not needed between every batch of reconditioned lubrication oil, e.g. if the same type of oil is treated. Another important aspect of the DST Stand-alone process is the need for transporting the oil between the customer and the SKF facility. It is also shown that the transportation of industrial oil has a relatively high impact.

3   Conclusion and Recommendations

According to the findings in this LCA, the DST processes have a significantly lower climate impact and lower fossil resource use than a conventional industrial oil life cycle. DST stand-alone achieves 96% CO_2-eq reduction and DST integrated 99%. The main difference is the avoidance of oil production due to the regeneration by DST processes. The highest impact in all three studied scenarios originates from incineration of waste oil. For both DST systems, it is recommended to minimize the oil waste as much as possible to further reduce the environmental impact. This can be done either by increasing the oil recovery level in its application process, or by reusing the additional oil – for example the flushing oil – for as long as possible.

The main differences between the two DST systems are the additional transport of oil to the facility, the use of flushing oil for cleaning the tanks between different types of oil for the DST stand-alone system. The main contributor is the production and waste treatment of flushing oil. To reduce the environmental impact even further, it is recommended to reduce the need for flushing oil (if possible) or to reuse the flushing oil. To reduce the impact from transport, the transportation distance should be kept to a minimum by opening more facilities or offering transport based on renewable fuels/electricity.

In this LCA, it is assumed that the oil input to the DST processes is cut-off from the impact from earlier life cycles. An alternative method to cut-off is to split, or to allocate, the initial environmental impact of the oil production evenly onto all oil regeneration cycles. It is recommended to gather more data on the exact oil regeneration cycles on specific application as studied, to conclude the split or allocate method.

Acknowledgements. This paper is based on sustainability study of industrial lubricant, carried out in collaboration with IVL – Swedish Environmental Research Institute. We would like to thank our colleagues in SKF for the data collection. We would also like to express our gratitude to IVL for their dedicated technical support.

References

1. Machinery Lubrication. https://www.machinerylubrication.com/Read/29169/oil-degradation-cause. Last accessed 11 Aug 2023
2. Duda M, Shaw JS (1997) Life cycle assessment. Society 34(6):38–43. https://doi.org/10.1007/s12115-997-1022-5
3. SKF RecondOil. https://www.skf.com/group/services/reconl/the-values-of-a-circular-use-of-oil. Last accessed 29 Aug 2023
4. Kristin J, Fredrik T, Håkan S (2021) LCA of SKF recondoil double separation technology process. IVL
5. Report: Data collection LCA RecondOil. SKF RecondOil Team (2021)
6. Report: LCA Data collection from Airasca and Schweinfurt. SKF RecondOil Team (2021)

Student Sessions

Planning Multiproduct Assembly Lines Through a Continuous-Time Model Accounting for Carbon Footprint

Nélida B. Camussi[1] and Diego C. Cafaro[1,2]([✉]) [iD]

[1] Intec (UNL-CONICET), Güemes 3450, 3000 Santa Fe, Argentina
dcafaro@fiq.unl.edu.ar
[2] Centro Interuniversitario de Investigaciones en Gestión Analítica de Procesos (GAP)
ITBA-UNL, Facultad de Ing. Química, Santiago del Estero 2829, 3000 Santa Fe, Argentina

Abstract. This work presents an efficient mathematical formulation for the optimal planning of multiproduct assembly lines. The aim is to establish a cyclic production agenda that minimizes the sum of inventory holding and transition costs per unit time. Besides, carbon footprint of assembled products is tracked along the line while optimally determining task times and operating modes. Batch sizing and sequencing in synchronous assembly lines have been typically addressed through discrete-time approaches that imply significant computational burden. In contrast, our novel representation resembles continuous-time pipeline scheduling models that permit to obtain optimal solutions in reasonable times. Results show the capabilities of the optimization approach to solve large instances of the problem, also demonstrating how task times and operating modes can be handled to reduce carbon footprint with no loss of productivity.

Keywords: Assembly lines · Sequencing · Batch sizing · Optimization

1 Introduction

Assembly lines are extremely effective flow-line production systems that consist of a number of workstations arranged along a unidirectional conveyor. In multiproduct assembly lines, campaigns or lots of different products are launched down, and every item in a lot moves from station to station such that in each of them, pieces of work or tasks are performed to manufacture the product. Every task usually requires material and/or energy consumption, which contribute to carbon (CO_2) emissions. There are usually alternative modes to perform these tasks (equipment, energy source, duration) yielding different carbon emissions per unit of product (i.e., carbon intensities). This has motivated the need to achieve systematic ways of measuring the carbon footprint of the products along their value chain [1]. Manufacturing tasks increase the carbon footprint of the products along the assembly line according to how they are performed. Operating modes affect not only costs and yield rates but also the magnitude of the carbon emissions, often being greener at the expense of longer times. In modern value chains it is becoming critical to plan operations sustainably and efficiently, meeting demand under carbon footprint specifications, at minimum total cost.

© The Author(s) 2025
H. Kohl et al. (Eds.): GCSM 2023, LNME, pp. 837–845, 2025.
https://doi.org/10.1007/978-3-031-77429-4_93

In turn, multiproduct pipelines are the most reliable and efficient mode of transportation of fluid products over land. A multiproduct pipeline moves different products (typically oil derivatives) over long distances in batches, which are pumped one after the other into the same duct, usually with no physical separation between them. Although the aim of pipelines and assembly lines is totally different, the operational planning of both systems share common features that deserve a deeper study. The first similarity is that both involve very expensive capital investment, from which a rapid return based on massive usage and high performance is expected. Moreover, predictable batch tracking in pipelines avoids environmental issues by strict control on volume balances, while synchronous assembly lines (simultaneously moving individual items between adjacent stations) reach similar standards by keeping the workforce balanced and work-in-process at low values.

Similar to pipeline scheduling models in which several batches of oil products are conveniently arranged to flow through the system one after the other, the movement of different products along the assembly line needs to be optimally planned (see Fig. 1). In a multiproduct assembly line, each product may have its own cycle time (time between two subsequent movements) which makes the problem challenging. Recent works have addressed the optimal planning of multiproduct assembly lines, which seek for the optimal sizing of lots and the most suitable sequence of products to be processed along the line. At the same time, new developments in process planning and production scheduling prove to be important for environmental performance [2].

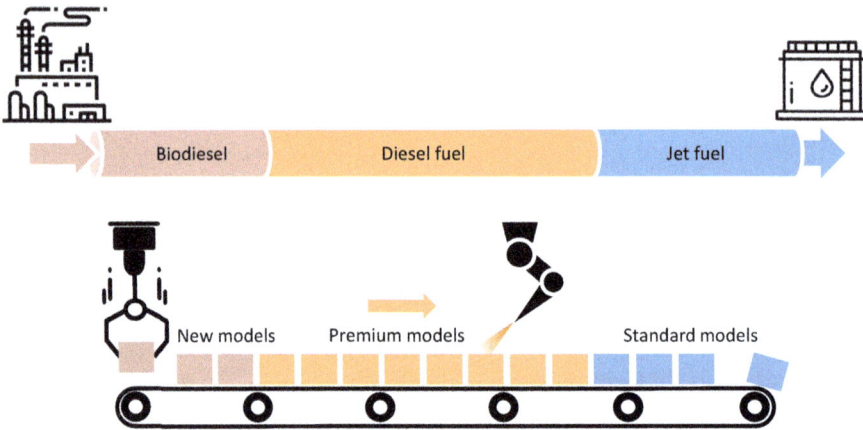

Fig. 1. A simple example illustrating the analogy of multiproduct pipelines and assembly lines

2   Problem Definition

In unidirectional pipelines carrying multiple products, different batches with different volumes move from one extreme to the other, as illustrated on Fig. 1. It is a simple, efficient, safe and economical way for the massive transportation of fluids. After the installation of the pipeline, the sequencing, timing and sizing of batch injections is critical, so as to timely meet demands while satisfying pipeline operational constraints. Many authors have proposed efficient mathematical models with those purposes [3, 4]. A schematic representation of this problem is shown on Fig. 1. Three batches, move from the beginning to the end of the pipeline following a certain chronological order or sequence (jet fuel, diesel fuel, biodiesel). By the incompressibility assumption, if the diesel fuel is denser and more viscous than the other products flowing through the pipeline, it controls the speed of the flow stream.

On the other hand, an assembly line is a production system focused on partitioning the whole work into smaller parts which are assigned to different workstations so as to maximize the throughput and the use of resources (workers, tools, equipment, etc.). In addition to balancing multiproduct assembly lines problems [5], research has been directed towards the organization of the production agenda in order to minimize inventory holding and changeover costs at the same time [6]. Figure 2 shows a paced assembly line with several workstations processing different products where different items synchronically move from one station to the next downwards. Analogously to pipelines, the slowest product controls the speed of the assembly line.

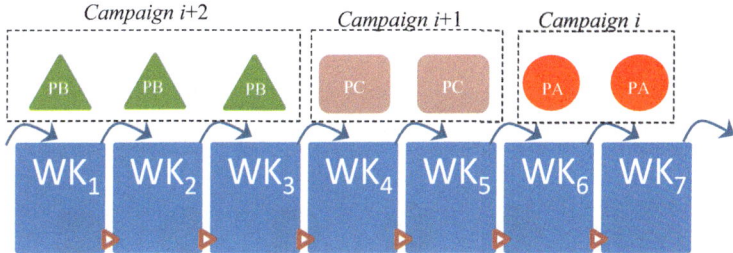

Fig. 2. Schematic outline of production items moving along an assembly line

In this case, instead of seeking for the length, duration and sequence of product batches sent through a pipeline, the aim is to find the optimal number of items comprised by a production campaign which must be processed along the assembly line so as to minimize, simultaneously, inventory holding and changeover costs. Furthermore, average carbon emissions per unit of product should be kept within admissible ranges.

3 Main Decision Variables and Constraints

3.1 Time Control

Every balanced assembly line has a cycle time, denoted as ct, which represents the time spent by the line for processing a new item of final product. It is clear that the cycle time is inversely proportional to the production rate p, i.e., $ct = 1/p$. The cycle time stands for the rhythm at which the line moves synchronously. If the line processes several models $j \in J$ with different cycle times ct_j, the simultaneous presence of two or more different products on the workstations makes the cycle time become equal to the maximum value among all of them, which limits the speed of the line. Figure 3 illustrates such condition, with three products (A, B, C) being simultaneously processed by the line. If cycle times satisfy the inequalities $ct_B > ct_C > ct_A$, then the input of at least one piece of model C (indicating the beginning of a new campaign) imposes the reduction of the line speed even though A has a faster cycle time. Similarly, the input of at least one piece of model B imposes the reduction of the production rate even though there are faster products along the line. Finally, and due to environmental constraints, cycle times for different products may need to be optimally managed by adopting "greener" (slower) or "grayer" (faster) processing modes $m \in MD$ during different runs. As a result, the cycle times in our optimization model are given by the parameter $ct_{j,m}$, indicating the time to process a single unit of j when the line runs in mode m.

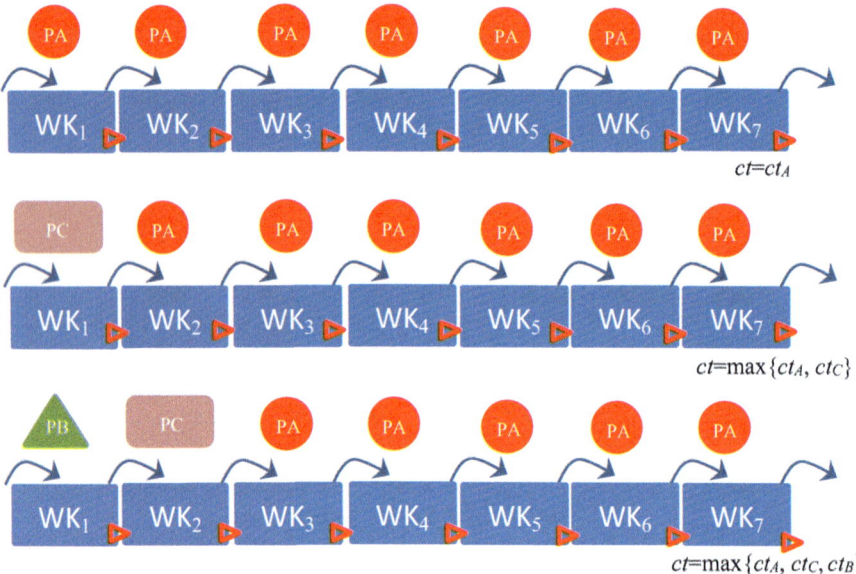

Fig. 3. Illustrative example comparing different cycle times, controlled by the slowest product

It is clear that each time a piece of any product enters the line in workstation 1, a new item of (the same or other) final product leaves the last workstation and is ready for fulfilling demand. There is a synchronic displacement downwards that makes each piece move from workstation k to workstation $k + 1$. If the precedent idea is now thought as a decision variable involving a group of products of the same kind instead of a single element, the notion of campaign is introduced. When a campaign i' displaces another group of items called campaign i (already in transit and exiting the line) we can use the variable $DP_{i,i',j}$ to account for the number of elements in the campaign i leaving the assembly line during the input of campaign i'. Through a simple mass balance equation, the total number of elements from previous campaigns i exiting the line while i' is entering should be equal to $N_{i'}$ and is imposed by Eq. (1). Note that $N_{i'}$ represents the number of elements in campaign i'.

$$N_{i'} = \sum_{i \le i'} \sum_{j:products} DP_{i,i',j} \quad \forall i' \in Runs \tag{1}$$

The elapsed time between the initial and final elements of the campaign i' is governed by $N_{i'}$ times the maximum cycle time of the products laying into the assembly line, according to the selected operating mode m, that is $ct_{j,m}$. This allows to calculate the start and end times, $S_{i'}$ and $C_{i'}$, respectively, expressed mathematically by Eq. (2). In that constraint, variables $x_{i,i'}$, $y_{i'',j}$ and $w_{i'',i',m}$ are binary variables which take value one when campaign i' makes some items in the previous campaign i be finished at the other extreme of the line ($x_{i,i'} = 1$); when product j is associated with campaign i'', moving between i and i' ($y_{i'',j} = 1$); and when campaign i'' runs in mode m during i' ($w_{i'',i',m} = 1$), respectively. MT is a large enough constant, in time units. Figure 4 shows how three items of PB in the new run i' displace the three items of PA in the last three workstations of the assembly line, belonging to campaign i.

$$C_{i'} \ge S_{i'} + N_{i'} ct_{j,m} - MT \left(3 - y_{i'',j} - x_{i,i'} - w_{i'',i',m} \right),$$
$$\forall i'' \in Runs : i \le i'' \le i', \forall j \in Prod, m \in Modes \tag{2}$$

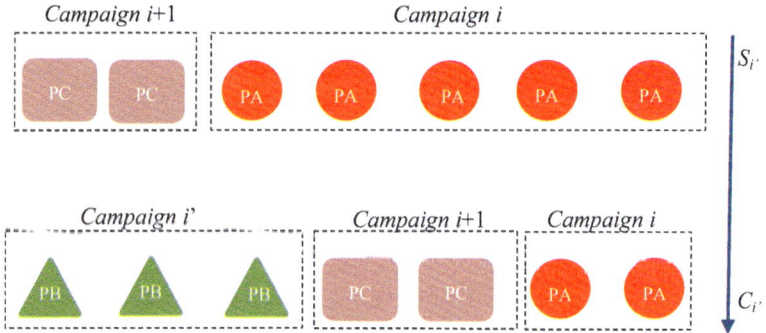

Fig. 4. Assembly line arrangement before and after the input of run i' with product PB

3.2 Inventory Control

We now focus on the calculation of the inventory level of the final product j after the input of production run i', $Inv_{i',j}$, which basically consists of the stock of j previous to i' plus the production of product j during campaign i', minus the consumption of j in the elapsed time interval (given by the demand rate r_j), as it is expressed by Eq. (3).

$$Inv_{i',j} = Inv_{i'-1,j} + \sum_{i'':1 \le i'' \le i'} DP_{i'',i',j} - r_j \left(C_{i'} - C_{i'-1} \right), \quad \forall i' \in Runs, j \in Prod \tag{3}$$

3.3 Carbon Footprint

The carbon footprint from the manufacturing of a campaign i during run i' can be obtained from the operating mode selected for processing the products in i that are in transit along the line. If $W_{i,i'}$ is the number of elements of run i into the assembly line at the end of run i' then the carbon footprint of tasks performed during run i' depends on the selected mode m, as shown in the block of Eqs. (4). Note that the parameter $cf_{j,m}$ is the carbon footprint (total emissions) in kg of CO_2 per unit of product j, per station and cycle, when the assembly line runs in mode m for product j. MF is a large enough constant, in carbon emissions units.

$$CF_{i,i'} \geq N_{i'} W_{i,i'} cf_{j,m} - MF\left(2 - y_{i,j} - w_{i,i',m} + x_{i,i'}\right)$$
$$CF_{i,i'} \geq \left[N_{i'} W_{i,i'-1} - D_{i,i'}(D_{i,i'} + 1)/2\right]cf_{j,m} - MF\left(3 - y_{i,j} - w_{i,i',m} + x_{i,i'}\right),$$
$$CF_{i',i'} \geq (N_{i'}(N_{i'} + 1)/2)cf_{j,m} - MF\left(2 - y_{i',j} - w_{i',i',m}\right)$$
$$\forall i, i' \in Runs : i < i', j \in Prod, m \in Modes \tag{4}$$

Finally, the total carbon emissions from a production cycle are computed as in the left-hand-side of Eq. (5). In one of its simplest versions, carbon footprint control is imposed as a maximum carbon intensity (emissions per unit of product), in an aggregate form. More specifically, total emissions are evenly distributed among all product units finished in a production cycle, and the maximum carbon intensity target is imposed as in the right-hand-side of constraint (5).

$$\sum_{i,i':i\leq i'} CF_{i,i'} \leq \sum_{i'} N_{i'} \, maxci \tag{5}$$

3.4 Objective Function

We seek to optimize the total cost per unit time, derived from the average inventory holding and changeover costs, as expressed in Eq. (6). In the first term, the inventory holding cost for product j per unit time is given by ic_j while the other term is related with the transition cost incurred between two consecutive campaigns processing different products ($TCost_i$), as captured by Eq. (7). Parameter $tc_{j,j'}$ is the transition cost when model j in a production run i'-1 changes to another model j' in the next campaign i'. T is the length of the whole production cycle and is defined by Eq. (8).

$$z = \left[\sum_{j\in P}\sum_{i:i>1} ic_j \frac{(Inv_{i,j} + Inv_{i-1,j})}{2}(C_i - C_{i-1}) + \sum_{i} TCost_i\right]\bigg/ T \tag{6}$$

$$TCost_{i'} \geq tc_{j,j'}\left(y_{i',j'} + y_{i'-1,j} - 1\right) , \forall j, j' \in Prod : j \neq j' \tag{7}$$

$$T \geq C_i \quad \forall i \in Runs \tag{8}$$

4 Case Study

In order to evaluate the performance of the proposed mathematical model, a motivation example from a truck trailers manufacturing industry in Argentina is addressed in this section. The aim is to find the optimal sizing and sequencing of a synchronous assembly line with five workstations that processes four products: P_1, P_2, P_3 and P_4. The sequence of products found by the MINLP solver comprises 6 successive campaigns i_1, i_2, i_3, i_4, i_5 and i_6 processing items of P_1, P_3, P_4, P_2, P_2, and P_1, respectively, along a total production cycle of 13.33 h. The lot sizes of each campaign are 1, 2, 4, 2, 2 and 5, respectively. On Table 1 we report the different cycle times and operating modes that can be adopted for each campaign. Figure 5 shows how the different production runs progress through the assembly line.

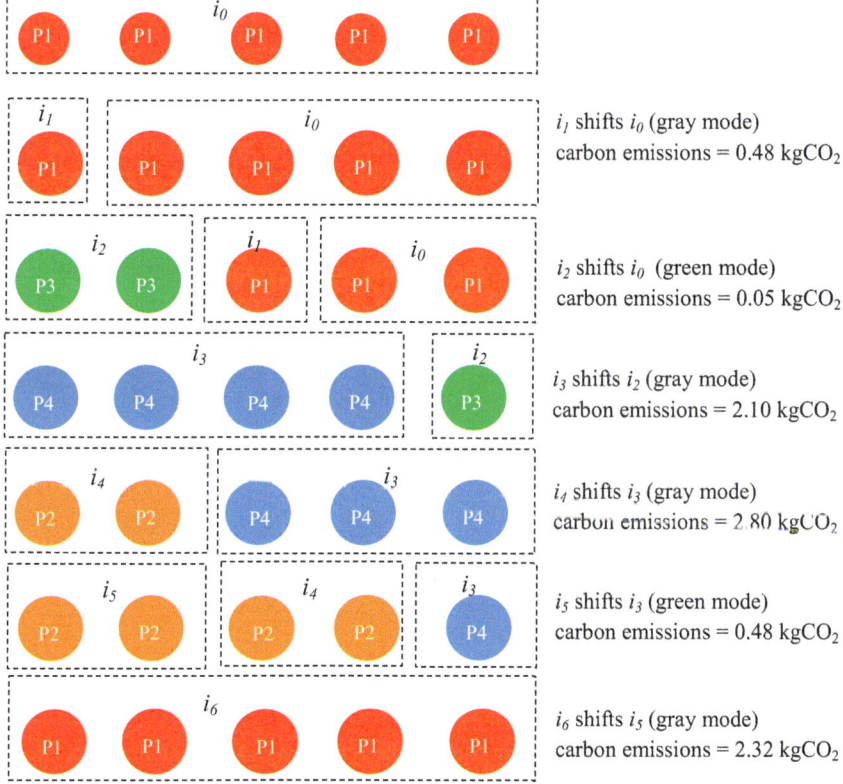

Fig. 5. Evolution of production runs along the assembly line, with illustrative carbon emissions.

Two out of the six runs are accomplished in green mode, with the illustrative carbon emissions given at the right of each line of Fig. 5. For confidentiality reasons, actual emissions are not disclosed. Note that a maximum of 1 kg of CO_2 (average) per unit of product is admitted as carbon footprint, thus making the model to favor production on green mode when other products of slower cycle times are in transit along the line.

Table 1. Cycle times (in hours) for each product and operating mode

	P_1	P_2	P_3	P_4
"green"	0.888	0.833	0.833	0.750
"gray"	0.800	0.750	0.750	0.675

5 Conclusions

As previous authors have emphasized [2], performance improvements can be obtained along with energy consumption and emissions reduction by properly managing batch sizes and product sequences. In that sense, we have developed a novel mixed-integer non-linear programming (MINLP) formulation for the optimal sequencing and sizing of runs in multiproduct, synchronous assembly lines. By means of a continuous representation in both time and spatial domains, which has been adapted from pipeline scheduling optimization models, the problem complexity can be tackled in reasonable computational times. Furthermore, the optimization model has been extended to assess the impacts of operating modes ("greener" or "grayer" modes, emitting different amounts of CO_2). Operating modes affect not only costs and yield rates but also the magnitude of the carbon emissions, often being greener at the expense of slower cycle times. This fact is especially relevant in assembly lines which process multiple products. Planning operations sustainably and efficiently, meeting demand under carbon footprint specifications at minimum total cost is a worthwhile effort. In fact, optimally planning operations to solve the tradeoff between emissions and costs is a key goal of modern production systems. Future research will focus on solving larger instances of the problem, with more workstations and a wider set of products for which continuous time models can be even more competitive in comparison to discrete counterparts.

References

1. Jeswiet J, Nava P (2009) Applying CES to assembly and comparing carbon footprints. Int J Sustain Eng 2:232–240
2. Shuterland JW, Skerlos SJ, Haapala KR, Cooper D, Zhao F, Huang A (2020) Industrial sustainability: reviewing the past and envisioning the future. J Manuf Sci Eng 142:110806
3. Rejowski R, Pinto JM (2003) Scheduling of a multiproduct pipeline system. Comput Chem Eng 27:1229–1246
4. Cafaro DC, Cerda J (2008) Efficient tool for the scheduling of multiproduct pipelines and terminal operations. Ind Eng Chem Res 47:9941–9956
5. Scholl A (1999) Balancing and sequencing of assembly lines (2nd ed) Physica-Verlag, Heidelberg, Germany
6. Camussi NB, Cerdá J, Cafaro DC (2021) Mathematical formulations for the optimal sequencing and lot sizing in multiproduct synchrpnous assembly lines. Comput Ind Eng 152:107006

An Approach to Quantifying the Lifecycle Sustainability Impacts of Tungsten Carbide Tools

Gatewood Arnold[1,2](✉), Julian Polizzi[1,2], Syed Ibn Mohsin[1,2], Daniel Carnesi[1,2], Julius Schoop[1,2], and Fazleena Badurdeen[1,2]

[1] Institute for Sustainable Manufacturing, University of Kentucky, Lexington, KY 40506, USA
gjar222@uky.edu
[2] Department of Mechanical and Aerospace Engineering, University of Kentucky, Lexington, KY 40506, USA

Abstract. Tungsten carbide cutting tools are widely used for their high machining performance and low cost. However, their cost does not currently reflect the material rarity, environmental costs, nor societal costs associated with their production. Critically, embodied energy makes a sizeable contribution to the total lifecycle energy consumption. This study quantifies the true triple bottom line cost of the tools with respect to established sustainability metrics by leveraging cross-disciplinary insights from the product, process, and system levels. Based on the adjusted cost, a comparative analysis of the cutting tool sustainability on the profitability of a nickel-based superalloy machining process was carried out for different tools. Tool wear rates were experimentally measured across a range of industrial feeds and speeds for milling of Inconel 718. Using a Taylorian cost optimization model, the relative influence of the true cost of different tungsten carbide tools was compared. In all cases considered, changes in tool cost had a minimal impact on process profitability, suggesting that more sustainably produced cutting tools may readily justify increased tooling costs. Thus, identification and further development of more sustainable cutting tools may offer a meaningful opportunity to enhance sustainability performance of tungsten carbide cutting tools enabling more sustainable manufacturing.

Keywords: Tungsten carbide · embodied energy · cutting tool · tool cost · sustainable manufacturing · machining

1 Introduction

Metal cutting operations such as milling, drilling, and turning consume tooling inserts as part of their normal operations. The rate of consumption of these tooling inserts depends on several factors, including the workpiece material, tool material, cutting speed, depth of cut, and tool geometry, as well as other process-specific parameters [1]. As a result, unnecessarily conservative cutting parameters are typically used in practice to ensure workpiece quality [2]. This practice is common because accurate tool wear prediction

© The Author(s) 2025
H. Kohl et al. (Eds.): GCSM 2023, LNME, pp. 846–855, 2025.
https://doi.org/10.1007/978-3-031-77429-4_94

can be expensive and cutting tools have a comparatively low cost [1]. The prohibitive cost of scrap and rework also encourages conservative tool use since a worn tool can lead to poor surface properties and part rejection [3].

Tungsten is a rare earth element that will be depleted within 40 years at the current rate of consumption if recycling rates do not increase [4]. In the modern day, tungsten carbide is the preferred material for tooling inserts due to its exceptional abrasion and heat resistance, and moderate manufacturing cost when compared to ceramic or super hard tool materials. However, this material comes with many sustainability problems associated with its production and use [3]. While there are well-established procedures to optimize cutting processes, they often neglect the additional social and environmental costs associated with the cutting tool.

2 Background

Tungsten carbide tools have remarkably high hardness, low porosity, and high strength. The primary elements in these tools are tungsten, cobalt, carbon, and nickel. The two most significant materials by mass are tungsten and cobalt; each has its own sustainability concerns. Tungsten requires substantial amounts of energy to extract, and cobalt mining can be dangerous due to a lack of regulations. While tool composition can vary, an average tungsten carbide tool consists of approximately 90% tungsten carbide and up to 10% binder, with the most common being cobalt. Because tungsten carbide is too brittle by itself for industrial use, a binder metal is added to the mix to improve toughness, resulting in an ideal cutting tool material [5].

Environmental problems associated with the production and disposal of tungsten carbide include dangerous mining practices, the use of harmful chemicals, and energy inefficient production methods [6]. While tungsten itself is not dangerous in small concentrations, other heavy metals found in tungsten mine tailings are resulting in workers and nearby ecosystems being exposed to toxins [6].

According to Ma et al. [7], China is the main producer of tungsten carbide, accounting for 82% of the global output in 2015, yet it has an immature tungsten industry. China does not properly regulate pollution levels, has depression of resource superiority, and has a low recycling rate [7]. Additionally, the electricity needed for post processing comes from coal-based power plants that are poorly regulated. The production of 1 cm^3 of tungsten carbide material can require as much as 8,590–9,724 kJ while the same amount of high-speed steel requires between 755 and 856 kJ of energy [1] This results in the release of 33.7 kg of CO_2 per kg of Carbide [3], which is the primary environmental burden associated with tungsten carbide production. The use of cleaner energy sources could significantly mitigate this concern [7].

Cobalt demand around the world is at an all-time high and is currently projected to double between 2021 and 2035 due to its use in battery technology [8]. This level of demand has caused the cobalt mining industry to outpace government regulations [8]. Most cobalt is extracted as a byproduct of other mining processes [1]. In these types of mines, worker safety, and environmental impacts are acceptable [11]; however, small artisanal mines can unfortunately be very dangerous [9]. Artisanal mines in the Congo account for 3% of all cobalt production, the lack of oversight results in regular cave collapses and poorly documented fatality rates [9]. This results in a significant, but hard to measure societal impact.

All of this leads to the production of a tool that has environmental and societal costs that are not currently reflected in the price of tooling. Instead, this price is paid later through negative externalities such as increased cancer rates, birth defects, and wildlife devastation [10]. While there is some research analyzing the environmental impacts of the elements released into the surrounding soil, these studies focus only on the environment impact and do not consider the impact on workers' health nor the economy.

The work of Henckens et al. [11] indicates that the free-market price mechanism fails to incorporate the potential depletion of mineral resources, thus failing to reflect future resource scarcity through price signals [11]. A price increase could promote sustainable utilization of resources and incentivize research into substitute materials. However, according to De Bruyn et al., the raw material makes up a small portion of the final cost of the product [12]. As such, a moderate increase in the price of a particular raw material is unlikely to result in a significant increase in the final product's price. Thus, a more effective method to quantify the triple bottom line (TBL) cost of tungsten carbide cutting tools must be established to curtail resource depletion and promote sustainable practices.

3 Research Methods

In an effort to assess the TBL impact of tungsten carbide cutting tools, a two-step approach is followed in this study. The first step, metrics-based analysis, serves to determine tool sustainability. Next, experimental analysis was conducted to determine the economic viability of using a more sustainable, and therefore more expensive, tool.

3.1 Experimental Analysis

An experiment was conducted on a 9.525 mm 2-flute tungsten carbide endmill and Inconel 718 using a Tormach1100MX. The input values for the variables of interest— feed and cutting speed—for the nine trials are shown in Table 1. For each trial, cuts were taken along the 50.8 mm by 6.35 mm face of the Inconel bar and a photo was taken of each flute using a 1.5× magnifying lens after a predetermined number of cuts using the setup in Fig. 2. The collected images, like those in Fig. 1 (a) and (b), were then used to measure the tool wear with MATLAB image processing software to manually measure the number of pixels at three different points along the uniform flank wear. Areas with larger wear were favored because these regions typically initiate further destruction of the tool. To calibrate the number of pixels to microns, images of calibration slides were taken with the same camera and processed in the same software to find the conversion of pixels to microns. The amount of time that it takes to reach the 50 μm cutoff for the wear was recorded from this data and used to make the Taylor plots in Fig. 1. The error bars for this plot use standard error. A wear cutoff threshold of 50 μm was used for this experiment based on industry partner feedback for machining Inconel 718 in aerospace applications requiring good surface integrity.

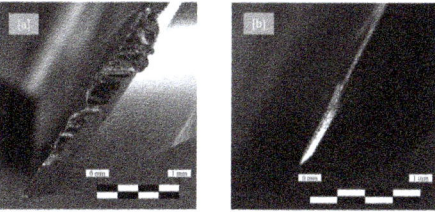

Fig. 1. (a) Tool wear under 2× magnification & additional lighting. (b) Tool wear under 1.5× magnification for simplified tool wear measurements.

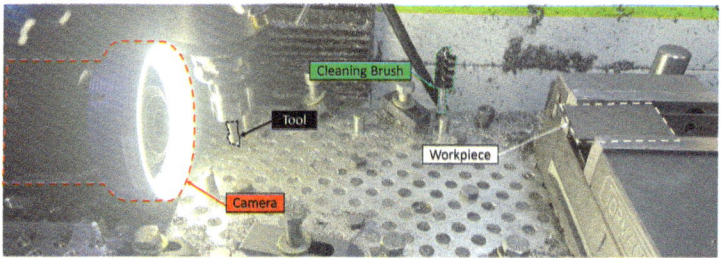

Fig. 2. Experimental setup showing the camera, tool, workpiece, and cleaning brush.

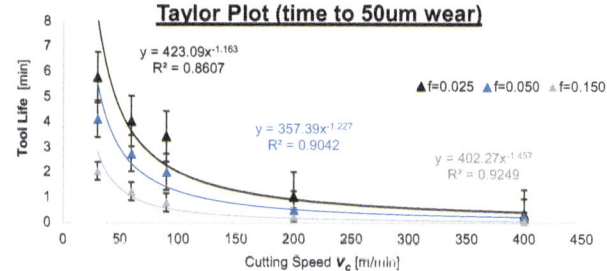

Fig. 3. Taylor plots of remaining tool life vs. cutting in the 9 trials.

Regression fitting techniques were applied to the experimental data and the general form of the Taylor equation $T = \frac{k}{V_c^n}$, where k and n are the constants found from the regression and V_c is the cutting speed. Three equations of best fit are derived from the three feeds tested and are shown in Fig. 3. The third equation was found to give the best R-squared value. The moderate speeds used in the trials—that match industry practices—meant that the regressions in Fig. 3 predicted unrealistic values for high cutting speeds, so to reflect the near immediate failure of cutting tools with Inconel 718 at high cutting speeds, points were manually added at 200 and 400 m/min to improve the predictive performance of the model in the upper limit of the cutting speed regime. Tools would fail at these high cutting speeds more quickly than data could be collected to make meaningful tool wear curves.

Table 1. Experimental inputs for the nine trials conducted.

Trial	v_c [m/min]	F [mm/rev]
1	30	0.025
2	30	0.05
3	30	0.15
4	60	0.025
5	60	0.05
6	60	0.15
7	90	0.025
8	90	0.05
9	90	0.15

3.2 Metrics Based Approach to Tool Sustainability

To evaluate a tool's sustainability performance, an unbiased, metrics-based process must be applied. In this study, the Product Sustainability Index (ProdSI) [13], was adapted and applied to evaluate tungsten carbide cutting tools. Similar processes such as LCA would also be appropriate for future studies with more available data. ProdSI is an established process in which a set of sustainability metrics is identified, data gathered, and then normalized such that each metric score ranges from zero to ten with a score of ten being the most sustainable. The scores are then weighted and aggregated across several levels from metrics to sub-clusters, clusters, sub-indices and then the ProdSI. For this study, the ProdSI method is adapted and simplified, focusing only on the metrics relevant to each TBL category and summed using Eq. 1. Normalization can be done using a variety of techniques as described in Shuaib et al. [13], and the weights for each metric are often determined by an industry professional. In this example, equal weights were used for the metrics of the environmental, economic, and societal categories, and benchmark normalization was used [13]. Scores that fell outside of the set range were capped at zero or ten to enable a fair comparison between different tools presented later. For this study, typical non-Chinese based manufacturers (nCBM) were compared to Chinese based manufacturers (CBM) due to a lack of data available that differentiated between different non-Chinese based manufacturers. To establish a datum, non-Chinese based manufacturers were given a score of 5. The final score was then produced using the following equation:

$$\text{ProdSI} = \frac{1}{3}(E_c + E_v + S_o) = \frac{1}{3}\left(\sum w_i^f C_i + \sum w_i^f C_i + \sum w_i^f C_i\right) \quad (1)$$

where E_c is economic, E_v is environmental, S_o is societal, w is weight, and C_i is the normalized scores for the different metrics in each category.

4   Results and Discussion

With an established tool wear rate, the rate of tool consumption for a given feed and speed can be calculated and used for determining the profitability across the range of possible inputs The profitability is calculated as the number of parts that can be made per tool with the addition of general shop overheads. Tools must have enough remaining tool life to complete an entire part because switching tools can lead to irregularities that initiate premature cracking and fatigue, meaning that a tool with five minutes of remaining tool life would be rejected for an operation taking six minutes. Truncating tool life in this way can lead to non-linear relationships between cutting speed and profitability as seen in Fig. 4 (a). Feed rate, however, follows an approximately linear relationship with the profitability as seen in Fig. 4 (b) because of the linear relationship between the feed rate and both the tool life and the productivity, which are key factors in the profitability.

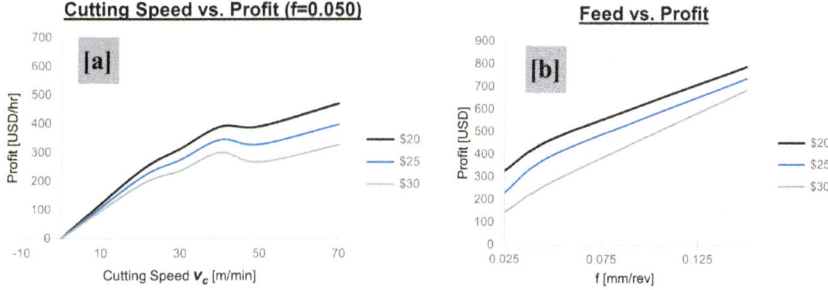

Fig. 4. A Profitability curve for a feed of 0.050 mm/rev, and (b) feed vs. profit plot derived from the three speed vs. profit plots collected data points for the trials.

By varying the tool cost in different scenarios using these calculations, the impact of the tooling cost can be determined and compared against a wide range of scenarios. For each given set of parameters, some optimum feed and cutting speed will give the maximum profitability. For example, using the parameters within the bounds of Table 2 that matches the experimental setup will give the cutting speed vs. profitability curves in. Figure 4 (a) and by getting data for all the three feeds, the feed curve vs. profitability in Fig. 4 (b) can be obtained. All other parameters are held constant between the three curves in Fig. 3 (a) and (b), except for the tooling cost, which increases from $20 to $30. This in turn makes a new maximum along the range of acceptable feeds for the given set of parameters, but as the other parameters besides the tooling cost change, the maximum of this curve can vary significantly. The equation used for finding the percentage change for the profit of an operation with a more expensive tool is displayed in Fig. 5.

To address the infinite range of combinations that can be made with the parameters examined and how they might impact the profitability, an acceptable range and step size was selected for each parameter so that the entire space of workable parameter combinations could be explored. The tooling cost was increased by 100% to demonstrate the behavior for any range of tooling cost increase that might be taken on and the diminishing impacts of increasing the tooling cost.

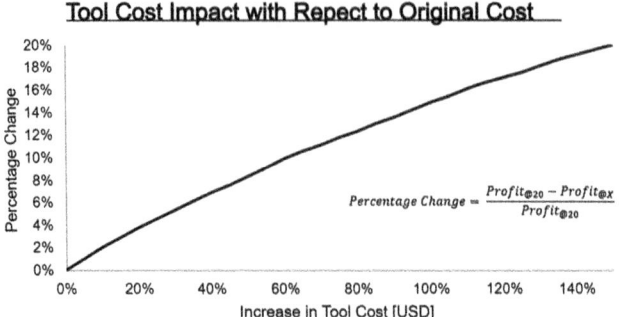

Fig. 5. Percentage change in the average profitability across the range of tool prices

Table 2. Range of parameters used for profitability calculations.

Parameter	Starting Value	Ending Value	Units
Overhead/hour	80	120	[$/h]
Material cost/part	20	60	[$/part]
Income/part	70	100	[$/part]
Cost/tool	20	50	[$/tool]
Volume to cut/part	500	10,000	[mm^3]
Tool diameter	6	10	[mm]
Flutes	4	6	[flutes]
Tool change time	0.5	0.5	[min]
Part changeover time	0.5	0.5	[min]

The maximum profitability across the entire range of possible parameters for each tool cost tested was averaged to determine the overall impact of the tooling cost as a single data point. Next, the percentage change in the profitability compared to the original tool cost was calculated across the entire range of tool prices using Eq. 2. These values were then plotted in Fig. 4 to show the impact of increasing the tool cost on the profitability of the machining conditions.

The results of this study show that large increases in tooling cost can be taken on without incurring a large profit loss and that the impact of increasing the cost of the tool has diminishing effects as the price increases. For instance, this study found that increasing the tooling cost from $20 to $30 only decreased the profitability by 8.39% for the entire range of the other parameters considered. Similarly, increasing the tool cost from $20 to $22, or by 10%, results in a decrease in average profitability of 2.05%. It should be noted, however, that the small number of trials in this test results in uncertainty that warrants future research.

Using the adapted ProdSI method, tungsten carbide cutting tools produced by two types of manufacturers, Chinese-based manufacturer (CBM) and non-Chinese based manufacturer (nCBM) were computed. While limited data was available, the preliminary results for ProdSI are shown in Table 3. A collection of metrics was used for TBL evaluation and sources of data are also shown. Results indicate that nCBM tools are more sustainable than those of CBM. According to Furberg et al., nCBM suppliers are often more sustainable than CBM suppliers due to their reduced energy use [14]. Thus, when selecting a tool supplier, each metric can be discussed with them to gather the appropriate data and analyzed to identify those with highest ProdSI score(s) [14].

Table 3. Values used for the ProdSI analysis.

Metric	Weight	Units	CBM	ProdSI Score		nCBM	ProdSI Score	
				Unweighted	Weighted		Unweighted	Weighted
[7] Climate change	0.17	kg CO_2 eq	69	0.00	0.00	14	5	0.83
[7] Photochemical oxidant Formation	0.17	kg NMCOC eq	0.18	0.00	0.00	0.049	5	0.83
[7] Terrestrial acidification	0.17	kg SO_2 eq	0.21	8.19	1.36	0.58	5	0.83
[7] Ozone depletion	0.17	kg CFC-11 eq	4.8E-07	9.08	1.51	2.6E-06	5	0.83
[7] Water depletion	0.17	m^3	0.6	0.00	0.00	0.24	5	0.83
[7] Freshwater eutrophication	0.17	kg P eq	2.80E-04	9.97	1.66	5.40E-02	5	0.83
[15] Energy cost	0.50	USD/kWh	0.076	7.82	3.91	0.174	5	2.5
[16] Worker compensation	0.50	USD	5.00	1.00	0.50	35.00	5	2.5
Fatality rate	1.00	Fatalities/100,000	[17] 2.13	1.78	3.50	[18] 3.5	5	5.00
Total ProdSI Score					4.15			5.00

5 Conclusion

A small increase in tool cost, which is associated with responsibly sourcing tools, has a marginal impact on profitability. The environmental and human rights issues tied to some tool manufacturers justifies a closer look into the complete lifecycle impact of carbide tools. These hidden costs should be represented in the actual cost of the tool, as is the case when tools are purchased from sustainable manufacturers. This actual cost should be calculated using the ProdSI method. In this study, the ProdSI method was inconclusive in its comparison of CBM and nCBM tools due to the lack of available information. Buyers should attempt to gain more data about a supplier and conduct their own analysis using the methods outlined in this paper.

6 Future Research

In this preliminary study, only a generic Inconel machining operation with a tungsten carbide tool was considered. A comprehensive analysis should include additional cutting tools and work piece materials. More specific data should be collected to improve accuracy for specific machining operations. Further testing could also be done to produce more data points leading to Taylor curve with a greater predictive power. Existing literature lacks comprehensive data collected from material suppliers. To improve this, data should be collected from the supplier through a field survey. Currently, end of life data collection is scarce, more information is required for analysis. Finally, a comprehensive sustainability analysis should be conducted. Overall, a more rigorous and data-driven implementation of a metrics bases sustainability analysis is envisioned to enable significantly more sustainable cutting tool utilization.

References

1. Ullah AMMS, Kitajima K, Akamatsu T, Furuno M, Tamaki JI, Kubo A (2011) On some eco-indicators of cutting tools. In: ASME 2011 MSEC
2. Gutnichenko O et al (2021) Improvement of tool utilization when hard turning with CBN tools at varying process parameters. Wear 477:203900. https://doi.org/10.1016/j.wear.2021.203900
3. Sun H, Liu Y, Pan J, Zhang J, Ji W (2020) Enhancing cutting tool sustainability based on remaining useful life prediction. J Clean Prod 244:118794
4. IshIda T, Itakura T, Moriguchi H, Ikegaya A (2012) Development of technologies for recycling cemented carbide scrap and reducing tungsten use in cemented carbide tools. SEI Tech Rev 75:30–46
5. Sekulić DP (2013) Advances in brazing: science, technology and applications. Elsevier
6. Han Z, Golev A, Edraki M (2021) A review of tungsten resources and potential extraction from mine waste. Minerals 11(7):701
7. Ma X, Qi C, Ye L, Yang D, Hong J (2017) Life cycle assessment of tungsten carbide powder production: a case study in China. J Clean Prod 149:936–944
8. Al Barazi S, Näher U, Vetter S, Schütte P, Liedtke M, Baier M, Franken G (2017) Cobalt from the DR Congo–potential risks and significance for the global cobalt market. Bundesanstalt für Geowissenschaften und Rohstoffe, Hannover
9. Prevention, C.f.d.C.a. (2023) Toxicological profile for cobalt: health effects [cited 2023]. Available from: https://www.atsdr.cdc.gov/toxprofiledocs/index.html
10. Chung AP et al (2019) Tailings microbial community profile and prediction of its functionality in basins of tungsten mine. Sci Rep 9(1):1–13
11. Henckens M, Van Ierland E, Driessen P, Worrell E (2016) Mineral resources: Geological scarcity, market price trends, and future generations. Resour Policy 49:102–111
12. De Bruyn S, Markowska A, de Jong F, Blom M (2009) Resource productivity, competitiveness and environmental policies. CE Delft, pp 1–72
13. Shuaib M, Seevers D, Zhang X, Badurdeen F, Rouch KE, Jawahir IS (2014) Product sustainability index (prodsi). J Ind Ecol 18(4):491–507. https://doi.org/10.1111/jiec.12179
14. Furberg A, Arvidsson R, Molander S (2019) Environmental life cycle assessment of cemented carbide (WC-Co) production. J Clean Prod 209:1126–1138
15. Valev NT (2023) Electricity prices around the world. GlobalPetrolPrices.com. https://www.globalpetrolprices.com/electricity_prices/#hl48

16. U.S., Congress, Bureau of Labor Statistics. Charting International Labor Comparisons
17. Herbert A, Ran B, Machida S, Shengli N, Kawakami T, Changyou Z, Yanyun W, Hongyuan Z (2012) National. Profile Report on Occupational Safety and Health in China
18. Bureau of Labor Statistics, National Census of Fatal Occupational Injuries in 2010 (preliminary results)

Spaceship Earth: A Sustainable Manufacturing Game

Shivangi Paliwal[1,2]([✉]), Patrick Gannon[1], Ethan Lauricella[3], Kyle E. Riddett[1], Fazleena Badurdeen[1,2], and Anthony Elam[4]

[1] Department of Mechanical & Aerospace Engineering, University of Kentucky, Lexington, KY 40506, USA
shivangi.paliwal@uky.edu
[2] Institute for Sustainable Manufacturing, University of Kentucky, Lexington, KY 40506, USA
[3] Department of Materials Engineering, University of Kentucky, Lexington, KY 40506, USA
[4] Center for Computational Science, University of Kentucky, Lexington, KY 40506, USA

Abstract. Developing an educational game about sustainable manufacturing can be an effective way to teach people about the importance of sustainable practices in the manufacturing industry. Through educational sustainable manufacturing games, users can learn about the impact of sustainable manufacturing on the environment, along with strategies and techniques that can be used to reduce waste and methods of sustainable production. A minimal number of articles have focused on board game development for sustainable manufacturing, and of the few, most focused on educational simulation of machines/systems, catering to specific profession. This paper chronicles the creation of "Spaceship Earth," a sustainable manufacturing board game from game design to testing. It explores the entire process of developing an educational game, from game theory to final testing. The game incorporates various elements such as resource management, decision-making, and problem-solving to simulate the real-world challenges faced by manufacturers when implementing sustainable practices. By examining the design of educational games, this paper aims to gain insight into the constraints of engineering education gamification through development of a board game. The paper demonstrates the potential of educational games as a means of promoting sustainability and provides useful framework for designing and developing similar games in the future.

Keywords: Sustainable Manufacturing · Gamified Learning · Board Games

1 Introduction

As innovative educational approaches continue to take shape, one concept has been gaining remarkable popularity, captivating educators, and learners alike: the transformative power of gamification. Education gamification refers to the incorporation of game-like components into educational settings, providing an enhanced learning experience. The theory of learning that considers the social and cultural aspects (Vygotsky 1978) along with flow theory (Csikszentmihalyi 1990) agree with effective game designs and positive

© The Author(s) 2025
H. Kohl et al. (Eds.): GCSM 2023, LNME, pp. 856–863, 2025.
https://doi.org/10.1007/978-3-031-77429-4_95

educational results. According to Vygotsky (1978) learning occurs through social inter-action, active participation, and contextual relevance. Additionally, Vygotsky (1978) asserted that play facilitates the process of learning. The integration of gamification in education has emerged as a transformative approach, utilizing game design principles and mechanics to cultivate captivating learning experiences.

Gamification has gained attention in diverse fields like business, workforce development, healthcare, and environmental studies (Roodt and Ryklief 2019). (Saleem et al. 2022) Consequently, integrating gamification into e-learning has led to several benefits like enhanced learning skills such as decision making, teamwork and communication (Nand et al. 2019). Therefore, to facilitate development of these skills gamification incorporates game elements to improve user engagement and satisfaction. In practical implementation, gamification applications commonly utilize a restricted range of game elements, namely point systems, leader boards and badges (Rapp et al. 2019). Therefore, there exists a requirement for the investigation of alternative game mechanics and their associated elements that can yield distinct effects within gamified environments. In addition, visualizing players' actions and their impact on the board enhances engagement with the game's purpose (Fjællingsdal and Klöckner 2020). Accordingly, to design games for educational purposes, it is crucial to prioritize motivation as the key element. This can be accomplished by utilizing effective computing techniques that emphasize emotional aspects within the design interfaces (Hakak et al. 2019). The various game elements, such as team-based activities, competitive dynamics and collaborative efforts are often overlooked (Rapp et al. 2019). However, these factors serve as the bedrock for establishing effective goal structures within academic settings.

A review of gamification papers in the field of engineering reveals a predominant emphasis on theoretical aspects derived from game theory. However, there is a notable scarcity of experimental studies investigating the outcomes experienced by participants engaging with engineering gamification (Marlppoulos et al. 2015).

Morelock et al. (2018) conducted an extensive review of papers from Scopus, ScienceDirect, Web of Science, ERIC, EBSCO, and Engineering Village, to identify a total of 10,825 papers describing games and gamification in engineering education They removed papers having similarity to game designs in other papers, lacked detail of the game, or based on theory of gamification instead of a designed game. Of these papers, 112 papers were most relevant and separated into digital vs non-digital. Digital games involved the use of computers to play the game, either cooperatively or by oneself, while non-digital describes games with only physical designs with players playing in the same room as one another. Each paper was evaluated based on RAT, Replacement, Amplification or Transformative framework. Replacement describes the games as merely another means to convey the same material without any additional benefit. Amplification enhances the learning experience using gamification concepts. Games considered transformative introduce concepts/ideas that are not feasible in traditional settings, allowing for new concepts to be taught in a more efficient manner. Of the non-digital games analyzed, 75% were deemed transformative, indicating their effectiveness in learning and skill development.

In recent years, a larger focus has been placed on sustainable practices as also outlined in the 17 Sustainable Development Goals by the United Nations established in 2015. Sustainable manufacturing is focused on enhancing closed-loop material flow, through the application of novel methodologies such as the 6Rs (Reduce, Reuse, Recycle, Recover, Redesign and Remanufacture (Jawahir et al. 2006) to enhance economic, environmental, and societal performance across product, process and system levels. While there are numerous simulations and games presented in literature for other related domains such as lean manufacturing (e.g., see Badurdeen et al. 2010 for an extensive review), nothing similar is available to increase awareness of sustainable manufacturing. A board game focused on sustainable manufacturing can enable more effectively teaching the relevant concepts to the next generation of manufacturing engineers. They can offer unique advantages in terms of fostering systemic understanding, promoting behavioral change, and encouraging active learning, making them a valuable addition to the conventional pedagogical tools.

2 Methodology

In this study, a sustainable manufacturing game was designed by leveraging the key findings from previous literature. The first step in the process was determining what steps were taken in the prior papers and how successful they were in accomplishing the goal of an educational game. Then, by adapting their processes in the most successful games, such as Be Blessed Taiwan (Tsai et al. 2021), a concept map was developed; some main topics necessary for a sustainable manufacturing game to be successful were included in the concept map. This allowed focusing efforts on critical topics and speeding up the game development process. One of the authors of this paper is an avid board game enthusiast with a large collection and knowledge of board games. Their expertise was also leveraged to design an approach to develop a sustainable manufacturing boardgame with engaging features to more ensure effective learning. The team utilized engineering design tools such as Pugh charts to develop the game. The Pugh charts compared prototypes based on desired outcomes, helping the team select a core concept. Each member contributed a game concept, which was anonymously ranked using the Pugh chart. This allowed the integration of beneficial aspects from each design into the game board's development. Once the game reached a point where it was playable, playtesting and evaluation was initiated. This consisted of the team playing key parts of the game very slowly and taking notes as it progressed. After a certain time, the game was stopped and analyzed to discern what gameplay elements worked and which did not. Particular emphasis was placed on whether the elements served to encourage the player to use the 6R's (Jawahir et al. 2006). The elements were reworked, and the cycle was reiterated until desired results were obtained. Figure 1 illustrates the game development and playtesting process.

After learning objectives were established and a general design path was determined, the framework of the physical game was constructed. A prebuilt kit was used for building a boardgame to quicken the process.

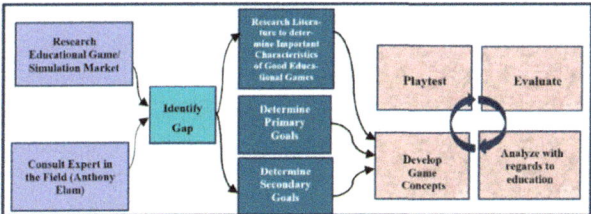

Fig. 1. Flow Chart of Game Development.

2.1 Spaceship Earth: A Sustainable Manufacturing Game

The Spaceship Earth concept was chosen because the environment is truly isolated, which helps convey the message within Ken Boulding's article, "Spaceship Earth." In the paper, Boulding (1966) urges readers to shift their consumption ideology from that of a cowboy mentality, where resources are plentiful and waste is ignored, to a spaceship mentality, where the circular economy and thoughtful resource use is key. In Spaceship Earth, four players are engaged in sustainable manufacturing by producing, recycling, and upgrading components and machinery. Crew members are designated specific roles to promote specialization, which allows individuals unique insights into the world of sustainable manufacturing through the lens of the crew member. It also allows individuals to replay the game a few times, each with a new perspective and set of overall goals. The goal of the group is to reach Earth in under 2 h (playing time), while surviving challenging internal supply chain issues and managing dwindling resources.

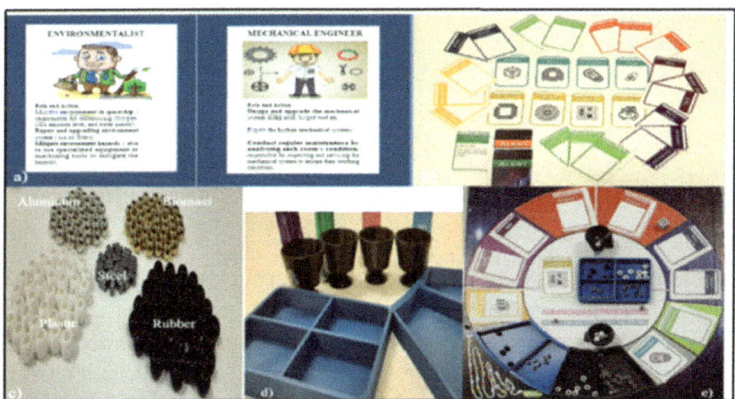

Fig. 2. a) Player cards, b) Cards, c) Raw materials, d) 3D printed player cups and four cornered board to serve as reclamation center, e) Final "Prototype" of game designed for play testing.

Figure 2 shows visuals of different parts of the board game developed. Figure 2 (a) displays two examples of player cards which introduce specialties within the team, as each crew member has strengths, weaknesses, and priorities within the ship. Figure 2 (e) includes all 6 rooms and the centralized recycling and reclamation center found within the board game. The color of each room corresponds with the systems found within them. Each area of the board is vital to team success. Steel, aluminum, plastic, and rubber (represented by small plastic beads) are available as resources for the crew to create many components necessary for fabricating machines and systems, as seen in Fig. 2 (c). Spaceship Earth has 6 rooms (Engine Room, Workshop, Generator Room, Storage, Greenhouse, and Laboratory) and a centralized center for recycling and resource reclamation. Each player has a "crew role" (e.g., Mechanical Engineer, Chemical Engineer, Researcher, and Botanist). All fabrication occurs in the Workshop. The highest priority machine to build and upgrade is the engine, which runs on biofuel produced in the Laboratory. Biofuel is created by the Chemical Engineer in the Laboratory from excess biomass grown in the Greenhouse. This biomass can also be used to create plastic, the only renewable resource on the ship. As players upgrade and build new machines, materials dwindle and the air is polluted, forcing players to act. To survive, the crew must use the 6R's to build and reuse existing equipment.

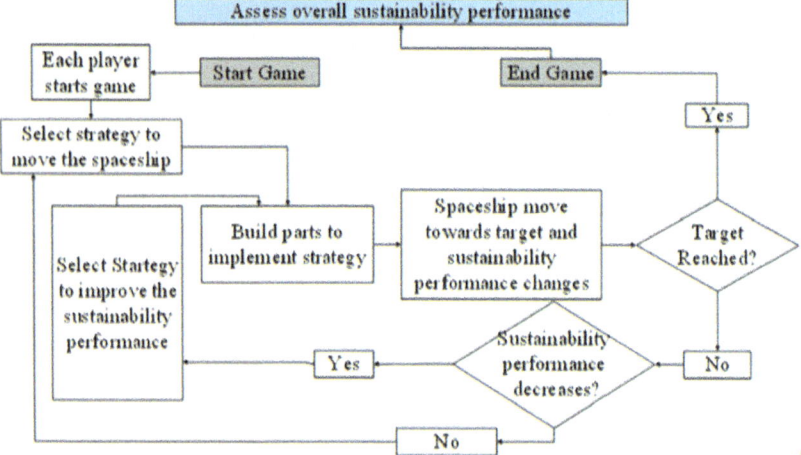

Fig. 3. In game decision making flow chart for Spaceship Earth

To win the game, the new crew (players) of Spaceship Earth must pilot the ship back home within 2 h. To accomplish this mission, the crew must be conscious of material waste, energy consumption, and air quality. There are several critical decisions the crew must make to become successful. The results of good and bad decisions are shown in Fig. 3. After each round certain metrics like material waste, energy consumption, and air quality are recorded along with ship position and overall time. These "per turn wastes" are explained further in the rules and are dependent on the systems or processes in question. After the game is completed, the team performance is evaluated based on the sustainability of their production. A scorecard (shown later in Fig. 4) is filled out with the resulting outcome values of the game, including waste, recycled materials, and systems built. A weighted value of each aspect, based on the difficulty of success determined through play testing, allows for an accurate assessment of overall team performance.

3   Results and Discussion

This section presents a discussion of the results derived from the empirical evaluation of the board game incorporating the Likert Scale. A playtesting session was conducted to assess the game's effectiveness. A sample of four players, specifically graduate-level students with no background in sustainable manufacturing, were selected for evaluation. The focus of this evaluation was gauging improvement in knowledge of sustainable manufacturing concepts after playing the game. It involved pre- and post-game evaluation using the Likert Scale with a 5-point rating system as a means of assessment. Five response options, ranging from "Strongly Disagree" to "Strongly Agree", were used with a set of questions to collect participant responses, and analyze different aspects of game. The pre-game questions asked participants about the impact of playing board games on their attitude towards sustainable manufacturing, their level of awareness and their familiarity with the concept of 6Rs. In contrast, the post-game questions focused on factors such as stimulating, challenging, informativeness, and goal clarity of game to understand these concepts after the game being played. The main objective of this board game is to teach the importance of 6Rs and sustainable manufacturing. Therefore, the cumulative response from each player was collected separately to assess and further improve the mechanics of the game. It was observed that knowledge and awareness of 6Rs increased after the participants played the game. Although the game was found to be challenging, feedback from players revealed that the game helped improve their knowledge about sustainable manufacturing. Additionally, an assessment rubric was used to evaluate player's choices and actions in terms of impact on sustainability metrics. The rubric assesses players' decisions in terms of the extent of applying the 6Rs for sustainable manufacturing. The scorecard (in Fig. 4) is used to determine the final score for the team to convey the effectiveness of 6R-informed decision making in steering Spaceship Earth.

Metric	Value	
Waste created	62	Enter the number of turns taken into the yellow box
Recycled Components	25	
Products created Level 1	29	
Products created Level 2	15	Enter true/false within the blue box's at the end of the game.
Products created Level 3	10	
Rounds Taken	20	
Built the Level 2 generator	TRUE	
Built the Level 3 generator	FALSE	
Built the Level 2 greenhouse	TRUE	
Built the Level 3 greenhouse	TRUE	
Build the Air Filter Level 2	FALSE	
Built Engine 2	TRUE	
Built Engine 3	FALSE	
Have less than 20 Space waste	FALSE	
Have less than 10 Scrap in scrap bin	TRUE	
Air quality above 30% for 70% of game	TRUE	
Total Score	138.5	

Fig. 4. Scorecard for team performance assessment

In Fig. 4, the values in white cells indicate example amounts for each total of performance metrics i.e., waste created, air quality, energy consumption after the end of the game. Here, waste created shows that 62 units of waste were created, 25 components were recycled into base materials, 29 "level 1" components were made, 15 "level 2" and

10 "level 3". Each level is indicated in the game represents the difficulty of producing. "Level 1" is the easiest to create, "level 2" requires more advanced machines, and "level 3" requires all the machine shop systems. The rounds taken is the number of rounds taken to finish the game. TRUE or FALSE included in the blue cells indicate if the system was built during the game with the level indicating an advanced version of the same machine. The higher the level, the more efficient the system works, requiring less materials and producing less waste when used. Whenever creating a component, some material becomes space waste, i.e., waste that cannot be recycled. Here, an example of 20 space waste units is created. If this value is lower than 20, players put TRUE indicating the amount of space waste created was less than 20. The air quality is automatically determined in Excel based on players' input values in a table recording the air quality per round. The impact is assessed using the metrics of energy consumption, total amount of waste produced, and the air quality till the ship reaches the target. The values are then consolidated to establish a total score for the team, with a lower score indicating better performance. This score is used to frame a clear picture of knowledge gained by participants during the game. If multiple teams were playing the game, this score can be used to comparatively assess the performance of each team, creating a competitive gaming environment that motivates more effective application of the 6Rs and better learning.

4   Conclusion

There is currently a lack of board games focused on sustainable manufacturing. However, the development of these games presents an intriguing avenue for future research to explore their efficacy when compared to more traditional educational environments. This study presented the development of 'Spaceship Earth' to make a meaningful contribution to the field of gamified learning of sustainable manufacturing by fostering the development of innovative educational ideas. The game was tested with one team of players by gathering pre-and post-game responses from each participant. Through the analysis of survey results and extracted metric data, the results revealed that board game Spaceship Earth enabled increasing the players' understanding of 6Rs for sustainable manufacturing. Further validation and enhancement of the board game's effectiveness require thorough testing and analysis with various teams. This limitation necessitates an extensive evaluation process to ensure its efficacy. Moving forward, future studies can further enhance the survey questionnaire and expand its application to encompass other games, thereby enabling standardized data collection and facilitating confident comparisons between different learning methods. This comprehensive approach will not only enhance the accuracy of research findings but also drive the advancement of the gamification field. The creation of the Spaceship Earth sustainable manufacturing board game opens possibilities for using gamification to understand educational outcomes and reach a wider audience in the manufacturing sector. As current results are limited to one playtesting because of the time frame, however in the future more testing could be done, and techniques of data collection could be improved to enhance the effectiveness of the game.

References

1. Badurdeen F, Marksberry P, Hall A, Gregory B (2010) Teaching Lean Manufacturing With Simulations and Games: A Survey and Future Directions. Simul Gaming 41(4):465–486
2. Csikszentmihalyi, M. (1990). Flow. The Psychology of Optimal Experience. New York (HarperPerennial) 1990
3. Fjællingsdal KS, Klöckner CA (2020) Green Across the Board: Board Games as Tools for Dialogue and Simplified Environmental Communication. Simul Gaming 51(5):632–652
4. Hakak, S., Noor, N. F. M., Ayub, M. N., Affal, H., Hussin, N., ahmed, E., & Imran, M. (2019). Cloud-assisted gamification for education and learning – Recent advances and challenges. Comput. Electr. Eng., 74(C), 22–34
5. Jawahir, I.S., Dillon, O.W., Rouch, K.E., Joshi, K.J., Venkatachalam, A., & Jaafar, I.H. (2006). Total life-cycle considerations in product design for manufacture: a framework for comprehensive evaluation. 10th International Research/Expert Conference, Barcelona, Spain, September, 1–10
6. Markopoulos AP, Fragkou A, Kasidiaris PD, Davim JP (2015) Gamification in engineering education and professional training. Int J Mech Eng Educ 43(2):118–131
7. Morelock, J., & Matusovich, H. M. (2018). All Games Are Not Created Equally: How Different Games Contribute to Learning Differently in Engineering
8. Nand, K., Baghaei, N., Casey, J., Barmada, B., Mehdipour, F., & Liang, H.-N. (2019). Engaging children with educational content via Gamification. Smart Learning Environments,
9. Rapp A, Hopfgartner F, Hamari J, Linehan C, Cena F (2019) Strengthening gamification studies: Current trends and future opportunities of gamification research. Int J Hum Comput Stud 127:1–6
10. Roodt S, Ryklief Y (2019) Using Digital Game-Based Learning to Improve the Academic Efficiency of Vocational Education students. International Journal of Game-Based Learning (IJGBL) 9(4):45–69
11. Saleem AN, Noori NM, Ozdamli F (2022) Gamification Applications in E-learning: A Literature Review. Technol Knowl Learn 27(1):139–159
12. Tsai J-C, Liu S-Y, Chang C-Y, Chen S-Y (2021) Using a Board Game to Teach about Sustainable Development. Sustainability 13(9):4942
13. Vygotsky, L. S. (1978). Mind in society: The development of higher mental processes (E. Rice, Ed. & Trans.). In: Cambridge, MA: Harvard University Press

The manufacturer's authorised representative in the EU is Springer
Nature Customer Service Centre GmbH, Europaplatz 3, 69115 Heidelberg,
Germany. If you have any concerns regarding our products, please
contact ProductSafety@springernature.com

Printed and bound by CPI Group (UK) Ltd, Croydon, CR0 4YY
27/04/2026
02097580-0007